Springer Series in the Data Sciences

Springer Series in the Data Sciences focuses primarily on monographs and graduate level textbooks. The target audience includes students and researchers working in and across the fields of mathematics, theoretical computer science, and statistics. Data Analysis and Interpretation is a broad field encompassing some of the fastest-growing subjects in interdisciplinary statistics, mathematics and computer science. It encompasses a process of inspecting, cleaning, transforming, and modeling data with the goal of discovering useful information, suggesting conclusions, and supporting decision making. Data analysis has multiple facets and approaches, including diverse techniques under a variety of names, in different business, science, and social science domains. Springer Series in the Data Sciences addresses the needs of a broad spectrum of scientists and students who are utilizing quantitative methods in their daily research. The series is broad but structured, including topics within all core areas of the data sciences. The breadth of the series reflects the variation of scholarly projects currently underway in the field of machine learning.

More information about this series at http://www.springer.com/series/13852

Guanghui Lan

First-order and Stochastic Optimization Methods for Machine Learning

Guanghui Lan
Industrial and Systems Engineering
Georgia Institute of Technology
Atlanta, GA, USA

ISSN 2365-5674 ISSN 2365-5682 (electronic)
Springer Series in the Data Sciences
ISBN 978-3-030-39570-4 ISBN 978-3-030-39568-1 (eBook)
https://doi.org/10.1007/978-3-030-39568-1

Mathematics Subject Classification (2010): 90C15, 90C25, 90C30, 68Q32, 93E35

This Springer imprint is published by the registered company Springer Nature Switzerland AG
The registered company address is: Gewerbestrasse 11, 6330 Cham, Switzerland

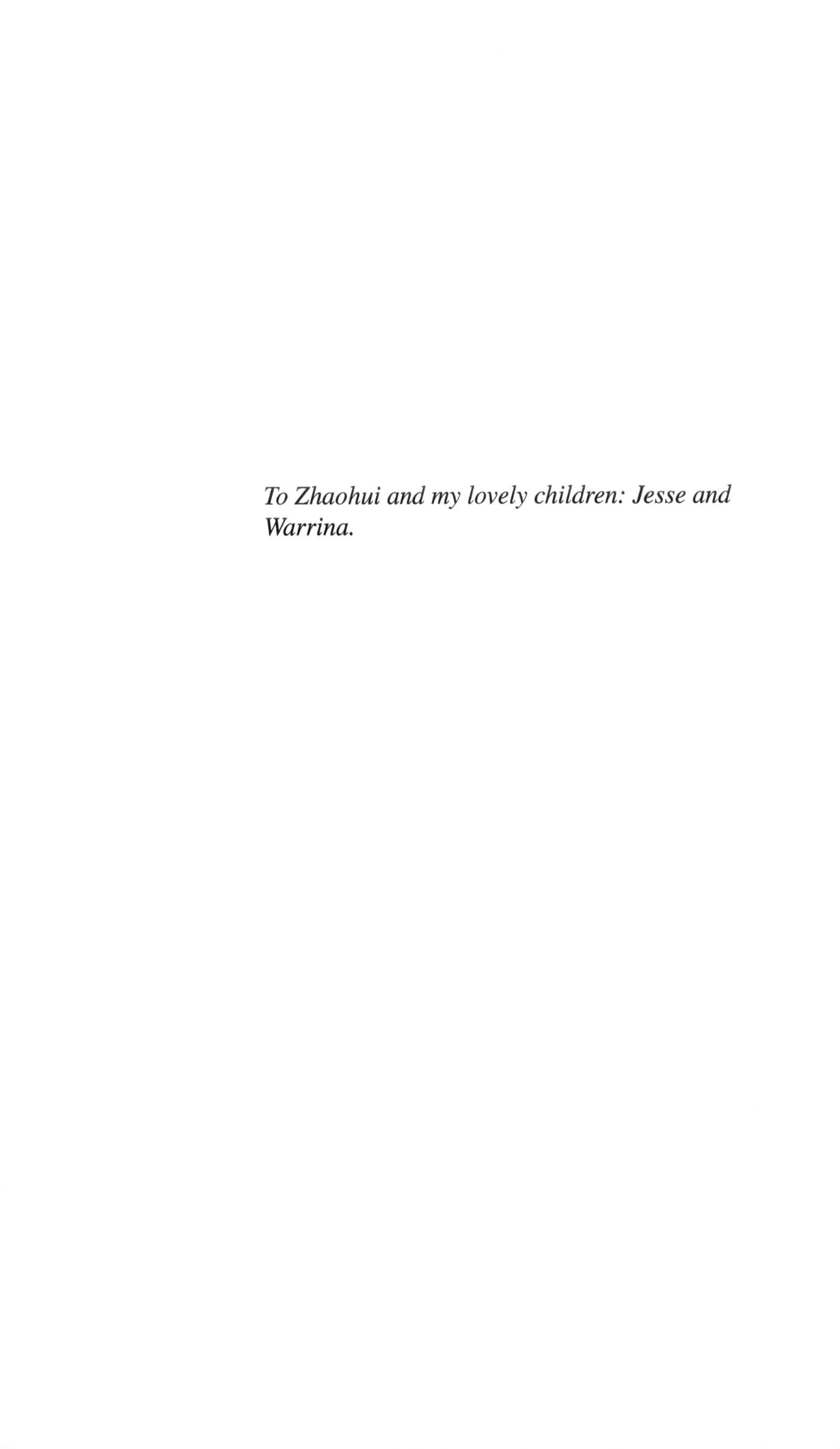

To Zhaohui and my lovely children: Jesse and Warrina.

Preface

Since its beginning, optimization has played a vital role in data science. The analysis and solution methods for many statistical and machine learning models rely on optimization. The recent surge of interest in optimization for computational data analysis also comes with a few significant challenges. The high problem dimensionality, large data volumes, inherent uncertainty, unavoidable nonconvexity, together with the increasing need to solve these problems in real time and sometimes under a distributed setting all contribute to create a considerable set of obstacles for existing optimization methodologies.

During the past 10 years or so, significant progresses have been made in the design and analysis of optimization algorithms to tackle some of these challenges. Nevertheless, they were scattered in a large body of literature across a few different disciplines. The lack of a systematic treatment for these progresses makes it more and more difficult for young researchers to step into this field, build up the necessary foundation, understand the current state of the art, and push forward the frontier of this exciting research area. In this book, I attempt to put some of these recent progresses into a slightly more organized manner. I mainly focus on the optimization algorithms that have been widely applied or may have the applied potential (from my perspective) to large-scale machine learning and data analysis. These include quite a few first-order methods, stochastic optimization methods, randomized and distributed methods, nonconvex stochastic optimization methods, projection-free methods, and operator sliding and decentralized methods. My goal is to introduce the basic algorithmic schemes that can provide the best performance guarantees under different settings. Before discussing these algorithms, I do provide a brief introduction to a few popular machine learning models to inspire the readers and also review some important optimization theory to equip the readers, especially the beginners, with a good theoretic foundation.

The target audience of this book includes the graduate students and senior undergraduate students who are interested in optimization methods and their applications in machine learning or machine intelligence. It can also be used as a reference book for more senior researchers. The initial draft of this book has been used as the text for a senior undergraduate class and a Ph.D. class here at Georgia Institute of Technology. For a one-semester senior undergraduate course, I would suggest to cover the following sections: 1.1, 1.2, 1.4, 1.5, 1.6, 1.7, 2.1, 2.2, 3.1, 3.2, 4.1, and 7.1 and encourage students to work on a course project. For a one-semester Ph.D. course, I would suggest to cover Sects. 1.1–1.7, 2.1–2.4, 3.1–3.6, 4.1–4.3, 5.1, 5.3, 5.4, 6.1–6.5, and 7.1–7.4 and encourage students to read and present those uncovered materials either in the book or from the literature.

Many of the materials that I selected to cover in this book originated from our research in the past few years. I am deeply indebted to my Ph.D. supervisors, former Ph.D. students, post-docs, and other collaborators. My foremost and sincerest appreciation will go to Arkadi Nemirovski, who guided me through different stages of my academic life and shaped me into whom I am now. Alex Shapiro provided much guidance to me regarding how to write the book and constantly reminded me of its status. Without his encouragement, I would probably have given up this effort. I am very thankful to Renato Monteiro for his kindness, support, and friendship. Working on this book often refreshes my pleasant memory of collaborating in this area with some very dedicated colleagues such as Yunmei Chen and Hongchao Zhang, and highly talented former students and post-docs including Cong Dang, Qi Deng, Saeed Ghadimi, Soomin Lee, Yuyuan Ouyang, Wei Zhang, and Yi Zhou. I am fortunate that my present students also work quite independently which helped me to spare some time to be dedicated to this book.

Atlanta, GA, USA
May 2019

Guanghui Lan

Contents

Chapter 1
Machine Learning Models

In this chapter, we introduce some widely used statistics and machine learning models in order to motivate our later discussion about optimization theory and algorithms.

1.1 Linear Regression

To motivate our discussion, let us start with a simple example. Julie needs to decide whether she should go to the restaurant "Bamboo Garden" for lunch or not. She went to ask for her friends Judy and Jim, who had been to this restaurant. Both of them gave a rating of 3 in the scale between 1 and 5 for the service in this restaurant. Given these ratings, it is a bit difficult for Julie to decide if she should pay a visit to "Bamboo Garden." Fortunately, she has kept a table of Judy and Jim's ratings for some other restaurants, as well as her own ratings in the past, as shown in Table 1.1.

Table 1.1 Historical ratings for the restaurants

Restaurant	Judy's rating	Jim's rating	Julie's ratings?
Goodfellas	1	5	2.5
Hakkasan	4.5	4	5
...	...	...	...
Bamboo Garden	3	3	?

To fix notation, let us use $u^{(i)}$ to denote the "input" variables (ratings of Judy and Jim in this example), also called input features, and $v^{(i)}$ to denote the "output" or target variable (rating of Julie's) to predict. A pair $(u^{(i)}, v^{(i)})$ is called a training example, and the dataset—a list of N training examples $\{(u^{(i)}, v^{(i)}\}$, $i = 1,\dots,N$—is called a training set. We will also use U to denote the space of input values, and V

G. Lan, *First-Order and Stochastic Optimization Methods for Machine Learning*, Springer Series in the Data Sciences, https://doi.org/10.1007/978-3-030-39568-1_1

the space of output values. In this example, $U = \mathbb{R}^2$ and $V = \mathbb{R}$. Specifically, $u_1^{(1)}$ and $u_2^{(1)}$ are the Judy and Jim's ratings for Goodfellas, respectively, and $v^{(1)}$ represents Julie's rating for Goodfellas.

Our goal is, given a training set, to learn a function $h : U \to V$ so that $h(u)$ is a "good" predictor for the corresponding value of v. This function h is usually called a hypothesis or decision function. Machine learning tasks of these types are called *supervised learning*. When the output v is continuous, we call the learning task *regression*. Otherwise, if v takes values on a discrete set of values, the learning task is called *classification*. Regression and classification are the two main tasks in supervised learning.

One simple idea is to approximate v by a linear function of u:

$$h(u) \equiv h_\theta(u) = \theta_0 + \theta_1 u_1 + \ldots + \theta_n u_n.$$

In our example, n simply equals 2. For notational convenience, we introduce the convention of $u_0 = 1$ so that

$$h(u) = \Sigma_{i=0}^n \theta_i u_i = \theta^T u,$$

where $\theta = (\theta_0; \ldots; \theta_n)$ and $u = (u_0; \ldots; u_n)$. In order to find the parameters $\theta \in \mathbb{R}^{n+1}$, we formulate an optimization problem of

$$\min_\theta \left\{ f(\theta) := \Sigma_{i=1}^N (h_\theta(u^{(i)}) - v^{(i)})^2 \right\}, \tag{1.1.1}$$

which gives rise to the ordinary least square regression model.

To derive a solution of θ for (1.1.1), let

$$U = \begin{bmatrix} {u^{(1)}}^T \\ {u^{(2)}}^T \\ \vdots \\ {u^{(N)}}^T \end{bmatrix}.$$

U is sometimes called the design matrix and it consists of all the input variables. Then, $f(\theta)$ can be written as:

$$\begin{aligned} f(\theta) &= \Sigma_{i=1}^N ({u^{(i)}}^T \theta - v^{(i)})^2 \\ &= (U\theta - v)^T (U\theta - v) \\ &= \theta^T U^T U \theta - 2\theta^T U^T v - v^T v. \end{aligned}$$

Taking the derivative of $f(\theta)$ and setting it to zero, we obtain the normal equation

$$U^T U \theta - U^T v = 0.$$

Thus the minimizer of (1.1.1) is given by

$$\theta^* = (U^T U)^{-1} U^T v.$$

The ordinary least square regression is among very few machine learning models that has an explicit solution. Note, however, that to compute θ^*, one needs to compute the inverse of an $(n+1)\times(n+1)$ matrix $(U^T U)$. If the dimension of n is big, computing the inverse of a large matrix can still be computationally expensive.

The formulation of the optimization problem in (1.1.1) follows a rather intuitive approach. In the sequel, we provide some statistical reasoning about this formulation. Let us denote

$$\varepsilon^{(i)} = v^{(i)} - \theta^T u^{(i)}, i = 1, \ldots, N. \tag{1.1.2}$$

In other words, $\varepsilon^{(i)}$ denotes the error associated with approximating $v^{(i)}$ by $\theta^T u^{(i)}$. Moreover, assume that $\varepsilon^{(i)}$, $i = 1, \ldots, N$, are i.i.d. (independently and identically distributed) according to a Gaussian (or Normal) distribution with mean 0 and variance σ^2. Then, the density of $\varepsilon^{(i)}$ is then given by

$$p(\varepsilon^{(i)}) = \tfrac{1}{\sqrt{2\pi}\sigma} \exp\left(-\tfrac{(\varepsilon^{(i)})^2}{2\sigma^2}\right).$$

Using (1.1.2) in the above equation, we have

$$p(v^{(i)}|u^{(i)};\theta) = \tfrac{1}{\sqrt{2\pi}\sigma} \exp\left(-\tfrac{(v^{(i)} - \theta^T u^{(i)})^2}{2\sigma^2}\right). \tag{1.1.3}$$

Here, $p(v^{(i)}|u^{(i)};\theta)$ denotes the distribution of the output $v^{(i)}$ given input $u^{(i)}$ and parameterized by θ.

Given the input variables $u^{(i)}$ and output $v^{(i)}$, $i = 1, \ldots, N$, the likelihood function with respect to (w.r.t.) the parameters θ is defined as

$$L(\theta) := \prod_{i=1}^{N} p(v^{(i)}|u^{(i)};\theta) = \prod_{i=1}^{N} \tfrac{1}{\sqrt{2\pi}\sigma} \exp\left(-\tfrac{(v^{(i)} - \theta^T u^{(i)})^2}{2\sigma^2}\right).$$

The principle of *maximum likelihood* tells us that we should choose θ to maximize the likelihood $L(\theta)$, or equivalently, the *log likelihood*

$$\begin{aligned} l(\theta) &:= \log L(\theta) \\ &= \Sigma_{i=1}^{N} \log\left[\tfrac{1}{\sqrt{2\pi}\sigma} \exp\left(-\tfrac{(v^{(i)} - \theta^T u^{(i)})^2}{2\sigma^2}\right)\right] \\ &= N \log \tfrac{1}{\sqrt{2\pi}\sigma} - \tfrac{1}{2\sigma^2} \Sigma_{i=1}^{N} (v^{(i)} - \theta^T u^{(i)})^2. \end{aligned}$$

This is exactly the ordinary least square regression problem, i.e., to minimize $\Sigma_{i=1}^{N}(v^{(i)} - \theta^T u^{(i)})^2$ w.r.t. θ. The above reasoning tells us that under certain probabilistic assumptions, the ordinary least square regression is the same as maximum likelihood estimation. It should be noted, however, that the probabilistic assumptions are by no means necessary for least-squares to be a rational procedure for regression.

1.2 Logistic Regression

Let us come back to the previous example. Suppose that Julie only cares about whether she will like the restaurant "Bamboo Garden" or not, rather her own ratings. Moreover, she only recorded some historical data indicating whether she likes or dislikes some restaurants, as shown in Table 1.2. These records are also visualized in Fig. 1.1, where each restaurant is represented by a green "O" or a red "X," corresponding to whether Julie liked or disliked the restaurant, respectively. The question is: with the rating of 3 from both of her friends, will Julie like Bamboo Garden? Can she use the past data to come up with a reasonable decision?

Table 1.2 Historical ratings for the restaurants

Restaurant	Judy's rating	Jim's rating	Julie likes?
Goodfellas	1	5	No
Hakkasan	4.5	4	Yes
...	...	...	...
Bamboo Garden	3	3	?

Similar to the regression model, the input values are still denoted by $U = ({u^{(1)}}^T; \ldots; {u^{(N)}}^T)$, i.e., the ratings given by Judy and Jim. But the output values are now binary, i.e., $v^{(i)} \in \{0, 1\}$, $i = 1, \ldots, N$. Here $v^{(i)} = 1$ means that Julie likes the i-th restaurant and $v^{(i)} = 0$ means that she dislikes the restaurant. Julie's goal is to come up with a decision function $h(u)$ to approximate these binary variables v. This type of machine learning task is called *binary classification.*

Julie's decision function can be as simple as a weighted linear combination of her friends' ratings:

$$h_\theta(u) = \theta_0 + \theta_1 \, u_1 + \ldots + \theta_n u_n \tag{1.2.1}$$

with $n = 2$. One obvious problem with the decision function in (1.2.1) is that its values can be arbitrarily large or small. On the other hand, Julie wishes its values to fall between 0 and 1 because those represent the range of v. A simple way to force h to fall within 0 and 1 is to map the linear decision function $\theta^T u$ through another function called the sigmoid (or logistic) function

$$g(z) = \frac{1}{1+\exp(-z)} \tag{1.2.2}$$

and define the decision function as

$$h_\theta(u) = g(\theta^T u) = \frac{1}{1+\exp(-\theta^T u)}. \tag{1.2.3}$$

Note that the range of the sigmoid function is given by $(0, 1)$, as shown in Fig. 1.2.

Now the question is how to determine the parameters θ for the decision function in (1.2.3). We have seen the derivation of the ordinary least square regression model as the consequence of maximum likelihood estimation under certain probabilistic assumptions. We will follow a similar approach for the classification problem.

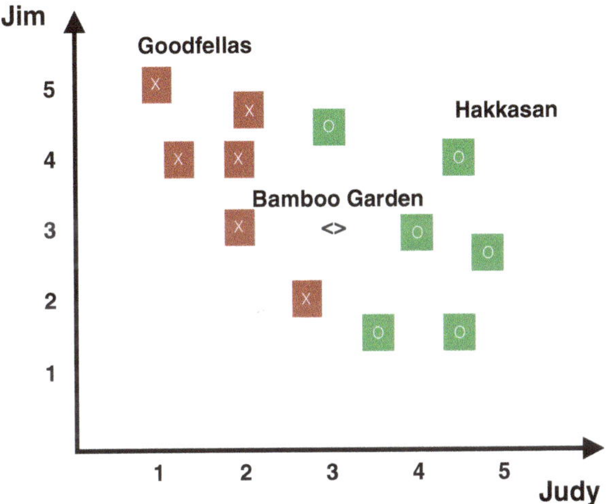

Fig. 1.1 Visualizing ratings of the restaurants

We assume that $v^{(i)}$, $i = 1, \ldots, N$, are independent Bernoulli random variables with success probability (or mean) of $h_\theta(u^{(i)})$. Thus their probability mass functions are given by

$$p(v^{(i)}|u^{(i)};\theta) = [h_\theta(u^{(i)})]^{v^{(i)}} [1 - h_\theta(u^{(i)})]^{1-v^{(i)}}, v^{(i)} \in \{0,1\},$$

and the associated likelihood function $L(\theta)$ is defined as

$$L(\theta) = \prod_{i=1}^{N} \left\{ [h_\theta(u^{(i)})]^{v^{(i)}} [1 - h_\theta(u^{(i)})]^{1-v^{(i)}} \right\}.$$

In view of the principle of maximum likelihood, we intend to maximize $L(\theta)$, or equivalently, the log likelihood

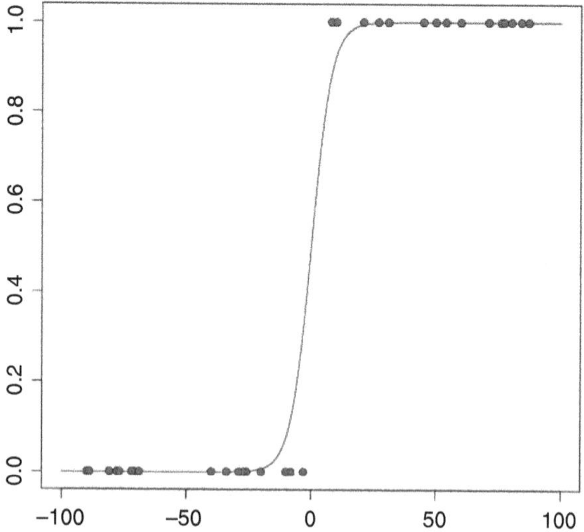

Fig. 1.2 The sigmoid (logistic) function

$$
\begin{aligned}
l(\theta) &= \Sigma_{i=1}^N \log\left\{[h_\theta(u^{(i)})]^{v^{(i)}}\,[1-h_\theta(u^{(i)})]^{1-v^{(i)}}\right\}\\
&= \Sigma_{i=1}^N \left\{v^{(i)}\log h_\theta(u^{(i)}) + [1-v^{(i)}]\log[1-h_\theta(u^{(i)})]\right\}.
\end{aligned}
$$

Accordingly, we formulate an optimization problem of

$$
\max_\theta \Sigma_{i=1}^N \left\{-\log[1+\exp(-\theta^T u^{(i)})] - [1-v^{(i)}]\theta^T u^{(i)}\right\}. \tag{1.2.4}
$$

Even though this model is used for binary classification, it is often called logistic regression for historical reasons.

Unlike linear regression, (1.2.4) does not have an explicit solution. Instead, we need to develop some numerical procedures to find its approximate solutions. These procedures are called optimization algorithms, a subject to be studied intensively later in our lectures.

Suppose that Julie can solve the above problem and find at least one of its optimal solutions θ^*. She then obtained a decision function $h_{\theta^*}(u)$ which can be used to predict whether she likes a new restaurant (say "Bamboo Garden") or not. More specifically, recall that the example corresponding to "Bamboo Garden" is $u = (1,3,3)$ (recall $u_1 = 1$). If $h_{\theta^*}((1,3,3)) > 0.5$, then Julie thinks she will like the restaurant, otherwise she will not. The values of u's that cause $h_{\theta^*}(u)$ to be 0.5 are called the "decision boundary" as shown in Fig. 1.3. The black line is the "decision boundary." Any point lying above the decision boundary represents a restaurant that Julie likes, while any point lying below the decision boundary is a restaurant that she does not

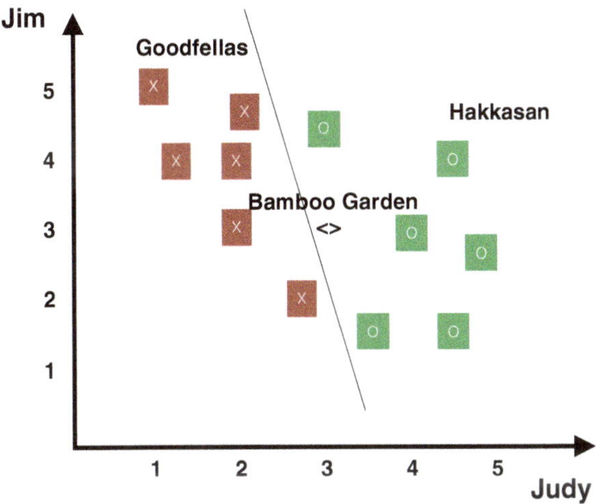

Fig. 1.3 Decision boundary

like. With this decision boundary, it seems that Bamboo Garden is slightly on the positive side, which means she may like this restaurant.

1.3 Generalized Linear Models

In the previous two sections, we have introduced two supervised machine learning models: the ordinary least square regression and the logistic regression. In the former model, we had assumed that $v|u;\theta \sim \mathcal{N}(\mu,\sigma^2)$, and in the latter model, $v|u;\theta \sim \text{Bernoulli(q)}$, for some appropriate definitions of μ and q as functions of u and θ. We will show in this section that both of them are certain special cases of a broader family of models, called Generalized Linear Models (GLMs), which essentially apply maximum likelihood estimation for exponential distribution family.

1.3.1 Exponential Family

An exponential family is a set of probability distributions given in the form of

$$p(v;\eta) = b(v)\exp(\eta^T T(v) - a(\eta)) \tag{1.3.1}$$

for a fixed choice of T, a, and b. This family is parameterized by η in the sense that different distributions can be obtained by varying η.

Let us first check that the normal distribution can indeed be written in the form of (1.3.1). For simplicity, we consider random variable $v \sim \mathcal{N}(\mu, 1)$, i.e.,

$$\begin{aligned} p(v;\mu) &= \tfrac{1}{\sqrt{2\pi}} \exp\left(-\tfrac{(v-\mu)^2}{2}\right) \\ &= \tfrac{1}{\sqrt{2\pi}} \exp\left(-\tfrac{v^2}{2}\right) \exp\left(\mu^T v - \tfrac{\mu^2}{2}\right). \end{aligned}$$

Clearly, $p(v;\mu)$ is a special case of (1.3.1) with $\eta = \mu$,

$$b(v) = \tfrac{1}{\sqrt{2\pi}} \exp\left(-\tfrac{v^2}{2}\right), T(v) = v \text{ and } a(\eta) = \tfrac{\mu^2}{2} = \tfrac{\eta^2}{2}.$$

To check that Bernoulli is a special exponential distribution, we first rewrite its density function

$$\begin{aligned} p(v;q) &= q^v (1-q)^{1-v} \\ &= \exp\left(v \log q + (1-v)\log(1-q)\right) \\ &= \exp\left(v \log \tfrac{q}{1-q} + \log(1-q)\right). \end{aligned}$$

Clearly, in view of (1.3.1), we have

$$\eta = \log \tfrac{q}{1-q}, b(v) = 1, T(v) = v \text{ and } a(\eta) = -\log(1-q).$$

It is interesting to note that by the first identity, we have

$$q = \tfrac{1}{1+\exp(-\eta)},$$

which gives exactly the logistic (sigmoid) function.

The exponential family covers a broad class of distributions, including normal, exponential, gamma, chi-squared, beta, Dirichlet, Bernoulli, categorical, Poisson, Wishart and Inverse Wishart distributions, etc.

1.3.2 Model Construction

Following how we construct the ordinary least square and logistic regression models, we can summarize the basic elements of the GLM model as follows.

- We need to know the distribution of the response (output) v, given input u and parameters θ to be estimated. More specifically, we assume that $v|u;\theta$ satisfies a family of exponential distribution parameterized by η.
- Given u, we need to construct a decision function (or hypothesis) $h(u)$ to predict the outcome $T(v)$ (in most cases $T(v) = v$ as in the ordinary least square and logistic regression). $T(v)$ is random and we expect that $h(u) = \mathbb{E}[T(v)|u]$. For

example, in logistic regression, we have chosen h in a way such that $h_\theta(u) = \mathbb{E}[v|u]$.
- We assume that η linearly depends on the input values u, i.e., $\eta = \theta^T u$. If η is a vector, we assume that $\eta_i = \theta_i^T u$.

While the first two elements are assumptions we make, the last element is more related to the design of the model. In particular, with this type of design, the resulting models are most likely easier to fit.

Let us check these elements have indeed been used in our development for the ordinary least square model. Recall we assume that $\varepsilon = v - \theta^T u$ is normally distributed according to $\mathcal{N}(0,\sigma^2)$. Hence, $v|u;\theta \sim \mathcal{N}(\eta,\sigma^2)$ with $\eta = \theta^T u$. This implies that all the above three elements hold since $v|u;\theta$ is normally distributed, $\eta = \mathbb{E}[v|u;\theta]$ and $h(u) = \theta^T u = \eta$.

Next we can check that these elements also hold for the logistic regression. Recall that we assume that $v|u;\theta$ satisfies a family of Bernoulli distribution with mean

$$q = h(u) = \tfrac{1}{1+\exp(-\theta^T u)}.$$

Denoting $\eta = \theta^T u$, and using it in the above identity, we obtain

$$q = \tfrac{1}{1+\exp(-\eta)},$$

or equivalently,

$$\eta = \log \tfrac{q}{1-q},$$

which is exactly the parameter we used to write Bernoulli distribution in the form of exponential family. These discussions imply that all the aforementioned elements of GLM hold.

Let us look at one more example of a GLM. Consider a classification problem in which the response variable v can take on any one of the k values, i.e., $v \in \{1,2,\ldots,k\}$. The response variable is still discrete, but can now take on more than two values. This type of machine learning task is called *multi-class classification.*

In order to derive a GLM for modeling this type of task, we first assume that v is distributed according to a multinomial distribution and show that multinomial is an exponential family distribution.

To parameterize a multinomial over k possible outcomes, we could use k parameters $q_1,\ldots,q_k$ to specify the probability of each outcome. However, these parameters would not be independent because any $k-1$ of the q_i's uniquely determines the last one due to the fact that $\sum_{i=1}^k q_i = 1$. So, we will instead parameterize the multinomial with only $k-1$ parameters, $q_1,\ldots,q_{k-1}$, where $q_i = p(y=i;q)$. For notational convenience, we also let $q_k = p(y=k;q) = 1-\sum_{i=1}^{k-1} q_i$.

To show that multinomial is an exponential distribution, let us define $T(v) \in \mathbb{R}^{k-1}$ as follows:

$$T(1) = \begin{bmatrix} 1 \\ 0 \\ 0 \\ \vdots \\ 0 \end{bmatrix}, T(2) = \begin{bmatrix} 0 \\ 1 \\ 0 \\ \vdots \\ 0 \end{bmatrix}, \ldots, T(k-1) = \begin{bmatrix} 0 \\ 0 \\ 0 \\ \vdots \\ k-1 \end{bmatrix}, T(k) = \begin{bmatrix} 0 \\ 0 \\ 0 \\ \vdots \\ 0 \end{bmatrix}.$$

Unlike our previous examples, here we do not have $T(v) = v$. Moreover, $T(v)$ is now in $\mathbb{R}^{k-1}$, instead of a real number. We will write $(T(v))_i$ to denote the i-th element of the vector $T(v)$.

We introduce one more useful piece of notation. An indicator function $I\{\cdot\}$ takes on a value of 1 if its argument is true, and 0 otherwise. So, we can also write the relationship between $T(v)$ and v as $(T(v))_i = I\{v = i\}$. Further, we have that $\mathbb{E}[(T(v))_i] = p(v = i) = q_i$.

We are now ready to show that multinomial distribution is a member of the exponential family. We have:

$$\begin{aligned} p(v;q) &= q_1^{I\{v=1\}} q_2^{I\{v=2\}} \ldots q_k^{I\{v=k\}} \\ &= q_1^{I\{v=1\}} q_2^{I\{v=2\}} \ldots q_k^{1-\sum_{i=1}^{k-1} I\{v=i\}} \\ &= q_1^{(T(v))_1} q_2^{(T(v))_2} \ldots q_k^{1-\sum_{i=1}^{k-1}(T(v))_i} \\ &= \exp\left(\sum_{i=1}^{k-1}(T(v))_i \log q_i + (1 - \sum_{i=1}^{k-1}(T(v))_i) \log q_k\right) \\ &= \exp\left(\sum_{=1}^{k-1}(T(v))_i \log \tfrac{q_i}{q_k} + \log q_k\right). \end{aligned}$$

This is an exponential distribution with

$$\eta_i = \log \tfrac{q_i}{q_k}, i = 1, \ldots, k-1,\ a(\eta) = -\log q_k, \text{ and } b = 1. \tag{1.3.2}$$

In order to define the decision function, we first represent q_i's in terms of η_i's, since $\mathbb{E}[(T(y))_i] = p(y = i) = q_i$ and we would like $h_i(u) = q_i$. By (1.3.2), we have

$$\tfrac{q_i}{q_k} = \exp(\eta_i), i = 1, \ldots, k-1. \tag{1.3.3}$$

For convenience, let us also denote $\eta_k = 0$ and

$$\tfrac{q_k}{q_k} = \exp(\eta_k).$$

Summing up these identities and using the fact that $\sum_{i=1}^k q_i = 1$, we have $\frac{1}{q_k} = \sum_{i=1}^k \exp(\eta_i)$, and hence

$$q_i = \frac{\exp(\eta_i)}{\sum_{i=1}^k \exp(\eta_i)}, i = 1, \ldots, k-1. \tag{1.3.4}$$

To finish the definition of the decision function, we assume $h_i(u) = q_i$ and set $\eta_i = \theta_i^T u$. Using these two relations together with (1.3.4), we obtain

$$h_i(u) = \frac{\exp(\theta_i^T u)}{\sum_{i=1}^k \exp(\theta_i^T u)}, i = 1, \ldots, k-1.$$

Finally, these parameters θ_i, $i = 1, \ldots, k-1$, used in the definition of $h_i(u)$ can be estimated by maximizing the log likelihood

$$l(\theta) = \sum_{i=1}^N \log p(v^{(i)}|u^{(i)};\theta) = \sum_{i=1}^N \log \prod_{j=1}^k \left(\frac{\exp(\eta_j)}{\sum_{j=1}^k \exp(\eta_j)}\right)^{I(v^{(i)}=j)}.$$

1.4 Support Vector Machines

In this section, we will provide a brief introduction to support vector machine, which has been considered as one of the most successful classification models.

Consider the binary classification problem with N training examples $(u^{(i)}, v^{(i)})$, $i = 1, \ldots, N$. For convenience, we assume throughout this section that the output $v^{(i)}$ is given by either 1 or -1, i.e., $v^{(i)} \in \{-1, 1\}$, rather than $v^{(i)} \in \{0, 1\}$ as in the previous sections. Observe that this is just a change of label for the class, but does not affect at all the fact that one particular example belongs to one class or the other.

We assume for this moment that these observed examples are separable. Formally speaking, we assume that there exists a linear function of u, denoted by

$$h_{w,b}(u) = b + w_1 u_1 + w_2 u_2 + \ldots w_n u_n,$$

such that for all $i = 1, \ldots, N$,

$$h_{w,b}(u^{(i)}) \begin{cases} > 0, & \text{if } v^{(i)} = 1, \\ < 0, & \text{Otherwise(i.e., v}^{(i)} = -1). \end{cases}$$

In particular, $h_{w,b}(u) = 0$ defines a hyperplane that separates the observed examples into two different classes. The examples that fall above the hyperplane are labeled by $v^{(i)} = 1$, while those below the hyperplane are labeled by $v^{(i)} = -1$. Notice that our notations for the decision function $h_{w,b}$ here also slightly differ from the previous sections. First of all, we get rid of u_0 which was assumed to be 1. Second, we denote the normal vector $w := (w_1, \ldots, w_n)$ and the intercept b by different notations, rather than a single $n+1$ vector $(\theta_0, \theta_1, \ldots, \theta_n)$. The main reason is that we will investigate the geometric meaning of the normal vector w for the separating hyperplane $h_{w,b}(u) = 0$.

By using logistic regression, we can possibly define a separating hyperplane. Recall that we use a decision function $h_\theta(u) = g(\theta^T u)$ to approximate the probability $p(y = 1|x;\theta)$. Given an u, we predict its output to be either 1 or 0 depending on whether $h_\theta(u) \geq 0.5$ or $h_\theta(u) < 0.5$, or equivalently, whether $\theta^T u \geq 0$ or $\theta^T u < 0$. Therefore, this vector θ gives rise to a possible separating hyperplane as denoted by $H1$ in Fig. 1.4.

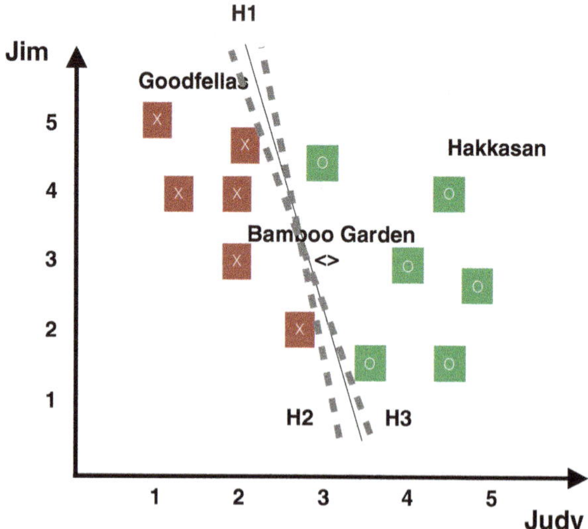

Fig. 1.4 Inspiration of SVM

However, there exist quite many other hyperplanes separating these observed examples, e.g., $H2$ and $H3$ as shown in Fig. 1.4. Given potentially an infinite number of separating hyperplanes, how should we evaluate their strength and thus choose the strongest separating hyperplane?

In order to answer this question, let us examine the so-called "margin" associated with a separating hyperplane $w^T u + b = 0$. For a given example $u^{(i)}$, i.e., the point $A = (u_1^{(i)}, \ldots, u_n^{(i)})$ in Fig. 1.5, we first compute its distance to the separating hyperplane. Let B be the projection of A to the separating hyperplane, it suffices to compute the length of the line segment $\overrightarrow{BA}$, denoted by $d^{(i)} = |\overrightarrow{BA}|$. Note that the unit direction of $\overrightarrow{BA}$ is given by $w/\|w\|$, and hence the coordinates of B are given by $u^{(i)} - d^{(i)} w/\|w\|$. Meanwhile, since B belongs to the separating hyperplane, we have

$$w^T \left(u^{(i)} - d^{(i)} \frac{w}{\|w\|} \right) + b = 0.$$

Solving the above equation for $d^{(i)}$, we have

$$d^{(i)} = \frac{w^T u^{(i)} + b}{\|w\|}. \tag{1.4.1}$$

In the above derivation, we have implicitly assumed that the point A sits above the separating hyperplane (i.e., $v^{(i)} = 1$). In case the point A sits below the hyperplane ($v^{(i)} = -1$), then the point B should be written as $u^{(i)} + d^{(i)} w/\|w\|$, and hence

$$d^{(i)} = -\frac{w^T u^{(i)} + b}{\|w\|}. \tag{1.4.2}$$

Putting (1.4.1) and (1.4.2) together, we can represent the distance $d^{(i)}$ by

$$d^{(i)} = \frac{v^{(i)}[w^T u^{(i)} + b]}{\|w\|} \tag{1.4.3}$$

for any $i = 1, \ldots, N$.

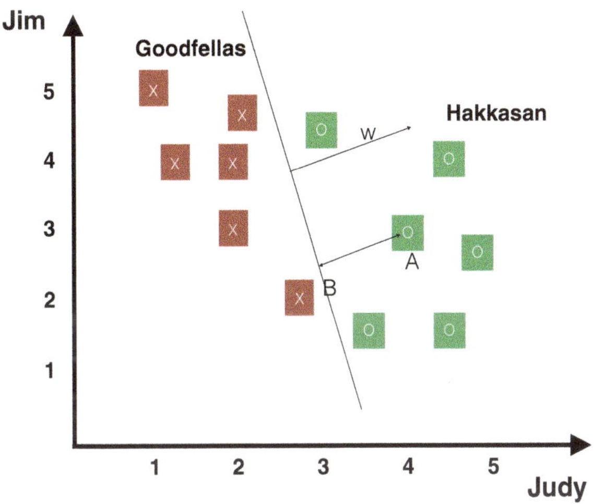

Fig. 1.5 Inspiration of SVM

With the above computation of $d^{(i)}$'s, we can now define the margin associated with the separating hyperplane $w^T u + b$ by

$$d(w,b) := \min_{i=1,\ldots,N} d^i \equiv \frac{\min_{i=1,\ldots,N} v^{(i)}[w^T u^{(i)} + b]}{\|w\|}. \tag{1.4.4}$$

The margin $d(w,b)$ provides a way to evaluate the strength of a separating hyperplane. Intuitively, a larger margin implies that the separating hyperplane can distinguish these two different classes of examples more significantly.

Therefore, a reasonable goal is to find (w,b) to maximize the margin $d(w,b)$, i.e.,

$$\max_{w,b} \frac{\min_{i=1,\ldots,N} v^{(i)}[w^T u^{(i)} + b]}{\|w\|}.$$

Specifically, this will result in a classifier that separates the positive and the negative training examples with a large "gap." The above optimization problem can be written equivalently as

$$\begin{array}{ll} \max_{w,b,r} & \frac{r}{\|w\|} \\ \text{s.t.} & v^{(i)}[w^T u^{(i)} + b] \geq r, i = 1, \ldots, N. \end{array}$$

For many reasons, most importantly for tractability, we wish the formulated optimization problem for machine learning to be convex (see definitions in Chap. 2). However, neither of these two formulations is convex as their objective functions are nonconvex. Fortunately, observing that multiplying w, b, and r by a scaling factor does not change the optimal value of the above problem, we can then assume $r = 1$ and reformulate it as

$$\begin{array}{ll} \max_{w,b} & \frac{1}{\|w\|} \\ \text{s.t.} & v^{(i)}[w^T u^{(i)} + b] \geq 1, i = 1, \ldots, N, \end{array} \tag{1.4.5}$$

or equivalently,

$$\begin{array}{ll} \min_{w,b} & \frac{1}{2}\|w\|^2 \\ \text{s.t.} & v^{(i)}[w^T u^{(i)} + b] \geq 1, i = 1, \ldots, N. \end{array}$$

The latter is a convex optimization problem as we will see shortly in Chap. 2.

Now we should provide some explanation regarding why the above model is called *support vector machine*. Suppose that we have a very large number of examples, i.e., N is very large. Once we solve (1.4.5) to optimality by identifying the optimal (w^*, b^*), we will find out that only a small number (out of N) of constraints of (1.4.5) are active at (w^*, b^*), i.e., only a small number of constraints are satisfied with equality. The corresponding $u^{(i)}$'s are then called *support vectors*. Geometrically, the support vectors gave the shortest distance to the optimal separating hyperplane $(w^*)^T u + b^* = 0$ among all training examples. If we move the optimal separating hyperplane along the direction of $w^*/\|w^*\|$ and stop until some vectors from the training examples are encountered, we will find the first set of support vectors. Similarly, moving the optimal separating hyperplane along $-w^*/\|w^*\|$, we will obtain another set of support vectors. These two set of support vectors resides on two hyperplanes parallel to the optimal separating hyperplane. They define the gap between these two different classes of training examples, which is exactly twice the objective value of (1.4.5).

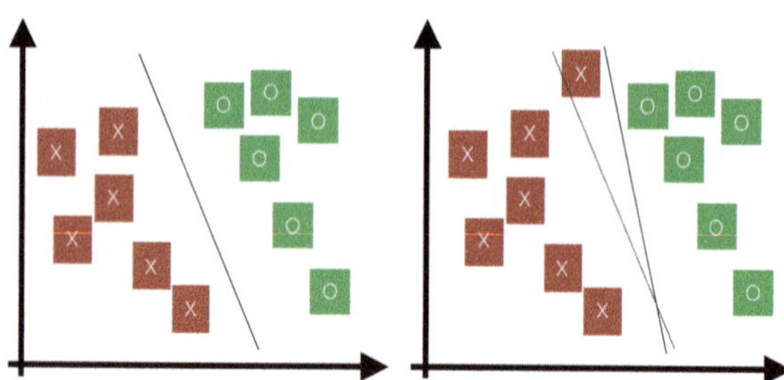

Fig. 1.6 SVM outliners

The derivation of the SVM as presented so far assumed that the training examples are linearly separable. However, this might not be the case in practice. Moreover, in some cases it is not clear that finding a separating hyperplane is exactly what we want, since that might be susceptible to outliers. To see this, Fig. 1.6a shows an optimal margin classifier. However, when a single outlier is added in Fig. 1.6b, the optimal separating hyperplane has to make a dramatic swing, leading to a much smaller margin for the resulting classifier.

In order to address these issues, we reformulate (1.4.5) as

$$\begin{array}{ll} \min_{w,b} & \frac{1}{2}\|w\|^2 + \lambda \sum_{i=1}^N \xi_i \\ \text{s.t.} & v^{(i)}[w^T u^{(i)} + b] \geq 1 - \xi_i, i = 1, \ldots, N, \\ & \xi_i \geq 0, i = 1, \ldots, N, \end{array} \tag{1.4.6}$$

for some $\lambda > 0$. In the above formulation, we allow the constraints in (1.4.5) to be violated and then penalize the total amount of violations. Observe that problem (1.4.6) can be written equivalently as

$$\min_{w,b} \frac{1}{2}\|w\|^2 + \lambda \sum_{i=1}^N \max\{0, 1 - v^{(i)}[w^T u^{(i)} + b]\}. \tag{1.4.7}$$

These formulations are called *soft-margin support vector machine.*

1.5 Regularization, Lasso, and Ridge Regression

Many supervised machine learning models, including a few problems discussed above, can be written in the following form:

$$f^* := \min_{x \in \mathbb{R}^n} \left\{ f(x) := \sum_{i=1}^N L(x^T u_i, v_i) + \lambda r(x) \right\}, \tag{1.5.1}$$

for some $\lambda \geq 0$, where $L(\cdot,\cdot)$ and $r(\cdot)$ are called the loss and regularization functions, respectively.

For instance, in SVM, we have $x = (w,b)$, $L(z,v) = \max\{0, 1 - vz\}$ and $r(x) = \|w\|^2$. In the ordinary least square regression, we have $x = \theta$, $L(z,v) = (z-v)^2$, and $r(\theta) = 0$. In fact, we can add many different types of regularization to the square loss function to avoid over-fitting, reduce variance of the prediction error, and handle correlated predictors.

The two most commonly used penalized models are ridge regression and Lasso regression. Ridge regression is given in the form of (1.5.1) with $L(z,v) = (z-v)^2$ and $r(x) = \|x\|_2^2$, while Lasso regression is represented by (1.5.1) with $L(z,v) = (z-v)^2$ and $r(x) = \|x\|_1$. Here $\|x\|_2^2 = \sum_{i=1}^n x_i^2$ and $\|x\|_1 = \sum_{i=1}^n |x_i|$. The elastic net combines both l_1 and l_2 penalty and is defined as

$$\lambda P_\alpha(\beta) = \lambda \left(\alpha \|\beta\|_1 + \tfrac{1}{2}(1-\alpha)\|\beta\|_2^2 \right).$$

Lasso regression leads to a sparse solution when the tuning parameter is sufficiently large. As the tuning parameter value λ is increased, all coefficients are set to zero. Since reducing parameters to zero removes them from the model, Lasso is a good selection tool. Ridge regression penalizes the l_2 norm of the model coefficients. It provides greater numerical stability and is easier and faster to compute than Lasso. Ridge regression reduces coefficient values simultaneously as the penalty is increased without however setting any of them to zero.

Variable selection is important in numerous modern applications with many features where the l_1 penalty has proven to be successful. Therefore, if the number of variables is large or if the solution is known to be sparse, we recommend using Lasso, which will select a small number of variables for sufficiently high λ that could be crucial to the interpretability of the model. The l_2 penalty does not have this effect: it shrinks the coefficients, but does not set them exactly to zero.

The two penalties also differ in the presence of correlated predictors. The l_2 penalty shrinks coefficients for correlated columns towards each other, while the l_1 penalty tends to select only one of them and set the other coefficients to zero. Using the elastic net argument α combines these two behaviors. The elastic net both selects variables and preserves the grouping effect (shrinking coefficients of correlated columns together). Moreover, while the number of predictors that can enter a Lasso model saturates at $\min(N,n)$, where N is the number of observations and n is the number of variables in the model, the elastic net does not have this limitation and can fit models with a larger number of predictors.

1.6 Population Risk Minimization

Let us come back to our motivating example. When Julie examines her decision process, she observes that her true objective was to design the decision function to predict her judgment about the restaurant "Bamboo Garden," rather than just fit the historical data she collected. In other words, the problem she intended to solve was not really the one in (1.1.1). Rather, her problem could be better formulated as

$$\min_{w,b} \mathbb{E}_{u,v}[h(u;w,b)-v]^2, \tag{1.6.1}$$

where u and v are random variables denoting Judy's and Jim's ratings, and her own judgment for a restaurant. During the process, she has implicitly assumed that the optimal solution of (1.1.1) will be a good approximate solution to (1.6.1).

In fact, Julie's intuition can be proved more rigorously. In stochastic optimization, (1.1.1) is called the sample average approximation (SAA) problem of (1.6.1). It can be shown that as N increases, an optimal solution of (1.1.1) will approximately solve problem (1.6.1). It is worth noting that in machine learning, problems (1.6.1) and (1.1.1) are called *population* and *empirical risk minimization*, respectively.

A few problems remain. First, how should we solve problem (1.1.1) efficiently if both the dimension of examples and the sample size N are very large? Second, should we really need to solve the empirical risk minimization problem? Why should not we design an algorithm to solve the population risk minimization problem directly? These are the problems that we will deal with mostly in this book.

1.7 Neural Networks

In the past few years, deep learning has generated much excitement in machine learning, especially in industry, due to many breakthrough results in speech recognition, computer vision, and text processing. For many researchers, deep learning is another name for a set of algorithms that use a neural network as an architecture. Even though neural networks have a long history, they became more successful in recent years due to the availability of inexpensive, parallel hardware (GPUs, computer clusters), massive amounts of data, and the recent development of efficient optimization algorithms, especially those designed for population risk minimization. In this section, we will start with the concept of a linear classifier and use that to develop the concept of neural networks.

Let us continue our discussion about the restaurant example. In the previous case, Julie was lucky because the examples are linearly separable which means that she can draw a linear decision function to separate the positive from negative instances. Her friend Jenny has different food tastes. If we plot Jenny's data, the graph will look rather different (see Fig. 1.7). Jenny likes some of the restaurants that Judy and Jim rated poor. The question is how we can come up with a decision function for Jenny. By looking at the data, the decision function must be more complex than the decision we saw before. Our heuristic approach to solve a complex problem is to

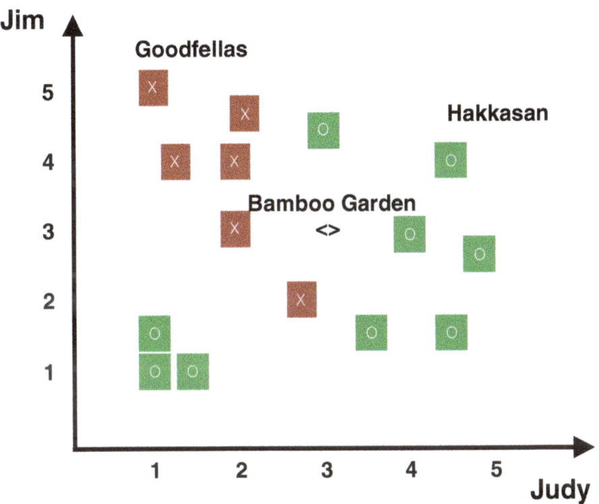

Fig. 1.7 Visualizing ratings of the restaurants without linear separability

decompose it into smaller problems that we can solve. For our particular situation, we know that if we throw away the "weird" examples from the bottom left corner of the figure, then the problem becomes simpler. Similarly, if we throw the "weird" examples on the top right figure, the problem is again also simpler. We solve for each case using our presumed stochastic optimization algorithm and the decision functions look like Fig. 1.8.

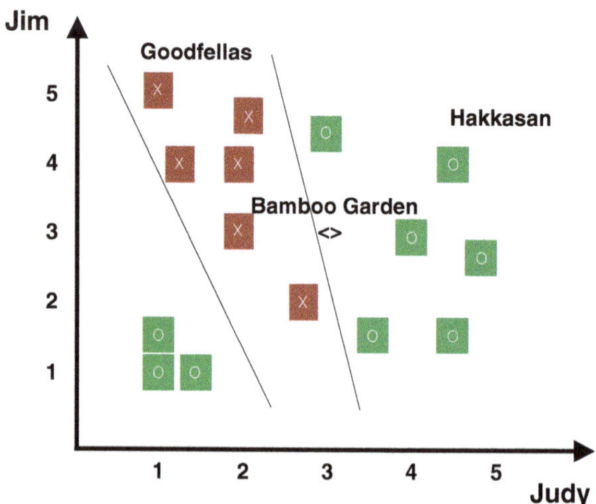

Fig. 1.8 Visualizing ratings of the restaurants without linear separability

Is it possible to combine these two decision functions into one final decision function for the original data? Let us suppose, as stated above, that the two decision functions are $h_1(u) := g(w_1u_1 + w_2u_2 + b_1)$ and $h_2(u) := g(w_3u_1 + w_4u_2 + b_2)$, where g denotes the sigmoid function defined in (1.2.2). For every example $u^{(i)}$, we can then compute $h_1(u^{(i)})$ and $h_2(u^{(i)})$, respectively. Table 1.3 lays out the data associated with these decision functions.

Table 1.3 Two separate linear decision functions

Restaurant	Judy's rating	Jim's rating	Jenny likes?
Goodfellas	$h_1(u^{(1)})$	$h_2(u^{(1)})$	No
Hakkasan	$h_1(u^{(2)})$	$h_2(u^{(2)})$	Yes
...	...	...	...
Bamboo Garden	$h_1(u^{(n+1)})$	$h_2(u^{(n+1)})$	?

Now the problem reduces to find a new parameter set to weight these two decision functions to approximate v. Let's call these parameters ω, c, and we intend to find them such that $h(u; w, b, \omega, c) := g(\omega_1 h_1(u) + \omega_2 h_2(u) + c)$ can approximate the label v. This can be formulated, again, as an optimization problem. In summary, we can find the decision function for Jenny by using the following two steps:

(a) Partition the data into two sets. Each set can be simply classified by a linear decision. Then use the previous sections to find the decision function for each set;

(b) Use the newly found decision functions and compute the decision values for each example. Then treat these values as input to another decision function. Use optimization methods to find the final decision function.

Figure 1.9 provides a graphical way to visualize the above process.

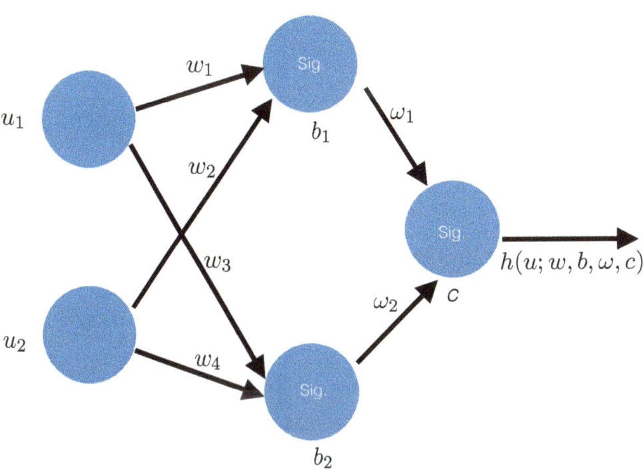

Fig. 1.9 A simple neural network

What we have just discussed is a special architecture in machine learning known as "neural networks." This instance of neural networks has one hidden layer, which has two "neurons." The first neuron computes values for function h_1 and the second neuron computes values for function h_2. The sigmoid function that maps real value to bounded values between 0 and 1 is also known as "the nonlinearity" or the "activation function." Since we are using sigmoid, the activation function is also called "sigmoid activation function." There exist many other types of activation functions. The parameters inside the network, such as w, ω are called "weights" whereas b, c are called "biases." If we have a more complex function to approximate, we may need to have a deeper network, i.e., one with more hidden layers and each layer with more than two neurons.

Let us get back to our problem of finding a good decision function for Jenny. It seems that in the above steps, we cheated a little bit when we divided the dataset into two sets because we looked at the data and decided that the two sets should be partitioned that way. Is there any way that such a step can be automated? It turns out that the natural way is to find the parameters ω, c, w, and b all at once on the complex dataset rather than doing the aforementioned two steps sequentially. To see this more clearly, let us write down how we will compute $h(u)$:

$$\begin{aligned} h(u) &= g(\omega_1 h_1(u; w_1, w_2, b_1) + \omega_2 h_2(u, w_3, w_4, b_2) + c) \\ &= g(\omega_1 g(w_1 u_1 + w_2 u_2 + b_1) + \omega_2 g(w_3 u_1 + w_4 u_2 + b_2) + c). \end{aligned}$$

We will find all these parameters $\omega_1, \omega_2, c, w_1, w_2, w_3, w_4, b_1$ and b_2 at the same time by solving

$$\min_{\omega_1,\omega_2,c,w_1,w_2,w_3,w_4,b_1,b_2} \mathbb{E}_{u,v}[(h(u)-v)^2].$$

This problem turns out to be a nonconvex stochastic optimization problem. The big question remains: how we solve this problem efficiently and whether we can provide any guarantee to our solution procedure?

1.8 Exercises and Notes

Exercises.

1. Consider the following loss function for logistic regression:

$$l(\theta) = \Sigma_{i=1}^{N} \left\{ -\log[1+\exp(-\theta^T u^{(i)})] - [1-v^{(i)}]\theta^T u^{(i)} \right\}.$$

Find the Hessian H for this function and show that l is a concave function.

2. Consider the Poisson distribution parameterized by λ:

$$p(v;\lambda) = \frac{e^{-\lambda}\lambda^y}{y!}.$$

Please show that the Poisson distribution is in the exponential family. If you would like to design a generalized linear model, what would be the decision function? How would you formulate the maximum log-likelihood for a given set of training examples?

3. Let u and v be given and denote $x \equiv (w_1, w_2, b_1, w_3, w_4, b_2, \omega_1, \omega_2, c)$. Also let us denote

$$\begin{aligned} h(u;x) &:= g(\omega_1 h_1(u;w_1,w_2,b_1) + \omega_2 h_2(u,w_3,w_4,b_2) + c) \\ &= g(\omega_1 g(w_1 u_1 + w_2 u_2 + b_1) + \omega_2 g(w_3 u_1 + w_4 u_2 + b_2) + c), \end{aligned}$$

where

$$g(z) = \frac{1}{1+\exp(-z)}.$$

and define $f(x) = [h(u;x) - v]^2$

(a) Compute the Hessian of f and show that f is not necessarily convex.
(b) Compute the gradient of f with respect to x.
(c) Discuss how to evaluate the gradient of f efficiently in computer.
(d) Derive the conditions under which the gradients of f are Lipschitz continuous, i.e.,

$$\|\nabla f(x_1) - \nabla f(x_2)\| \le L\|x_1 - x_2\|, \ \forall x_1, x_2 \in \mathbb{R}^9,$$

for some $L > 0$.

Notes. Further reading on statistical learning models and deep learning architectures can be found in [46] and [8], respectively. Some recent online course materials for machine learning can be found, e.g., in [30].

Chapter 2
Convex Optimization Theory

Many machine learning tasks can be formulated as an optimization problem given in the form of

$$\min_{x \in X} f(x), \tag{2.0.1}$$

where f, x, and X denote the objective function, decision variables, and feasible set, respectively. Unfortunately, solving an optimization problem is challenging. In general, we cannot guarantee whether one can find an optimal solution, and if so, how much computational effort one needs. However, it turns out that we can provide such guarantees for a special but broad class optimization problems, namely convex optimization, where X is a convex set and f is a convex function. In fact, many machine learning models we formulated so far, such as least square linear regression, logistic regression, and support vector machine, are convex optimization problems.

Our goal in this chapter is to provide a brief introduction to the basic convex optimization theory, including convex sets, convex functions, strong duality, and KKT conditions. We will also briefly discuss some consequences of these theoretic results in machine learning, e.g., the representer theorem, Kernel trick, and dual support vector machine. We include proofs for some important results but the readers can choose to skip them for the first pass through the text.

2.1 Convex Sets

2.1.1 Definition and Examples

We begin with the definition of the notion of a convex set.

Definition 2.1. A set $X \subseteq \mathbb{R}^n$ is said to be convex if it contains all of its segments, that is

$$\lambda x + (1-\lambda)y \in X, \quad \forall (x,y,\lambda) \in X \times X \times [0,1].$$

G. Lan, *First-Order and Stochastic Optimization Methods for Machine Learning*, Springer Series in the Data Sciences, https://doi.org/10.1007/978-3-030-39568-1_2

Note that the point $\lambda x + (1-\lambda)y$ is called a convex combination of x and y. Figure 2.1 show the examples of a convex set (left) and a nonconvex set (right).

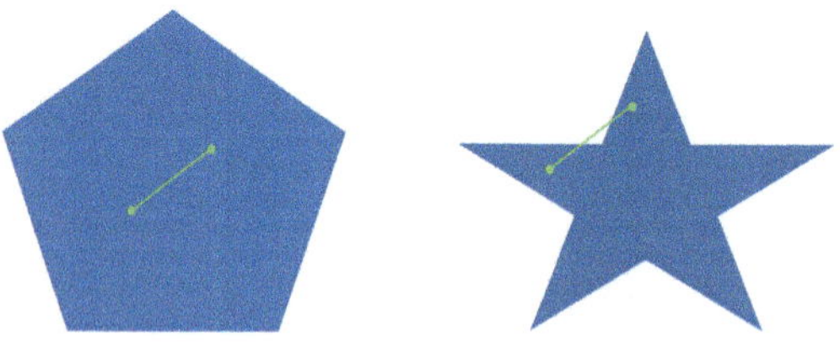

Fig. 2.1 Convex vs. nonconvex sets

It is easy to check that the following sets are convex.

(a) n-dimensional Euclidean space, $\mathbb{R}^n$. Given $x, y \in \mathbb{R}^n$, we must have $\lambda x + (1-\lambda)y \in \mathbb{R}^n$.

(b) Nonnegative orthant, $\mathbb{R}^n_+ := \{x \in \mathbb{R}^n : x_i \geq 0, i = 1, \ldots, n\}$. Let $x, y \in \mathbb{R}^n_+$ be given. Then for any $\lambda \in [0,1]$,

$$(\lambda x + (1-\lambda)y)_i = \lambda x_i + (1-\lambda)y_i \geq 0.$$

(c) Balls defined by an arbitrary norm, $\{x \in \mathbb{R}^n | \|x\| \leq 1\}$ (e.g., the l_2 norm $\|x\|_2 = \sqrt{\sum_{i=1}^n x_i^2}$ or l_1 norm $\|x\|_1 = \sum_{i=1}^n |x_i|$ balls). To show this set is convex, it suffices to apply the Triangular inequality and the positive homogeneity associated with a norm. Suppose that $\|x\| \leq 1, \|y\| \leq 1$ and $\lambda \in [0,1]$. Then

$$\|\lambda x + (1-\lambda)y\| \leq \|\lambda x\| + \|(1-\lambda)y\| = \lambda\|x\| + (1-\lambda)\|y\| \leq 1.$$

(d) Affine subspace, $\{x \in \mathbb{R}^n | Ax = b\}$. Suppose $x, y \in \mathbb{R}^n$, $Ax = b$, and $Ay = b$. Then

$$A(\lambda x + (1-\lambda)y) = \lambda Ax + (1-\lambda)Ay = b.$$

(e) Polyhedron, $\{x \in \mathbb{R}^n | Ax \leq b\}$. For any $x, y \in \mathbb{R}^n$ such that $Ax \leq b$ and $Ay \leq b$, we have

$$A(\lambda x + (1-\lambda)y) = \lambda Ax + (1-\lambda)Ay \leq b$$

for any $\lambda \in [0,1]$.

(f) The set of all positive semidefinite matrices S^n_+. S^n_+ consists of all matrices $A \in \mathbb{R}^{n\times n}$ such that $A = A^T$ and $x^T Ax \geq 0$ for all $x \in \mathbb{R}^n$. Now consider $A, B \in \mathrm{S}^n_+$ and $\lambda \in [0,1]$. Then we must have

$$[\lambda A + (1-\lambda)B]^T = \lambda A^T + (1-\lambda)B^T = \lambda A + (1-\lambda)B.$$

Moreover, for any $x \in \mathbb{R}^n$,

$$x^T(\lambda A + (1-\lambda)B)x = \lambda x^T Ax + (1-\lambda)x^T Bx \geq 0.$$

(g) Intersections of convex sets. Let $X_i, i = 1, \ldots, k$, be convex sets. Assume that $x, y \in \cap_{i=1}^k X_i$, i.e., $x, y \in X_i$ for all $i = 1, \ldots, k$. Then for any $\lambda \in [0,1]$, we have $\lambda x + (1-\lambda)y \in X_i$ by the convexity of X_i, $i = 1, \ldots, k$, whence $\lambda x + (1-\lambda)y \in \cap_{i=1}^k X_i$.

(h) Weighted sums of convex sets. Let $X_1, \ldots, X_k \subseteq \mathbb{R}^n$ be nonempty convex subsets and $\lambda_1, \ldots, \lambda_k$ be reals. Then the set

$$\begin{aligned}&\lambda_1 X_1 + \ldots + \lambda_k X_k \\ &\equiv \{x = \lambda_1 x_1 + \ldots + \lambda_k x_k : x_i \in X_i, 1 \le i \le k\}\end{aligned}$$

is convex. The proof also follows directly from the definition of convex sets.

2.1.2 Projection onto Convex Sets

In this subsection we define the notion of projection over a convex set, which is important to the theory and computation of convex optimization.

Definition 2.2. Let $X \subset \mathbb{R}^n$ be a closed convex set. For any $y \in \mathbb{R}^n$, we define the closest point to y in X as:

$$\mathrm{Proj}_X(y) = \operatorname*{argmin}_{x \in X} \|y - x\|_2^2. \tag{2.1.2}$$

$\mathrm{Proj}_X(y)$ is called the projection of y onto X.

In the above definition, we require the set X to be closed in order to guarantee the existence of projection. On the other hand, if X is not closed, then the projection over X is not well defined. As an example, the projection of the point $\{2\}$ onto the interval $(0,1)$ does not exist. The existence of the projection over a closed convex set is formally stated as follows.

Proposition 2.1. *Let $X \subset \mathbb{R}^n$ be a closed convex set, and $y \in \mathbb{R}^n$ be given. Then $\mathrm{Proj}_X(y)$ must exist.*

Proof. Let $\{x_i\} \subseteq X$ be a sequence such that

$$\|y - x_i\|_2 \to \inf_{x \in X} \|y - x\|_2, \ i \to \infty.$$

The sequence $\{x_i\}$ clearly is bounded. Passing to a subsequence, we may assume that $x_i \to \bar{x}$ as $i \to \infty$. Since X is closed, we have $\bar{x} \in X$, and

$$\|y - \bar{x}\|_2 = \lim_{i \to \infty} \|y - x_i\|_2 = \inf_{x \in X} \|y - x\|_2.$$

■

The following result further shows that the projection onto a closed convex set X is unique.

Proposition 2.2. *Let X be a closed convex set, and $y \in \mathbb{R}^n$ be given. Then* $\mathrm{Proj}_X(y)$ *is unique.*

Proof. Let a and b be the two closet points in X to the given point y, so that $\|y-a\|_2 = \|y-b\|_2 = d$. Since X is convex, the point $z = (a+b)/2 \in X$. Therefore $\|y-z\|_2 \geq d$. We now have

$$\underbrace{\|(y-a)+(y-b)\|_2^2}_{=\|2(y-z)\|_2^2 \geq 4d^2} + \underbrace{\|(y-a)-(y-b)\|_2^2}_{=\|a-b\|^2} = \underbrace{2\|y-a\|_2^2 + 2\|y-b\|_2^2}_{4d^2},$$

whence $\|a-b\|_2 = 0$. Thus, the closest to y point in X is unique. ∎

In many cases when the set X is relatively simple, we can compute $\mathrm{Proj}_X(y)$ explicitly. In fact, in Sect. 1.4, we computed the distance from a given point $y \in \mathbb{R}^n$ to a given hyperplane $H := \{x \in \mathbb{R}^n | w^T x + b = 0\}$ by using projection. Following the same reasoning (see Fig. 2.2a), we can write down

$$\mathrm{Proj}_H(y) = y - \frac{(w^T y + b)w}{\|w\|_2^2}.$$

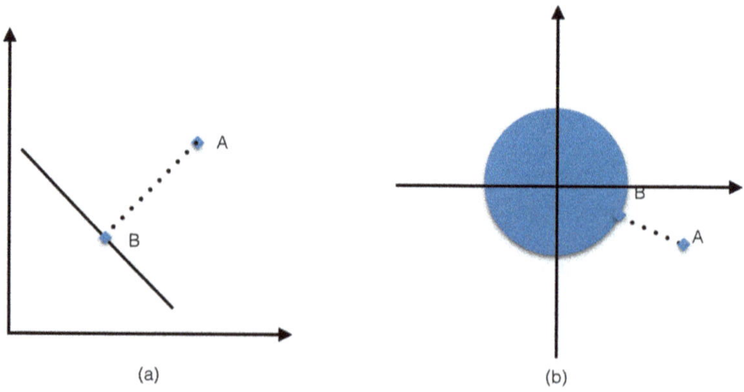

Fig. 2.2 Projection over convex sets

As another example, let us consider the projection of $y \in \mathbb{R}^n$ onto the standard Euclidean ball defined as $\mathsf{B} := \{x \in \mathbb{R}^n | \|x\|_2 \leq 1\}$ (see Fig. 2.2b). We can easily see that $\mathrm{Proj}_{\mathsf{B}}(y) = \frac{y}{\|y\|_2}$.

Projection over a convex set will be used extensively as a subroutine for solving more complicated optimization problems later in this book.

2.1.3 Separation Theorem

One fundamental result in convex analysis is the Separation Theorem. In this section, we will prove this theorem based on the projection onto a closed convex set and discuss some of its consequences.

We first discuss the separation of a point from a closed convex set.

Theorem 2.1. *Let $X \subseteq \mathbb{R}^n$ be a nonempty closed convex set, and a point $y \notin X$ be given. Then there exists $w \in \mathbb{R}^n$, $w \neq 0$ such that*

$$\langle w, y \rangle < \langle w, x \rangle, \quad \forall x \in X.$$

Proof. Our proof is based on projecting y onto the set X. In particular, let $\text{Proj}_X(y)$ be defined in (2.1.2), we show that the vector $w = y - \text{Proj}_X(y)$ separates y and X. Note that $w \neq 0$ since $y \notin X$. Also let $x \in X$ be given and denote $z = tx + (1-t)\text{Proj}_X(y)$ for any $t \in [0,1]$. Then we must have $z \in X$ and hence

$$\begin{aligned}
\|y - \text{Proj}_X(y)\|_2^2 \le \|y - z\|^2 &= \|y - [tx + (1-t)\text{Proj}_X(y)]\|_2^2 \\
&= \|y - \text{Proj}_X(y) - t(x - \text{Proj}_X(y))\|_2^2 \\
&= \|w - t(x - \text{Proj}_X(y))\|_2^2.
\end{aligned}$$

Define $\phi(t) := \|y - \text{Proj}_X(y) - t(x - \text{Proj}_X(y))\|_2^2$. It then follows from the above inequality that $\phi(0) \le \phi(t)$ for any $t \in [0,1]$. We have

$$0 \le \phi'(0) = -2w^T(x - \text{Proj}_X(y)),$$

which implies that

$$\forall x \in X : w^T x \le w^T \text{Proj}_X(y) = w^T(y - w) = w^T y - \|w\|_2^2.$$

■

We can generalize the above theorem to separate a closed convex set from another compact convex set.

Corollary 2.1. *Let X_1, X_2 be two nonempty closed convex sets and $X_1 \cap X_2 = \emptyset$. If X_2 is bounded, there exists $w \in \mathbb{R}^n$ such that*

$$\sup_{x \in X_1} w^T x < \inf_{x \in X_2} w^T x. \tag{2.1.3}$$

Proof. The set $X_1 - X_2$ is convex (the weighted sum of convex sets) and closed (the difference of a closed with a compact set). Moreover, $X_1 \cap X_2 = \emptyset$ implies $0 \notin X_1 - X_2$. So by Theorem 2.1, there exists w such that

$$\sup_{y \in X_1 - X_2} w^T y < w^T 0 = 0.$$

Or equivalently,

$$\begin{aligned} 0 &> \sup_{x_1 \in X_1, x_2 \in X_2} w^T(x_1 - x_2) \\ &= \sup_{x_1 \in X_1} w^T x_1 + \sup_{x_2 \in X_2} w^T(-x_2) \\ &= \sup_{x_1 \in X_1} w^T x_1 - \inf_{x_2 \in X_2} w^T x_2. \end{aligned}$$

Since X_2 is bounded, the last infimum becomes a min. Moreover, it is finite and can be moved to the left-hand side. ■

When X_2 is unbounded, Corollary 2.1 may fail. One possible fix is to replace the strict inequality in (2.1.3) by an inequality. However, this might cause some problems. For example, consider the line segment connecting $(-1,0)$ and $(0,0)$, and the other one connecting $(-1,0)$ and $(2,0)$. Observe that the inner product between $(0,1)$ and any point from these line segments equals 0. The vector $w = (0,1)$ appears to "separate" these two line segments, while apparently they are not separable.

To address this issue, we say that a linear form $w^T x$ properly separates nonempty sets S and T if and only if

$$\begin{aligned} \sup_{x \in S} w^T x &\le \inf_{y \in T} w^T y \\ \inf_{x \in S} w^T x &< \sup_{y \in T} w^T y. \end{aligned} \tag{2.1.4}$$

In this case, the hyperplanes associated with w that separate S and T are exactly the hyperplanes

$$\{x : w^T x - b = 0\} \text{ with } \sup_{x \in S} w^T x \le b \le \inf_{y \in T} w^T y.$$

The proper separation property holds under quite general assumptions on the intersection $X_1 \cap X_2$. To state this more general result, we need to introduce the notion of *relative interior* $\text{ri}(X)$, defined as the interior of X when we view it as subset of the affine subspace it generates. Without specific mention, we assume that the set X is full dimensional so that $\text{int}(X) = \text{ri}(X)$.

Theorem 2.2. *If the two nonempty convex sets* X_1 *and* X_2 *satisfy* $\text{ri}(X_1) \cap \text{ri}(X_2) = \emptyset$, *they can be properly separated.*

The above separation theorem can be derived from Theorem 2.1, but requiring us to establish a few technical results. We will first prove the result about the separability of a set in $\mathbb{R}^n$.

Lemma 2.1. *Every nonempty subset* $S \subseteq \mathbb{R}^n$ *is separable: one can find a sequence* $\{x_i\}$ *of points from* S *which is dense in* S *such that every point* $x \in S$ *is the limit of an appropriate subsequence of the sequence.*

Proof. Let $r_1, r_2, \ldots$ be the countable set of all rational vectors in $\mathbb{R}^n$. For every positive integer t, let $X_t \subset S$ be the countable set given by the following construction: we examine, one after another, at the points $r_1, r_2, \ldots$ and for every point r_s check whether there is a point $z \in S$ which is at most at the distance $1/t$ away from r_s. If points z with this property exist, we take one of them and add it to X_t and then pass to r_{s+1}, otherwise directly pass to r_{s+1}.

It is clear that every point $x \in S$ is at the distance at most $2/t$ from certain point of X_t. Indeed, since the rational vectors are dense in $\mathbb{R}^n$, there exists s such that r_s is at the distance $\leq \frac{1}{t}$ from x. Therefore, when processing r_s, we definitely add to X_t a point z which is at the distance $\leq 1/t$ from r_s and thus is at the distance $\leq 2/t$ from x. By construction, the countable union $\cup_{t=1}^{\infty} X_t$ of countable sets $X_t \subset S$ is a countable set in S, and by the fact that every point $x \in S$ is at most $2/t$ from X_t, this set is dense in S. ∎

With the help of Lemma 2.1, we can refine the basic separation result stated in Theorem 2.1 by removing the "closedness" assumption, and using the notion of proper separation.

Proposition 2.3. *Let $X \subseteq \mathbb{R}^n$ be a nonempty convex set and $y \in \mathbb{R}^n, y \notin X$ be given. Then there exists $w \in \mathbb{R}^n$, $w \neq 0$ such that*

$$\sup_{x \in X} w^T x \leq w^T y,$$
$$\inf_{x \in X} w^T x < w^T y.$$

Proof. First note that we can perform the following simplification.

- Shifting X and $\{y\}$ by $-y$ (which clearly does not affect the possibility of separating the sets), we can assume that $\{0\} \not\subset X$.
- Replacing, if necessary, $\mathbb{R}^n$ with $\mathrm{Lin}(X)$, we may further assume that $\mathbb{R}^n = \mathrm{Lin}(X)$, i.e., the linear subspace generated by X.

In view of Lemma 2.1, let $\{x_i \in X\}$ be a sequence which is dense in X. Since X is convex and does not contain 0, we have

$$0 \notin \mathrm{Conv}(\{x_1, \ldots, x_i\}) \; \forall i.$$

Noting that $\mathrm{Conv}(\{x_1, \ldots, x_i\})$ are closed convex sets, we conclude from Theorem 2.1 that

$$\exists w_i : 0 = w_i^T 0 > \max_{1 \leq j \leq i} w_i^T x_j. \tag{2.1.5}$$

By scaling, we may assume that $\|w_i\|_2 = 1$. The sequence $\{w_i\}$ of unit vectors possesses a converging subsequence $\{w_{i_s}\}_{s=1}^{\infty}$ and the limit w of this subsequence is also a unit vector. By (2.1.5), for every fixed j and all large enough s we have $w_{i_s}^T x_j < 0$, whence

$$w^T x_j \leq 0 \; \forall j. \tag{2.1.6}$$

Since $\{x_j\}$ is dense in X, (2.1.6) implies that $w^T x \leq 0$ for all $x \in X$, and hence that

$$\sup_{x \in X} w^T x \leq 0 = w^T 0. \tag{2.1.7}$$

Now, it remains to verify that

$$\inf_{x \in X} w^T x < w^T 0 = 0.$$

Assuming the opposite, (2.1.7) would imply that $w^T x = 0$ for all $x \in X$, which is impossible, since $\mathrm{Lin}(X) = \mathbb{R}^n$ and w is nonzero. ■

We can now further show that two nonempty convex sets (not necessarily bounded or closed) can be properly separated.

Proposition 2.4. *If the two nonempty convex sets X_1 and X_2 satisfy $X_1 \cap X_2 = \emptyset$, they can be properly separated.*

Proof. Let $\widehat{X} = X_1 - X_2$. The set $\widehat{X}$ clearly is convex and does not contain 0 (since $X_1 \cap X_2 = \emptyset$). By Proposition 2.3, $\widehat{X}$ and $\{0\}$ can be separated: there exists f such that

$$\begin{aligned} \sup_{x \in X_1} w^T s - \inf_{y \in X_2} w^T y = \sup_{x \in X_1, y \in X_2} [w^T x - w^T y] \le 0 = \inf_{z \in \{0\}} w^T z, \\ \inf_{x \in X_1} w^T x - \sup_{y \in X_2} w^T y = \inf_{x \in X_1, y \in X_2} [w^T x - w^T y] < 0 = \sup_{z \in \{0\}} w^T z, \end{aligned}$$

whence

$$\begin{aligned} \sup_{x \in X_1} w^T x \le \inf_{y \in X_2} w^T y, \\ \inf_{x \in X_1} w^T x < \sup_{y \in X_2} w^T y. \end{aligned}$$

■

We are now ready to prove Theorem 2.2, which is even stronger than Proposition 2.4 in the sense that we only need $\mathrm{ri}(X_1) \cap \mathrm{ri}(X_2) = \emptyset$. In other words, these two sets can possibly intersect on their boundaries.

Proof of Theorem 2.2. The sets $X_1' = \mathrm{ri}(X_1)$ and $X_2' = \mathrm{ri}(X_2)$ are convex and nonempty, and these sets do not intersect. By Proposition 2.4, X_1' and X_2' can be separated: for properly chosen w, one has

$$\begin{aligned} \sup_{x \in X_1'} w^T x \le \inf_{y \in X_2'} w^T y, \\ \inf_{x \in X_1'} w^T x < \sup_{y \in X_2'} w^T y. \end{aligned}$$

Since X_1' is dense in X_1 and X_2' is dense in X_2, inf's and sup's in the above relations remain the same when replacing X_1' with X_1 and X_2' with X_2. Thus, w separates X_1 and X_2. ■

In fact, we can show the reverse statement of Theorem 2.2 also holds.

Theorem 2.3. *If the two nonempty convex sets X_1 and X_2 can be properly separated, then* $\mathrm{ri}(X_1) \cap \mathrm{ri}(X_2) = \emptyset$.

Proof. We will first need to prove the following claim.

Claim. Let X be a convex set, $f(x) = w^T x$ be a linear form, and $a \in \mathrm{ri}(X)$. Then

$$w^T a = \max_{x \in X} w^T x \Leftrightarrow f(x) = \text{const}\ \forall x \in X.$$

Indeed, shifting X, we may assume $a = 0$. Let, on the contrary to what should be proved, $w^T x$ be nonconstant on X, so that there exists $y \in X$ with $w^T y \neq w^T a = 0$. The case of $w^T y > 0$ is impossible, since $w^T a = 0$ is the maximum of $w^T x$ on X. Thus, $w^T y < 0$. Since $0 \in \mathrm{ri}(X)$, all points $z = -\varepsilon y$ belong to X, provided that $\varepsilon > 0$ is small enough. At every point of this type, $w^T z > 0$, which contradicts the fact that $\max_{x \in X} w^T x = w^T a = 0$.

Now let us use the above claim to prove our main result. Let $a \in \mathrm{ri}X_1 \cap \mathrm{ri}X_2$. Assume, on contrary to what should be proved, that $w^T x$ separates X_1 and X_2, so that

$$\sup_{x \in X_1} w^T x \leq \inf_{y \in X_2} w^T y.$$

Since $a \in X_2$, we get $w^T a \geq \sup_{x \in X_1} w^T x$, that is, $w^T a = \max_{x \in X_1} w^T x$. By the above claim, $w^T x = w^T a$ for all $x \in X_1$. Moreover, since $a \in X_1$, we get $w^T a \leq \inf_{y \in X_2} w^T y$, that is, $w^T a = \min_{y \in T} w^T y$. By the above claim, $w^T y = w^T a$ for all $y \in X_2$. Thus,

$$z \in X_1 \cup X_2 \Rightarrow w^T z \equiv w^T a,$$

so that w does not properly separate X_1 and X_2, which is a contradiction. ∎

As a consequence of Theorem 2.2, we have the following supporting hyperplane theorem.

Corollary 2.2. *Let $X \subseteq \mathbb{R}^n$ be a convex set, and y be a point from its relative boundary. Then there exists $w \in \mathbb{R}^n$ and $w \neq 0$ such that*

$$\langle w, y \rangle \geq \sup_{x \in X} \langle w, x \rangle, \text{ and } \langle w, y \rangle > \inf_{x \in X} \langle w, x \rangle.$$

The hyperplane $\{x | \langle w, x \rangle = \langle w, y \rangle\}$ is called a supporting hyperplane of X at y.

Proof. Since y is a point from the relative boundary of X, it is outside the relative interior of X and therefore $\{x\}$ and $\mathrm{ri}X$ can be separated by the Separation Theorem. The separating hyperplane is exactly the desired supporting hyperplane to X at y. ∎

2.2 Convex Functions

2.2.1 Definition and Examples

Let $X \subseteq \mathbb{R}^n$ be a given convex set. A function $f : X \to \mathbb{R}$ is said to be convex if it always lies below its chords (Fig. 2.3), that is

$$f(\lambda x + (1-\lambda)y) \leq \lambda f(x) + (1-\lambda) f(y), \quad \forall (x, y, \lambda) \in X \times X \times [0,1]. \quad (2.2.8)$$

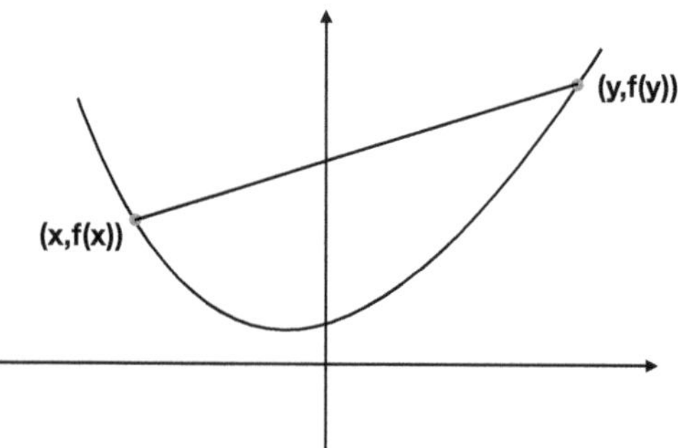

Fig. 2.3 The graph of a convex function

We say a function is strictly convex if (2.2.8) holds with strict inequality for any $x \neq y$ and $\lambda \in (0,1)$. We say that f is concave if $-f$ is convex, and similarly that f is strictly concave if $-f$ is strictly concave.

Some examples of convex functions are given as follows.

(a) Exponential, $f(x) = \exp(ax)$ for any $a \in \mathbb{R}$.
(b) Negative logarithm, $f(x) = -\log x$ with $x > 0$.
(c) Affine functions, $f(x) = w^T x + b$.
(d) Quadratic functions, $f(x) = \frac{1}{2}x^T Ax + b^T x$ with $A \in \mathbb{S}_+^n$ or $A \succeq 0$.
(e) Norms, $f(x) = \|x\|$.
(f) Nonnegative weighted sums of convex functions. Let $f_1, f_2, \ldots, f_k$ be convex functions and $w_1, w_2, \ldots, w_k$ be nonnegative real numbers. Then $f(x) = \sum_{i=1}^k w_i f_i(x)$ is a convex function.

2.2.2 *Differentiable Convex Functions*

Suppose that a function $f : X \to \mathbb{R}$ is differentiable over its domain. Then f is convex if and only if

$$f(y) \geq f(x) + \langle \nabla f(x), y - x \rangle$$

for any $x, y \in X$, where ∇f denotes the gradients of f. The function $f(x) + \langle \nabla f(x), y - x \rangle$ is the first-order Taylor approximation of f at the point x. The above first-order condition for convexity says that f is convex if and only if the tangent line underestimates f everywhere in its domain. Similar to the definition of convexity, f will be strictly convex if this condition holds with strict inequality, concave if the inequality is reversed, and strictly concave if the reverse inequality is strict.

Suppose that a function $f : X \to \mathbb{R}$ is twice differentiable. Then f is convex if and only if its Hessian is positive semidefinite, i.e.,

$$\nabla^2 f(x) \succeq 0.$$

In one dimension, this is equivalent to the condition that the second-order derivative $f''(x)$ is nonnegative. Again analogous to both the definition and the first-order conditions for convexity, f is strictly convex if its Hessian is positive definite, concave if the Hessian is negative semidefinite, and strictly concave if the Hessian is negative definite. The function f is said to be strongly convex modulus μ with respect to the norm $\|\cdot\|$ if

$$f(y) \geq f(x) + \langle g, y-x \rangle + \tfrac{\mu}{2}\|y-x\|^2$$

for some $\mu > 0$. Clearly, strong convexity implies strict convexity.

2.2.3 Non-differentiable Convex Functions

Note that convex functions are not always differentiable everywhere over its domain. For example, the absolute value function $f(x) = |x|$ is not differentiable when $x = 0$. In this subsection, we will introduce an important notion about convex functions, i.e., subgradients, to generalize the gradients for differentiable convex functions.

Definition 2.3. $g \in \mathbb{R}^n$ is a subgradient of f at $x \in X$ if for any $y \in X$

$$f(y) \geq f(x) + \langle g, y-x \rangle.$$

The set of subgradients of f at x is called the subdifferential, denoted by $\partial f(x)$.

In order to show the existence of the subgradients for a convex function, we need to use the epigraph of a function $f : X \to \mathbb{R}$ given by

$$\text{epi}(f) = \{(x,t) \in X \times \mathbb{R} : f(x) \leq t\}.$$

It can be easily shown that f is convex if and only if epi(f) is a convex set.

The next result establishes the existence of subgradients for convex functions.

Proposition 2.5. *Let $X \subseteq \mathbb{R}^n$ be convex and $f : X \to \mathbb{R}$. If $\forall x \in X$, $\partial f(x) \neq \emptyset$, then f is convex. Moreover, if f is convex, then for any $x \in \text{ri}(X)$, $\partial f(x) \neq \emptyset$.*

Proof. The first claim is obvious. Let $g \in \partial f(\lambda x + (1-\lambda)y)$ for some $\lambda \in [0,1]$. Then by definition we have

$$f(y) \geq f(\lambda x + (1-\lambda)y) + \lambda \langle g, y-x \rangle,$$
$$f(x) \geq f(\lambda x + (1-\lambda)y) + (1-\lambda)\langle g, x-y \rangle.$$

Multiplying the first inequality by $1-\lambda$ and the second one by λ, and then summing them up, we show the convexity of f.

We now show that f has subgradients in the interior of X. We will construct such a subgradient by using a supporting hyperplane to the epigraph of f. Let $x \in X$. Then $(x, f(x)) \in \text{epi}(f)$. By the convexity of epi(f) and the separating hyperplane theorem, there exists $(w, v) \in \mathbb{R}^n \times \mathbb{R}$ ($(w,v) \neq 0$) such that

$$\langle w,x\rangle + vf(x) \geq \langle w,y\rangle + vt, \;\; \forall (y,t) \in \text{epi}(f). \tag{2.2.9}$$

Clearly, by tending t to infinity, we can see that $v \leq 0$. Now let us assume that x is in the interior of X. Then for $\varepsilon > 0$ small enough, $y = x + \varepsilon w \in X$, which implies that $v \neq 0$, since otherwise, we have $0 \geq \varepsilon \|w\|_2^2$ and hence $w = 0$, contradicting with the fact that $(w,v) \neq 0$. Letting $t = f(y)$ in (2.2.9), we obtain

$$f(y) \geq f(x) + \tfrac{1}{v}\langle w, y - x\rangle,$$

which implies that w/v is a subgradient of f at x. ■

Let f be a convex and differentiable function. Then by definition,

$$\begin{aligned} f(y) &\geq \tfrac{1}{\lambda}\left[f((1-\lambda)x + \lambda y) - (1-\lambda)f(x)\right] \\ &= f(x) + \tfrac{1}{\lambda}\left[f((1-\lambda)x + \lambda y) - f(x)\right]. \end{aligned}$$

Tending λ to 0, we show that $\nabla f(x) \in \partial f(x)$.

Below we provide some basic subgradient calculus for convex functions. Observe that many of them mimic the calculus for gradient computation.

(a) Scaling: $\partial(af) = a\partial f$ provided $a > 0$. The condition $a > 0$ makes function f remain convex.
(b) Addition: $\partial(f_1 + f_2) = \partial(f_1) + \partial(f_2)$.
(c) Affine composition: if $g(x) = f(Ax + b)$, then $\partial g(x) = A^T \partial f(Ax + b)$.
(d) Finite pointwise maximum: if $f(x) = \max_{i=1,\ldots,m} f_i(x)$, then

$$\partial f(x) = \text{conv}\left\{\cup_{i: f_i(x) = f(x)} \partial f_i(x)\right\},$$

which is the convex hull of union of subdifferentials of all active $i : f_i(x) = f(x)$ functions at x.
(e) General pointwise maximum: if $f(x) = \max_{s \in S} f_s(x)$, then under some regularity conditions (on S and f_s),

$$\partial f(x) = \text{cl}\left\{\text{conv}\left(\cup_{s: f_s(x) = f(x)} \partial f_s(x)\right)\right\}.$$

(f) Norms: important special case, $f(x) = \|x\|_p$. Let q be such that $1/p + 1/q = 1$, then

$$\partial f(x) = \{y : \|y\|_q \leq 1 \, \text{and} \, y^T x = \max\{z^T x : \|z\|_q \leq 1\}.$$

Other notions of convex analysis will prove to be useful. In particular the notion of closed convex functions is convenient to exclude pathological cases: these are convex functions with closed epigraphs (see Sect. 2.4 for more details).

2.2.4 Lipschitz Continuity of Convex Functions

Our goal in this section is to show that convex functions are Lipschitz continuous inside the interior of its domain.

We will first show that a convex function is locally bounded.

Lemma 2.2. *Let f be convex and $x_0 \in \operatorname{int}\operatorname{dom} f$. Then f is locally bounded, i.e., $\exists \varepsilon > 0$ and $M(x_0, \varepsilon) > 0$ such that*

$$f(x) \le M(x_0,\varepsilon)\ \forall\, x \in B_\varepsilon(x_0) := \{x \in \mathbb{R}^n : \|x - x_0\|_2 \le \varepsilon\}.$$

Proof. Since $x_0 \in \operatorname{int}\operatorname{dom} f$, $\exists \varepsilon > 0$ such that the vectors $x_0 \pm \varepsilon e_i \in \operatorname{int}\operatorname{dom} f$ for $i = 1,\ldots,n$, where e_i denotes the unit vector along coordinate i. Also let $H_\varepsilon(x_0) := \{x \in \mathbb{R}^n : \|x - x_0\|_\infty \le \varepsilon\}$ denote the hypercube formed by the vectors $x_0 \pm \varepsilon e_i$. It can be easily seen that $B_\varepsilon(x_0) \subseteq H_\varepsilon(x_0)$ and hence that

$$\max_{x \in B_\varepsilon(x_0)} f(x) \le \max_{x \in H_\varepsilon(x_0)} f(x) \le \max_{i=1,\ldots,n} f(x_0 \pm \varepsilon e_i) =: M(x_0, \varepsilon).$$

■

Next we show that f is locally Lipschitz continuous.

Lemma 2.3. *Let f be convex and $x_0 \in \operatorname{int}\operatorname{dom} f$. Then f is locally Lipschitz, i.e., $\exists \varepsilon > 0$ and $\bar{M}(x_0, \varepsilon) > 0$ such that*

$$|f(y) - f(x_0)| \le \bar{M}(x_0,\varepsilon)\|x - y\|,\ \forall y \in B_\varepsilon(x_0) := \{x \in \mathbb{R}^n : \|x - x_0\|_2 \le \varepsilon\}. \tag{2.2.10}$$

Proof. We assume that $y \ne x_0$ (otherwise, the result is obvious). Let $\alpha = \|y - x_0\|_2/\varepsilon$. We extend the line segment connecting x_0 and y so that it intersects the ball $B_\varepsilon(x_0)$, and then obtain two intersection points z and u (see Fig. 2.4). It can be easily seen that

$$y = (1-\alpha)x_0 + \alpha z, \tag{2.2.11}$$
$$x_0 = [y + \alpha u]/(1+\alpha). \tag{2.2.12}$$

It then follows from the convexity of f and (2.2.11) that

$$\begin{aligned} f(y) - f(x_0) &\le \alpha[f(z) - f(x_0)] = \tfrac{f(z)-f(x_0)}{\varepsilon}\|y - x_0\|_2 \\ &\le \tfrac{M(x_0,\varepsilon)-f(x_0)}{\varepsilon}\|y - x_0\|_2, \end{aligned}$$

where the last inequality follows from Lemma 2.2. Similarly, by the convexity f, (2.2.11), and Lemma 2.2, we have

$$f(x_0) - f(y) \le \|y - x_0\|_2 \tfrac{M(x_0,\varepsilon)-f(x_0)}{\varepsilon}.$$

Combining the previous two inequalities, we show (2.2.10) holds with $\bar{M}(x_0,\varepsilon) = [M(x_0,\varepsilon) - f(x_0)]/\varepsilon$. ■

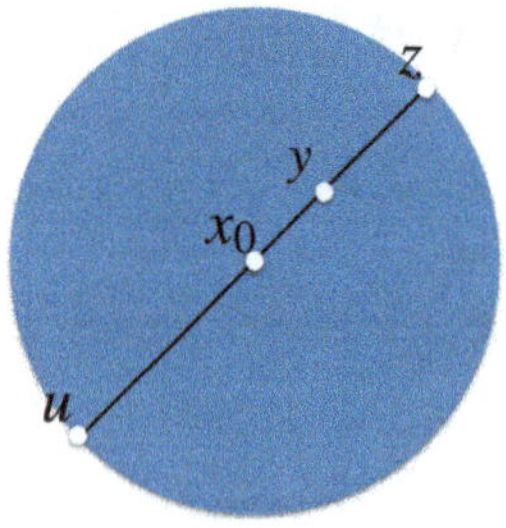

Fig. 2.4 Local Lipschitz continuity of a convex function

The following simple result shows the relation between the Lipschitz continuity of f and the boundedness of subgradients.

Lemma 2.4. *The following statements hold for a convex function f.*

(a) If $x_0 \in \operatorname{int}\operatorname{dom} f$ and f is locally Lipschitz (i.e., (2.2.10) holds), then $\|g(x_0)\| \le \bar{M}(x_0,\varepsilon)$ for any $g(x_0) \in \partial f(x_0)$.
(b) If $\exists g(x_0) \in \partial f(x_0)$ and $\|g(x_0)\|_2 \le \bar{M}(x_0,\varepsilon)$, then $f(x_0) - f(y) \le \bar{M}(x_0,\varepsilon)\|x_0 - y\|_2$.

Proof. We first show part (a). Let $y = x_0 + \varepsilon g(x_0)/\|g(x_0)\|_2$. By the convexity of f and (2.2.10), we have

$$\varepsilon\|g(x_0)\|_2 = \langle g(x_0), y - x_0\rangle \le f(y) - f(x_0) \le \bar{M}(x_0,\varepsilon)\|y - x_0\| = \varepsilon\bar{M}(x_0,\varepsilon),$$

which implies part (a). Part (b) simply follows the convexity of f, i.e.,

$$f(x_0) - f(y) \le \langle g(x_0), x_0 - y\rangle \le \bar{M}(x_0,\varepsilon)\|x_0 - y\|_2.$$

■

Below we state the global Lipschitz continuity of a convex function in its interior of domain.

Theorem 2.4. *Let f be a convex function and let K be a closed and bounded set contained in the relative interior of the domain* $\operatorname{dom} f$ *of f. Then f is Lipschitz continuous on K, i.e., there exists constant M such that*

$$|f(x) - f(y)| \le M_K\|x - y\|_2 \quad \forall x, y \in K. \tag{2.2.13}$$

Proof. The result directly follows from the local Lipschitz continuity of a convex function (see Lemmas 2.3 and 2.4) and the boundedness of K. ■

Remark 2.1. All three assumptions on K—i.e., (a) closedness, (b) boundedness, and (c) $K \subset \operatorname{ri}\operatorname{dom} f$—are essential, as it is seen from the following three examples:

- $f(x) = 1/x$, $\mathrm{dom} f = (0, +\infty)$, $K = (0, 1]$. We have (b), (c) but not (a); f is neither bounded, nor Lipschitz continuous on K.
- $f(x) = x^2$, $\mathrm{dom} f = \mathbb{R}$, $K = \mathbb{R}$. We have (a), (c) and not (b); f is neither bounded nor Lipschitz continuous on K.
- $f(x) = -\sqrt{x}$, $\mathrm{dom} f = [0, +\infty)$, $K = [0, 1]$. We have (a), (b) and not (c); f is not Lipschitz continuous on K although is bounded. Indeed, we have $\lim_{t \to +0} \frac{f(0)-f(t)}{t} = \lim_{t \to +0} t^{-1/2} = +\infty$, while for a Lipschitz continuous f the ratios $t^{-1}(f(0) - f(t))$ should be bounded.

2.2.5 Optimality Conditions for Convex Optimization

The following results state the basic optimality conditions for convex optimization.

Proposition 2.6. *Let f be convex. If x is a local minimum of f, then x is a global minimum of f. Furthermore this happens if and only if $0 \in \partial f(x)$.*

Proof. It can be easily seen that $0 \in \partial f(x)$ if and only if x is a global minimum of f. Now assume that x is a local minimum of f. Then for $\lambda > 0$ small enough one has for any y,

$$f(x) \le f((1-\lambda)x + \lambda y) \le (1-\lambda) f(x) + \lambda f(y),$$

which implies that $f(x) \le f(y)$ and thus that x is a global minimum of f. ■

The above result can be easily generalized to the constrained case. Given a convex set $X \subseteq \mathbb{R}^n$ and a convex function $f : X \to \mathbb{R}$, we intend to

$$\min_{x \in X} f(x).$$

We first define the indicator function of the convex set X, i.e.,

$$I_X(x) := \begin{cases} 0, & x \in X, \\ \infty, & Otherwise. \end{cases}$$

By definition of subgradients, we can see that the subdifferential of I_X is given by the normal cone of X, i.e.,

$$\partial I_X(x) = \{w \in \mathbb{R}^n | \langle w, y - x \rangle \le 0, \forall y \in X\}. \tag{2.2.14}$$

Proposition 2.7. *Let $f : X \to \mathbb{R}$ be a convex function and X be a convex set. Then x^* is an optimal solution of $\min_{x \in X} f(x)$ if and only if there exists $g^* \in \partial f(x^*)$ such that*

$$\langle g^*, y - x^* \rangle \ge 0, \forall y \in X.$$

Proof. Clearly the problem is equivalent to $\min_{x\in\mathbb{R}^n} f(x)+I_X(x)$, where the I_X denotes the indicator function of X. The results then immediately follow from (2.2.14) and Proposition 2.6. ∎

In particular, if $X=\mathbb{R}^n$, then we must have $0\in\partial f(x)$, which reduces to the case in Proposition 2.6.

2.2.6 Representer Theorem and Kernel

In this subsection, we introduce a very important application of the optimality condition for convex optimization in machine learning.

Recall that many supervised machine learning models can be written in the following form:

$$f^* := \min_{x\in\mathbb{R}^n} \left\{ f(x) := \textstyle\sum_{i=1}^N L(x^T u_i, v_i) + \lambda r(x) \right\}, \tag{2.2.15}$$

for some $\lambda \geq 0$. For the sake of simplicity we assume that $r(x)=\|x\|_2^2/2$. It turns out that under these assumptions, we can always write the solutions to problem (2.2.15) as a linear combination of the input variables u_i's as shown in the following statement.

Theorem 2.5. *The optimal solution of (2.2.15) with $r(x)=\|x\|_2^2/2$ can be written as*

$$x^* = \textstyle\sum_{i=1}^N \alpha_i u^{(i)}$$

for some real-valued weights α_i.

Proof. Let $L'(z,v)$ denote a subgradient of L w.r.t. z. Then by the chain rule of subgradient computation, the subgradients of f can be written in the form of

$$f'(x) = \textstyle\sum_{i=1}^N L'(x^T u^{(i)}) u^{(i)} + \lambda x.$$

Noting that $0\in\partial f(x^*)$ and letting $w_i = L'(x^T u^{(i)})$, there must exist w_i's such that

$$x = -\tfrac{1}{\lambda}\textstyle\sum_{i=1}^N w_i u^{(i)}.$$

The result then follows by setting $\alpha_i = -1/(\lambda w_i)$. ∎

This result has some important consequence in machine learning. For any inner product of $x^T u$ in machine learning models, we can replace it with

$$x^T u = u^T x = \textstyle\sum_{i=1}^N \alpha_i (u^{(i)})^T u^{(i)},$$

and then view these α_i, $i=1,\ldots,N$, as unknown variables (or parameters).

More generally, we may consider a nonlinear transformation of our original input variables u. Recall in our regression example in Chap. 1, we have an input variable u, i.e., the rating of a friend (say Judy), and we can consider regression using the features u, u^2, and u^3 to obtain a cubic function. We can use $\phi(u)$ to define such a nonlinear mapping from the original input to a new feature space.

Rather than learning the parameters associated with the original input variables u, we may instead learn using these expanded features $\phi(u)$. To do so, we simply need to go over our previous models, and replace u everywhere in it with $\phi(u)$.

Since the model can be written entirely in terms of the inner products $\langle u, z\rangle$, we can replace all those inner products with $\langle \phi(u), \phi(z)\rangle$. Given a feature mapping ϕ, let us define the so-called kernel

$$K(u,z) = \phi(u)^T \phi(z).$$

Then, we replace everywhere we previously had $\langle u, z\rangle$ with $K(u,z)$. In particular, we can write the new objective function as

$$\begin{aligned}
\Phi(\alpha) = f(x) &= \Sigma_{i=1}^N L(x^T u^{(i)}, v^{(i)}) + \tfrac{\lambda}{2}\|x\|_2^2 \\
&= \Sigma_{i=1}^N L\left(\phi(u^{(i)})^T \Sigma_{j=1}^N \alpha_j \phi(u^{(j)}), v_i\right) + \tfrac{\lambda}{2}\|\Sigma_{j=1}^N \alpha_j \phi(u^{(j)})\|_2^2 \\
&= \Sigma_{i=1}^N L\left(\phi(u^{(i)})^T \Sigma_{j=1}^N \alpha_j \phi(u^{(j)}), v_i\right) \\
&\quad + \tfrac{\lambda}{2}\Sigma_{i=1}^N \Sigma_{j=1}^N \alpha_i \alpha_j \phi(u^{(i)})^T \phi(u^{(j)}) \\
&= \Sigma_{i=1}^N L\left(\Sigma_{j=1}^N \alpha_j K(u^{(i)}, u^{(j)}), v_i\right) \\
&\quad + \tfrac{\lambda}{2}\Sigma_{i=1}^N \Sigma_{j=1}^N \alpha_i \alpha_j K(u^{(i)}, u^{(j)}).
\end{aligned}$$

In this way, we can write the objective function in terms of the Kernel matrix

$$K = \{K(u^{(i)}, v^{(j)})\}_{i,j=1}^N.$$

Even more interestingly, in many cases, we do not need to compute the nonlinear mapping $\phi(u)$ explicitly for every u, since the Kernel might be easier to compute than ϕ. One commonly used Kernel is the Gaussian or Radial Basis Function (RBF) kernel given by

$$K(u,z) = \exp\left(-\tfrac{1}{2\tau^2}\|u-z\|_2^2\right)$$

applicable to data in any dimension and the other one is the min-kernel given by $K(x,z) = \min(x,z)$ applicable to data in $\mathbb{R}$.

2.3 Lagrange Duality

In this section, we consider differentiable convex optimization problems of the form

$$\begin{aligned}
f^* \equiv \min_{x\in X} &\ f(x) \\
\text{s.t.} &\ g_i(x) \le 0, i = 1,\ldots,m, \\
&\ h_j(x) = 0, j = 1,\ldots,p,
\end{aligned} \tag{2.3.16}$$

where $X \subseteq \mathbb{R}^n$ is a closed convex set, $f : X \to \mathbb{R}$ and $g_i : X \to \mathbb{R}$ are differentiable convex functions, and $h_j : \mathbb{R}^n \to \mathbb{R}$ are affine functions. Our goal is to introduce Lagrange duality and a few optimality conditions for these convex optimization problems with function constraints.

2.3.1 Lagrange Function and Duality

We define the Lagrangian function L for (2.3.16) as

$$\mathrm{L}(x,\lambda,y) = f(x) + \Sigma_{i=1}^m \lambda_i g_i(x) + \Sigma_{j=1}^p y_j h_j(x),$$

for some $\lambda_i \in \mathbb{R}_+$, $i = 1,\ldots,m$, and $y_j \in \mathbb{R}, j = 1,\ldots,p$. These λ_i's and y_j's are called dual variables or Lagrange multipliers.

Intuitively, the Lagrangian function L can be viewed as a relaxed version of the objective function in the original problem (2.3.16) by allowing violation of the constraints ($g_i(x) \le 0$ and $h_j(x) \le 0$).

Let us consider the minimization of $\mathrm{L}(x,\lambda,y)$ w.r.t. x. Suppose that $\lambda \ge 0$ and $y \in \mathbb{R}^p$ are given. Let us define

$$\phi(\lambda,y) := \min_{x\in X} \mathrm{L}(x,\lambda,y).$$

Clearly, for any feasible point x to (2.3.16) (i.e., $x \in X$, $g_i(x) \le 0$, and $h_j(x) = 0$), we have

$$\begin{aligned}\mathrm{L}(x,\lambda,y) &= f(x) + \Sigma_{i=1}^m \lambda_i g_i(x) + \Sigma_{j=1}^p y_j h_j(x)\\ &= f(x) + \Sigma_{i=1}^m \lambda_i g_i(x) \le f(x).\end{aligned}$$

In particular, letting $x = x^*$ be the optimal solution of (2.3.16), we must have

$$\phi(\lambda,y) \le f^*.$$

In other words, $\phi(\lambda,y)$ gives us a lower bound on the optimal value f^*. In order to obtain the strongest lower bound, we intend to maximize $\phi(\lambda,y)$ w.r.t. $\lambda \ge 0$ and $y \in \mathbb{R}^p$, and thus define the Lagrange dual as

$$\phi^* \equiv \max_{\lambda\ge 0,y} \left\{ \phi(\lambda,y) := \min_{x\in X} \mathrm{L}(x,\lambda,y) \right\}. \tag{2.3.17}$$

By construction, we must have

$$\phi^* \le f^*.$$

This relation is the so-called *weak duality*. What is more interesting is that under certain conditions, we have $\phi^* = f^*$ as stated in Theorem 2.6. The proof of this result, however, is more involved. Hence, we provide this proof separately in Sect. 2.3.2.

Theorem 2.6. *Suppose that (2.3.16) is below bounded and that there exists $\bar{x} \in \text{int}X$ s.t. $g(\bar{x}) < 0$ and $h(\bar{x}) = 0$. Then the Lagrange dual is solvable and we must have*

$$\phi^* = f^*.$$

The above theorem says that as long as the primal problem (2.3.16) has a strictly feasible solution (called *Slater condition*), the optimal value for the Lagrange dual must be equal to the optimal value of the primal. This result is called *strong duality*. In practice, nearly all convex problems satisfy this type of constraint qualification, and hence the primal and dual problems have the same optimal value.

2.3.2 Proof of Strong Duality

In this subsection, we provide a proof for the strong duality for convex optimization. The proof follows from the separation theorem and the consequent convex theorem on alternatives. For the sake of simplicity, we focus on the case when there exist only nonlinear inequality constraints. The readers can easily adapt the proof to the case when affine constraints do exist, or even further refine the results if there only exist affine constraints.

Before proving Theorem 2.6, we will first establish the Convex Theorem on Alternative (CTA). Consider a system of constraints on x

$$\begin{aligned} f(x) &< c, \\ g_j(x) &\le 0,\ j = 1,\ldots,m, \\ x &\in X, \end{aligned} \tag{I}$$

along with system of constraints on λ:

$$\begin{aligned} \inf_{x\in X}[f(x) + \textstyle\sum_{j=1}^m \lambda_j g_j(x)] &\ge c, \\ \lambda_j \ge 0,\ j = 1,\ldots,m. \end{aligned} \tag{II}$$

We first discuss the trivial part of the CTA.

Proposition 2.8. *If (II) is solvable, then (I) is insolvable.*

What is more interesting is that the reverse statement is also true under the slater condition.

Proposition 2.9. *If (I) is insolvable and the subsystem*

$$\begin{aligned} g_j(x) &< 0,\ j = 1,\ldots,m, \\ x &\in X \end{aligned}$$

is solvable, then (II) is solvable.

Proof. Assume that (I) has no solutions. Consider two sets S and T in $\mathbb{R}^{m+1}$:

$$S := \left\{u \in \mathbb{R}^{m+1} : u_0 < c, u_1 \leq 0, \ldots, u_m \leq 0\right\},$$

$$T := \left\{ u \in \mathbb{R}^{m+1} : \exists x \in X : \begin{array}{c} f(x) \leq u_0 \\ g_1(x) \leq u_1 \\ \cdots\cdots\cdots \\ g_m(x) \leq u_m \end{array} \right\}.$$

First observe that S and T are nonempty convex sets. Moreover, S and T do not intersect (otherwise (I) would have a solution). By Theorem 2.2, S and T can be separated: $\exists (a_0, \ldots, a_m) \neq 0$ such that

$$\inf_{u \in T} a^T u \geq \sup_{u \in S} a^T u$$

or equivalently,

$$\inf_{x \in X} \inf_{\substack{u_0 \geq f(x) \\ u_1 \geq g_1(x) \\ \vdots \\ u_m \geq g_m(x)}} [a_0 u_0 + a_1 u_1 + \ldots + a_m u_m] \geq \sup_{\substack{u_0 < c \\ u_1 \leq 0 \\ \vdots \\ u_m \leq 0}} [a_0 u_0 + a_1 u_1 + \ldots + a_m u_m].$$

In order to bound the RHS, we must have $a \geq 0$, whence

$$\inf_{x \in X} [a_0 f(x) + a_1 g_1(x) + \ldots + a_m g_m(x)] \geq a_0 c. \tag{2.3.18}$$

Finally, we observe that $a_0 > 0$. Indeed, otherwise $0 \neq (a_1, \ldots, a_m) \geq 0$ and

$$\inf_{x \in X} [a_1 g_1(x) + \ldots + a_m g_m(x)] \geq 0,$$

while $\exists \bar{x} \in X : g_j(\bar{x}) < 0$ for all j. Now, dividing both sides of (2.3.18) by a_0, we have

$$\inf_{x \in X} \left[f(x) + \Sigma_{j=1}^m \left(\frac{a_j}{a_0}\right) g_j(x) \right] \geq c.$$

By setting $\lambda_j = a_j / a_0$ we obtain the result. ■

We are now ready to prove the strong duality.

Proof of Theorem 2.6. The system

$$f(x) < f^*, \ g_j(x) \leq 0, \ j = 1, \ldots, m, \ x \in X$$

has no solutions, while the system

$$g_j(x) < 0, \ j = 1, \ldots, m, \ x \in X$$

has a solution. By CTA,

$$\exists \lambda^* \geq 0 : f(x) + \sum_j \lambda_j^* g_j(x) \geq f^* \ \forall x \in X,$$

whence

$$\phi(\lambda^*) \geq f^*.$$

Combined with Weak Duality, the above inequality says that

$$f^* = \phi(\lambda^*) = \phi^*.$$

■

2.3.3 *Saddle Points*

Now let us examine some interesting consequences of strong duality. In particular, we can derive a few optimality conditions for convex optimization in order to check whether an $x^* \in X$ is optimal to (2.3.16) or not.

The first one is given in the form of a pair of *saddle points.*

Theorem 2.7. *Let $x^* \in X$ be given.*

(a) If x^* can be extended, by a $\lambda^* \geq 0$ and $y^* \in \mathbb{R}^p$, to a saddle point of the Lagrange function on $X \times \{\lambda \geq 0\}$:

$$\mathrm{L}(x, \lambda^*, y^*) \geq \mathrm{L}(x^*, \lambda^*, y^*) \geq \mathrm{L}(x^*, \lambda, y) \ \forall (x \in X, \lambda \geq 0, y \in \mathbb{R}^p),$$

then x^* is optimal for (2.3.16).

(b) If x^* is optimal for (2.3.16) which is convex and satisfies the Slater condition, then x^* can be extended, by a $\lambda^* \geq 0$ and $y^* \in \mathbb{R}^p$, to a saddle point of the Lagrange function on $X \times \{\lambda \geq 0\} \times \mathbb{R}^p$.

Proof. We first prove part (a). Clearly,

$$\sup_{\lambda \geq 0, y} \mathrm{L}(x^*, \lambda, y) = \begin{cases} +\infty, & x^* \text{ is infeasible} \\ f(x^*), & \text{otherwise} \end{cases}$$

Thus, $\lambda^* \geq 0$, $\mathrm{L}(x^*, \lambda^*, y^*) \geq \mathrm{L}(x^*, \lambda, y) \ \forall \lambda \geq 0 \forall y$ is equivalent to

$$g_j(x^*) \leq 0 \forall j, \ \lambda_i^* g_i(x^*) = 0 \forall i, \ h_j(x^*) = 0 \forall j.$$

Consequently, $\mathrm{L}(x^*, \lambda^*, y^*) = f(x^*)$, hence

$$\mathrm{L}(x, \lambda^*, y^*) \geq \mathrm{L}(x^*, \lambda^*, y^*) \ \forall x \in X$$

reduces to

$$\mathrm{L}(x,\lambda^*,y^*) \geq f(x^*)\ \forall x.$$

Since for $\lambda \geq 0$ and y, one has $f(x) \geq \mathrm{L}(x,\lambda,y)$ for all feasible x, the above inequality then implies that

$$x \text{ is feasible } \Rightarrow f(x) \geq f(x^*).$$

We now show part (b). By Lagrange Duality, $\exists \lambda^* \geq 0$, y^*:

$$f(x^*) = \phi(\lambda^*,y^*) \equiv \inf_{x\in X}\left[f(x)+\Sigma_i\lambda_i^* g_i(x)+\Sigma_j y_j^* h_j(x)\right]. \tag{2.3.19}$$

Since x^* is feasible, we have

$$\inf_{x\in X}\left[f(x)+\Sigma_i\lambda_i^* g_i(x)+\Sigma_j y_j^* h_j(x)\right] \leq f(x^*)+\Sigma_i\lambda_i^* g_i(x^*) \leq f(x^*).$$

By (2.3.19), the last " $\leq$ " here should be " $=$ ". This identity, in view of the fact that $\lambda^* \geq 0$, is possible if and only if $\lambda_j^* g_j(x^*) = 0 \forall j. Therefore, we have$

$$f(x^*) = \mathrm{L}(x^*,\lambda^*,y^*) \geq \mathrm{L}(x^*,\lambda,y)\ \forall \lambda \geq 0, \forall y \in \mathbb{R}^p,$$

where the last inequality follows from the definition of L (or weak duality). Now (2.3.19) reads $\mathrm{L}(x,\lambda^*,y^*) \geq f(x^*) = \mathrm{L}(x^*,\lambda^*,y^*)$. ■

2.3.4 Karush–Kuhn–Tucker Conditions

We are now ready to derive the Karush–Kuhn–Tucker (KKT) optimality conditions for convex programming.

Theorem 2.8. *Let (2.3.16) be a convex program, let x^* be its feasible solution, and let the functions f, $g_1,\ldots,g_m$ be differentiable at x^*. Then*

(a) Exist Lagrange multipliers $\lambda^* \geq 0$ and y^* such that the following KKT conditions

$$\begin{array}{l}\nabla f(x^*)+\Sigma_{i=1}^m \lambda_i^* \nabla g_i(x^*)+\Sigma_{i=1}^p y_j^* \nabla h_j(x^*) \in N_X^*(x^*)\ [\text{stationarity}]\\ \lambda_i^* g_i(x^*) = 0,\ 1 \leq i \leq m\ [\text{complementary slackness}]\\ g_i(x^*) \leq 0, 1 \leq i \leq m; h_j(x^*) = 0,\ 1 \leq j \leq p\ [\text{primal feasibility}]\end{array}$$

are sufficient for x^* to be optimal.

(b) If (2.3.16) satisfies restricted Slater condition: $\exists \bar{x} \in \mathrm{rint}X : g_i(\bar{x}) \leq 0, h_j(\bar{x}) = 0$ for all constraints and $g_j(\bar{x}) < 0$ for all nonlinear constraints, then the KKT conditions are necessary and sufficient for x^* to be optimal.

Proof. We first prove part (a). Indeed, complementary slackness plus $\lambda^* \geq 0$ ensure that

$$\mathrm{L}(x^*,\lambda^*,y^*) \geq \mathrm{L}(x^*,\lambda,y) \quad \forall \lambda \geq 0, \forall y \in \mathbb{R}^p.$$

Further, $\mathrm{L}(x,\lambda^*,y^*)$ is convex in $x \in X$ and differentiable at $x^* \in X$, so that the stationarity condition implies that

$$\mathrm{L}(x,\lambda^*,y^*) \geq \mathrm{L}(x^*,\lambda^*,y^*) \quad \forall x \in X.$$

Thus, x^* can be extended to a saddle point of the Lagrange function and therefore is optimal for (2.3.16).

We now show that part (b) holds. By Saddle Point Optimality condition, from optimality of x^* it follows that $\exists \lambda^* \geq 0$ and $y^* \in \mathbb{R}^p$ such that (x^*,λ^*,y^*) is a saddle point of $\mathrm{L}(x,\lambda,y)$ on $X \times \{\lambda \geq 0\} \times \mathbb{R}^p$. This is equivalent to $h_j(x^*) = 0$,

$$\lambda_i^* g_i(x^*) = 0 \; \forall i,$$

and

$$\min_{x \in X} \mathrm{L}(x,\lambda^*,y^*) = \mathrm{L}(x^*,\lambda^*,y^*).$$

Since the function $\mathrm{L}(x,\lambda^*,y^*)$ is convex in $x \in X$ and differentiable at $x^* \in X$, the last identity implies the stationarity condition. ■

Let us look at one example.

Example 2.1. Assuming $a_i > 0$, $p \geq 1$, show that the solution of the problem

$$\min_x \left\{ \Sigma_i \tfrac{a_i}{x_i} : x > 0, \Sigma_i x_i^p \leq 1 \right\}$$

is given by

$$x_i^* = \frac{a_i^{1/(p+1)}}{\left(\Sigma_j a_j^{p/(p+1)}\right)^{1/p}}.$$

Proof. Assuming $x^* > 0$ is a solution such that $\Sigma_i (x_i^*)^p = 1$, the KKT stationarity condition reads

$$\nabla_x \left\{ \Sigma_i \tfrac{a_i}{x_i} + \lambda(\Sigma_i x_i^p - 1) \right\} = 0 \Leftrightarrow \tfrac{a_i}{x_i^2} = p\lambda x_i^{p-1}$$

whence $x_i = [a_i/(p\lambda)]^{1/(p+1)}$. Since $\Sigma_i x_i^p$ should be 1, we get

$$x_i^* = \frac{a_i^{1/(p+1)}}{\left(\Sigma_j a_j^{p/(p+1)}\right)^{1/p}}.$$

This x^* is optimal because the problem is convex and the KKT conditions are satisfied at this point. ■

By examining the KKT conditions, we can obtain explicit solutions for many simple convex optimization problems, which can be used as subproblems in iterative algorithms for solving more complicated convex or even nonconvex optimization problems.

2.3.5 Dual Support Vector Machine

In this subsection, we discuss one interesting application of the optimality conditions for convex programming in support vector machines.

Recall that the support vector machine can be formulated as

$$\begin{array}{ll}\min_{w,b} & \frac{1}{2}\|w\|^2 \\ \text{s.t.} & v^{(i)}(w^T u^{(i)} + b) \ge 1, i = 1,\ldots,m.\end{array}$$

We can write the constraints equivalently as

$$g_i(w,b) = -v^{(i)}(w^T u^{(i)} + b) + 1 \le 0.$$

Thus Lagrangian function L for our problem is given by

$$\mathrm{L}(w,b,\lambda) = \tfrac{1}{2}\|w\|^2 - \Sigma_{i=1}^N \lambda_i [v^{(i)}(w^T u^{(i)} + b) - 1].$$

For fixed λ_i's, the problem is unconstrained. Let us minimize $\mathrm{L}(w,b,\lambda)$ w.r.t. w and b. Setting the derivatives of L w.r.t. w and b to zero, i.e.,

$$\nabla_w \mathrm{L}(w,b,\lambda) = w - \Sigma_{i=1}^N \lambda_i v^{(i)} u^{(i)} = 0,$$

we have

$$w = \Sigma_{i=1}^m \lambda_i v^{(i)} u^{(i)}. \tag{2.3.20}$$

Moreover, we have

$$\nabla_b \mathrm{L}(w,b,\lambda) = \Sigma_{i=1}^m \lambda_i v^{(i)} = 0. \tag{2.3.21}$$

Plugging the above definition of w into $\mathrm{L}(w,b,\lambda)$, we obtain

$$\mathrm{L}(w,b,\lambda) = \Sigma_{i=1}^N \lambda_i - \tfrac{1}{2}\Sigma_{i,j=1}^N v^{(i)} v^{(j)} \lambda_i \lambda_j (u^{(i)})^T u^{(j)} - b\Sigma_{i=1}^m \lambda_i v^{(i)}.$$

Since by (2.3.21), the last term must be zero, we have

$$\mathrm{L}(w,b,\lambda) = \Sigma_{i=1}^N \lambda_i - \tfrac{1}{2}\Sigma_{i,j=1}^N v^{(i)} v^{(j)} \lambda_i \lambda_j (u^{(i)})^T u^{(j)}.$$

Therefore, we can write the dual SVM problem as

$$\begin{array}{ll}\max_\lambda & \Sigma_{i=1}^N \lambda_i - \frac{1}{2}\Sigma_{i,j=1}^N v^{(i)} v^{(j)} \lambda_i \lambda_j (u^{(i)})^T u^{(j)} \\ \text{s.t.} & \lambda_i \ge 0, i = 1,\ldots,m \\ & \Sigma_{i=1}^N \lambda_i v^{(i)} = 0.\end{array}$$

Once we find the optimal λ^*, we can use (2.3.20) to compute optimal w^*. Moreover, with optimal w^*, we can easily solve the primal problem to find the intercept term b as

$$b^* = -\frac{\max_{i:v^{(i)}=-1} w^{*T} v^{(i)} + \min_{i:v^{(i)}=1} w^{*T} v^{(i)}}{2}.$$

It is also interesting to observe that the dual problem only depends on the inner product and we can generalize it easily by using the Kernel trick (Sect. 2.2.6).

2.4 Legendre–Fenchel Conjugate Duality

2.4.1 Closure of Convex Functions

We can extend the domain of a convex function $f : X \to \mathbb{R}$ to the whole space $\mathbb{R}^n$ by setting $f(x) = +\infty$ for any $x \notin X$. In view of the definition of a convex function in (2.2.8), and our discussion about the epigraphs in Sect. 2.2, a function $f : \mathbb{R}^n \to \mathbb{R} \cup \{+\infty\}$ is convex if and only if its epigraph

$$\mathrm{epi}(f) = \{(x,t) \in \mathbb{R}^{n+1} : f(x) \leq t\}$$

is a nonempty convex set.

As we know, closed convex sets possess many nice topological properties. For example, a closed convex set is comprised of the limits of all converging sequences of elements. Moreover, by the Separation Theorem, a closed and nonempty convex set X is the intersection of all closed half-spaces containing X. Among these half-spaces, the most interesting ones are the supporting hyperplanes touching X on the relative boundary.

In functional Language, the "closedness" of epigraph corresponds to a special type of continuity, i.e., the lower semicontinuity. Let $f : \mathbb{R}^n \to \mathbb{R} \cup \{+\infty\}$ be a given function (not necessarily convex). We say that f is lower semicontinuous (l.s.c.) at a point $\bar{x}$, if for every sequence of points $\{x_i\}$ converging to $\bar{x}$ one has

$$f(\bar{x}) \leq \liminf_{i \to \infty} f(x_i).$$

Of course, lim inf of a sequence with all terms equal to $+\infty$ is $+\infty$. f is called lower semicontinuous, if it is lower semicontinuous at every point.

A trivial example of a lower semicontinuous function is a continuous one. Note, however, that a semicontinuous function is not necessarily continuous. What it is obliged is to make only "jumps down." For example, the function

$$f(x) = \begin{cases} 0, & x \neq 0 \\ a, & x = 0 \end{cases}$$

is lower semicontinuous if $a \leq 0$ ("jump down at $x = 0$ or no jump at all"), and is not lower semicontinuous if $a > 0$ ("jump up").

The following statement links lower semicontinuity with the geometry of the epigraph.

Proposition 2.10. *A function f defined on $\mathbb{R}^n$ and taking values from $\mathbb{R} \cup \{+\infty\}$ is lower semicontinuous if and only if its epigraph is closed (e.g., due to its emptiness).*

Proof. We first prove the "only if" part (from lower semicontinuity to closed epigraph). Let (x,t) be the limit of the sequence $\{x_i,t_i\} \subset \text{epi} f$. Then we have $f(x_i) \leq t_i$. Thus the following relation holds: $t = \lim_{i\to\infty} t_i \geq \lim_{i\to\infty} f(x_i) \geq f(x)$.

We now show the "if" part (from closed epigraph to lower semicontinuity). Suppose for contradiction that $f(x) > \gamma > \lim_{i\to\infty} f(x_i)$ for some constant γ, where x_i converges to x. Then there exists a subsequence $\{x_{i_k}\}$ such that $f(x_{i_k}) \leq \gamma$ for all i_k. Since the epigraph is closed, then x must belong to this set, which implies that $f(x) \leq \gamma$, which is a contradiction. ■

As an immediate consequence of Proposition 2.10, the upper bound

$$f(x) = \sup_{\alpha \in \mathscr{A}} f_\alpha(x)$$

of arbitrary family of lower semicontinuous functions is lower semicontinuous. Indeed, the epigraph of the upper bound is the intersection of the epigraphs of the functions forming the bound, and the intersection of closed sets always is closed.

Now let us look at proper lower semicontinuous (l.s.c.) convex functions. According to our general convention, a convex function f is proper if $f(x) < +\infty$ for at least one x and $f(x) > -\infty$ for every x. This implies that proper convex functions have convex nonempty epigraphs. Also as we just have seen, "lower semicontinuous" means "with closed epigraph." Hence proper l.s.c. convex functions always have closed convex nonempty epigraphs.

Similar to the fact that a closed convex set is intersection of closed half-spaces, we can provide an outer description of a proper l.s.c. convex function. More specifically, we can show that a proper l.s.c. convex function f is the upper bound of all its affine minorants given in the form of $t \geq d^T x - a$. Moreover, at every point $\bar{x} \in \text{ri}\,\text{dom} f$ from the relative interior of the domain f, f is even not the upper bound, but simply the maximum of its minorants: there exists an affine function $f_{\bar{x}}(x)$ which underestimates $f(x)$ everywhere in $\mathbb{R}^n$ and is equal to f at $x = \bar{x}$. This is exactly the first-order approximation $f(\bar{x}) + \langle g(\bar{x}), x - \bar{x} \rangle$ given by the definition of subgradients.

Now, what if the convex function is not lower semicontinuous (see Fig. 2.5)? A similar question also arises about convex sets—what to do with a convex set which is not closed? To deal with these convex sets, we can pass from the set to its closure and thus get a "normal" object which is very "close" to the original one. Specifically, while the "main part" of the original set—its relative interior—remains unchanged, we add a relatively small "correction," i.e., the relative boundary, to the set. The same approach works for convex functions: if a proper convex function f is not l.s.c. (i.e., its epigraph, being convex and nonempty, is not closed), we can "correct" the function—replace it with a new function with the epigraph being the closure of $\text{epi}(f)$. To justify this approach, we should make sure that the closure of the epigraph of a convex function is also an epigraph of such a function.

Thus, we conclude that the closure of the epigraph of a convex function f is the epigraph of certain function, referred to as *the closure* $\text{cl} f$ *of* f. Of course, this

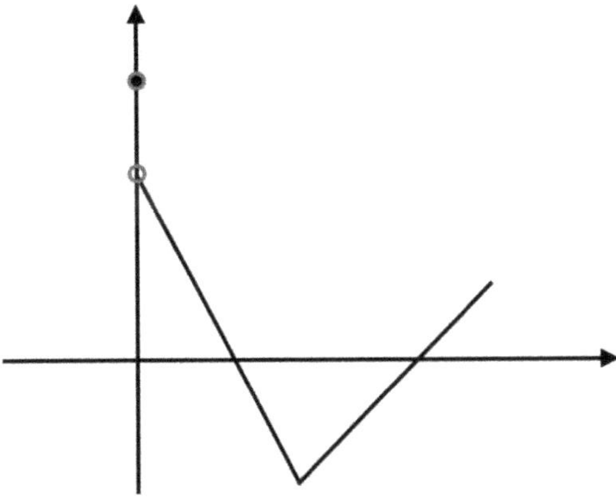

Fig. 2.5 Example for an upper semicontinuous function. The domain of this function is $[0,+\infty)$, and it "jumps up" at 0. However, the function is still convex

latter function is convex (its epigraph is convex—it is the closure of a convex set), and since its epigraph is closed, $\mathrm{cl}f$ is proper. The following statement gives direct description of $\mathrm{cl}f$ in terms of f:

(i) For every x one has $\mathrm{cl}f(x) = \lim\limits_{r\to+0} \inf\limits_{x':\|x'-x\|_2\le r} f(x')$. In particular,

$$f(x) \ge \mathrm{cl}f(x)$$

for all x, and

$$f(x) = \mathrm{cl}f(x)$$

whenever $x \in \mathrm{ri}\,\mathrm{dom}f$, or equivalently whenever $x \notin \mathrm{cl}\,\mathrm{dom}f$. Thus, the "correction" $f \mapsto \mathrm{cl}f$ may vary f only at the points from the relative boundary of $\mathrm{dom}f$,

$$\mathrm{dom}f \subset \mathrm{dom}\,\mathrm{cl}f \subset \mathrm{cl}\,\mathrm{dom}f,$$

hence

$$\mathrm{ri}\,\mathrm{dom}f = \mathrm{ri}\,\mathrm{dom}\,\mathrm{cl}f.$$

(ii) The family of affine minorants of $\mathrm{cl}f$ is exactly the family of affine minorants of f, so that

$$\mathrm{cl}f(x) = \sup\{\phi(x) : \phi \text{ is an affine minorant of } f\},$$

due to the fact that $\mathrm{cl}f$ is l.s.c. and is therefore the upper bound of its affine minorants, and the sup in the right-hand side can be replaced with max whenever $x \in \mathrm{ri}\,\mathrm{dom}\,\mathrm{cl}f = \mathrm{ri}\,\mathrm{dom}f$.

2.4.2 Conjugate Functions

Let f be a convex function. We know that f "basically" is the upper bound of all its affine minorants. This is exactly the case when f is proper, otherwise the corresponding equality takes place everywhere except, perhaps, some points from the relative boundary of $\mathrm{dom} f$. Now, when an affine function $d^T x - a$ is an affine minorant of f? It is the case if and only if

$$f(x) \geq d^T x - a$$

for all x or, which is the same, if and only if

$$a \geq d^T x - f(x)$$

for all x. We see that if the slope d of an affine function $d^T x - a$ is fixed, then in order for the function to be a minorant of f we should have

$$a \geq \sup_{x \in \mathbb{R}^n} [d^T x - f(x)].$$

The supremum in the right-hand side of the latter relation is certain function of d and this function, denoted by f^*, is called the *Legendre–Fenchel conjugate* of f:

$$f^*(d) = \sup_{x \in \mathbb{R}^n} [d^T x - f(x)].$$

Geometrically, the Legendre–Fenchel transformation answers the following question: given a slope d of an affine function, i.e., given the hyperplane $t = d^T x$ in $\mathbb{R}^{n+1}$, what is the minimal "shift down" of the hyperplane which places it below the graph of f?

From the definition of the conjugate it follows that this is a proper l.s.c. convex function. Indeed, we lose nothing when replacing $\sup_{x \in \mathbb{R}^n} [d^T x - f(x)]$ by $\sup_{x \in \mathrm{dom} f} [d^T x - f(x)]$, so that the conjugate function is the upper bound of a family of affine functions. This bound is finite at least at one point, namely, at every d coming from affine minorant of f, and we know that such a minorant exists. Therefore, f^* must be a proper l.s.c. convex function, as claimed.

The most elementary (and the most fundamental) fact about the conjugate function is its symmetry.

Proposition 2.11. *Let f be a convex function. Then $(f^*)^* = \mathrm{cl} f$. In particular, if f is l.s.c., then $(f^*)^* = f$.*

Proof. The conjugate function of f^* at the point x is, by definition,

$$\sup_{d \in \mathbb{R}^n} [x^T d - f^*(d)] = \sup_{d \in \mathbb{R}^n, a \geq f^*(d)} [d^T x - a].$$

The second sup here is exactly the supremum of all affine minorants of f due to the origin of the Legendre–Fenchel transformation: $a \geq f^*(d)$ if and only if the affine form $d^T x - a$ is a minorant of f. The result follows since we already know that the upper bound of all affine minorants of f is the closure of f. ■

The Legendre–Fenchel transformation is a very powerful tool—this is a "global" transformation, so that *local* properties of f^* correspond to *global* properties of f.

- $d = 0$ belongs to the domain of f^* if and only if f is below bounded, and if it is the case, then $f^*(0) = -\inf f$;
- if f is proper and l.s.c., then the subgradients of f^* at $d = 0$ are exactly the minimizers of f on $\mathbb{R}^n$;
- $\mathrm{dom} f^*$ is the entire $\mathbb{R}^n$ if and only if $f(x)$ grows, as $\|x\|_2 \to \infty$, faster than $\|x\|_2$: there exists a function $r(t) \to \infty$, as $t \to \infty$ such that

$$f(x) \geq r(\|x\|_2) \quad \forall x.$$

Thus, whenever we can compute explicitly the Legendre–Fenchel transformation of f, we get a lot of "global" information on f.

Unfortunately, the more detailed investigation of the properties of Legendre–Fenchel transformation is beyond our scope. Below we simply list some facts and examples.

- From the definition of Legendre transformation,

$$f(x) + f^*(d) \geq x^T d \quad \forall x, d.$$

Specifying here f and f^*, we get certain inequality, e.g., the following one: [Young's Inequality] *if p and q are positive reals such that $\frac{1}{p} + \frac{1}{q} = 1$, then*

$$\frac{|x|^p}{p} + \frac{|d|^q}{q} \geq xd \quad \forall x, d \in \mathbb{R}$$

- The Legendre–Fenchel transformation of the function

$$f(x) \equiv -a$$

is the function which is equal to a at the origin and is $+\infty$ outside the origin; similarly, the Legendre–Fenchel transformation of an affine function $\bar{d}^T x - a$ is equal to a at $d = \bar{d}$ and is $+\infty$ when $d \neq \bar{d}$;
- The Legendre–Fenchel transformation of the strictly convex quadratic form

$$f(x) = \frac{1}{2} x^T A x$$

($A \succeq 0$) is the quadratic form

$$f^*(d) = \frac{1}{2} d^T A^{-1} d$$

- The Legendre–Fenchel transformation of the Euclidean norm

$$f(x) = \|x\|_2$$

is the function which is equal to 0 in the closed unit ball centered at the origin and is $+\infty$ outside the ball.

2.5 Exercises and Notes

Exercises.

1. *Determine whether the following sets are convex or not.*

   (a) $\{x \in \mathbb{R}^2 : x_1 + i^2 x_2 \le 1, i = 1, \ldots, 10\}$.
   (b) $\{x \in \mathbb{R}^2 : x_1^2 + 2ix_1x_2 + i^2x_2^2 \le 1, i = 1, \ldots, 10\}$.
   (c) $\{x \in \mathbb{R}^2 : x_1^2 + ix_1x_2 + i^2x_2^2 \le 1, i = 1, \ldots, 10\}$.
   (d) $\{x \in \mathbb{R}^2 : x_1^2 + 5x_1x_2 + 4x_2^2 \le 1\}$.
   (e) $\{x \in \mathbb{R}^2 : \exp\{x_1\} \le x_2\}$.
   (f) $\{x \in \mathbb{R}^2 : \exp\{x_1\} \ge x_2\}$.
   (g) $\{x \in \mathbb{R}^n : \sum_{i=1}^n x_i^2 = 1\}$.

2. Assume that $X = \{x_1, \ldots, x_k\}$ and $Y = \{y_1, \ldots, y_m\}$ are finite sets in $\mathbb{R}^n$, with $k + m \ge n + 2$, and all the points $x_1, \ldots, x_k, y_1, \ldots, y_m$ are distinct. Assume that for any subset $S \subset X \cup Y$ comprised of $n+2$ points the convex hulls of the sets $X \cap S$ and $Y \cap S$ do not intersect. Then the convex hulls of X and Y also do not intersect. (Hint: Assume on contrary that the convex hulls of X and Y intersect, so that

$$\sum_{i=1}^k \lambda_i x_i = \sum_{j=1}^m \mu_j y_j$$

   for certain nonnegative λ_i, $\sum_i \lambda_i = 1$, and certain nonnegative μ_j, $\sum_j \mu_j = 1$, and look at the expression of this type with the minimum possible total number of nonzero coefficients λ_i, μ_j.)

3. Prove that the following functions are convex on the indicated domains:

   (a) $\frac{x^2}{y}$ on $\{(x, y) \in \mathbb{R}^2 \mid y > 0\}$,
   (b) $\ln(\exp\{x\} + \exp\{y\})$ on the 2D plane.

4. A function f defined on a convex set Q is called log-convex on Q, if it takes real positive values on Q and the function $\ln f$ is convex on Q. Prove that

   (a) a log-convex on Q function is convex on Q,
   (b) the sum (more generally, linear combination with positive coefficients) of two log-convex functions on Q also is log-convex on the set.

5. Show the following statements related to the computation of subgradients.

   (a) The subgradient of $f(x) = \sqrt{x}$ does not exist at $x = 0$.
   (b) The subdifferential of $f(x) = |x|$ is given by $[-1, 1]$.
   (c) Let u and v be given. What is the subdifferential of $f(w, b) = \max\{0, v(w^T u + b)\} + \rho \|w\|_2^2$ at w and b.
   (d) The subdifferential of $f(x) = \|x\|$ is given by $\partial f(0) = \{x \in \mathbb{R}^n \| \|x\| \le 1\}$ and $\partial f(x) = \{x / \|x\|\}$ for $x \ne 0$.

6. Find the minimizer of a linear function

$$f(x) = c^T x$$

on the set

$$V_p = \{x \in \mathbb{R}^n \mid \Sigma_{i=1}^n |x_i|^p \leq 1\},$$

where p, $1 < p < \infty$, is a parameter. What happens with the solution when the parameter becomes 0.5?

7. Let $a_1, \ldots, a_n > 0$, $\alpha, \beta > 0$. Solve the optimization problem

$$\min_x \left\{ \Sigma_{i=1}^n \frac{a_i}{x_i^\alpha} : x > 0, \Sigma_i x_i^\beta \leq 1 \right\}.$$

8. Consider the optimization problem

$$\max_{x,y} \{ f(x,y) = ax + by + \ln(\ln y - x) + \ln(y) : (x,y) \in X = \{y > \exp\{x\}\} \},$$

where $a, b \in \mathbb{R}$ are parameters. Is the problem convex? What is the domain in space of parameters where the problem is solvable? What is the optimal value? Is it convex in the parameters?

9. Let $a_1, \ldots, a_n$ be positive reals, and let $0 < s < r$ be two reals. Find maximum and minimum of the function

$$\Sigma_{i=1}^n a_i |x_i|^r$$

on the surface

$$\Sigma_{i=1}^n |x_i|^s = 1.$$

Notes. Further readings on convex analysis and convex optimization theory can be found on the monographs [50, 118], classic textbooks [12, 16, 79, 97, 104, 107, 120], and online course materials [92].

Chapter 3
Deterministic Convex Optimization

In this chapter, we study algorithms for solving convex optimization problems. We will focus on algorithms that have been applied or have the potential to be applied for solving machine learning and other data analysis problems. More specifically, we will discuss first-order methods which have been proven effective for large-scale optimization. These methods also form the basis for other computationally efficient methods, e.g., stochastic and randomized methods to be discussed in later chapters.

3.1 Subgradient Descent

We start with the simplest gradient descent method to minimize a differentiable convex function f. Starting from an initial point $x_1 \in \mathbb{R}^n$, the gradient descent method updates the search point x_t according to

$$x_{t+1} = x_t - \gamma_t \nabla f(x_t), t = 1, 2, \ldots, \tag{3.1.1}$$

where $\gamma_t > 0$ is a certain stepsize at the t-th iteration. The rationale behind (3.1.1) is to move a small step along the direction (also known as the steepest descent direction) that minimizes the local first-order Taylor approximation of f.

We need to make two essential modifications to the gradient descent method in order to solve a general convex optimization problem given by

$$f^* := \min_{x \in X} f(x). \tag{3.1.2}$$

Here $X \subseteq \mathbb{R}^n$ is a closed convex set and $f : X \to \mathbb{R}$ is a proper l.s.c. convex function. Without specific mention, we assume that the set of optimal solutions of (3.1.2) is nonempty and x^* is an arbitrary solution of (3.1.2). First, since the objective function f is not necessarily differentiable, it makes sense to replace $\nabla f(x_t)$ in (3.1.1) with a subgradient $g(x_t) \in \partial f(x_t)$. Second, the recursion in (3.1.1) applies only to

G. Lan, *First-Order and Stochastic Optimization Methods for Machine Learning*,
Springer Series in the Data Sciences, https://doi.org/10.1007/978-3-030-39568-1_3

unconstrained problems. For the constrained case when $X \neq \mathbb{R}^n$, the search point x_{t+1} defined in (3.1.1) may fall outside the feasible set X. Hence, it is necessary to "push" x_{t+1} back to X by using projection. Incorporating these enhancements, we update x_t according to

$$x_{t+1} := \text{argmin}_{x \in X} \|x - (x_t - \gamma_t g(x_t))\|_2, t = 1, 2, \ldots, \tag{3.1.3}$$

for some $g(x_t) \in \partial f(x_t)$ and $\gamma_t > 0$.

The projected subgradient iteration in (3.1.3) admits some natural explanation from the proximity control point of view. Indeed, (3.1.3) can be written equivalently as

$$\begin{aligned} x_{t+1} &= \text{argmin}_{y \in X} \tfrac{1}{2} \|x - (x_t - \gamma_t g(x_t))\|_2^2 \\ &= \text{argmin}_{x \in X} \gamma_t \langle g(x_t), x - x_t \rangle + \tfrac{1}{2} \|x - x_t\|_2^2 \\ &= \text{argmin}_{x \in X} \gamma_t \left[f(x_t) + \langle g(x_t), x - x_t \rangle \right] + \tfrac{1}{2} \|x - x_t\|_2^2 \\ &= \text{argmin}_{x \in X} \gamma_t \langle g(x_t), x \rangle + \tfrac{1}{2} \|x - x_t\|_2^2. \end{aligned} \tag{3.1.4}$$

This implies that we would like to minimize the linear approximation $f(x_t) + \langle g(x_t), x - x_t \rangle$ of $f(x)$ over X, without moving too far away from x_t so as to have $\|x - x_t\|_2^2$ small. The parameter $\gamma_t > 0$ balances these two terms, and its selection will depend on the properties of the objective function f, e.g., the differentiability of f, the Lipschitz continuity of its gradients, and so forth.

3.1.1 General Nonsmooth Convex Problems

We will first consider a general convex function f which is Lipschitz continuous over X, i.e., $\exists M > 0$ such that

$$|f(x) - f(y)| \leq M \|x - y\|_2 \quad \forall x, y \in X. \tag{3.1.5}$$

Observe that this assumption is not too restrictive in view of Theorem 2.4.

The following lemma provides an important characterization for x_{t+1} by using the representation in (3.1.4).

Lemma 3.1. *Let x_{t+1} be defined in (3.1.3). For any $y \in X$, we have*

$$\gamma_t \langle g(x_t), x_{t+1} - x \rangle + \tfrac{1}{2} \|x_{t+1} - x_t\|_2^2 \leq \tfrac{1}{2} \|x - x_t\|_2^2 - \tfrac{1}{2} \|x - x_{t+1}\|_2^2.$$

Proof. Denote $\phi(x) = \gamma_t \langle g(x_t), x \rangle + \frac{1}{2} \|x - x_t\|_2^2$. By the strong convexity of ϕ, we have

$$\phi(x) \geq \phi(x_{t+1}) + \langle \phi'(x_{t+1}), x - x_{t+1} \rangle + \tfrac{1}{2} \|x - x_{t+1}\|_2^2.$$

Moreover, by the first-order optimality condition of (3.1.4), we have $\langle \phi'(x_{t+1}), x - x_{t+1}\rangle \geq 0$ for any $x \in X$. The result immediately follows by combining these two inequalities. ■

The following theorem describes some general convergence properties for the subgradient descent method. Note that in our convergence analysis for first-order methods, we often use the following simple inequality

$$bt - \tfrac{at^2}{2} \leq \tfrac{b^2}{2a}, \forall a > 0, b \in \mathbb{R}, t \in \mathbb{R}. \tag{3.1.6}$$

Theorem 3.1. *Let $x_t, t = 1, \ldots, k$, be generated by (3.1.3). Under Assumption (3.1.5), we have*

$$\textstyle\sum_{t=s}^k \gamma_t [f(x_t) - f(x)] \leq \tfrac{1}{2}\left[\|x - x_s\|_2^2 + M^2 \sum_{t=s}^k \gamma_t^2\right], \forall x \in X. \tag{3.1.7}$$

Proof. By the convexity of f and Lemma 3.1,

$$\begin{aligned}
\gamma_t[f(x_t) - f(x)] &\leq \gamma_t \langle g(x_t), x_t - x\rangle \\
&\leq \tfrac{1}{2}\|x - x_t\|_2^2 - \tfrac{1}{2}\|x - x_{t+1}\|_2^2 \\
&\quad + \gamma_t \langle g(x_t), x_t - x_{t+1}\rangle - \tfrac{1}{2}\|x_{t+1} - x_t\|_2^2 \\
&\leq \tfrac{1}{2}\|x - x_t\|_2^2 - \tfrac{1}{2}\|x - x_{t+1}\|_2^2 + \tfrac{\gamma_t^2}{2}\|g(x_t)\|_2^2 \\
&\leq \tfrac{1}{2}\|x - x_t\|_2^2 - \tfrac{1}{2}\|x - x_{t+1}\|_2^2 + \tfrac{\gamma_t^2}{2} M^2,
\end{aligned}$$

where the third inequality follows from the Cauchy–Schwarz inequality and (3.1.6). The result then immediately follows by summing up the above inequality from $t = s$ to k. ■

We now provide a simple specific choice for the stepsizes γ_t.

Corollary 3.1. *Let us denote*

$$D_X^2 \equiv D_{X, \|\cdot\|_2^2/2} := \max_{x_1, x_2 \in X} \tfrac{\|x_1 - x_2\|_2^2}{2}. \tag{3.1.8}$$

Suppose that the number of iterations k is fixed and

$$\gamma_t = \sqrt{\tfrac{2D_X^2}{kM^2}}, t = 1, \ldots, k.$$

Then

$$f(\bar{x}_1^k) - f^* \leq \tfrac{\sqrt{2}MD_X}{2\sqrt{k}}, \quad \forall k \geq 1,$$

where

$$\bar{x}_s^k = \left(\textstyle\sum_{t=s}^k \gamma_t\right)^{-1} \textstyle\sum_{t=s}^k (\gamma_t x_t). \tag{3.1.9}$$

Proof. By Theorem 3.1 and the fact that $f(\bar{x}_s^k) \le (\Sigma_{t=s}^k \gamma_t)^{-1} \Sigma_{t=s}^k f(x_t)$, we have

$$f(\bar{x}_s^k) \le (2\Sigma_{t=s}^k \gamma_t)^{-1} \left[\|x^* - x_s\|_2^2 + M^2 \Sigma_{t=s}^k \gamma_t^2 \right]. \tag{3.1.10}$$

If $\gamma_t = \gamma$, $t = 1, \ldots, k$, we have

$$f(\bar{x}_1^k) - f^* \le \tfrac{1}{2} \left(\tfrac{2}{k\gamma} D_X^2 + M^2 \gamma \right).$$

The result follows by minimizing the right-hand side (RHS) of above inequality w.r.t. γ. ■

We can also use variable stepsizes without fixing the number of iterations k a priori.

Corollary 3.2. *If*

$$\gamma_t = \sqrt{\tfrac{2D_X^2}{tM^2}}, t = 1, \ldots, k,$$

then

$$f(\bar{x}_{\lceil k/2 \rceil}^k) - f^* \le \tfrac{(1+2\ln 3)MD_X}{2(\sqrt{2}-1)\sqrt{k+1}}$$

for any $k \ge 3$, where $\bar{x}_{\lceil k/2 \rceil}^k$ is defined in (3.1.9).

Proof. It suffices to bound the RHS of (3.1.10) by using the following bounds:

$$\begin{aligned}
\Sigma_{t=\lceil k/2 \rceil}^k \gamma_t &= \Sigma_{t=\lceil k/2 \rceil}^k \sqrt{\tfrac{2D_X^2}{tM^2}} \ge \tfrac{\sqrt{2}D_X}{M} \int_{(k+1)/2}^{k+1} t^{-1/2} dt \\
&\ge \tfrac{\sqrt{2}D_X}{M} (1 - \tfrac{1}{\sqrt{2}})(k+1)^{1/2}, \forall k \ge 1. \\
\Sigma_{t=\lceil k/2 \rceil}^k \gamma_t^2 &= \tfrac{2D_X^2}{M^2} \Sigma_{t=\lceil k/2 \rceil}^k \tfrac{1}{t} \le \tfrac{2D_X^2}{M^2} \int_{\lceil k/2 \rceil - 1}^k \tfrac{1}{t} \le \tfrac{2D_X^2}{M^2} \ln \tfrac{k}{\lceil k/2 \rceil - 1} \\
&\le \tfrac{2D_X^2}{M^2} \ln 3, \forall k \ge 3.
\end{aligned}$$

■

Observe that in (3.1.9), we define the output solution as the weighted average of the iterates $\{x_k\}$. However, we can define the output solution $\hat{x}_k \in \{x_1, \ldots, x_k\}$ as the best solution found so far in the trajectory, i.e.,

$$f(\hat{x}_k) = \min_{i=1,\ldots,k} f(x_i). \tag{3.1.11}$$

We can easily see that all the results stated in Theorem 3.1 and Corollaries 3.1 and 3.2 still hold with this different selection of output solution. Moreover, it is worth noting that the definition of the diameter D_X in (3.1.8) depends on the norm $\|\cdot\|_2$. We will discuss later how to generalize such a characteristic of the feasible set X.

3.1.2 Nonsmooth Strongly Convex Problems

In this subsection, we assume that f, in addition to (3.1.5), is strongly convex, i.e., $\exists \mu > 0$ s.t.

$$f(y) \geq f(x) + \langle g(x), y-x\rangle + \tfrac{\mu}{2}\|y-x\|_2^2, \forall x, y \in X, \tag{3.1.12}$$

where $g(x) \in \partial f(x)$.

Theorem 3.2. *Let $x_t, t = 1,\ldots,k$, be generated by (3.1.3). Under Assumptions (3.1.5) and (3.1.12), if for some $w_t \geq 0$,*

$$\tfrac{w_t(1-\mu\gamma_t)}{\gamma_t} \leq \tfrac{w_{t-1}}{\gamma_{t-1}}, \tag{3.1.13}$$

then

$$\textstyle\sum_{t=1}^k w_t[f(x_t) - f(x)] \leq \tfrac{w_1(1-\mu\gamma_1)}{2\gamma_1}\|x^* - x_1\|_2^2 - \tfrac{w_k}{2\gamma_k}\|x - x_{k+1}\|_2^2 + M^2 \textstyle\sum_{t=1}^k w_t\gamma_t. \tag{3.1.14}$$

Proof. By the strong convexity of f and Lemma 3.1,

$$\begin{aligned} f(x_t) - f(x) &\leq \langle g(x_t), x_t - x\rangle - \tfrac{\mu}{2}\|x - x_t\|_2^2 \\ &\leq \tfrac{1-\mu\gamma_t}{2\gamma_t}\|x - x_t\|_2^2 - \tfrac{1}{2\gamma_t}\|x - x_{t+1}\|_2^2 \\ &\quad + \langle g(x_t), x_t - x_{t+1}\rangle - \tfrac{1}{2\gamma_t}\|x_{t+1} - x_t\|_2^2 \\ &\leq \tfrac{1-\mu\gamma_t}{2\gamma_t}\|x - x_t\|_2^2 - \tfrac{1}{2\gamma_t}\|x - x_{t+1}\|_2^2 + \tfrac{\gamma_t}{2}\|g(x_t)\|_2^2 \\ &\leq \tfrac{1-\mu\gamma_t}{2\gamma_t}\|x - x_t\|_2^2 - \tfrac{1}{2\gamma_t}\|x - x_{t+1}\|_2^2 + \tfrac{\gamma_t}{2}M^2, \end{aligned}$$

where the last inequality follows from the Cauchy–Schwarz inequality and (3.1.6). Summing up these inequalities with weight w_t, we obtain (3.1.14). ∎

Below we provide a specific selection of $\{\gamma_k\}$ and $\{w_k\}$.

Corollary 3.3. *If*

$$\gamma_t = \tfrac{2}{\mu t} \text{ and } w_t = t, \forall t \geq 1, \tag{3.1.15}$$

then

$$f(\bar{x}_1^k) - f(x) + \tfrac{\mu k}{2(k+1)}\|x_{k+1} - x\|^2 \leq \tfrac{4M^2}{\mu(k+1)}, \ \forall x \in X.$$

where $\bar{x}_1^k$ is defined in (3.1.9).

Proof. It can be easily seen that

$$\tfrac{w_t(1-\mu\gamma_t)}{\gamma_t} = \tfrac{\mu t(t-2)}{2} \text{ and } \tfrac{w_{t-1}}{\gamma_{t-1}} = \tfrac{\mu(t-1)^2}{2}$$

and hence that (3.1.13) holds. It then follows from (3.1.14) and (3.1.15) that

$$\textstyle\sum_{t=1}^k t[f(x_t) - f(x)] \leq \tfrac{\mu}{4}\|x_1 - x\|^2 - \tfrac{\mu k^2}{4}\|x_{k+1} - x\|^2 + \tfrac{2kM^2}{\mu}.$$

Using the definition of $\bar{x}_k$ in (3.1.9) and the convexity of f, we conclude that

$$f(\bar{x}_1^k) - f(x) \le \tfrac{2}{k(k+1)}\left(\tfrac{\mu}{4}\|x_1 - x\|^2 - \tfrac{\mu k^2}{4}\|x_{k+1} - x\|^2 + \tfrac{2kM^2}{\mu}\right).$$

■

In view of Corollary 3.3, we can bound both the function optimality gap $f(\bar{x}_1^k) - f(x^*)$ and the distance to the optimal solution $\|x_{k+1} - x\|^2$ by $\mathcal{O}(1/k)$. Here $\mathcal{O}(1/k)$ means "in the order of $1/k$". Similar to the general convex case, we can use the $\hat{x}_k$ in (3.1.11) in place of $\bar{x}_1^k$ as the output solution.

3.1.3 Smooth Convex Problems

In this subsection, we consider differentiable convex functions f with Lipschitz continuous gradients, i.e.,

$$\|\nabla f(x) - \nabla f(y)\|_2 \le L\|x - y\|_2, \;\; \forall x, y \in X. \tag{3.1.16}$$

These functions are referred to as smooth convex function in this text. Since f is differentiable, we can set the subgradient $g(x_t) = \nabla f(x_t)$ in (3.1.3), and the resulting algorithm is often called projected gradient method.

We first prove a convenient representation about smoothness. Observe that this result does not depend on the convexity of f.

Lemma 3.2. *For any $x, y \in X$, we have*

$$f(y) - f(x) - \langle f'(x), y - x\rangle \le \tfrac{L}{2}\|y - x\|_2^2. \tag{3.1.17}$$

Proof. For all $x, y \in X$, we have

$$\begin{aligned} f(y) &= f(x) + \int_0^1 \langle f'(x + \tau(y - x)), y - x\rangle d\tau \\ &= f(x) + \langle f'(x), y - x\rangle + \int_0^1 \langle f'(x + \tau(y - x)) - f'(x), y - x\rangle d\tau. \end{aligned}$$

Therefore,

$$\begin{aligned} & f(y) - f(x) - \langle f'(x), y - x\rangle \\ &= \int_0^1 \langle f'(x + \tau(y - x)) - f'(x), y - x\rangle d\tau \\ &\le \int_0^1 \|f'(x + \tau(y - x)) - f'(x)\|_2 \|y - x\|_2 d\tau \\ &\le \int_0^1 \tau L\|y - x\|_2^2 d\tau = \tfrac{L}{2}\|y - x\|_2^2. \end{aligned}$$

■

Our next result shows that the function values at the iterates x_t, $t \ge 1$, are monotonically nonincreasing.

Lemma 3.3. *Let $\{x_t\}$ be generated by (3.1.3). If (3.1.16) holds and*

$$\gamma_t \leq \tfrac{2}{L}, \tag{3.1.18}$$

then

$$f(x_{t+1}) \leq f(x_t), \ \forall t \geq 1.$$

Proof. By the optimality condition of (3.1.3), we have

$$\langle \gamma_t g(x_t) + x_{t+1} - x_t, x - x_{t+1} \rangle \geq 0, \ \forall x \in X.$$

Letting $x = x_t$ in the above relation, we obtain

$$\gamma_t \langle g(x_t), x_{t+1} - x_t \rangle \leq -\|x_{t+1} - x_t\|_2^2. \tag{3.1.19}$$

It then follows from (3.1.17) and the above relation that

$$\begin{aligned} f(x_{t+1}) &\leq f(x_t) + \langle g(x_t), x_{t+1} - x_t \rangle + \tfrac{L}{2}\|x_{t+1} - x_t\|_2^2 \\ &\leq f(x_t) - \left(\tfrac{1}{\gamma_t} - \tfrac{L}{2}\right)\|x_{t+1} - x_t\|_2^2 \leq f(x_t). \end{aligned}$$

■

We are now ready to establish the main convergence properties for the projected gradient method applied to smooth convex optimization problems.

Theorem 3.3. *Let $\{x_t\}$ be generated by (3.1.3). If (3.1.16) holds and*

$$\gamma_t = \gamma \leq \tfrac{1}{L}, \forall t \geq 1, \tag{3.1.20}$$

then

$$f(x_{k+1}) - f(x) \leq \tfrac{1}{2\gamma k}\|x - x_1\|_2^2, \ \forall x \in X.$$

Proof. By (3.1.17), we have

$$\begin{aligned} f(x_{t+1}) &\leq f(x_t) + \langle g(x_t), x_{t+1} - x_t \rangle + \tfrac{L}{2}\|x_{t+1} - x_t\|_2^2 \\ &\leq f(x_t) + \langle g(x_t), x - x_t \rangle + \langle g(x_t), x_{t+1} - x \rangle + \tfrac{L}{2}\|x_{t+1} - x_t\|_2^2. \end{aligned} \tag{3.1.21}$$

It then follows from the above inequality, the convexity of f, and Lemma 3.1 that

$$\begin{aligned} f(x_{t+1}) &\leq f(x) + \tfrac{1}{2\gamma_t}\left(\|x - x_t\|_2^2 - \|x - x_{t+1}\|_2^2 - \|x_t - x_{t+1}\|_2^2\right) \\ &\quad + \tfrac{L}{2}\|x_{t+1} - x_t\|_2^2 \\ &\leq f(x) + \tfrac{1}{2\gamma}\left(\|x - x_t\|_2^2 - \|x - x_{t+1}\|_2^2\right), \end{aligned}$$

where the last inequality follows from (3.1.20). Summing up the above inequalities from $t = 1$ to k, and using Lemma 3.3, we have

$$k[f(x_{k+1}) - f(x)] \leq \textstyle\sum_{t=1}^k [f(x_{t+1}) - f(x)] \leq \tfrac{1}{2\gamma}\|x - x_1\|_2^2.$$

■

In view of Theorem 3.3, one may choose $\gamma = 1/L$ and then the rate of convergence of the projected gradient method becomes $f(x_{k+1}) - f^* \leq L/(2k)$.

3.1.4 Smooth and Strongly Convex Problems

In this section, we discuss the convergence properties of the projected gradient method when the objective function f is smooth and strongly convex.

Theorem 3.4. *Let $\{x_t\}$ be generated by (3.1.3). Assume (3.1.12) and (3.1.16) hold, and let $\gamma_t = \gamma = 1/L$, $t = 1,\ldots,k$. Then,*

$$\|x - x_{k+1}\|_2^2 \le (1 - \tfrac{\mu}{L})^k \|x - x_1\|_2^2. \tag{3.1.22}$$

Proof. It follows from (3.1.21), the strong convexity of f, and Lemma 3.1 that

$$\begin{aligned} f(x_{t+1}) &\le f(x) - \tfrac{\mu}{2}\|x - x_t\|_2^2 + \tfrac{1}{2\gamma_t}\left(\|x - x_t\|_2^2 - \|x - x_{t+1}\|_2^2 - \|x_t - x_{t+1}\|_2^2\right) \\ &\quad + \tfrac{L}{2}\|x_{t+1} - x_t\|_2^2 \\ &\le f(x) + \tfrac{1-\mu\gamma}{2\gamma}\|x - x_t\|_2^2 - \tfrac{1}{2\gamma}\|x - x_{t+1}\|_2^2. \end{aligned}$$

Using the above relation, the facts $\gamma = 1/L$ and $f(x_t) - f(x^*) \ge 0$, we have

$$\|x_{t+1} - x^*\|_2^2 \le (1 - \tfrac{\mu}{L})\|x_t - x^*\|_2^2,$$

which clearly implies (3.1.22). ■

In order to find a solution $\bar{x} \in X$ such that $\|\bar{x} - x^*\|^2 \le \varepsilon$, it suffices to have

$$\begin{aligned} (1 - \tfrac{\mu}{L})^k \|x - x_1\|_2^2 \le \varepsilon &\Longleftrightarrow k \log(1 - \tfrac{\mu}{L}) \le \log \tfrac{\varepsilon}{\|x - x_1\|_2^2} \\ &\Longleftrightarrow k \ge \tfrac{1}{-\log(1 - \frac{\mu}{L})} \log \tfrac{\|x - x_1\|_2^2}{\varepsilon} \\ &\Longleftarrow k \ge \tfrac{L}{\mu} \log \tfrac{\|x - x_1\|_2^2}{\varepsilon}, \end{aligned} \tag{3.1.23}$$

where the last inequality follows from the fact that $-\log(1 - \alpha) \ge \alpha$ for any $\alpha \in [0, 1)$.

3.2 Mirror Descent

The subgradient descent method is intrinsically linked to the Euclidean structure of $\mathbb{R}^n$. More specifically, the construction of the method relies on the Euclidean projection (see (3.1.3)), and the quantities D_X and M used in the efficiency estimate (see Corollary 3.1) are defined in terms of the Euclidean norm. In this section we develop a substantial generalization of the subgradient descent method allowing to adjust, to some extent, the method to the possibly non-Euclidean geometry of the problem in question. We shall see in the mean time that we can gain a lot, both theoretically and numerically, from such an adjustment.

Let $\|\cdot\|$ be a (general) norm on $\mathbb{R}^n$ and $\|x\|_* = \sup_{\|y\|\le 1}\langle x,y\rangle$ be its dual norm. We say that a function $\nu : X \to \mathbb{R}$ is a *distance generating function* modulus $\sigma_\nu > 0$ with respect to $\|\cdot\|$, if ν is convex and continuous on X, the set

$$X^o = \left\{x \in X : \text{there exists } p \in \mathbb{R}^n \text{ such that } x \in \arg\min_{u\in X}[p^T u + \nu(u)]\right\}$$

is convex (note that X^o always contains the relative interior of X), and restricted to X^o, ν is continuously differentiable and strongly convex with parameter σ_ν with respect to $\|\cdot\|$, i.e.,

$$\langle x' - x, \nabla\nu(x') - \nabla\nu(x)\rangle \ge \sigma_\nu \|x' - x\|^2, \ \forall x', x \in X^o. \tag{3.2.1}$$

The simplest example of a distance generating function is $\nu(x) = \|x\|_2^2/2$ (modulus 1 with respect to $\|\cdot\|_2$, $X^o = X$). Associated with the distance generating function, we define the prox-function (Bregman's distance or divergence) $V : X^o \times X \to \mathbb{R}_+$ as follows

$$V(x,z) = \nu(z) - [\nu(x) + \langle \nabla\nu(x), z - x\rangle. \tag{3.2.2}$$

Note that $V(x,\cdot)$ is nonnegative and is strongly convex modulus σ_ν with respect to the norm $\|\cdot\|$. Also by the strong convexity of ν, we have

$$V(x,z) \ge \tfrac{\sigma_\nu}{2}\|x - z\|^2. \tag{3.2.3}$$

In case $\nu(x) = \|x\|_2^2/2$, we have $V(x,z) = \|z - x\|_2^2/2$.

Without loss of generality, we assume that the strong convexity modulus σ_ν of ν is given by 1. Indeed, if $\sigma_\nu \ne 1$, we can always choose ν/σ_ν as the distance generating function and define its associated prox-function. The following quantity $D_X > 0$ will be used frequently in the complexity analysis of first-order and stochastic algorithms.

$$D_X^2 \equiv D_{X,\nu}^2 := \max_{x_1, x \in X} V(x_1, x). \tag{3.2.4}$$

Clearly, the definition of $D_{X,\nu}$ generalizes the definition of $D_{X,\|\cdot\|_2^2/2}$ in (3.1.8) with $\nu(x) = \|x\|_2^2/2$.

With the definition of the prox-function, we modify the subgradient iteration in (3.1.4) to

$$x_{t+1} = \mathrm{argmin}_{x\in X}\, \gamma_t\langle g(x_t), x\rangle + V(x_t, x), t = 1, 2, \ldots. \tag{3.2.5}$$

This implies that we would like to minimize a linear approximation of f, but do not move too far away from x_t in terms of $V(x_t, x)$. It can be easily seen that (3.1.4) is a special case of (3.2.5) with $V(x_t, x) = \|x - x_t\|_2^2/2$. The following lemma characterizes the updated solution x_{t+1} in (3.2.5).

Lemma 3.4. *Let x_{t+1} be defined in (3.2.5). For any $y \in X$, we have*

$$\gamma_t\langle g(x_t), x_{t+1} - x\rangle + V(x_t, x_{t+1}) \le V(x_t, x) - V(x_{t+1}, x).$$

Proof. By the optimality condition of (3.2.5),

$$\langle \gamma_t g(x_t) + \nabla V(x_t, x_{t+1}), x - x_t \rangle \geq 0, \ \forall x \in X,$$

where $\nabla V(x_t, x_{t+1})$ denotes the gradient of $V(x_t, \cdot)$ at x_{t+1}. Using the definition of the prox-function (3.2.2), it is easy to verify that

$$V(x_t, x) = V(x_t, x_{t+1}) + \langle \nabla V(x_t, x_{t+1}), x - x_{t+1} \rangle + V(x_{t+1}, x), \ \forall x \in X. \tag{3.2.6}$$

The result then immediately follows by combining the above two relations. ∎

With the help of Lemma 3.4, we can easily establish some general convergence properties for the mirror descent method. In order to provide an efficiency estimate that does not rely on the Euclidean structure, we assume that the subgradients of f satisfy

$$\|g(x_t)\|_* \leq M, \ \forall t \geq 1.$$

Theorem 3.5. *Let $x_t, t = 1, \ldots, k$, be generated by (3.2.5) and define $\bar{x}_s^k$ as in (3.1.9). Then*

$$f(\bar{x}_s^k) - f^* \leq \left(\sum_{t=s}^k \gamma_t \right)^{-1} \left[V(x_s, x^*) + \tfrac{1}{2} M^2 \sum_{t=s}^k \gamma_t^2 \right],$$

where x^* denotes an arbitrary solution of (3.1.2).

Proof. By the convexity of f and Lemma 3.4,

$$\begin{aligned} \gamma_t [f(x_t) - f(x)] &\leq \gamma_t \langle g(x_t), x_t - x \rangle \\ &\leq V(x_t, x) - V(x_{t+1}, x) + \gamma_t \langle g(x_t), x_t - x_{t+1} \rangle - V(x_t, x_{t+1}). \end{aligned}$$

Note that by the strong convexity of ν, the Cauchy–Schwarz inequality, and the fact that $bt - at^2/2 \leq b^2/(2a)$ for any $a > 0$, we have

$$\begin{aligned} \gamma_t \langle g(x_t), x_t - x_{t+1} \rangle - V(x_t, x_{t+1}) &\leq \gamma_t \langle g(x_t), x_t - x_{t+1} \rangle - \tfrac{1}{2} \|x_{t+1} - x_t\|^2 \\ &\leq \gamma_t^2 \|g(x_t)\|_*^2 \leq \tfrac{1}{2} \gamma_t^2 M^2. \end{aligned}$$

Combining the above two relations, we conclude

$$\gamma_t [f(x_t) - f(x)] \leq V(x_t, x) - V(x_{t+1}, x) + \tfrac{1}{2} \gamma_t^2 M^2.$$

The result then immediately follows by summing up the above inequality from $t = s$ to k, and using the fact that $f(\bar{x}_s^k) \leq (\sum_{t=s}^k \gamma_t)^{-1} \sum_{t=s}^k f(x_t)$. ∎

We now provide a simple specific choice for the stepsizes γ_t.

Corollary 3.4. *Let the number of iterations k be fixed and assume that*

$$\gamma_t = \sqrt{\tfrac{2D_X^2}{kM^2}}, t = 1, \ldots, k.$$

Then

$$f(\bar{x}_{1,k}) - f^* \leq \tfrac{\sqrt{2} M D_X}{\sqrt{k}}, \ \forall k \geq 1.$$

Proof. The proof is almost identical to that for Corollary 3.1 and hence the details are skipped. ■

We can also use variable stepsizes without fixing the number of iterations k a priori.

Excercise 1 *Show that if*

$$\gamma_t = \frac{D_X}{M\sqrt{t}}, \quad t = 1, 2, \ldots,$$

then $f(\bar{x}_{\lceil k/2 \rceil, k}) - f^* \leq \mathcal{O}(1)(\frac{MD_X}{\sqrt{k}})$*, where* $\mathcal{O}(1)$ *denotes an absolute constant.*

Comparing the results obtained in Corollaries 3.1 and 3.4, we see that for both the subgradient and mirror descent methods, the inaccuracy in terms of the objective of the approximate solution is bounded by $\mathcal{O}(k^{-1/2})$. A benefit of the mirror descent over the subgradient descent algorithm is its potential possibility to reduce the constant factor hidden in $\mathcal{O}(\cdot)$ by adjusting the norm $\|\cdot\|$ and the distance generating function $\nu(\cdot)$ to the geometry of the problem.

Example 3.1. Let $X = \{x \in \mathbb{R}^n : \sum_{i=1}^n x_i = 1,\ x \geq 0\}$ be a standard simplex. Consider two setups for the mirror descent method:

– *Euclidean setup*, where $\|\cdot\| = \|\cdot\|_2$ and $\nu(x) = \frac{1}{2}\|x\|_2^2$
– ℓ_1*-setup*, where $\|x\| = \|x\|_1 := \sum_{i=1}^n |x_i|$ and ν is the *entropy*

$$\nu(x) = \sum_{i=1}^n x_i \ln x_i. \tag{3.2.7}$$

The Euclidean setup leads to the subgradient descent method which is easily implementable (computing the projection subproblem in (3.1.4) requires $O(n \ln n)$ operations) and guarantees that

$$f(\bar{x}_1^k) - f(x_*) \leq \mathcal{O}(1)\bar{M}k^{-1/2}, \tag{3.2.8}$$

provided that the constant $\bar{M} = \max_{x \in X} \|g(x)\|$ is known and the stepsizes in Corollary 3.1 are used (note that the Euclidean diameter of X is of order of 1). The ℓ_1-setup corresponds to $X^o = \{x \in X : x > 0\}$, $D_X = \sqrt{\ln n}$, $x_1 = \mathrm{argmin}_X \omega = n^{-1}(1, \ldots, 1)^T$, $\sigma_\nu = 1$, and $\|x\|_* = \|x\|_\infty \equiv \max_i |x_i|$. The associated mirror descent is easily implementable: the prox-function here is $V(x, z) = \sum_{i=1}^n z_i \ln \frac{z_i}{x_i}$, and the subproblem $x^+ = \mathrm{argmin}_{z \in X} \left[y^T(z - x) + V(x, z)\right]$ can be computed in $\mathcal{O}(n)$ operations according to the explicit formula:

$$x_i^+ = \frac{x_i e^{-y_i}}{\sum_{k=1}^n x_k e^{-y_k}}, \quad i = 1, \ldots, n.$$

The efficiency estimate guaranteed with the ℓ_1-setup is

$$f(\bar{x}_1^k) - f(x_*) \leq \mathcal{O}(1)\sqrt{\ln n}\bar{M}_* k^{-1/2}, \tag{3.2.9}$$

provided that the constant $\bar{M}_* = \max_{x\in X} \|g(x)\|_*$ is known and the constant stepsizes in Corollary 3.4 are used. To compare (3.2.9) and (3.2.8), observe that $\bar{M}_* \le \bar{M}$, and the ratio $\bar{M}_*/\bar{M}$ can be as small as $n^{-1/2}$. Thus, the efficiency estimate for the ℓ_1-setup never is much worse than the estimate for the Euclidean setup, and for large n can be *far better* than the latter estimate:

$$\sqrt{\tfrac{1}{\ln n}} \le \tfrac{\bar{M}}{\sqrt{\ln n}\bar{M}_*} \le \sqrt{\tfrac{n}{\ln n}},\, k = 1,2,\ldots,$$

both the upper and the lower bounds being achievable. Thus, when X is a standard simplex of large dimension, we have strong reasons to prefer the ℓ_1-setup to the usual Euclidean one.

It should be noted that the mirror descent method will exhibit stronger rate of convergence when applied to strongly convex or smooth problems. We leave the development of these results as an exercise (see Sect. 3.10).

3.3 Accelerated Gradient Descent

In this subsection, we discuss an important improvement to the gradient descent method, namely the accelerated (or fast) gradient method applied to smooth convex optimization problems. Note that in this discussion, we incorporate the idea of mirror-descent method into the accelerated gradient method by using the prox-function discussed in the previous subsection.

In particular, we assume that given an arbitrary norm $\|\cdot\|$ ($\|\cdot\|_*$ denotes its conjugate),

$$\|\nabla f(x) - \nabla f(y)\|_* \le L\|x-y\|, \;\; \forall x,y \in X. \tag{3.3.1}$$

Similar to (3.1.17), we can show that for any $x,y \in X$, we have

$$f(y) - f(x) - \langle f'(x), y-x\rangle \le \tfrac{L}{2}\|y-x\|^2. \tag{3.3.2}$$

Moreover, we assume that for some $\mu \ge 0$,

$$f(y) \ge f(x) + \langle f'(x), y-x\rangle + \mu V(x,y), \;\; \forall x,y \in X. \tag{3.3.3}$$

If $\mu = 0$, (3.3.3) is implied by the convexity of f. If $\mu > 0$, it generalizes the definition of strong convexity in terms of Bregman's distance.

In fact, there exist many variants of the accelerated gradient method. Below we study one of its simplest variants. Given $(x_{t-1}, \bar{x}_{t-1}) \in X \times X$, we set

$$\underline{x}_t = (1-q_t)\bar{x}_{t-1} + q_t x_{t-1}, \tag{3.3.4}$$

$$x_t = \operatorname{argmin}_{x\in X}\{\gamma_t[\langle f'(\underline{x}_t), x\rangle + \mu V(\underline{x}_t, x)] + V(x_{t-1}, x)\}, \tag{3.3.5}$$

$$\bar{x}_t = (1-\alpha_t)\bar{x}_{t-1} + \alpha_t x_t, \tag{3.3.6}$$

for some $q_t \in [0,1]$, $\gamma_t \geq 0$, and $\alpha_t \in [0,1]$. In comparison with the gradient descent method, the accelerated gradient descent method builds up a lower approximation of the objective function, defined by

$$f(\underline{x}_t) + \langle f'(\underline{x}_t), x - \underline{x}_t \rangle + \mu V(\underline{x}_t, x),$$

at the search point $\underline{x}_t$, which is different from the other search point x_t used for proximity control (see (3.3.5)). Moreover, we compute the output solution $\bar{x}_t$ as a convex combination of the sequence $\{x_t\}$. Note that we have not specified the parameters $\{q_t\}$, $\{\gamma_t\}$, and $\{\alpha_t\}$ yet. In fact, the selection of these parameters will depend on the problem classes to be solved. We will discuss this issue after establishing some generic convergence properties of this method.

The first technical result below characterizes the solution of the projection (or prox-mapping) step (3.3.5). It is worth noting that the function ν is not necessarily strongly convex.

Lemma 3.5. *Let the convex function $p : X \to \mathbb{R}$, the points $\tilde{x}, \tilde{y} \in X$, and the scalars $\mu_1, \mu_2 \geq 0$ be given. Let $\nu : X \to \mathbb{R}$ be a differentiable convex function and $V(x,z)$ be defined in (3.2.2). If*

$$\hat{u} \in \text{Argmin}\{p(u) + \mu_1 V(\tilde{x}, u) + \mu_2 V(\tilde{y}, u) : u \in X\},$$

then for any $u \in X$, we have

$$p(\hat{u}) + \mu_1 V(\tilde{x}, \hat{u}) + \mu_2 V(\tilde{y}, \hat{u}) \leq p(u) + \mu_1 V(\tilde{x}, u) + \mu_2 V(\tilde{y}, u) - (\mu_1 + \mu_2) V(\hat{u}, u).$$

Proof. The definition of $\hat{u}$ and the fact $V(\tilde{x}, \cdot)$ is a differentiable convex function imply that, for some $p'(\hat{u}) \in \partial p(\hat{u})$, we have

$$\langle p'(\hat{u}) + \mu_1 \nabla V(\tilde{x}, \hat{u}) + \mu_2 \nabla V(\tilde{y}, \hat{u}), u - \hat{u} \rangle \geq 0, \quad \forall u \in X,$$

where $\nabla V(\tilde{x}, \hat{u})$ denotes the gradient of $V(\tilde{x}, \cdot)$ at $\hat{u}$. Using the definition of $V(x,z)$ in (3.2.2), it is easy to verify that

$$V(\tilde{x}, u) = V(\tilde{x}, \hat{u}) + \langle \nabla V(\tilde{x}, \hat{u}), u - \hat{u} \rangle + V(\hat{u}, u), \quad \forall u \in X.$$

Using the above two relations and the assumption that p is convex, we then conclude that

$$\begin{aligned}
&p(u) + \mu_1 V(\tilde{x}, u) + \mu_2 V(\tilde{y}, u) \\
&\quad = p(u) + \mu_1 [V(\tilde{x}, \hat{u}) + \langle \nabla V(\tilde{x}, \hat{u}), u - \hat{u} \rangle + V(\hat{u}, u)] \\
&\qquad + \mu_2 [V(\tilde{y}, \hat{u}) + \langle \nabla V(\tilde{y}, \hat{u}), u - \hat{u} \rangle + V(\hat{u}, u)] \\
&\quad \geq p(\hat{u}) + \mu_1 V(\tilde{x}, \hat{u}) + \mu_2 V(\tilde{y}, \hat{u}) \\
&\qquad + \langle p'(\hat{u}) + \mu_1 \nabla V(\tilde{x}, \hat{u}) + \mu_2 \nabla V(\tilde{y}, \hat{u}), u - \hat{u} \rangle \\
&\qquad + (\mu_1 + \mu_2) V(\hat{u}, u) \\
&\quad \geq [p(\hat{u}) + \mu_1 V(\tilde{x}, \hat{u}) + \mu_2 V(\tilde{y}, \hat{u})] + (\mu_1 + \mu_2) V(\hat{u}, u).
\end{aligned}$$

■

Proposition 3.1 below describes some important recursion of the accelerated gradient descent method.

Proposition 3.1. *Let* $(\underline{x}_t, x_t, \bar{x}_t) \in X \times X \times X$ *be generated by the accelerated gradient method in (3.3.4)–(3.3.6). If*

$$\alpha_t \geq q_t, \tag{3.3.7}$$

$$\tfrac{L(\alpha_t - q_t)}{1-q_t} \leq \mu, \tag{3.3.8}$$

$$\tfrac{Lq_t(1-\alpha_t)}{1-q_t} \leq \tfrac{1}{\gamma_t}, \tag{3.3.9}$$

then for any $x \in X$,

$$f(\bar{x}_t) - f(x) + \alpha_t(\mu + \tfrac{1}{\gamma_t})V(x_t, x) \leq (1-\alpha_t)[f(\bar{x}_{t-1}) - f(x)] + \tfrac{\alpha_t}{\gamma_t}V(x_{t-1}, x). \tag{3.3.10}$$

Proof. Denote $d_t = \bar{x}_t - \underline{x}_t$. It follows from (3.3.4) and (3.3.6) that

$$\begin{aligned} d_t &= (q_t - \alpha_t)\bar{x}_{t-1} + \alpha_t x_t - q_t x_{t-1} \\ &= \alpha_t \left[x_t - \tfrac{\alpha_t - q_t}{\alpha_t(1-q_t)}\underline{x}_t - \tfrac{q_t(1-\alpha_t)}{\alpha_t(1-q_t)}x_{t-1} \right], \end{aligned} \tag{3.3.11}$$

which, in view of the convexity of $\|\cdot\|^2$ and (3.3.7), implies that

$$\|d_t\|^2 \leq \alpha_t \left[\tfrac{\alpha_t - q_t}{1-q_t}\|x_t - \underline{x}_t\|^2 + \tfrac{q_t(1-\alpha_t)}{1-q_t}\|x_t - x_{t-1}\|^2 \right].$$

Using the above relation and (3.3.2), we have

$$\begin{aligned} f(\bar{x}_t) &\leq f(\underline{x}_t) + \langle f'(\underline{x}_t), \bar{x}_t - \underline{x}_t \rangle + \tfrac{L}{2}\|d_t\|^2 \\ &= (1-\alpha_t)[f(\underline{x}_t) + \langle f'(\underline{x}_t), \bar{x}_{t-1} - \underline{x}_t \rangle] + \alpha_t[f(\underline{x}_t) + \langle f'(\underline{x}_t), x_t - \underline{x}_t \rangle] + \tfrac{L}{2}\|d_t\|^2 \\ &\leq (1-\alpha_t)f(\bar{x}_{t-1}) \\ &\quad + \alpha_t \left[f(\underline{x}_t) + \langle f'(\underline{x}_t), x_t - \underline{x}_t \rangle + \tfrac{L(\alpha_t - q_t)}{2(1-q_t)}\|x_t - \underline{x}_t\|^2 + \tfrac{Lq_t(1-\alpha_t)}{2(1-q_t)}\|x_t - x_{t-1}\|^2 \right] \\ &\leq (1-\alpha_t)f(\bar{x}_{t-1}) \\ &\quad + \alpha_t \left[f(\underline{x}_t) + \langle f'(\underline{x}_t), x_t - \underline{x}_t \rangle + \mu V(\underline{x}_t, x_t) + \tfrac{1}{\gamma_t}V(x_{t-1}, x_t) \right], \end{aligned} \tag{3.3.12}$$

where the last inequality follows from (3.2.3), (3.3.8), and (3.3.9). Now using the above inequality, the definition of x_t in (3.3.5), and Lemma 3.5, we conclude

$$\begin{aligned} f(\bar{x}_t) &\leq (1-\alpha_t)f(\bar{x}_{t-1}) + \alpha_t[f(\underline{x}_t) + \langle f'(\underline{x}_t), x - \underline{x}_t \rangle + \mu V(\underline{x}_t, x)] \\ &\quad + \tfrac{\alpha_t}{\gamma_t}V(x_{t-1}, x) - \alpha_t(\mu + \tfrac{1}{\gamma_t})V(x_t, x) \\ &\leq (1-\alpha_t)f(\bar{x}_{t-1}) + \alpha_t f(x) + \tfrac{\alpha_t}{\gamma_t}V(x_{t-1}, x) - \alpha_t(\mu + \tfrac{1}{\gamma_t})V(x_t, x), \end{aligned}$$

where the last inequality follows from (3.3.3). Subtracting $f(x)$ from both sides of the above inequality and rearranging the terms, we obtain the result. ∎

Below we discuss the convergence of the accelerated gradient descent method for smooth convex function which are not necessarily strongly convex (i.e., $\mu = 0$).

Theorem 3.6. *Let $(\underline{x}_t, x_t, \bar{x}_t) \in X \times X \times X$ be generated by the accelerated gradient descent method in (3.3.4)–(3.3.6). If*

$$\alpha_t = q_t, \tag{3.3.13}$$

$$L\alpha_t \leq \tfrac{1}{\gamma_t}, \tag{3.3.14}$$

$$\tfrac{\gamma_t(1-\alpha_t)}{\alpha_t} \leq \tfrac{\gamma_{t-1}}{\alpha_{t-1}}, \tag{3.3.15}$$

for any $t = 1, \ldots, k$, then we have

$$f(\bar{x}_k) - f(x^*) + \tfrac{\alpha_k}{\gamma_k} V(x_k, x^*) \leq \tfrac{\alpha_k \gamma_1 (1-\alpha_1)}{\gamma_k \alpha_1}[f(\bar{x}_0) - f(x^*)] + \tfrac{\alpha_k}{\gamma_k} V(x_0, x^*). \tag{3.3.16}$$

In particular, if

$$q_t = \alpha_t = \tfrac{2}{t+1} \text{ and } \gamma_t = \tfrac{t}{2L},$$

for any $t = 1, \ldots, k$, then

$$f(\bar{x}_k) - f(x^*) \leq \tfrac{4L}{k(k+1)} V(x_0, x^*).$$

Proof. Using the fact that $\mu = 0$, (3.3.13) and (3.3.14), we can easily see that (3.3.7)–(3.3.9) hold. It then follows from (3.3.10) that

$$\begin{aligned}
\tfrac{\gamma_t}{\alpha_t}[f(\bar{x}_t) - f(x^*)] + V(x_t, x^*) &\leq \tfrac{\gamma_t(1-\alpha_t)}{\alpha_t}[f(\bar{x}_{t-1}) - f(x^*)] + V(x_{t-1}, x^*) \\
&\leq \tfrac{\gamma_{t-1}}{\alpha_{t-1}}[f(\bar{x}_{t-1}) - f(x^*)] + V(x_{t-1}, x^*),
\end{aligned}$$

where the last inequality follows from (3.3.15) and the fact that $f(\bar{x}_{t-1}) - f(x^*) \geq 0$. Summing up these inequalities and rearranging the terms, we obtain (3.3.16). ∎

It follows from the previous result that in order to find a solution $\bar{x} \in X$ such that $f(\bar{x}) - f(x^*) \leq \varepsilon$, the number of iteration performed by the accelerated gradient method can be bounded by $\mathcal{O}(1/\sqrt{\varepsilon})$. This bound turns out to be optimal for solving a general class of large-scale smooth convex optimization problems. One way to improve this complexity bound is to consider more specialized problems. Below we introduce a substantially improved result for the smooth and strongly convex case, i.e., $\mu > 0$.

Theorem 3.7. *Let $(\underline{x}_t, x_t, \bar{x}_t) \in X \times X \times X$ be generated by the accelerated gradient descent method in (3.3.4)–(3.3.6). If $\alpha_t = \alpha$, $\gamma_t = \gamma$, and $q_t = q$, $t = 1, \ldots, k$, satisfy (3.3.7)–(3.3.9) and*

$$\tfrac{1}{\gamma(1-\alpha)} \leq \mu + \tfrac{1}{\gamma}, \tag{3.3.17}$$

then for any $x \in X$,

$$f(\bar{x}_k) - f(x) + \alpha(\mu + \tfrac{1}{\gamma})V(x_{k-1},x) \le (1-\alpha)^k [f(\bar{x}_0) - f(x) + \alpha(\mu + \tfrac{1}{\gamma})V(x_1,x)]. \tag{3.3.18}$$

In particular, if

$$\alpha = \sqrt{\tfrac{\mu}{L}}, q = \tfrac{\alpha - \mu/L}{1-\mu/L}, \text{ and } \gamma = \tfrac{\alpha}{\mu(1-\alpha)}, \tag{3.3.19}$$

then for any $x \in X$,

$$f(\bar{x}_k) - f(x) + \alpha(\mu + \tfrac{1}{\gamma})V(x_{k-1},x) \le \left(1 - \sqrt{\tfrac{\mu}{L}}\right)^k [f(\bar{x}_0) - f(x) + \alpha(\mu + \tfrac{1}{\gamma})V(x_1,x)]. \tag{3.3.20}$$

Proof. The result in (3.3.18) follows directly from (3.3.10) and (3.3.17). Moreover, we can easily check that the parameters in (3.3.19) satisfy (3.3.7)–(3.3.9) and (3.3.17) with equality, which implies (3.3.20). ■

Using a similar reasoning to (3.1.23), we can see that the total number of iterations performed by the accelerated gradient method to solve strongly convex problems can be bounded by $\mathcal{O}(\sqrt{L/\mu}\log 1/\varepsilon)$ in order to find a point $\bar{x} \in X$ such that $f(\bar{x}) - f(x^*) \le \varepsilon$.

We now turn our attention to a relatively easy extension of the accelerated gradient descent method for solving a certain class of nonsmooth optimization problems given by

$$\min_{x \in X} \{ f(x) := \hat{f}(x) + F(x) \}. \tag{3.3.21}$$

Here $\hat{f}$ is a simple (not necessarily smooth) convex function and F is a smooth convex function with Lipschitz continuous gradients. Moreover, we assume that the Lipschitz constant of ∇F is given by L and that for some $\mu \ge 0$,

$$F(y) \ge F(x) + \langle F'(x), y - x\rangle + \mu V(x,y), \ \forall x,y \in X. \tag{3.3.22}$$

In order to solve the above composite problem, we only need to slightly modify (3.3.4)–(3.3.6) as follows.

$$\underline{x}_t = (1-q_t)\bar{x}_{t-1} + q_t x_{t-1}, \tag{3.3.23}$$

$$x_t = \operatorname{argmin}_{x \in X}\{\gamma_t[\langle f'(\underline{x}_t), x\rangle + \mu V(\underline{x}_t, x) + \hat{f}(x)] + V(x_{t-1},x)\}, \tag{3.3.24}$$

$$\bar{x}_t = (1-\alpha_t)\bar{x}_{t-1} + \alpha_t x_t. \tag{3.3.25}$$

Hence, the difference exists in that we keep $\hat{f}$ inside the subproblem (3.3.24).

Corollary 3.5. *The performance guarantees stated in Theorems 3.6 and 3.7 still hold for the above variant of accelerated gradient descent method applied to (3.3.21).*

Proof. It suffices to show that Proposition 3.1 holds. First note that relation (3.3.12) still holds with f replaced by F, i.e.,

$$\begin{aligned}F(\bar{x}_t) &\le (1-\alpha_t)F(\bar{x}_{t-1})\\&\quad+\alpha_t\left[F(\underline{x}_t)+\langle F'(\underline{x}_t),x_t-\underline{x}_t\rangle+\mu V(\underline{x}_t,x_t)+\tfrac{1}{\gamma_t}V(x_{t-1},x_t)\right].\end{aligned}$$

Moreover, by convexity of $\hat{f}$, we have $\hat{f}(\bar{x}_t)\le(1-\alpha_t)\hat{f}(\bar{x}_{t-1})+\alpha_t\hat{f}(x_t)$. Adding up the previous two relations, using the definition of x_t in (3.3.24), and Lemma 3.5, we conclude

$$\begin{aligned}f(\bar{x}_t) &\le (1-\alpha_t)f(\bar{x}_{t-1})\\&\quad+\alpha_t\left[F(\underline{x}_t)+\langle F'(\underline{x}_t),x_t-\underline{x}_t\rangle+\mu V(\underline{x}_t,x_t)+\hat{f}(x_t)+\tfrac{1}{\gamma_t}V(x_{t-1},x_t)\right]\\&\le(1-\alpha_t)f(\bar{x}_{t-1})+\alpha_t[F(\underline{x}_t)+\langle F'(\underline{x}_t),x-\underline{x}_t\rangle+\mu V(\underline{x}_t,x)+\hat{f}(x)]\\&\quad+\tfrac{\alpha_t}{\gamma_t}V(x_{t-1},x)-\alpha_t(\mu+\tfrac{1}{\gamma_t})V(x_t,x)\\&\le(1-\alpha_t)f(\bar{x}_{t-1})+\alpha_t f(x)+\tfrac{\alpha_t}{\gamma_t}V(x_{t-1},x)-\alpha_t(\mu+\tfrac{1}{\gamma_t})V(x_t,x),\end{aligned}$$

where the last inequality follows from (3.3.22). Subtracting $f(x)$ from both sides of the above inequality and rearranging the terms, we obtain the result. ∎

3.4 Game Interpretation for Accelerated Gradient Descent

In this subsection, we intend to provide some intuition that might help us to better understand the accelerated gradient descent method.

First let us consider the smooth case without the strong convexity assumption, i.e., $\mu=0$ in (3.3.3). Let J_f be the conjugate function of f, i.e., $J_f(y)=\max\langle y,x\rangle-f(x)$. Since f is convex and smooth, it is proper and the conjugate of J_f is given by $(J_f)^*=f$. We can then rewrite (3.1.2) equivalently as

$$\min_{x\in X}\max_{y}\left\{\langle x,y\rangle-J_f(y)\right\}. \tag{3.4.1}$$

This saddle point reformulation admits some natural buyer-supplier game interpretation. In particular, the dual variables y can be viewed as the prices for a list of products, and x are the order quantities for the buyer. The supplier's goal is to specify the prices to maximize the profit $\langle x,y\rangle-J_f(y)$ while the buyer intends to minimize the cost $\langle x,y\rangle$ by determining the order quantities x. Given an initial pair of order quantities and product prices $(x_0,y_0)\in X\times\mathbb{R}^n$, we need to design an iterative algorithm to play this game so that the buyer and supplier can achieve the equilibrium as soon as possible.

Next we describe the supplier and buyer's strategies to play this game iteratively, and then demonstrate that the accelerated gradient descent method can be viewed as a special case of this procedure. Let V be the prox-function defined in (3.2.2) and define

$$W(y_1,y_2):=J_f(y_2)-J_f(y_1)-\langle J_f'(y_1),y_2-y_1\rangle. \tag{3.4.2}$$

The supplier and buyer will iteratively perform the following three steps.

$$\tilde{x}_t = x_{t-1} + \lambda_t(x_{t-1} - x_{t-2}), \tag{3.4.3}$$

$$y_t = \text{argmin}_y \langle -\tilde{x}_t, y \rangle + J_f(y) + \tau_t W(y_{t-1}, y), \tag{3.4.4}$$

$$x_t = \text{argmin}_{x \in X} \langle y_t, x \rangle + \eta_t V(x_{t-1}, x). \tag{3.4.5}$$

In (3.4.3), the supplier predicts the demand by using historical data x_{t-1} and x_{t-2}. In (3.4.4), the supplier intends to maximize the profit without moving too far away from y_{t-1} in terms of $W(y_{t-1}, y)$. Then in (3.4.5) the buyer determines the order quantity by minimizing the cost without moving too far away from x_{t-1} in terms of $V(x_{t-1}, x)$.

It is interesting to notice that problem (3.4.4) is equivalent to the computation of gradients.

Lemma 3.6. *Let $\tilde{x} \in X$ and y_0 be given. Also for any $\tau > 0$, let us denote $z = [\tilde{x} + \tau J_f'(y_0)]/(1+\tau)$. Then we have*

$$\nabla f(z) = \text{argmin}_y \left\{ \langle -\tilde{x}, y \rangle + J_f(y) + \tau_t W(y_0, y) \right\}.$$

Proof. In view of the definition of $W(y_0, y)$, we have

$$\begin{aligned}
&\text{argmin}_y \left\{ \langle -\tilde{x}, y \rangle + J_f(y) + \tau_t W(y_0, y) \right\} \\
&= \arg\min_y \left\{ -\langle \tilde{x} + \tau J_f'(y_0), y \rangle + (1+\tau) J_f(y) \right\} \\
&= \arg\max_y \left\{ \langle z, y \rangle - J_f(y) \right\} = \nabla f(z).
\end{aligned}$$

■

In view of the above result, if

$$J_f'(y_{t-1}) = \underline{x}_{t-1}, \tag{3.4.6}$$

$$\underline{x}_t = \tfrac{1}{1+\tau_t} \left(\tilde{x}_t + \tau_t \underline{x}_{t-1} \right), \tag{3.4.7}$$

then

$$y_t = \text{argmin}_y \left\{ -\langle \underline{x}_t, y \rangle + J_f(y) \right\} = \nabla f(\underline{x}_t). \tag{3.4.8}$$

Moreover, by the optimality condition of (3.4.8), we must have $\underline{x}_t = J_f'(y_t)$ for some $J_f'(y_t) \in \partial J_f(y_t)$. Therefore, we can show that (3.4.6), (3.4.7), and (3.4.8) hold by induction, under the assumption that $J_f'(y_0) = \underline{x}_0$. The latter assumption can be satisfied by setting $y_0 = \nabla f(\underline{x}_0)$. Using these observations, we can reformulate (3.4.3)–(3.4.5) as

$$\begin{aligned}
\underline{x}_t &= \tfrac{1}{1+\tau_t} \left(\tilde{x}_t + \tau_t \underline{x}_{t-1} \right) \\
&= \tfrac{1}{1+\tau_t} \left[\tau_t \underline{x}_{t-1} + (1+\lambda_t) x_{t-1} - \lambda_t x_{t-2} \right],
\end{aligned} \tag{3.4.9}$$

$$x_t = \text{argmin}_{x \in X} \langle \nabla f(\underline{x}_t), x \rangle + \eta_t V(x_{t-1}, x). \tag{3.4.10}$$

Now we will show that the above definition of $\underline{x}_t$ and x_t will be equivalent to those in the accelerated gradient descent method (3.3.4)–(3.3.6). It can be seen from (3.3.4) and (3.3.6) that

$$\begin{aligned}\underline{x}_t &= (1-q_t)\left[(1-\alpha_{t-1})\bar{x}_{t-1}\right] + q_t x_{t-1}\\ &= (1-q_t)\left[\tfrac{1-\alpha_{t-1}}{1-q_{t-1}}(\underline{x}_{t-1} - q_{t-1}x_{t-1}) + \alpha_{t-1}x_{t-1}\right] + q_t x_{t-1}\\ &= \tfrac{(1-q_t)(1-\alpha_{t-1})}{1-q_{t-1}}\underline{x}_{t-1} + \left[(1-q_t)\alpha_{t-1} + q_t\right]x_{t-1} - \tfrac{q_{t-1}(1-q_t)(1-\alpha_{t-1})}{1-q_{t-1}}x_{t-2}. \quad (3.4.11)\end{aligned}$$

In particular, if $q_t = \alpha_t$ as in the smooth case, the above relation can be further simplified to

$$\underline{x}_t = (1-\alpha_t)\underline{x}_{t-1} + [(1-\alpha_t)\alpha_{t-1} + \alpha_t]x_{t-1} - \alpha_{t-1}(1-\alpha_t)x_{t-2},$$

which is equivalent to (3.4.9) if

$$\tau_t = \tfrac{1-\alpha_t}{\alpha_t} \text{ and } \lambda_t = \tfrac{\alpha_{t-1}(1-\alpha_t)}{\alpha_t}.$$

Now let us consider the strongly convex case with $\mu > 0$. In order to provide a game interpretation, we define

$$\tilde{f}(x) = f(x) - \mu\nu(x).$$

By (3.3.3), the function $\tilde{f}(x)$ must be a convex function. Indeed, for any $x, y \in X$,

$$\tilde{f}(y) - \tilde{f}(x) - \langle \nabla \tilde{f}(x), y - x\rangle = f(y) - f(x) - \langle \nabla f(x), y - x\rangle - \mu V(x,y) \geq 0.$$

We can then rewrite (3.1.2) as

$$\min_{x\in X} \mu\nu(x) + \tilde{f}(x),$$

or equivalently,

$$\min_{x\in X} \mu\nu(x) + \max_y \left\{\langle x,y\rangle - J_{\tilde{f}}(y)\right\}, \quad (3.4.12)$$

where $J_{\tilde{f}}$ denotes the conjugate function of $\tilde{f}$. Accordingly, we define the iterative game between the supplier and buyer as follows.

$$\tilde{x}_t = x_{t-1} + \lambda_t(x_{t-1} - x_{t-2}), \quad (3.4.13)$$

$$y_t = \mathrm{argmin}_y \langle -\tilde{x}_t, y\rangle + J_{\tilde{f}}(y) + \tau_t W(y_{t-1}, y), \quad (3.4.14)$$

$$x_t = \mathrm{argmin}_{x\in X} \langle y_t, x\rangle + \mu\nu(x) + \eta_t V(x_{t-1}, x). \quad (3.4.15)$$

In (3.4.13), the supplier predicts the demand by using historical data x_{t-1} and x_{t-2}. In (3.4.14), the supplier still intends to maximize the profit without moving too far away from y_{t-1} in terms of $W(y_{t-1}, y)$, but the local cost function changes to $J_{\tilde{f}}(y)$. Then in (3.4.15) the buyer determines the order quantity by minimizing the local cost $\langle y_t, x\rangle + \mu\nu(x)$ without moving too far away from x_{t-1} in terms of $V(x_{t-1}, x)$.

Similar to (3.4.9) and (3.4.10), we can show that

$$\underline{x}_t = \tfrac{1}{1+\tau_t}\left[\tau_t \underline{x}_{t-1} + (1+\lambda_t)x_{t-1} - \lambda_t x_{t-2}\right], \tag{3.4.16}$$

$$\begin{aligned} x_t &= \mathrm{argmin}_{x\in X}\langle \nabla \tilde{f}(\underline{x}_t), x\rangle + \mu \nu(x) + \eta_t V(x_{t-1}, x) \\ &= \mathrm{argmin}_{x\in X}\langle \nabla f(\underline{x}_t) - \mu \nabla \nu(\underline{x}_t), x\rangle + \mu w(x) + \eta_t V(x_{t-1}, x) \\ &= \mathrm{argmin}_{x\in X}\langle \nabla f(\underline{x}_t), x\rangle + \mu V(\underline{x}_t, x) + \eta_t V(x_{t-1}, x). \end{aligned} \tag{3.4.17}$$

By properly choosing τ_t, λ_t, q_t, and α_t, we can show that these steps are equivalent to (3.3.4) and (3.3.5).

3.5 Smoothing Scheme for Nonsmooth Problems

In this section, we consider the convex programming problem of

$$f^* \equiv \min_{x\in X}\left\{f(x) := \hat{f}(x) + F(x)\right\}, \tag{3.5.1}$$

where $\hat{f} : X \to \mathbb{R}$ is a simple Lipschitz continuous convex function and

$$F(x) := \max_{y\in Y}\left\{\langle Ax, y\rangle - \hat{g}(y)\right\}. \tag{3.5.2}$$

Here, $Y \subseteq \mathbb{R}^m$ is a compact convex set, $\hat{g} : Y \to \mathbb{R}$ is a continuous convex function on Y, and A denotes a linear operator from $\mathbb{R}^n$ to $\mathbb{R}^m$. Observe also that problem (3.5.1)–(3.5.2) can be written in an adjoint form:

$$\max_{y\in Y}\left\{g(y) := -\hat{g}(y) + G(y)\right\}, \quad G(y) := \min_{x\in X}\left\{\langle Ax, y\rangle + \hat{f}(x)\right\}. \tag{3.5.3}$$

While the function F given by (3.5.2) is a nonsmooth convex function in general, it can be closely approximated by a class of smooth convex functions defined as follows. Let $\omega(y)$ be a distance-generating function of Y with modulus 1 and prox-center $c_\omega = \mathrm{argmin}_{y\in Y}\omega(y)$. Also let us denote

$$W(y) \equiv W(c_\omega, y) := \omega(y) - \omega(c_\omega) - \langle \nabla\omega(c_\omega), y - c_\omega\rangle,$$

and, for some $\eta > 0$,

$$F_\eta(x) := \max_y\left\{\langle Ax, y\rangle - \hat{g}(y) - \eta\, W(y) : \; y \in Y\right\}, \tag{3.5.4}$$

$$f_\eta(x) := \hat{f}(x) + F_\eta(x). \tag{3.5.5}$$

Then we have, for every $x \in X$,

$$F_\eta(x) \le F(x) \le F_\eta(x) + \eta\, D_Y^2, \tag{3.5.6}$$

and, as a consequence,

$$f_\eta(x) \le f(x) \le f_\eta(x) + \eta\, D_Y^2, \tag{3.5.7}$$

where $D_Y \equiv D_{Y,\omega}$ is defined in (3.2.4). Moreover, we can show that F_η is a smooth convex function.

Lemma 3.7. *$F_\eta(\cdot)$ has Lipschitz-continuous gradient with constant*

$$\mathscr{L}_\eta \equiv \mathscr{L}(F_\eta) := \tfrac{\|A\|^2}{\eta}, \tag{3.5.8}$$

where $\|A\|$ denotes the operator norm of A.

Proof. Let $x_1, x_2 \in Y$ be given, and denote

$$y_1 = \mathrm{argmax}_y \{\langle Ax_1, y\rangle - \hat{g}(y) - \eta\, W(y) :\ y \in Y\}, \tag{3.5.9}$$
$$y_2 = \mathrm{argmax}_y \{\langle Ax_2, y\rangle - \hat{g}(y) - \eta\, W(y) :\ y \in Y\}. \tag{3.5.10}$$

The gradients of F_η at x_1 and x_2, by implicit function theorem, are given by $A^T y_1$ and $A^T y_2$, respectively. By the optimality conditions for (3.5.9) and (3.5.10), we have

$$\langle Ax_1 - \hat{g}'(y_1) - \eta[\nabla\omega(y_1) - \nabla\omega(c_\omega)], y - y_1\rangle \le 0, \tag{3.5.11}$$
$$\langle Ax_2 - \hat{g}'(y_2) - \eta[\nabla\omega(y_2) - \nabla\omega(c_\omega)], y - y_2\rangle \le 0, \tag{3.5.12}$$

for any $y \in Y$ and some $\hat{g}'(y_1) \in \partial\hat{g}(y_1)$ and $\hat{g}'(y_2) \in \partial\hat{g}(y_2)$. Letting $y = y_2$ and $y = y_1$ in (3.5.11) and (3.5.12) respectively, and summing up the resulting inequalities, we obtain

$$-\langle A(x_1 - x_2), y_1 - y_2\rangle + \langle \eta\nabla\omega(y_1) - \nabla\omega(y_2), y_1 - y_2\rangle \le 0.$$

Using the above inequality and the strong convexity of ω, we conclude

$$\begin{aligned}\eta\|y_1 - y_2\|^2 &\le \eta\langle\nabla\omega(y_1) - \nabla\omega(y_2), y_1 - y_2\rangle\\ &\le -\langle A(x_1 - x_2), y_1 - y_2\rangle \le \|A\|\|x_1 - x_2\|\|y_1 - y_2\|,\end{aligned}$$

which implies that $\|y_1 - y_2\| \le \|A\|\|x_1 - x_2\|/\eta$ and hence that

$$\|\nabla F_\eta(x_1) - \nabla F_\eta(x_2)\| = \|A^T(y_1 - y_2)\| \le \|A\|^2\|x_1 - x_2\|/\eta.$$

■

Since F_η is a smooth convex function and $\hat{f}$ is a simple convex function, we can apply the accelerated gradient descent method in (3.3.23)–(3.3.25) to solve

$$f_\eta^* := \min_{x\in X} f_\eta(x). \tag{3.5.13}$$

Then, in view of Theorem 3.6 and Corollary 3.5, after performing the algorithm for at most

$$\left\lceil 2\sqrt{\frac{2\mathscr{L}_\eta D_X^2}{\varepsilon}} \right\rceil \tag{3.5.14}$$

iterations, we will be able to find a solution $\bar{x} \in X$ such that $f_\eta(\bar{x}) - f_\eta^* \leq \frac{\varepsilon}{2}$. Noting that by (3.5.7), $f(\bar{x}) \leq f_\eta(\bar{x}) + \eta D_Y^2$ and $f^* \geq f_\eta^*$, we then conclude that

$$f(\bar{x}) - f^* \leq \tfrac{\varepsilon}{2} + \eta D_Y^2.$$

If one chooses $\eta > 0$ small enough, e.g.,

$$\eta = \tfrac{\varepsilon}{2D_Y^2},$$

then we will have

$$f(\bar{x}) - f^* \leq \varepsilon$$

and the total number of iterations in (3.5.14) will reduce to

$$\left\lceil \tfrac{4\|A\| D_X D_Y}{\varepsilon} \right\rceil.$$

Observe that in the above discussion, we have assumed $\hat{f}$ to be relatively simple. Of course this does not have to be true. For example, if $\hat{f}$ itself is a smooth convex function and its gradients are $\hat{L}$-Lipschitz continuous, then we can apply the accelerated gradient descent method in (3.3.23)–(3.3.25) (rather than (3.3.23)–(3.3.25)) to solve (3.5.13). This method will require us to compute the gradient of both $\hat{f}$ and F_η at each iteration. Then it is not difficult to show that the total number of iterations performed by this method for finding a solution $\bar{x} \in X$ such that $f(\bar{x}) - f^*$ can be bounded by

$$\mathscr{O}\left\{ \sqrt{\tfrac{\hat{L}D_X^2}{\varepsilon}} + \tfrac{\|A\| D_X D_Y}{\varepsilon} \right\}.$$

It is worth noting that this bound can be substantially improved, in terms of the total number of gradient computations of $\nabla \hat{f}$. We will have some further discussions for this type of improvement in Sect. 8.2.

3.6 Primal–Dual Method for Saddle-Point Optimization

In this subsection, we continue our discussion about the following more general (than (3.5.1)) bilinear saddle point optimization problem:

$$\min_{x \in X} \left\{ \hat{f}(x) + \max_{y \in Y} \langle Ax, y \rangle - \hat{g}(y) \right\}. \tag{3.6.1}$$

Here $X \subseteq \mathbb{R}^n$ and $Y \subseteq \mathbb{R}^m$ are closed convex sets, $A \in \mathbb{R}^{n \times m}$ denotes a linear mapping from $\mathbb{R}^n$ to $\mathbb{R}^m$, and $\hat{f} : X \to \mathbb{R}$ and $\hat{g} : X \to \mathbb{R}$ are convex functions satisfying

$$\hat{f}(y) - \hat{f}(x) - \langle \hat{f}'(x), y - x \rangle \geq \mu_p V(x,y), \quad \forall x, y \in X, \tag{3.6.2}$$
$$\hat{g}(y) - \hat{g}(x) - \langle \hat{g}'(x), y - x \rangle \geq \mu_d W(x,y), \quad \forall x, y \in Y, \tag{3.6.3}$$

for some $\mu_p \geq 0$ and $\mu_d \geq 0$, where V and W, respectively, denote the prox-functions associated with distances generating functions ν and ω in the primal and dual spaces, i.e.,

$$V(x,y) := \nu(y) - \nu(x) - \langle \nu'(x,y), y - x \rangle, \tag{3.6.4}$$
$$W(x,y) := \omega(y) - \omega(x) - \langle \omega'(x,y), y - x \rangle. \tag{3.6.5}$$

For the sake of simplicity, we assume that both ν and ω have modulus 1 w.r.t. the respective norms in the primal and dual spaces. Without specific mention, we assume that the set of optimal primal and dual solutions of (3.6.1) is nonempty, and that $z^* := (x^*, y^*)$ is an optimal pair of primal and dual solutions of (3.6.1).

As discussed in the previous subsection, problem (3.6.1) covers the optimization problem (3.1.2) as a special case. As another example, we can write the Lagrangian dual of

$$\min_{x \in X} \{f(x) : Ax = b\} \tag{3.6.6}$$

in the form of (3.6.1) by dualizing the linear constraint $Ax = b$.

Denote $z \equiv (x,y)$ and $\bar{z} \equiv (\bar{x}, \bar{y})$ and define

$$Q(\bar{z}, z) := \hat{f}(\bar{x}) + \langle A\bar{x}, y \rangle - \hat{g}(y) - [\hat{f}(x) + \langle Ax, \bar{y} \rangle - \hat{g}(\bar{y})]. \tag{3.6.7}$$

Note that by definition $\bar{z}$ is a saddle point if and only if $\forall z = (x,y) \in Z$,

$$\hat{f}(\bar{x}) + \langle A\bar{x}, y \rangle - \hat{g}(y) \leq \hat{f}(\bar{x}) + \langle A\bar{x}, \bar{y} \rangle - \hat{g}(\bar{y}) \leq [\hat{f}(x) + \langle Ax, \bar{y} \rangle - \hat{g}(\bar{y})].$$

It then follows that $\bar{z} \in Z \equiv X \times Y$ is a pair of saddle points if and only if $Q(\bar{z}, z) \leq 0$ for any $z \in Z$, or equivalent, $Q(z, \bar{z}) \geq 0$ for any $z \in Z$.

The smoothing scheme described in Sect. 3.5 can be viewed as an indirect approach for solving bilinear saddle point problems. Below we describe a direct primal–dual method for solving the saddle point problem (3.6.1). This method generalizes the game interpretation of the accelerated gradient descent method discussed in Sect. 3.4. Given $(x_{t-2}, x_{t-1}, y_{t-1}) \in X \times X \times Y$, this algorithm updates x_t and y_t according to

$$\tilde{x}_t = x_{t-1} + \lambda_t (x_{t-1} - x_{t-2}), \tag{3.6.8}$$
$$y_t = \text{argmin}_y \langle -A\tilde{x}_t, y \rangle + \hat{g}(y) + \tau_t W(y_{t-1}, y), \tag{3.6.9}$$
$$x_t = \text{argmin}_{x \in X} \langle y_t, Ax \rangle + \hat{f}(x) + \eta_t V(x_{t-1}, x). \tag{3.6.10}$$

Clearly, the accelerated gradient descent method in (3.4.3)–(3.4.5) can be viewed as a special case of the above primal–dual method applied to problem (3.4.12).

We first state a simple relation based on the optimality conditions for (3.6.9) and (3.6.10).

Lemma 3.8. *Assume that* $\phi : X \to \mathbb{R}$ *satisfies*

$$\phi(y) \geq \phi(x) + \langle \phi'(x), y - x \rangle + \mu V(x,y), \ \forall x,y \in X \tag{3.6.11}$$

for some $\mu \geq 0$. *If*

$$\bar{x} = \text{argmin}\{\phi(x) + V(\tilde{x},x)\}, \tag{3.6.12}$$

then

$$\phi(\bar{x}) + V(\tilde{x},\bar{x}) + (\mu+1)V(\bar{x},x) \leq \phi(x) + V(\tilde{x},x), \forall x \in X.$$

Proof. It follows from the definition of V that $V(\tilde{x},x) = V(\tilde{x},\bar{x}) + \langle \nabla V(\tilde{x},\bar{x}), x - \bar{x}\rangle + V(\bar{x},x)$. Using this relation, (3.6.11), and the optimality condition for problem (3.6.12), we have

$$\begin{aligned} \phi(x) + V(\tilde{x},x) &= \phi(x) + [V(\tilde{x},\bar{x}) + \langle \nabla V(\tilde{x},\bar{x}), x - \bar{x}\rangle + V(\bar{x},x)] \\ &\geq \phi(\bar{x}) + \langle \phi'(\bar{x}), x - \bar{x}\rangle + \mu V(\bar{x},x) \\ &\quad + [V(\tilde{x},\bar{x}) + \langle \nabla V(\tilde{x},\bar{x}), x - \bar{x}\rangle + V(\bar{x},x)] \\ &\geq \phi(\bar{x}) + V(\tilde{x},\bar{x}) + (\mu+1)V(\bar{x},x), \forall x \in X. \end{aligned}$$

■

We now prove some general convergence properties of the primal–dual method.

Theorem 3.8. *If*

$$\gamma_t \lambda_t = \gamma_{t-1}, \tag{3.6.13}$$

$$\gamma_t \tau_t \leq \gamma_{t-1}(\tau_{t-1} + \mu_d), \tag{3.6.14}$$

$$\gamma_t \eta_t \leq \gamma_{t-1}(\eta_{t-1} + \mu_p), \tag{3.6.15}$$

$$\tau_t \eta_{t-1} \geq \lambda_t \|A\|^2, \tag{3.6.16}$$

for any $t \geq 2$, *then*

$$\begin{aligned} \textstyle\sum_{t=1}^k \gamma_t Q(z_t,z) &\leq \gamma_1 \eta_1 V(x_0,x) - \gamma_k(\eta_k + \mu_p)V(x_k,x) \\ &\quad + \gamma_1 \tau_1 W(y_0,y) - \gamma_k \left(\tau_k + \mu_d - \tfrac{\|A\|^2}{\eta_k}\right) W(y_k,y). \end{aligned} \tag{3.6.17}$$

Moreover, we have

$$\gamma_k \left(\tau_k + \mu_d - \tfrac{\|A\|^2}{\eta_k}\right) W(y_k,y^*) \leq \gamma_1 \eta_1 V(x_0,x^*) + \gamma_1 \tau_1 W(y_0,y^*), \tag{3.6.18}$$

$$\gamma_k \left(\eta_k - \tfrac{\|A\|^2}{\tau_k + \mu_d}\right) V(x_{k-1},x_k) \leq \gamma_1 \eta_1 V(x_0,x^*) + \gamma_1 \tau_1 W(y_0,y^*). \tag{3.6.19}$$

Proof. It follows from Lemma 3.8 applied to (3.6.9) and (3.6.10) that

$$\langle -A\tilde{x}_t, y_t - y\rangle + \hat{g}(y_t) - \hat{g}(y) \leq \tau_t[W(y_{t-1},y) - W(y_{t-1},y_t)] - (\tau_t + \mu_d)W(y_t,y),$$

$$\langle A(x_t - x), y_t\rangle + \hat{f}(x_t) - \hat{f}(x) \leq \eta_t[V(x_{t-1},x) - V(x_{t-1},x_t)] - (\eta_t + \mu_p)V(x_t,x).$$

Summing up these inequalities and using the definition of the gap function Q, we have

$$Q(z_t,z)+\langle A(x_t-\tilde{x}_t),y_t-y\rangle \le \eta_t[V(x_{t-1},x)-V(x_{t-1},x_t)]-(\eta_t+\mu_p)V(x_t,x)$$
$$+\tau_t[W(y_{t-1},y)-W(y_{t-1},y_t)]-(\tau_t+\mu_d)W(y_t,y).$$

Noting that by (3.4.13), we have

$$\langle A(x_t-\tilde{x}_t),y_t-y\rangle = \langle A(x_t-x_{t-1}),y_t-y\rangle-\lambda_t\langle A(x_{t-1}-x_{t-2}),y_t-y\rangle$$
$$=\langle A(x_t-x_{t-1}),y_t-y\rangle-\lambda_t\langle A(x_{t-1}-x_{t-2}),y_{t-1}-y\rangle$$
$$+\lambda_t\langle A(x_{t-1}-x_{t-2}),y_{t-1}-y_t\rangle.$$

Combining the above two relations, multiplying both sides by $\gamma_t \ge 0$, and summing up the resulting inequality, we have

$$\begin{aligned}
&\textstyle\sum_{t=1}^k \gamma_t Q(z_t,z)\\
&\le \textstyle\sum_{t=1}^k \gamma_t[\eta_t V(x_{t-1},x)-(\eta_t+\mu_p)V(x_t,x)]\\
&\quad+\textstyle\sum_{t=1}^k \gamma_t[\tau_t W(y_{t-1},y)-(\tau_t+\mu_d)W(y_t,y)]\\
&\quad+\textstyle\sum_{t=1}^k \gamma_t[\langle A(x_t-x_{t-1}),y_t-y\rangle-\lambda_t\langle A(x_{t-1}-x_{t-2}),y_{t-1}-y\rangle]\\
&\quad-\textstyle\sum_{t=1}^k \gamma_t[\tau_t W(y_{t-1},y_t)+\eta_t V(x_{t-1},x_t)+\lambda_t\langle A(x_{t-1}-x_{t-2}),y_{t-1}-y_t\rangle].
\end{aligned}$$

The above inequality, in view of (3.6.13)–(3.6.15) and the fact that $x_0 = x_{-1}$, then implies that

$$\begin{aligned}
&\textstyle\sum_{t=1}^k \gamma_t Q(z_t,z) \le \gamma_1\eta_1 V(x_0,x)\\
&\quad-\gamma_k(\eta_k+\mu_p)V(x_k,x)+\gamma_1\tau_1 W(y_0,y)-\gamma_k(\tau_k+\mu_d)W(y_k,y)\\
&\quad-\textstyle\sum_{t=1}^k \gamma_t[\tau_t W(y_{t-1},y_t)+\eta_t V(x_{t-1},x_t)+\lambda_t\langle A(x_{t-1}-x_{t-2}),y_{t-1}-y_t\rangle]\\
&\quad+\gamma_k\langle A(x_k-x_{k-1}),y_k-y\rangle.
\end{aligned}$$

Also note that by (3.6.13) and (3.6.16), we have

$$\begin{aligned}
&-\textstyle\sum_{t=1}^k \gamma_t[\tau_t W(y_{t-1},y_t)+\eta_t V(x_{t-1},x_t)+\lambda_t\langle A(x_{t-1}-x_{t-2}),y_{t-1}-y_t\rangle]\\
&\le -\textstyle\sum_{t=2}^k[\frac{\gamma_t\tau_t}{2}\|y_{t-1}-y_t\|^2+\frac{\gamma_{t-1}\eta_{t-1}}{2}\|x_{t-2}-x_{t-1}\|^2\\
&\qquad-\gamma_t\lambda_t\|A\|\|x_{t-1}-x_{t-2}\|\|y_{t-1}-y_t\|]\\
&\qquad-\gamma_k\eta_k V(x_{k-1},x_k)\\
&\le -\gamma_k\eta_k V(x_{k-1},x_k).
\end{aligned}$$

Combining the above two inequalities, we obtain

$$\begin{aligned}
\textstyle\sum_{t=1}^k \gamma_t Q(z_t,z) &\le \gamma_1\eta_1 V(x_0,x)-\gamma_k(\eta_k+\mu_p)V(x_k,x)+\gamma_1\tau_1 W(y_0,y)\\
&\quad-\gamma_k(\tau_k+\mu_d)W(y_k,y)-\gamma_k\eta_k V(x_{k-1},x_k)\\
&\quad+\gamma_k\langle A(x_k-x_{k-1}),y_k-y\rangle.
\end{aligned} \tag{3.6.20}$$

The result in (3.6.17) then follows from the above inequality and the fact that by (3.6.16),

$$
\begin{aligned}
&-(\tau_k+\mu_d)W(y_k,y)-\eta_k V(x_{k-1},x_k)+\langle A(x_k-x_{k-1}),y_k-y\rangle \\
&\le -(\tau_k+\mu_d)W(y_k,y)-\tfrac{\eta_k}{2}\|x_k-x_{k-1}\|^2+\|A\|\|x_k-x_{k-1}\|\|y_k-y\| \\
&\le -(\tau_k+\mu_d)W(y_k,y)+\tfrac{\|A\|^2}{2\eta_k}\|y_k-y\|^2 \\
&\le -\left(\tau_k+\mu_d-\tfrac{\|A\|^2}{\eta_k}\right)W(y_k,y).
\end{aligned}
$$

Fixing $z=z^*$ in the above inequality and using the fact that $Q(z_t,z^*)\ge 0$, we obtain (3.6.18). Finally, (3.6.19) follows from similar ideas and a different bound

$$
\begin{aligned}
&-(\tau_k+\mu_d)W(y_k,y)-\eta_k V(x_{k-1},x_k)+\langle A(x_k-x_{k-1}),y_k-y\rangle \\
&\le -\tfrac{\tau_k+\mu_d}{2}\|y_k-y\|_2^2-\eta_k V(x_{k-1},x_k)+\|A\|\|x_k-x_{k-1}\|\|y_k-y\|_2 \\
&\le -\eta_k V(x_{k-1},x_k)+\tfrac{\|A\|^2}{2(\tau_k+\mu_d)}\|x_k-x_{k-1}\|^2 \\
&\le -\left(\eta_k-\tfrac{\|A\|^2}{\tau_k+\mu_d}\right)V(x_{k-1},x_k).
\end{aligned}
$$

■

Based on Theorem 3.8, we will provide different ways to specify the algorithmic parameters $\{\tau_t\}$, $\{\eta_t\}$, and $\{\gamma_t\}$ for solving different classes of problems.

3.6.1 General Bilinear Saddle Point Problems

In this subsection, we assume that the parameters $\mu_p=\mu_d=0$ in (3.6.2). Moreover, for the sake of simplicity, we assume that both the primal and dual feasible sets X and Y are bounded. Discussions about primal–dual type method for solving problems with unbounded feasible sets can be found later in Sects. 3.6.4 and 4.4.

Given $\bar z\in Z$, we define the primal–dual gap as

$$
\max_{z\in Z} Q(\bar z,z). \tag{3.6.21}
$$

Our goal is to show that the primal–dual gap evaluated at the output solution

$$
\bar z_k=\frac{\sum_{t=1}^k(\gamma_t z_k)}{\sum_{t=1}^k\gamma_t} \tag{3.6.22}
$$

will converge to zero by properly specifying the algorithmic parameters. We use the following quantities in the convergence analysis of the primal dual method:

$$
D_X^2\equiv D_{X,\nu}^2:=\max_{x_0,x\in X}V(x_0,x)\ \text{ and }\ D_Y^2\equiv D_{Y,\omega}^2=\max_{y_0,y\in Y}W(y_0,y). \tag{3.6.23}
$$

Corollary 3.6. *If $\gamma_t = 1$, $\tau_t = \tau$, $\eta_t = \eta$, $\lambda_t = 1$, and $\tau\eta \ge \|A\|^2$ for any $t = 1,\ldots,k$, then*

$$\max_{z\in Z} Q(\bar{z}_k, z) \le \tfrac{1}{k}(\eta D_X^2 + \tau D_Y^2).$$

Proof. We can easily check that all the conditions (3.6.13)–(3.6.16) hold. It then follows from (3.6.17) that

$$\Sigma_{t=1}^k Q(z_t, z) \le \eta V(x_0, x) + \tau W(y_0, y). \tag{3.6.24}$$

Dividing both sides by k and using the convexity of $Q(\bar{z}_k, z)$ w.r.t. $\bar{z}_k$, we have

$$Q(\bar{z}_k, z) \le \tfrac{1}{k}[\eta V(x_0, x) + \tau W(y_0, y)].$$

The result then follows by maximizing both sides of the above inequality w.r.t. $z \in Z$. ■

It can be seen from the above result that the best selection of τ and η is given by

$$\eta = \tfrac{\|A\| D_Y}{D_X} \text{ and } \tau = \tfrac{\|A\| D_X}{D_Y}.$$

With such a selection, the rate of convergence of the primal–dual method reduces to

$$\max_{z\in Z} Q(\bar{z}_k, z) \le \tfrac{\|A\| D_X D_Y}{k}.$$

3.6.2 Smooth Bilinear Saddle Point Problems

In this subsection, we first assume that $\mu_p = 0$ but $\mu_d > 0$. We call these problems smooth bilinear saddle point problems because the objective function in (3.6.1) is a differentiable convex function with Lipschitz continuous gradient. We will show that by properly specifying algorithmic parameters, the primal–dual method will exhibit an $\mathcal{O}(1/k^2)$ rate of convergence. For the sake of simplicity, we assume that the primal feasible region X is bounded.

Corollary 3.7. *If $\gamma_t = t$, $\tau_t = \mu_d(t-1)/2$, $\eta_t = 2\|A\|^2/(\mu_d t)$, and $\lambda_t = (t-1)/t$, then*

$$\max_{z\in Z} Q(\bar{z}_k, z) \le \tfrac{4\|A\|^2 D_X^2}{\mu_d k(k+1)}.$$

where $\bar{z}_k$ is defined in (3.6.22).

Proof. Observe that all the conditions (3.6.13)–(3.6.16) hold. It then follows from (3.6.17) that

$$\Sigma_{t=1}^k \gamma_t Q(z_t, z) \le \gamma_1 \eta_1 V(x_0, x) = \tfrac{2\|A\|^2 V(x_0, x)}{\mu_d}. \tag{3.6.25}$$

Dividing both sides by $\sum_t \gamma_t$ and using the convexity of $Q(\bar{z}_k, z)$ w.r.t. $\bar{z}_k$, we have

$$Q(\bar{z}_k, z) \leq \frac{4\|A\|^2 V(x_0,x)}{\mu_d k(k+1)}.$$

The result then follows by maximizing both sides of the above inequality w.r.t. $z \in X$. ■

Next, we assume that $\mu_p > 0$ but $\mu_d = 0$. Moreover, for the sake of simplicity, we assume that the dual feasible set Y is bounded. Similar to the previous corollary, we can show the following result.

Corollary 3.8. *If* $\gamma_t = t+1$, $\tau_t = 4\|A\|^2/[\mu_p(t+1)]$, $\eta_t = \mu_p t/2$, *and* $\lambda_t = t/(t+1)$, *then*

$$\max_{z \in Z} Q(\bar{z}_k, z) \leq \frac{2}{k(k+3)}\left[\mu_p D_X^2 + \frac{4\|A\|^2 D_Y^2}{\mu_p}\right].$$

Proof. Observe that all the conditions (3.6.13)–(3.6.16) hold. It then follows from (3.6.17) that

$$\begin{aligned} \sum_{t=1}^k \gamma_t Q(z_t, z) &\leq \gamma_1 \eta_1 V(x_0,x) + \gamma_1 \tau_1 W(y_0,y) \\ &\leq \mu_p V(x_0,x) + \frac{4\|A\|^2 W(y_0,y)}{\mu_p}. \end{aligned} \tag{3.6.26}$$

Dividing both sides by $\sum_t \gamma_t$ and using the convexity of $Q(\bar{z}_k, z)$ w.r.t. $\bar{z}_k$, we have

$$Q(\bar{z}_k, z) \leq \frac{2}{k(k+3)}\left[\mu_p V(x_0,x) + \frac{4\|A\|^2 W(y_0,y)}{\mu_p}\right].$$

The result then follows by maximizing both sides of the above inequality w.r.t. $z \in X$. ■

3.6.3 Smooth and Strongly Convex Bilinear Saddle Point Problems

In this subsection, we assume that $\mu_p > 0$ and $\mu_d > 0$. We call these problems smooth and strongly convex bilinear saddle point problems because the objective function in (3.6.1) is both smooth and strongly convex.

Corollary 3.9. *Assume that*

$$\lambda = \frac{1}{2}\left[2 + \frac{\mu_p \mu_d}{\|A\|^2} - \sqrt{(2 + \frac{\mu_p \mu_d}{\|A\|^2})^2 - 4}\right].$$

If $\gamma_t = \lambda^{-t}$, $\tau_t = \mu_d \lambda/(1-\lambda)$, $\eta_t = \mu_p \lambda/(1-\lambda)$ *and* $\lambda_t = \lambda$, *then*

$$\frac{\mu_p}{1-\lambda} V(x_k, x^*) + \mu_d W(y_k, y^*) \leq \frac{\lambda^k}{1-\lambda}\left[\mu_p V(x_0, x^*) + \mu_d W(y_0, y^*)\right].$$

Proof. Observe that all the conditions (3.6.13)–(3.6.16) hold. It then follows from (3.6.17) (with $z = z^*$) and $Q(z_t, z^*) \geq 0$ that

$$0 \leq \textstyle\sum_{t=1}^k \gamma_t Q(z_t, z^*) \leq \gamma_1 \tau_1 W(y_0, y^*) - \gamma_k \mu_d W(y_k, y^*) \\ + \gamma_1 \eta_1 V(x_0, x^*) - \gamma_k(\eta_k + \mu_p) V(x_k, x^*),$$

which clearly implies the result. ■

3.6.4 Linearly Constrained Problems

In this subsection, we analyze the convergence of the primal–dual method applied to the linear constrained problem (3.6.6). For the sake of simplicity, let us assume that $b = 0$. We can then apply this algorithm with $\hat{f}(x) = f(x)$ and $\hat{g}(x) = 0$ to the saddle point reformulation of (3.6.6) given by

$$\min_{x \in X} \max_{y} \{f(x) + \langle Ax, y \rangle\}. \tag{3.6.27}$$

Since the dual feasible set is unbounded, we set the dual prox-function to $W(y_{t-1}, y) = \|y - y_{t-1}\|_2^2/2$ and restate the algorithm as follows.

$$\tilde{x}_t = x_{t-1} + \lambda_t(x_{t-1} - x_{t-2}), \tag{3.6.28}$$

$$y_t = \mathrm{argmin}_y\{\langle -A\tilde{x}_t, y\rangle + \tfrac{\tau_t}{2}\|y - y_{t-1}\|_2^2\} = y_{t-1} + \tfrac{1}{\tau_t} A\tilde{x}_t, \tag{3.6.29}$$

$$x_t = \mathrm{argmin}_{x \in X} \langle y_t, Ax\rangle + f(x) + \eta_t V(x_{t-1}, x). \tag{3.6.30}$$

Moreover, to deal with the unbounded dual feasible set of (3.6.27) we have to modify the convergence analysis given in the previous subsections.

We first prove a general convergence result for the above algorithm without specifying any algorithmic parameters.

Theorem 3.9. *Suppose that the conditions in (3.6.13)–(3.6.16) hold (with $\mu_d = 0$). If, in addition,*

$$\gamma_1 \tau_1 = \gamma_k \tau_k, \tag{3.6.31}$$

$$\tau_k \eta_k \geq \|A\|^2, \tag{3.6.32}$$

then

$$f(\bar{x}_k) - f(x^*) \leq \tfrac{1}{\sum_{t=1}^k \gamma_t}\left[\gamma_1 \eta_1 V(x_0, x^*) + \tfrac{\gamma_1 \tau_1}{2}\|y_0\|_2^2\right], \tag{3.6.33}$$

$$\|A\bar{x}_k\|_2 \leq \tfrac{1}{\sum_{t=1}^k \gamma_t}\left\{\tfrac{\gamma_1 \tau_1 \sqrt{\eta_k} + \gamma_k \|A\| \sqrt{\tau_k}}{\sqrt{\gamma_k(\eta_k \tau_k - \|A\|^2)}}\sqrt{2\gamma_1 \eta_1 V(x_0, x^*)}\right. \\ \left. + \left[\tfrac{\gamma_1 \tau_1 \sqrt{\eta_k} + \gamma_k \|A\| \sqrt{\tau_k}}{\sqrt{\gamma_k(\eta_k \tau_k - \|A\|^2)}}\sqrt{\gamma_1 \tau_1} + \gamma_1 \tau_1\right]\|y_0 - y^*\|_2\right\}. \tag{3.6.34}$$

Proof. It follows from (3.6.31) and (3.6.32) that

$$\begin{aligned}
&\gamma_1\tau_1 W(y_0,y) - \gamma_k\tau_k W(y_k,y) - \gamma_k\eta_k V(x_{k-1},x_k) - \gamma_k\langle A(x_k - x_{k-1}), y_k - y\rangle \\
&\le \tfrac{\gamma_1\tau_1}{2}\|y-y_0\|_2^2 - \tfrac{\gamma_k\tau_k}{2}\|y-y_k\|_2^2 - \tfrac{\gamma_k\eta_k}{2}\|x_k-x_{k-1}\|^2 - \gamma_k\langle A(x_k - x_{k-1}), y_k - y\rangle \\
&= \tfrac{\gamma_1\tau_1}{2}\|y_0\|_2^2 - \tfrac{\gamma_k\tau_k}{2}\|y_k\|_2^2 - \tfrac{\gamma_k\eta_k}{2}\|x_k - x_{k-1}\|^2 - \gamma_k\langle A(x_k - x_{k-1}), y_k\rangle \\
&\quad + \langle \gamma_1\tau_1(y_k - y_0) + \gamma_k A(x_k - x_{k-1}), y\rangle \\
&\le \tfrac{\gamma_1\tau_1}{2}\|y_0\|_2^2 + \langle \gamma_1\tau_1(y_k - y_0) + \gamma_k A(x_k - x_{k-1}), y\rangle,
\end{aligned}$$

which, in view of (3.6.20), then implies that

$$\begin{aligned}
\textstyle\sum_{t=1}^k \gamma_t Q(z_t,z) &\le \gamma_1\eta_1 V(x_0,x) - \gamma_k(\eta_k + \mu_p)V(x_k,x) \\
&\quad + \tfrac{\gamma_1\tau_1}{2}\|y_0\|_2^2 + \langle \gamma_1\tau_1(y_k - y_0) + \gamma_k A(x_k - x_{k-1}), y\rangle.
\end{aligned}$$

Fixing $x = x^*$ and noting that $Ax^* = 0$, we have $Q(z_t,(x^*,y)) = f(x_t) - f(x^*) + \langle Ax_t, y\rangle$. Using the previous two relations, we obtain

$$\begin{aligned}
&\textstyle\sum_{t=1}^k \gamma_t[f(x_t) - f(x^*)] + \langle \sum_{t=1}^k(\gamma_t A x_t) - \gamma_1\tau_1(y_k - y_0) - \gamma_k A(x_k - x_{k-1}), y\rangle \\
&\le \gamma_1\eta_1 V(x_0,x^*) - \gamma_k(\eta_k + \mu_p)V(x_k,x^*) + \tfrac{\gamma_1\tau_1}{2}\|y_0\|_2^2, \qquad (3.6.35)
\end{aligned}$$

for any y, which implies

$$\textstyle\sum_{t=1}^k(\gamma_t A x_t) - \gamma_1\tau_1(y_k - y_0) - \gamma_k A(x_k - x_{k-1}) = 0,$$

since otherwise the left-hand side of (3.6.35) can be unbounded. Using the above two observations and the convexity of f, we have

$$\begin{aligned}
f(\bar{x}_k) - f(x^*) &\le \tfrac{1}{\sum_{t=1}^k \gamma_t}\textstyle\sum_{t=1}^k \gamma_t[f(x_t) - f(x^*)] \\
&\le \tfrac{1}{\sum_{t=1}^k \gamma_t}\left[\gamma_1\eta_1 V(x_0,x^*) + \tfrac{\gamma_1\tau_1}{2}\|y_0\|_2^2\right], \\
\|A\bar{x}_k\|_2 &= \tfrac{1}{\sum_{t=1}^k \gamma_t}\textstyle\sum_{t=1}^k \|\gamma_t A x_t\|_2 \\
&= \tfrac{1}{\sum_{t=1}^k \gamma_t}\left[\|\gamma_1\tau_1(y_k - y_0) + \gamma_k A(x_k - x_{k-1})\|_2\right] \\
&\le \tfrac{1}{\sum_{t=1}^k \gamma_t}\left[\gamma_1\tau_1\|y_k - y_0\|_2 + \gamma_k\|A\|\|x_k - x_{k-1}\|\right]. \qquad (3.6.36)
\end{aligned}$$

Also by (3.6.18) and (3.6.19), we have

$$\begin{aligned}
\|y_k - y^*\|_2^2 &\le \tfrac{\eta_k}{\gamma_k(\eta_k\tau_k - \|A\|^2)}\left[2\gamma_1\eta_1 V(x_0,x^*) + \gamma_1\tau_1\|y_0 - y^*\|_2^2\right], \\
\|x_k - x_{k-1}\|^2 &\le \tfrac{\tau_k}{\gamma_k(\eta_k\tau_k - \|A\|^2)}\left[2\gamma_1\eta_1 V(x_0,x^*) + \gamma_1\tau_1\|y_0 - y^*\|_2^2\right],
\end{aligned}$$

which implies that

$$
\begin{aligned}
&\gamma_1\tau_1\|y_k-y_0\|_2+\gamma_k\|A\|\|x_k-x_{k-1}\| \\
&\le \gamma_1\tau_1(\|y_k-y^*\|_2+\|y_0-y^*\|_2)+\gamma_k\|A\|\|x_k-x_{k-1}\| \\
&\le \gamma_1\tau_1\left[\sqrt{\tfrac{\eta_k}{\gamma_k(\eta_k\tau_k-\|A\|^2)}}\left(\sqrt{2\gamma_1\eta_1V(x_0,x^*)}+\sqrt{\gamma_1\tau_1}\|y_0-y^*\|_2\right)+\|y_0-y^*\|_2\right] \\
&\quad+\gamma_k\|A\|\sqrt{\tfrac{\tau_k}{\gamma_k(\eta_k\tau_k-\|A\|^2)}}\left(\sqrt{2\gamma_1\eta_1V(x_0,x^*)}+\sqrt{\gamma_1\tau_1}\|y_0-y^*\|_2\right) \\
&=\tfrac{\gamma_1\tau_1\sqrt{\eta_k}+\gamma_k\|A\|\sqrt{\tau_k}}{\sqrt{\gamma_k(\eta_k\tau_k-\|A\|^2)}}\left[\sqrt{2\gamma_1\eta_1V(x_0,x^*)}+\sqrt{\gamma_1\tau_1}\|y_0-y^*\|_2\right]+\gamma_1\tau_1\|y_0-y^*\|_2.
\end{aligned}
$$

Using this relation in (3.6.36), we obtain (3.6.34). ■

Let us first consider the case when f is a general convex function which is not necessarily strongly convex. The proof of this result directly follows from Theorem 3.9.

Corollary 3.10. *If $\gamma_t=1$, $\tau_t=\tau$, $\eta_t=\eta$, $\lambda_t=1$, and $\tau\eta\ge\|A\|^2$ for any $t=1,\ldots,k$, then*

$$
\begin{aligned}
f(\bar{x}_k)-f(x^*)&\le\tfrac{1}{k}\left[\eta V(x_0,x^*)+\tfrac{\tau}{2}\|y_0\|_2^2\right], \\
\|A\bar{x}_k\|_2&\le\tfrac{1}{k}\left[\tfrac{\tau\sqrt{\eta}+\|A\|\sqrt{\tau}}{\eta\tau-\|A\|^2}\sqrt{2\eta V(x_0,x^*)}+\left(\tfrac{\tau\sqrt{\eta\tau}+\|A\|\tau}{\eta\tau-\|A\|^2}+\tau\right)\|y_0-y^*\|_2\right].
\end{aligned}
$$

In particular, if $\eta=2\|A\|$ and $\tau=\|A\|$, then

$$
\begin{aligned}
f(\bar{x}_k)-f(x^*)&\le\tfrac{\|A\|}{k}\left[2V(x_0,x^*)+\tfrac{1}{2}\|y_0\|^2\right], \\
\|A\bar{x}_k\|_2&\le\tfrac{2(\sqrt{2}+1)}{k}\sqrt{V(x_0,x^*)}+\tfrac{\|A\|+\sqrt{2}+1}{k}\|y_0-y^*\|.
\end{aligned}
$$

Next we consider the case when f is strongly convex with modulus μ_p.

Corollary 3.11. *If $\gamma_t=t+1$, $\tau_t=4\|A\|^2/[\mu_p(t+1)]$, $\eta_t=\mu_p t/2$, and $\lambda_t=t/(t+1)$, then for any $k\ge2$,*

$$
\begin{aligned}
f(\bar{x}_k)-f(x^*)&\le\tfrac{2}{k(k+3)}\left[\mu_pV(x_0,x^*)+\tfrac{2\|A\|^2}{\mu_p}\|y_0\|_2^2\right], \\
\|A\bar{x}_k\|_2&\le\tfrac{2}{k(k+3)}\left[2(2+\sqrt{3})\|A\|\sqrt{2V(x^0,x^*)}+\tfrac{4(3+\sqrt{3})\|A\|^2\|y_0-y^*\|_2}{\mu_p}\right].
\end{aligned}
$$

Proof. Note that $\sum_{t=1}^k\gamma_t=k(k+3)/2$. Also observe that

$$\eta_k\tau_k-\|A\|^2=\tfrac{2k\|A\|^2}{k+1}-\|A\|^2=\tfrac{(k-1)\|A\|^2}{k+1}$$

and hence that

$$\tfrac{\gamma_1\tau_1\sqrt{\eta_k}+\gamma_k\|A\|\sqrt{\tau_k}}{\sqrt{\gamma_k(\eta_k\tau_k-\|A\|^2)}}=\tfrac{4\|A\|\sqrt{k/2}+2\|A\|\sqrt{k+1}}{\sqrt{\mu_p(k-1)}}\le\tfrac{2\|A\|}{\sqrt{\mu_p}}(2+\sqrt{3}),\forall k\ge2.$$

The results then follows from (3.6.33) and (3.6.34), the selection of γ_t, τ_t and η_t, and the previous bounds. ■

3.7 Alternating Direction Method of Multipliers

In this section, we discuss a popular primal–dual type method called the alternating direction method of multipliers (ADMM) for solving a special class of linearly constrained convex optimization problems.

Consider the problem of

$$\min_{x\in X, y\in Y} \{f(x)+g(y) : Ax+By=b\}, \tag{3.7.1}$$

where $X \subseteq \mathbb{R}^p$ and $Y \subseteq \mathbb{R}^q$ are closed convex sets, the vector $b \in \mathbb{R}^m$ and matrices $A \in \mathbb{R}^{m\times p}$ and $B \in \mathbb{R}^{m\times q}$ are given. Moreover, we assume that the optimal solution (x^*, y^*) exists, along with an arbitrary dual multiplier $\lambda^* \in \mathbb{R}^m$ associated with the linear constraint $Ax+By=b$. This problem can be viewed as a special case of problem (3.6.6).

ADMM updates both the primal and dual variables (x_t, y_t, λ_t). Given $(y_{t-1}, \lambda_{t-1}) \in Y \times \mathbb{R}^m$ and some penalty parameter $\rho > 0$, this algorithm computes (x_t, y_t, λ_t) according to

$$x_t = \text{argmin}_{x\in X} f(x) + \langle \lambda_{t-1}, Ax+By_{t-1}-b\rangle + \tfrac{\rho}{2}\|Ax+By_{t-1}-b\|^2, \tag{3.7.2}$$

$$y_t = \text{argmin}_{y\in Y} g(y) + \langle \lambda_{t-1}, Ax_t+By-b\rangle + \tfrac{\rho}{2}\|Ax_t+By-b\|^2, \tag{3.7.3}$$

$$\lambda_t = \lambda_{t-1} + \rho(Ax_t+By_t-b). \tag{3.7.4}$$

For notational convenience, let us denote $z_t \equiv (x_t, y_t, \lambda_t)$, $z \equiv (x, y, \lambda)$, and let the primal–dual gap function Q be defined in (3.6.7), i.e.,

$$\begin{aligned} Q(z_t, z) = {} & f(x_t) + g(y_t) + \langle \lambda, Ax_t + By_t - b\rangle \\ & - [f(x) + g(y) + \langle \lambda_t, Ax+By-b\rangle]. \end{aligned} \tag{3.7.5}$$

Moreover, we denote an arbitrary pair of primal–dual solution of (3.7.1) by $z^* = (x^*, y^*, \lambda^*)$.

Theorem 3.10. *Let z_t, $t = 1,\ldots,k$, be the sequence generated by ADMM with some $\rho > 0$, and define $\bar z_k = \sum_{t=1}^k z_t/k$. Then we have*

$$f(\bar x_k) + g(\bar y_k) - f(x^*) - g(y^*) \le \tfrac{1}{2k}\left[\tfrac{1}{\rho}\|\lambda_0\|_2^2 + \rho\|B(y_0-y^*)\|_2^2\right], \tag{3.7.6}$$

$$\|A\bar x_k + B\bar y_k - b\|_2 \le \tfrac{1}{k}\left(\tfrac{2}{\rho}\|\lambda_0 - \lambda^*\|_2 + \|B(y_0-y^*)\|_2\right). \tag{3.7.7}$$

Proof. In view of the optimality conditions of (3.7.2) and (3.7.3), we have

$$f(x_t) - f(x) + \langle \lambda_{t-1} + \rho(Ax_t + By_{t-1} - b), A(x_t - x)\rangle \le 0,$$

$$g(y_t) - g(y) + \langle \lambda_{t-1} + \rho(Ax_t + By_t - b), B(y_t - y)\rangle \le 0,$$

which together with (3.7.4) then imply that

$$f(x_t) - f(x) \le -\langle \lambda_t + \rho B(y_{t-1} - y_t), A(x_t - x)\rangle,$$
$$g(y_t) - g(y) \le -\langle \lambda_t, B(y_t - y)\rangle.$$

Using these two relations and (3.7.5), we have

$$\begin{aligned} Q(z_t,z) &\le -\langle \lambda_t + \rho B(y_{t-1} - y_t), A(x_t - x)\rangle - \langle \lambda_t, B(y_t - y)\rangle \\ &\quad + \langle \lambda, Ax_t + By_t - b\rangle - \langle \lambda_t, Ax + By - b\rangle \\ &= \langle \lambda - \lambda_t, Ax_t + By_t - b\rangle + \rho \langle B(y_t - y_{t-1}, A(x_t - x)\rangle \\ &= \langle \lambda - \lambda_t, \tfrac{1}{\rho}(\lambda_t - \lambda_{t-1})\rangle + \rho \langle B(y_t - y_{t-1}), A(x_t - x)\rangle. \end{aligned}$$

Noting that

$$2\langle \lambda - \lambda_t, \lambda_t - \lambda_{t-1}\rangle = \|\lambda - \lambda_{t-1}\|_2^2 - \|\lambda - \lambda_t\|_2^2 - \|\lambda_{t-1} - \lambda_t\|_2^2,$$

and

$$\begin{aligned} &2\langle B(y_t - y_{t-1}), A(x_t - x)\rangle \\ &= \|Ax + By_{t-1} - b\|_2^2 - \|Ax + By_t - b\|_2^2 + \|Ax_t + By_t - b\|_2^2 - \|Ax_t + By_{t-1} - b\|_2^2 \\ &= \|Ax + By_{t-1} - b\|_2^2 - \|Ax + By_t - b\|_2^2 + \tfrac{1}{\rho^2}\|\lambda_{t-1} - \lambda_t\|_2^2 - \|Ax_t + By_{t-1} - b\|_2^2, \end{aligned}$$

we conclude that

$$\begin{aligned} Q(z_t,z) &\le \tfrac{1}{2\rho}\left(\|\lambda_{t-1} - \lambda\|_2^2 - \|\lambda_t - \lambda\|_2^2\right) \\ &\quad + \tfrac{\rho}{2}\left(\|Ax + By_{t-1} - b\|_2^2 - \|Ax + By_t - b\|_2^2 - \|Ax_t + By_{t-1} - b\|_2^2\right). \end{aligned}$$

Summing up the above inequality from $t = 1, \ldots, k$, we obtain

$$\begin{aligned} \Sigma_{t=1}^k Q(z_t, z) &\le \tfrac{1}{2\rho}\left(\|\lambda_0 - \lambda\|_2^2 - \|\lambda_k - \lambda\|_2^2\right) \\ &\quad + \tfrac{\rho}{2}\left(\|B(y_0 - y)\|_2^2 - \|B(y_k - y)\|_2^2\right). \end{aligned} \tag{3.7.8}$$

Setting $z = z^*$ in the above inequality and using the fact that $Q(\bar{z}_k, z^*) \ge 0$, we can see that

$$\|\lambda_k - \lambda^*\|_2^2 \le \|\lambda_0 - \lambda^*\|_2^2 + \rho^2\|B(y_0 - y^*)\|_2^2$$

and hence that

$$\|\lambda_k - \lambda_0\|_2 \le \|\lambda_0 - \lambda^*\|_2 + \|\lambda_k - \lambda^*\|_2 \le 2\|\lambda_0 - \lambda^*\|_2 + \rho\|B(y_0 - y^*)\|_2. \tag{3.7.9}$$

Moreover, letting $z = (x^*, y^*, \lambda)$ in (3.7.8), and noting that

$$\begin{aligned} \tfrac{1}{k}\Sigma_{t=1}^k Q(z_t, (x^*, y^*, \lambda)) &\ge Q(\bar{z}_k, (x^*, y^*, \lambda)) \\ &= f(\bar{x}_k) + g(\bar{y}_k) - f(x^*) - g(y^*) + \langle \lambda, A\bar{x}_k + B\bar{y}_k - b\rangle, \end{aligned}$$

and that

$$\tfrac{1}{2}\left(\|\lambda_0-\lambda\|_2^2-\|\lambda_k-\lambda\|_2^2\right)=\tfrac{1}{2}\left(\|\lambda_0\|_2^2-\|\lambda_k\|_2^2\right)-\langle\lambda,\lambda_0-\lambda_k\rangle,$$

we have

$$\begin{aligned}&f(\bar{x}_k)+g(\bar{y}_k)-f(x^*)-g(y^*)+\langle\lambda,A\bar{x}_k+B\bar{y}_k-b+\tfrac{1}{\rho k}(\lambda_0-\lambda_k)\rangle\\&\le\tfrac{1}{2k}\left[\tfrac{1}{\rho}\left(\|\lambda_0\|_2^2-\|\lambda_k\|_2^2\right)+\rho\left(\|B(y_0-y^*)\|_2^2-\|B(y_k-y^*)\|_2^2\right)\right]\\&\le\tfrac{1}{2k}\left[\tfrac{1}{\rho}\|\lambda_0\|_2^2+\rho\|B(y_0-y^*)\|_2^2\right].\end{aligned}$$

for any $\lambda\in\mathbb{R}^m$. This relation implies that $A\bar{x}_k+B\bar{y}_k-b+\frac{1}{\rho k}(\lambda_0-\lambda_k)=0$, and hence that (3.7.6) holds. (3.7.7) also directly follows from the previous observation and (3.7.9). ■

In comparison with the primal–dual method, the selection of the algorithmic parameter in the ADMM method seems to be simpler. Moreover, the rate of convergence of ADMM depends only on the norm of one part of the constraint matrix, i.e., $\|B\|$ rather than $\|[A,B]\|$. However, this method requires the solution of more complicated subproblems and it is not straightforward to generalize this algorithm for solving problems with more than two blocks of variables.

3.8 Mirror-Prox Method for Variational Inequalities

This section focuses on variational inequality (VI) that can be used to model a wide range of optimization, equilibrium, and complementarity problems. Given a nonempty closed convex set $X\subseteq\mathbb{R}^n$ and a continuous map $F:X\to\mathbb{R}^n$, the variational inequality problem, denoted by $\mathrm{VI}(X,F)$, is to find $x^*\in X$ satisfying

$$\langle F(x^*),x-x^*\rangle\ge 0\quad\forall x\in X. \tag{3.8.1}$$

Such a point x^* is often called a *strong solution* of $\mathrm{VI}(X,F)$. In particular, if F is given by the gradient of f, then (3.8.1) is exactly the first-order necessary optimality condition of $\min_{x\in X}f(x)$.

One important class of VI problems is called monotone VI for which the operator $F(\cdot)$ is monotone, i.e.,

$$\langle F(x)-F(y),x-y\rangle\ge 0\quad\forall x,y\in X. \tag{3.8.2}$$

These monotone VIs cover convex optimization problems as a special case. They also cover the following saddle point problems

$$\min_{x\in X}\max_{y\in Y}F(x,y),$$

where F is convex w.r.t. x and concave w.r.t. y.

A related notion is a *weak solution* of VI(X,F), i.e., a point $x^* \in X$ such that

$$\langle F(x), x - x^* \rangle \geq 0 \quad \forall x \in X. \tag{3.8.3}$$

Note that if $F(\cdot)$ is monotone and continuous, a weak solution of VI(X,F) must be a strong solution and vice versa.

In this section, we focus on the mirror-prox method for solving these VI problems. The mirror-prox method evolves from Korpelevich's extragradient method. We assume that a distance generating function ν with modulus 1 and its associated *prox-function* $V : X^o \times X \to \mathbb{R}_+$ are given. The basic scheme of the mirror-prox method can be described as follows.

Input: Initial point $x_1 \in X$ and stepsizes $\{\gamma_k\}_{k \geq 1}$.
(0) Set $k = 1$.
(1) Compute

$$y_k = \text{argmin}_{x \in X} \{ \langle \gamma_k F(x_k), x \rangle + V(x_k, x) \}, \tag{3.8.4}$$
$$x_{k+1} = \text{argmin}_{x \in X} \{ \langle \gamma_k F(y_k), x \rangle + V(x_k, x) \}. \tag{3.8.5}$$

(2) Set $k = k+1$ and go to Step 1.

We now add a few remarks about the above mirror-prox method. First, observe that under the Euclidean case when $\|\cdot\| = \|\cdot\|_2$ and $\nu(x) = \|x\|^2/2$, the computation of (y_t, x_t), $t \geq 1$, is the same as Korpelevich's extragradient or Euclidean extragradient method. Second, the above method is slightly different that Nemirovski's mirror-prox method for solving monotone VI problems which can possibly skip the extragradient step (3.8.5) depending on the progress of the algorithm. We will establish the convergence of the mirror-prox method for solving different types of VIs.

3.8.1 Monotone Variational Inequalities

In this subsection, we discuss the convergence properties of the mirror-prox method for solving monotone VIs.

We first show an important recursion of the mirror-prox method for VI(X,F), which holds for VI problems not necessarily monotone.

Lemma 3.9. *Let $x_1 \in X$ be given and the pair $(y_k, x_{k+1}) \in X \times X$ be computed according to (3.8.4)–(3.8.5). Then for any $x \in X$, we have*

$$\gamma_k \langle F(y_k), y_k - x \rangle - \tfrac{\gamma_k^2}{2} \|F(x_k) - F(y_k)\|_*^2 + V(x_k, y_k) \leq V(x_k, x) - V(x_{k+1}, x). \tag{3.8.6}$$

Proof. By (3.8.4) and Lemma 3.5 (with $p(\cdot) = \gamma_k \langle F(x_k), \cdot \rangle$, $\tilde{x} = x_k$, and $\hat{u} = y_k$), we have

$$\gamma_k\langle F(x_k), y_k - x\rangle + V(x_k, y_k) + V(y_k, x) \le V(x_k, x), \ \forall x \in X.$$

Letting $x = x_{k+1}$ in the above inequality, we obtain

$$\gamma_k\langle F(x_k), y_k - x_{k+1}\rangle + V(x_k, y_k) + V(y_k, x_{k+1}) \le V(x_k, x_{k+1}). \tag{3.8.7}$$

Moreover, by (3.8.5) and Lemma 3.5 (with $p(\cdot) = \gamma_k\langle F(y_k), \cdot\rangle$, $\tilde{x} = x_k$, and $\hat{u} = x_{k+1}$), we have

$$\gamma_k\langle F(y_k), x_{k+1} - x\rangle + V(x_k, x_{k+1}) + V(x_{k+1}, x) \le V(x_k, x), \ \forall x \in X.$$

Replacing $V(x_k, x_{k+1})$ in the above inequality with the bound in (3.8.7) and noting that $\langle F(y_k), x_{k+1} - x\rangle = \langle F(y_k), y_k - x\rangle - \langle F(y_k), y_k - x_{k+1}\rangle$, we have

$$\begin{aligned}&\gamma_k\langle F(y_k), y_k - x\rangle + \gamma_k\langle F(x_k) - F(y_k), y_k - x_{k+1}\rangle \\ &+ V(x_k, y_k) + V(y_k, x_{k+1}) + V(x_{k+1}, x) \le V(x_k, x).\end{aligned}$$

Moreover, by using the Cauchy–Schwarz inequality and the strong convexity of ν, we have

$$\begin{aligned}&\gamma_k\langle F(x_k) - F(y_k), y_k - x_{k+1}\rangle + V(x_k, y_k) + V(y_k, x_{k+1}) \\ &\ge -\gamma_k\|F(x_k) - F(y_k)\|_*\|y_k - x_{k+1}\| + V(x_k, y_k) + V(y_k, x_{k+1}) \\ &\ge -\gamma_k\|F(x_k) - F(y_k)\|_*\left[2V(y_k, x_{k+1})\right]^{1/2} + V(x_k, y_k) + V(y_k, x_{k+1}) \\ &\ge -\tfrac{\gamma_k^2}{2}\|F(x_k) - F(y_k)\|_*^2 + V(x_k, y_k),\end{aligned}$$

where the last inequality follows from (3.1.6). Combining the above two conclusions we arrive at relation (3.8.20). ■

We define a termination criterion to characterize the weak solutions of VI problems as follows:

$$g(\bar{x}) := \sup_{x \in X}\langle F(x), \bar{x} - x\rangle. \tag{3.8.8}$$

Note that we must have $g(\bar{x}) \ge 0$ by setting $x = \bar{x}$ in the right-hand side of the above definition. Hence, $g(\bar{x})$ measures how much the condition in (3.8.3) is violated. It should be noted that the mirror-prox method can also generate a solution that approximately satisfies the stronger criterion in (3.8.1). We will discuss the computation of strong solutions in next subsection.

For the sake of simplicity, we focus on the case when F is Lipschitz continuous, i.e.,

$$\|F(x) - F(y)\|_* \le L\|x - y\|, \quad \forall x, y \in X. \tag{3.8.9}$$

Note, however, that this algorithm can also be applied to more general problems for which F is Hölder continuous, locally Lipschitz continuous, continuous, or bounded. We leave the extension to these cases as an exercise.

Theorem 3.11. *Let $x_1 \in X$ be given and the pair $(y_k, x_{k+1}) \in X \times X$ be computed according to (3.8.4)–(3.8.5). Also let us denote*

$$\bar{y}_k := \frac{\sum_{t=1}^{k}(\gamma_t y_t)}{\sum_{t=1}^{k}\gamma_t}. \tag{3.8.10}$$

If (3.8.9) holds and

$$\gamma_k = \tfrac{1}{L},\ k = 1,2,\ldots, \tag{3.8.11}$$

then we have

$$g(\bar{y}_k) \leq \tfrac{LD_X^2}{k}, k \geq 1, \tag{3.8.12}$$

where D_X and g are defined in (3.2.4) and (3.8.8), respectively.

Proof. Note that by the monotonicity of F, i.e., (3.8.2), we have $\langle F(y_k), y_k - x\rangle \geq \langle F(x), y_k - x\rangle$ for any $x \in X$. Moreover, by the Lipschitz continuity of F, i.e., (3.8.9), and the strong convexity of v we have

$$-\tfrac{\gamma_k^2}{2}\|F(x_k) - F(y_k)\|_*^2 + V(x_k, y_k) \geq -\tfrac{L\gamma_k^2}{2}\|x_k - y_k\|^2 + \|x_k - y_k\|^2 \geq 0.$$

Using these relations in (3.8.6), we conclude that

$$\gamma_k\langle F(x), y_k - x\rangle \leq V(x_k, x) - V(x_{k+1}, x), \forall x \in X \text{ and } k \geq 1.$$

Summing up these relations and then maximizing both sides w.r.t. x, we obtain (3.8.12). ■

Theorem 3.11 establishes the rate of convergence for the mirror-prox method for solving monotone VIs. In the next subsection we will discuss its convergence for solving a class of VI problems which are not necessarily monotone.

3.8.2 Generalized Monotone Variational Inequalities

We study in this subsection a class of generalized monotone VI (GMVI) problems which satisfy for any $x^* \in X^*$

$$\langle F(x), x - x^*\rangle \geq 0 \quad \forall x \in X. \tag{3.8.13}$$

Clearly, condition (3.8.13) is satisfied if $F(\cdot)$ is monotone Moreover, this assumption holds if $F(\cdot)$ is pseudo-monotone, i.e.,

$$\langle F(y), x - y\rangle \geq 0 \implies \langle F(x), x - y\rangle \geq 0. \tag{3.8.14}$$

As an example, $F(\cdot)$ is pseudo-monotone if it is the gradient of a real-valued differentiable pseudo-convex function. It is also not difficult to construct VI problems that satisfy (3.8.13), but their operator $F(\cdot)$ is neither monotone nor pseudo-monotone anywhere. One set of simple examples is given by all the functions $F : \mathbb{R} \to \mathbb{R}$ satisfying

$$F(x)\begin{cases} = 0, \; x = x_0; \\ \geq 0, \; x \geq x_0; \\ \leq 0, \; x \leq x_0. \end{cases} \tag{3.8.15}$$

These problems, although satisfying (3.8.13) with $x^* = x_0$, can be neither monotone nor pseudo-monotone.

In this subsection, we introduce a termination criterion for solving VI associated with the prox-mapping by first providing a simple characterization of a strong solution to $\mathrm{VI}(X,F)$.

Lemma 3.10. *A point $x \in X$ is a strong solution of* $\mathrm{VI}(X,F)$ *if and only if*

$$x = \mathrm{argmin}_{z\in X}\langle \gamma F(x), z\rangle + V(x,z) \tag{3.8.16}$$

for some $\gamma > 0$.

Proof. (3.8.16) holds if and only if

$$\langle \gamma F(x) + \nabla \nu(x) - \nabla \nu(x), z - x\rangle \geq 0, \quad \forall z \in X, \tag{3.8.17}$$

or equivalently, $\langle \gamma F(x), z - x\rangle \geq 0$ for any $z \in X$, which, in view of the fact that $\gamma > 0$ and definition (3.8.1), implies that x is a strong solution of $\mathrm{VI}(X,F)$. ■

Motivated by Lemma 3.10, we can define the residual function for a given $x \in X$ as follows.

Definition 3.1. Let $\|\cdot\|$ be a given norm in $\mathbb{R}^n$, $\nu(\cdot)$ be a distance generating function modulus 1 w.r.t. $\|\cdot\|$, and

$$x^+ := \mathrm{argmin}_{z\in X}\{\langle \gamma F(x), z\rangle + V(x,z)\}$$

for some positive constant γ. Then we define the residual $R_\gamma(\cdot)$ at the point $x \in X$ as

$$R_\gamma(x) \equiv P_X(x, F(x), \gamma) := \tfrac{1}{\gamma}\left[x - x^+\right]. \tag{3.8.18}$$

Observe that in the Euclidean setup where $\|\cdot\| = \|\cdot\|_2$ and $\nu(x) = \|x\|_2^2/2$, the residual $R_\gamma(\cdot)$ in (3.8.18) reduces to

$$R_\gamma(x) = \tfrac{1}{\gamma}\left[x - \Pi_X(x - \gamma F(x))\right], \tag{3.8.19}$$

where $\Pi_X(\cdot)$ denotes the Euclidean projection over X. In particular, if $F(\cdot)$ is the gradient of a real-valued differentiable function $f(\cdot)$, the residual $R_\gamma(\cdot)$ in (3.8.19) is often called the projected gradient of $f(\cdot)$ at x.

The following two results are immediate consequences of Lemma 3.10 and Definition 3.1.

Lemma 3.11. *A point $x \in X$ is a strong solution of* $\mathrm{VI}(X,F)$ *if and only if* $\|R_\gamma(x)\| = 0$ *for some $\gamma > 0$.*

Lemma 3.12. *Suppose that* $x_k \in X$, $\gamma_k \in (0,\infty)$, $k = 1,2,\ldots$, *and that*

$$y_k = \mathrm{argmin}_{z\in X}\left\{\langle \gamma_k F(x_k), z\rangle + V(x_k,z)\right\}$$

satisfy the following conditions:

(i) $\lim_{k\to\infty} V(x_k,y_k) = 0$;
(ii) There exists $K \in \mathbb{N}$ *and* $\gamma^* > 0$ *such that* $\gamma_k \geq \gamma^*$ *for any* $k \geq K$.

Then we have $\lim_{k\to\infty}\|R_{\gamma_k}(x_k)\| = 0$. *If in addition, the sequence* $\{x_k\}$ *is bounded, there exists an accumulation point* $\tilde{x}$ *of* $\{x_k\}$ *such that* $\tilde{x} \in X^*$, *where* X^* *denotes the solution set of* VI(X,F).

Proof. It follows from the strong convexity of ν and condition (i) that $\lim_{k\to\infty}\|x_k - y_k\| = 0$. This observation, in view of Condition (ii) and Definition 3.1, then implies that $\lim_{k\to\infty}\|R_{\gamma_k}(x_k)\| = 0$. Moreover, if $\{x_k\}$ is bounded, there exist a subsequence $\{\tilde{x}_i\}$ of $\{x_k\}$ obtained by setting $\tilde{x}_i = x_{n_i}$ for $n_1 \leq n_2 \leq \ldots$, such that $\lim_{i\to\infty}\|\tilde{x}_i - \tilde{x}\| = 0$. Let $\{\tilde{y}_i\}$ be the corresponding subsequence in $\{y_k\}$, i.e., $y_i = \mathrm{argmin}_{z\in X}\left\{\langle \gamma_{n_i} F(x_{n_i}), z\rangle + V(x_{n_i},z)\right\}$, and $\tilde{\gamma}_i = \gamma_{n_i}$. We have $\lim_{i\to\infty}\|\tilde{x}_i - \tilde{y}_i\| = 0$. Moreover, by (3.8.17), we have

$$\langle F(\tilde{x}_i) + \tfrac{1}{\tilde{\gamma}_i}\left[\nabla\nu(\tilde{y}_i) - \nabla\nu(\tilde{x}_i)\right], z - \tilde{y}_i\rangle \geq 0, \;\; \forall z \in X, \forall i \geq 1.$$

Tending i to $+\infty$ in the above inequality, and using the continuity of $F(\cdot)$ and $\nabla\nu(\cdot)$, and condition (ii), we conclude that $\langle F(\tilde{x}), z - \tilde{x}\rangle \geq 0$ for any $z \in X$. ∎

We now specialize Lemma 3.9 for solving GMVIs.

Lemma 3.13. *Let* $x_1 \in X$ *be given and the pair* $(y_k, x_{k+1}) \in X \times X$ *be computed according to (3.8.4)–(3.8.5). Also let* X^* *denote the solution set of* VI(X,F). *Then, the following statements hold:*

(a) There exists $x^* \in X^*$ *such that*

$$-\tfrac{\gamma_k^2}{2}\|F(x_k) - F(y_k)\|_*^2 + V(x_k,y_k) \leq V(x_k,x^*) - V(x_{k+1},x^*); \qquad (3.8.20)$$

(b) If $F(\cdot)$ *is Lipschitz continuous (i.e., condition (3.8.9) holds), then we have*

$$\left(1 - L^2\gamma_k^2\right)V(x_k,y_k) \leq V(x_k,x^*) - V(x_{k+1},x^*). \qquad (3.8.21)$$

Proof. We first show part (a). Fixing $x = x^*$ in (3.8.6) and using the fact that $\langle F(y_k), y_k - x^*\rangle \geq 0$ due to (3.8.13), we obtain the result.

Now, it follows from the assumption (3.8.9) and the strong convexity of ν that

$$\|F(x_k) - F(y_k)\|_*^2 \leq L^2\|x_k - y_k\|^2 \leq 2L^2 V(y_k,x_k).$$

Combining the previous observation with (3.8.20), we obtain (3.8.21). ∎

We are now ready to establish the complexity of the mirror descent method for solving GMVI problems.

Theorem 3.12. *Suppose that $F(\cdot)$ is Lipschitz continuous (i.e., condition (3.8.9) holds) and that the stepsizes γ_k are set to*

$$\gamma_k = \tfrac{1}{\sqrt{2}L}, \quad k \geq 1. \tag{3.8.22}$$

Also let R_γ be defined in (3.8.18). For any $k \in \mathbb{N}$, there exists $i \leq k$ such that

$$\|R_{\gamma_i}(x_i)\|^2 \leq \tfrac{8L^2}{k} V(x_1, x^*), \quad k \geq 1. \tag{3.8.23}$$

Proof. Using (3.8.21) and (3.8.22), we have

$$\tfrac{1}{2} V(x_k, y_k) \leq V(x_k, x^*) - V(x_{k+1}, x^*), \;\; k \geq 1.$$

Also it follows from the strong convexity of ν and definition (3.8.18) that

$$V(x_k, y_k) \geq \tfrac{1}{2}\|x_k - y_k\|^2 = \tfrac{\gamma_k^2}{2}\|R_{\gamma_k}(x_k)\|^2. \tag{3.8.24}$$

Combining the above two observations, we obtain

$$\gamma_k^2 \|R_{\gamma_k}(x_k)\|^2 \leq 4\left[V(x_k, x^*) - V(x_{k+1}, x^*)\right], k \geq 1.$$

By summing up these inequalities we arrive at

$$\sum_{i=1}^{k} \gamma_i^2 \min_{i=1,\ldots,k} \|R_{\gamma_i}(x_i)\|^2 \leq \sum_{i=1}^{k} \gamma_i^2 \|R_{\gamma_i}(x_i)\|^2 \leq 4V(x_1, x^*), \;\; k \geq 1,$$

which implies that

$$\min_{i=1,\ldots,k} \|R_{\gamma_i}(x_i)\|^2 \leq \tfrac{4}{\sum_{i=1}^k \gamma_i^2} \gamma_i^2 V(x_1, x^*). \tag{3.8.25}$$

Using the above inequality and (3.8.22), we obtain the bound in (3.8.23). ■

In view of Theorem 3.12, the mirror-prox method can be applied to solve the GMVI problems which are not necessarily monotone. Moreover, the rate of convergence for computing an approximate strong solution, in terms of $\min_{i=1,\ldots,k} R_{\gamma_i}$, can be bounded by $\mathcal{O}(1/\sqrt{k})$.

3.9 Accelerated Level Method

In this section, we discuss an important class of first-order methods, i.e., bundle-level methods for large-scale convex optimization. These methods have been regarded as one of the most efficient first-order optimization methods in practice as they can utilize historical first-order information through the cutting plane model. While other first-order methods discussed in the previous sections usually require us to estimate quite a few problem parameters (e.g., L and D_X), the bundle-level type methods we present in this section do not require much problem information, but can still attain the best performance guarantees for solving a few different classes of convex optimization problems.

3.9.1 Nonsmooth, Smooth, and Weakly Smooth Problems

Consider the convex programming

$$f^* := \min_{x \in X} f(x), \tag{3.9.1}$$

where X is a convex compact set and $f : X \to \mathbb{R}$ is a closed convex function. In the classic black-box setting, f is represented by a first-order oracle which, given an input point $x \in X$, returns $f(x)$ and $f'(x) \in \partial f(x)$, where $\partial f(x)$ denotes the subdifferential of f at $x \in X$. Moreover, we assume

$$f(y) - f(x) - \langle f'(x), y - x\rangle \le \tfrac{M}{1+\rho}\|y - x\|^{1+\rho}, \ \ \forall x, y \in X \tag{3.9.2}$$

for some $M > 0$, $\rho \in [0,1]$ and $f'(x) \in \partial f(x)$. Clearly, this class of problems cover nonsmooth ($\rho = 0$), smooth ($\rho = 1$), as well as weakly smooth problems, i.e., problems with Hölder continuous gradient when $\rho \in (0,1)$.

Let us first provide a brief introduction to the classical cutting plane method for solving (3.9.1). The basic idea of this type of method is to construct lower and upper bounds on f^* and to guarantee that the gap between these bounds converges to 0.

While the upper bound of f^* is given by the objective value of a feasible solution, we need to discuss how to compute lower bounds on f^*. Given a sequence of search points $x_1, x_2, \ldots, x_k \in X$, an important construct, the cutting plane model, of the objective function f of problem (3.9.1) is given by

$$m_k(x) := \max\{h(x_i, x) : 1 \le i \le k\}, \tag{3.9.3}$$

where

$$h(z, x) := f(z) + \langle f'(z), x - z\rangle. \tag{3.9.4}$$

In the cutting plane method, we approximate f by m_k and update the search points according to

$$x_{k+1} \in \mathrm{Argmin}_{x \in X} m_k(x). \tag{3.9.5}$$

Hence, x_{k+1} defines a new feasible solution in the next iteration, and the updated model $m_k(x_{k+1})$ provides an improved lower bound on f^*. However, this scheme converges slowly, both theoretically and practically.

To improve the basic scheme of the cutting plane method, we need to introduce a new way to construct the lower bounds of f^* by utilizing the (approximate) level sets of f. Let $\mathcal{E}_f(l)$ denote the level set of f given by

$$\mathscr{E}_f(l) := \{x \in X : f(x) \le l\}. \tag{3.9.6}$$

Also for some $z \in X$, let $h(z,x)$ be the cutting plane defined in (3.9.4) and denote

$$\bar{h} := \min\left\{h(z,x) : x \in \mathscr{E}_f(l)\right\}. \tag{3.9.7}$$

Then, it is easy to verify that

$$\min\{l, \bar{h}\} \le f(x), \quad \forall x \in X. \tag{3.9.8}$$

Indeed, if $l \le f^*$, then $\mathscr{E}_f(l) = \emptyset$, $\bar{h} = +\infty$ and $\min\{l,\bar{h}\} = l$. Hence (3.9.8) is obviously true. Now consider the case $l > f^*$. Clearly, for an arbitrary optimal solution x^* of (3.9.1), we have $x^* \in \mathscr{E}_f(l)$. Moreover, by (3.9.4), (3.9.7) and the convexity of f, we have $\bar{h} \le h(z,x) \le f(x)$ for any $x \in \mathscr{E}_f(l)$. Hence, $\bar{h} \le f(x^*) = f^*$ and thus (3.9.8) holds.

Note, however, that solving problem (3.9.7) is usually as difficult as solving the original problem (3.9.1). To compute a convenient lower bound of f^*, we replace $\mathscr{E}_f(l)$ in (3.9.7) with a convex and compact set X' satisfying

$$\mathscr{E}_f(l) \subseteq X' \subseteq X. \tag{3.9.9}$$

The set X' will be referred to as a *localizer* of the level set $\mathscr{E}_f(l)$. The following result shows the computation of a lower bound on f^* by solving such a relaxation of (3.9.7).

Lemma 3.14. *Let X' be a localizer of the level set $\mathscr{E}_f(l)$ for some $l \in \mathbb{R}$ and $h(z,x)$ be defined in (3.9.4). Denote*

$$\underline{h} := \min\left\{h(z,x) : x \in X'\right\}. \tag{3.9.10}$$

We have

$$\min\{l, \underline{h}\} \le f(x), \quad \forall x \in X. \tag{3.9.11}$$

Proof. Note that if $X' = \emptyset$ (i.e., (3.9.10) is infeasible), then $\underline{h} = +\infty$. In this case, we have $\mathscr{E}_f(l) = \emptyset$ and $f(x) \ge l$ for any $x \in X$. Now assume that $X' \ne \emptyset$. By (3.9.7), (3.9.9), and (3.9.10), we have $\underline{h} \le \bar{h}$, which together with (3.9.8) then clearly imply (3.9.11). ∎

We will also employ some ideas from the accelerated gradient descent method to guarantee fast convergence of level methods. In particular, we will use three different sequences, i.e., $\{x_k^l\}, \{x_k\}$, and $\{x_k^u\}$ for updating the lower bound, search points, and upper bound respectively. Similar to the mirror descent method, we assume that a distance generating function ν with modulus 1 is given.

We are now ready to describe a gap reduction procedure, denoted by $\mathscr{G}_{APL}$, which, for a given search point p and a lower bound lb on f^*, computes a new search point p^+ and a new lower bound lb^+ satisfying $f(p^+) - \mathrm{lb}^+ \le q[f(p) - \mathrm{lb}]$ for some $q \in (0,1)$. Note that the value of q will depend on the two algorithmic input parameters: $\beta, \theta \in (0,1)$.

The APL gap reduction procedure: $(p^+, \mathrm{lb}^+) = \mathscr{G}_{APL}(p, \mathrm{lb}, \beta, \theta)$

(0) Set $x_0^u = p$, $\overline{f}_0 = f(x_0^u)$, $\underline{f}_0 = \mathrm{lb}$, and $l = \beta \underline{f}_0 + (1-\beta)\overline{f}_0$. Also let $x_0 \in X$ and the initial localizer X_0' be arbitrarily chosen, say $x_0 = p$ and $X_0' = X$. Set the prox-function $V(x_0, x) = \nu(x) - [\nu(x_0) + \langle \nu'(x_0), x - x_0\rangle]$. Also let $k = 1$.

(1) *Update lower bound:* set $x_k^l = (1-\alpha_k)x_{k-1}^u + \alpha_k x_{k-1}$, $h(x_k^l, x) = f(x_k^l) + \langle f'(x_k^l), x - x_k^l\rangle$,

$$\underline{h}_k := \min_{x \in X_{k-1}'} \left\{ h(x_k^l, x) \right\}, \text{ and } \underline{f}_k := \max \left\{ \underline{f}_{k-1}, \min\{l, \underline{h}_k\} \right\}. \tag{3.9.12}$$

If $\underline{f}_k \geq l - \theta(l - \underline{f}_0)$, then **terminate** the procedure with $p^+ = x_{k-1}^u$ and $\mathrm{lb}^+ = \underline{f}_k$.

(2) *Update prox-center:* set

$$x_k := \mathrm{argmin}_{x \in X_{k-1}'} \left\{ V(x_0, x) : h(x_k^l, x) \leq l \right\}. \tag{3.9.13}$$

(3) *Update upper bound:* set $\bar{f}_k = \min\{\bar{f}_{k-1}, f(\alpha_k x_k + (1-\alpha_k)x_{k-1}^u)\}$, and choose x_k^u such that $f(x_k^u) = \bar{f}_k$. If $\overline{f}_k \leq l + \theta(\overline{f}_0 - l)$, then **terminate** the procedure with $p^+ = x_k^u$ and $\mathrm{lb}^+ = \underline{f}_k$.

(4) *Update localizer:* choose an arbitrary X_k' such that $\underline{X}_k \subseteq X_k' \subseteq \overline{X}_k$, where

$$\begin{aligned} \underline{X}_k &:= \left\{ x \in X_{k-1}' : h(x_k^l, x) \leq l \right\}, \\ \overline{X}_k &:= \left\{ x \in X : \langle \nu'(x_k) - \nu'(x_0), x - x_k\rangle \geq 0 \right\}. \end{aligned} \tag{3.9.14}$$

(6) Set $k = k+1$ and go to step 1.

We now add a few comments about procedure $\mathscr{G}_{APL}$ described above. First, note that the level l used in (3.9.13) is fixed throughout the procedure. Also, the two parameters (i.e., β and θ) are fixed a priori, say, $\beta = \theta = 0.5$.

Second, procedure $\mathscr{G}_{APL}$ can be terminated in either step 1 or 3. If it terminates in step 1, then we say that significant progress has been made on the lower bound $\underline{f}_k$. Otherwise, if it terminates in step 3, then significant progress has been made on the upper bound $\overline{f}_k$.

Third, observe that in step 4 of procedure $\mathscr{G}_{APL}$, we can choose any set X_k' satisfying $\underline{X}_k \subseteq X_k' \subseteq \overline{X}_k$ (the simplest way is to set $X_k' = \underline{X}_k$ or $X_k' = \overline{X}_k$). While the number of constraints in $\underline{X}_k$ increases with k, the set $\overline{X}_k$ has only one more constraint than X. By choosing X_k' between these two extremes, we can control the number of constraints in subproblems (3.9.12) and (3.9.13). Hence, even though the iteration cost of procedure $\mathscr{G}_{APL}$ can be higher than projected gradient descent type methods, it is still controllable to a certain extent.

We summarize below a few more observations regarding the execution of procedure $\mathscr{G}_{APL}$.

Lemma 3.15. *The following statements hold for procedure $\mathscr{G}_{APL}$.*

(a) $\{X'_k\}_{k\geq 0}$ *is a sequence of localizers of the level set* $\mathscr{E}_f(l)$.
(b) $\underline{f}_0 \leq \underline{f}_1 \leq \ldots \leq \underline{f}_k \leq f^*$ *and* $\overline{f}_0 \geq \overline{f}_1 \geq \ldots \geq \overline{f}_k \geq f^*$ *for any* $k \geq 1$.
(c) Problem (3.9.13) is always feasible unless the procedure terminates.
(d) $\emptyset \neq \underline{X}_k \subseteq \overline{X}_k$ *for any* $k \geq 1$ *and hence step 4 is always feasible unless the procedure terminates.*
(e) Whenever the procedure terminates, we have $f(p^+) - \text{lb}^+ \leq q[f(p) - \text{lb}]$, *where*

$$q \equiv q(\beta, \theta) := 1 - (1-\theta)\min\{\beta, 1-\beta\}. \tag{3.9.15}$$

Proof. We first show part (a). First, noting that $\mathscr{E}_f(l) \subseteq X'_0$, we can show that $\mathscr{E}_f(l) \subseteq X'_k$, $k \geq 1$, by using induction. Suppose that X'_{k-1} is a localizer of the level set $\mathscr{E}_f(l)$. Then, for any $x \in \mathscr{E}_f(l)$, we have $x \in X'_{k-1}$. Moreover, by the definition of h, we have $h(x^l_k, x) \leq f(x) \leq l$ for any $x \in \mathscr{E}_f(l)$. Using these two observations and the definition of $\underline{X}_k$ in (3.9.14), we have $\mathscr{E}_f(l) \subseteq \underline{X}_k$, which, in view of the fact that $\underline{X}_k \subseteq X'_k$, implies $\mathscr{E}_f(l) \subseteq X'_k$, i.e., X'_k is a localizer of $\mathscr{E}_f(l)$.

We now show part (b). The first relation follows from Lemma 5.7, (3.9.12), and the fact that X'_k, $k \geq 0$, are localizers of $\mathscr{E}_f(l)$ due to part (a). The second relation of part (b) follows immediately from the definition of $\overline{f}_k$, $k \geq 0$.

To show part (c), suppose that problem (3.9.13) is infeasible. Then, by the definition of $\underline{h}_k$ in (3.9.12), we have $\underline{h}_k > l$, which implies $\underline{f}_k \geq l$, which in turn implies that the procedure should have terminated in step 1 at iteration k.

To show part (d), note that by part (c), the set $\underline{X}_k$ is nonempty. Moreover, by the optimality condition of (3.9.13) and the definition of $\underline{X}_k$ in (3.9.14), we have $\langle \nabla v(x_k), x - x_k \rangle \geq 0$ for any $x \in \underline{X}_k$, which then implies that $\underline{X}_k \subseteq \overline{X}_k$.

We now provide the proof of part (e). Suppose first that the procedure terminates in step 1 of the k-th iteration. We must have $\underline{f}_k \geq l - \theta(l - \underline{f}_0)$. By using this condition, and the facts that $f(p^+) \leq \overline{f}_0$ (see part (b)) and $l = \beta \underline{f}_0 + (1-\beta)\overline{f}_0$, we obtain

$$\begin{aligned} f(p^+) - \text{lb}^+ &= f(p^+) - \underline{f}_k \leq \overline{f}_0 - [l - \theta(l - \underline{f}_0)] \\ &= [1 - (1-\beta)(1-\theta)](\overline{f}_0 - \underline{f}_0). \end{aligned} \tag{3.9.16}$$

Now suppose that the procedure terminates in step 3 of the k-th iteration. We must have $\overline{f}_k \leq l + \theta(\overline{f}_0 - l)$. By using this condition, and the facts that $\text{lb}^+ \geq \underline{f}_0$ (see part b) and $l = \beta \underline{f}_0 + (1-\beta)\overline{f}_0$, we have

$$f(p^+) - \text{lb}^+ = \overline{f}_k - \text{lb}^+ \leq l + \theta(\overline{f}_0 - l) - \underline{f}_0 = [1 - (1-\theta)\beta](\overline{f}_0 - \underline{f}_0).$$

Part (e) then follows by combining the above two relations. ■

By showing how the gap between the upper bound (i.e., $f(x^u_k)$) and the level l decreases with respect to k, we will establish in Theorem 3.13 some important convergence properties of procedure $\mathscr{G}_{APL}$. Before that we first need to show two technical results.

Lemma 3.16. *Let $(x_{k-1}, x^u_{k-1}) \in X \times X$ be given at the k-th iteration, $k \geq 1$, of an iterative scheme and denote $x^l_k = \alpha_k x_{k-1} + (1-\alpha_k)x^u_{k-1}$. Also let $h(z,\cdot)$ be defined in (3.9.4) and suppose that the pair of new search points $(x_k, \tilde{x}^u_k) \in X \times X$ satisfy that, for some $l \in \mathbb{R}$ and $\alpha_k \in (0,1]$,*

$$h(x^l_k, x_k) \leq l, \tag{3.9.17}$$

$$\tilde{x}^u_k = \alpha_t x_k + (1-\alpha_k)x^u_{k-1}. \tag{3.9.18}$$

Then,

$$f(\tilde{x}^u_k) \leq (1-\alpha_k)f(x^u_{k-1}) + \alpha_k l + \tfrac{M}{1+\rho}\|\alpha_k(x_k - x_{k-1})\|^{1+\rho}. \tag{3.9.19}$$

Proof. It can be easily seen from (3.9.18) and the definition of x^l_k that

$$\tilde{x}^u_k - x^l_k = \alpha_k(x_k - x_{k-1}). \tag{3.9.20}$$

Using this observation, (3.9.2), (3.9.4), (3.9.17), (3.9.18), and the convexity of f, we have

$$\begin{aligned}
f(\tilde{x}^u_k) &\leq h(x^l_k, \tilde{x}^u_k) + \tfrac{M}{1+\rho}\|\tilde{x}^u_k - x^l_k\|^{1+\rho} \\
&= (1-\alpha_k)h(x^l_k, x^u_{k-1}) + \alpha_k h(x^l_k, x_k) + \tfrac{M}{1+\rho}\|\tilde{x}^u_k - x^l_k\|^{1+\rho} \\
&= (1-\alpha_k)h(x^l_k, x^u_{k-1}) + \alpha_k h(x^l_k, x_k) + \tfrac{M}{1+\rho}\|\alpha_k(x_k - x_{k-1})\|^{1+\rho} \\
&\leq (1-\alpha_k)f(x^u_{k-1}) + \alpha_k h(x^l_k, x_k) + \tfrac{M}{1+\rho}\|\alpha_k(x_k - x_{k-1})\|^{1+\rho} \\
&\leq (1-\alpha_k)f(x^u_{k-1}) + \alpha_k l + \tfrac{M}{1+\rho}\|\alpha_k(x_k - x_{k-1})\|^{1+\rho},
\end{aligned}$$

where the three inequalities follow from (3.9.2) and (3.9.4), the convexity, and (3.9.17), respectively, while the two identities follow from (3.9.18) and (3.9.20), respectively. ■

Lemma 3.17. *Let $w_k \in (0,1]$, $k = 1,2,\ldots$, be given. Also let us denote*

$$W_k := \begin{cases} 1, & k = 1, \\ (1-w_k)W_{k-1}, & k \geq 2. \end{cases} \tag{3.9.21}$$

Suppose that $W_k > 0$ for all $k \geq 2$ and that the sequence $\{\delta_k\}_{k\geq 0}$ satisfies

$$\delta_k \leq (1-w_k)\delta_{k-1} + B_k, \quad k = 1,2,\ldots. \tag{3.9.22}$$

Then, we have $\delta_k \leq W_k(1-w_1)\delta_0 + W_k\Sigma^k_{i=1}(B_i/W_i)$.

Proof. Dividing both sides of (3.9.22) by W_k, we obtain

$$\tfrac{\delta_1}{W_1} \leq \tfrac{(1-w_1)\delta_0}{W_1} + \tfrac{B_1}{W_1}$$

and

$$\frac{\delta_k}{W_k} \le \frac{\delta_{k-1}}{W_{k-1}} + \frac{B_k}{W_k}, \quad \forall k \ge 2.$$

The result then immediately follows by summing up the above inequalities and rearranging the terms. ■

We are now ready to analyze the convergence behavior of the APL gap reduction procedure. Note that the following quantities will be used in our analysis:

$$\gamma_k := \begin{cases} 1 & k = 1, \\ (1-\alpha_k)\,\gamma_{k-1} & k \ge 2, \end{cases} \tag{3.9.23}$$

$$\Gamma_k := \left\{\gamma_1^{-1}\alpha_1^{1+\rho}, \gamma_2^{-1}\alpha_2^{1+\rho}, \ldots, \gamma_k^{-1}\alpha_k^{1+\rho}\right\}. \tag{3.9.24}$$

Theorem 3.13. *Let $\alpha_k \in (0,1]$, $k = 1,2,\ldots$, be given. Also let $(x_k^l, x_k, x_k^u) \in X \times X \times X$, $k \ge 1$, be the search points, l be the level, and V be the prox-function in procedure $\mathscr{G}_{APL}$. Then, we have*

$$\begin{aligned} f(x_k^u) - l &\le (1-\alpha_1)\gamma_k\,[f(x_0^u) - l] \\ &\quad + \tfrac{M}{1+\rho}\,[2V(x_0,x_k)]^{(1+\rho)/2}\,\gamma_k\,\|\Gamma_k\|_{2/(1-\rho)} \end{aligned} \tag{3.9.25}$$

for any $k \ge 1$, where $\|\cdot\|_p$ denotes the l_p norm, γ_k and Γ_k, respectively, are defined in (3.9.23) and (3.9.24). In particular, if $\alpha_k \in (0,1]$, $k = 1,2,\ldots$, are chosen such that for some $c > 0$,

$$\alpha_1 = 1 \quad \text{and} \quad \gamma_k\,\|\Gamma_k\|_{2/(1-\rho)} \le c\,k^{-(1+3\rho)/2}, \tag{3.9.26}$$

then the number of iterations performed by procedure $\mathscr{G}_{APL}$ can be bounded by

$$K_{APL}(\Delta_0) := \left\lceil \left(\frac{2cMD_X^{1+\rho}}{\beta\theta(1+\rho)\Delta_0}\right)^{2/(1+3\rho)} \right\rceil, \tag{3.9.27}$$

where $\Delta_0 = \overline{f}_0 - \underline{f}_0$ and D_X is defined in (3.2.4).

Proof. We first show that the prox-centers $\{x_k\}$ in procedure $\mathscr{G}_{APL}$ are "close" to each other in terms of $\sum_{i=1}^k \|x_{i-1} - x_i\|^2$. Observe that the function $V(x_0,x)$ is strongly convex with modulus 1, $x_0 = \arg\min_{x\in X} V(x_0,x)$ and $V(x_0,x_0) = 0$. Hence, we have

$$\tfrac{1}{2}\|x_1 - x_0\|^2 \le V(x_0,x_1). \tag{3.9.28}$$

Moreover, by (3.9.14), we have $\langle \nabla V(x_0,x_k), x - x_k\rangle \ge 0$ for any $x \in \overline{X}_k$, which together with the fact that $X_k' \subseteq \overline{X}_k$ then imply that $\langle \nabla V(x_0,x_k), x - x_k\rangle \ge 0$ for any $x \in X_k'$. Using this observation, the fact that $x_{k+1} \in X_k'$ due to (3.9.13), and the strong convexity of $V(x_0,\cdot)$, we have

$$\begin{aligned} \tfrac{1}{2}\|x_{k+1} - x_k\|^2 &\le V(x_0,x_{k+1}) - V(x_0,x_k) - \langle \nabla V(x_0,x_k), x_{k+1} - x_k\rangle \\ &\le V(x_0,x_{k+1}) - V(x_0,x_k) \end{aligned}$$

for any $k \ge 1$. Summing up the above inequalities with (3.9.28), we arrive at

$$\tfrac{1}{2}\textstyle\sum_{i=1}^{k}\|x_i - x_{i-1}\|^2 \le V(x_0, x_k). \tag{3.9.29}$$

Next, we establish a recursion for procedure $\mathscr{G}_{APL}$. Let us denote $\tilde{x}_k^u \equiv \alpha_k x_k + (1-\alpha_k)x_{k-1}^u$. By the definitions of x_k^u and $\tilde{x}_k^u$, we have $f(x_k^u) \le f(\tilde{x}_k^u)$. Also by (3.9.13), we have $h(x_k^l, x) \le l$. Using these observations and Lemma 3.16, we have

$$f(x_k^u) \le f(\tilde{x}_k^u) \le (1-\alpha_k) f(x_{k-1}^u) + \alpha_k l + \tfrac{M}{1+\rho}\|\alpha_k(x_k - x_{k-1})\|^{1+\rho}$$

for any $k \ge 1$. Subtracting l from both sides of the above inequality, we obtain

$$f(x_k^u) - l \le (1-\alpha_k)[f(x_{k-1}^u) - l] + \tfrac{M}{1+\rho}\|\alpha_k(x_k - x_{k-1})\|^{1+\rho}, \tag{3.9.30}$$

for any $k \ge 1$. Using the above inequality and Lemma 3.17 (with $\delta_k = f(x_k^u) - l$, $w_k = 1 - \alpha_k$, $W_k = \gamma_k$, and $B_k = M\|\alpha_k(x_k - x_{k-1})\|^{1+\rho}/(1+\rho)$), we obtain

$$\begin{aligned} f(x_k^u) - l &\le (1-\alpha_1)\gamma_k[f(x_0^u) - l] + \tfrac{M}{1+\rho}\gamma_k \textstyle\sum_{i=1}^{k} \gamma_i^{-1}\|\alpha_i(x_i - x_{i-1})\|^{1+\rho} \\ &\le (1-\alpha_1)\gamma_k[f(x_0^u) - l] + \tfrac{M}{1+\rho}\|\Gamma_k\|_{2/(1-\rho)} \left[\textstyle\sum_{i=1}^{k}\|x_i - x_{i-1}\|^2\right]^{(1+\rho)/2}, \end{aligned}$$

for any $k \ge 1$, where the last inequality follows from Hölder's inequality. The above conclusion together with (3.9.29) then imply that (3.9.25) holds.

Now, denote $K = K_{APL}(\varepsilon)$ and suppose that condition (3.9.26) holds. Then by (3.9.25), (3.9.26), and (3.2.4), we have

$$\begin{aligned} f(x_K^u) - l &\le \tfrac{cM}{1+\rho}[2V(x_0, x_K)]^{(1+\rho)/2} K^{-(1+3\rho)/2} \le \tfrac{2cM}{1+\rho} D_X^{1+\rho} K^{-(1+3\rho)/2} \\ &\le \theta\beta\Delta_0 = \theta(\overline{f}_0 - l), \end{aligned}$$

where the last equality follows from the fact that $l = \beta \underline{f}_0 + (1-\beta)\overline{f}_0 = \overline{f}_0 - \beta\Delta_0$. Hence, procedure $\mathscr{G}_{APL}$ must terminate in step 3 of the K-th iteration. ∎

In view of Theorem 3.13, we discuss below a few possible selections of $\{\alpha_k\}$, which satisfy condition (3.9.26) and thus guarantee the termination of procedure $\mathscr{G}_{APL}$. It is worth noting that these selections of $\{\alpha_k\}$ do not rely on any problem parameters, including M, ρ, and D_X, nor on any other algorithmic parameters, such as β and θ.

Proposition 3.2. *Let γ_k and Γ_k, respectively, be defined in (3.9.23) and (3.9.24).*

(a) If $\alpha_k = 2/(k+1)$, $k = 1, 2, \ldots$, then $\alpha_k \in (0,1]$ and relation (3.9.26) holds with $c = 2^{1+\rho}3^{-(1-\rho)/2}$.

(b) If α_k, $k = 1, 2, \ldots$, are recursively defined by

$$\alpha_1 = \gamma_1 = 1, \quad \gamma_k = \alpha_k^2 = (1-\alpha_k)\gamma_{k-1}, \quad \forall k \ge 2, \tag{3.9.31}$$

then we have $\alpha_k \in (0,1]$ *for any* $k \geq 1$. *Moreover, condition (3.9.26) is satisfied with* $c = \frac{4}{3^{(1-\rho)/2}}$.

Proof. We first show part (a). By (3.9.23) and (3.9.24), we have

$$\gamma_k = \tfrac{2}{k(k+1)} \text{ and } \gamma_k^{-1}\alpha_k^{1+\rho} = \left(\tfrac{2}{k+1}\right)^{\rho} k \leq 2^{\rho} k^{1-\rho}. \tag{3.9.32}$$

Using (3.9.32) and the simple observation that $\sum_{i=1}^k i^2 = k(k+1)(2k+1)/6 \leq k(k+1)^2/3$, we have

$$\begin{aligned}
\gamma_k \|\Gamma_k\|_{2/(1-\rho)} &\leq \gamma_k \left[\textstyle\sum_{i=1}^k \left(2^{\rho} i^{1-\rho}\right)^{2/(1-\rho)}\right]^{(1-\rho)/2} = 2^{\rho}\gamma_k \left(\textstyle\sum_{i=1}^k i^2\right)^{(1-\rho)/2} \\
&\leq 2^{\rho}\gamma_k \left[\tfrac{k(k+1)^2}{3}\right]^{(1-\rho)/2} = \left(2^{1+\rho}3^{-(1-\rho)/2}\right)\left[k^{-(1+\rho)/2}(k+1)^{-\rho}\right] \\
&\leq \left(2^{1+\rho}3^{-(1-\rho)/2}\right) k^{-(1+3\rho)/2}.
\end{aligned}$$

We now show that part (b) holds. Note that by (3.9.31), we have

$$\alpha_k = \tfrac{1}{2}\left(-\gamma_{k-1} + \sqrt{\gamma_{k-1}^2 + 4\gamma_{k-1}}\right), \quad k \geq 2, \tag{3.9.33}$$

which clearly implies that $\alpha_k > 0$, $k \geq 2$. We now show that $\alpha_k \leq 1$ and $\gamma_k \leq 1$ by induction. Indeed, if $\gamma_{k-1} \leq 1$, then by (3.9.33), we have

$$\alpha_k \leq \tfrac{1}{2}\left(-\gamma_{k-1} + \sqrt{\gamma_{k-1}^2 + 4\gamma_{k-1} + 4}\right) = 1.$$

The previous conclusion, together with the fact that $\alpha_k^2 = \gamma_k$ due to (3.9.31), then also imply that $\gamma_k \leq 1$. Now let us bound $1/\sqrt{\gamma_k}$ for any $k \geq 2$. First observe that by (3.9.31) we have, for any $k \geq 2$,

$$\frac{1}{\sqrt{\gamma_k}} - \frac{1}{\sqrt{\gamma_{k-1}}} = \frac{\sqrt{\gamma_{k-1}} - \sqrt{\gamma_k}}{\sqrt{\gamma_{k-1}\gamma_k}} = \frac{\gamma_{k-1} - \gamma_k}{\sqrt{\gamma_{k-1}\gamma_k}\left(\sqrt{\gamma_{k-1}} + \sqrt{\gamma_k}\right)} = \frac{\alpha_k \gamma_{k-1}}{\gamma_{k-1}\sqrt{\gamma_k} + \gamma_k\sqrt{\gamma_{k-1}}}.$$

Using the above identity, (3.9.31), and the fact that $\gamma_k \leq \gamma_{k-1}$ due to (3.9.31), we conclude that

$$\frac{1}{\sqrt{\gamma_k}} - \frac{1}{\sqrt{\gamma_{k-1}}} \geq \frac{\alpha_k}{2\sqrt{\gamma_k}} = \frac{1}{2} \text{ and } \frac{1}{\sqrt{\gamma_k}} - \frac{1}{\sqrt{\gamma_{k-1}}} \leq \frac{\alpha_k}{\sqrt{\gamma_k}} = 1,$$

which, in view of the fact that $\gamma_1 = 1$, then implies that $(k+1)/2 \leq 1/\sqrt{\gamma_k} \leq k$. Using the previous inequality and (3.9.31), we conclude that

$$\gamma_k \leq \tfrac{4}{(k+1)^2}, \quad \gamma_k^{-1}\alpha_k^{1+\rho} = \left(\sqrt{\gamma_k}\right)^{-(1-\rho)} \leq k^{1-\rho},$$

and

$$\gamma_k \|\Gamma_k\|_{2/(1-\rho)} \le \gamma_k \left[\sum_{i=1}^k i^2\right]^{(1-\rho)/2} \le \gamma_k \left(\int_0^{k+1} u^2 \, du\right)^{(1-\rho)/2}$$
$$\le \tfrac{4}{3^{(1-\rho)/2}} (k+1)^{-(1+3\rho)/2} \le \tfrac{4}{3^{(1-\rho)/2}} k^{-(1+3\rho)/2}.$$

■

In view of Lemma 3.15(e) and the termination criterion of procedure $\mathcal{G}_{APL}$, each call to this procedure can reduce the gap between a given upper and lower bound on f^* by a constant factor q (see (3.9.15)). In the following APL method, we will iteratively call procedure $\mathcal{G}_{APL}$ until a certain accurate solution of problem (3.9.1) is found.

The APL Method

Input: initial point $p_0 \in X$, tolerance $\varepsilon > 0$ and algorithmic parameters $\beta, \theta \in (0,1)$.

(0) Set $p_1 \in \text{Argmin}_{x \in X} h(p_0, x)$, $\text{lb}_1 = h(p_0, p_1)$ and $\text{ub}_1 = f(p_1)$. Let $s = 1$.
(1) If $\text{ub}_s - \text{lb}_s \le \varepsilon$, **terminate**;
(2) Set $(p_{s+1}, \text{lb}_{s+1}) = \mathcal{G}_{APL}(p_s, \text{lb}_s, \beta, \theta)$ and $\text{ub}_{s+1} = f(p_{s+1})$;
(3) Set $s = s+1$ and go to step 1.

Whenever s increments by 1, we say that a phase of the APL method occurs. Unless explicitly mentioned otherwise, an iteration of procedure $\mathcal{G}_{APL}$ is also referred to as an iteration of the APL method. The main convergence properties of the above APL method are summarized as follows.

Theorem 3.14. *Let M, ρ, D_X, and q be defined in (3.9.2), (3.2.4), and (3.9.15). Suppose that $\alpha_k \in (0,1]$, $k = 1, 2, \ldots$, in procedure $\mathcal{G}_{APL}$ are chosen such that condition (3.9.26) holds for some $c > 0$.*

(a) The number of phases performed by the APL method does not exceed

$$\bar{S}(\varepsilon) := \left\lceil \max\left\{0, \log_{1/q} \tfrac{2MD_X^{1+\rho}}{(1+\rho)\varepsilon}\right\}\right\rceil. \tag{3.9.34}$$

(b) The total number of iterations performed by the APL method can be bounded by

$$\bar{S}(\varepsilon) + \tfrac{1}{1-q^{2/(1+3\rho)}} \left(\tfrac{2cMD_X^{1+\rho}}{\beta\theta(1+\rho)\varepsilon}\right)^{2/(1+3\rho)}. \tag{3.9.35}$$

Proof. Denote $\delta_s \equiv \text{ub}_s - \text{lb}_s$, $s \ge 1$. Without loss of generality, we assume that $\delta_1 > \varepsilon$, since otherwise the statements are obviously true. By Lemma 3.15(e) and the origin of ub_s and lb_s, we have

$$\delta_{s+1} \le q\delta_s, \quad s \ge 1. \tag{3.9.36}$$

Also note that, by (3.9.2), (3.2.4), and the definition of p_1 in the APL method, we have

$$\delta_1 = f(p_1) - h(p_0, p_1) = f(p_1) - \left[f(p_0) + \langle f'(p_0), p_1 - p_0\rangle\right]$$
$$\leq \tfrac{M\|p_1 - p_0\|^{1+\rho}}{1+\rho} \leq \tfrac{2MD_X^{1+\rho}}{1+\rho}. \quad (3.9.37)$$

The previous two observations then clearly imply that the number of phases performed by the APL method is bounded by (3.9.34).

We now bound the total number of iterations performed by the APL method. Suppose that procedure $\mathscr{G}_{APL}$ has been called $\bar{s}$ times for some $1 \leq \bar{s} \leq \bar{S}(\varepsilon)$. It follows from (3.9.36) that $\delta_s > \varepsilon q^{s-\bar{s}}$, $s = 1, \ldots, \bar{s}$, since $\delta_{\bar{s}} > \varepsilon$ due to the origin of $\bar{s}$. Using this observation, we obtain

$$\textstyle\sum_{s=1}^{\bar{s}} \delta_s^{-2/(1+3\rho)} < \sum_{s=1}^{\bar{s}} \frac{q^{2(\bar{s}-s)/(1+3\rho)}}{\varepsilon^{2/(1+3\rho)}} = \sum_{t=0}^{\bar{s}-1} \frac{q^{2t/(1+3\rho)}}{\varepsilon^{2/(1+3\rho)}} \leq \frac{1}{(1-q^{2/(1+3\rho)})\varepsilon^{2/(1+3\rho)}}.$$

Moreover, by Theorem 3.13, the total number of iterations performed by the APL method is bounded by

$$\textstyle\sum_{s=1}^{\bar{s}} K_{APL}(\delta_s) \leq \bar{s} + \sum_{s=1}^{\bar{s}} \left(\frac{cMD_X^{1+\rho}}{\beta\theta(1+\rho)\delta_s}\right)^{2/(1+3\rho)}.$$

Our result then immediately follows by combining the above two inequalities. ■

Clearly, in view of Theorem 3.14, the APL method can achieve the $\mathscr{O}(1/\varepsilon^2)$ and $\mathscr{O}(1/\sqrt{\varepsilon})$ iteration complexity for nonsmooth and smooth convex optimization, respectively. It also achieves the best possible complexity bounds for weakly smooth problems. What is interesting is that this algorithm does not require the input of any smoothness information, such as whether the problem is smooth, nonsmooth, or weakly smooth, as well as the specific values of Lipschitz constant and smoothness level. In addition, its iteration cost is more or less controllable depending on how much historical first-order information to be used.

3.9.2 Saddle Point Problems

In this subsection, we consider the bilinear saddle point problem (3.5.1) that we have studied in Sect. 3.5. As shown in Sect. 3.5, the nonsmooth function F in (3.5.2) can be approximated by the smooth convex function F_η given by (3.5.4).

Since F_η is a smooth convex function, we can apply a smooth convex optimization method, e.g., the accelerated gradient method to $\min_{x \in X} f_\eta(x)$, for a properly chosen $\eta > 0$. It has been shown that one can obtain an ε-solution of problem (3.5.1)–(3.5.2) in at most $\mathscr{O}(1/\varepsilon)$ iterations. However, this approach would require us to input a number of problem parameters (e.g., $\|A\|$ and D_Y) or algorithmic parameters (e.g., the number of iterations or target accuracy).

Our goal in this subsection is to present a completely problem parameter-free smoothing technique, namely: the *uniform smoothing level (USL) method*, obtained by properly modifying the APL method in Sect. 3.9.1. In the USL method, the smoothing parameter η is adjusted dynamically during their execution rather than being fixed in advance. Moreover, an estimate on the value of D_Y can be provided automatically. We start by describing the USL gap reduction procedure, denoted by $\mathcal{G}_{USL}$, which will be iteratively called by the USL method. Specifically, for a given search point p, a lower bound lb on f^*, and an initial estimate $\tilde{D}$ on D_Y, procedure $\mathcal{G}_{USL}$ will either compute a new search point p^+ and a new lower bound lb^+ satisfying $f(p^+) - \text{lb}^+ \leq q[f(p) - \text{lb}]$ for some $q \in (0,1)$, or provide an updated estimate $\tilde{D}^+$ on D_Y in case the current estimate $\tilde{D}$ is not accurate enough.

The USL Gap Reduction Procedure: $(p^+, \text{lb}^+, \tilde{D}^+) = \mathcal{G}_{USL}(p, \text{lb}, \tilde{D}, \beta, \theta)$

(0) Set $x_0^u = p$, $\overline{f}_0 = f(x_0^u)$, $\underline{f}_0 = \text{lb}$, $l = \beta \underline{f}_0 + (1-\beta)\overline{f}_0$, and

$$\eta := \theta(\overline{f}_0 - l)/(2\tilde{D}). \tag{3.9.38}$$

Also let $x_0 \in X$ and the initial localizer X_0' be arbitrarily chosen, say $x_0 = p$ and $X_0' = X$. Set the prox-function $d(x) = \nu(x) - [\nu(x_0) + \langle \nu'(x_0), x - x_0 \rangle]$. Set $k = 1$.

(1) *Update lower bound:* set $x_k^l = (1-\alpha_k)x_{k-1}^u + \alpha_k x_{k-1}$ and

$$h(x_k^l, x) = h_\eta(x_k^l, x) := \hat{f}(x) + F_\eta(x_k^l) + \langle \nabla F_\eta(x_k^l), x - x_k^l \rangle. \tag{3.9.39}$$

Compute $\underline{f}_k$ according to (3.9.12). If $\underline{f}_k \geq l - \theta(l - \underline{f}_0)$, then **terminate** the procedure with $p^+ = x_{k-1}^u$, $\text{lb}^+ = \underline{f}_k$, and $\tilde{D}^+ = \tilde{D}$;

(2) *Update prox-center:* set x_k according to (3.9.13);

(3) *Update upper bound:* set $\bar{f}_k = \min\{\bar{f}_{k-1}, f(\alpha_k x_k + (1-\alpha_k)x_{k-1}^u)\}$, and choose x_k^u such that $f(x_k^u) = \bar{f}_k$. Check the following two possible termination criterions:

(3a) if $\overline{f}_k \leq l + \theta(\overline{f}_0 - l)$, **terminate** the procedure with $p^+ = x_k^u$, $\text{lb}^+ = \underline{f}_k$, and $\tilde{D}^+ = \tilde{D}$,

(3b) Otherwise, if $f_\eta(x_k^u) \leq l + \frac{\theta}{2}(\overline{f}_0 - l)$, **terminate** the procedure with $p^+ = x_k^u$, $\text{lb}^+ = \underline{f}_k$, and $\tilde{D}^+ = 2\tilde{D}$;

(4) *Update localizer:* choose an arbitrary X_k' such that $\underline{X}_k \subseteq X_k' \subseteq \overline{X}_k$, where $\underline{X}_k$ and $\overline{X}_k$ are defined in (3.9.14);

(5) Set $k = k+1$ and go to Step 1.

We notice that there are a few essential differences between procedure $\mathcal{G}_{USL}$ described above and procedure $\mathcal{G}_{APL}$ in Sect. 3.9.1. First, in comparison with procedure $\mathcal{G}_{APL}$, procedure $\mathcal{G}_{USL}$ needs to use one additional input parameter, namely $\tilde{D}$, to define η (see (3.9.38)) and hence the approximation function f_η in (3.5.5).

Second, we use the support functions $h_\eta(x_k^l, x)$ of $f_\eta(x)$ defined in (3.9.39) procedure $\mathcal{G}_{USL}$ rather than the cutting planes of $f(x)$ in procedure $\mathcal{G}_{APL}$. Notice that by (3.9.39), the convexity of F_η, and the first relation in (3.5.6), we have

$$h_\eta(x_k^l, x) \leq \hat{f}(x) + F_\eta(x) \leq \hat{f}(x) + F(x) = f(x), \tag{3.9.40}$$

which implies that the functions $h_\eta(x_k^l, x)$ underestimate f everywhere on X. Hence, $\underline{f}_k$ computed in step 1 of this procedure are indeed lower bounds of f^*.

Third, there are three possible ways to terminate procedure $\mathscr{G}_{USL}$. Similar to procedure $\mathscr{G}_{APL}$, if it terminates in step 1 and step 3a, then we say that significant progress has been made on the lower and upper bounds on f^*, respectively. The new added termination criterion in step 3b will be used only if the value of $\tilde{D}$ is not properly specified. We formalize these observations in the following simple result.

Lemma 3.18. *The following statements hold for procedure $\mathscr{G}_{USL}$.*

(a) If the procedure terminates in step 1 or step 3a, we have $f(p^+) - \mathrm{lb}^+ \le q[f(p) - \mathrm{lb}]$, where q is defined in (3.9.15);
(b) If the procedure terminates in step 3b, then $\tilde{D} < D_Y$.

Proof. The proof of part (a) is the same as that of Lemma 3.15(e) and we only need to show part (b). Observe that whenever step 3b occurs, we have $\overline{f}_k > l + \theta(\overline{f}_0 - l)$ and $f_\eta(x_k^u) \le l + \frac{\theta}{2}(\overline{f}_0 - l)$. Hence,

$$f(x_k^u) - f_\eta(x_k^u) = \overline{f}_k - f_\eta(x_k^u) > \tfrac{\theta}{2}(\overline{f}_0 - l),$$

which, in view of the second relation in (3.5.7), then implies that $\eta D_Y > \theta(\overline{f}_0 - l)/2$. Using this observation and (3.9.38), we conclude that $\tilde{D} < D_Y$. ■

We observe that all the results in Lemma 3.15(a)–(d) regarding the execution of procedure $\mathscr{G}_{APL}$ also hold for procedure $\mathscr{G}_{USL}$. In addition, similar to Theorem 3.13, we establish below some important convergence properties of procedure $\mathscr{G}_{USL}$ by showing how the gap between $f(x_k^u)$ and the level l decreases.

Theorem 3.15. *Let $\alpha_k \in (0,1]$, $k = 1,2,\ldots$, be given. Also let $(x_k^l, x_k, x_k^u) \in X \times X \times X$, $k \ge 1$, be the search points, l be the level, $V(x_0,\cdot)$ be the prox-function, and η be the smoothing parameter (see (3.9.38)) in procedure $\mathscr{G}_{USL}$. Then, we have*

$$f_\eta(x_k^u) - l \le (1-\alpha_1)\gamma_k [f_\eta(x_0^u) - l] + \tfrac{\|A\|^2 V(x_0,x_k)}{\eta} \gamma_k \|\Gamma_k\|_\infty, \tag{3.9.41}$$

for any $k \ge 1$, where $\|\cdot\|_\infty$ denotes the l_∞ norm, γ_k and Γ_k, respectively, are defined in (3.9.23) and (3.9.24). In particular, if $\alpha_k \in (0,1]$, $k = 1,2,\ldots$, are chosen such that condition (3.9.26) holds with $\rho = 1$ for some $c > 0$, then the number of iterations performed by procedure $\mathscr{G}_{APL}$ can be bounded by

$$K_{USL}(\Delta_0, \tilde{D}) := \left\lceil \frac{2\|A\|\sqrt{cD_X \tilde{D}}}{\beta\theta\Delta_0} \right\rceil, \tag{3.9.42}$$

where D_X is defined in (3.2.4).

Proof. Note that by (3.9.40) and (3.5.5), we have $h_\eta(z,x) \le f_\eta(x)$ for any $z, x \in X$. Moreover, by (3.5.5), (3.9.39), and the fact that F_η has Lipschitz continuous gradients with constant $\mathscr{L}_\eta$, we obtain

$$f_\eta(x) - h_\eta(z,x) = F_\eta(x) - [F_\eta(z) + \langle \nabla F_\eta(z), x - z\rangle] \le \tfrac{\mathscr{L}_\eta}{2}\|x-z\|^2 = \tfrac{\|A\|^2}{2\eta}\|x-z\|^2,$$

for any $z, x \in X$, where the last inequality follows from the smoothness of F_η. In view of these observations, (3.9.41) follows from an argument similar to the one used in the proof of (3.9.25) with $f = f_\eta$, $M = \mathscr{L}_\eta$, and $\rho = 1$.

Now using (3.2.4), (3.9.26) (with $\rho = 1$), (3.9.38), and (3.9.41), we obtain

$$\begin{aligned} f_\eta(x_k^u) - l &\le \tfrac{\|A\|^2 V(x_0, x_k)}{\eta} \gamma_k(1) \, \|\Gamma_k(1,\rho)\|_\infty \le \tfrac{c\|A\|^2 V(x_0, x_k)}{\eta k^2} \\ &\le \tfrac{c\|A\|^2 D_X}{\eta k^2} = \tfrac{2c\|A\|^2 D_X \tilde{D}}{\theta(\overline{f}_0 - l) k^2}. \end{aligned}$$

Denoting $K = K_{USL}(\Delta_0, \tilde{D})$ and noting that $\Delta_0 = \overline{f}_0 - \underline{f}_0 = (\overline{f}_0 - l)/\beta$, we conclude from the previous inequality that $f_\eta(x_K^u) - l \le \theta(\overline{f}_0 - l)/2$. This result together with (3.5.7) imply that if $\tilde{D} \ge D_Y$, then $f(x_K^u) - l \le f_\eta(x_K^u) - l + \eta D_Y \le \theta(\overline{f}_0 - l)$. In view of these two observations and the termination criterions used in step 3, procedure $\mathscr{G}_{USL}$ must terminate in at most $K_{APL}(\Delta_0, \tilde{D})$ iterations. ■

In view of Lemma 3.18, each call to procedure $\mathscr{G}_{USL}$ can reduce the gap between a given upper and lower bound on f^* by a constant factor q, or update the estimate on D_Y by a factor of 2. In the following USL method, we will iteratively call procedure $\mathscr{G}_{USL}$ until a certain accurate solution is found.

The USL Method

Input: $p_0 \in X$, tolerance $\varepsilon > 0$, initial estimate $Q_1 \in (0, D_Y]$, and algorithmic parameters $\beta, \theta \in (0,1)$.

(1) Set

$$p_1 \in \text{Argmin}_{x \in X} \left\{ h_0(p_0, x) := \hat{f}(x) + F(p_0) + \langle F'(p_0), x - p_0 \rangle \right\}, \qquad (3.9.43)$$

$\text{lb}_1 = h_0(p_0, p_1)$ and $\text{ub}_1 := \min\{f(p_1), f(\tilde{p}_1)\}$. Let $s = 1$.

(2) If $\text{ub}_s - \text{lb}_s \le \varepsilon$, **terminate**;

(3) Set $(p_{s+1}, \text{lb}_{s+1}, Q_{s+1}) = \mathscr{G}_{USL}(p_s, \text{lb}_s, Q_s, \beta, \theta)$, and $\text{ub}_{s+1} = f(p_{s+1})$;

(4) Set $s = s + 1$ and go to step 1.

We now make a few remarks about the USL method described above. First, each phase s, $s \ge 1$, of the USL method is associated with an estimation Q_s on D_Y, and $Q_1 \in (0, D_Y]$ is a given input parameter. Note that such a Q_1 can be easily obtained by the definition of D_Y. Second, we differentiate two types of phases: a phase is called *significant* if procedure $\mathscr{G}_{USL}$ terminates in step 1 or step 3a, otherwise, it is called *nonsignificant*. Third, in view of Lemma 3.18(b), if phase s is nonsignificant, then we must have $Q_s \le D_Y$. In addition, using the previous observation, and the facts that $Q_1 \le D_Y$ and that Q_s can be increased by a factor of 2 only in the nonsignificant phases, we must have $Q_s \le 2D_Y$ for all significant phases.

Before establishing the complexity of the above USL method, we first present a technical result which will be used to provide a convenient estimate on the gap between the initial lower and upper bounds on f^*.

Proposition 3.3. *Let F be defined in (3.5.2) and v be a prox-function of Y with modulus 1. We have*

$$F(x_0) - F(x_1) - \langle F'(x_1), x_0 - x_1 \rangle \le 2\left(2\|A\|^2 D_Y\right)^{1/2} \|x_0 - x_1\|, \quad \forall x_0, x_1 \in \mathbb{R}^n, \tag{3.9.44}$$

where $F'(x_1) \in \partial F(x_1)$ and D_Y is defined in (3.2.4).

Proof. We first need to provide some characterization for the subgradients of F. Let F and F_η be defined in (3.5.2) and (3.5.4), respectively. Also let us denote, for any $\eta > 0$ and $x \in X$,

$$\psi_x(z) := F_\eta(x) + \langle \nabla F_\eta(x), z - x \rangle + \tfrac{\mathscr{L}_\eta}{2}\|z - x\|^2 + \eta D_Y, \tag{3.9.45}$$

where D_Y and $\mathscr{L}_\eta$ are defined in (3.2.4) and (3.5.8), respectively. Clearly, in view of (3.9.2) and (3.5.6), ψ_x is a majorant of both F_η and F. For a given $x \in X$, let Z_x be the set

$$Z_x := \{z \in \mathbb{R}^n | \psi_x(z) + \langle \nabla \psi_x(z), x - z \rangle = F(x)\}, \tag{3.9.46}$$

where $\nabla \psi_x(z) = \nabla F_\eta(x) + \mathscr{L}_\eta(z - x)$. Equivalently, we have

$$Z_x := \left\{ z \in \mathbb{R}^n : \|z - x\|^2 = \tfrac{2}{\mathscr{L}_\eta}\left[\eta D_Y + F_\eta(x) - F(x)\right] \right\}. \tag{3.9.47}$$

Clearly, by the first relation in (3.5.6)

$$\|z - x\|^2 \le \tfrac{2\eta D_Y}{\mathscr{L}_\eta}, \quad \forall z \in Z_x. \tag{3.9.48}$$

Moreover, for any given $x \in \mathbb{R}^n$ and $p \in \mathbb{R}^n$, there exists $z \in Z_x$ such that

$$\langle F'(x), p \rangle \le \langle \nabla \psi_x(z), p \rangle = \langle \nabla F_\eta(x) + \mathscr{L}_\eta(z - x), p \rangle, \tag{3.9.49}$$

where $F'(x) \in \partial F(x)$. Indeed, let us denote

$$t = \tfrac{1}{\|p\|}\left\{ \tfrac{2}{\mathscr{L}_\eta}\left[\eta D_Y + F_\eta(x) - F(x)\right] \right\}^{1/2}$$

and $z_0 = x + tp$. Clearly, in view of (3.9.47), we have $z_0 \in Z_x$. By convexity of F, the fact that $F(z_0) \le \psi_x(z_0)$, and (3.9.46), we have

$$\begin{aligned} F(x) + \langle F'(x), tp \rangle \le F(x + tp) \le \psi_x(z_0) &= F(x) + \langle \nabla \psi_x(z_0), z_0 - x \rangle \\ &= F(x) + t\langle \nabla \psi_x(z_0), p \rangle, \end{aligned}$$

which clearly implies the result in (3.9.49).

Now we are ready to prove our main conclusion. First note that by the convexity of F, we have

$$F(x_0) - \left[F(x_1) + \langle F'(x_1), x_0 - x_1 \right]\rangle \le \langle F'(x_0), x_0 - x_1 \rangle + \langle F'(x_1), x_1 - x_0 \rangle.$$

Moreover, by (3.9.49), $\exists z_0 \in Z_{x_0}$, and $z_1 \in Z_{x_1}$ s.t.

$$\begin{aligned} &\langle F'(x_0), x_0 - x_1 \rangle + \langle F'(x_1), x_1 - x_0 \rangle \\ &\le \langle \nabla F_\eta(x_0) - \nabla F_\eta(x_1), x_0 - x_1 \rangle + \mathscr{L}_\eta \langle z_0 - x_0 - (z_1 - x_1), x_0 - x_1 \rangle \end{aligned}$$

$$\leq \mathcal{L}_\eta \|x_0 - x_1\|^2 + \mathcal{L}_\eta (\|z_0 - x_0\| + \|z_1 - x_1\|) \|x_0 - x_1\|$$
$$\leq \mathcal{L}_\eta \|x_0 - x_1\|^2 + 2\mathcal{L}_\eta \left(\tfrac{2\eta D_Y}{\mathcal{L}_\eta}\right)^{1/2} \|x_0 - x_1\|$$
$$= \tfrac{\|A\|^2}{\eta} \|x_0 - x_1\|^2 + 2\left(2\|A\|^2 D_Y\right)^{1/2} \|x_0 - x_1\|,$$

where the last inequality and equality follow from (3.9.48) and (3.5.8), respectively. Combining the above two relations, we have

$$F(x_0) - \left[F(x_1) + \langle F'(x_1), x_0 - x_1 \rangle\right] \leq \tfrac{\|A\|^2}{\eta} \|x_0 - x_1\|^2 + 2\left(2\|A\|^2 D_Y\right)^{1/2} \|x_0 - x_1\|.$$

The result now follows by tending η to $+\infty$ in the above relation. ■

We are now ready to show the main convergence results for the USL method.

Theorem 3.16. *Suppose that $\alpha_k \in (0,1]$, $k = 1, 2, \ldots$, in procedure $\mathcal{G}_{USL}$ are chosen such that condition (3.9.26) holds with $\rho = 1$ for some $c > 0$. The following statements hold for the USL method applied to problem (3.5.1)–(3.5.2).*

(a) The number of nonsignificant phases is bounded by $\tilde{\mathrm{S}}_F(Q_1) := \lceil \log D_Y / Q_1 \rceil$, and the number of significant phases is bounded by

$$\mathrm{S}_F(\varepsilon) := \left\lceil \max\left\{0, \log_{1/q}\left(\tfrac{4\|A\|\sqrt{D_X D_Y}}{\varepsilon}\right)\right\}\right\rceil. \tag{3.9.50}$$

(b) The total number of gap reduction iterations performed by the USL method does not exceed

$$\mathrm{S}_F(\varepsilon) + \tilde{\mathrm{S}}_F(Q_1) + \tfrac{\tilde{c}\bar{\Delta}_F}{\varepsilon}, \tag{3.9.51}$$

where $\tilde{c} := 2[\sqrt{2}/(1-q) + \sqrt{2} + 1]\sqrt{c}/\beta\theta$.

Proof. Denote $\delta_s \equiv \mathrm{ub}_s - \mathrm{lb}_s$, $s \geq 1$. Without loss of generality, we assume that $\delta_1 > \varepsilon$, since otherwise the statements are obviously true. The first claim in part (a) immediately follows from the facts that a nonsignificant phase can occur only if $Q_1 \leq D_Y$ due to Lemma 3.18(b) and that Q_s, $s \geq 1$, is increased by a factor of 2 in each nonsignificant phase. In order to show the second claim in part (a), we first bound the initial optimality gap $\mathrm{ub}_1 - \mathrm{lb}_1$. By the convexity of F, (3.5.5), and (3.9.43), we can easily see that $\mathrm{lb}_1 \leq f^*$. Moreover, we conclude from (3.5.5), (3.9.44), and (3.9.43) that

$$\mathrm{ub}_1 - \mathrm{lb}_1 \leq f(p_1) - \mathrm{lb}_1 = F(p_1) - F(p_0) - \langle F'(p_0), p_1 - p_0 \rangle$$
$$\leq 2\left(2\|A\|^2 D_Y\right)^{1/2} \|p_1 - p_0\| \leq 4\|A\|\sqrt{D_X D_Y},$$

where the last inequality follows from the fact that $\|p_1 - p_0\| \leq \sqrt{2} D_X$. Using this observation and Lemma 3.18(a), we can easily see that the number of significant phases is bounded by $\mathrm{S}_F(\varepsilon)$.

We now show that part (b) holds. Let $B = \{b_1, b_2, \ldots, b_k\}$ and $N = \{n_1, n_2, \ldots, n_m\}$, respectively, denote the set of indices of the significant and nonsignificant phases.

Note that $\delta_{b_{t+1}} \le q\,\delta_{b_t}, t \ge 1$, and hence that $\delta_{b_t} \ge q^{t-k}\delta_{b_k} > \varepsilon q^{t-k}$, $1 \le t \le k$. Also observe that $Q_{b_t} \le 2D_Y$ (see the remarks right after the statement of the USL method). Using these observations and Theorem 3.15, we conclude that the total number of iterations performed in the significant phases is bounded by

$$
\begin{aligned}
\Sigma_{t=1}^{k} K_{USL}(\delta_{b_t}, Q_{b_t}) &\le \Sigma_{t=1}^{k} K_{USL}(\varepsilon q^{t-k}, 2D_Y) \\
&\le k + \tfrac{2\|A\|}{\beta\theta\varepsilon}\sqrt{C_1 D_X D_Y}\,\Sigma_{t=1}^{k} q^{k-t} \\
&\le \mathrm{S}_F + \tfrac{2\|A\|}{\beta\theta(1-q)\varepsilon}\sqrt{2C_1 D_X D_Y}, \qquad (3.9.52)
\end{aligned}
$$

where the last inequality follows from part (a) and the observation that $\Sigma_{t=1}^{k} q^{k-t} \le 1/(1-q)$. Moreover, note that $\Delta_{n_r} > \varepsilon$ for any $1 \le r \le m$ and that $Q_{n_{r+1}} = 2Q_{n_r}$ for any $1 \le r \le m$. Using these observations and Theorem 3.15, we conclude that the total number of iterations performed in the nonsignificant phases is bounded by

$$
\begin{aligned}
\Sigma_{r=1}^{m} K_{USL}(\delta_{n_r}, Q_{n_r}) &\le \Sigma_{r=1}^{m} K_{USL}(\varepsilon, Q_{n_r}) \\
&\le m + \tfrac{2\|A\|}{\beta\theta\varepsilon}\sqrt{C_1 D_X Q_1}\,\Sigma_{r=1}^{m} 2^{(r-1)/2} \\
&\le \tilde{\mathrm{S}}_F + \tfrac{2\|A\|}{\beta\theta\varepsilon}\sqrt{C_1 D_X Q_1}\sum_{r=1}^{\tilde{\mathrm{S}}_F} 2^{(r-1)/2} \\
&\le \tilde{\mathrm{S}}_F + \tfrac{2\|A\|}{(\sqrt{2}-1)\beta\theta\varepsilon}\sqrt{C_1 D_X D_Y}. \qquad (3.9.53)
\end{aligned}
$$

Combining (3.9.52) and (3.9.53), we obtain (3.9.51). ■

It is interesting to observe that if $Q_1 = D_Y$, then there are no nonsignificant phases and the number of iterations performed by the USL method is simply bounded optimally by (3.9.52). In this case, we do not need to compute the value of $f_\eta(x_k^u)$ in step 3b. It is interesting to note that, in view of Theorem 3.16, the USL method still achieves the optimal complexity bound in (3.9.51) even without a good initial estimate on D_Y.

3.10 Exercises and Notes

1. Let $\{x_t\}$ be the sequence generated by the mirror descent method (3.2.5) applied to (3.1.2).

   (a) Assume the gradient $\|g(x_t)\|_* \le M$,

   $$f(y) \ge f(x) + \langle g(x), y - x\rangle + \mu V(x,y), \forall x, y \in X \qquad (3.10.1)$$

   and $\gamma_t = 2/(\mu t)$ for some $\mu > 0$. Provide a bound on $\|x_t - x^*\|^2$ and $f(\bar{x}_1^t) - f^*$ where x^* is the optimal solution of (3.1.2) and $\bar{x}_1^k$ is defined in (3.1.9).

(b) Please derive the selection of stepsizes $\{\gamma_t\}$ and the rate of convergence for mirror descent method applied to smooth convex optimization problems for which

$$\|\nabla f(x) - \nabla f(y)\|_* \le L\|x - y\|, \forall x, y \in X.$$

What if the problem is also strongly convex, i.e., (3.10.1) holds?

2. Show that if we replace step (3.3.6) in the accelerated gradient descent method ((3.3.4)–(3.3.6)) by

$$\bar{x}_t = \text{argmin}_{x \in X}\{f(\underline{x}_t) + \langle f(\underline{x}_t), x - \underline{x}_t\rangle + \tfrac{L}{2}\|x - \underline{x}_t\|^2\},$$

the algorithm still converges with similar performance guarantees as in Theorems 3.6 and 3.7.

3. In the game interpretation of accelerated gradient descent method, please specify the selection of τ_t, λ_t, q_t, and α_t so that (3.4.16) and (3.4.17) are equivalent to (3.3.5) and (3.3.6)
4. In the smoothing scheme for solving saddle point problems, let us define

$$F_\eta(x) := \max_y\{\langle Ax, y\rangle - \hat{g}(y) + \eta(D_Y^2 - W(y))\}.$$

Moreover, let us modify the accelerated gradient descent method in (3.3.23)–(3.3.25) as follows:

$$\underline{x}_t = (1 - q_t)\bar{x}_{t-1} + q_t x_{t-1}, \tag{3.10.2}$$
$$x_t = \text{argmin}_{x \in X}\{\gamma_t[\langle \nabla F_{\eta_t}(\underline{x}_t), x\rangle + \mu V(\underline{x}_t, x) + \hat{f}(x)] + V(x_{t-1}, x)\}, \tag{3.10.3}$$
$$\bar{x}_t = (1 - \alpha_t)\bar{x}_{t-1} + \alpha_t x_t. \tag{3.10.4}$$

Show that if η_t in (3.10.3) is set to

$$\eta = \tfrac{\|A\|D_X}{t}, t = 1, 2\ldots,$$

then by using the above algorithm one can find an ε-solution of (3.5.1), i.e., a point $\bar{x} \in X$ such that $f(\bar{x}) - f^* \le \varepsilon$, in at most

$$\mathcal{O}(\tfrac{\|A\|D_X D_Y}{\varepsilon})$$

iterations.

5. Consider the saddle point problem in (3.6.1), i.e., $f^* = \min_{x \in X} f(x)$ with

$$f(x) = \hat{f}(x) + \max_{y \in Y}\langle Ax, y\rangle - \hat{g}(y).$$

Also let $Q(\bar{z}, z)$ be defined in (3.6.7).

(a) Show that $\bar{z} \in Z$ is a saddle point of (3.6.1) if and only if $Q(\bar{z}, z) \le 0$ for any $z \in Z$.
(b) Show that $f(\bar{x}) - f^* \le \max_{z \in Z} Q(\bar{z}, z)$.

(c) Show that $0 \le f^* - \min_{x\in X}\{\hat{f}(x) + \langle Ax, \bar{y}\rangle - \hat{g}(\bar{y})\} \le \max_{z\in Z} Q(\bar{z}, z)$.

6. Consider the linear constrained convex optimization problem in (3.7.1). Establish the rate of convergence of the following preconditioned ADMM method obtained by replacing (3.7.2) with

$$x_t = \text{argmin}_{x\in X}\,\{f(x) + \langle \lambda_{t-1}, Ax + By_{t-1} - b\rangle + \rho\,\langle Ax_{t-1} + By_{t-1} - b, Ax\rangle \\ + \tfrac{\eta}{2}\|x - x_{t-1}\|^2\}$$

for some $\eta > 0$.

7. Consider the variational inequality problem in (3.8.1). Establish the rate of convergence of the mirror-prox method in (3.8.4) and (3.8.5) under the following situation.

(a) The operator is monotone, i.e., (3.8.2) holds, and Hölder continuous, i.e.,

$$\|F(x) - F(y)\|_* \le L\|x - y\|^\nu, \forall x, y \in X \tag{3.10.5}$$

for some $\nu \in [0, 1]$ and $L \ge 0$.

(b) The operator is not monotone, but satisfies (3.8.3). Moreover, it satisfies (3.10.5) for some $\nu \in (0, 1]$ and $L \ge 0$.

8. Describe a variant of the APL method discussed in Sect. 3.9.1 with the parameter $\alpha_k = 1$ in the APL gap reduction procedure, and establish its rate of convergence when applied to solve problem 3.9.1.

Notes. The rate of convergence for subgradient descent for solving was first established in [94]. The mirror descent method was first introduced by Nemirovski and Yudin in [94] and later simplified in [5]. Nesterov first introduced the accelerated gradient descent method in [96] and different variants of this method can be found, e.g., in [7, 72, 97, 99, 100]. The game interpretation of the accelerated gradient descent method and the relationship between this method and the primal–dual method were introduced in Lan and Zhou [71]. Nesterov [99] first presented the smoothing scheme for solving bilinear saddle point problems and the primal–dual method for solving these problems was first introduced by Chambolle and Pock in [18]. Chen, Lan, and Ouyang [23], and Dang and Lan [26] presented the generalization of the primal–dual method, e.g., to the non-Euclidean setting and smooth and/or strongly convex bilinear saddle point problem. Boyd et al. provide a comprehensive survey on ADMM in [17]. The rate of convergence analysis of ADMM was first presented by [48, 86], and the rate of convergence for different variants of ADMM, in terms of their primal optimality gap and feasibility violation, was established in [105]. Inspired by the smoothing scheme for solving bilinear saddle point problem, Nemirovski developed the mirror-prox method for solving variation inequalities in [93]. This method evolves from Korpelevish's extragradient method [61], see also [33] for a comprehensive treatment for variational inequalities and complementarity problems. The complexity of the mirror-prox method for solving the generalized monotone variational inequalities problems was established in [27]. Lan [66] first presented the accelerated prox-level and uniform smoothing level method for solving different classes of convex optimization problems, see [9, 10, 60, 76] for earlier developments of bundle-level type methods.

Chapter 4
Stochastic Convex Optimization

In this chapter, we focus on stochastic convex optimization problems which have found wide applications in machine learning. We will first study two classic methods, i.e., stochastic mirror descent and accelerated stochastic gradient descent methods. We will then present stochastic optimization methods for solving general convex concave saddle point, stochastic bilinear saddle point, and stochastic variational inequality problems. Finally, we discuss how to incorporate randomized block decomposition into stochastic optimization methods.

4.1 Stochastic Mirror Descent

We consider the following optimization problem

$$f^* \equiv \min_{x \in X} \{ f(x) := \mathbb{E}[F(x,\xi)] \}, \tag{4.1.1}$$

where $X \subset \mathbb{R}^m$ is a nonempty bounded closed convex set, and ξ is a random vector whose probability distribution P is supported on set $\Xi \subset \mathbb{R}^d$ and $F : X \times \Xi \to \mathbb{R}$. We assume that for every $\xi \in \Xi$ the function $F(\cdot,\xi)$ is convex on X, and that the expectation

$$\mathbb{E}[F(x,\xi)] = \textstyle\int_{\Xi} F(x,\xi)dP(\xi) \tag{4.1.2}$$

is well defined and finite valued for every $x \in X$. It follows that function $f(\cdot)$ is convex and finite valued on X. Moreover, we assume that $f(\cdot)$ is continuous on X. Clearly, continuity of $f(\cdot)$ follows from convexity if $f(\cdot)$ is finite valued and convex on a neighborhood of X. With these assumptions, (4.1.1) becomes a convex programming problem. A basic difficulty of solving stochastic optimization problem (4.1.1) is that the multidimensional integral (expectation) (4.1.2) cannot be computed with a high accuracy for dimension d, say, greater than 5. The aim of this

G. Lan, *First-Order and Stochastic Optimization Methods for Machine Learning*,
Springer Series in the Data Sciences, https://doi.org/10.1007/978-3-030-39568-1_4

section is to introduce one computational approach based on Monte Carlo sampling techniques. To this end we make the following assumptions.

Assumption 1 *It is possible to generate an independent and identically distributed (i.i.d.) sample* $\xi_1, \xi_2, \ldots,$ *of realizations of the random vector* ξ.

Assumption 2 *There is a mechanism, a stochastic first-order oracle* (SFO) *which for every given* $x \in X$ *and* $\xi \in \Xi$ *returns a stochastic subgradient—a vector* $G(x,\xi)$ *such that* $\mathrm{g}(x) := \mathbb{E}[G(x,\xi)]$ *is well defined.*

Throughout this section, we assume that the stochastic subgradient G satisfies the following assumptions.

Assumption 3 *For any* $x \in X$, *we have*

$$a)\ \mathbb{E}[G(x,\xi_t)] \equiv f'(x) \in \partial\Psi(x), \tag{4.1.3}$$

$$b)\ \mathbb{E}\left[\|G(x,\xi_t) - f'(x)\|_*^2\right] \le \sigma^2. \tag{4.1.4}$$

Recall that if $F(\cdot,\xi)$, $\xi \in \Xi$, is convex and $f(\cdot)$ is finite valued in a neighborhood of a point x, then

$$\partial f(x) = \mathbb{E}\left[\partial_x F(x,\xi)\right]. \tag{4.1.5}$$

In that case we can employ a measurable selection $\mathrm{G}(x,\xi) \in \partial_x F(x,\xi)$ as a stochastic subgradient.

The stochastic mirror descent method, also referred to as mirror descent stochastic approximation, is obtained by replacing the exact subgradient $g(x_t)$ in (3.2.5) with a stochastic subgradient $G_t := \mathrm{G}(x_t,\xi_t)$ returned by the stochastic oracle. More specifically, it updates x_t according to

$$x_{t+1} = \mathrm{argmin}_{x \in X} \gamma_t \langle G_t, x\rangle + V(x_t,x), t = 1,2,\ldots. \tag{4.1.6}$$

Here V denotes the prox-function associated with the distance generating function ω. For the sake of simplicity, we assume that the modulus σ_ν of ν is given by 1 (see Sect. 3.2). We will establish the convergence of this stochastic optimization under different assumptions about the objective function f.

4.1.1 General Nonsmooth Convex Functions

In this subsection, we assume that the objective function f has bounded subgradients such that

$$\|g(x)\|_* \le M, \ \ \forall x \in X. \tag{4.1.7}$$

This assumption implies that f is Lipschitz continuous over X in view of Lemma 2.4.

It can be easily seen that the result in Lemma 3.4 holds with g_t replaced by G_t. By using this result, we can establish the convergence properties for the above stochastic mirror descent method.

Theorem 4.1. *Let $x_t, t = 1,\ldots,k$, be generated by (4.1.6) and define $\bar{x}_s^k$ as in (3.1.9). Then*

$$\mathbb{E}[f(\bar{x}_s^k)] - f^* \le \left(\sum_{t=s}^k \gamma_t\right)^{-1} \left[\mathbb{E}[V(x_s, x^*)] + (M^2 + \sigma^2)\sum_{t=s}^k \gamma_t^2\right], \tag{4.1.8}$$

where x^* denotes an arbitrary solution of (4.1.1) and the expectation is taken w.r.t. $\xi_1,\ldots,\xi_k$.

Proof. Let us denote $\delta_t = G_t - g_t$, $t = 1,\ldots,k$. By the convexity of f and Lemma 3.4,

$$\begin{aligned}\gamma_t[f(x_t) - f(x)] &\le \gamma_t\langle G_t, x_t - x\rangle - \gamma_t\langle \delta_t, x_t - x\rangle\\ &\le V(x_t, x) - V(x_{t+1}, x) + \gamma_t\langle G_t, x_t - x_{t+1}\rangle - V(x_t, x_{t+1})\\ &\quad - \gamma_t\langle \delta_t, x_t - x\rangle\\ &\le V(x_t, x) - V(x_{t+1}, x) + \gamma_t^2\|G_t\|_*^2 - \gamma_t\langle \delta_t, x_t - x\rangle,\end{aligned}$$

where the last inequality follows from the strong convexity of ν, the Cauchy–Schwarz inequality, and the fact that $bt - at^2/2 \le b^2/(2a)$ for any $a > 0$. The previous conclusion together with the fact that

$$\|G_t\|_*^2 \le 2(\|g_t\|_*^2 + \|\delta\|_*^2) \le 2(M^2 + \|\delta\|_*^2) \tag{4.1.9}$$

due to (4.1.7) then imply that

$$\gamma_t[f(x_t) - f(x)] \le V(x_t, x) - V(x_{t+1}, x) + 2\gamma_t^2(M^2 + \|\delta\|_*^2) - \gamma_t\langle \delta_t, x_t - x\rangle.$$

Summing up these inequalities and using the fact that $f(\bar{x}_s^k) \le (\sum_{t=s}^k \gamma_t)^{-1} \sum_{t=s}^k f(x_t)$, we obtain

$$\begin{aligned}f(\bar{x}_s^k) - f^* \le \left(\sum_{t=s}^k \gamma_t\right)^{-1} &\Big[V(x_s, x^*) - V(x_{k+1}, x^*)\\ &+ 2\sum_{t=s}^k \gamma_t^2(M^2 + \|\delta_t\|_*^2) - \sum_{t=s}^k \gamma_t\langle \delta_t, x_t - x\rangle\Big].\end{aligned} \tag{4.1.10}$$

The result then follows by taking expectation on both sides of the above inequality. ■

Assuming that the total number of steps k is given in advance and optimizing the right-hand side of (4.1.8), we arrive at the constant stepsize policy

$$\gamma_t = \frac{D_X}{\sqrt{k(M^2+\sigma^2)}}, \quad t = 1,\ldots,k, \tag{4.1.11}$$

where D_X is defined in (3.2.4), and the associated efficiency estimate

$$\mathbb{E}\left[f(\tilde{x}_1^k) - f(x_*)\right] \le 2D_X\sqrt{\tfrac{M^2+\sigma^2}{k}}. \tag{4.1.12}$$

Certainly we can allow different stepsize strategy similar to the deterministic mirror descent method discussed in the previous chapter.

So far, all our efficiency estimates were upper bounds on the expected non-optimality, in terms of the objective, of approximate solutions generated by the algorithms. Here we complement these results with bounds on probabilities of large deviations. Observe that by Markov inequality, (4.1.12) implies that

$$\text{Prob}\left\{f(\tilde{x}_1^k) - f(x_*) > \varepsilon\right\} \le \tfrac{2D_X\sqrt{M^2+\sigma^2}}{\varepsilon\sqrt{k}}, \ \forall \varepsilon > 0. \tag{4.1.13}$$

This implies that in order to find an (ε, Λ)-solution of (4.1.1), i.e., a point $\bar{x} \in X$ s.t. $\text{Prob}\{f(\tilde{x}_1^k) - f(x_*) > \varepsilon)\} < \Lambda$ for some $\Lambda \in (0,1)$, one needs to run the stochastic mirror descent method for

$$\mathscr{O}\left\{\tfrac{D_X^2(M^2+\sigma^2)}{\Lambda^2\varepsilon^2}\right\} \tag{4.1.14}$$

iterations. It is possible, however, to obtain much finer bounds on deviation probabilities when imposing more restrictive assumptions on the distribution of $\mathsf{G}(x,\xi)$. Specifically, assume the following "light-tail" assumption.

Assumption 4 *For any $x \in X$, we have*

$$\mathbb{E}\left[\exp\{\|G(x,\xi_t) - g(x)\|_*^2/\sigma^2\}\right] \le \exp\{1\}. \tag{4.1.15}$$

It can be easily seen that Assumption 4 implies Assumption 3(b). Indeed, if a random variable Y satisfies $\mathbb{E}[\exp\{Y/a\}] \le \exp\{1\}$ for some $a > 0$, then by Jensen's inequality $\exp\{\mathbb{E}[Y/a]\} \le \mathbb{E}[\exp\{Y/a\}] \le \exp\{1\}$, and therefore $\mathbb{E}[Y] \le a$. Of course, condition (4.1.15) holds if $\|\mathsf{G}(x,\xi) - g(x)\|_* \le \sigma$ for all $(x,\xi) \in X \times \Xi$.

Assumption 4 is sometimes called the sub-Gaussian assumption. Many different random variables, such as Gaussian, uniform, and any random variables with a bounded support, will satisfy this assumption.

Now let us state the following well-known result for the martingale-difference sequence.

Lemma 4.1. *Let $\xi_{[t]} \equiv \{\xi_1, \xi_2, \ldots, \xi_t\}$ be a sequence of iid random variables, and $\zeta_t = \zeta_t(\xi_{[t]})$ be deterministic Borel functions of $\xi_{[t]}$ such that $\mathbb{E}_{|\xi_{[t-1]}}[\zeta_t] = 0$ a.s. and $\mathbb{E}_{|\xi_{[t-1]}}[\exp\{\zeta_t^2/\sigma_t^2\}] \le \exp\{1\}$ a.s., where $\sigma_t > 0$ are deterministic. Then*

$$\forall \lambda \ge 0 : \text{Prob}\left\{\textstyle\sum_{t=1}^N \zeta_t > \lambda\sqrt{\sum_{t=1}^N \sigma_t^2}\right\} \le \exp\{-\lambda^2/3\}.$$

Proof. For simplicity, let us denote the conditional expectation $\mathbb{E}_{|\xi_{[t-1]}}$ by $\mathbb{E}_{|t-1}$. Let us set $\bar{\zeta}_t = \zeta_t/\sigma_t$. We have $\exp\{x\} \le x + \exp\{9x^2/16\}$ for all x, so that $\mathbb{E}_{|t-1}[\bar{\zeta}_t] = 0$ and $\mathbb{E}_{|t-1}[\exp\{\bar{\zeta}_t^2\}] \le \exp\{1\}$ a.s. In view of these relations and the moment inequality, we have

$$\forall \lambda \in [0,4/3] : \mathbb{E}_{|t-1}\left[\exp\{\lambda \bar{\zeta}_t\}\right] \le \mathbb{E}_{|t-1}\left[\exp\{(9\lambda^2/16)\bar{\zeta}_t^2\}\right] \le \exp\{9\lambda^2/16\}. \tag{4.1.16}$$

Besides this, we have $\lambda x \le \frac{3}{8}\lambda^2 + \frac{2}{3}x^2$, whence

$$\mathbb{E}_{|t-1}\left[\exp\{\lambda \bar{\zeta}_t\}\right] \le \exp\{3\lambda^2/8\}\mathbb{E}_{|t-1}\left[\exp\{2\bar{\zeta}_t^2/3\}\right] \le \exp\{\frac{2}{3} + 3\lambda^2/8\}.$$

Combining the latter inequality with (4.1.16), we get

$$\forall \lambda \ge 0 : \mathbb{E}_{|t-1}\left[\exp\{\lambda \bar{\zeta}_t\}\right] \le \exp\{3\lambda^2/4\},$$

or, which is the same,

$$\forall \nu \ge 0 : \mathbb{E}_{|t-1}\left[\exp\{\nu \zeta_t\}\right] \le \exp\{3\nu^2\sigma_t^2/4\}.$$

Now, since ζ_τ is a deterministic function of ξ^τ, we have the recurrence

$$\begin{aligned}&\forall \nu \ge 0 : \mathbb{E}\left[\exp\{\nu \textstyle\sum_{\tau=1}^t \zeta_\tau\}\right] = \mathbb{E}\left[\exp\{\nu \textstyle\sum_{\tau=1}^{t-1} \zeta_\tau\}\mathbb{E}_{|t-1}\exp\{\nu\zeta_\tau\}\right]\\ &\le \exp\{3\nu^2\sigma_\tau^2/4\}\mathbb{E}\left[\exp\{\nu \textstyle\sum_{\tau=1}^{t-1}\zeta_\tau\}\right],\end{aligned}$$

whence

$$\forall \nu \ge 0 : \mathbb{E}\left[\exp\{\nu \textstyle\sum_{t=1}^N \zeta_t\}\right] \le \exp\{3\nu^2 \textstyle\sum_{t=1}^N \sigma_t^2/4\}.$$

Applying Chebyshev inequality, we get for a positive λ

$$\begin{aligned}&\text{Prob}\left\{\textstyle\sum_{t=1}^N \zeta_t > \lambda\sqrt{\textstyle\sum_{t=1}^N \sigma_t^2}\right\}\\ &\le \inf_{\nu\ge 0}\exp\left\{3\nu^2\textstyle\sum_{t=1}^N\sigma_t^2/4\right\}\exp\left\{-\lambda\nu\sqrt{\textstyle\sum_{t=1}^N\sigma_t^2}\right\}\\ &= \exp\left\{-\lambda^2/3\right\}.\end{aligned}$$

■

Proposition 4.1. *In the case of Assumption 4, for the constant stepsizes* (4.1.11) *one has that for any $\lambda \ge 0$ the following holds*

$$\text{Prob}\left\{f(\tilde{x}_1^k) - f(x_*) > \tfrac{3D_X}{\sqrt{k}}\left(\sqrt{M^2+\sigma^2} + \lambda\sigma\right)\right\} \le \exp\{-\lambda\} + \exp\{-\lambda^2/3\}.$$

Proof. Let $\zeta_t = \gamma_t\langle\delta_t, x^* - x_t\rangle$. Clearly, $\{\zeta_t\}_{t\ge 1}$ is a martingale-difference sequence. Moreover, it follows from the definition of D_X and (4.1.15) that

$$\begin{aligned}\mathbb{E}_{|\xi_{[t-1]}}\left[\exp\{\zeta_t^2/(\gamma_t D_X\sigma)^2\}\right] &\le \mathbb{E}_{|\xi_{[t-1]}}\left[\exp\{(\gamma_t D_X\|\delta_t\|_*)^2/(\gamma_t D_X\sigma)^2\}\right]\\ &\le \exp(1),\end{aligned}$$

The previous two observations, in view of Lemma 4.1, then imply that

$$\forall \lambda \ge 0 : \text{Prob}\left\{\textstyle\sum_{t=s}^k \zeta_t > \lambda D_X\sigma\sqrt{\textstyle\sum_{t=s}^k \gamma_t^2}\right\} \le \exp\{-\lambda^2/3\}. \tag{4.1.17}$$

Now observe that under Assumption 4,

$$\mathbb{E}_{|\xi_{t-1}}\left[\exp\{\|\delta_t\|_*^2/\sigma^2\}\right] \le \exp\{1\}.$$

Setting $\theta_t = \gamma_t^2/\sum_{t=s}^k \gamma_t^2$, we have

$$\exp\left\{\sum_{t=s}^k \theta_t(\|\delta_t\|_*^2/\sigma^2)\right\} \le \sum_{t=s}^k \theta_t \exp\{\|\delta_t\|_*^2/\sigma^2\},$$

whence, taking expectations,

$$\mathbb{E}\left[\exp\left\{\sum_{t=s}^k \gamma_t^2\|\delta_t\|_*^2/\left(\sigma^2\sum_{t=s}^k \gamma_t^2\right)\right\}\right] \le \exp\{1\}.$$

It then follows from Markov's inequality that

$$\forall \lambda \ge 0 : \text{Prob}\left\{\sum_{t=s}^k \gamma_t^2\|\delta_t\|_*^2 > (1+\lambda)\sigma^2\sum_{t=s}^k \gamma_t^2\right\} \le \exp\{-\lambda\}. \tag{4.1.18}$$

Using (4.1.17) and (4.1.18) in (4.1.10), we conclude that

$$\begin{aligned}\text{Prob}\Bigg\{f(\bar{x}_s^k)] - f^* > \left(\sum_{t=s}^k \gamma_t\right)^{-1}&\left[D_X^2 + 2\sum_{t=s}^k \gamma_t^2[M^2 + (1+\lambda)\sigma^2]\right.\\ &\left.+\lambda D_X\sigma\sqrt{\sum_{t=s}^k \gamma_t^2}\right]\Bigg\} \le \exp\{-\lambda\} + \exp\{-\lambda^2/3\}.\end{aligned} \tag{4.1.19}$$

The result immediately follows from the above inequality and (4.1.11). ■

In view of Proposition 4.1, if Assumption 4 holds, then the number of iterations performed by the stochastic mirror descent method to find an (ε, Λ)-solution of (4.1.1) can be bounded by

$$\mathcal{O}\left\{\tfrac{D_X^2(M^2+\sigma^2)\log(1/\Lambda)}{\varepsilon^2}\right\}.$$

In this subsection, we focus on nonsmooth problems which are not necessarily strongly convex. We will discuss the convergence of the stochastic mirror descent method for solving strongly convex nonsmooth problems in Sect. 4.6 under a more general setting.

4.1.2 Smooth Convex Problems

In this section, we still consider problem (4.1.1), but assume that $f : X \to \mathbb{R}$ is a convex function with Lipschitz continuous gradient, that is,

$$\|\nabla f(x) - \nabla f(x')\|_* \le L\|x - x'\|, \; \forall x, x' \in X. \tag{4.1.20}$$

We first derive the rate of convergence for a direct application of the stochastic mirror descent to smooth stochastic convex optimization problem mentioned above. Note that

$$\begin{aligned}\|\nabla f(x_t)\|_*^2 &\le (\|\nabla f(x_1)+\nabla f(x_t)-\nabla f(x_1)\|_*^2)\\ &\le 2\|\nabla f(x_1)\|_*^2+2\|\nabla f(x_t)-\nabla f(x_1)\|_*^2\\ &\le 2\|\nabla f(x_1)\|_*^2+2L^2\|x_t-x_1\|^2\\ &\le 2\|\nabla f(x_1)\|_*^2+2L^2D_X^2. \end{aligned} \quad (4.1.21)$$

We can easily see from the above inequality and (4.1.12) that the rate of convergence for a direct application of the stochastic mirror descent algorithm is bounded by

$$\mathcal{O}(1)\left[\frac{D_X(\|\nabla f(x_1)\|_*+LD_X+\sigma)}{\sqrt{k}}\right]. \quad (4.1.22)$$

One problem associated with the above method exists in that it does not explore the smoothness properties of f. We will see that a sharper rate of convergence can be obtained by exploiting the smoothness of f, coupled with a different convergence analysis. We will also slightly modify the way to compute the output solution as follows:

$$x_{t+1}^{av}=\left(\sum_{\tau=1}^{t}\gamma_\tau\right)^{-1}\sum_{\tau=1}^{t}\gamma_\tau x_{\tau+1}. \quad (4.1.23)$$

More specifically, the sequence $\{x_t^{av}\}_{t\ge 2}$ is obtained by averaging the iterates $x_t, t\ge 2$ with their corresponding weights γ_{t-1}, while the one in the original stochastic mirror descent is obtained by taking the average of the whole trajectory $x_t, t\ge 1$ with weights γ_t. Note, however, that if the constant stepsizes are used, i.e., $\gamma_t=\gamma, \forall t\ge 1$, then the averaging step stated above is exactly the same as the one stated in the original stochastic mirror descent method up to shifting one iterate.

The following lemma establishes an important recursion for the above stochastic mirror descent algorithm for smooth optimization problems.

Lemma 4.2. *Assume that the stepsizes γ_t satisfy $L\gamma_t<1$, $t\ge 1$. Also let $\delta_t := G(x_t,\xi_t)-g(x_t)$, where $g(x_t)=\mathbb{E}[G(x_t,\xi_t)]=\nabla f(x_t)$. Then, we have*

$$\gamma_t[f(x_{t+1})-f(x)]+V(x_{t+1},x)\le V(x_t,x)+\Delta_t(x),\ \forall x\in X, \quad (4.1.24)$$

where

$$\Delta_t(x):=\gamma_t\langle\delta_t,x-x_t\rangle+\frac{\|\delta_t\|_*^2\gamma_t^2}{2(1-L\gamma_t)}. \quad (4.1.25)$$

Proof. Denoting $d_t := x_{t+1}-x_t$, due to the strong convexity of ν, we have $\|d_t\|^2/2\le V(x_t,x_{t+1})$, which together with (3.1.17), then imply that

$$\begin{aligned}\gamma_t f(x_{t+1}) &\le \gamma_t[f(x_t)+\langle g(x_t),d_t\rangle+\tfrac{L}{2}\|d_t\|^2]\\ &= \gamma_t[f(x_t)+\langle g(x_t),d_t\rangle]+\tfrac{1}{2}\|d_t\|^2-\tfrac{1-L\gamma_t}{2}\|d_t\|^2\\ &\le \gamma_t[f(x_t)+\langle g(x_t),d_t\rangle]+V(x_t,x_{t+1})-\tfrac{1-L\gamma_t}{2}\|d_t\|^2\end{aligned}$$

$$
\begin{aligned}
&= \gamma_t[f(x_t) + \langle G_t, d_t\rangle] - \gamma_t\langle \delta_t, d_t\rangle + V(x_t, x_{t+1}) - \tfrac{1-L\gamma_t}{2}\|d_t\|^2 \\
&\le \gamma_t[f(x_t) + \langle G_t, d_t\rangle] + V(x_t, x_{t+1}) - \tfrac{1-L\gamma_t}{2}\|d_t\|^2 + \|\delta_t\|_* \gamma_t \|d_t\| \\
&\le \gamma_t[f(x_t) + \langle G_t, d_t\rangle] + V(x_t, x_{t+1}) + \tfrac{\|\delta_t\|_*^2 \gamma_t^2}{2(1-L\gamma_t)}.
\end{aligned} \tag{4.1.26}
$$

Moreover, it follows from Lemma 3.4 with g_t replaced by G_t that

$$
\begin{aligned}
&\gamma_t f(x_t) + [\gamma_t\langle G_t, x_{t+1} - x_t\rangle + V(x_t, x_{t+1})] \\
&\le \gamma_t f(x_t) + [\gamma_t\langle G_t, x - x_t\rangle + V(x_t, x) - V(x_{t+1}, x)] \\
&= \gamma_t[f(x_t) + \langle g(x_t), x - x_t\rangle] + \gamma_t\langle \delta_t, x - x_t\rangle + V(x_t, x) - V(x_{t+1}, x) \\
&\le \gamma_t f(x) + \gamma_t\langle \delta_t, x - x_t\rangle + V(x_t, x) - V(x_{t+1}, x),
\end{aligned}
$$

where the last inequality follows from the convexity of $f(\cdot)$. Combining the above two conclusions and rearranging the terms, we obtain (4.1.24). ■

We are now ready to describe the general convergence properties of the above stochastic mirror descent algorithm without specifying the stepsizes γ_t.

Theorem 4.2. *Assume that the stepsizes γ_t satisfy $0 < \gamma_t \le 1/(2L)$, $\forall t \ge 1$. Let $\{x_{t+1}^{av}\}_{t\ge 1}$ be the sequence computed according to (4.1.23) by the modified stochastic mirror descent algorithm.*

(a) Under Assumption 3,

$$
\mathbb{E}\left[f(x_{k+1}^{av}) - f^*\right] \le K_0(k),\ \forall k \ge 1, \tag{4.1.27}
$$

where

$$
K_0(k) := \left(\textstyle\sum_{t=1}^k \gamma_t\right)^{-1}\left[D_X^2 + \sigma^2 \textstyle\sum_{t=1}^k \gamma_t^2\right].
$$

(b) Under Assumptions 3 and 4, $\forall \lambda > 0, k \ge 1$,

$$
\text{Prob}\left\{f(x_{k+1}^{av}) - f^* > K_0(k) + \lambda K_1(k)\right\} \le \exp\{-\lambda^2/3\} + \exp\{-\lambda\}, \tag{4.1.28}
$$

where

$$
K_1(k) := \left(\textstyle\sum_{t=1}^k \gamma_t\right)^{-1}\left[D_X \sigma \sqrt{\textstyle\sum_{t=1}^k \gamma_t^2} + \sigma^2 \textstyle\sum_{t=1}^k \gamma_t^2\right].
$$

Proof. Summing up (4.1.24) from $t = 1$ to k, we have

$$
\begin{aligned}
\textstyle\sum_{t=1}^k [\gamma_t(f(x_{t+1}) - f^*)] &\le V(x_1, x^*) - V(x_{t+1}, x^*) + \textstyle\sum_{t=1}^k \Delta_t(x^*) \\
&\le V(x_1, x^*) + \textstyle\sum_{t=1}^k \Delta_t(x^*) \le D_X^2 + \textstyle\sum_{t=1}^k \Delta_t(x^*),
\end{aligned}
$$

which, in view of the fact that

$$
f(x_{t+1}^{av}) \le (\textstyle\sum_{t=1}^k \gamma_t)^{-1} \textstyle\sum_{t=1}^k \gamma_t f(x_{t+1}),
$$

then implies that

$$
\left(\textstyle\sum_{t=1}^k \gamma_t\right)\left[f(x_{t+1}^{av}) - f^*\right] \le D_X^2 + \textstyle\sum_{t=1}^k \Delta_t(x^*). \tag{4.1.29}
$$

Denoting $\zeta_t := \gamma_t \langle \delta_t, x^* - x_t \rangle$ and observing that

$$\Delta_t(x^*) = \zeta_t + \frac{\gamma_t^2 \|\delta_t\|_*^2}{2(1-L\gamma_t)},$$

we then conclude from (4.1.29) that

$$\begin{aligned}\left(\sum_{t=1}^k \gamma_t\right)\left[f(x_{t+1}^{av}) - f^*\right] &\le D_X^2 + \sum_{t=1}^k \left[\zeta_t + \frac{\gamma_t^2 \|\delta_t\|_*^2}{2(1-L\gamma_t)}\right] \\ &\le D_X^2 + \sum_{t=1}^k \left(\zeta_t + \gamma_t^2 \|\delta_t\|_*^2\right), \qquad (4.1.30)\end{aligned}$$

where the last inequality follows from the assumption that $\gamma_t \le 1/(2L)$.

Note that the pair (x_t, x_t^{av}) is a function of the history $\xi_{[t-1]} := (\xi_1, \ldots, \xi_{t-1})$ of the generated random process and hence is random. Taking expectations of both sides of (4.1.30) and noting that under Assumption 3, $\mathbb{E}[\|\delta_t\|_*^2] \le \sigma^2$, and

$$\mathbb{E}_{|\xi_{[\tau-1]}}[\zeta_t] = 0, \qquad (4.1.31)$$

we obtain

$$\left(\sum_{t=1}^k \gamma_t\right)\mathbb{E}\left[f(x_{t+1}^{av}) - f^*\right] \le D_X^2 + \sigma^2 \sum_{t=1}^k \gamma_t^2,$$

which clearly implies part (a).

The proof of part (b) is similar to that of Proposition 4.1 and hence the details are skipped. ■

We now describe the selection of the stepsizes for the modified stochastic mirror descent. For the sake of simplicity, let us suppose that the number of iterations for the above algorithm is fixed in advance, say equal to k, and that the *constant stepsize policy* is applied, i.e., $\gamma_t = \gamma, t = 1, \cdots, k$, for some $\gamma < 1/(2L)$ (note that the assumption of constant stepsizes does not hurt the efficiency estimate). We then conclude from Theorem 4.2 that the obtained solution $x_{k+1}^{av} = k^{-1}\sum_{t=1}^k x_{t+1}$ satisfies

$$\mathbb{E}\left[f(x_{k+1}^{av}) - f^*\right] \le \frac{D_X^2}{k\gamma} + \gamma\sigma^2.$$

Minimizing the right-hand side of the above inequality with respect to γ over the interval $(0, 1/(2L)]$, we conclude that

$$\mathbb{E}\left[f(x_{k+1}^{av}) - f^*\right] \le K_0^*(k) := \frac{2LD_X^2}{k} + \frac{2D_X\sigma}{\sqrt{k}}, \qquad (4.1.32)$$

by choosing γ as

$$\gamma = \min\left\{\frac{1}{2L}, \sqrt{\frac{D_X^2}{k\sigma^2}}\right\}.$$

Moreover, with this choice of γ, we have

$$K_1(k) = \frac{D_X\sigma}{\sqrt{k}} + \gamma\sigma^2 \le \frac{2D_X\sigma}{\sqrt{k}},$$

hence, bound (4.1.28) implies that

$$\text{Prob}\left\{f(x_{k+1}^{av}) - f^* > \frac{2LD_X^2}{k} + \frac{2(1+\lambda)D_X\sigma}{\sqrt{k}}\right\} \le \exp\{-\lambda^2/3\} + \exp\{-\lambda\}$$

for any $\lambda > 0$.

It is interesting to compare the rate of convergence (4.1.32) obtained for the modified stochastic mirror descent and the one stated in (4.1.22) for a direction application of the original stochastic mirror descent. Clearly, the latter one is always worse than the former one. Moreover, in the range

$$L \le \frac{\sqrt{k}\sigma^2}{D_X}, \tag{4.1.33}$$

the first component in (4.1.32) (for abbreviation, the L-component) merely does not affect the error estimate (4.1.32). Note that the range stated in (4.1.33) extends as N increases, meaning that if k is large, the Lipschitz constant of f does not affect the complexity of finding good approximate solutions. In contrast, this phenomenon does not appear in the error estimate (4.1.22) derived for the original stochastic mirror descent algorithm which employs a simple stepsizes strategy without taking into account the structure of the objective function f.

In this subsection, we focus on smooth problems which are not necessarily strongly convex. We will discuss the convergence of the stochastic mirror descent method for solving smooth and strongly convex smooth problems in Section 4.6 under a more general setting.

It should be noted that the stochastic mirror descent is a direct descendant of the mirror descent algorithm. It is well known that algorithms of these types are not optimal for smooth convex optimization. We will study a stochastic version of the optimal methods for smooth convex optimization in Sect. 4.2. In fact, we will show that these methods are also optimal for solving nonsmooth problem by properly specifying stepsizes.

4.1.3 Accuracy Certificates

In this subsection, we discuss one way to estimate lower and upper bounds for the optimal value of problem (4.1.1) when running the stochastic mirror descent algorithm. For the sake of simplicity, we focus on general nonsmooth convex programming problems with bounded subgradients, i.e., (4.1.7) holds. Discussions about accuracy certificates for smooth problems will be presented in Sect. 4.2.4.

Let k be the total number of steps and denote

$$\nu_t := \frac{\gamma_t}{\sum_{i=1}^k \gamma_i},\ t = 1,\ldots,k, \text{ and } \tilde{x}_k := \textstyle\sum_{t=1}^k \nu_t x_t. \tag{4.1.34}$$

Consider the functions

$$f^k(x) := \sum_{t=1}^k \nu_t \left[f(x_t) + g(x_t)^T(x - x_t)\right],$$
$$\hat{f}^k(x) := \sum_{t=1}^k \nu_t [F(x_t, \xi_t) + G(x_t, \xi_t)^T(x - x_t)],$$

and define

$$f_*^k := \min_{x \in X} f^k(x) \text{ and } f^{*k} := \sum_{t=1}^k \nu_t f(x_t). \tag{4.1.35}$$

Since $\nu_t > 0$ and $\sum_{t=1}^k \nu_t = 1$, it follows by convexity of $f(\cdot)$ that the function $f^k(\cdot)$ underestimates $f(\cdot)$ everywhere on X, and hence $f_*^k \le f^*$. Since $\tilde{x}_k \in X$ we also have that $f^* \le f(\tilde{x}_k)$, and by convexity of $f(\cdot)$ that $f(\tilde{x}_k) \le f^{*k}$. That is, for *any realization* of the random sample $\xi_1, \ldots, \xi_k$ we have that

$$f_*^k \le f^* \le f(\tilde{x}_k) \le f^{*k}. \tag{4.1.36}$$

It follows from (4.1.36) that $\mathbb{E}[f_*^k] \le f^* \le \mathbb{E}[f^{*k}]$ as well.

The bounds f_*^k and f^{*k} are unobservable since the values $f(x_t)$ are not known exactly. Therefore we consider their computable counterparts

$$\underline{f}^k = \min_{x \in X} \hat{f}^k(x) \;\text{ and }\; \overline{f}^k = \sum_{t=1}^k \nu_t F(x_t, \xi_t). \tag{4.1.37}$$

The bound $\overline{f}^k$ can be easily calculated while running the stochastic mirror descent procedure. The bound $\underline{f}^k$ involves solving the optimization problem of minimizing a linear objective function over set X.

Since x_t is a function of $\xi_{[t-1]} = (\xi_1, \ldots, \xi_{t-1})$, and ξ_t is independent of $\xi_{[t-1]}$, we have that

$$\mathbb{E}\left[\overline{f}^k\right] = \sum_{t=1}^k \nu_t \mathbb{E}\left\{\mathbb{E}[F(x_t, \xi_t)|\xi_{[t-1]}]\right\} = \sum_{t=1}^k \nu_t \mathbb{E}\left[f(x_t)\right] = \mathbb{E}[f^{*k}]$$

and

$$\begin{aligned}
\mathbb{E}\left[\underline{f}^k\right] &= \mathbb{E}\left[\mathbb{E}\left\{\min_{x \in X}\left[\sum_{t=1}^k \nu_t [F(x_t, \xi_t) + G(x_t, \xi_t)^T(x - x_t)]\right] \Big| \xi_{[t-1]}\right\}\right] \\
&\le \mathbb{E}\left[\min_{x \in X}\left\{\mathbb{E}\left[\sum_{t=1}^k \nu_t [F(x_t, \xi_t) + G(x_t, \xi_t)^T(x - x_t)]\right] \Big| \xi_{[t-1]}\right\}\right] \\
&= \mathbb{E}\left[\min_{x \in X} f^k(x)\right] = \mathbb{E}\left[f_*^k\right].
\end{aligned}$$

It follows that

$$\mathbb{E}\left[\underline{f}^k\right] \le f^* \le \mathbb{E}\left[\overline{f}^k\right]. \tag{4.1.38}$$

That is, on average $\underline{f}^k$ and $\overline{f}^k$ give, respectively, a lower and an upper bound for the optimal value of problem (4.1.1).

Our goal in the remaining part of this subsection is to understand how good the bounds $\underline{f}^k$ and $\overline{f}^k$ are. In the sequel, we denote $\Delta_t := F(x_t, \xi_t) - f(x_t)$ and $\delta_t := G(x_t, \xi_t) - g(x_t)$. Since x_t is a function of $\xi_{[t-1]}$ and ξ_t is independent of $\xi_{[t-1]}$, we have that the conditional expectations

$$\mathbb{E}_{|t-1}[\Delta_t] = 0 \;\text{ and }\; \mathbb{E}_{|t-1}[\delta_t] = 0, \tag{4.1.39}$$

and hence the unconditional expectations $\mathbb{E}[\Delta_t] = 0$ and $\mathbb{E}[\delta_t] = 0$ as well.

We make the following assumptions about the Δ_t.

Assumption 5 *There exists a positive constant Q such that for any $t \geq 0$:*

$$\mathbb{E}\left[\Delta_t^2\right] \leq Q^2. \tag{4.1.40}$$

We first need to show the following simple result.

Lemma 4.3. *Let $\zeta_1, \ldots, \zeta_j$ be a sequence of elements of $\mathbb{R}^n$. Define the sequence v_t, $t = 1, 2, \ldots$ in X^o as follows: $v_1 \in X^o$ and*

$$v_{t+1} = \mathrm{argmin}_{x \in X} \left\{ \langle \zeta_t, x \rangle + V(v_t, x) \right\}.$$

Then for any $x \in X$ the following inequalities hold

$$\langle \zeta_t, v_t - x \rangle \leq V(v_t, x) - V(v_{t+1}, x) + \tfrac{\|\zeta_t\|_*^2}{2}, \tag{4.1.41}$$

$$\textstyle\sum_{t=1}^{j} \langle \zeta_t, v_t - x \rangle \leq V(v_1, x) + \frac{1}{2} \sum_{t=1}^{j} \|\zeta_t\|_*^2. \tag{4.1.42}$$

Proof. By Lemma 3.4, we have

$$\langle \zeta_t, v_{t+1} - x \rangle + V(v_t, v_{t+1}) \leq V(v_t, x) - V(v_{t+1}, x),$$

which in view of the fact that

$$\langle \zeta_t, v_t - v_{t+1} \rangle - V(v_t, v_{t+1}) \leq \langle \zeta_t, v_t - v_{t+1} \rangle - \tfrac{1}{2} \|v_{t+1} - v_t\|^2 \leq \tfrac{1}{2} \|\zeta_t\|_*^2,$$

then implies (4.1.41). Summing up (4.1.41) from $t = 1$ to j, we conclude (4.1.42) due to $V(v, x) \geq 0$ for any $v \in Z^o, x \in Z$. ■

We are now ready to bound the expected gap between the aforementioned upper and lower bounds for the optimal value of problem (4.1.1).

Theorem 4.3. *Suppose that Assumptions 1-3 and Assumption 5 hold. Then*

$$\mathbb{E}\left[f^{*k} - f_*^k\right] \leq \tfrac{4D_X^2 + (2M^2 + 3\sigma^2) \sum_{t=1}^{k} \gamma_t^2}{2 \sum_{t=1}^{k} \gamma_t}, \tag{4.1.43}$$

$$\mathbb{E}\left[\left|\overline{f}^k - f^{*k}\right|\right] \leq Q \sqrt{\textstyle\sum_{t=1}^{k} v_t^2}, \tag{4.1.44}$$

$$\begin{aligned}\mathbb{E}\left[\left|\underline{f}^k - f_*^k\right|\right] &\leq \tfrac{D_X^2 + (M^2 + \sigma^2) \sum_{t=1}^{k} \gamma_t^2}{\sum_{t=1}^{k} \gamma_t} \\ &\quad + \left(Q + 2\sqrt{2} D_X \sigma\right) \sqrt{\textstyle\sum_{t=1}^{k} v_t^2}.\end{aligned} \tag{4.1.45}$$

In particular, in the case of constant stepsize policy (4.1.11) *we have*

$$\begin{aligned}&\mathbb{E}\left[f^{*k} - f_*^k\right] \leq \tfrac{7D_X \sqrt{M^2 + \sigma^2}}{2\sqrt{k}}, \\ &\mathbb{E}\left[\left|\overline{f}^k - f^{*k}\right|\right] \leq Q k^{-1/2}, \\ &\mathbb{E}\left[\left|\underline{f}^k - f_*^k\right|\right] \leq \tfrac{2D_X \sqrt{M^2 + \sigma^2}}{\sqrt{k}} + \left(Q + 2\sqrt{2} D_X \sigma\right) k^{-1/2}.\end{aligned} \tag{4.1.46}$$

Proof. If in Lemma 4.3 we take $v_1 := x_1$ and $\zeta_t := \gamma_t \mathsf{G}(x_t, \xi_t)$, then the corresponding iterates v_t coincide with x_t. Therefore, we have by (4.1.41) and since $V(x_1, u) \le D_X^2$ that

$$\sum_{t=1}^k \gamma_t (x_t - u)^T \mathsf{G}(x_t, \xi_t) \le D_X^2 + 2^{-1} \sum_{t=1}^k \gamma_t^2 \|\mathsf{G}(x_t, \xi_t)\|_*^2, \ \forall u \in X. \tag{4.1.47}$$

It follows that for any $u \in X$:

$$\begin{aligned} &\sum_{t=1}^k \nu_t \left[-f(x_t) + (x_t - u)^T \mathrm{g}(x_t) \right] + \sum_{t=1}^k \nu_t f(x_t) \\ &\le \frac{D_X^2 + 2^{-1} \sum_{t=1}^k \gamma_t^2 \|\mathsf{G}(x_t, \xi_t)\|_*^2}{\sum_{t=1}^k \gamma_t} + \sum_{t=1}^k \nu_t \delta_t^T (x_t - u). \end{aligned}$$

Since

$$f^{*k} - f_*^k = \sum_{t=1}^k \nu_t f(x_t) + \max_{u \in X} \sum_{t=1}^k \nu_t \left[-f(x_t) + (x_t - u)^T \mathrm{g}(x_t) \right],$$

it follows that

$$f^{*k} - f_*^k \le \frac{D_X^2 + 2^{-1} \sum_{t=1}^k \gamma_t^2 \|\mathsf{G}(x_t, \xi_t)\|_*^2}{\sum_{t=1}^k \gamma_t} + \max_{u \in X} \sum_{t=1}^k \nu_t \delta_t^T (x_t - u). \tag{4.1.48}$$

Let us estimate the second term in the right-hand side of (4.1.48). Let

$$\begin{aligned} &u_1 = v_1 = x_1; \\ &u_{t+1} = \operatorname{argmin}_{x \in X} \{ \langle -\gamma_t \delta_t, x \rangle + V(u_t, x) \}, t = 1, 2, \ldots, k; \\ &v_{t+1} = \operatorname{argmin}_{x \in X} \{ \langle \gamma_t \delta_t, x \rangle + V(v_t, x) \}, t = 1, 2, \ldots k. \end{aligned} \tag{4.1.49}$$

Observe that δ_t is a deterministic function of $\xi_{[t]}$, whence u_t and v_t are deterministic functions of $\xi_{[t-1]}$. By using Lemma 4.3 we obtain

$$\sum_{t=1}^k \gamma_t \delta_t^T (v_t - u) \le D_X^2 + 2^{-1} \sum_{t=1}^k \gamma_t^2 \|\delta_t\|_*^2, \ \forall u \in X. \tag{4.1.50}$$

Moreover,

$$\delta_t^T (v_t - u) = \delta_t^T (x_t - u) + \delta_t^T (v_t - x_t),$$

and hence it follows by (4.1.50) that

$$\max_{u \in X} \sum_{t=1}^k \nu_t \delta_t^T (x_t - u) \le \sum_{t=1}^k \nu_t \delta_t^T (x_t - v_t) + \frac{D_X^2 + 2^{-1} \sum_{t=1}^k \gamma_t^2 \|\delta_t\|_*^2}{\sum_{t=1}^k \gamma_t}. \tag{4.1.51}$$

Observe that by similar reasoning applied to $-\delta_t$ in the role of δ_t we get

$$\max_{u \in X} \left[-\sum_{t=1}^k \nu_t \delta_t^T (x_t - u) \right] \le \left[-\sum_{t=1}^k \nu_t \delta_t^T (x_t - u_t) \right] + \frac{D_X^2 + 2^{-1} \sum_{t=1}^k \gamma_t^2 \|\delta_t\|_*^2}{\sum_{t=1}^k \gamma_t}. \tag{4.1.52}$$

Moreover, $\mathbb{E}_{|t-1} [\delta_t] = 0$ and u_t,v_t and x_t are functions of $\xi_{[t-1]}$, while $\mathbb{E}_{|t-1} \delta_t = 0$ and hence

$$\mathbb{E}_{|t-1} \left[(x_t - v_t)^T \delta_t \right] = \mathbb{E}_{|t-1} \left[(x_t - u_t)^T \delta_t \right] = 0. \tag{4.1.53}$$

We also have that $\mathbb{E}_{|t-1}\left[\|\delta_t\|_*^2\right] \le \sigma^2$ by (4.1.4), it follows from (4.1.51) and (4.1.53) that

$$\mathbb{E}\left[\max_{u\in X}\sum_{t=1}^k \nu_t \delta_t^T(x_t-u)\right] \le \frac{D_X^2+2^{-1}\sigma^2\sum_{t=1}^k \gamma_t^2}{\sum_{t=1}^k \gamma_t}. \tag{4.1.54}$$

Therefore, by taking expectation of both sides of (4.1.48) and using (4.1.4), (4.1.9) together with (4.1.54) we obtain the estimate (4.1.43).

In order to prove (4.1.44) let us observe that $\overline{f}^k - f^{*k} = \sum_{t=1}^k \nu_t \Delta_t$, and that for $1 \le s < t \le k$,

$$\mathbb{E}[\Delta_s\Delta_t] = \mathbb{E}\{\mathbb{E}_{|t-1}[\Delta_s\Delta_t]\} = \mathbb{E}\{\Delta_s\mathbb{E}_{|t-1}[\Delta_t]\} = 0.$$

Therefore

$$\begin{aligned}\mathbb{E}\left[\left(\overline{f}^k - f^{*k}\right)^2\right] = \mathbb{E}\left[\left(\sum_{t=1}^k \nu_t\Delta_t\right)^2\right] &= \sum_{t=1}^k \nu_t^2\mathbb{E}\left[\Delta_t^2\right]\\ &= \sum_{t=1}^k \nu_t^2\mathbb{E}\left\{\mathbb{E}_{|t-1}\left[\Delta_t^2\right]\right\}.\end{aligned}$$

Moreover, by condition (4.1.40) of assumption (A1) we have that $\mathbb{E}_{|t-1}\left[\Delta_t^2\right] \le Q^2$, and hence

$$\mathbb{E}\left[\left(\overline{f}^k - f^{*k}\right)^2\right] \le Q^2\sum_{t=1}^k \nu_t^2. \tag{4.1.55}$$

Since $\sqrt{\mathbb{E}[Y^2]} \ge \mathbb{E}|Y|$ for any random variable Y, inequality (4.1.44) follows from (4.1.55).

Let us now look at (4.1.45). We have

$$\begin{aligned}\left|\underline{f}^k - f_*^k\right| &= \left|\min_{x\in X}\hat{f}^k(x) - \min_{x\in X} f^k(x)\right| \le \max_{x\in X}\left|\hat{f}^k(x) - f^k(x)\right|\\ &\le \left|\sum_{t=1}^k \nu_t\Delta_t\right| + \max_{x\in X}\left|\sum_{t=1}^k \nu_t\delta_t^T(x_t-x)\right|.\end{aligned} \tag{4.1.56}$$

We already showed above (see (4.1.55)) that

$$\mathbb{E}\left[\left|\sum_{t=1}^k \nu_t\Delta_t\right|\right] \le Q\sqrt{\sum_{t=1}^k \nu_t^2}. \tag{4.1.57}$$

Invoking (4.1.51), (4.1.52), we get

$$\begin{aligned}\max_{x\in X}\left|\sum_{t=1}^k \nu_t\delta_t^T(x_t-x)\right| \le \left|\sum_{t=1}^k \nu_t\delta_t^T(x_t-v_t)\right| + \left|\sum_{t=1}^k \nu_t\delta_t^T(x_t-u_t)\right|\\ + \frac{D_X^2+2^{-1}\sum_{t=1}^k \gamma_t^2\|\delta_t\|_*^2}{\sum_{t=1}^k \gamma_t}.\end{aligned} \tag{4.1.58}$$

Moreover, for $1 \le s < t \le k$ we have that $\mathbb{E}\left[\left(\delta_s^T(x_s-v_s)\right)\left(\delta_t^T(x_t-v_t)\right)\right] = 0$, and hence

$$\begin{aligned}\mathbb{E}\left[\left|\sum_{t=1}^k \nu_t\delta_t^T(x_t-v_t)\right|^2\right] &= \sum_{t=1}^k \nu_t^2\mathbb{E}\left[\left|\delta_t^T(x_t-v_t)\right|^2\right]\\ &\le \sigma^2\sum_{t=1}^k \nu_t^2\mathbb{E}\left[\|x_t-v_t\|^2\right]\\ &\le 2\sigma^2D_X^2\sum_{t=1}^k \nu_t^2,\end{aligned}$$

where the last inequality follows from the strong convex of ν. It follows that

$$\mathbb{E}\left[\left|\sum_{t=1}^k \nu_t\delta_t^T(x_t-v_t)\right|\right] \le \sqrt{2}D_X\sigma\sqrt{\sum_{t=1}^k \nu_t^2}.$$

By similar reasons,

$$\mathbb{E}\left[\left|\sum_{t=1}^{k} \nu_t \delta_t^T (x_t - u_t)\right|\right] \le \sqrt{2} D_X \sigma \sqrt{\sum_{t=1}^{k} \nu_t^2}.$$

These two inequalities combine with (4.1.57), (4.1.58), and (4.1.56) to imply (4.1.45). ■

Theorem 4.3 shows that for large k the online observable random quantities $\overline{f}^k$ and $\underline{f}^k$ are close to the upper bound f^{*k} and lower bound f_*^k, respectively. Besides this, on average, $\overline{f}^k$ indeed overestimates f^*, and $\underline{f}_k$ indeed underestimates f^*. More specifically, for the constant stepsize policy (4.1.11), we have that all estimates given in the right-hand side of (4.1.46) are of order $O(k^{-1/2})$. It follows that for the constant stepsize policy, difference between the upper $\overline{f}^k$ and lower $\underline{f}^k$ bounds converges on average to zero, with increase of the sample size k, at a rate of $O(k^{-1/2})$. It is possible to derive and refine the large-deviation properties of the gap between the lower and upper bounds, especially if we augmented (4.1.40) with

$$\mathbb{E}\left[\exp\left\{\delta_t^2/Q^2\right)\right\}\right] \le \exp\{1\}. \tag{4.1.59}$$

The development will be similar to Theorem 4.2(b) (see Sect. 4.2.4 for some related discussions).

Notice that the sample average approximation (SAA) approach (Sect. 1.6) also provides a lower on average bound—the random quantity $\hat{f}_{\text{SAA}}^k$, which is the optimal value of the sample average problem (see (4.1.60) below). Suppose the same sample ξ_t, $t = 1, \ldots, k$, is applied for both stochastic mirror descent and SAA methods. Besides this, assume that the constant stepsize policy is used in the stochastic mirror descent method, and hence $\nu_t = 1/k$, $t = 1, .., k$. Finally, assume (as it often is the case) that $\mathsf{G}(x, \xi)$ is a subgradient of $F(x, \xi)$ in x. By convexity of $F(\cdot, \xi)$ and since $\underline{f}^k = \min_{x \in X} \hat{f}^k(x)$, we have

$$\begin{aligned}\hat{f}_{\text{SAA}}^k &:= \min_{x \in X} k^{-1} \sum_{t=1}^{k} F(x, \xi_t) \\ &\ge \min_{x \in X} \sum_{t=1}^{k} \nu_t \left(F(x_t, \xi_t) + \mathsf{G}(x_t, \xi_t)^T (x - x_t)\right) = \underline{f}^k.\end{aligned} \tag{4.1.60}$$

That is, for the same sample the lower bound $\underline{f}^k$ is smaller than the lower bound obtained by the SAA method. However, it should be noted that the lower bound $\underline{f}^k$ is computed much faster than $\hat{f}_{\text{SAA}}^k$, since computing the latter one amounts to solving the sample average optimization problem associated with the generated sample.

Similar to the SAA method, in order to estimate the variability of the lower bound $\underline{f}^k$, one can run the stochastic mirror descent method M times, with independent samples, each of size k, and consequently compute the average and sample variance of M realizations of the random quantity $\underline{f}^k$. Alternatively, one can run the stochastic mirror descent procedure once but with kM iterations, then partition the obtained trajectory into M consecutive parts, each of size k, for each of these parts calculate the corresponding stochastic mirror descent lower bound and consequently compute the average and sample variance of the M obtained numbers. The latter approach is similar, in spirit, to the batch means method used in simulation output analysis. One advantage of this approach is that as more iterations being run, the stochastic mirror-descent method can output a solution $\tilde{x}_{kM}$ with much better objective value than $\tilde{x}_k$. However, this method has the same shortcoming as the batch means method, that is, the correlation among consecutive blocks will result in a biased estimation for the sample variance.

4.2 Stochastic Accelerated Gradient Descent

In this section, we study a class of stochastic composite optimization problems given by

$$\Psi^* := \min_{x \in X} \{\Psi(x) := f(x) + h(x)\}, \tag{4.2.1}$$

where X is a closed convex set in $\mathbb{R}^m$, $h(x)$ is a simple convex function with known structure (e.g., $h(x) = 0$ or $h(x) = \|x\|_1$), and $f : X \to \mathbb{R}$ is a general convex function such that for some $L \geq 0$, $M \geq 0$ and $\mu \geq 0$,

$$\mu V(x,y) \leq f(y) - f(x) - \langle f'(x), y - x\rangle \leq \tfrac{L}{2}\|y - x\|^2 + M\|y - x\|, \quad \forall x, y \in X, \tag{4.2.2}$$

where $f'(x) \in \partial f(x)$ and $\partial f(x)$ denotes the subdifferential of f at x. Moreover, we only have access to stochastic first-order information about f. More specifically, at the t-th iteration, for a given $x_t \in X$, the stochastic first-order oracle (SFO) returns $F(x_t, \xi_t)$ and $G(x_t, \xi_t)$ satisfying $\mathbb{E}[F(x_t, \xi_t)] = f(x_t)$ and $\mathbb{E}[G(x_t, \xi_t)] \equiv g(x_t) \in \partial f(x_t)$, where $\{\xi_t\}_{t \geq 1}$ is a sequence of independently and identically distributed random variables.

Since the parameters L, M, μ, and σ can be zero, problem (4.2.1) described above covers a wide range of convex programming problems. In particular, if f is a general Lipschitz continuous function with constant M, then relation (4.2.2) holds with $L = 0$. If f is a smooth convex function with L-Lipschitz continuous gradient, then (4.2.2) holds with $M = 0$. Clearly, relation (4.2.2) also holds if f is given as the summation of smooth and nonsmooth convex functions. Moreover, f is strongly convex if $\mu > 0$ and problem (4.2.1) covers different classes of deterministic convex programming problems if $\sigma = 0$.

If $\mu > 0$ in (4.2.2), then, by the classic complexity theory for convex programming, to find *an ε-solution of (4.2.1)*, i.e., a point $\bar{x} \in X$ s.t. $\mathbb{E}[\Psi(\bar{x}) - \Psi^*] \leq \varepsilon$, the number of calls (or iterations) to SFO cannot be smaller than

$$\mathcal{O}(1)\left\{\sqrt{\tfrac{L}{\mu}} \log \tfrac{L\|x_0 - x^*\|^2}{\varepsilon} + \tfrac{(M+\sigma)^2}{\mu\varepsilon}\right\}, \tag{4.2.3}$$

where x_0 denotes an initial point, x^* is the optimal solution of problem (4.2.1), and $\mathcal{O}(1)$ represents an absolute constant. Moreover, if $\mu = 0$ in (4.2.2), then by the complexity theory of convex programming, the number of calls to the SFO cannot be smaller than

$$\mathscr{O}(1)\left\{\sqrt{\tfrac{L\|x_0-x^*\|^2}{\varepsilon}}+\tfrac{\sigma^2}{\varepsilon^2}\right\}. \tag{4.2.4}$$

While the terms without involving σ in (4.2.3) and (4.2.4) come from deterministic convex programming, we briefly discuss how the critical terms involving σ is derived. Let us focus on the term $\sigma^2/(\mu\varepsilon)$ in (4.2.3). Consider the problem of $\min_x\{\Psi(x)=\mu(x-\alpha)^2\}$ with unknown α. Also suppose that the stochastic gradient is given by $2\mu(x-\alpha-\xi/\mu)$ with $\xi\sim\mathbb{N}(0,\sigma^2)$. Under this setting, our optimization problem is equivalent to the estimation of the unknown mean α from the observations of $\zeta=\alpha+\xi/\mu\sim\mathbb{N}(\alpha,\sigma^2/\mu^2)$, and the residual is μ times the expected squared error of recovery of the mean α. By standard statistical reasons, when the initial range for α is larger than σ/μ, to make this expected squared error smaller than $\delta^2\equiv\varepsilon/\mu$, or equivalently, $\mathbb{E}[\Psi(\bar{x})-\Psi_*]\le\varepsilon$, the number of observations we need is at least $N=\mathscr{O}(1)\left((\sigma^2/\mu^2)/\delta^2\right)=\mathscr{O}(1)\left(\sigma^2/(\mu\varepsilon)\right)$.

Our goal in this section is to present an optimal stochastic gradient descent type algorithm, namely the stochastic accelerated gradient descent method, which can achieve the lower complexity bounds stated in (4.2.3) and (4.2.4). The stochastic accelerated gradient descent method, also called accelerated stochastic approximation (AC-SA), is obtained by replacing exact gradients with stochastic gradients in the accelerated gradient descent method. The basic scheme of this algorithm is described as follows.

$$\underline{x}_t=(1-q_t)\bar{x}_{t-1}+q_t x_{t-1}, \tag{4.2.5}$$
$$x_t=\arg\min_{x\in X}\left\{\gamma_t\left[\langle G(\underline{x}_t,\xi_t),x\rangle+h(x)+\mu V(\underline{x}_t,x)\right]+V(x_{t-1},x)\right\}, \tag{4.2.6}$$
$$\bar{x}_t=(1-\alpha_t)\bar{x}_{t-1}+\alpha_t x_t. \tag{4.2.7}$$

Observe that while the original accelerated gradient descent method was designed for solving deterministic convex optimization problems only, by using a novel convergence analysis, we will demonstrate that this algorithm is optimal for not only smooth, but also general nonsmooth and stochastic optimization problems.

The following result describes some properties of the composite function Ψ.

Lemma 4.4. *Let $\bar{x}_t:=(1-\alpha_t)\bar{x}_{t-1}+\alpha_t x_t$ for some $\alpha_t\in[0,1]$ and $(\bar{x}_{t-1},x_t)\in X\times X$. We have*

$$\begin{aligned}\Psi(\bar{x}_t)\le(1-\alpha_t)\Psi(\bar{x}_{t-1})+\alpha_t[f(z)+\langle f'(z),x_t-z\rangle+h(x_t)]\\ +\tfrac{L}{2}\|\bar{x}_t-z\|^2+M\|\bar{x}_t-z\|,\end{aligned}$$

for any $z\in X$.

Proof. First observe that by the definition of $\bar{x}_t$ and the convexity of f, we have

$$\begin{aligned}f(z)+\langle f'(z),\bar{x}_t-z\rangle&=f(z)+\langle f'(z),\alpha_t x_t+(1-\alpha_t)\bar{x}_{t-1}-z\rangle\\&=(1-\alpha_t)[f(z)+\langle f'(z),\bar{x}_{t-1}-z\rangle]\\&\quad+\alpha_t[f(z)+\langle f'(z),x_t-z\rangle]\\&\le(1-\alpha_t)f(\bar{x}_{t-1})+\alpha_t[f(z)+\langle f'(z),x_t-z\rangle].\end{aligned}$$

Using this observation and (4.2.2), we have

$$f(\bar{x}_t) \le f(z) + \langle f'(z), \bar{x}_t - z\rangle + \tfrac{L}{2}\|\bar{x}_t - z\|^2 + M\|\bar{x}_t - z\|$$
$$\le (1-\alpha_t) f(\bar{x}_{t-1}) + \alpha_t [f(z) + \langle f'(z), x_t - z\rangle] + \tfrac{L}{2}\|\bar{x}_t - z\|^2 + M\|\bar{x}_t - z\|.$$

Using the convexity of h, we have $h(\bar{x}_t) \le (1-\alpha_t)h(\bar{x}_{t-1}) + \alpha_t h(x_t)$. Adding up the above two inequalities and using the definition of Ψ in (4.2.1), we obtain the result. ■

In the sequel, we still use δ_t, $t \ge 1$, to denote the error for the computation of the subgradient of f, i.e.,

$$\delta_t \equiv G(\underline{x}_t, \xi_t) - f'(\underline{x}_t), \quad \forall t \ge 1, \tag{4.2.8}$$

where $f'(\underline{x}_t)$ represents an arbitrary element of $\partial f(\underline{x}_t)$ wherever it appears.

The following proposition establishes a basic recursion for the generic stochastic accelerated gradient descent method.

Proposition 4.2. *Let $(x_{t-1}, \bar{x}_{t-1}) \in X \times X$ be given. Also let $(\underline{x}_t, x_t, \bar{x}_t) \in X \times X \times X$ be computed according to (4.2.5), (4.2.6), and (4.2.7). If*

$$\frac{q_t(1-\alpha_t)}{\alpha_t(1-q_t)} = \frac{1}{1+\mu\gamma_t}, \tag{4.2.9}$$

$$1 + \mu\gamma_t > L\alpha_t\gamma_t, \tag{4.2.10}$$

then for any $x \in X$, we have

$$\Psi(\bar{x}_t) \le (1-\alpha_t)\Psi(\bar{x}_{t-1}) + \alpha_t l_\Psi(\underline{x}_t, x) + \tfrac{\alpha_t}{\gamma_t}[V(x_{t-1}, x) - (1+\mu\gamma_t)V(x_t, x)] + \Delta_t(x), \tag{4.2.11}$$

where

$$l_\Psi(\underline{x}_t, x) := f(\underline{x}_t) + \langle f'(\underline{x}_t), x - \underline{x}_t\rangle + h(x) + \mu V(\underline{x}_t, x), \tag{4.2.12}$$

$$\Delta_t(x) := \frac{\alpha_t\gamma_t(M + \|\delta_t\|_*)^2}{2[1+\mu\gamma_t - L\alpha_t\gamma_t]} + \alpha_t\langle \delta_t, x - x_{t-1}^+\rangle, \tag{4.2.13}$$

$$x_{t-1}^+ := \frac{\mu\gamma_t}{1+\mu\gamma_t}\underline{x}_t + \frac{1}{1+\mu\gamma_t}x_{t-1}. \tag{4.2.14}$$

Proof. Denote $d_t := \bar{x}_t - \underline{x}_t$. By Lemma 4.4 (with $z = \underline{x}_t$), we have

$$\Psi(\bar{x}_t) \le (1-\alpha_t)\Psi(\bar{x}_{t-1}) + \alpha_t[f(\underline{x}_t) + \langle f'(\underline{x}_t), x_t - \underline{x}_t\rangle + h(x_t)] + \tfrac{L}{2}\|d_t\|^2 + M\|d_t\|.$$

Moreover, by (4.2.6) and Lemma 3.5, we have

$$\gamma_t[\langle G(\underline{x}_t, \xi_t), x_t - \underline{x}_t\rangle + h(x_t) + \mu V(\underline{x}_t, x_t)] + V(x_{t-1}, x_t)$$
$$\le \gamma_t[\langle G(\underline{x}_t, \xi_t), x - \underline{x}_t\rangle + h(x) + \mu V(\underline{x}_t, x)] + V(x_{t-1}, x) - (1+\mu\gamma_t)V(x_t, x)$$

for any $x \in X$. Using the fact that $G(\underline{x}_t, \xi_t) = f'(\underline{x}_t) + \delta_t$ and combining the above two inequalities, we obtain

$$\begin{aligned}\Psi(\bar{x}_t) &\le (1-\alpha_t)\Psi(\bar{x}_{t-1}) + \alpha_t[f(\underline{x}_t) + \langle f'(\underline{x}_t), x - \underline{x}_t\rangle + h(x) + \mu V(\underline{x}_t, x)]\\ &\quad + \tfrac{\alpha_t}{\gamma_t}[V(x_{t-1},x) - (1+\mu\gamma_t)V(x_t,x) - V(x_{t-1},x_t) - \mu\gamma_t V(\underline{x}_t, x_t)]\\ &\quad + \tfrac{L}{2}\|d_t\|^2 + M\|d_t\| + \alpha_t\langle\delta_t, x - x_t\rangle. \end{aligned} \tag{4.2.15}$$

Observing that by (3.3.11), (4.2.9), and (4.2.14), we have

$$\begin{aligned} d_t &= \alpha_t[x_t - \tfrac{\alpha_t - q_t}{\alpha_t(1-q_t)}\underline{x}_t - \tfrac{q_t(1-\alpha_t)}{\alpha_t(1-q_t)}x_{t-1}]\\ &= \alpha_t[x_t - \tfrac{\mu\gamma_t}{1+\mu\gamma_t}\underline{x}_t - \tfrac{1}{1+\mu\gamma_t}x_{t-1}]\\ &= \alpha_t[x_t - x_{t-1}^+]. \end{aligned} \tag{4.2.16}$$

It then follows from this observation, the strong convexity of V, the convexity of $\|\cdot\|^2$, and (4.2.14) that

$$\begin{aligned} V(x_{t-1},x_t) + \mu\gamma_t V(\underline{x}_t, x_t) &\ge \tfrac{1}{2}\left[\|x_t - x_{t-1}\|^2 + \mu\gamma_t\|x_t - \underline{x}_t\|^2\right]\\ &\ge \tfrac{1+\mu\gamma_t}{2}\|x_t - \tfrac{1}{1+\mu\gamma_t}x_{t-1} - \tfrac{\mu\gamma_t}{1+\mu\gamma_t}\underline{x}_t\|^2 \qquad (4.2.17)\\ &= \tfrac{1+\mu\gamma_t}{2}\|x_t - x_{t-1}^+\|^2\\ &= \tfrac{1+\mu\gamma_t}{2\alpha_t^2}\|d_t\|^2. \qquad (4.2.18) \end{aligned}$$

It also follows from (4.2.16) that

$$\alpha_t\langle\delta_t, x - x_t\rangle = \langle\delta_t, d_t\rangle + \alpha_t\langle\delta_t, x - x_{t-1}^+\rangle. \tag{4.2.19}$$

Using the above two relations in (4.2.15), we have

$$\begin{aligned}\Psi(\bar{x}_t) &\le (1-\alpha_t)\Psi(\bar{x}_{t-1}) + \alpha_t[f(\underline{x}_t) + \langle f'(\underline{x}_t), x - \underline{x}_t\rangle + h(x) + \mu V(\underline{x}_t, x)]\\ &\quad + \tfrac{\alpha_t}{\gamma_t}[V(x_{t-1},x) - (1+\mu\gamma_t)V(x_t,x)]\\ &\quad - \tfrac{1+\mu\gamma_t - L\alpha_t\gamma_t}{2\alpha_t\gamma_t}\|d_t\|^2 + (M + \|\delta_t\|_*)\|d_t\| + \alpha_t\langle\delta_t, x - x_{t-1}^+\rangle. \end{aligned}$$

The result then immediately follows from the above inequality, the definition of l_Ψ, and the simple inequality in (3.1.6). ■

Proposition 4.3 below follows from Proposition 4.2 by taking summation over the relations in (4.2.11).

Proposition 4.3. *Let $\{\bar{x}_t\}_{t\ge1}$ be computed by the stochastic accelerated gradient descent algorithm. Also assume that $\{\alpha_t\}_{t\ge1}$ and $\{\gamma_t\}_{t\ge1}$ are chosen such that relations (4.2.9) and (4.2.10) hold. We have*

$$\begin{aligned}\Psi(\bar{x}_k) - \Gamma_k\textstyle\sum_{t=1}^k[\frac{\alpha_t}{\Gamma_t}l_\Psi(\underline{x}_t, x)] &\le \Gamma_k(1-\alpha_1)\Psi(\bar{x}_1) + \Gamma_k\textstyle\sum_{t=1}^k\frac{\alpha_t}{\gamma_t\Gamma_t}[V(x_{t-1},x)\\ &\quad - (1+\mu\gamma_t)V(x_t,x)] + \Gamma_k\textstyle\sum_{t=1}^k\frac{\Delta_t(x)}{\Gamma_t}, \end{aligned} \tag{4.2.20}$$

for any $x \in X$ and any $t \ge 1$, where $l_\Psi(z,x)$ and $\Delta_t(x)$ are defined in (4.2.12) and (4.2.13), respectively, and

$$\Gamma_t := \begin{cases} 1, & t = 1, \\ (1-\alpha_t)\Gamma_{t-1}, & t \geq 2. \end{cases} \tag{4.2.21}$$

Proof. Dividing both sides of relation (4.2.11) by Γ_t, and using the definition of Γ_t in (4.2.21), we have

$$\begin{aligned}\tfrac{1}{\Gamma_t}\Psi(\bar{x}_t) \leq{}& \tfrac{1}{\Gamma_{t-1}}\Psi(\bar{x}_{t-1}) + \tfrac{\alpha_t}{\Gamma_t} l_\Psi(\underline{x}_t, x) \\ &+ \tfrac{\alpha_t}{\Gamma_t\gamma_t}[V(x_{t-1},x) - (1+\mu\gamma_t)V(x_t,x)] + \tfrac{\Delta_t(x)}{\Gamma_t}.\end{aligned}$$

We obtain the result by summing up over the above inequalities. ■

Theorem 4.4 below summarizes the main convergence properties of the generic stochastic accelerated gradient descent method.

Theorem 4.4. *Assume that $\{q_t\}$, $\{\alpha_t\}$, and $\{\gamma_t\}$ are chosen such that $\alpha_1 = 1$ and relations (4.2.9) and (4.2.10) hold. Also assume that $\{\alpha_t\}_{t\geq 1}$ and $\{\gamma_t\}_{t\geq 1}$ are chosen such that*

$$\tfrac{\alpha_t}{\gamma_t\Gamma_t} \leq \tfrac{\alpha_{t-1}(1+\mu\gamma_{t-1})}{\gamma_{t-1}\Gamma_{t-1}}, \tag{4.2.22}$$

where Γ_t is defined in (4.2.21).

(a) Under Assumption 3, we have

$$\mathbb{E}[\Psi(\bar{x}_k) - \Psi^*] \leq \mathrm{B}_e(k) := \tfrac{\Gamma_k}{\gamma_1}V(x_0,x^*) + \Gamma_k\textstyle\sum_{t=1}^k \tfrac{\alpha_t\gamma_t(M^2+\sigma^2)}{\Gamma_t(1+\mu\gamma_t - L\alpha_t\gamma_t)}, \tag{4.2.23}$$

for any $t \geq 1$, where x^* is an arbitrary optimal solution of (4.2.1).

(b) Under Assumption 4, we have

$$\mathrm{Prob}\left\{\Psi(\bar{x}_k) - \Psi^* \geq \mathrm{B}_e(k) + \lambda\,\mathrm{B}_p(k)\right\} \leq \exp\{-\lambda^2/3\} + \exp\{-\lambda\}, \tag{4.2.24}$$

for any $\lambda > 0$ and $k \geq 1$, where

$$\mathrm{B}_p(k) := \sigma\Gamma_k R_X(x^*)\left(\textstyle\sum_{t=1}^k \tfrac{\alpha_t^2}{\Gamma_t^2}\right)^{1/2} + \Gamma_k\textstyle\sum_{t=1}^k \tfrac{\alpha_t\gamma_t\sigma^2}{\Gamma_t(1+\mu\gamma_t - L\alpha_t\gamma_t)}, \tag{4.2.25}$$

$$R_X(x^*) := \max_{x\in X}\|x - x^*\|. \tag{4.2.26}$$

(c) If X is compact and the condition (4.2.22) is replaced by

$$\tfrac{\alpha_t}{\gamma_t\Gamma_t} \geq \tfrac{\alpha_{t-1}}{\gamma_{t-1}\Gamma_{t-1}}, \tag{4.2.27}$$

then parts (a) and (b) still hold by simply replacing the first term in the definition of $\mathrm{B}_e(k)$ with $\alpha_k D_X/\gamma_k$, where D_X is defined in (3.2.4).

Proof. We first show part (a). Observe that by the definition of Γ_t in (4.2.21) and the fact that $\alpha_1 = 1$, we have

$$\textstyle\sum_{t=1}^k \tfrac{\alpha_t}{\Gamma_t} = \tfrac{\alpha_1}{\Gamma_1} + \sum_{t=2}^k \tfrac{1}{\Gamma_t}\left(1 - \tfrac{\Gamma_t}{\Gamma_{t-1}}\right) = \tfrac{1}{\Gamma_1} + \sum_{t=2}^k\left(\tfrac{1}{\Gamma_t} - \tfrac{1}{\Gamma_{t-1}}\right) = \tfrac{1}{\Gamma_k}. \tag{4.2.28}$$

Using the previous observation and (4.2.2), we obtain

$$\Gamma_k \sum_{t=1}^k \left[\tfrac{\alpha_t}{\Gamma_t} l_\Psi(\underline{x}_t, x)\right] \le \Gamma_k \sum_{t=1}^k \left[\tfrac{\alpha_t}{\Gamma_t} \Psi(x)\right] = \Psi(x), \quad \forall x \in X. \tag{4.2.29}$$

Moreover, it follows from the condition (4.2.22) that

$$\Gamma_k \sum_{t=1}^k \tfrac{\alpha_t}{\gamma_t \Gamma_t} [V(x_{t-1}, x) - (1+\mu\gamma_t) V(x_t, x)] \le \Gamma_k \tfrac{\alpha_1}{\gamma_1 \Gamma_1} V(x_0, x) = \tfrac{\Gamma_k}{\gamma_1} V(x_0, x), \tag{4.2.30}$$

where the last inequality follows from the facts that $\Gamma_1 = 1$ and that $V(x_t, x) \ge 0$. Using the fact that $V(x_t, x) \ge 0$ and replacing the above two bounds into (4.2.20), we have

$$\Psi(\bar{x}_k) - \Psi(x) \le \tfrac{\Gamma_k}{\gamma_1} V(x_0, x) - \tfrac{\alpha_k (1+\mu\gamma_k)}{\gamma_k} V(x_k, x) + \Gamma_k \sum_{t=1}^k \tfrac{\Delta_t(x)}{\Gamma_t}, \quad \forall x \in X, \tag{4.2.31}$$

where $\Delta_t(x)$ is defined in (4.2.13). Observe that the triple $(\underline{x}_t, x_{t-1}, \bar{x}_{t-1})$ is a function of the history $\xi_{[t-1]} := (\xi_1, \ldots, \xi_{t-1})$ of the generated random process and hence is random. Taking expectations on both sides of (4.2.31) and noting that under Assumption 3, $\mathbb{E}[\|\delta_t\|_*^2] \le \sigma^2$, and

$$\mathbb{E}_{|\xi_{[t-1]}}[\langle \delta_t, x^* - x_{t-1}^+ \rangle] = 0, \tag{4.2.32}$$

we have

$$\begin{aligned}
\mathbb{E}\left[\Psi(\bar{x}_k) - \Psi^*\right] &\le \tfrac{\Gamma_k}{\gamma_1} V(x_0, x^*) + \Gamma_k \sum_{t=1}^k \tfrac{\alpha_t \gamma_t \mathbb{E}\left[(M + \|\delta_t\|_*)^2\right]}{2\Gamma_t (1+\mu\gamma_t - L\alpha_t\gamma_t)} \\
&\le \tfrac{\Gamma_k}{\gamma_1} V(x_0, x^*) + \Gamma_k \sum_{t=1}^k \tfrac{\alpha_t \gamma_t (M^2 + \sigma^2)}{\Gamma_t (1+\mu\gamma_t - L\alpha_t\gamma_t)}.
\end{aligned}$$

To show part (b), let us denote $\zeta_t := \Gamma_t^{-1} \alpha_t \langle \delta_t, x^* - x_{t-1}^+ \rangle$. Clearly, from the definition of $R_X(x^*)$ given by (4.2.26), we have $\|x^* - x_{t-1}^+\| \le R_X(x^*)$, which together with Assumption 4 imply that

$$\begin{aligned}
&\mathbb{E}_{|\xi_{[t-1]}}\left[\exp\{\zeta_t^2 / [\Gamma_t^{-1} \alpha_t \sigma R_X(x^*)]^2\}\right] \\
&\le \mathbb{E}_{|\xi_{[t-1]}}\left[\exp\{(\|\delta_t\|_* \|x^* - x_{t-1}^+\|)^2 / [\sigma R_X(x^*)]^2\}\right] \\
&\le \mathbb{E}_{|\xi_{[t-1]}}\left[\exp\{(\|\delta_t\|_*)^2 / \sigma^2\}\right] \le \exp(1).
\end{aligned}$$

Moreover, observe that $\{\zeta_t\}_{t\ge1}$ is a martingale-difference. Using the previous two observations and Lemma 4.1, we have

$$\forall \lambda \ge 0 : \text{Prob}\left\{\sum_{t=1}^k \zeta_t > \lambda\, \sigma R_X(x^*) \left[\sum_{t=1}^k (\Gamma_t^{-1} \alpha_t)^2\right]^{1/2}\right\} \le \exp\{-\lambda^2/3\}. \tag{4.2.33}$$

Also observe that under Assumption 4, $\mathbb{E}_{|\xi_{[t-1]}}\left[\exp\{\|\delta_t\|_*^2/\sigma^2\}\right] \le \exp\{1\}$. Setting

$$\pi_t^2 = \frac{\alpha_t^2}{\Gamma_t(\mu + \gamma_t - L\alpha_t^2)} \quad \text{and} \quad \theta_t = \frac{\pi_t^2}{\sum_{t=1}^k \pi_t^2},$$

we have

$$\exp\left\{\sum_{t=1}^k \theta_t (\|\delta_t\|_*^2/\sigma^2)\right\} \le \sum_{t=1}^k \theta_t \exp\{\|\delta_t\|_*^2/\sigma^2\},$$

whence, taking expectations,

$$\mathbb{E}\left[\exp\left\{\sum_{t=1}^{k}\pi_t^2\|\delta_t\|_*^2/\left(\sigma^2\sum_{t=1}^{t}\pi_t^2\right)\right\}\right] \le \exp\{1\}.$$

It then follows from Markov's inequality that

$$\forall\lambda \ge 0 : \text{Prob}\left\{\sum_{t=1}^{t}\pi_t^2\|\delta_t\|_*^2 > (1+\lambda)\sigma^2\sum_{t=1}^{k}\pi_t^2\right\} \le \exp\{-\lambda\}. \tag{4.2.34}$$

Combining (4.2.31), (4.2.33), and (4.2.34), and rearranging the terms, we obtain (4.2.24).

Finally, observing that by the condition (4.2.27), the fact that $V(u,x) \ge 0$ and the definition of D_X,

$$\begin{aligned}
&\Gamma_k \sum_{t=1}^{k} \tfrac{\alpha_t}{\gamma_t\Gamma_t}[V(x_{t-1},x) - V(x_t,x)] \\
&\le \Gamma_k\left[\tfrac{\alpha_1}{\gamma_1\Gamma_1}D_X + \sum_{t=2}^{k}\left(\tfrac{\alpha_t}{\gamma_t\Gamma_t} - \tfrac{\alpha_{t-1}}{\gamma_{t-1}\Gamma_{t-1}}\right)D_X - \tfrac{\alpha_k}{\gamma_k\Gamma_k}V(x_k,x)\right] \\
&\le \tfrac{\alpha_k}{\gamma_k}D_X - \tfrac{\alpha_k}{\gamma_k}V(x_k,x) \le \tfrac{\alpha_k}{\gamma_k}D_X,
\end{aligned} \tag{4.2.35}$$

we can show part (c) similar to parts (a) and (b) by replacing the bound in (4.2.30) with the one given above. ■

4.2.1 Problems Without Strong Convexity

In this subsection, we consider problem (4.2.1), but now the objective function f is not necessarily strongly convex. We present the stochastic accelerated gradient descent methods for solving these problems by setting $\mu = 0$ and properly choosing the stepsize parameters $\{\alpha_t\}_{t\ge1}$ and $\{\gamma_t\}_{t\ge1}$ in the generic algorithmic framework.

Observe that if μ is set to 0, then by (4.2.9) we have $q_t = \alpha_t$. Hence the identities (4.2.5) and (4.2.6), respectively, reduce to

$$\underline{x}_t = (1-\alpha_t)\bar{x}_{t-1} + \alpha_t x_{t-1}, \tag{4.2.36}$$

$$x_t = \arg\min_{x\in X}\{\gamma_t[\langle G(\underline{x}_t,\xi_t),x\rangle + h(x)] + V(x_{t-1},x)\}. \tag{4.2.37}$$

We will study and compare two stochastic accelerated gradient descent algorithms, each of them employed with a different *stepsize policy to choose* $\{\alpha_t\}_{t\ge1}$ *and* $\{\gamma_t\}_{t\ge1}$.

The first stepsize policy and its associated convergence results stated below follows as an immediate consequence of Theorem 4.4.

Proposition 4.4. *Let*

$$\alpha_t = \tfrac{2}{t+1} \quad \textit{and} \quad \gamma_t = \gamma t, \ \forall t \ge 1, \tag{4.2.38}$$

for some $\gamma \le 1/(4L)$. *Then, under Assumption 3, we have* $\mathbb{E}[\Psi(\bar{x}_t) - \Psi^*] \le \mathrm{C}_{e,1}(t)$, $\forall t \ge 1$, *where*

$$\mathrm{C}_{e,1}(k) \equiv \mathrm{C}_{e,1}(x_0, \gamma, k) := \tfrac{2V(x_0,x^*)}{\gamma k(k+1)} + \tfrac{4\gamma(M^2+\sigma^2)(k+1)}{3}. \tag{4.2.39}$$

If in addition, Assumption 4 holds, then, $\forall \lambda > 0, \forall k \ge 1$,

$$\mathrm{Prob}\left\{\Psi(\bar{x}_k) - \Psi^* > \mathrm{C}_{e,1}(k) + \lambda\, \mathrm{C}_{p,1}(k)\right\} \le \exp\{-\lambda^2/3\} + \exp\{-\lambda\}, \tag{4.2.40}$$

where

$$\mathrm{C}_{p,1}(k) \equiv \mathrm{C}_{p,1}(\gamma, k) := \tfrac{2\sigma R_X(x^*)}{\sqrt{3k}} + \tfrac{4\sigma^2\gamma(k+1)}{3}. \tag{4.2.41}$$

Proof. Clearly, by the definition of Γ_t in (4.2.21), the stepsize policy (4.2.38), and the facts that $\gamma \le 1/(4L)$ and $\mu = 0$, we have

$$\Gamma_t = \tfrac{2}{t(t+1)}, \quad \tfrac{\alpha_t}{\Gamma_t} = t, \quad 1 + \mu\gamma_t - L\alpha_t\gamma_t = 1 - \tfrac{2\gamma L t}{t+1} \ge \tfrac{1}{2}, \quad \forall t \ge 1, \tag{4.2.42}$$

and hence the specification of α_t and γ_t in (4.2.38) satisfies conditions (4.2.10) and (4.2.22). It can also be easily seen from the previous result and (4.2.38) that

$$\textstyle\sum_{t=1}^k \tfrac{\alpha_t\gamma_t}{\Gamma_t(1-L\alpha_t\gamma_t)} \le 2\sum_{t=1}^k \gamma t^2 = \tfrac{\gamma k(k+1)(2k+1)}{3} \le \tfrac{2\gamma k(k+1)^2}{3}, \tag{4.2.43}$$

$$\textstyle\sum_{t=1}^k (\Gamma_t^{-1}\alpha_t)^2 = \sum_{t=1}^k t^2 = \tfrac{k(k+1)(2k+1)}{6} \le \tfrac{k(k+1)^2}{3}. \tag{4.2.44}$$

Now let $\mathrm{B}_e(k)$ and $\mathrm{B}_p(k)$ be defined in (4.2.23) and (4.2.25) respectively. By (4.2.42), (4.2.43), and (4.2.44), we have

$$\mathrm{B}_e(k) \le \tfrac{\Gamma_k}{\gamma} V(x_0, x^*) + \tfrac{2(M^2+\sigma^2)\Gamma_k \gamma k(k+1)^2}{3} = \mathrm{C}_{e,1}(k),$$

$$\mathrm{B}_p(k) \le \Gamma_k[\sigma R_X(x^*)\left(\tfrac{k(k+1)^2}{3}\right)^{1/2} + \tfrac{2\sigma^2\gamma k(k+1)^2}{3}] = \mathrm{C}_{p,1}(k),$$

which, in view of Theorem 4.4, clearly imply our results. ■

We now briefly discuss how to derive the optimal rate of convergence. Given a fixed in advance number of iterations k, let us suppose that the stepsize parameters $\{\alpha_t\}_{t=1}^k$ and $\{\gamma_t\}_{t=1}^k$ are set to (4.2.38) with

$$\gamma = \gamma_k^* = \min\left\{\tfrac{1}{4L}, \left[\tfrac{3V(x_0,x^*)}{2(M^2+\sigma^2)k(k+1)^2}\right]^{1/2}\right\}. \tag{4.2.45}$$

Note that γ_N^* in (4.2.45) is obtained by minimizing $\mathrm{C}_{e,1}(N)$ (cf. (4.2.39)) with respect to γ over the interval $[0, 1/(4L)]$. Then, it can be shown from (4.2.39) and (4.2.41) that

$$\mathrm{C}_{e,1}(x_0, \gamma_k^*, k) \le \tfrac{8LV(x_0,x^*)}{k(k+1)} + \tfrac{4\sqrt{2(M^2+\sigma^2)V(x_0,x^*)}}{\sqrt{3k}} =: \mathrm{C}_{e,1}^*(N), \tag{4.2.46}$$

$$\mathrm{C}_{p,1}(\gamma_N^*, N) \le \tfrac{2\sigma R_X(x^*)}{\sqrt{3k}} + \tfrac{2\sigma\sqrt{6V(x_0,x^*)}}{3\sqrt{k}} =: \mathrm{C}_{p,1}^*(k). \tag{4.2.47}$$

Indeed, let

$$\bar{\gamma} := \left[\frac{3V(x_0,x^*)}{2(M^2+\sigma^2)k(k+1)^2}\right]^{1/2}.$$

According to the relation (4.2.45), we have $\gamma_k^* \le \min\{1/(4L), \bar{\gamma}\}$. Using these facts and (4.2.39) we obtain

$$\begin{aligned} \mathrm{C}_{e,1}(x_0, \gamma_k^*, k) &\le \tfrac{8LV(x_0,x^*)}{k(k+1)} + \tfrac{2V(x_0,x^*)}{\bar{\gamma}k(k+1)} + \tfrac{4\bar{\gamma}(M^2+\sigma^2)(k+1)}{3} \\ &= \tfrac{8LV(x_0,x^*)}{k(k+1)} + \tfrac{4\sqrt{2(M^2+\sigma^2)V(x_0,x^*)}}{\sqrt{3k}}. \end{aligned}$$

Also by (4.2.41), we have

$$\mathrm{C}_{p,1}(k) \le \tfrac{2\sigma R_X(x^*)}{\sqrt{3k}} + \tfrac{4\sigma^2\bar{\gamma}(k+1)}{3},$$

which leads to (4.2.47).

Hence, by Proposition 4.4, we have, under Assumption 3, $\mathbb{E}[\Psi(\bar{x}_k) - \Psi^*] \le \mathrm{C}_{e,1}^*(k)$, which gives us an optimal expected rate of convergence for solving problems without strong convexity. Moreover, if Assumption 4 holds, then $\mathrm{Prob}\{\Psi(\bar{x}_k) - \Psi^* \ge \mathrm{C}_{e,1}^*(k) + \lambda \mathrm{C}_{p,1}^*(k)\} \le \exp(-\lambda^2/3) + \exp(-\lambda)$. It is worth noting that both $\mathrm{C}_{p,1}^*$ and $\mathrm{C}_{e,1}^*$ are in the same order of magnitude, i.e., $\mathcal{O}(1/\sqrt{k})$. Observe that we need to estimate a bound on $V(x_0,x^*)$ to implement this stepsize policy since $V(x_0,x^*)$ is usually unknown.

One possible drawback of the stepsize policy (4.2.38) with $\gamma = \gamma_k^*$ is the need of fixing k in advance. In Proposition 4.5, we propose an alternative stepsize policy which does not require to fix the number of iterations k. Note that, to apply this stepsize policy properly, we need to assume that all the iterates $\{x_k\}_{k\ge1}$ stay in a bounded set.

Proposition 4.5. *Assume that X is compact. Let $\{\bar{x}_t\}_{t\ge1}$ be computed by the stochastic accelerated gradient descent method with*

$$\alpha_t = \tfrac{2}{t+1} \text{ and } \tfrac{1}{\gamma_t} = \tfrac{2L}{t} + \gamma\sqrt{t}, \ \forall t \ge 1, \tag{4.2.48}$$

for some $\gamma > 0$. Then, under Assumption 3, we have $\mathbb{E}[\Psi(\bar{x}_k) - \Psi^*] \le \mathrm{C}_{e,2}(k)$, $\forall k \ge 1$, where

$$\mathrm{C}_{e,2}(k) \equiv \mathrm{C}_{e,2}(\gamma,k) := \tfrac{4LD_X}{k(k+1)} + \tfrac{2\gamma D_X}{\sqrt{k}} + \tfrac{4\sqrt{2}}{3\gamma\sqrt{k}}(M^2+\sigma^2), \tag{4.2.49}$$

and D_X is defined in (3.2.4). If in addition, Assumption 4 holds, then, $\forall \lambda > 0$, $\forall k \ge 1$,

$$\mathrm{Prob}\left\{\Psi(\bar{x}_k) - \Psi^* > \mathrm{C}_{e,2}(k) + \lambda\,\mathrm{C}_{p,2}(k)\right\} \le \exp\{-\lambda^2/3\} + \exp\{-\lambda\}, \tag{4.2.50}$$

where

$$\mathrm{C}_{p,2}(k) \equiv \mathrm{C}_{p,2}(\gamma,k) := \tfrac{2\sigma D_X}{\sqrt{3k}} + \tfrac{4\sqrt{2}\sigma^2}{3\gamma\sqrt{k}}. \tag{4.2.51}$$

Proof. Clearly, by the definition of Γ_t in (4.2.21), the stepsize policy (4.2.48), and the fact that $\mu = 0$, we have

$$\Gamma_t = \tfrac{2}{t(t+1)}, \ \tfrac{\alpha_t}{\gamma_t \Gamma_t} = \tfrac{2L}{t} + \gamma\sqrt{t}, \ \tfrac{1}{\gamma_t} - L\alpha_t = \tfrac{2L}{t} + \gamma\sqrt{t} - \tfrac{2L}{t+1} \geq \gamma\sqrt{t}, \tag{4.2.52}$$

and hence the specification of α_t and γ_t in (4.2.48) satisfies conditions (4.2.10) and (4.2.27). It can also be easily seen from the previous observations and (4.2.48) that

$$\textstyle\sum_{t=1}^k (\Gamma_t^{-1}\alpha_t)^2 = \sum_{t=1}^k t^2 = \tfrac{k(k+1)(2k+1)}{6} \leq \tfrac{k(k+1)^2}{3}, \tag{4.2.53}$$

$$\textstyle\sum_{t=1}^k \tfrac{\alpha_t\gamma_t}{\Gamma_t(1-L\alpha_t\gamma_t)} = \sum_{t=1}^k \tfrac{t}{1/\gamma_t - L\alpha_t} \leq \tfrac{1}{\gamma}\sum_{t=1}^k \sqrt{t} \leq \tfrac{1}{\gamma}\int_1^{k+1}\sqrt{x}dx \leq \tfrac{2}{3\gamma}(k+1)^{3/2}. \tag{4.2.54}$$

Now let $\mathrm{B}_e'(k)$ be obtained by replacing the first term in the definition of $\mathrm{B}_e(k)$ in (4.2.23) with $\alpha_k D_X/\gamma_k$ and $\mathrm{B}_p(k)$ be defined in (4.2.25). By (4.2.48), (4.2.52), (4.2.53), and (4.2.54), we have

$$\mathrm{B}_e'(k) \leq \tfrac{\alpha_k D_X}{\gamma_k} + \tfrac{2\Gamma_k(M^2+\sigma^2)}{3\gamma}(k+1)^{3/2} \leq \mathrm{C}_{e,2}(k),$$

$$\mathrm{B}_p(k) \leq \Gamma_k\left[\sigma R_X(x^*)\left(\tfrac{k(k+1)^2}{3}\right)^{1/2} + \tfrac{2\Gamma_k\sigma^2}{3\gamma}(k+1)^{3/2}\right] \leq \mathrm{C}_{p,2}(t),$$

which, in view of Theorem 4.4(c), then clearly imply our results. ■

Clearly, if we set γ in the stepsize policy (4.2.48) as

$$\gamma = \tilde{\gamma}^* := \left[\tfrac{2\sqrt{2}(M^2+\sigma^2)}{3D_X}\right]^{1/2},$$

then by (4.2.49), we have

$$\mathbb{E}[\Psi(\bar{x}_k) - \Psi^*] \leq \tfrac{4L\bar{V}(x^*)}{k(k+1)} + 4\left[\tfrac{2\sqrt{2}D_X(M^2+\sigma^2)}{3k}\right]^{1/2} =: \mathrm{C}_{e,2}^*,$$

which also gives an optimal expected rate of convergence for problems without strong convexity. As discussed before, one obvious advantage of the stepsize policy (4.2.48) with $\gamma = \tilde{\gamma}^*$ over the one in (4.2.38) with $\gamma = \gamma_k^*$ is that the former one does not require the knowledge of k. Hence, it allows possibly earlier termination of the algorithm, especially when coupled with the validation to be discussed in Subsection 4.2.4. Note, however, that the convergence rate $\mathrm{C}_{e,1}^*$ depends on $V(x_0, x^*)$, which can be significantly smaller than D_X in $\mathrm{C}_{e,2}^*$ given a good starting point $x_0 \in X$.

4.2.2 Nonsmooth Strongly Convex Problems

The objective of this subsection is to present a stochastic accelerated gradient descent algorithm for solving strongly convex problems with $\mu > 0$. We start by presenting this algorithm with a simple stepsize policy and discussing its convergence properties. It is worth noting that this stepsize policy does not depend on σ, M and $V(x_0,x^*)$, and hence it is quite convenient for implementation.

Proposition 4.6. *Let $\{\bar{x}_t\}_{t\geq 1}$ be computed by the stochastic accelerated gradient descent algorithm with*

$$\alpha_t = \tfrac{2}{t+1},\ \tfrac{1}{\gamma_t} = \tfrac{\mu(t-1)}{2} + \tfrac{2L}{t},\ \text{and}\ q_t = \tfrac{\alpha_t}{\alpha_t+(1-\alpha_t)(1+\mu\gamma_t)}\ \ \forall t\geq 1. \tag{4.2.55}$$

Then under Assumption 3, we have

$$\mathbb{E}[\Psi(\bar{x}_k) - \Psi^*] \leq \mathrm{D}_e(k) := \tfrac{4LV(x_0,x^*)}{k(k+1)} + \tfrac{4(M^2+\sigma^2)}{\mu(k+1)},\ \ \forall t\geq 1. \tag{4.2.56}$$

If in addition, Assumption 4 holds, then, $\forall\lambda > 0, \forall t\geq 1$,

$$\mathrm{Prob}\left\{\Psi(\bar{x}_k) - \Psi^* \geq \mathrm{D}_e(k) + \lambda\mathrm{D}_p(k)\right\} \leq \exp\{-\lambda^2/3\} + \exp\{-\lambda\}, \tag{4.2.57}$$

where

$$\mathrm{D}_p(k) := \tfrac{2\sigma R_X(x^*)}{\sqrt{3k}} + \tfrac{4\sigma^2}{\mu(k+1)}. \tag{4.2.58}$$

Proof. Clearly, by the definition of Γ_t in (4.2.21) and the stepsize policy (4.2.55), we have

$$\Gamma_t = \tfrac{2}{t(t+1)},\ \ \tfrac{\alpha_t}{\Gamma_t} = t, \tag{4.2.59}$$

$$\begin{aligned}
\tfrac{\alpha_t}{\gamma_t\Gamma_t} &= t\left[\tfrac{\mu(t-1)}{2} + \tfrac{2L}{t}\right] = \tfrac{\mu t(t-1)}{2} + 2L,\\
\tfrac{\alpha_{t-1}(1+\mu\gamma_{t-1}}{\gamma_{t-1}\Gamma_{t-1}} &= \tfrac{\alpha_{t-1}}{\gamma_{t-1}\Gamma_{t-1}} + \tfrac{\alpha_{t-1}\mu}{\Gamma_{t-1}} = \tfrac{\mu t(t-1)}{2} + 2L,\\
\tfrac{1}{\gamma_t} + \mu - L\alpha_t &= \tfrac{\mu(t+1)}{2} + \tfrac{2L}{t} - \tfrac{2L}{t} > \tfrac{\mu(t+1)}{2},
\end{aligned}$$

and hence that the specification of q_t, α_t, and γ_t in (4.2.55) satisfies conditions (4.2.9), (4.2.10), and (4.2.22). It can also be easily seen from the previous results and (4.2.55) that (4.2.44) holds and that

$$\textstyle\sum_{t=1}^k \tfrac{\alpha_t\gamma_t}{\Gamma_t[1+\mu\gamma_t - L\alpha_t\gamma_t]} = \sum_{t=1}^k \tfrac{t}{1/\gamma_t+\mu-L\alpha_t} \leq \sum_{\tau=1}^t \tfrac{2}{\mu} \leq \tfrac{2k}{\mu}. \tag{4.2.60}$$

Let $\mathrm{B}_e(k)$ and $\mathrm{B}_p(k)$ be defined in (4.2.23) and (4.2.25), respectively. By (4.2.44), (4.2.59), and (4.2.60), we have

$$\begin{aligned}
\mathrm{B}_e(k) &\leq \Gamma_k\left[\tfrac{V(x_0,x^*)}{\gamma_1} + \tfrac{2k(M^2+\sigma^2)}{\mu}\right] = \mathrm{D}_e(t),\\
\mathrm{B}_p(k) &\leq \Gamma_k\left[\sigma R_X(x^*)\left(\tfrac{k(k+1)^2}{3}\right)^{1/2} + \tfrac{2k\sigma^2}{\mu}\right] = \mathrm{D}_p(k),
\end{aligned}$$

which, in view of Theorem 4.4, clearly imply our results. ∎

We now make a few remarks about the results obtained in Proposition 4.6. First, in view of (4.2.3), the stochastic accelerated gradient descent method with the stepsize policy (4.2.55) achieves the optimal rate of convergence for solving nonsmooth strongly convex problems, i.e., for those problems without a smooth component ($L = 0$). It is also nearly optimal for solving smooth and strongly convex problems, in the sense that the second term $4(M^2+\sigma^2)/[\mu(k+1)]$ of $\mathrm{D}_e(k)$ in (4.2.56) is unimprovable. The first term of $\mathrm{D}_e(k)$ (for abbreviation, L-component) depends on the product of L and $V(x_0,x^*)$, which can be as big as $LV(x_0,x^*) \le 2(k+1)(M^2+\sigma^2)/\mu$ without affecting the rate of convergence (up to a constant factor 2). Note that in comparison with (4.2.3), it seems that it is possible to improve the L-component of $\mathrm{D}_e(k)$. We will show in the next subsection an optimal multi-epoch stochastic accelerated gradient descent algorithm for solving smooth and strongly convex problems which can substantially reduce the L-component in $\mathrm{D}_e(t)$. Another possible approach is to use a batch of samples of ξ (with increasing batch size) so that the variance to estimate the gradients will decrease exponentially fast at every iteration. We will discuss this type of approach in Sect. 5.2.3.

Second, observe that the bounds $\mathrm{D}_e(k)$ and $\mathrm{D}_p(k)$, defined in (4.2.56) and (4.2.58), respectively, are not in the same order of magnitude, that is, $\mathrm{D}_e(k) = \mathcal{O}(1/k)$ and $\mathrm{D}_p(k) = \mathcal{O}(1/\sqrt{k})$. We now discuss some consequences of this fact. By (4.2.56) and Markov's inequality, under Assumption 3, we have

$$\mathrm{Prob}\{\Psi(\bar{x}_k) - \Psi^* \ge \lambda \mathrm{D}_e(k)\} \le 1/\lambda$$

for any $\lambda > 0$ and $k \ge 1$. Hence, for a given confidence level $\Lambda \in (0,1)$, one can easily see that the number of iterations for finding an (ε,Λ)-solution $\bar{x} \in X$ such that $\mathrm{Prob}\{\Psi(\bar{x}) - \Psi^* < \varepsilon\} \ge 1-\Lambda$ can be bounded by

$$\mathcal{O}\left\{\tfrac{1}{\Lambda}\left(\sqrt{\tfrac{LV(x_0,x^*)}{\varepsilon}} + \tfrac{M^2+\sigma^2}{\mu\varepsilon}\right)\right\}. \tag{4.2.61}$$

Moreover, if Assumption 4 holds, then by setting the value of λ in (4.2.57) such that $\exp(-\lambda^2/3)+\exp(-\lambda) \le \Lambda$ and using definitions of D_e and D_p in (4.2.56) and (4.2.58), we conclude that the number of iterations for finding an (ε,Λ)-solution of (4.2.1) can be bounded by

$$\mathcal{O}\left\{\sqrt{\tfrac{LV(x_0,x^*)}{\varepsilon}} + \tfrac{M^2+\sigma^2}{\mu\varepsilon} + \tfrac{\sigma^2}{\mu\varepsilon}\log\tfrac{1}{\Lambda} + \left(\tfrac{\sigma R_X(x^*)}{\varepsilon}\log\tfrac{1}{\Lambda}\right)^2\right\}. \tag{4.2.62}$$

Note that the above iteration-complexity bound has a significantly worse dependence on ε than the one in (4.2.61), although it depends only logarithmically on $1/\Lambda$. A certain domain shrinking procedure will be incorporated in next subsection to address this issue.

4.2.3 Smooth and Strongly Convex Problems

In this subsection, we show that the generic stochastic accelerated gradient descent method can yield an optimal algorithm for solving strongly convex problems even if the problems are smooth. More specifically, we present an optimal algorithm obtained by properly restarting the algorithm presented in Sect. 4.2.1 for solving problems without strong convexity. We also discuss how to improve the large-deviation properties associated with the optimal expected rate of convergence for solving these strongly convex problems.

A Multi-Epoch Stochastic Accelerated Gradient Descent Method

(0) Let a point $p_0 \in X$, and a bound Δ_0 such that $\Psi(p_0) - \Psi(x^*) \le \Delta_0$ be given.
(1) For $s = 1, 2, \dots$

(a) Run N_s iterations of the stochastic accelerated gradient method with $x_0 = p_{k-1}$, $\alpha_t = 2/(t+1)$, $q_t = \alpha_t$, and $\gamma_t = \gamma_s t$, where

$$N_s = \left\lceil \max\left\{4\sqrt{\tfrac{2L}{\mu}}, \tfrac{64(M^2+\sigma^2)}{3\mu\Delta_0 2^{-(s)}}\right\}\right\rceil, \tag{4.2.63}$$

$$\gamma_s = \min\left\{\tfrac{1}{4L}, \left[\tfrac{3\Delta_0 2^{-(s-1)}}{2\mu(M^2+\sigma^2)N_s(N_s+1)^2}\right]^{1/2}\right\}; \tag{4.2.64}$$

(b) Set $p_s = \bar{x}_{N_s}$, where $\bar{x}_{N_s}$ is the solution obtained in Step 1.a).

We say that an epoch of the algorithm described above, referred to as *the multi-epoch stochastic accelerated gradient descent method*, occurs whenever s increments by 1. Clearly, the sth epoch of this algorithm consists of N_s iterations of the stochastic accelerated gradient descent method, which are also called iterations of the multi-epoch stochastic accelerated gradient descent method for the sake of notational convenience. The following proposition summarizes the convergence properties of the multi-epoch algorithm.

Proposition 4.7. *Let $\{p_s\}_{s\ge1}$ be computed by the multi-epoch stochastic accelerated gradient descent method. Then under Assumption 3,*

$$\mathbb{E}[\Psi(p_s) - \Psi^*] \le \Delta_s \equiv \Delta_0 2^{-s}, \quad \forall s \ge 0. \tag{4.2.65}$$

As a consequence, this multi-epoch algorithm will find a solution $\bar{x} \in X$ of (4.2.1) such that $\mathbb{E}[\Psi(\bar{x}) - \Psi^*] \le \varepsilon$ for any $\varepsilon \in (0, \Delta_0)$ in at most $S := \lceil \log \Delta_0/\varepsilon \rceil$ epochs. Moreover, the total number of iterations performed by this algorithm to find such a solution is bounded by $\mathcal{O}(\mathrm{T}_1(\varepsilon))$, where

$$\mathrm{T}_1(\varepsilon) := \sqrt{\tfrac{L}{\mu}} \max\left(1, \log \tfrac{\Delta_0}{\varepsilon}\right) + \tfrac{M^2+\sigma^2}{\mu\varepsilon}. \tag{4.2.66}$$

Proof. We first show that (4.2.65) holds by using induction. Clearly (4.2.65) holds for $s = 0$. Assume that for some $s \ge 1$, $\mathbb{E}[\Psi(p_{s-1}) - \Psi^*] \le \Delta_{s-1} = \Delta_0 2^{-(s-1)}$. This assumption together with (4.2.2) clearly imply that

$$\mathbb{E}[V(p_{s-1}, x^*)] \le \mathbb{E}\left[\tfrac{\Psi(p_{s-1}) - \Psi^*}{\mu}\right] \le \tfrac{\Delta_{s-1}}{\mu}. \tag{4.2.67}$$

Also note that by the definitions of N_s and Δ_s, respectively, in (4.2.63) and (4.2.65), we have

$$Q_1(N_s) \equiv \frac{8L\Delta_{s-1}}{\mu N_s(N_s+1)} \le \frac{8L\Delta_{s-1}}{\mu N_s^2} = \frac{16L\Delta_s}{\mu N_s^2} \le \tfrac{1}{2}\Delta_s, \tag{4.2.68}$$

$$Q_2(N_s) \equiv \frac{(M^2+\sigma^2)\Delta_{s-1}}{6\mu N_s} \le \frac{\Delta_s^2}{64}. \tag{4.2.69}$$

We then conclude from Proposition 4.4, (4.2.64), (4.2.67), (4.2.68), and (4.2.69) that

$$\begin{aligned}
\mathbb{E}[\Psi(p_s) - \Psi^*] &\le \frac{2\mathbb{E}[V(p_{s-1},x^*)]}{\gamma_s N_s(N_s+1)} + \frac{4\gamma_s(M^2+\sigma^2)(N_s+1)}{3} \\
&\le \frac{2\Delta_{s-1}}{\mu\gamma_s N_s(N_s+1)} + \frac{4\gamma_s(M^2+\sigma^2)(N_s+1)}{3} \\
&\le \max\left\{Q_1(N_s), 4\sqrt{Q_2(N_s)}\right\} + 4\sqrt{Q_2(N_s)} \le \Delta_s.
\end{aligned}$$

We have thus shown that (4.2.65) holds. Now suppose that the multi-epoch stochastic accelerated gradient descent algorithm is run for S epochs. By (4.2.65), we have $\mathbb{E}[\Psi(p_S) - \Psi^*] \le \Delta_0 2^{-S} \le \Delta_0 2^{\log \frac{\varepsilon}{\Delta_0}} = \varepsilon$. Moreover, it follows from (4.2.63) that the total number of iterations can be bounded by

$$\begin{aligned}
\textstyle\sum_{s=1}^S N_s &\le \textstyle\sum_{s=1}^S \left[4\sqrt{\frac{2L}{\mu}} + \frac{64(M^2+\sigma^2)}{3\mu\Delta_0 2^{-s}} + 1\right] \\
&= S\left(4\sqrt{\tfrac{2L}{\mu}} + 1\right) + \frac{64(M^2+\sigma^2)}{3\mu\Delta_0}\textstyle\sum_{s=1}^S 2^s \\
&\le S\left(4\sqrt{\tfrac{2L}{\mu}} + 1\right) + \frac{64(M^2+\sigma^2)}{3\mu\Delta_0}2^{S+1} \\
&\le \left(4\sqrt{\tfrac{2L}{\mu}} + 1\right)\left\lceil \log\tfrac{\Delta_0}{\varepsilon}\right\rceil + \frac{86(M^2+\sigma^2)}{\mu\varepsilon},
\end{aligned}$$

which clearly implies bound (4.2.66). ■

A few remarks about the results in Proposition 4.7 are in place. First, in view of (4.2.3), the multi-epoch stochastic accelerated gradient descent method achieves the optimal expected rate of convergence for solving strongly convex problems. Note that since Δ_0 only appears inside the logarithmic term of (4.2.66), the selection of the initial point $p_0 \in X$ has little effect on the efficiency of this algorithm. Second, suppose that we run the multi-epoch stochastic accelerated gradient descent method for $K_\Lambda := \lceil \log \Delta_0/(\Lambda\varepsilon)\rceil$ epochs for a given confidence level $\Lambda \in (0,1)$. Then, by (4.2.65) and Markov's inequality, we have $\text{Prob}[\Psi(p_{K_\Lambda}) - \Psi^* > \varepsilon] \le \mathbb{E}[\Psi(p_{K_\Lambda}) - \Psi^*]/\varepsilon \le \Lambda$, which implies that the total number of iterations performed by the multi-epoch stochastic accelerated gradient descent method for finding an (ε,Λ)-solution of (4.2.1) can be bounded by $\mathcal{O}(T_1(\Lambda\varepsilon))$. Third, similar to the single-epoch stochastic accelerated gradient descent method, under the stronger Assumption 4, we can improve the iteration complexity of the multi-epoch stochastic accelerated gradient descent method for finding an (ε,Λ)-solution of (4.2.1), so that it will depend on $\log(1/\Lambda)$ rather than $1/\Lambda$. However, such an iteration complexity will have a worse dependence on ε in the sense that it will be in the order of $1/\varepsilon^2$ rather than $1/\varepsilon$.

Let us suppose now that Assumption 4 holds. We introduce a shrinking multi-epoch stochastic accelerated gradient descent algorithm which possesses an iteration-complexity bound linearly dependent on both $\log(1/\Lambda)$ and $1/\varepsilon$, for finding an (ε,Λ)-solution of (4.2.1). It is worth noting that the stochastic accelerated gradient descent algorithms with stepsize policy either (4.2.38) or (4.2.38) can be used here to update iterate p_s of the shrinking multi-epoch stochastic accelerated gradient descent, although we focus on the former one in this chapter.

The Shrinking Multi-Epoch Stochastic Accelerated Gradient Descent

(0) Let a point $p_0 \in X$, and a bound Δ_0 such that $\Psi(p_0) - \Psi(x^*) \leq \Delta_0$ be given. Set $\bar{S} := \lceil \log(\Delta_0/\varepsilon) \rceil$ and $\lambda := \lambda(\bar{S}) > 0$ such that $\exp\{-\lambda^2/3\} + \exp\{-\lambda\} \leq \Lambda/\bar{S}$.
(1) For $s = 1,\ldots,\bar{S}$

(a) Run $\hat{N}_s$ iterations of the stochastic accelerated gradient descent algorithm for applied to $\min_{x\in\hat{X}_s}\{\Psi(x)\}$, with input $x_0 = p_{s-1}$, $\alpha_t = 2/(t+1)$, $q_t = \alpha_t$, and $\gamma_t = \hat{\gamma}_s t$, where,

$$\hat{X}_s := \left\{x \in X : V(p_{s-1},x) \leq \hat{R}^2_{s-1} := \tfrac{\Delta_0}{\mu 2^{s-1}}\right\}, \tag{4.2.70}$$

$$\hat{N}_s = \left\lceil \max\left\{4\sqrt{\tfrac{2L}{\mu}}, \tfrac{\max\{256(M^2+\sigma^2),288\lambda^2\sigma^2\}}{3\mu\Delta_0 2^{-(s+1)}}\right\}\right\rceil, \tag{4.2.71}$$

$$\hat{\gamma}_s = \min\left\{\tfrac{1}{4L}, \left[\tfrac{3\Delta_0 2^{-(s-1)}}{2\mu(M^2+\sigma^2)N_s(N_s+1)^2}\right]^{1/2}\right\}; \tag{4.2.72}$$

(b) Set $p_s = \bar{x}_{\hat{N}_s}$, where $\bar{x}_{\hat{N}_s}$ is the solution obtained in Step 1.a).

Note that in the shrinking multi-epoch stochastic accelerated gradient descent algorithm, the epoch limit $\bar{S}$ is computed for a given accuracy ε. The value of $\bar{S}$ is then used in the computation of $\lambda(\bar{S})$ and subsequently in $\hat{N}_s$ and $\hat{\gamma}_s$ (cf. (4.2.71) and (4.2.72)). This is in contrast to the multi-epoch stochastic accelerated gradient descent algorithm without shrinkage, in which the definitions of N_s and γ_s in (4.2.63) and (4.2.64) do not depend on the target accuracy ε.

The following result shows some convergence properties of the shrinking multi-epoch stochastic accelerated gradient descent algorithm.

Lemma 4.5. *Let $\{p_s\}_{s\geq 1}$ be computed by the shrinking multi-epoch stochastic accelerated gradient descent algorithm. Also for any $s \geq 0$, let $\Delta_s \equiv \Delta_0 2^{-s}$ and denote the event $A_s := \{\Psi(p_s) - \Psi^* \leq \Delta_s\}$. Then under Assumption 2,*

$$\text{Prob}[\Psi(p_s) - \Psi^* \geq \Delta_s | A_{s-1}] \leq \tfrac{\Lambda}{\bar{S}}, \;\; \forall 1 \leq s \leq \bar{S}. \tag{4.2.73}$$

Proof. By the conditional assumption in (4.2.73), we have $\Psi(p_{s-1}) - \Psi^* \leq \Delta_{s-1}$, which together with the strong convexity of f and the definition of $\hat{R}_{s-1}$ in (4.2.70) imply that

$$V(p_{s-1},x^*) \leq \tfrac{[\Psi(p_{s-1})-\Psi^*]}{\mu} \leq \tfrac{\Delta_{s-1}}{\mu} = \hat{R}^2_{s-1}. \tag{4.2.74}$$

Hence, the restricted problem $\min_{x\in\hat{X}_s}\{\Psi(x)\}$ has the same solution as (4.2.1). We then conclude from Proposition 4.4 applied to the previous restricted problem that

$$\begin{aligned}&\text{Prob}[\Psi(p_s)-\Psi^*>\hat{C}_{e,1}(\hat{N}_s)+\lambda\hat{C}_{p,1}(\hat{N}_s)|A_{s-1}]\\&\leq\exp\{-\lambda^2/3\}+\exp\{-\lambda\}\leq\tfrac{\Lambda}{\bar{S}},\end{aligned}\tag{4.2.75}$$

where

$$\hat{C}_{e,1}(\hat{N}_s):=\tfrac{2V(p_{s-1},x^*)}{\hat{\gamma}_s\hat{N}_s(\hat{N}_s+1)}+\tfrac{4\hat{\gamma}_s(M^2+\sigma^2)(\hat{N}_s+1)}{3}\quad\text{and}$$
$$\hat{C}_{p,1}(\hat{N}_s):=\frac{2\sigma R_{\hat{X}_s}(x^*)}{\sqrt{3\hat{N}_s}}+\tfrac{4\sigma^2\hat{\gamma}_s(\hat{N}_s+1)}{3}.$$

Let $Q_1(\cdot)$ and $Q_2(\cdot)$ be defined in (4.2.68) and (4.2.69), respectively. Note that by the definition of $\hat{N}_s$ in (4.2.71), we have $Q_1(\hat{N}_s)\leq\Delta_s/4$ and $Q_2(\hat{N}_s)\leq\Delta_s^2/256$. Using the previous observations, (4.2.74), and the definition of $\hat{\gamma}_s$ in (4.2.72), we obtain

$$\begin{aligned}\hat{C}_{e,1}(\hat{N}_s)&\leq\tfrac{2\Delta_{s-1}}{\mu\hat{\gamma}_s\hat{N}_s(\hat{N}_s+1)}+\tfrac{4\hat{\gamma}_s(M^2+\sigma^2)(\hat{N}_s+1)}{3}\\&\leq\max\left\{Q_1(\hat{N}_s),4\sqrt{Q_2(\hat{N}_s)}\right\}+4\sqrt{Q_2(\hat{N}_s)}\leq\tfrac{\Delta_s}{2}.\end{aligned}\tag{4.2.76}$$

Moreover, note that by the strong convexity of ν, (4.2.70), and (4.2.74), we have for any $x\in\hat{X}_s$,

$$\|x-p_{s-1}\|\leq\sqrt{2V(p_{s-1},x)}\leq\sqrt{\tfrac{2\Delta_{s-1}}{\mu}}$$

and

$$\|x-x^*\|\leq\|x-p_{s-1}\|+\|p_{s-1}-x^*\|\leq2\sqrt{\tfrac{2\Delta_{s-1}}{\mu}},$$

and hence that $R_{\hat{X}_s}(x^*)\leq2\sqrt{2\Delta_{s-1}/\mu}$, which together with (4.2.71) and (4.2.72) then imply that

$$\begin{aligned}\hat{C}_{p,1}(\hat{N}_s)&\leq4\sigma\sqrt{\tfrac{2\Delta_{s-1}}{3\mu\hat{N}_s}}+\tfrac{4\sigma^2(\hat{N}_s+1)}{3}\left[\tfrac{3\Delta_0 2^{-(s-1)}}{2\mu(M^2+\sigma^2)N_s(N_s+1)^2}\right]^{1/2}\\&\leq\tfrac{4\sigma}{3}\sqrt{\tfrac{6\Delta_{s-1}}{\mu\hat{N}_s}}+\tfrac{2\sigma}{3}\sqrt{\tfrac{6\Delta_{s-1}}{\mu\hat{N}_s}}\\&=2\sigma\sqrt{\tfrac{6\Delta_{s-1}}{\mu\hat{N}_k}}\leq\tfrac{\sqrt{\Delta_{s-1}\Delta_0 2^{-(s+1)}}}{2\lambda}=\tfrac{\Delta_s}{2\lambda}.\end{aligned}\tag{4.2.77}$$

Combining (4.2.75), (4.2.76), and (4.2.77), we obtain (4.2.73). ■

The following proposition establishes the iteration complexity of the shrinking multi-epoch stochastic accelerated gradient descent algorithm.

Proposition 4.8. *Let $\{p_s\}_{s\geq1}$ be computed by the shrinking multi-epoch accelerated stochastic gradient descent algorithm. Then under Assumption 4, we have*

$$\text{Prob}[\Psi(p_{\bar{S}})-\Psi^*>\varepsilon]\leq\Lambda.\tag{4.2.78}$$

Moreover, the total number of iterations performed by the algorithm to find such a solution is bounded by $\mathscr{O}(\mathrm{T}_2(\varepsilon,\Lambda))$, *where*

$$\mathrm{T}_2(\varepsilon,\Lambda) := \sqrt{\tfrac{L}{\mu}}\max\left(1,\log\tfrac{\Delta_0}{\varepsilon}\right) + \tfrac{M^2+\sigma^2}{\mu\varepsilon} + \left[\ln\tfrac{\log(\Delta_0/\varepsilon)}{\Lambda}\right]^2\tfrac{\sigma^2}{\mu\varepsilon}. \tag{4.2.79}$$

Proof. Denote $\Delta_s = \Delta_0 2^{-s}$. Let A_s denote the event of $\{\Psi(p_s) - \Psi^* \le \Delta_s\}$ and $\bar{A}_s$ be its complement. Clearly, we have $\mathrm{Prob}(A_0) = 1$. It can also be easily seen that

$$\begin{aligned}\mathrm{Prob}[\Psi(p_s) - > \Delta_s] &\le \mathrm{Prob}[\Psi(p_s) - \Psi^* > \Delta_s | A_{s-1}] + \mathrm{Prob}[\bar{A}_{s-1}]\\ &\le \tfrac{\Lambda}{\bar{S}} + \mathrm{Prob}[\Psi(p_{s-1}) - \Psi^* > \Delta_{s-1}],\ \forall 1 \le s \le \bar{S}\end{aligned}$$

where the last inequality follows from Lemma 4.5 and the definition of $\bar{A}_{s-1}$. Summing up both sides of the above inequality from $s=1$ to $\bar{S}$, we obtain (4.2.78). Now, by (4.2.71), the total number of stochastic accelerated gradient descent iterations can be bounded by

$$\begin{aligned}\textstyle\sum_{s=1}^{\bar{S}}\hat{N}_s &\le \textstyle\sum_{s=1}^{\bar{S}}\left\{8\sqrt{\tfrac{L}{\mu}} + \tfrac{\max\{256(M^2+\sigma^2),288\lambda^2\sigma^2\}}{3\mu\Delta_0 2^{-(s+1)}} + 1\right\}\\ &= \bar{S}\left(8\sqrt{\tfrac{L}{\mu}}+1\right) + \tfrac{\max\{256(M^2+\sigma^2),288\lambda^2\sigma^2\}}{3\mu\Delta_0}\textstyle\sum_{s=1}^{\bar{S}}2^{s+1}\\ &\le \bar{S}\left(8\sqrt{\tfrac{L}{\mu}}+1\right) + \tfrac{\max\{256(M^2+\sigma^2),288\lambda^2\sigma^2\}}{3\mu\Delta_0}2^{\bar{S}+2}.\end{aligned}$$

Using the above conclusion, the fact that $\bar{S} = \lceil\log(\Delta_0/\varepsilon)\rceil$, the observation that $\lambda = \mathscr{O}\{\ln(\bar{S}/\Lambda)\}$ and (4.2.79), we conclude that the total number of stochastic accelerated gradient descent iterations is bounded by $\mathscr{O}(\mathrm{T}_2(\varepsilon,\lambda))$. ■

While Proposition 4.8 shows the large-deviation properties of the shrinking multi-epoch stochastic accelerated gradient descent method, we can also derive the expected rate of convergence for this algorithm. For the sake of simplicity, we only consider the case when $M=0$ and $\varepsilon > 0$ is small enough such that $\hat{N}_s \ge \lambda^2 N_s, s = 1,\ldots,\bar{S}$, where N_s is defined in (4.2.63). Then by using an argument similar to the one used in the proof of Proposition 4.7, we can show that

$$\mathbb{E}[\Psi(p_s) - \Psi^*] = \mathscr{O}\left(\tfrac{\Delta_0 2^{-s}}{\lambda^{2-2^{-s}}}\right),\ s = 1,\ldots,\bar{S}.$$

Using this result and the definition of $\bar{S}$, we conclude that $\mathbb{E}[\Psi(p_{\bar{S}}) - \Psi^*] = \mathscr{O}\left(\varepsilon/\lambda^{2-\varepsilon/\Delta_0}\right)$.

4.2.4 Accuracy Certificates

In this subsection, we show that one can compute, with little additional computational effort, certain stochastic lower bounds of the optimal value of (4.2.1) during the execution of the accelerated stochastic gradient descent algorithms. These stochastic lower bounds, when grouped with certain stochastic upper bounds on the optimal value, can provide online accuracy certificates for the generated solutions.

We start by discussing the accuracy certificates for the generic stochastic accelerated gradient descent algorithm. Let $l_\Psi(z,x)$ be defined in (4.2.12) and denote

$$\text{lb}_t := \min_{x\in X}\left\{\underline{\Psi}_t(x) := \Gamma_t\Sigma_{\tau=1}^t\left[\tfrac{\alpha_\tau}{\Gamma_\tau}l_\Psi(\underline{x}_\tau,x)\right]\right\}. \tag{4.2.80}$$

By (4.2.29), the function $\underline{\Psi}_t(\cdot)$ underestimates $\Psi(\cdot)$ everywhere on X. Note, however, that lb_t is unobservable since $\underline{\Psi}_t(\cdot)$ is not known exactly. Along with lb_t, let us define

$$\tilde{\text{lb}}_t = \min_{x\in X}\left\{\tilde{\underline{\Psi}}_t(x) := \Gamma_t\Sigma_{\tau=1}^t\tfrac{\alpha_\tau}{\Gamma_\tau}\tilde{l}_\Psi(\underline{x}_\tau,\xi_\tau,x)\right\}, \tag{4.2.81}$$

where

$$\tilde{l}_\Psi(z,\xi,x) := F(z,\xi)+\langle G(z,\xi),x-z\rangle+\mu V(z,x)+h(x).$$

In view of the assumption that problem (4.2.6) is easy to solve, the bound $\tilde{\text{lb}}_t$ is easily computable. Moreover, since $\underline{x}_t$ is a function of $\xi_{[t-1]}$, and ξ_t is independent of $\xi_{[t-1]}$, we have that

$$\begin{aligned}\mathbb{E}\left[\tilde{\text{lb}}_t\right] &= \mathbb{E}\left[\mathbb{E}_{\xi_{[t-1]}}\left[\min_{x\in X}\left(\Gamma_t\Sigma_{\tau=1}^t\tilde{l}_\Psi(\underline{x}_\tau,\xi_\tau,x)\right)\right]\right]\\ &\le \mathbb{E}\left[\min_{x\in X}\mathbb{E}_{\xi_{[t-1]}}\left[\left(\Gamma_t\Sigma_{\tau=1}^t\tilde{l}_\Psi(\underline{x}_\tau,\xi_\tau,x)\right)\right]\right]\\ &= \mathbb{E}\left[\min_{x\in X}\underline{\Psi}_t(x)\right] = \mathbb{E}\left[\text{lb}_t\right] \le \Psi^*. \end{aligned} \tag{4.2.82}$$

That is, on average, $\tilde{\text{lb}}_t$ gives a lower bound for the optimal value of (4.2.1). In order to see how good the lower bound $\tilde{\text{lb}}_t$ is, we estimate the expectations and probabilities of the corresponding errors in Theorem 4.5 below. To establish the large-deviation results for $\tilde{\text{lb}}_t$, we also need the following assumption for the SFO.

Assumption 6 *For any $x\in X$ and $t\ge 1$, we have $\mathbb{E}\left[\exp\{\|F(x,\xi_t)\ -f(x)\|_*^2/Q^2\}\right] \le \exp\{1\}$ for some $Q>0$.*

Note that while Assumption 4 describes certain "light-tail" assumption about the stochastic gradients $G(x,\xi)$, Assumption 6 imposes a similar restriction on the function values $F(x,\xi)$. Such an additional assumption is needed to establish the large deviation properties for the derived stochastic online lower and upper bounds on Ψ^*, both of which involve the estimation of function values, i.e., $F(\underline{x}_t,\xi_t)$ in (4.2.81) and $F(\bar{x}_t,\xi_t)$ in (4.2.91). On the other hand, we do not need to use the estimation of function values in the stochastic accelerated gradient descent algorithm in (4.2.5)–(4.2.7).

Theorem 4.5. *Consider the generic stochastic accelerated gradient descent algorithm applied to problem (4.2.1). Also assume that* $\{\alpha_t\}_{t\geq 1}$, $\{q_t\}_{t\geq 1}$, *and* $\{\gamma_t\}_{t\geq 1}$ *are chosen such that* $\alpha_1 = 1$ *and relations (4.2.9), (4.2.10) and (4.2.22) hold. Let* $\tilde{\text{lb}}_t$ *be defined in (4.2.81). Then,*

(a) under Assumption 3, we have, for any $t \geq 2$,

$$\mathbb{E}[\Psi(\bar{x}_t) - \tilde{\text{lb}}_t] \leq \tilde{\text{B}}_e(t) := \tfrac{\Gamma_t}{\gamma_1} \max_{x\in X} V(x_0, x) + \Gamma_t \textstyle\sum_{\tau=1}^{t} \tfrac{\alpha_\tau \gamma_\tau (M^2+\sigma^2)}{\Gamma_\tau (1+\mu\gamma_\tau - L\alpha_\tau\gamma_\tau)}; \quad (4.2.83)$$

(b) if Assumptions 4 and 6 hold, then for any $t \geq 1$ *and* $\lambda > 0$,

$$\text{Prob}\left\{\Psi(\bar{x}_t) - \tilde{\text{lb}}_t > \tilde{\text{B}}_e(t) + \lambda \tilde{\text{B}}_p(t)\right\} \leq 2\exp(-\lambda^2/3) + \exp(-\lambda), \quad (4.2.84)$$

where

$$\begin{aligned} \tilde{\text{B}}_p(t) := Q\Gamma_t \left(\textstyle\sum_{\tau=1}^{t} \tfrac{\alpha_\tau^2}{\Gamma_\tau^2}\right)^{1/2} + \sigma \Gamma_t R_X(x^*) \left(\textstyle\sum_{\tau=1}^{t} \tfrac{\alpha_\tau^2}{\Gamma_\tau^2}\right)^{1/2} \\ + \sigma^2 \Gamma_t \textstyle\sum_{\tau=1}^{t} \tfrac{\alpha_\tau\gamma_\tau}{\Gamma_\tau(1+\mu\gamma_\tau - L\alpha_\tau\gamma_\tau)}, \end{aligned} \quad (4.2.85)$$

and $R_X(x^*)$ *is defined in (4.2.26);*

(c) If X *is compact and the condition (4.2.22) is replaced by (4.2.27), then parts (a) and (b) still hold by simply replacing the first term in the definition of* $\tilde{\text{B}}_e(t)$ *with* $\alpha_t D_X/\gamma_t$, *where* D_X *is defined in (3.2.4).*

Proof. Let $\zeta_t := F(\underline{x}_t, \xi_t) - f(\underline{x}_t)$, $t \geq 1$, and δ_t be defined in (4.2.8). Noting that by (4.2.20) and (4.2.30), relation (4.2.81), and the fact that $V(x_t, x) \geq 0$ due to (5.1.15), we have

$$\begin{aligned} \Psi(\bar{x}_t) - \tilde{\underline{\Psi}}_t(x) &= \Psi(\bar{x}_t) - \Gamma_t \textstyle\sum_{\tau=1}^{t} \tfrac{\alpha_\tau}{\Gamma_\tau} \left[l_\Psi(\underline{x}_\tau, x) + \zeta_\tau + \langle \delta_\tau, x - \underline{x}_\tau\rangle\right] \\ &\leq \Gamma_t \textstyle\sum_{\tau=1}^{t} \tfrac{\alpha_\tau}{\gamma_\tau \Gamma_\tau} \left[V(x_{\tau-1}, x) - V(x_\tau, x)\right] \\ &\quad + \Gamma_t \textstyle\sum_{\tau=1}^{t} \tfrac{1}{\Gamma_\tau} \left[\Delta_\tau(x) - \alpha_\tau(\zeta_\tau + \langle \delta_\tau, x - \underline{x}_\tau\rangle)\right] \\ &\leq \tfrac{\Gamma_t}{\gamma_1} V(x_0, x) + \Gamma_t \textstyle\sum_{\tau=1}^{t} \tfrac{1}{\Gamma_\tau} \left[\Delta_\tau(x) - \alpha_\tau(\zeta_\tau + \langle \delta_\tau, x - \underline{x}_\tau\rangle)\right] \\ &= \tfrac{\Gamma_t}{\gamma_1} V(x_0, x) + \Gamma_t \textstyle\sum_{\tau=1}^{t} \tfrac{1}{\Gamma_\tau} \left[\alpha_\tau \langle \delta_\tau, \underline{x}_\tau - x_{\tau-1}^+\rangle + \tfrac{\alpha_\tau \gamma_t (M + \|\delta_\tau\|_*)^2}{2(1+\mu\gamma_\tau - L\alpha_\tau\gamma_\tau)} - \alpha_\tau \zeta_\tau\right], \end{aligned} \quad (4.2.86)$$

where the last identity follows from (4.2.13). Note that $\underline{x}_t$ and x_{t-1}^+ are functions of $\xi_{[t-1]} = (\xi_1, \ldots, \xi_{t-1})$ and that ξ_t is independent of $\xi_{[t-1]}$. The rest of the proof is similar to the one used in using arguments similar to the ones in the proof of Theorem 4.4 and hence the details are skipped. ■

We now add a few comments about the results obtained in Theorem 4.5. First, note that relations (4.2.83) and (4.2.84) tell us how the gap between $\Psi(\bar{x}_t)$ and $\tilde{\text{lb}}_t$ converges to zero. By comparing these two relations with (4.2.23) and (4.2.24), we can easily see that both $\Psi(\bar{x}_t) - \tilde{\text{lb}}_t$ and $\Psi(\bar{x}_t) - \Psi^*$ converge to zero in the same order of magnitude.

Second, it is possible to specialize the results in Theorem 4.5 for solving different classes of stochastic convex optimization problems. In particular, Proposition 4.9 below discusses the lower bounds $\tilde{\text{lb}}_t^*$ for solving strongly convex problems. The proof of this result is similar to that of Proposition 4.6 and hence the details are skipped.

Proposition 4.9. *Let $\bar{x}_t$ be computed by the accelerated stochastic gradient descent algorithm for solving strongly convex problems with stepsize policy (4.2.55). Also let $\tilde{\text{lb}}_t$ be defined as in (4.2.81). If $\mu > 0$ in condition (4.2.2), then under Assumption 3, we have, $\forall t \geq 1$,*

$$\mathbb{E}[\Psi(\bar{x}_t) - \tilde{\text{lb}}_t] \leq \tilde{D}_e(t) := \tfrac{4L\max_{x\in X} V(x_0,x)}{\nu t(t+1)} + \tfrac{4(M^2+\sigma^2)}{\nu\mu(t+1)}. \tag{4.2.87}$$

If Assumptions 4 and 6 hold, then, $\forall \lambda > 0, \forall t \geq 1$,

$$\text{Prob}\left\{\Psi(\bar{x}_t) - \tilde{\text{lb}}_t > \tilde{D}_e(t) + \lambda \tilde{D}_p(t)\right\} \leq 2\exp(-\lambda^2/3) + \exp(-\lambda), \tag{4.2.88}$$

where

$$\tilde{D}_p(t) := \tfrac{Q}{(t+1)^{1/2}} + \tfrac{2\sigma R_X(x^*)}{\sqrt{3t}} + \tfrac{4\sigma^2}{\nu\mu(t+1)}, \tag{4.2.89}$$

$R_X(x^*)$ is defined in (4.2.26), and Q is from Assumption 6.

Theorem 4.5 presents a way to assess the quality of the solutions $\bar{x}_t$, $t \geq 1$, by computing the gap between $\Psi(\bar{x}_t)$ and $\tilde{\text{lb}}_t$ (cf. (4.2.81)). While $\tilde{\text{lb}}_t$ can be computed easily, the estimation of of $\Psi(\bar{x}_t)$ can be time consuming, requiring a large number of samples for ξ. In the remaining part of this section, we will briefly discuss how to enhance these lower bounds with efficiently computable upper bounds on the optimal value Ψ^* so that one can assess the quality of the generated solutions in an online manner. More specifically, for any $t \geq 1$, let us denote

$$\beta_t := \textstyle\sum_{\tau=\lceil t/2 \rceil}^t \tau,$$

$$\text{ub}_t := \beta_t^{-1} \textstyle\sum_{\tau=\lceil t/2 \rceil}^t \tau \Psi(\bar{x}_\tau) \quad \text{and} \quad \bar{x}_t^{ag} := \beta_t^{-1} \textstyle\sum_{\tau=\lceil t/2 \rceil}^t \tau \bar{x}_\tau. \tag{4.2.90}$$

Clearly, we have $\text{ub}_t \geq \Psi(\bar{x}_t^{ag}) \geq \Psi^*$ due to the convexity of Ψ. Also let us define

$$\bar{\text{ub}}_t = \beta_t^{-1} \textstyle\sum_{\tau=\lceil t/2 \rceil}^t \tau \{F(\bar{x}_\tau, \xi_\tau) + h(\bar{x}_\tau)\}, \ \forall t \geq 1. \tag{4.2.91}$$

Since $\mathbb{E}_{\xi_\tau}[F(\bar{x}_\tau, \xi_\tau)] = f(\bar{x}_\tau)$, we have $\mathbb{E}[\bar{\text{ub}}_t] = \text{ub}_t \geq \Psi^*$. That is, $\bar{\text{ub}}_t$, $t \geq 1$, on average, provide online upper bounds on Ψ^*. Accordingly, we define the new online lower bounds as

$$\bar{\text{lb}}_t = \beta_t^{-1} \textstyle\sum_{\tau=\lceil t/2 \rceil}^t \tau \tilde{\text{lb}}_\tau, \ \forall t \geq 1, \tag{4.2.92}$$

where $\tilde{\text{lb}}_\tau$ is defined in (4.2.81).

To bound the gap between these lower and upper bounds, let $\tilde{\text{B}}_e(\tau)$ be defined in (4.2.83) and suppose that $\tilde{\text{B}}_e(t) = \mathcal{O}(t^{-q})$ for some $q \in [1/2, 1]$. In view of Theorem 4.5(a), (4.2.90), and (4.2.92), we have

$$\mathbb{E}[\bar{\text{ub}}_t - \bar{\text{lb}}_t] = \beta_t^{-1}\Sigma_{\tau=\lceil t/2\rceil}^t \tau[\Psi(\bar{x}_\tau) - \bar{\text{lb}}_\tau] \le \beta_t^{-1}\Sigma_{\tau=\lceil t/2\rceil}^t [\tau\tilde{\text{B}}_e(\tau)]$$
$$= \mathcal{O}\left(\beta_t^{-1}\Sigma_{\tau=\lceil t/2\rceil}^t \tau^{1-q}\right) = \mathcal{O}(t^{-q}), \ \ t \ge 3,$$

where the last identity follows from the facts that $\Sigma_{\tau=\lceil t/2\rceil}^t \tau^{1-q} = \mathcal{O}(t^{2-q})$ and that

$$\beta_t \ge \tfrac{1}{2}\left[t(t+1) - \left(\tfrac{t}{2}+1\right)\left(\tfrac{t}{2}+2\right)\right] \ge \tfrac{1}{8}\left(3t^2 - 2t - 8\right).$$

Therefore, the gap between the online upper bound $\bar{\text{ub}}_t$ and lower bound $\bar{\text{lb}}_t$ converges to 0 in the same order of magnitude as the one between $\Psi(\bar{x}_t)$ and $\bar{\text{lb}}_t$. It should be mentioned that the stochastic upper bound $\bar{\text{ub}}_t$, on average, overestimates the value of $\Psi(\bar{x}_t^{ag})$ (cf. (4.2.90)), indicating that one can also use $\bar{x}_t^{ag}$, $t \ge 1$, as the output of the stochastic accelerated gradient descent algorithm.

4.3 Stochastic Convex–Concave Saddle Point Problems

We show in this section how the stochastic mirror descent algorithm can be modified to solve a convex–concave stochastic saddle point problem. Consider the following minimax (saddle point) problem

$$\min_{x\in X}\max_{y\in Y} \left\{\phi(x,y) := \mathbb{E}[\Phi(x,y,\xi)]\right\}. \tag{4.3.1}$$

Here $X \subset \mathbb{R}^n$ and $Y \subset \mathbb{R}^m$ are nonempty bounded closed convex sets, ξ is a random vector whose probability distribution P is supported on set $\Xi \subset \mathbb{R}^d$ and $\Phi : X \times Y \times \Xi \to \mathbb{R}$. We assume that for every $\xi \in \Xi$, function $\Phi(x,y,\xi)$ is convex in $x \in X$ and concave in $y \in Y$, and for all $x \in X$, $y \in Y$ the expectation

$$\mathbb{E}[\Phi(x,y,\xi)] = \int_\Xi \Phi(x,y,\xi)dP(\xi)$$

is well defined and finite valued. It follows that $\phi(x,y)$ is convex in $x \in X$ and concave in $y \in Y$, finite valued, and hence (4.3.1) is a convex–concave saddle point problem. In addition, we assume that $\phi(\cdot,\cdot)$ is Lipschitz continuous on $X \times Y$. It is well known that in the above setting the problem (4.3.1) is solvable, i.e., the corresponding "primal" and "dual" optimization problems $\min_{x\in X}\left[\max_{y\in Y}\phi(x,y)\right]$ and $\max_{y\in Y}\left[\min_{x\in X}\phi(x,y)\right]$, respectively, have optimal solutions and equal optimal values, denoted ϕ^*, and the pairs (x^*,y^*) of optimal solutions to the respective problems form the set of saddle points of $\phi(x,y)$ on $X \times Y$.

As in the case of the minimization problem (4.1.1) we assume that neither the function $\phi(x,y)$ nor its sub/supergradients in x and y are available explicitly. However, we make the following assumption.

Assumption 7 *There exists a stochastic first-order oracle which for every given* $x \in X$, $y \in Y$ *and* $\xi \in \Xi$ *returns value* $\Phi(x,y,\xi)$ *and a stochastic subgradient, that is,* $(n+m)$*-dimensional vector*

$$\mathsf{G}(x,y,\xi) = \begin{bmatrix} \mathsf{G}_x(x,y,\xi) \\ -\mathsf{G}_y(x,y,\xi) \end{bmatrix}$$

such that vector

$$\mathsf{g}(x,y) = \begin{bmatrix} \mathsf{g}_x(x,y) \\ -\mathsf{g}_y(x,y) \end{bmatrix} := \begin{bmatrix} \mathbb{E}[\mathsf{G}_x(x,y,\xi)] \\ -\mathbb{E}[\mathsf{G}_y(x,y,\xi)] \end{bmatrix}$$

is well defined, and $\mathsf{g}_x(x,y) \in \partial_x \phi(x,y)$ *and* $-\mathsf{g}_y(x,y) \in \partial_y(-\phi(x,y))$.

For example, under mild assumptions we can set

$$\mathsf{G}(x,y,\xi) = \begin{bmatrix} \mathsf{G}_x(x,y,\xi) \\ -\mathsf{G}_y(x,y,\xi) \end{bmatrix} \in \begin{bmatrix} \partial_x \Phi(x,y,\xi) \\ \partial_y(-\Phi(x,y,\xi)) \end{bmatrix}.$$

Let $\|\cdot\|_X$ be a norm on $\mathbb{R}^n$ and $\|\cdot\|_Y$ be a norm on $\mathbb{R}^m$, and let $\|\cdot\|_{*,X}$ and $\|\cdot\|_{*,Y}$ stand for the corresponding dual norms. As in Sect. 4.1, the basic assumption we make about the stochastic oracle (aside of its unbiasedness which we have already postulated) is that there exist positive constants M_X^2 and M_Y^2 such that

$$\mathbb{E}\left[\|\mathsf{G}_x(u,v,\xi)\|_{*,X}^2\right] \le M_X^2 \text{ and } \mathbb{E}\left[\|\mathsf{G}_y(u,v,\xi)\|_{*,Y}^2\right] \le M_Y^2, \ \forall (u,v) \in X \times Y. \tag{4.3.2}$$

4.3.1 General Algorithmic Framework

We equip X and Y with distance generating functions $v_X : X \to \mathbb{R}$ modulus 1 with respect to $\|\cdot\|_X$, and $v_Y : Y \to \mathbb{R}$ modulus 1 with respect to $\|\cdot\|_Y$. Let $D_X \equiv D_{X,v_X}$ and $D_Y \equiv D_{Y,v_Y}$ be the respective constants (see Sect. 3.2). We equip $\mathbb{R}^n \times \mathbb{R}^m$ with the norm

$$\|(x,y)\| := \sqrt{\tfrac{1}{2D_X^2}\|x\|_X^2 + \tfrac{1}{2D_Y^2}\|y\|_Y^2}, \tag{4.3.3}$$

so that the dual norm is

$$\|(\zeta,\eta)\|_* = \sqrt{2D_X^2\|\zeta\|_{*,X}^2 + 2D_Y^2\|\eta\|_{*,Y}^2}. \tag{4.3.4}$$

It follows by (4.3.2) that

$$\mathbb{E}\left[\|\mathsf{G}(x,y,\xi)\|_*^2\right] \le 2D_X^2 M_X^2 + 2D_Y^2 M_Y^2 =: M^2. \tag{4.3.5}$$

We use notation $z = (x,y)$ and equip $Z := X \times Y$ with the distance generating function as follows:

$$\nu(z) := \tfrac{\nu_X(x)}{2D_X^2} + \tfrac{\nu_Y(y)}{2D_Y^2}.$$

It is immediately seen that ν indeed is a distance generating function for Z modulus 1 with respect to the norm $\|\cdot\|$, and that $Z^o = X^o \times Y^o$ and $D_Z \equiv D_{Z,\nu} = 1$. In what follows, $V(z,u) : Z^o \times Z \to \mathbb{R}$ is the prox-function associated with ν and Z (see Sect. 3.2).

We are ready now to present the stochastic mirror descent algorithm for solving general saddle point problems. This is the iterative procedure

$$z_{t+1} := \operatorname{argmin}_{z\in X} \{\gamma_t\{\langle G(z_t,\xi_t),z\rangle + V(z_t,z)\},\tag{4.3.6}$$

where the initial point $z_1 \in Z$ is chosen to be the minimizer of $\nu(z)$ on Z. Moreover, we define the approximate solution $\tilde{z}_j$ of (4.3.1) after j iterations as

$$\tilde{z}_j = (\tilde{x}_j,\tilde{y}_j) := \left(\textstyle\sum_{t=1}^j \gamma_t\right)^{-1} \textstyle\sum_{t=1}^j \gamma_t z_t.\tag{4.3.7}$$

Let us analyze the convergence properties of the algorithm. We measure quality of an approximate solution $\tilde{z} = (\tilde{x},\tilde{y})$ by the error

$$\varepsilon_\phi(\tilde{z}) := \left[\max_{y\in Y} \phi(\tilde{x},y) - \phi_*\right] + \left[\phi_* - \min_{x\in X} \phi(x,\tilde{y})\right] = \max_{y\in Y} \phi(\tilde{x},y) - \min_{x\in X} \phi(x,\tilde{y}).$$

By convexity of $\phi(\cdot,y)$ we have

$$\phi(x_t,y_t) - \phi(x,y_t) \le g_x(x_t,y_t)^T(x_t - x),\ \ \forall x \in X,$$

and by concavity of $\phi(x,\cdot)$,

$$\phi(x_t,y) - \phi(x_t,y_t) \le g_y(x_t,y_t)^T(y - y_t),\ \ \forall y \in Y,$$

so that for all $z = (x,y) \in Z$,

$$\phi(x_t,y) - \phi(x,y_t) \le g_x(x_t,y_t)^T(x_t - x) + g_y(x_t,y_t)^T(y - y_t) = g(z_t)^T(z_t - z).$$

Using once again the convexity-concavity of ϕ we write

$$\begin{aligned}
\varepsilon_\phi(\tilde{z}_j) &= \max_{y\in Y} \phi(\tilde{x}_j,y) - \min_{x\in X} \phi(x,\tilde{y}_j)\\
&\le \left[\textstyle\sum_{t=1}^j \gamma_t\right]^{-1} \left[\max_{y\in Y} \textstyle\sum_{t=1}^j \gamma_t \phi(x_t,y) - \min_{x\in X} \textstyle\sum_{t=1}^j \gamma_t \phi(x,y_t)\right]\\
&\le \left(\textstyle\sum_{t=1}^j \gamma_t\right)^{-1} \max_{z\in Z} \textstyle\sum_{t=1}^j \gamma_t g(z_t)^T(z_t - z).
\end{aligned}\tag{4.3.8}$$

We now provide a bound on the right-hand side of (4.3.8).

Lemma 4.6. *For any $j \ge 1$ the following inequality holds*

$$\mathbb{E}\left[\max_{z\in Z}\Sigma_{t=1}^{j}\gamma_t \mathrm{g}(z_t)^T(z_t-z)\right] \le 2+\tfrac{5}{2}M^2\Sigma_{t=1}^{j}\gamma_t^2. \tag{4.3.9}$$

Proof. Using (4.1.41) with $\zeta_t = \gamma_t \mathsf{G}(z_t,\xi_t)$, we have for any $u\in Z$

$$\gamma_t(z_t-u)^T\mathsf{G}(z_t,\xi_t) \le V(z_t,u)-V(z_{t+1},u)+\tfrac{\gamma_t^2}{2}\|\mathsf{G}(z_t,\xi_t)\|_*^2 \tag{4.3.10}$$

This relation implies that for every $u\in Z$ one has

$$\begin{aligned}\gamma_t(z_t-u)^T\mathrm{g}(z_t) &\le V(z_t,u)-V(z_{t+1},u)\\ &\quad+\tfrac{\gamma_t^2}{2}\|\mathsf{G}(z_t,\xi_t)\|_*^2-\gamma_t(z_t-u)^T\Delta_t,\end{aligned} \tag{4.3.11}$$

where $\Delta_t := \mathsf{G}(z_t,\xi_t)-\mathrm{g}(z_t)$. Summing up these inequalities over $t=1,\ldots,j$, we get

$$\begin{aligned}\Sigma_{t=1}^{j}\gamma_t(z_t-u)^T\mathrm{g}(z_t) &\le V(z_1,u)-V(z_{t+1},u)\\ &\quad+\Sigma_{t=1}^{j}\tfrac{\gamma_t^2}{2}\|\mathsf{G}(z_t,\xi_t)\|_*^2-\Sigma_{t=1}^{j}\gamma_t(z_t-u)^T\Delta_t.\end{aligned}$$

Let us also apply Lemma 4.3 to an auxiliary sequence with $v_1=z_1$ and $\zeta_t=-\gamma_t\Delta_t$:

$$\forall u\in Z:\quad \Sigma_{t=1}^{j}\gamma_t\Delta_t^T(u-v_t)\le V(z_1,u)+\tfrac{1}{2}\Sigma_{t=1}^{j}\gamma_t^2\|\Delta_t\|_*^2. \tag{4.3.12}$$

Observe that

$$\mathbb{E}\|\Delta_t\|_*^2 \le 4\mathbb{E}\|\mathsf{G}(z_t,\xi_t)\|_*^2 \le 4\left(2D_X^2M_X^2+2D_Y^2M_Y^2\right)=4M^2,$$

so that when taking the expectation of both sides of (4.3.12) we get

$$\mathbb{E}\left[\sup_{u\in Z}\left(\Sigma_{t=1}^{j}\gamma_t\Delta_t^T(u-v_t)\right)\right]\le 1+2M^2\Sigma_{t=1}^{j}\gamma_t^2 \tag{4.3.13}$$

(recall that $V(z_1,\cdot)$ is bounded by 1 on Z). Now we sum up (4.3.11) from $t=1$ to j to obtain

$$\begin{aligned}\Sigma_{t=1}^{j}\gamma_t(z_t-u)^T\mathrm{g}(z_t) &\le V(z_1,u)+\Sigma_{t=1}^{j}\tfrac{\gamma_t^2}{2}\|\mathsf{G}(z_t,\xi_t)\|_*^2-\Sigma_{t=1}^{j}\gamma_t(z_t-u)^T\Delta_t\\ &=V(z_1,u)+\Sigma_{t=1}^{j}\tfrac{\gamma_t^2}{2}\|\mathsf{G}(z_t,\xi_t)\|_*^2-\Sigma_{t=1}^{j}\gamma_t(z_t-v_t)^T\Delta_t+\Sigma_{t=1}^{j}\gamma_t(u-v_t)^T\Delta_t.\end{aligned} \tag{4.3.14}$$

When taking into account that z_t and v_t are deterministic functions of $\xi_{[t-1]}=(\xi_1,\ldots,\xi_{t-1})$ and that the conditional expectation of Δ_t, $\xi_{[t-1]}$ being given, vanishes, we conclude that $\mathbb{E}[(z_t-v_t)^T\Delta_t]=0$. We take now suprema in $u\in Z$ and then expectations on both sides of (4.3.14):

$$\begin{aligned}\mathbb{E}\left[\sup_{u\in Z}\Sigma_{t=1}^{j}\gamma_t(z_t-u)^T\mathrm{g}(z_t)\right] &\le \sup_{u\in Z}V(z_1,u)+\Sigma_{t=1}^{j}\tfrac{\gamma_t^2}{2}\mathbb{E}\|\mathsf{G}(z_t,\xi_t)\|_*^2\\ &\quad+\sup_{u\in Z}\Sigma_{t=1}^{j}\gamma_t(u-v_t)^T\Delta_t\end{aligned}$$

$$\begin{aligned}[\text{by (4.3.13)}] \quad &\le 1+\tfrac{M^2}{2}\textstyle\sum_{t=1}^{j}\gamma_t^2+\left[1+2M^2\sum_{t=1}^{j}\gamma_t^2\right]\\ &= 2+\tfrac{5}{2}M^2\textstyle\sum_{t=1}^{j}\gamma_t^2.\end{aligned}$$

and we arrive at (4.3.9). ∎

In order to obtain an error bound for the solution $\tilde{z}_j$ it suffices to substitute inequality (4.3.9) into (4.3.8) to obtain

$$\mathbb{E}[\varepsilon_\phi(\tilde{z}_j)] \le \left(\textstyle\sum_{t=1}^{j}\gamma_t\right)^{-1}\left[2+\tfrac{5}{2}M^2\textstyle\sum_{t=1}^{j}\gamma_t^2\right].$$

Let us use the constant stepsize strategy

$$\gamma_t = \tfrac{2}{M\sqrt{5N}},\ t=1,\ldots,N. \tag{4.3.15}$$

Then $\varepsilon_\phi(\tilde{z}_N) \le 2M\sqrt{\frac{5}{N}}$, and hence (see definition (4.3.5) of M) we obtain

$$\varepsilon_\phi(\tilde{z}_N) \le 2\sqrt{\frac{10\left[\alpha_y D_X^2 M_X^2+\alpha_x D_Y^2 M_Y^2\right]}{\alpha_x\alpha_y N}}. \tag{4.3.16}$$

Same as in the minimization case discussed in Sect. 3.2, we can pass from constant stepsizes on a fixed "time horizon" to decreasing stepsize policy

$$\gamma_t := \tfrac{1}{(M\sqrt{t})},\quad t=1,2,\ldots,$$

and from the averaging of all iterates to the "sliding averaging"

$$\tilde{z}_j = \left(\textstyle\sum_{t=j-\lfloor j/\ell\rfloor}^{j}\gamma_t\right)^{-1}\textstyle\sum_{t=j-\lfloor j/\ell\rfloor}^{j}\gamma_t z_t,$$

arriving at the efficiency estimate

$$\varepsilon_\phi(\tilde{z}_j) \le O(1)\frac{\ell\overline{D}_{Z,v}M}{\sqrt{j}}, \tag{4.3.17}$$

where the quantity $\overline{D}_{Z,v}=\left[2\sup_{z\in Z^o,w\in Z}V(z,w)\right]^{1/2}$ is assumed to be finite.

We give below a bound on the probabilities of large deviations of the error $\varepsilon_\phi(\tilde{z}_N)$. The proof of this result is similar to that of Proposition 4.10 and hence details are skipped.

Proposition 4.10. *Suppose that conditions of the bound* (4.3.16) *are verified and, further, it holds for all* $(u,v)\in Z$ *that*

$$\begin{aligned}&\mathbb{E}\left[\exp\left\{\|G_x(u,v,\xi)\|_{*,X}^2/M_X^2\right\}\right] \le \exp\{1\},\\ &\mathbb{E}\left[\exp\left\{\|G_y(x,y,\xi)\|_{*,Y}^2/M_Y^2\right\}\right] \le \exp\{1\}.\end{aligned} \tag{4.3.18}$$

Then for the stepsizes (4.3.15) *one has for any* $\lambda \geq 1$ *that*

$$\text{Prob}\left\{\varepsilon_\phi(\tilde{z}_N) > \frac{(8+2\lambda)\sqrt{5}M}{\sqrt{N}}\right\} \leq 2\exp\{-\lambda\}. \tag{4.3.19}$$

4.3.2 Minimax Stochastic Problems

Consider the following minimax stochastic problem

$$\min_{x\in X}\max_{1\leq i\leq m}\left\{f_i(x) := \mathbb{E}[F_i(x,\xi)]\right\}, \tag{4.3.20}$$

where $X \subset \mathbb{R}^n$ is a nonempty bounded closed convex set, ξ is a random vector whose probability distribution P is supported on set $\Xi \subset \mathbb{R}^d$ and $F_i : X \times \Xi \to \mathbb{R}$, $i = 1,\ldots,m$. We assume that for a.e. ξ the functions $F_i(\cdot,\xi)$ are convex and for every $x \in \mathbb{R}^n$, $F_i(x,\cdot)$ are integrable, i.e., the expectations

$$\mathbb{E}[F_i(x,\xi)] = \int_\Xi F_i(x,\xi)dP(\xi),\ \ i = 1,\ldots,m, \tag{4.3.21}$$

are well defined and finite valued. To find a solution to the minimax problem (4.3.20) is exactly the same as to solve the saddle point problem

$$\min_{x\in X}\max_{y\in Y}\left\{\phi(x,y) := \Sigma_{i=1}^m y_i f_i(x)\right\}, \tag{4.3.22}$$

with $Y := \{y \in \mathbb{R}^m : y \geq 0, \Sigma_{i=1}^m y_i = 1\}$.

Similar to Assumptions 1 and 2, assume that we cannot compute $f_i(x)$ (and thus $\phi(x,y)$) explicitly, but are able to generate independent realizations $\xi_1,\xi_2,\ldots$ distributed according to P, and for given $x \in X$ and $\xi \in \Xi$ we can compute $F_i(x,\xi)$ and its *stochastic subgradient* $\mathsf{G}_i(x,\xi)$, i.e., such that $\mathsf{g}_i(x) = \mathbb{E}[\mathsf{G}_i(x,\xi)]$ is well defined and $\mathsf{g}_i(x) \in \partial f_i(x)$, $x \in X$, $i = 1,\ldots,m$. In other words we have a stochastic oracle for the problem (4.3.22) such that Assumption 7 holds, with

$$\mathsf{G}(x,y,\xi) := \begin{bmatrix} \Sigma_{i=1}^m y_i \mathsf{G}_i(x,\xi) \\ (-F_1(x,\xi),\ldots,-F_m(x,\xi)) \end{bmatrix}, \tag{4.3.23}$$

and

$$\mathsf{g}(x,y) := \mathbb{E}[\mathsf{G}(x,y,\xi)] = \begin{bmatrix} \Sigma_{i=1}^m y_i \mathsf{g}_i(x) \\ (-f_1(x),\ldots,-f_m(x)) \end{bmatrix} \in \begin{bmatrix} \partial_x\phi(x,y) \\ -\partial_y\phi(x,y) \end{bmatrix}. \tag{4.3.24}$$

Suppose that the set X is equipped with norm $\|\cdot\|_X$, whose dual norm is $\|\cdot\|_{*,X}$, and a distance generating function ν modulus 1 with respect to $\|\cdot\|_X$. We equip the set Y with norm $\|\cdot\|_Y := \|\cdot\|_1$, so that $\|\cdot\|_{*,Y} = \|\cdot\|_\infty$, and with the distance generating function

$$v_Y(y) := \sum_{i=1}^m y_i \ln y_i,$$

and hence $D_Y^2 = \ln m$. Next, following (4.3.3) we set

$$\|(x,y)\| := \sqrt{\frac{\|x\|_X^2}{2D_X^2} + \frac{\|y\|_1^2}{2D_Y^2}},$$

and hence

$$\|(\zeta,\eta)\|_* = \sqrt{2D_X^2\|\zeta\|_{*,X}^2 + 2D_Y^2\|\eta\|_\infty^2}.$$

Let us assume uniform bounds:

$$\mathbb{E}\left[\max_{1\le i\le m} \|\mathsf{G}_i(x,\xi)\|_{*,X}^2\right] \le M_X^2, \quad \mathbb{E}\left[\max_{1\le i\le m} |F_i(x,\xi)|^2\right] \le M_Y^2, \quad i=1,\ldots,m.$$

Note that

$$\begin{aligned}\mathbb{E}\left[\|\mathsf{G}(x,y,\xi)\|_*^2\right] &= 2D_X^2\,\mathbb{E}\left[\left\|\sum_{i=1}^m y_i \mathsf{G}_i(x,\xi)\right\|_{*,X}^2\right] + 2D_Y^2\,\mathbb{E}\left[\|F(x,\xi)\|_\infty^2\right]\\ &\le 2D_X^2 M_X^2 + 2D_Y^2 M_Y^2 = 2D_X^2 M_X^2 + 2M_Y^2 \ln m =: M^2.\end{aligned} \tag{4.3.25}$$

Let us now use the stochastic mirror descent algorithm (4.3.6) and (4.3.7) with the constant stepsize strategy

$$\gamma_t = \frac{2}{M\sqrt{5N}}, \quad t = 1,2,\ldots,N.$$

When substituting the value of M, we obtain from (4.3.16):

$$\begin{aligned}\mathbb{E}\left[\varepsilon_\phi(\tilde{z}_N)\right] &= \mathbb{E}\left[\max_{y\in Y}\phi(\hat{x}_N,y) - \min_{x\in X}\phi(x,\hat{y}_N)\right] \le 2M\sqrt{\tfrac{5}{N}}\\ &\le 2\sqrt{\frac{10\left[D_X^2M_X^2 + M_Y^2\ln m\right]}{N}}.\end{aligned} \tag{4.3.26}$$

Looking at the bound (4.3.26) one can make the following important observation. The error of the stochastic mirror descent algorithm in this case is "almost independent" of the number m of constraints (it grows as $O(\sqrt{\ln m})$ as m increases). The interested reader can easily verify that if a stochastic gradient descent (with Euclidean distance generating function) was used in the same setting (i.e., the algorithm tuned to the norm $\|\cdot\|_y := \|\cdot\|_2$), the corresponding bound would grow with m much faster (in fact, our error bound would be $O(\sqrt{m})$ in that case).

Note that properties of the stochastic mirror descent can be used to reduce significantly the arithmetic cost of the algorithm implementation. To this end let us look at the definition (4.3.23) of the stochastic oracle: in order to obtain a realization $\mathsf{G}(x,y,\xi)$ one has to compute m random subgradients $\mathsf{G}_i(x,\xi)$, $i=1,\ldots,m$, and then the convex combination $\sum_{i=1}^m y_i\mathsf{G}_i(x,\xi)$. Now let η be an independent of ξ and uniformly distributed in $[0,1]$ random variable, and let $\iota(\eta,y):[0,1]\times Y\to\{1,\ldots,m\}$ equals to i when $\sum_{s=1}^{i-1}y_s < \eta \le \sum_{s=1}^{i}y_s$. That is, random variable $\hat{\imath} = \iota(\eta,y)$ takes values $1,\ldots,m$ with probabilities $y_1,\ldots,y_m$. Consider random vector

$$\mathsf{G}(x,y,(\xi,\eta)) := \begin{bmatrix} \mathsf{G}_{\imath(\eta,y)}(x,\xi) \\ (-F_1(x,\xi),\ldots,-F_m(x,\xi)) \end{bmatrix}. \tag{4.3.27}$$

We refer to $\mathsf{G}(x,y,(\xi,\eta))$ as a randomized oracle for problem (4.3.22), the corresponding random parameter being (ξ,η). By construction we still have $\mathbb{E}\big[\mathsf{G}(x,y,(\xi,\eta))\big] = \mathsf{g}(x,y)$, where g is defined in (4.3.24), and, moreover, the same bound (4.3.25) holds for $\mathbb{E}\big[\|\mathsf{G}(x,y,(\xi,\eta))\|_*^2\big]$. We conclude that the accuracy bound (4.3.26) holds for the error of the stochastic mirror descent algorithm with randomized oracle. On the other hand, in the latter procedure only one randomized subgradient $\mathsf{G}_{\hat\imath}(x,\xi)$ per iteration is to be computed. This simple idea is further developed in another interesting application of the stochastic mirror descent algorithm to bilinear matrix games which we discuss next.

4.3.3 Bilinear Matrix Games

Consider the standard matrix game problem, that is, problem (4.3.1) with

$$\phi(x,y) := y^T Ax + b^T x + c^T y,$$

where $A \in \mathbb{R}^{m\times n}$, and X and Y are the standard simplices, i.e.,

$$X := \left\{x \in \mathbb{R}^n : x \ge 0, \textstyle\sum_{j=1}^n x_j = 1\right\},\ Y := \left\{y \in \mathbb{R}^m : y \ge 0, \textstyle\sum_{i=1}^m y_i = 1\right\}.$$

In the case in question it is natural to equip X (respectively, Y) with the usual $\|\cdot\|_1$-norm on $\mathbb{R}^n$ (respectively, $\mathbb{R}^m$). We choose entropies as the corresponding distance generating functions:

$$\nu_X(x) := \textstyle\sum_{i=1}^n x_i \ln x_i, \quad \nu_y(x) := \textstyle\sum_{i=1}^m y_i \ln y_i.$$

As we already have seen, this choice results in $D_X^2 = \ln n$ and $D_Y^2 = \ln m$. According to (4.3.3) we set

$$\|(x,y)\| := \sqrt{\frac{\|x\|_1^2}{2\ln n} + \frac{\|y\|_1^2}{2\ln m}},$$

and thus

$$\|(\zeta,\eta)\|_* = \sqrt{2\|\zeta\|_\infty^2 \ln n + 2\|\eta\|_\infty^2 \ln m}. \tag{4.3.28}$$

In order to compute the estimates $\Phi(x,y,\xi)$ of $\phi(x,y)$ and $\mathsf{G}(x,y,\xi)$ of $\mathsf{g}(x,y) = (b + A^T y, -c - Ax)$ to be used in the stochastic mirror descent iterations (4.3.6), we use the randomized oracle

$$\begin{aligned}\Phi(x,y,\xi) &= c^T x + b^T y + A_{\iota(\xi_1,y)\iota(\xi_2,x)},\\ \mathsf{G}(x,y,\xi) &= \begin{bmatrix} c + A^{\iota(\xi_1,y)} \\ -b - A_{\iota(\xi_2,x)} \end{bmatrix},\end{aligned}$$

where ξ_1 and ξ_2 are independent uniformly distributed on $[0,1]$ random variables, $\hat{j} = \iota(\xi_1, y)$ and $\hat{\imath} = \iota(\xi_2, x)$ are defined as in (4.3.27), i.e., $\hat{j}$ can take values $1,\ldots,m$ with probabilities $y_1,\ldots,y_m$ and $\hat{\imath}$ can take values $1,\ldots,n$ with probabilities $x_1,\ldots,x_n$, and A_j, $[A^i]^T$ are j-th column and i-th row in A, respectively.

Note that $\mathsf{g}(x,y) := \mathbb{E}\big[\mathsf{G}(x,y,(\hat{j},\hat{\imath}))\big] \in \begin{bmatrix} \partial_x \phi(x,y) \\ \partial_y(-\phi(x,y)) \end{bmatrix}$. Besides this,

$$|\mathsf{G}(x,y,\xi)_i| \le \max_{1\le j\le m} \|A^j + b\|_\infty, \text{ for } i = 1,\ldots,n,$$

and

$$|\mathsf{G}(x,y,\xi)_i| \le \max_{1\le j\le n} \|A_j + c\|_\infty, \text{ for } i = n+1,\ldots,n+m.$$

Hence, by the definition (4.3.28) of $\|\cdot\|_*$,

$$\mathbb{E}\|\mathsf{G}(x,y,\xi)\|_*^2 \le M^2 := 2\ln n \max_{1\le j\le m} \|A^j + b\|_\infty^2 + 2\ln m \max_{1\le j\le n} \|A_j + c\|_\infty^2.$$

Therefore, the inputs of the stochastic mirror descent algorithm satisfy the conditions of validity of the bound (4.3.16) with M as above. Using the constant stepsize strategy with

$$\gamma_t = \tfrac{2}{M\sqrt{5N}}, \quad t = 1,\ldots,N,$$

we obtain from (4.3.16):

$$\mathbb{E}\big[\varepsilon_\phi(\tilde{z}_N)\big] = \mathbb{E}\left[\max_{y\in Y} \phi(\tilde{x}_N, y) - \min_{x\in X} \phi(x, \tilde{y}_N)\right] \le 2M\sqrt{\tfrac{5}{N}}. \tag{4.3.29}$$

We continue with the counterpart of Proposition 4.10 for the Saddle Point Mirror SA in the setting of bilinear matrix games.

Proposition 4.11. *For any $\Omega \ge 1$ it holds that*

$$\operatorname{Prob}\left\{\varepsilon_\phi(\tilde{z}_N) > 2M\sqrt{\tfrac{5}{N}} + \tfrac{4\overline{M}}{\sqrt{N}}\Omega\right\} \le \exp\{-\Omega^2/2\}, \tag{4.3.30}$$

where

$$\overline{M} := \max_{1\le j\le m} \|A^j + b\|_\infty + \max_{1\le j\le n} \|A_j + c\|_\infty. \tag{4.3.31}$$

Proof. As in the proof of Proposition 4.10, when setting $\Gamma_N = \sum_{t=1}^N \gamma_t$ and using the relations (4.3.8), (4.3.12), (4.3.14), combined with the fact that $\|\mathsf{G}(z,\xi_y)\|_* \le \overline{M}$, we obtain

$$\begin{aligned}\Gamma_N \varepsilon_\phi(\tilde{z}_N) &\le 2+\textstyle\sum_{t=1}^N \frac{\gamma_t^2}{2}\left[\|\mathsf{G}(z_t,\xi_t)\|_*^2+\|\Delta_t\|_*^2\right]+\sum_{t=1}^N \gamma_t(v_t-z_t)^T\Delta_t\\ &\le 2+\tfrac{5}{2}M^2\textstyle\sum_{t=1}^N\gamma_t^2+\underbrace{\textstyle\sum_{t=1}^N\gamma_t(v_t-z_t)^T\Delta_t}_{\alpha_N},\end{aligned} \tag{4.3.32}$$

where the second inequality follows from the definition of Δ_t and the fact that $\|\Delta_t\|_*=\|\mathsf{G}(z_t,\xi_t)-\mathsf{g}(z_t)\|_*\le\|\mathsf{G}(z_t,\xi_t)\|+\|\mathsf{g}(z_t)\|_*\le 2M$.

Note that $\zeta_t=\gamma_t(v_t-z_t)^T\Delta_t$ is a bounded martingale-difference, i.e., $\mathbb{E}(\zeta_t|\xi_{[t-1]})=0$, and $|\zeta_t|\le 4\gamma_t\overline{M}$ (here $\overline{M}$ is defined in (4.3.31)). Then by Lemma 4.1 for any $\Omega\ge 0$:

$$\mathrm{Prob}\left(\alpha_N>4\Omega\overline{M}\sqrt{\textstyle\sum_{t=1}^N\gamma_t^2}\right)\le e^{-\Omega^2/2}. \tag{4.3.33}$$

Indeed, let us denote $v_t=(v_t^{(x)},v_t^{(y)})$ and $\Delta_t=(\Delta_t^{(x)},\Delta_t^{(y)})$. When taking into account that $\|v_t^{(x)}\|_1\le 1$, $\|v_t^{(y)}\|_1\le 1$, and $\|x_t\|_1\le 1$, $\|y_t\|_1\le 1$, we conclude that

$$\begin{aligned}|(v_t-z_t)^T\Delta_t| &\le |(v_t^{(x)}-x_t)^T\Delta_t^{(x)}|+|(v_t^{(y)}-y_t)^T\Delta_t^{(y)}|\\ &\le 2\|\Delta_t^{(x)}\|_\infty+2\|\Delta_t^{(y)}\|_\infty\\ &\le 4\max_{1\le j\le m}\|A^j+b\|_\infty+4\max_{1\le j\le n}\|A_j+c\|_\infty\\ &=4\overline{M}.\end{aligned}$$

We conclude from (4.3.32) and (4.3.33) that

$$\mathrm{Prob}\left(\Gamma_N\varepsilon_\phi(\tilde{z}_N)>2+\tfrac{5}{2}M^2\textstyle\sum_{t=1}^N\gamma_t^2+4\Omega\overline{M}\sqrt{\textstyle\sum_{t=1}^N\gamma_t^2}\right)\le e^{-\Omega^2/2},$$

and the bound (4.3.30) of the proposition can be easily obtained by substituting the constant stepsizes γ_t as defined in (4.3.15). ■

Consider a bilinear matrix game with $m=n$ and $b=c=0$. Suppose that we are interested to solve it within a fixed relative accuracy ρ, that is, to ensure that a (perhaps random) approximate solution $\tilde{z}_N$, we get after N iterations, satisfies the error bound

$$\varepsilon_\phi(\tilde{z}_N)\le\rho\max_{1\le i,j\le n}|A_{ij}|$$

with probability at least $1-\delta$. According to (4.3.30), the number of stochastic mirror descent iterations (4.3.6) can be bounded by

$$N=O(1)\frac{\ln n+\ln(\delta^{-1})}{\rho^2}. \tag{4.3.34}$$

The computational cost of building $\tilde{z}_N$ with this approach is

$$O(1)\frac{\left[\ln n+\ln(\delta^{-1})\right]\mathscr{R}}{\rho^2}$$

arithmetic operations, where $\mathscr{R}$ is the arithmetic cost of extracting a column/row from A, given the index of this column/row. The total number of rows and columns visited by the algorithm does not exceed the sample size N, given in (4.3.34), so that the total number of entries in A used in course of the entire computation does not exceed

$$M = O(1)\frac{n(\ln n + \ln(\delta^{-1}))}{\rho^2}.$$

When ρ is fixed and n is large, this is incomparably less that the total number n^2 of entries of A. Thus, *the algorithm in question produces reliable solutions of prescribed quality to large-scale matrix games by inspecting a negligible, as $n \to \infty$, part of randomly selected data.* Note that randomization here is critical. It is easily seen that a deterministic algorithm which is capable to find a solution with (deterministic) relative accuracy $\rho \leq 0.1$, has to "see" in the worst case *at least* $O(1)n$ rows/columns of A.

4.4 Stochastic Accelerated Primal–Dual Method

Let $X \subseteq \mathbb{R}^n$, $Y \subseteq \mathbb{R}^m$ be given closed convex sets equipped with their respective inner product $\langle \cdot, \cdot \rangle$ and norm $\| \cdot \|$. The basic problem of interest in this section is the saddle-point problem (SPP) given in the form of:

$$\min_{x \in X} \left\{ f(x) := \max_{y \in Y} \hat{f}(x) + \langle Ax, y \rangle - \hat{g}(y) \right\}. \tag{4.4.1}$$

Here, $\hat{f}(x)$ is a general smooth convex function and A is a linear operator such that

$$\begin{aligned} \hat{f}(u) - \hat{f}(x) - \langle \nabla \hat{f}(x), u - x \rangle &\leq \tfrac{L_{\hat{f}}}{2} \|u - x\|^2, \ \forall x, u \in X, \\ \|Au - Ax\|_* &\leq \|A\| \|u - x\|, \ \forall x, u \in X, \end{aligned} \tag{4.4.2}$$

and $\hat{g} : Y \to \mathbb{R}$ is a relatively simple, proper, convex, lower semi-continuous (l.s.c.) function (i.e., problem (4.4.15) below is easy to solve). In particular, if $\hat{g}$ is the convex conjugate of some convex function F and $Y \equiv \mathbb{R}^m$, then (4.4.1) is equivalent to the primal problem:

$$\min_{x \in X} \hat{f}(x) + F(Ax). \tag{4.4.3}$$

Problems of these types have recently found many applications in data analysis, especially in imaging processing and machine learning. In many of these applications, $\hat{f}(x)$ is a convex data fidelity term, while $F(Ax)$ is a certain regularization, e.g., total variation, low rank tensor, overlapped group lasso, and graph regularization.

Since the objective function f defined in (4.4.1) is nonsmooth in general, traditional nonsmooth optimization methods, e.g., subgradient or mirror descent methods, would exhibit an $\mathscr{O}(1/\sqrt{N})$ rate of convergence when applied to (4.4.1), where N denotes the number of iterations. As discussed in Sect. 3.5, if X and Y are compact, then the rate of convergence of this smoothing scheme applied to (4.4.1) can be bounded by:

$$\mathcal{O}\left(\tfrac{L_{\hat{f}}}{N^2} + \tfrac{\|A\|}{N}\right), \tag{4.4.4}$$

which significantly improves the previous bound $\mathcal{O}(1/\sqrt{N})$.

While Nesterov's smoothing scheme or its variants rely on a smooth approximation to the original problem (4.4.1), primal–dual methods discussed in Sect. 3.6 work directly with the original saddle-point problem. In Sect. 3.6, we assume $\hat{f}$ to be relatively simple so that the subproblems can be solved efficiently. With little additional effort, one can show that, by linearizing $\hat{f}$ at each step, this method can also be applied for a general smooth convex function $\hat{f}$ and the rate of convergence of this modified algorithm is given by

$$\mathcal{O}\left(\tfrac{L_{\hat{f}}+\|A\|}{N}\right). \tag{4.4.5}$$

The rate of convergence in (4.4.4) has a significantly better dependence on $L_{\hat{f}}$ than that in (4.4.5). Therefore, Nesterov's smoothing scheme allows a very large Lipschitz constant $L_{\hat{f}}$ (as big as $\mathcal{O}(N)$) without affecting the rate of convergence (up to a constant factor of 2). This is desirable in many data analysis applications, where $L_{\hat{f}}$ is usually significantly bigger than $\|A\|$.

Similar to Assumptions 1 and 2 made for minimization problems, in the stochastic setting we assume that there exists a *stochastic first-order oracle* (SFO) that can provide unbiased estimators to the gradient operators $\nabla\hat{f}(x)$ and $(-Ax, A^T y)$. More specifically, at the i-th call to SFO, $(x_i, y_i) \in X \times Y$ being the input, the oracle will output the *stochastic gradient* $(\hat{\mathsf{G}}(x_i), \hat{\mathsf{A}}_x(x_i), \hat{\mathsf{A}}_y(y_i)) \equiv (\mathsf{G}(x_i,\xi_i), \mathsf{A}_x(x_i,\xi_i), \mathsf{A}_y(y_i,\xi_i))$ such that

$$\mathbb{E}[\hat{\mathsf{G}}(x_i)] = \nabla\hat{f}(x_i), \quad \mathbb{E}\left[\begin{pmatrix} -\hat{\mathsf{A}}_x(x_i) \\ \hat{\mathsf{A}}_y(y_i) \end{pmatrix}\right] = \begin{pmatrix} -Ax_i \\ A^T y_i \end{pmatrix}. \tag{4.4.6}$$

Here $\{\xi_i \in \mathbb{R}^d\}_{i=1}^{\infty}$ is a sequence of i.i.d. random variables. In addition, we assume that, for some $\sigma_{x,\hat{f}}, \sigma_y, \sigma_{x,A} \geq 0$, the following assumption holds for all $x_i \in X$ and $y_i \in Y$:

Assumption 8

$$\begin{aligned}
&\mathbb{E}[\|\hat{\mathsf{G}}(x_i) - \nabla\hat{f}(x_i)\|_*^2] \leq \sigma_{x,\hat{f}}^2, \\
&\mathbb{E}[\|\hat{\mathsf{A}}_x(x_i) - Ax_i\|_*^2] \leq \sigma_y^2, \\
&\mathbb{E}[\|\hat{\mathsf{A}}_y(y_i) - A^T y_i\|_*^2] \leq \sigma_{x,A}^2.
\end{aligned}$$

Sometimes we simply denote $\sigma_x := \sqrt{\sigma_{x,\hat{f}}^2 + \sigma_{x,A}^2}$ for the sake of notational convenience. It should also be noted that deterministic SPP is a special case of the above setting with $\sigma_x = \sigma_y = 0$.

We can apply a few stochastic optimization algorithms discussed earlier to solve the above stochastic SPP. More specifically, the stochastic mirror descent method, when applied to the stochastic SPP, will achieve a rate of convergence given by

$$\mathcal{O}\left\{(L_{\hat{f}}+\|A\|+\sigma_x+\sigma_y)\tfrac{1}{\sqrt{N}}\right\}. \tag{4.4.7}$$

Moreover, the accelerated stochastic gradient descent method, when applied to the aforementioned stochastic SPP, possesses a rate of convergence given by

$$\mathcal{O}\left\{\tfrac{L_{\hat{f}}}{N^2}+(\|A\|+\sigma_x+\sigma_y)\tfrac{1}{\sqrt{N}}\right\}, \tag{4.4.8}$$

which improves the bound in (4.4.55) in terms of its dependence on the Lipschitz constant $L_{\hat{f}}$.

Our goals in this section are to further accelerate the primal–dual method discussed in Sect. 3.6, to achieve the rate of convergence in (4.4.4) for deterministic SPP, and to deal with unbounded feasible sets X and Y. Moreover, we intend to develop a stochastic accelerated primal–dual method that can further improve the rate of convergence stated in (4.4.8). It is worth noting that further improvement of the complexity bound for SPP, in terms of the number of gradient computations of $\hat{f}$, will be discussed in Sect. 8.2.

4.4.1 Accelerated Primal–Dual Method

One possible limitation of primal–dual method in Sect. 3.6 when applied to problem (4.4.1) is that both $\hat{f}$ and $\hat{g}$ need to be simple enough. To make this algorithm applicable to more practical problems we consider more general cases, where $\hat{g}$ is simple, but $\hat{f}$ may not be so. In particular, we assume that $\hat{f}$ is a general smooth convex function satisfying (4.4.2). In this case, we can replace $\hat{f}$ by its linear approximation $\hat{f}(x_t)+\langle\nabla\hat{f}(x_t),x-x_t\rangle$ and obtain the so-called "linearized primal–dual method" in Algorithm 4.1. By some extra effort we can show that if for $t=1,\ldots,N$, $0<\theta_t=\tau_{t-1}/\tau_t=\eta_{t-1}/\eta_t\le 1$, and $L_{\hat{f}}\eta_t+\|A\|^2\eta_t\tau_t\le 1$, then (x^N,y^N) has an $\mathcal{O}((L_{\hat{f}}+\|A\|)/N)$ rate of convergence in the sense of the partial duality gap.

Algorithm 4.1 Linearized primal–dual method for solving deterministic SPP

1: Choose $x_1\in X$, $y_1\in Y$. Set $\bar{x}_1=x_1$.
2: For $t=1,\ldots,N$, calculate

$$y_{t+1}=\mathrm{argmin}_{y\in Y}\langle -A\bar{x}_t,y\rangle+\hat{g}(y)+\tfrac{1}{2\tau_t}\|y-y_t\|^2, \tag{4.4.9}$$

$$x_{t+1}=\mathrm{argmin}_{x\in X}\langle\nabla\hat{f}(x_t),x\rangle+\langle Ax,y_{t+1}\rangle+\tfrac{1}{2\eta_t}\|x-x_t\|^2, \tag{4.4.10}$$

$$\tilde{x}_{t+1}=\theta_t(x_{t+1}-x_t)+x_{t+1}. \tag{4.4.11}$$

3: Output $\bar{x}_N,\bar{y}_N$.

Algorithm 4.2 Accelerated primal–dual method for deterministic SPP

1: Choose $x_1 \in X, y_1 \in Y$. Set $\bar{x}_1 = x_1, \bar{y}_1 = y_1, \tilde{x}_1 = x_1$.
2: For $t = 1, 2, \ldots, N-1$, calculate

$$\underline{x}_t = (1-\beta_t^{-1})\bar{x}_t + \beta_t^{-1} x_t, \tag{4.4.14}$$

$$y_{t+1} = \mathrm{argmin}_{y\in Y} \langle -A\tilde{x}_t, y\rangle + \hat{g}(y) + \tfrac{1}{\tau_t} V_Y(y, y_t), \tag{4.4.15}$$

$$x_{t+1} = \mathrm{argmin}_{x\in X} \langle \nabla \hat{f}(\underline{x}_t), x\rangle + \langle x, A^T y_{t+1}\rangle + \tfrac{1}{\eta_t} V_X(x, x_t), \tag{4.4.16}$$

$$\tilde{x}_{t+1} = \theta_{t+1}(x_{t+1} - x_t) + x_{t+1}, \tag{4.4.17}$$

$$\bar{x}_{t+1} = (1-\beta_t^{-1})\bar{x}_t + \beta_t^{-1} x_{t+1}, \tag{4.4.18}$$

$$\bar{y}_{t+1} = (1-\beta_t^{-1})\bar{y}_t + \beta_t^{-1} y_{t+1}. \tag{4.4.19}$$

3: Output x_N^{ag}, y_N^{ag}.

In order to further improve the above rate of convergence of Algorithm 4.1, we propose an accelerated primal–dual (APD) method in Algorithm 4.2 which integrates the accelerated gradient descent algorithm into the linearized version of the primal dual method. For any $x, u \in X$ and $y, v \in Y$, the functions $V_X(\cdot,\cdot)$ and $V_Y(\cdot,\cdot)$ are Bregman divergences defined as

$$V_X(x,u) := \nu_X(x) - \nu_X(u) - \langle \nabla \nu_X(u), x-u\rangle, \tag{4.4.12}$$

$$V_Y(y,v) := \nu_Y(y) - \nu_Y(v) - \langle \nabla \nu_Y(v), y-v\rangle, \tag{4.4.13}$$

where $\nu_X(\cdot)$ and $\nu_Y(\cdot)$ are strongly convex functions with strong convexity parameters (modulus) 1. We assume that $\hat{g}(y)$ is a simple convex function, so that the optimization problem in (4.4.15) can be solved efficiently.

Note that if $\beta_t = 1$ for all $t \geq 1$, then $\underline{x}_t = x_t$, $\bar{x}_{t+1} = x_{t+1}$, and Algorithm 4.2 is the same as the linearized version of Algorithm 4.1. However, by specifying a different selection of β_t (e.g., $\beta_t = O(t)$), we can significantly improve the rate of convergence of Algorithm 4.2 in terms of its dependence on $L_{\hat{f}}$. It should be noted that the iteration cost for the APD algorithm is about the same as that for Algorithm 4.1.

In order to analyze the convergence of Algorithm 4.2, we use the same notion as the one we used in the analysis of the primal–dual method in Sect. 3.6 to characterize the solutions of (4.4.1). Specifically, denoting $Z = X \times Y$, for any $\tilde{z} = (\tilde{x}, \tilde{y}) \in Z$ and $z = (x, y) \in Z$, we define

$$Q(\tilde{z}, z) := \left[\hat{f}(\tilde{x}) + \langle A\tilde{x}, y\rangle - \hat{g}(y)\right] - \left[\hat{f}(x) + \langle Ax, \tilde{y}\rangle - \hat{g}(\tilde{y})\right]. \tag{4.4.20}$$

It can be easily seen that $\tilde{z}$ is a solution of problem (4.4.1), if and only if $Q(\tilde{z}, z) \leq 0$ for all $z \in Z$. Therefore, if Z is bounded, it is suggestive to use the gap function

$$g(\tilde{z}) := \max_{z\in Z} Q(\tilde{z}, z) \tag{4.4.21}$$

to assess the quality of a feasible solution $\tilde{z} \in Z$. In fact, we can show that $f(\tilde{x}) - f^* \leq g(\tilde{z})$ for all $\tilde{z} \in Z$, where f^* denotes the optimal value of problem (4.4.1).

However, if Z is unbounded, then $g(\tilde{z})$ is not well defined even for a nearly optimal solution $\tilde{z} \in Z$. Hence, in the sequel, we will consider the bounded and unbounded case separately, by employing a slightly different error measure for the latter situation. In particular, we establish the convergence of Algorithm 4.2 in Theorems 4.6 and 4.7 for the bounded and unbounded case, respectively.

We need to prove two technical results: Proposition 4.12 shows some important properties for the function $Q(\cdot,\cdot)$ in (4.4.20) and Lemma 4.7 establishes a bound on $Q(\bar{x}_t, z)$. Note that the following quantity will be used in the convergence analysis of the APD algorithm.

$$\gamma_t = \begin{cases} 1, & t = 1, \\ \theta_t^{-1}\gamma_{t-1}, & t \geq 2. \end{cases} \tag{4.4.22}$$

Proposition 4.12. *Assume that $\beta_t \geq 1$ for all t. If $\bar{z}_{t+1} = (\bar{x}_{t+1}, \bar{y}_{t+1})$ is generated by Algorithm 4.2, then for all $z = (x,y) \in Z$,*

$$\begin{aligned} &\beta_t Q(\bar{z}_{t+1}, z) - (\beta_t - 1)Q(\bar{z}_t, z) \\ &\leq \langle \nabla \hat{f}(\underline{x}_t), x_{t+1} - x\rangle + \tfrac{L_{\hat{f}}}{2\beta_t}\|x_{t+1} - x_t\|^2 \\ &\quad + [\hat{g}(y_{t+1}) - \hat{g}(y)] + \langle Ax_{t+1}, y\rangle - \langle Ax, y_{t+1}\rangle. \end{aligned} \tag{4.4.23}$$

Proof. By Eqs. (4.4.14) and (4.4.18), $\bar{x}_{t+1} - \underline{x}_t = \beta_t^{-1}(x_{t+1} - x_t)$. Using this observation and the convexity of $\hat{f}(\cdot)$, we have

$$\begin{aligned} \beta_t \hat{f}(\bar{x}_{t+1}) \leq\ & \beta_t \hat{f}(\underline{x}_t) + \beta_t\langle \nabla \hat{f}(\underline{x}_t), \bar{x}_{t+1} - \underline{x}_t\rangle + \tfrac{\beta_t L_{\hat{f}}}{2}\|\bar{x}_{t+1} - \underline{x}_t\|^2 \\ =\ & \beta_t \hat{f}(\underline{x}_t) + \beta_t\langle \nabla \hat{f}(\underline{x}_t), \bar{x}_{t+1} - \underline{x}_t\rangle + \tfrac{L_{\hat{f}}}{2\beta_t}\|x_{t+1} - x_t\|^2 \\ =\ & \beta_t \hat{f}(\underline{x}_t) + (\beta_t - 1)\langle \nabla \hat{f}(\underline{x}_t), \bar{x}_t - \underline{x}_t\rangle \\ & + \langle \nabla \hat{f}(\underline{x}_t), x_{t+1} - \underline{x}_t\rangle + \tfrac{L_{\hat{f}}}{2\beta_t}\|x_{t+1} - x_t\|^2 \\ =\ & (\beta_t - 1)\left[\hat{f}(\underline{x}_t) + \langle \nabla \hat{f}(\underline{x}_t), \bar{x}_t - \underline{x}_t\rangle\right] \\ & + \left[\hat{f}(\underline{x}_t) + \langle \nabla \hat{f}(\underline{x}_t), x_{t+1} - \underline{x}_t\rangle\right] + \tfrac{L_{\hat{f}}}{2\beta_t}\|x_{t+1} - x_t\|^2 \\ =\ & (\beta_t - 1)\left[\hat{f}(\underline{x}_t) + \langle \nabla \hat{f}(\underline{x}_t), \bar{x}_t - \underline{x}_t\rangle\right] + \left[\hat{f}(\underline{x}_t) + \langle \nabla \hat{f}(\underline{x}_t), x - \underline{x}_t\rangle\right] \\ & + \langle \nabla \hat{f}(\underline{x}_t), x_{t+1} - x\rangle \\ & + \tfrac{L_{\hat{f}}}{2\beta_t}\|x_{t+1} - x_t\|^2 \\ \leq\ & (\beta_t - 1)\hat{f}(\bar{x}_t) + \hat{f}(x) + \langle \nabla \hat{f}(\underline{x}_t), x_{t+1} - x\rangle + \tfrac{L_{\hat{f}}}{2\beta_t}\|x_{t+1} - x_t\|^2. \end{aligned}$$

Moreover, by (4.4.19) and the convexity of $\hat{g}(\cdot)$, we have

$$\begin{aligned} \beta_t \hat{g}(\bar{y}_{t+1}) - \beta_t \hat{g}(y) &\leq (\beta_t - 1)\hat{g}(\bar{y}_t) + \hat{g}(y_{t+1}) - \beta_t \hat{g}(y) \\ &= (\beta_t - 1)\left[\hat{g}(\bar{y}_t) - \hat{g}(y)\right] + \hat{g}(y_{t+1}) - \hat{g}(y). \end{aligned}$$

By (4.4.20), (4.4.18), (4.4.19), and the above two inequalities, we obtain

$$\begin{aligned}
&\beta_t Q(\bar{z}_{t+1},z) - (\beta_t - 1)Q(\bar{z}_t,z) \\
&= \beta_t\left\{\left[\hat{f}(\bar{x}_{t+1}) + \langle A\bar{x}_{t+1}, y\rangle - \hat{g}(y)\right] - \left[\hat{f}(x) + \langle Ax, \bar{y}_{t+1}\rangle - \hat{g}(\bar{y}_{t+1})\right]\right\} \\
&\quad -(\beta_t - 1)\left\{\left[\hat{f}(\bar{x}_t) + \langle A\bar{x}_t, y\rangle - \hat{g}(y)\right] - \left[\hat{f}(x) + \langle Ax, \bar{y}_t\rangle - \hat{g}(\bar{y}_t)\right]\right\} \\
&= \beta_t \hat{f}(\bar{x}_{t+1}) - (\beta_t - 1)\hat{f}(\bar{x}_t) - \hat{f}(x) + \beta_t\left[\hat{g}(\bar{y}_{t+1}) - \hat{g}(y)\right] \\
&\quad -(\beta_t - 1)\left[\hat{g}(\bar{y}_t) - \hat{g}(y)\right] + \langle A(\beta_t\bar{x}_{t+1} - (\beta_t - 1)\bar{x}_t), y\rangle \\
&\quad -\langle Ax, \beta_t\bar{y}_{t+1} - (\beta_t - 1)\bar{y}_t\rangle \\
&\le \langle \nabla\hat{f}(\underline{x}_t), x_{t+1} - x\rangle + \tfrac{L_{\hat{f}}}{2\beta_t}\|x_{t+1} - x_t\|^2 \\
&\quad +\hat{g}(y_{t+1}) - \hat{g}(y) + \langle Ax_{t+1}, y\rangle - \langle Ax, y_{t+1}\rangle.
\end{aligned}$$

■

Lemma 4.7 establishes a bound for $Q(\bar{z}_{t+1},z)$ for all $z \in Z$, which will be used in the proof of both Theorems 4.6 and 4.7.

Lemma 4.7. *Let $\bar{z}_{t+1} = (\bar{x}_{t+1}, \bar{y}_{t+1})$ be the iterates generated by Algorithm 4.2. Assume that the parameters $\beta_t, \theta_t, \eta_t$, and τ_t satisfy*

$$\beta_1 = 1,\ \beta_{t+1} - 1 = \beta_t\theta_{t+1}, \tag{4.4.24}$$

$$0 < \theta_t \le \min\{\tfrac{\eta_{t-1}}{\eta_t}, \tfrac{\tau_{t-1}}{\tau_t}\}, \tag{4.4.25}$$

$$\tfrac{1}{\eta_t} - \tfrac{L_{\hat{f}}}{\beta_t} - \|A\|^2\tau_t \ge 0. \tag{4.4.26}$$

Then, for any $z \in Z$, we have

$$\begin{aligned}
\beta_t\gamma_t Q(\bar{z}_{t+1},z) &\le \mathrm{B}_t(z, z_{[t]}) + \gamma_t\langle A(x_{t+1} - x_t), y - y_{t+1}\rangle \\
&\quad - \gamma_t\left(\tfrac{1}{2\eta_t} - \tfrac{L_{\hat{f}}}{2\beta_t}\right)\|x_{t+1} - x_t\|^2,
\end{aligned} \tag{4.4.27}$$

where γ_t is defined in (4.4.22), $z_{[t]} := \{(x_i, y_i)\}_{i=1}^{t+1}$, and

$$\mathrm{B}_t(z, z_{[t]}) := \textstyle\sum_{i=1}^{t}\left\{\tfrac{\gamma_i}{\eta_i}\left[V_X(x, x_i) - V_X(x, x_{i+1})\right] + \tfrac{\gamma_i}{\tau_i}\left[V_Y(y, y_i) - V_Y(y, y_{i+1})\right]\right\}. \tag{4.4.28}$$

Proof. First of all, we explore the optimality conditions of (4.4.15) and (4.4.16). Applying Lemma 3.5 to (4.4.15), we have

$$\begin{aligned}
&\langle -A\tilde{x}_t, y_{t+1} - y\rangle + \hat{g}(y_{t+1}) - \hat{g}(y) \\
&\le \tfrac{1}{\tau_t}V_Y(y, y_t) - \tfrac{1}{\tau_t}V_Y(y_{t+1}, y_t) - \tfrac{1}{\tau_t}V_Y(y, y_{t+1}) \\
&\le \tfrac{1}{\tau_t}V_Y(y, y_t) - \tfrac{1}{2\tau_t}\|y_{t+1} - y_t\|^2 - \tfrac{1}{\tau_t}V_Y(y, y_{t+1}),
\end{aligned} \tag{4.4.29}$$

where the last inequality follows from the fact that, by the strong convexity of $v_Y(\cdot)$ and (4.4.13),

$$V_Y(y_1, y_2) \ge \tfrac{1}{2}\|y_1 - y_2\|^2, \text{ for all } y_1, y_2 \in Y. \tag{4.4.30}$$

Similarly, from (4.4.16) we can derive that

$$\begin{aligned}&\langle \nabla \hat{f}(\underline{x}_t), x_{t+1} - x\rangle + \langle x_{t+1} - x, A^T y_{t+1}\rangle \\ &\quad \le \tfrac{1}{\eta_t} V_X(x, x_t) - \tfrac{1}{2\eta_t}\|x_{t+1} - x_t\|^2 - \tfrac{1}{\eta_t} V_X(x, x_{t+1}).\end{aligned} \tag{4.4.31}$$

Our next step is to establish a crucial recursion of Algorithm 4.2. It follows from (4.4.23), (4.4.29), and (4.4.31) that

$$\begin{aligned}&\beta_t Q(\bar{z}_{t+1}, z) - (\beta_t - 1) Q(\bar{z}_t, z) \\ &\le \langle \nabla \hat{f}(\underline{x}_t), x_{t+1} - x\rangle + \tfrac{L_{\hat{f}}}{2\beta_t}\|x_{t+1} - x_t\|^2 + [\hat{g}(y_{t+1}) - \hat{g}(y)] + \langle A x_{t+1}, y\rangle - \langle A x, y_{t+1}\rangle \\ &\le \tfrac{1}{\eta_t} V_X(x, x_t) - \tfrac{1}{\eta_t} V_X(x, x_{t+1}) - \left(\tfrac{1}{2\eta_t} - \tfrac{L_{\hat{f}}}{2\beta_t}\right)\|x_{t+1} - x_t\|^2 \\ &\quad + \tfrac{1}{\tau_t} V_Y(y, y_t) - \tfrac{1}{\tau_t} V_Y(y, y_{t+1}) - \tfrac{1}{2\tau_t}\|y_{t+1} - y_t\|^2 \\ &\quad - \langle x_{t+1} - x, A^T y_{t+1}\rangle + \langle A\tilde{x}_t, y_{t+1} - y\rangle + \langle A x_{t+1}, y\rangle - \langle A x, y_{t+1}\rangle.\end{aligned} \tag{4.4.32}$$

Also observe that by (4.4.17), we have

$$\begin{aligned}&- \langle x_{t+1} - x, A^T y_{t+1}\rangle + \langle A\tilde{x}_t, y_{t+1} - y\rangle + \langle A x_{t+1}, y\rangle - \langle A x, y_{t+1}\rangle \\ &= \langle A(x_{t+1} - x_t), y - y_{t+1}\rangle - \theta_t\langle A(x_t - x_{t-1}), y - y_{t+1}\rangle \\ &= \langle A(x_{t+1} - x_t), y - y_{t+1}\rangle - \theta_t\langle A(x_t - x_{t-1}), y - y_t\rangle - \theta_t\langle A(x_t - x_{t-1}), y_t - y_{t+1}\rangle.\end{aligned}$$

Multiplying both sides of (4.4.32) by γ_t, using the above identity and the fact that $\gamma_t\theta_t = \gamma_{t-1}$ due to (4.4.22), we obtain

$$\begin{aligned}&\beta_t\gamma_t Q(\bar{z}_{t+1}, z) - (\beta_t - 1)\gamma_t Q(\bar{z}_t, z) \\ &\le \tfrac{\gamma_t}{\eta_t} V_X(x, x_t) - \tfrac{\gamma_t}{\eta_t} V_X(x, x_{t+1}) + \tfrac{\gamma_t}{\tau_t} V_Y(y, y_t) - \tfrac{\gamma_t}{\tau_t} V_Y(y, y_{t+1}) \\ &\quad + \gamma_t\langle A(x_{t+1} - x_t), y - y_{t+1}\rangle - \gamma_{t-1}\langle A(x_t - x_{t-1}), y - y_t\rangle \\ &\quad - \gamma_t\left(\tfrac{1}{2\eta_t} - \tfrac{L_{\hat{f}}}{2\beta_t}\right)\|x_{t+1} - x_t\|^2 - \tfrac{\gamma_t}{2\tau_t}\|y_{t+1} - y_t\|^2 \\ &\quad - \gamma_{t-1}\langle A(x_t - x_{t-1}), y_t - y_{t+1}\rangle.\end{aligned} \tag{4.4.33}$$

Now, applying Cauchy–Schwartz inequality to the last term in (4.4.33), using the notation $\|A\|$ in (4.4.2) and noticing that $\gamma_{t-1}/\gamma_t = \theta_t \le \min\{\eta_{t-1}/\eta_t, \tau_{t-1}/\tau_t\}$ from (4.4.25), we have

$$\begin{aligned}&-\gamma_{t-1}\langle A(x_t - x_{t-1}), y_t - y_{t+1}\rangle \le \gamma_{t-1}\|A(x_t - x_{t-1})\|_*\|y_t - y_{t+1}\| \\ &\le \|A\|\gamma_{t-1}\|x_t - x_{t-1}\|\,\|y_t - y_{t+1}\| \le \tfrac{\|A\|^2\gamma_{t-1}^2\tau_t}{2\gamma_t}\|x_t - x_{t-1}\|^2 + \tfrac{\gamma_t}{2\tau_t}\|y_t - y_{t+1}\|^2 \\ &\le \tfrac{\|A\|^2\gamma_{t-1}\tau_{t-1}}{2}\|x_t - x_{t-1}\|^2 + \tfrac{\gamma_t}{2\tau_t}\|y_t - y_{t+1}\|^2.\end{aligned}$$

Noting that $\theta_{t+1} = \gamma_t/\gamma_{t+1}$, so by (4.4.24) we have $(\beta_{t+1} - 1)\gamma_{t+1} = \beta_t\gamma_t$. Combining the above two relations with inequality (4.4.33), we get the following recursion for Algorithm 4.2.

$$
\begin{aligned}
&(\beta_{t+1}-1)\gamma_{t+1}Q(\bar{z}_{t+1},z)-(\beta_t-1)\gamma_t Q(\bar{z}_t,z) = \beta_t\gamma_t Q(\bar{z}_{t+1},z)-(\beta_t-1)\gamma_t Q(\bar{z}_t,z)\\
\le\ & \tfrac{\gamma_t}{\eta_t}V_X(x,x_t)-\tfrac{\gamma_t}{\eta_t}V_X(x,x_{t+1})+\tfrac{\gamma_t}{\tau_t}V_Y(y,y_t)-\tfrac{\gamma_t}{\tau_t}V_Y(y,y_{t+1})\\
&+\gamma_t\langle A(x_{t+1}-x_t),y-y_{t+1}\rangle-\gamma_{t-1}\langle A(x_t-x_{t-1}),y-y_t\rangle\\
&-\gamma_t\left(\tfrac{1}{2\eta_t}-\tfrac{L_{\hat{f}}}{2\beta_t}\right)\|x_{t+1}-x_t\|^2+\tfrac{\|A\|^2\gamma_{t-1}\tau_{t-1}}{2}\|x_t-x_{t-1}\|^2, \forall t\ge 1.
\end{aligned}
$$

Applying the above inequality inductively and assuming that $x_0 = x_1$, we conclude that

$$
\begin{aligned}
&(\beta_{t+1}-1)\gamma_{t+1}Q(\bar{z}_{t+1},z)-(\beta_1-1)\gamma_1 Q(\bar{z}_1,z)\\
&\le \mathrm{B}_t(z,z_{[t]})+\gamma_t\langle A(x_{t+1}-x_t),y-y_{t+1}\rangle\\
&-\gamma_t\left(\tfrac{1}{2\eta_t}-\tfrac{L_{\hat{f}}}{2\beta_t}\right)\|x_{t+1}-x_t\|^2-\textstyle\sum_{i=1}^{t-1}\gamma_i\left(\tfrac{1}{2\eta_i}-\tfrac{L_{\hat{f}}}{2\beta_i}-\tfrac{\|A\|^2\tau_i}{2}\right)\|x_{i+1}-x_i\|^2,
\end{aligned}
$$

which, in view of (4.4.26) and the facts that $\beta_1 = 1$ and $(\beta_{t+1}-1)\gamma_{t+1} = \beta_t\gamma_t$ by (4.4.24), implies (4.4.27). ■

Theorem 4.6 below describes the convergence properties of Algorithm 4.2 when Z is bounded. This result follows as an immediate consequence of Lemma 4.7.

Theorem 4.6. *Assume that for some $D_X, D_Y > 0$:*

$$
\sup_{x_1,x_2\in X} V_X(x_1,x_2)\le D_X^2 \ \text{ and } \ \sup_{y_1,y_2\in Y} V_Y(y_1,y_2)\le D_Y^2. \tag{4.4.34}
$$

Also assume that the parameters $\beta_t, \theta_t, \eta_t, \tau_t$ in Algorithm 4.2 are chosen such that (4.4.24), (4.4.25), and (4.4.26) hold. Then for all $t \ge 1$,

$$
g(\bar{z}_{t+1})\le \tfrac{1}{\beta_t\eta_t}D_X^2+\tfrac{1}{\beta_t\tau_t}D_Y^2. \tag{4.4.35}
$$

Proof. Let $\mathrm{B}_t(z,z_{[t]})$ be defined in (4.4.28). First note that by the definition of γ_t in (4.4.22) and relation (4.4.25), we have $\theta_t = \gamma_{t-1}/\gamma_t \le \eta_{t-1}/\eta_t$ and hence $\gamma_{t-1}/\eta_{t-1} \le \gamma_t/\eta_t$. Using this observation and (4.4.34), we conclude that

$$
\begin{aligned}
\mathrm{B}_t(z,z_{[t]}) =\ & \tfrac{\gamma_1}{\eta_1}V_X(x,x_1)-\textstyle\sum_{i=1}^{t-1}\left(\tfrac{\gamma_i}{\eta_i}-\tfrac{\gamma_{i+1}}{\eta_{i+1}}\right)V_X(x,x_{i+1})-\tfrac{\gamma_t}{\eta_t}V_X(x,x_{t+1})\\
&+\tfrac{\gamma_1}{\tau_1}V_Y(y,y_1)-\textstyle\sum_{i=1}^{t-1}\left(\tfrac{\gamma_i}{\tau_i}-\tfrac{\gamma_{i+1}}{\tau_{i+1}}\right)V_Y(y,y_{i+1})-\tfrac{\gamma_t}{\tau_t}V_Y(y,y_{t+1})\\
\le\ & \tfrac{\gamma_1}{\eta_1}D_X^2-\textstyle\sum_{i=1}^{t-1}\left(\tfrac{\gamma_i}{\eta_i}-\tfrac{\gamma_{i+1}}{\eta_{i+1}}\right)D_X^2-\tfrac{\gamma_t}{\eta_t}V_X(x,x_{t+1})\\
&+\tfrac{\gamma_1}{\tau_1}D_Y^2-\textstyle\sum_{i=1}^{t-1}\left(\tfrac{\gamma_i}{\tau_i}-\tfrac{\gamma_{i+1}}{\tau_{i+1}}\right)D_Y^2-\tfrac{\gamma_t}{\tau_t}V_Y(y,y_{t+1})\\
=\ & \tfrac{\gamma_t}{\eta_t}D_X^2-\tfrac{\gamma_t}{\eta_t}V_X(x,x_{t+1})+\tfrac{\gamma_t}{\tau_t}D_Y^2-\tfrac{\gamma_t}{\tau_t}V_Y(y,y_{t+1}).
\end{aligned} \tag{4.4.36}
$$

Now applying the Cauchy–Schwartz inequality to the inner product term in (4.4.27), we get

$$
\begin{aligned}
&\gamma_t\langle A(x_{t+1}-x_t),y-y_{t+1}\rangle\le\|A\|\gamma_t\|x_{t+1}-x_t\|\|y-y_{t+1}\|\\
&\le\tfrac{\|A\|^2\gamma_t\tau_t}{2}\|x_{t+1}-x_t\|^2+\tfrac{\gamma_t}{2\tau_t}\|y-y_{t+1}\|^2.
\end{aligned} \tag{4.4.37}
$$

Using the above two relations, (4.4.26), (4.4.27), and (4.4.30), we have

$$\beta_t \gamma_t Q(\bar{z}_{t+1}, z) \leq \tfrac{\gamma_t}{\eta_t} D_X^2 - \tfrac{\gamma_t}{\eta_t} V_X(x, x_{t+1}) + \tfrac{\gamma_t}{\tau_t} D_Y^2 - \tfrac{\gamma_t}{\tau_t} \left(V_Y(y, y_{t+1}) - \tfrac{1}{2}\|y - y_{t+1}\|^2\right)$$
$$-\gamma_t \left(\tfrac{1}{2\eta_t} - \tfrac{L_{\hat{f}}}{2\beta_t} - \tfrac{\|A\|^2 \tau_t}{2}\right) \|x_{t+1} - x_t\|^2 \leq \tfrac{\gamma_t}{\eta_t} D_X^2 + \tfrac{\gamma_t}{\tau_t} D_Y^2, \ \forall z \in Z,$$

which together with (4.4.21) then clearly imply (4.4.35). ∎

There are various options for choosing the parameters β_t, η_t, τ_t, and θ_t such that (4.4.24)–(4.4.26) hold. Below we provide such an example.

Corollary 4.1. *Suppose that (4.4.34) holds. In Algorithm 4.2, if the parameters are set to*

$$\beta_t = \tfrac{t+1}{2}, \ \theta_t = \tfrac{t-1}{t}, \ \eta_t = \tfrac{t}{2L_{\hat{f}} + t\|A\| D_Y / D_X} \ \text{and} \ \tau_t = \tfrac{D_Y}{\|A\| D_X}, \tag{4.4.38}$$

then for all $t \geq 2$,

$$g(\bar{z}_t) \leq \tfrac{4L_{\hat{f}} D_X^2}{t(t-1)} + \tfrac{4\|A\| D_X D_Y}{t}. \tag{4.4.39}$$

Proof. It suffices to verify that the parameters in (4.4.38) satisfy (4.4.24)–(4.4.26) in Theorem 4.6. It is easy to check that (4.4.24) and (4.4.25) hold. Furthermore,

$$\tfrac{1}{\eta_t} - \tfrac{L_{\hat{f}}}{\beta_t} - \|A\|^2 \tau_t = \tfrac{2L_{\hat{f}} + t\|A\| D_Y / D_X}{t} - \tfrac{2L_{\hat{f}}}{t+1} - \tfrac{\|A\| D_Y}{D_X} \geq 0,$$

so (4.4.26) holds. Therefore, by (4.4.35), for all $t \geq 1$ we have

$$\begin{aligned} g(\bar{z}_t) &\leq \tfrac{1}{\beta_{t-1}\eta_{t-1}} D_X^2 + \tfrac{1}{\beta_{t-1}\tau_{t-1}} D_Y^2 = \tfrac{4L_{\hat{f}} + 2(t-1)\|A\| D_Y / D_X}{t(t-1)} \cdot D_X^2 + \tfrac{2\|A\| D_X / D_Y}{t} \cdot D_Y^2 \\ &= \tfrac{4L_{\hat{f}} D_X^2}{t(t-1)} + \tfrac{4\|A\| D_X D_Y}{t}. \end{aligned}$$

∎

Clearly, in view of (4.4.4), the rate of convergence of Algorithm 4.2 applied to problem (4.4.1) is optimal when the parameters are chosen according to (4.4.38). Also observe that we need to estimate D_Y / D_X to use these parameters. However, it should be pointed out that replacing the ratio D_Y / D_X in (4.4.38) by any positive constant only results in an increase in the RHS of (4.4.39) by a constant factor.

Now, we study the convergence properties of the APD algorithm for the case when $Z = X \times Y$ is unbounded, by using a perturbation-based termination criterion based on the enlargement of a maximal monotone operator. More specifically, it can be seen that there always exists a perturbation vector v such that

$$\tilde{g}(\tilde{z}, v) := \max_{z \in Z} Q(\tilde{z}, z) - \langle v, \tilde{z} - z \rangle \tag{4.4.40}$$

is well defined, although the value of $g(\tilde{z})$ in (4.4.21) may be unbounded if Z is unbounded. In the following result, we show that the APD algorithm can compute a nearly optimal solution $\tilde{z}$ with a small residue $\tilde{g}(\tilde{z}, v)$, for a small perturbation vector v (i.e., $\|v\|$ is small). In addition, our derived iteration complexity bounds are proportional to the distance from the initial point to the solution set. For the case when Z is unbounded, we assume that $V_X(x,x_t) = \|x-x_t\|^2/2$ and $V_Y(y,y_t) = \|y-y_t\|^2/2$ in Algorithm 4.2, where the norms are induced by the inner products.

We will first prove a technical result which specializes the results in Lemma 4.7 for the case when (4.4.24), (4.4.41), and (4.4.42) hold.

Lemma 4.8. *Let $\hat{z} = (\hat{x},\hat{y}) \in Z$ be a saddle point of* (4.4.1). *If the parameters $\beta_t, \theta_t, \eta_t$ and τ_t satisfy* (4.4.24),

$$\theta_t = \tfrac{\eta_{t-1}}{\eta_t} = \tfrac{\tau_{t-1}}{\tau_t}, \tag{4.4.41}$$

$$\tfrac{1}{\eta_t} - \tfrac{L_{\hat{f}}}{\beta_t} - \tfrac{\|A\|^2\tau_t}{p} \geq 0, \tag{4.4.42}$$

then for all $t \geq 1$,

$$\|\hat{x}-x_{t+1}\|^2 + \tfrac{\eta_t(1-p)}{\tau_t}\|\hat{y}-y_{t+1}\|^2 \leq \|\hat{x}-x_1\|^2 + \tfrac{\eta_t}{\tau_t}\|\hat{y}-y_1\|^2, \tag{4.4.43}$$

$$\tilde{g}(\bar{z}_{t+1}, v_{t+1}) \leq \tfrac{1}{2\beta_t\eta_t}\|\bar{x}_{t+1}-x_1\|^2 + \tfrac{1}{2\beta_t\tau_t}\|\bar{y}_{t+1}-y_1\|^2 =: \delta_{t+1}, \tag{4.4.44}$$

where $\tilde{g}(\cdot,\cdot)$ is defined in (4.4.40) and

$$v_{t+1} = \left(\tfrac{1}{\beta_t\eta_t}(x_1-x_{t+1}), \tfrac{1}{\beta_t\tau_t}(y_1-y_{t+1}) - \tfrac{1}{\beta_t}A(x_{t+1}-x_t)\right). \tag{4.4.45}$$

Proof. It is easy to check that the conditions in Lemma 4.7 are satisfied. By (4.4.41), (4.4.27) in Lemma 4.7 becomes

$$\begin{aligned}\beta_t Q(\bar{z}_{t+1},z) \leq\ & \tfrac{1}{2\eta_t}\|x-x_1\|^2 - \tfrac{1}{2\eta_t}\|x-x_{t+1}\|^2 + \tfrac{1}{2\tau_t}\|y-y_1\|^2 - \tfrac{1}{2\tau_t}\|y-y_{t+1}\|^2 \\ & + \langle A(x_{t+1}-x_t), y-y_{t+1}\rangle - \left(\tfrac{1}{2\eta_t} - \tfrac{L_{\hat{f}}}{2\beta_t}\right)\|x_{t+1}-x_t\|^2.\end{aligned} \tag{4.4.46}$$

To prove (4.4.43), observe that

$$\langle A(x_{t+1}-x_t), y-y_{t+1}\rangle \leq \tfrac{\|A\|^2\tau_t}{2p}\|x_{t+1}-x_t\|^2 + \tfrac{p}{2\tau_t}\|y-y_{t+1}\|^2 \tag{4.4.47}$$

where p is the constant in (4.4.42). By (4.4.42) and the above two inequalities, we get

$$\beta_t Q(\bar{z}_{t+1},z) \leq \tfrac{1}{2\eta_t}\|x-x_1\|^2 - \tfrac{1}{2\eta_t}\|x-x_{t+1}\|^2 + \tfrac{1}{2\tau_t}\|y-y_1\|^2 - \tfrac{1-p}{2\tau_t}\|y-y_{t+1}\|^2.$$

Letting $z = \hat{z}$ in the above, and using the fact that $Q(\bar{z}_{t+1},\hat{z}) \geq 0$, we obtain (4.4.43). Now we prove (4.4.44). Noting that

$$\begin{aligned}&\|x-x_1\|^2 - \|x-x_{t+1}\|^2 = 2\langle x_{t+1}-x_1, x\rangle + \|x_1\|^2 - \|x_{t+1}\|^2 \\ =&2\langle x_{t+1}-x_1, x-\bar{x}_{t+1}\rangle + 2\langle x_{t+1}-x_1, \bar{x}_{t+1}\rangle + \|x_1\|^2 - \|x_{t+1}\|^2 \\ =&2\langle x_{t+1}-x_1, x-\bar{x}_{t+1}\rangle + \|\bar{x}_{t+1}-x_1\|^2 - \|\bar{x}_{t+1}-x_{t+1}\|^2,\end{aligned} \tag{4.4.48}$$

we conclude from (4.4.42) and (4.4.46) that for any $z \in Z$,

$$\begin{aligned}
&\beta_t Q(\bar{z}_{t+1}, z) + \langle A(x_{t+1} - x_t), \bar{y}_{t+1} - y\rangle \\
&\quad - \tfrac{1}{\eta_t}\langle x_1 - x_{t+1}, \bar{x}_{t+1} - x\rangle - \tfrac{1}{\tau_t}\langle y_1 - y_{t+1}, \bar{y}_{t+1} - y\rangle \\
&\le \tfrac{1}{2\eta_t}\left(\|\bar{x}_{t+1} - x_1\|^2 - \|\bar{x}_{t+1} - x_{t+1}\|^2\right) + \tfrac{1}{2\tau_t}\left(\|\bar{y}_{t+1} - y_1\|^2 - \|\bar{y}_{t+1} - y_{t+1}\|^2\right) \\
&\quad + \langle A(x_{t+1} - x_t), \bar{y}_{t+1} - y_{t+1}\rangle - \left(\tfrac{1}{2\eta_t} - \tfrac{L_{\hat{f}}}{2\beta_t}\right)\|x_{t+1} - x_t\|^2 \\
&\le \tfrac{1}{2\eta_t}\left(\|\bar{x}_{t+1} - x_1\|^2 - \|\bar{x}_{t+1} - x_{t+1}\|^2\right) + \tfrac{1}{2\tau_t}\left(\|\bar{y}_{t+1} - y_1\|^2 - \|\bar{y}_{t+1} - y_{t+1}\|^2\right) \\
&\quad + \tfrac{p}{2\tau_t}\|\bar{y}_{t+1} - y_{t+1}\|^2 - \left(\tfrac{1}{2\eta_t} - \tfrac{L_{\hat{f}}}{2\beta_t} - \tfrac{\|A\|^2\tau_t}{2p}\right)\|x_{t+1} - x_t\|^2 \\
&\le \tfrac{1}{2\eta_t}\|\bar{x}_{t+1} - x_1\|^2 + \tfrac{1}{2\tau_t}\|\bar{y}_{t+1} - y_1\|^2.
\end{aligned}$$

The result in (4.4.44) and (4.4.45) immediately follows from the above inequality and (4.4.40).

■

We are now ready to establish the convergence of Algorithm 4.2 when X or Y is unbounded.

Theorem 4.7. *Let $\{\bar{z}_t\} = \{(\bar{x}_t, \bar{y}_t)\}$ be the iterates generated by Algorithm 4.2 with $V_X(x, x_t) = \|x - x_t\|^2/2$ and $V_Y(y, y_t) = \|y - y_t\|^2/2$. Assume that the parameters $\beta_t, \theta_t, \eta_t$, and τ_t satisfy* (4.4.24), *(4.4.41), and (4.4.42) for all $t \ge 1$ and for some $0 < p < 1$, then there exists a perturbation vector v_{t+1} such that*

$$\tilde{g}(\bar{z}_{t+1}, v_{t+1}) \le \tfrac{(2-p)D^2}{\beta_t \eta_t (1-p)} =: \varepsilon_{t+1} \tag{4.4.49}$$

for any $t \ge 1$. Moreover, we have

$$\|v_{t+1}\| \le \tfrac{1}{\beta_t \eta_t}\|\hat{x} - x_1\| + \tfrac{1}{\beta_t \tau_t}\|\hat{y} - y_1\| + \left[\tfrac{1}{\beta_t \eta_t}\left(1 + \sqrt{\tfrac{\eta_1}{\tau_1(1-p)}}\right) + \tfrac{2\|A\|}{\beta_t}\right] D, \tag{4.4.50}$$

where $(\hat{x}, \hat{y})$ is a pair of solutions for problem (4.4.1) and

$$D := \sqrt{\|\hat{x} - x_1\|^2 + \tfrac{\eta_1}{\tau_1}\|\hat{y} - y_1\|^2}. \tag{4.4.51}$$

Proof. We have established the expression of v_{t+1} and δ_{t+1} in Lemma 4.8. It suffices to estimate the bound on $\|v_{t+1}\|$ and δ_{t+1}. It follows from the definition of D, (4.4.41), and (4.4.43) that for all $t \ge 1$, $\|\hat{x} - x_{t+1}\| \le D$ and $\|\hat{y} - y_{t+1}\| \le D\sqrt{\tfrac{\tau_1}{\eta_1(1-p)}}$. Now by (4.4.45), we have

$$\begin{aligned}
\|v_{t+1}\| &\le \tfrac{1}{\beta_t \eta_t}\|x_1 - x_{t+1}\| + \tfrac{1}{\beta_t \tau_t}\|y_1 - y_{t+1}\| + \tfrac{\|A\|}{\beta_t}\|x_{t+1} - x_t\| \\
&\le \tfrac{1}{\beta_t \eta_t}\left(\|\hat{x} - x_1\| + \|\hat{x} - x_{t+1}\|\right) + \tfrac{1}{\beta_t \tau_t}\left(\|\hat{y} - y_1\| + \|\hat{y} - y_{t+1}\|\right) \\
&\quad + \tfrac{\|A\|}{\beta_t}\left(\|\hat{x} - x_{t+1}\| + \|\hat{x} - x_t\|\right) \\
&\le \tfrac{1}{\beta_t \eta_t}\left(\|\hat{x} - x_1\| + D\right) + \tfrac{1}{\beta_t \tau_t}\left(\|\hat{y} - y_1\| + D\sqrt{\tfrac{\tau_1}{\eta_1(1-p)}}\right) + \tfrac{2\|A\|}{\beta_t} D \\
&= \tfrac{1}{\beta_t \eta_t}\|\hat{x} - x_1\| + \tfrac{1}{\beta_t \tau_t}\|\hat{y} - y_1\| + D\left[\tfrac{1}{\beta_t \eta_t}\left(1 + \sqrt{\tfrac{\eta_1}{\tau_1(1-p)}}\right) + \tfrac{2\|A\|}{\beta_t}\right].
\end{aligned}$$

To estimate the bound of δ_{t+1}, consider the sequence $\{\gamma_t\}$ defined in (4.4.22). Using the fact that $(\beta_{t+1}-1)\gamma_{t+1}=\beta_t\gamma_t$ due to (4.4.24) and (4.4.22), and applying (4.4.18) and (4.4.19) inductively, we have

$$\bar{x}_{t+1}=\tfrac{1}{\beta_t\gamma_t}\textstyle\sum_{i=1}^t\gamma_i x_{i+1},\ \bar{y}_{t+1}=\tfrac{1}{\beta_t\gamma_t}\sum_{i=1}^t\gamma_i y_{i+1}\ \text{and}\ \tfrac{1}{\beta_t\gamma_t}\sum_{i=1}^t\gamma_i=1. \qquad (4.4.52)$$

Thus $\bar{x}_{t+1}$ and $\bar{y}_{t+1}$ are convex combinations of sequences $\{x_{i+1}\}_{i=1}^t$ and $\{y_{i+1}\}_{i=1}^t$. Using these relations and (4.4.43), we have

$$\begin{aligned}
\delta_{t+1} &= \tfrac{1}{2\beta_t\eta_t}\|\bar{x}_{t+1}-x_1\|^2+\tfrac{1}{2\beta_t\tau_t}\|\bar{y}_{t+1}-y_1\|^2\\
&\le \tfrac{1}{\beta_t\eta_t}\left(\|\hat{x}-\bar{x}_{t+1}\|^2+\|\hat{x}-x_1\|^2\right)+\tfrac{1}{\beta_t\tau_t}\left(\|\hat{y}-\bar{y}_{t+1}\|^2+\|\hat{y}-y_1\|^2\right)\\
&= \tfrac{1}{\beta_t\eta_t}\left(D^2+\|\hat{x}-\bar{x}_{t+1}\|^2+\tfrac{\eta_t(1-p)}{\tau_t}\|\hat{y}-\bar{y}_{t+1}\|^2+\tfrac{\eta_t p}{\tau_t}\|\hat{y}-\bar{y}_{t+1}\|^2\right)\\
&\le \tfrac{1}{\beta_t\eta_t}\left[D^2+\tfrac{1}{\beta_t\gamma_t}\textstyle\sum_{i=1}^t\gamma_i\left(\|\hat{x}-x_{i+1}\|^2+\tfrac{\eta_t(1-p)}{\tau_t}\|\hat{y}-y_{i+1}\|^2+\tfrac{\eta_t p}{\tau_t}\|\hat{y}-y_{i+1}\|^2\right)\right]\\
&\le \tfrac{1}{\beta_t\eta_t}\left[D^2+\tfrac{1}{\beta_t\gamma_t}\textstyle\sum_{i=1}^t\gamma_i\left(D^2+\tfrac{\eta_t p}{\tau_t}\cdot\tfrac{\tau_1}{\eta_1(1-p)}D^2\right)\right]=\tfrac{(2-p)D^2}{\beta_t\eta_t(1-p)}.
\end{aligned}$$

■

Below we suggest a specific parameter setting which satisfies (4.4.24), (4.4.41), and (4.4.42).

Corollary 4.2. *In Algorithm 4.2, if N is given and the parameters are set to*

$$\beta_t=\tfrac{t+1}{2},\ \theta_t=\tfrac{t-1}{t},\ \eta_t=\tfrac{t+1}{2(L_{\hat{f}}+N\|A\|)},\ \textit{and}\ \tau_t=\tfrac{t+1}{2N\|A\|} \qquad (4.4.53)$$

then there exists v_N that satisfies (4.4.49) *with*

$$\varepsilon_N\le\frac{10L_{\hat{f}}\hat{D}^2}{N^2}+\frac{10\|A\|\hat{D}^2}{N}\ \textit{and}\ \|v_N\|\le\frac{15L_{\hat{f}}\hat{D}}{N^2}+\frac{19\|A\|\hat{D}}{N}, \qquad (4.4.54)$$

where $\hat{D}=\sqrt{\|\hat{x}-x_1\|^2+\|\hat{y}-y_1\|^2}$.

Proof. For the parameters β_t, γ_t, η_t, and τ_t in (4.4.53), it is clear that (4.4.24), (4.4.41) hold. Furthermore, let $p=1/4$, for any $t=1,\ldots,N-1$, we have

$$\frac{1}{\eta_t}-\frac{L_{\hat{f}}}{\beta_t}-\frac{\|A\|^2\tau_t}{p}=\frac{2L_{\hat{f}}+2\|A\|N}{t+1}-\frac{2L_{\hat{f}}}{t+1}-\frac{2\|A\|^2(t+1)}{\|A\|N}\ge\frac{2\|A\|N}{t+1}-\frac{2\|A\|(t+1)}{N}\ge 0,$$

thus (4.4.42) holds. By Theorem 4.7, inequalities (4.4.49) and (4.4.50) hold. Noting that $\eta_t\le\tau_t$, in (4.4.49) and (4.4.50) we have $D\le\hat{D}$, $\|\hat{x}-x_1\|+\|\hat{y}-y_1\|\le\sqrt{2}\hat{D}$, hence

$$\|v_{t+1}\|\le\frac{\sqrt{2}\hat{D}}{\beta_t\eta_t}+\frac{(1+\sqrt{4/3})\hat{D}}{\beta_t\eta_t}+\frac{2\|A\|\hat{D}}{\beta_t}$$

and

$$\varepsilon_{t+1}\le\frac{(2-p)\hat{D}^2}{\beta_t\eta_t(1-p)}=\frac{7\hat{D}^2}{3\beta_t\eta_t}.$$

Also note that by (4.4.53), $\frac{1}{\beta_{N-1}\eta_{N-1}} = \frac{4(L_{\hat{f}}+\|A\|N)}{N^2} = \frac{4L_{\hat{f}}}{N^2} + \frac{4\|A\|}{N}$. Using these three relations and the definition of β_t in (4.4.53), we obtain (4.4.54) after simplifying the constants. ■

It is interesting to notice that if the parameters in Algorithm 4.2 are set to (4.4.53), then both residues ε_N and $\|v_N\|$ in (4.4.54) reduce to zero with approximately the same rate of convergence (up to a factor of $\hat{D}$).

4.4.2 Stochastic Bilinear Saddle Point Problems

In order to solve stochastic SPP, i.e., problem (4.4.1) with a stochastic first-order oracle, we develop a stochastic counterpart of the APD method, namely stochastic APD and demonstrate that it can actually achieve the rate of convergence given by

$$\mathcal{O}\left\{(L_{\hat{f}} + \|A\| + \sigma_x + \sigma_y)\frac{1}{\sqrt{N}}\right\}. \tag{4.4.55}$$

Therefore, this algorithm further improves the error bound in (4.4.8) in terms of its dependence on $\|A\|$. In fact it can be shown that such a rate of convergence is theoretically not improvable for the stochastic SPP described above unless certain special properties are assumed.

The stochastic APD method is obtained by replacing the gradient operators $-A\tilde{x}_t$, $\nabla\hat{f}(x_t^{md})$ and $A^T y_{t+1}$, used in (4.4.15) and (4.4.16), with the stochastic gradient operators computed by the SFO, i.e., $-\hat{\mathrm{A}}_x(\tilde{x}_t)$, $\hat{\mathrm{G}}(\underline{x}_t)$, and $\hat{\mathrm{A}}_y(y_{t+1})$, respectively. This algorithm is formally described as in Algorithm 4.3.

Algorithm 4.3 Stochastic APD method for stochastic SPP

Modify (4.4.15) and (4.4.16) in Algorithm 4.2 to

$$y_{t+1} = \mathrm{argmin}y \in Y\langle -\hat{\mathrm{A}}_x(\tilde{x}_t), y\rangle + \hat{g}(y) + \tfrac{1}{\tau_t}V_Y(y, y_t), \tag{4.4.56}$$

$$x_{t+1} = \mathrm{argmin}x \in X\langle \hat{\mathrm{G}}(\underline{x}_t), x\rangle + \langle x, \hat{\mathrm{A}}_y(y_{t+1})\rangle + \tfrac{1}{\eta_t}V_X(x, x_t). \tag{4.4.57}$$

As noted earlier, one possible way to solve stochastic SPP is to apply the accelerated stochastic gradient descent method to a certain smooth approximation of (4.4.1). However, the rate of convergence of this approach will depend on the variance of the stochastic gradients computed for the smooth approximation problem, which is usually unknown and difficult to characterize. On the other hand, the stochastic APD method described above works directly with the original problem without requiring the application of the smoothing technique, and its rate of convergence will depend on the variance of the stochastic gradient operators computed for the original problem, i.e., $\sigma_{x,\hat{f}}^2$, σ_y^2, and $\sigma_{x,A}^2$.

Similar to Sect. 4.4.1, we use the two gap functions $g(\cdot)$ and $\tilde{g}(\cdot,\cdot)$, respectively, defined in (4.4.21) and (4.4.40) as the termination criteria for the stochastic APD algorithm, depending on whether the feasible set $Z = X \times Y$ is bounded or not. More specifically, we establish the convergence of the stochastic APD method for case when Z is bounded or unbounded in Theorems 4.8 and 4.9, respectively. Since the algorithm is stochastic in nature, for both cases we establish its expected rate of convergence in terms of $g(\cdot)$ or $\tilde{g}(\cdot,\cdot)$, i.e., the "average" rate of convergence over many runs of the algorithm. In addition, we show that if Z is bounded, then the convergence of the APD algorithm can be strengthened under the following "light-tail" assumption on the stochastic first-order oracle.

Assumption 9

$$\begin{aligned}
&\mathbb{E}\left[\exp\{\|\nabla \hat{f}(x) - \hat{\mathsf{G}}(x)\|_*^2/\sigma_{x,\hat{f}}^2\}\right] \le \exp\{1\},\\
&\mathbb{E}\left[\exp\{\|Ax - \hat{\mathsf{A}}_x(x)\|_*^2/\sigma_y^2\}\right] \le \exp\{1\}\\
&\mathbb{E}\left[\exp\{\|A^T y - \hat{\mathsf{A}}_y(y)\|_*^2/\sigma_{x,A}^2\}\right] \le \exp\{1\}.
\end{aligned}$$

Let $\hat{\mathsf{G}}(\underline{x}_t)$, $\hat{\mathsf{A}}_x(\tilde{x}_t)$, and $\hat{\mathsf{A}}_y(y_{t+1})$ be the output from the SFO at the t-th iteration of Algorithm 4.3. We denote

$$\begin{aligned}
&\Delta_{x,\hat{f}}^t := \hat{\mathsf{G}}(\underline{x}_t) - \nabla \hat{f}(x_t^{md}),\ \Delta_{x,A}^t := \hat{\mathsf{A}}_y(y_{t+1}) - A^T y_{t+1},\ \Delta_y^t := -\hat{\mathsf{A}}_x(\tilde{x}_t) + A\tilde{x}_t,\\
&\Delta_x^t := \Delta_{x,\hat{f}}^t + \Delta_{x,A}^t \ \text{ and } \ \Delta^t := (\Delta_x^t, \Delta_y^t).
\end{aligned}$$

Moreover, for a given $z = (x,y) \in Z$, let us denote $\|z\|^2 = \|x\|^2 + \|y\|^2$ and its associate dual norm for $\Delta = (\Delta_x, \Delta_y)$ by $\|\Delta\|_*^2 = \|\Delta_x\|_*^2 + \|\Delta_y\|_*^2$. We also define the Bregman divergence $V(z,\tilde{z}) := V_X(x,\tilde{x}) + V_Y(y,\tilde{y})$ for $z = (x,y)$ and $\tilde{z} = (\tilde{x},\tilde{y})$.

We first need to estimate a bound on $Q(\bar{z}_{t+1}, z)$ for all $z \in Z$. This result is analogous to Lemma 4.7 for the deterministic APD method.

Lemma 4.9. *Let $\bar{z}_t = (\bar{x}_t, \bar{y}_t)$ be the iterates generated by Algorithm 4.3. Assume that the parameters $\beta_t, \theta_t, \eta_t$, and τ_t satisfy* (4.4.24), (4.4.25), *and*

$$\frac{q}{\eta_t} - \frac{L_{\hat{f}}}{\beta_t} - \frac{\|A\|^2\tau_t}{p} \ge 0 \tag{4.4.58}$$

for some $p,q \in (0,1)$. Then, for any $z \in Z$, we have

$$\begin{aligned}
\beta_t\gamma_t Q(\bar{z}_{t+1}, z) \le\ & \mathrm{B}_t(z, z_{[t]}) + \gamma_t\langle A(x_{t+1} - x_t), y - y_{t+1}\rangle\\
& - \gamma_t\left(\tfrac{q}{2\eta_t} - \tfrac{L_{\hat{f}}}{2\beta_t}\right)\|x_{t+1} - x_t\|^2 + \textstyle\sum_{i=1}^t \Lambda_i(z),
\end{aligned} \tag{4.4.59}$$

where γ_t and $\mathrm{B}_t(z, z_{[t]})$, respectively, are defined in (4.4.22) and (4.4.28), $z_{[t]} = \{(x_i, y_i)\}_{i=1}^{t+1}$ and

$$\Lambda_i(z) := -\tfrac{(1-q)\gamma_i}{2\eta_i}\|x_{i+1} - x_i\|^2 - \tfrac{(1-p)\gamma_i}{2\tau_i}\|y_{i+1} - y_i\|^2 - \gamma_i\langle \Delta^i, z_{i+1} - z\rangle. \tag{4.4.60}$$

Proof. Similar to (4.4.29) and (4.4.31), we conclude from the optimality conditions of (4.4.56) and (4.4.57) that

$$
\begin{aligned}
&\langle -\hat{A}_x(\tilde{x}_t), y_{t+1} - y\rangle + \hat{g}(y_{t+1}) - \hat{g}(y) \le \tfrac{1}{\tau_t} V_Y(y, y_t) - \tfrac{1}{2\tau_t}\|y_{t+1} - y_t\|^2 \\
&\quad - \tfrac{1}{\tau_t} V_Y(y, y_{t+1}), \\
&\langle \hat{G}(\underline{x}_t), x_{t+1} - x\rangle + \langle x_{t+1} - x, \hat{A}_y(y_{t+1})\rangle \le \tfrac{1}{\eta_t} V_X(x, x_t) - \tfrac{1}{2\eta_t}\|x_{t+1} - x_t\|^2 \\
&\quad - \tfrac{1}{\eta_t} V_X(x, x_{t+1}).
\end{aligned}
$$

Now we establish an important recursion for Algorithm 4.3. Observing that Proposition 4.12 also holds for Algorithm 4.3, and applying the above two inequalities to (4.4.23) in Proposition 4.12, similar to (4.4.33), we have

$$
\begin{aligned}
&\beta_t \gamma_t Q(\bar{z}_{t+1}, z) - (\beta_t - 1)\gamma_t Q(\bar{z}_t, z) \\
&\le \tfrac{\gamma_t}{\eta_t} V_X(x, x_t) - \tfrac{\gamma_t}{\eta_t} V_X(x, x_{t+1}) + \tfrac{\gamma_t}{\tau_t} V_Y(y, y_t) - \tfrac{\gamma_t}{\tau_t} V_Y(y, y_{t+1}) \\
&\quad + \gamma_t \langle A(x_{t+1} - x_t), y - y_{t+1}\rangle - \gamma_{t-1}\langle A(x_t - x_{t-1}), y - y_t\rangle \\
&\quad - \gamma_t\left(\tfrac{1}{2\eta_t} - \tfrac{L_{\hat{f}}}{2\beta_t}\right)\|x_{t+1} - x_t\|^2 - \tfrac{\gamma_t}{2\tau_t}\|y_{t+1} - y_t\|^2 - \gamma_{t-1}\langle A(x_t - x_{t-1}), y_t - y_{t+1}\rangle \\
&\quad - \gamma_t \langle \Delta^t_{x,\hat{f}} + \Delta^t_{x,A}, x_{t+1} - x\rangle - \gamma_t\langle \Delta^t_y, y_{t+1} - y\rangle, \ \forall z \in Z.
\end{aligned}
\tag{4.4.61}
$$

By the Cauchy–Schwartz inequality and (4.4.25), for all $p \in (0,1)$,

$$
\begin{aligned}
&-\gamma_{t-1}\langle A(x_t - x_{t-1}), y_t - y_{t+1}\rangle \le \gamma_{t-1}\|A(x_t - x_{t-1})\|_* \|y_t - y_{t+1}\| \\
&\le \|A\|\gamma_{t-1}\|x_t - x_{t-1}\|\|y_t - y_{t+1}\| \le \tfrac{\|A\|^2\gamma_{t-1}^2\tau_t}{2p\gamma_t}\|x_t - x_{t-1}\|^2 + \tfrac{p\gamma_t}{2\tau_t}\|y_t - y_{t+1}\|^2 \\
&\le \tfrac{\|A\|^2\gamma_{t-1}\tau_{t-1}}{2p}\|x_t - x_{t-1}\|^2 + \tfrac{p\gamma_t}{2\tau_t}\|y_t - y_{t+1}\|^2.
\end{aligned}
\tag{4.4.62}
$$

By (4.4.24), (4.4.60), (4.4.61), and (4.4.62), we can develop the following recursion for Algorithm 4.3:

$$
\begin{aligned}
&(\beta_{t+1} - 1)\gamma_{t+1} Q(\bar{z}_{t+1}, z) - (\beta_t - 1)\gamma_t Q(\bar{z}_t, z) = \beta_t\gamma_t Q(\bar{z}_{t+1}, z) \\
&\quad - (\beta_t - 1)\gamma_t Q(\bar{z}_t, z) \\
&\le \tfrac{\gamma_t}{\eta_t} V_X(x, x_t) - \tfrac{\gamma_t}{\eta_t} V_X(x, x_{t+1}) + \tfrac{\gamma_t}{\tau_t} V_Y(y, y_t) - \tfrac{\gamma_t}{\tau_t} V_Y(y, y_{t+1}) \\
&+ \gamma_t\langle A(x_{t+1} - x_t), y - y_{t+1}\rangle - \gamma_{t-1}\langle A(x_t - x_{t-1}), y - y_t\rangle \\
&- \gamma_t\left(\tfrac{q}{2\eta_t} - \tfrac{L_{\hat{f}}}{2\beta_t}\right)\|x_{t+1} - x_t\|^2 + \tfrac{\|A\|^2\gamma_{t-1}\tau_{t-1}}{2p}\|x_t - x_{t-1}\|^2 + \Lambda_t(x), \ \forall z \in Z.
\end{aligned}
$$

Applying the above inequality inductively and assuming that $x_0 = x_1$, we obtain

$$
\begin{aligned}
&(\beta_{t+1} - 1)\gamma_{t+1} Q(\bar{z}_{t+1}, z) - (\beta_1 - 1)\gamma_1 Q(\bar{z}_1, z) \\
&\le \mathcal{B}_t(z, z_{[t]}) + \gamma_t\langle A(x_{t+1} - x_t), y - y_{t+1}\rangle - \gamma_t\left(\tfrac{q}{2\eta_t} - \tfrac{L_{\hat{f}}}{2\beta_t}\right)\|x_{t+1} - x_t\|^2 \\
&\quad - \textstyle\sum_{i=1}^{t-1}\gamma_i\left(\tfrac{q}{2\eta_i} - \tfrac{L_{\hat{f}}}{2\beta_i} - \tfrac{\|A\|^2\tau_i}{2p}\right)\|x_{i+1} - x_i\|^2 + \sum_{i=1}^{t}\Lambda_i(x), \ \forall z \in Z.
\end{aligned}
$$

Relation (4.4.59) then follows immediately from the above inequality, (4.4.24) and (4.4.58). ■

We also need the following technical result whose proof is based on Lemma 4.3.

Lemma 4.10. *Let η_i, τ_i and γ_i, $i = 1,2,\ldots$, be given positive constants. For any $z_1 \in Z$, if we define $z_1^v = z_1$ and*

$$z_{i+1}^v = \mathrm{argmin}_{z=(x,y)\in Z} \left\{-\eta_i\langle \Delta_x^i, x\rangle - \tau_i\langle \Delta_y^i, y\rangle + V(z, z_i^v)\right\}, \tag{4.4.63}$$

then

$$\textstyle\sum_{i=1}^t \gamma_i\langle -\Delta^i, z_i^v - z\rangle \le \mathrm{B}_t(z, z_{[t]}^v) + \sum_{i=1}^t \frac{\eta_i\gamma_i}{2}\|\Delta_x^i\|_*^2 + \sum_{i=1}^t \frac{\tau_i\gamma_i}{2}\|\Delta_y^i\|_*^2, \tag{4.4.64}$$

where $z_{[t]}^v := \{z_i^v\}_{i=1}^t$ and $\mathrm{B}_t(z, z_{[t]}^v)$ is defined in (4.4.28).

Proof. Noting that (4.4.63) implies $z_{i+1}^v = (x_{i+1}^v, y_{i+1}^v)$ where

$$\begin{aligned} x_{i+1}^v &= \mathrm{argmin}_{x\in X}\left\{-\eta_i\langle \Delta_x^i, x\rangle + V_X(x, x_i^v)\right\} \\ y_{i+1}^v &= \mathrm{argmin}_{y\in Y}\left\{-\tau_i\langle \Delta_y^i, y\rangle + V(y, y_i^v)\right\}, \end{aligned}$$

from Lemma 4.3 we have

$$\begin{aligned} V_X(x, x_{i+1}^v) &\le V_X(x, x_i^v) - \eta_i\langle \Delta_x^i, x - x_i^v\rangle + \frac{\eta_i^2\|\Delta_x^i\|_*^2}{2}, \\ V_Y(y, y_{i+1}^v) &\le V_Y(y, y_i^v) - \tau_i\langle \Delta_y^i, y - y_i^v\rangle + \frac{\tau_i^2\|\Delta_y^i\|_*^2}{2} \end{aligned}$$

for all $i \ge 1$. Thus

$$\begin{aligned} \tfrac{\gamma_i}{\eta_i}V_X(x, x_{i+1}^v) &\le \tfrac{\gamma_i}{\eta_i}V_X(x, x_i^v) - \gamma_i\langle \Delta_x^i, x - x_i^v\rangle + \frac{\gamma_i\eta_i\|\Delta_x^i\|_*^2}{2}, \\ \tfrac{\gamma_i}{\tau_i}V_Y(y, y_{i+1}^v) &\le \tfrac{\gamma_i}{\tau_i}V_Y(y, y_i^v) - \gamma_i\langle \Delta_y^i, y - y_i^v\rangle + \frac{\gamma_i\tau_i\|\Delta_y^i\|_*^2}{2}. \end{aligned}$$

Adding the above two inequalities together, and summing up them from $i = 1$ to t we get

$$0 \le \mathrm{B}_t(z, z_{[t]}^v) - \textstyle\sum_{i=1}^t \gamma_i\langle \Delta^i, z - z_i^v\rangle + \sum_{i=1}^t \frac{\gamma_i\eta_i\|\Delta_x^i\|_*^2}{2} + \sum_{i=1}^t \frac{\gamma_i\tau_i\|\Delta_y^i\|_*^2}{2},$$

so (4.5.35) holds. ■

We are now ready to establish the convergence of stochastic APD for the case when Z is bounded.

Theorem 4.8. *Suppose that (4.4.34) holds for some $D_X, D_Y > 0$. Also assume that for all $t \ge 1$, the parameters $\beta_t, \theta_t, \eta_t$, and τ_t in Algorithm 4.3 satisfy (4.4.24), (4.4.25), and (4.4.58) for some $p, q \in (0,1)$. Then,*

(a) Under Assumption 8, for all $t \ge 1$,

$$\mathbb{E}[g(\bar{z}_{t+1})] \le \mathrm{Q}_0(t), \tag{4.4.65}$$

where

$$\mathrm{Q}_0(t) := \tfrac{1}{\beta_t\gamma_t}\left\{\tfrac{4\gamma_t}{\eta_t}D_X^2 + \tfrac{4\gamma_t}{\tau_t}D_Y^2\right\} + \tfrac{1}{2\beta_t\gamma_t}\textstyle\sum_{i=1}^t\left\{\tfrac{(2-q)\eta_i\gamma_i}{1-q}\sigma_x^2 + \tfrac{(2-p)\tau_i\gamma_i}{1-p}\sigma_y^2\right\}. \tag{4.4.66}$$

(b) Under Assumption 9, for all $\lambda > 0$ and $t \ge 1$,

$$\mathrm{Prob}\{g(\bar{z}_{t+1}) > \mathrm{Q}_0(t) + \lambda \mathrm{Q}_1(t)\} \le 3\exp\{-\lambda^2/3\} + 3\exp\{-\lambda\}, \tag{4.4.67}$$

where

$$\begin{aligned} \mathrm{Q}_1(t) := &\tfrac{1}{\beta_t\gamma_t}\left(2\sigma_x D_X + \sqrt{2}\sigma_y D_Y\right)\sqrt{2\textstyle\sum_{i=1}^t\gamma_i^2} \\ &+ \tfrac{1}{2\beta_t\gamma_t}\textstyle\sum_{i=1}^t\left[\tfrac{(2-q)\eta_i\gamma_i}{1-q}\sigma_x^2 + \tfrac{(2-p)\tau_i\gamma_i}{1-p}\sigma_y^2\right]. \end{aligned} \tag{4.4.68}$$

Proof. First, applying the bounds in (4.4.36) and (4.4.37) to (4.4.59), we get

$$\begin{aligned} \beta_t\gamma_t Q(\bar{z}_{t+1}, z) \le &\tfrac{\gamma_t}{\eta_t}D_X^2 - \tfrac{\gamma_t}{\eta_t}V_X(x, x_{t+1}) + \tfrac{\gamma_t}{\tau_t}D_Y^2 - \tfrac{\gamma_t}{\tau_t}V_Y(y, y_{t+1}) + \tfrac{\gamma_t}{2\tau_t}\|y - y_{t+1}\|^2 \\ &- \gamma_t\left(\tfrac{q}{2\eta_t} - \tfrac{L_{\hat{f}}}{2\beta_t} - \tfrac{\|A\|^2\tau_t}{2}\right)\|x_{t+1} - x_t\|^2 + \textstyle\sum_{i=1}^t\Lambda_i(z) \\ \le &\tfrac{\gamma_t}{\eta_t}D_X^2 + \tfrac{\gamma_t}{\tau_t}D_Y^2 + \textstyle\sum_{i=1}^t\Lambda_i(z), \ \ \forall z \in Z. \end{aligned} \tag{4.4.69}$$

By (4.4.60), we have

$$\begin{aligned} \Lambda_i(z) &= -\tfrac{(1-q)\gamma_i}{2\eta_i}\|x_{i+1} - x_i\|^2 - \tfrac{(1-p)\gamma_i}{2\tau_i}\|y_{i+1} - y_i\|^2 + \gamma_i\langle \Delta^i, z - z_{i+1}\rangle \\ &= -\tfrac{(1-q)\gamma_i}{2\eta_i}\|x_{i+1} - x_i\|^2 - \tfrac{(1-p)\gamma_i}{2\tau_i}\|y_{i+1} - y_i\|^2 + \gamma_i\langle \Delta^i, z_i - z_{i+1}\rangle + \gamma_i\langle \Delta^i, z - z_i\rangle \\ &\le \tfrac{\eta_i\gamma_i}{2(1-q)}\|\Delta_x^i\|_*^2 + \tfrac{\tau_i\gamma_i}{2(1-p)}\|\Delta_y^i\|_*^2 + \gamma_i\langle \Delta^i, z - z_i\rangle, \end{aligned} \tag{4.4.70}$$

where the last relation follows from (3.1.6). For all $i \ge 1$, letting $z_1^v = z_1$, and z_{i+1}^v as in (4.4.63), we conclude from (4.4.70) and Lemma 4.10 that, $\forall z \in Z$,

$$\begin{aligned} &\textstyle\sum_{i=1}^t\Lambda_i(z) \\ &\le \textstyle\sum_{i=1}^t\left\{\tfrac{\eta_i\gamma_i}{2(1-q)}\|\Delta_x^i\|_*^2 + \tfrac{\tau_i\gamma_i}{2(1-p)}\|\Delta_y^i\|_*^2 + \gamma_i\langle \Delta^i, z_i^v - z_i\rangle + \gamma_i\langle -\Delta^i, z_i^v - z\rangle\right\} \\ &\le \mathrm{B}_t(z, z_{[t]}^v) + \underbrace{\tfrac{1}{2}\textstyle\sum_{i=1}^t\left\{\tfrac{(2-q)\eta_i\gamma_i}{(1-q)}\|\Delta_x^i\|_*^2 + \tfrac{(2-p)\tau_i\gamma_i}{1-p}\|\Delta_y^i\|_*^2 + \gamma_i\langle \Delta^i, z_i^v - z_i\rangle\right\}}_{U_t}, \end{aligned} \tag{4.4.71}$$

where similar to (4.4.36) we have $\mathrm{B}_t(z, z_{[t]}^v) \le D_X^2\gamma_t/\eta_t + D_Y^2\gamma_t/\tau_t$. Using the above inequality, (4.4.21), (4.4.34), and (4.4.69), we obtain

$$\beta_t\gamma_t g(\bar{z}_{t+1}) \le \tfrac{2\gamma_t}{\eta_t}D_X^2 + \tfrac{2\gamma_t}{\tau_t}D_Y^2 + U_t. \tag{4.4.72}$$

Now it suffices to bound the above quantity U_t, both in expectation (part (a)) and in probability (part (b)).

We first show part (a). Note that by our assumptions on SFO, at iteration i of Algorithm 4.3, the random noises Δ^i are independent of z_i and hence $\mathbb{E}[\langle \Delta^i, z - z_i\rangle] = 0$. In addition, Assumption 8 implies that $\mathbb{E}[\|\Delta_x^i\|_*^2] \le \sigma_{x,\hat{f}}^2 + \sigma_{x,A}^2 = \sigma_x^2$ (noting that $\Delta_{x,\hat{f}}^i$ and $\Delta_{x,A}^i$ are independent at iteration i), and $\mathbb{E}[\|\Delta_y^i\|_*^2] \le \sigma_y^2$. Therefore,

$$\mathbb{E}[U_t] \le \tfrac{1}{2}\textstyle\sum_{i=1}^t \left\{ \frac{(2-q)\eta_i\gamma_i\sigma_x^2}{1-q} + \frac{(2-p)\tau_i\gamma_i\sigma_y^2}{1-p} \right\}. \tag{4.4.73}$$

Taking expectation on both sides of (4.4.72) and using the above inequality, we obtain (4.4.65).

We now show that part (b) holds. Note that by our assumptions on SFO and the definition of z_i^v, the sequences $\{\langle \Delta_{x,\hat{f}}^i, x_i^v - x_i\rangle\}_{i\ge 1}$ is a martingale-difference sequence. By the large-deviation theorem for martingale-difference sequence (see Lemma 4.1), and the fact that

$$\begin{aligned}
&\mathbb{E}[\exp\left\{\gamma_i^2 \langle \Delta_{x,\hat{f}}^i, x_i^v - x_i\rangle^2 / \left(2\gamma_i^2 D_X^2 \sigma_{x,\hat{f}}^2\right)\right\}] \\
&\le \mathbb{E}[\exp\left\{\|\Delta_{x,\hat{f}}^i\|_*^2 \|x_i^v - x_i\|^2 / \left(2 D_X^2 \sigma_{x,\hat{f}}^2\right)\right\}] \\
&\le \mathbb{E}[\exp\left\{\|\Delta_{x,\hat{f}}^i\|_*^2 V_X(x_i^v, x_i) / \left(D_X^2 \sigma_{x,\hat{f}}^2\right)\right\}] \\
&\le \mathbb{E}[\exp\left\{\|\Delta_{x,\hat{f}}^i\|_*^2 / \sigma_{x,\hat{f}}^2\right\}] \le \exp\{1\},
\end{aligned}$$

we conclude that

$$\text{Prob}\left\{\textstyle\sum_{i=1}^t \gamma_i \langle \Delta_{x,\hat{f}}^i, x_i^v - x_i\rangle > \lambda \cdot \sigma_{x,\hat{f}} D_X \sqrt{2\sum_{i=1}^t \gamma_i^2}\right\} \le \exp\{-\lambda^2/3\}, \forall \lambda > 0.$$

By using a similar argument, we can show that, $\forall \lambda > 0$,

$$\begin{aligned}
&\text{Prob}\left\{\textstyle\sum_{i=1}^t \gamma_i \langle \Delta_y^i, y_i^v - y_i\rangle > \lambda \cdot \sigma_y D_Y \sqrt{2\sum_{i=1}^t \gamma_i^2}\right\} \le \exp\{-\lambda^2/3\}, \\
&\text{Prob}\left\{\textstyle\sum_{i=1}^t \gamma_i \langle \Delta_{x,A}^i, x - x_i\rangle > \lambda \cdot \sigma_{x,A} D_X \sqrt{2\sum_{i=1}^t \gamma_i^2}\right\} \\
&\le \exp\{-\lambda^2/3\}.
\end{aligned}$$

Using the previous three inequalities and the fact that $\sigma_{x,\hat{f}} + \sigma_{x,A} \le \sqrt{2}\sigma_x$, we have, $\forall \lambda > 0$,

$$\begin{aligned}
&\text{Prob}\left\{\textstyle\sum_{i=1}^t \gamma_i \langle \Delta^i, z_i^v - z_i\rangle > \lambda\left[\sqrt{2}\sigma_x D_X + \sigma_y D_Y\right]\sqrt{2\sum_{i=1}^t \gamma_i^2}\right\} \\
&\le \text{Prob}\left\{\textstyle\sum_{i=1}^t \gamma_i \langle \Delta^i, z_i^v - z_i\rangle > \lambda\left[(\sigma_{x,\hat{f}} + \sigma_{x,A}) D_X + \sigma_y D_Y\right]\sqrt{2\sum_{i=1}^t \gamma_i^2}\right\} \\
&\le 3\exp\{-\lambda^2/3\}.
\end{aligned} \tag{4.4.74}$$

Now let $S_i := (2-q)\eta_i\gamma_i/(1-q)$ and $S := \sum_{i=1}^t S_i$. By the convexity of exponential function, we have

$$\mathbb{E}\left[\exp\left\{\tfrac{1}{S}\sum_{i=1}^t S_i\|\Delta^i_{x,\hat f}\|_*^2/\sigma^2_{x,\hat f}\right\}\right] \le \mathbb{E}\left[\tfrac{1}{S}\sum_{i=1}^t S_i \exp\left\{\|\Delta^i_{x,\hat f}\|_*^2/\sigma^2_{x,\hat f}\right\}\right] \le \exp\{1\},$$

where the last inequality follows from Assumption 9. Therefore, by Markov's inequality, for all $\lambda > 0$,

$$\begin{aligned}&\mathrm{Prob}\left\{\sum_{i=1}^t \tfrac{(2-q)\eta_i\gamma_i}{1-q}\|\Delta^i_{x,\hat f}\|_*^2 > (1+\lambda)\sigma^2_{x,\hat f}\sum_{i=1}^t \tfrac{(2-q)\eta_i\gamma_i}{1-q}\right\}\\ &= \mathrm{Prob}\left\{\exp\left\{\tfrac{1}{S}\sum_{i=1}^t S_i\|\Delta^i_{x,\hat f}\|_*^2/\sigma^2_{x,\hat f}\right\} \ge \exp\{1+\lambda\}\right\} \le \exp\{-\lambda\}.\end{aligned}$$

Using a similar argument, we can show that

$$\begin{aligned}&\mathrm{Prob}\left\{\sum_{i=1}^t \tfrac{(2-q)\eta_i\gamma_i}{1-q}\|\Delta^i_{x,A}\|_*^2 > (1+\lambda)\sigma^2_{x,A}\sum_{i=1}^t \tfrac{(2-q)\eta_i\gamma_i}{1-q}\right\} \le \exp\{-\lambda\},\\ &\mathrm{Prob}\left\{\sum_{i=1}^t \tfrac{(2-p)\tau_i\gamma_i}{1-p}\|\Delta^i_y\|_*^2 > (1+\lambda)\sigma_y^2\sum_{i=1}^t \tfrac{(2-p)\tau_i\gamma_i}{1-p}\right\} \le \exp\{-\lambda\}.\end{aligned}$$

Combining the previous three inequalities, we obtain

$$\begin{aligned}&\mathrm{Prob}\left\{\sum_{i=1}^t \tfrac{(2-q)\eta_i\gamma_i}{1-q}\|\Delta^i_x\|_*^2 + \sum_{i=1}^t \tfrac{(2-p)\tau_i\gamma_i}{1-p}\|\Delta^i_y\|_*^2 >\right.\\ &\left.(1+\lambda)\left[\sigma_x^2\sum_{i=1}^t \tfrac{(2-q)\eta_i\gamma_i}{1-q} + \sigma_y^2\sum_{i=1}^t \tfrac{(2-p)\tau_i\gamma_i}{1-p}\right]\right\} \le 3\exp\{-\lambda\}.\end{aligned} \tag{4.4.75}$$

Our result now follows directly from (4.4.71), (4.4.72), (4.4.74), and (4.4.75). ∎

We provide below a specific choice of the parameters β_t, θ_t, η_t, and τ_t for the stochastic APD method for the case when Z is bounded.

Corollary 4.3. *Suppose that (4.4.34) holds. In Algorithm 4.3, if the parameters are set to*

$$\beta_t = \tfrac{t+1}{2},\ \theta_t = \tfrac{t-1}{t},\ \eta_t = \frac{2\sqrt{2}D_X t}{6\sqrt{2}L_{\hat f}D_X + 3\sqrt{2}\|A\|D_Y t + 3\sigma_x t^{3/2}}\ \text{and}\ \tau_t = \frac{2\sqrt{2}D_Y}{3\sqrt{2}\|A\|D_X + 3\sigma_y\sqrt{t}}. \tag{4.4.76}$$

Then under Assumption 8, (4.4.65) *holds, and*

$$Q_0(t) \le \frac{12L_{\hat f}D_X^2}{t(t+1)} + \frac{12\|A\|D_XD_Y}{t} + \frac{6\sqrt{2}(\sigma_xD_X + \sigma_yD_Y)}{\sqrt{t}}. \tag{4.4.77}$$

If in addition, Assumption 9 holds, then for all $\lambda > 0$, (4.4.67) *holds, and*

$$Q_1(t) \le \frac{5\sqrt{2}\sigma_xD_X + 4\sqrt{2}\sigma_yD_Y}{\sqrt{t}}. \tag{4.4.78}$$

Proof. First we check that the parameters in (4.4.76) satisfy the conditions in Theorem 4.8. The inequalities (4.4.24) and (4.4.25) can be checked easily. Furthermore, setting $p = q = 2/3$ we have for all $t \ge 1$,

$$\frac{q}{\eta_t} - \frac{L_{\hat{f}}}{\beta_t} - \frac{\|A\|^2 \tau_t}{p} \geq \frac{2L_{\hat{f}}D_X + \|A\|D_Y t}{D_X t} - \frac{2L_{\hat{f}}}{t+1} - \frac{\|A\|^2 D_Y t}{\|A\| D_X t} \geq 0,$$

thus (4.4.58) hold, and hence Theorem 4.8 holds. Now it suffice to show that (4.4.77) and (4.4.78) hold.

Observe that by (4.4.22) and (4.4.76), we have $\gamma_t = t$. Also, observe that $\sum_{i=1}^t \sqrt{i} \leq \int_1^{t+1} \sqrt{u}du \leq \frac{2}{3}(t+1)^{3/2} \leq \frac{2\sqrt{2}}{3}(t+1)\sqrt{t}$, thus

$$\frac{1}{\gamma_t}\sum_{i=1}^t \eta_i \gamma_i \leq \frac{2\sqrt{2}D_X}{3\sigma_x t}\sum_{i=1}^t \sqrt{i} \leq \frac{8D_X(t+1)\sqrt{t}}{9\sigma_x t}$$
$$\frac{1}{\gamma_t}\sum_{i=1}^t \tau_i \gamma_i \leq \frac{2\sqrt{2}D_Y}{3\sigma_y t}\sum_{i=1}^t \sqrt{i} \leq \frac{8D_Y(t+1)\sqrt{t}}{9\sigma_y t}.$$

Applying the above bounds to (4.4.66) and (4.5.40), we get

$$\begin{aligned}
Q_0(t) \leq\ & \frac{2}{t+1}\left(\frac{6\sqrt{2}L_{\hat{f}}D_X + 3\sqrt{2}\|A\|D_Y t + 3\sigma_x t^{3/2}}{\sqrt{2}D_X t} \cdot D_X^2 + \frac{3\sqrt{2}\|A\|D_X + 3\sigma_y\sqrt{t}}{\sqrt{2}D_Y} \cdot D_Y^2\right. \\
& \left. + 2\sigma_x^2 \cdot \frac{8D_X(t+1)\sqrt{t}}{9\sigma_x t} + 2\sigma_y^2 \cdot \frac{4D_Y(t+1)\sqrt{t}}{9\sigma_y t}\right), \\
Q_1(t) \leq\ & \frac{2}{t(t+1)}\left(\sqrt{2}\sigma_x D_X + \sigma_y D_Y\right)\sqrt{\frac{2t(t+1)^2}{3}} + \frac{4\sigma_x^2}{t+1} \cdot \frac{8D_X(t+1)\sqrt{t}}{9\sigma_x t} \\
& + \frac{4\sigma_y^2}{t+1} \cdot \frac{8D_Y(t+1)\sqrt{t}}{9\sigma_y t}.
\end{aligned}$$

Simplifying the above inequalities, we see that (4.4.77) and (4.4.78) hold. ■

In view of (4.4.77), the stochastic APD method allows us to have very large Lipschitz constants $L_{\hat{f}}$ (as big as $\mathcal{O}(N^{3/2})$) and $\|A\|$ (as big as $\mathcal{O}(\sqrt{N})$) without significantly affecting its rate of convergence.

We now present the convergence results for the stochastic APD method applied to stochastic saddle-point problems with possibly unbounded feasible set Z. Similar to the proof of Theorem 4.7, first we specialize the result of Lemma 4.14 under (4.4.24), (4.4.41), and (4.4.58). The following lemma is analogous to Lemma 4.8.

Lemma 4.11. *Let $\hat{z} = (\hat{x}, \hat{y}) \in Z$ be a saddle point of* (4.4.1). *If $V_X(x, x_t) = \|x - x_t\|^2/2$ and $V_Y(y, y_t) = \|y - y_t\|^2/2$ in Algorithm 4.3, and the parameters $\beta_t, \theta_t, \eta_t$ and τ_t satisfy* (4.4.24), (4.4.41), *and* (4.4.58), *then*

$$\begin{aligned}
& \|\hat{x} - x_{t+1}\|^2 + \|\hat{x} - x_{t+1}^v\|^2 + \frac{\eta_t(1-p)}{\tau_t}\|\hat{y} - y_{t+1}\|^2 + \frac{\eta_t}{\tau_t}\|\hat{y} - y_{t+1}^v\|^2 \\
& \leq 2\|\hat{x} - x_1\|^2 + \frac{2\eta_t}{\tau_t}\|\hat{y} - y_1\|^2 + \frac{2\eta_t}{\gamma_t}U_t,
\end{aligned} \tag{4.4.79}$$

$$\tilde{g}(\bar{z}_{t+1}, v_{t+1}) \leq \frac{1}{\beta_t \eta_t}\|\bar{x}_{t+1} - x_1\|^2 + \frac{1}{\beta_t \tau_t}\|\bar{y}_{t+1} - y_1\|^2 + \frac{1}{\beta_t \gamma_t}U_t =: \delta_{t+1}, \tag{4.4.80}$$

for all $t \geq 1$, where (x_{t+1}^v, y_{t+1}^v), U_t, and $\tilde{g}(\cdot,\cdot)$ are defined in (4.4.63), (4.4.71), *and* (4.4.40), *respectively, and*

$$v_{t+1} = \left(\frac{1}{\beta_t \eta_t}(2x_1 - x_{t+1} - x_{t+1}^v), \frac{1}{\beta_t \tau_t}(2y_1 - y_{t+1} - y_{t+1}^v) + \frac{1}{\beta_t}A(x_{t+1} - x_t)\right). \tag{4.4.81}$$

Proof. Applying (4.4.58), (4.4.47), and (4.4.71) to (4.4.59) in Lemma 4.14, we get

$$\beta_t\gamma_t Q(\bar{z}_{t+1},z) \le \bar{\mathrm{B}}(z,z_{[t]}) + \tfrac{p\gamma_t}{2\tau_t}\|y-y_{t+1}\|^2 + \bar{\mathrm{B}}(z,z^v_{[t]}) + U_t,$$

where $\bar{\mathrm{B}}(\cdot,\cdot)$ is defined as

$$\bar{\mathrm{B}}(z,\tilde{z}_{[t]}) := \tfrac{\gamma_t}{2\eta_t}\|x-\tilde{x}_1\|^2 - \tfrac{\gamma_t}{2\eta_t}\|x-\tilde{x}_{t+1}\|^2 + \tfrac{\gamma_t}{2\tau_t}\|y-\tilde{y}_1\|^2 - \tfrac{\gamma_t}{2\tau_t}\|y-\tilde{y}_{t+1}\|^2,$$

for all $z \in Z$ and $\tilde{z}_{[t]} \subset Z$ thanks to (4.4.41). Now letting $z = \hat{z}$, and noting that $Q(\bar{z}_{t+1},\hat{z}) \ge 0$, we get (4.4.79). On the other hand, if we only apply (4.4.58) and (4.4.71) to (4.4.59) in Lemma 4.14, then we get

$$\beta_t\gamma_t Q(\bar{z}_{t+1},z) \le \bar{\mathrm{B}}(z,z_{[t]}) + \gamma_t\langle A(x_{t+1}-x_t), y-y_{t+1}\rangle + \bar{\mathrm{B}}(z,z^v_{[t]}) + U_t.$$

Applying (4.4.41) and (4.4.48) to $\bar{\mathrm{B}}(z,z_{[t]})$ and $\bar{\mathrm{B}}(z,z^v_{[t]})$ in the above inequality, we get (4.4.80). ■

With the help of Lemma 4.11, we are ready to prove Theorem 4.9, which summarizes the convergence properties of Algorithm 4.2 when X or Y is unbounded.

Theorem 4.9. *Let $\{\bar{z}_t\} = \{(\bar{x}_t,\bar{y}_t)\}$ be the iterates generated by Algorithm 4.2 with $V_X(x,x_t) = \|x-x_t\|^2/2$ and $V_Y(y,y_t) = \|y-y_t\|^2/2$. Assume that the parameters β_t,θ_t,η_t, and τ_t in Algorithm 4.3 satisfy* (4.4.24), (4.4.41), *and (4.4.58) for all $t \ge 1$ and some $p,q \in (0,1)$, then there exists a perturbation vector v_{t+1} such that*

$$\mathbb{E}[\tilde{g}(\bar{z}_{t+1},v_{t+1})] \le \tfrac{1}{\beta_t\eta_t}\left(\tfrac{6-4p}{1-p}D^2 + \tfrac{5-3p}{1-p}C^2\right) =: \varepsilon_{t+1} \tag{4.4.82}$$

for any $t \ge 1$. Moreover, we have

$$\begin{aligned}\mathbb{E}[\|v_{t+1}\|] \le{}& \tfrac{2\|\hat{x}-x_1\|}{\beta_t\eta_t} + \tfrac{2\|\hat{y}-y_1\|}{\beta_t\tau_t}\\ &+ \sqrt{2D^2+2C^2}\left[\tfrac{2}{\beta_t\eta_t} + \tfrac{1}{\beta_t\tau_t}\sqrt{\tfrac{\tau_1}{\eta_1}}\left(\sqrt{\tfrac{1}{1-p}}+1\right) + \tfrac{2\|A\|}{\beta_t}\right],\end{aligned} \tag{4.4.83}$$

where $(\hat{x},\hat{y})$ is a pair of solutions for problem (4.4.1)*, D is defined in (4.4.51), and*

$$C := \sqrt{\textstyle\sum_{i=1}^t \tfrac{\eta_i^2\sigma_x^2}{1-q} + \sum_{i=1}^t \tfrac{\eta_i\tau_i\sigma_y^2}{1-p}}. \tag{4.4.84}$$

Proof. Let δ_{t+1} and v_{t+1} be defined in (4.4.80) and (4.4.81), respectively. Also let C and D, respectively, be defined in (4.4.84) and (4.4.51). It suffices to estimate $\mathbb{E}[\|v_{t+1}\|]$ and $\mathbb{E}[\delta_{t+1}]$. First it follows from (4.4.41), (4.4.84), and (4.4.73) that

$$\mathbb{E}[U_t] \le \tfrac{\gamma_t}{\eta_t}C^2. \tag{4.4.85}$$

Using the above inequality, (4.4.41), (4.4.51), and (4.4.79), we have $\mathbb{E}[\|\hat{x}-x_{t+1}\|^2] \le 2D^2+2C^2$ and $\mathbb{E}[\|\hat{y}-y_{t+1}\|^2] \le (2D^2+2C^2)\tau_1/[\eta_1(1-p)]$, which, by Jensen's inequality, then imply that $\mathbb{E}[\|\hat{x}-x_{t+1}\|] \le \sqrt{2D^2+2C^2}$ and

$\mathbb{E}[\|\hat{y}-y_{t+1}\|] \le \sqrt{2D^2+2C^2}\sqrt{\tau_1/[\eta_1(1-p)]}$. Similarly, we can show that $\mathbb{E}[\|\hat{x}-x^v_{t+1}\|] \le \sqrt{2D^2+2C^2}$ and $\mathbb{E}[\|\hat{y}-y^v_{t+1}\|] \le \sqrt{2D^2+2C^2}\sqrt{\tau_1/\eta_1}$. Therefore, by (4.4.81) and the above four inequalities, we have

$$\begin{aligned}
&\mathbb{E}[\|v_{t+1}\|]\\
&\le \mathbb{E}\Big[\tfrac{1}{\beta_t\eta_t}\left(\|x_1-x_{t+1}\|+\|x_1-x^v_{t+1}\|\right)+\tfrac{1}{\beta_t\tau_t}\left(\|y_1-y_{t+1}\|+\|y_1-y^v_{t+1}\|\right)\\
&\quad+\tfrac{\|A\|}{\beta_t}\|x_{t+1}-x_t\|\Big]\\
&\le \mathbb{E}\Big[\tfrac{1}{\beta_t\eta_t}\left(2\|\hat{x}-x_1\|+\|\hat{x}-x_{t+1}\|+\|\hat{x}-x^v_{t+1}\|\right)\\
&\quad+\tfrac{1}{\beta_t\tau_t}\left(2\|\hat{y}-y_1\|+\|\hat{y}-y_{t+1}\|+\|\hat{y}-y^v_{t+1}\|\right)+\tfrac{\|A\|}{\beta_t}\left(\|\hat{x}-x_{t+1}\|+\|\hat{x}-x_t\|\right)\Big]\\
&\le \tfrac{2\|\hat{x}-x_1\|}{\beta_t\eta_t}+\tfrac{2\|\hat{y}-y_1\|}{\beta_t\tau_t}+\sqrt{2D^2+2C^2}\left[\tfrac{2}{\beta_t\eta_t}+\tfrac{1}{\beta_t\tau_t}\sqrt{\tfrac{\tau_1}{\eta_1}}\left(\sqrt{\tfrac{1}{1-p}}+1\right)+\tfrac{2\|A\|}{\beta_t}\right],
\end{aligned}$$

thus (4.4.83) holds.

Now let us estimate a bound on δ_{t+1}. By (4.4.52), (4.4.73), (4.4.79), and (4.4.85), we have

$$\begin{aligned}
&\mathbb{E}[\delta_{t+1}] = \mathbb{E}\left[\tfrac{1}{\beta_t\eta_t}\|\bar{x}_{t+1}-x_1\|^2+\tfrac{1}{\beta_t\tau_t}\|\bar{y}_{t+1}-y_1\|^2\right]+\tfrac{1}{\beta_t\gamma_t}\mathbb{E}[U_t]\\
&\le \mathbb{E}\left[\tfrac{2}{\beta_t\eta_t}\left(\|\hat{x}-\bar{x}_{t+1}\|^2+\|\hat{x}-x_1\|^2\right)+\tfrac{2}{\beta_t\tau_t}\left(\|\hat{y}-\bar{y}_{t+1}\|^2+\|\hat{y}-y_1\|^2\right)\right]+\tfrac{1}{\beta_t\eta_t}C^2\\
&= \mathbb{E}\left[\tfrac{1}{\beta_t\eta_t}\left(2D^2+2\|\hat{x}-\bar{x}_{t+1}\|^2+\tfrac{2\eta_t(1-p)}{\tau_t}\|\hat{y}-\bar{y}_{t+1}\|^2+\tfrac{2\eta_t p}{\tau_t}\|\hat{y}-\bar{y}_{t+1}\|^2\right)\right]+\tfrac{1}{\beta_t\eta_t}C^2\\
&\le \tfrac{1}{\beta_t\eta_t}\Big[2D^2+C^2+\tfrac{2}{\beta_t\gamma_t}\textstyle\sum_{i=1}^t\gamma_i\Big(\mathbb{E}\left[\|\hat{x}-x_{i+1}\|^2\right]+\tfrac{\eta_t(1-p)}{\tau_t}\mathbb{E}\left[\|\hat{y}-y_{i+1}\|^2\right]\\
&\quad+\tfrac{\eta_t p}{\tau_t}\mathbb{E}\left[\|\hat{y}-y_{i+1}\|^2\right]\Big)\Big]\\
&\le \tfrac{1}{\beta_t\eta_t}\left[2D^2+C^2+\tfrac{2}{\beta_t\gamma_t}\textstyle\sum_{i=1}^t\gamma_i\left(2D^2+C^2+\tfrac{\eta_t p}{\tau_t}\cdot\tfrac{\tau_1}{\eta_1(1-p)}(2D^2+C^2)\right)\right]\\
&= \tfrac{1}{\beta_t\eta_t}\left(\tfrac{6-4p}{1-p}D^2+\tfrac{5-3p}{1-p}C^2\right).
\end{aligned}$$

Therefore (4.4.82) holds. ■

Below we specialize the results in Theorem 4.9 by choosing a set of parameters satisfying (4.4.24), (4.4.41), and (4.4.58).

Corollary 4.4. *In Algorithm 4.3, if N is given and the parameters are set to*

$$\beta_t = \tfrac{t+1}{2},\ \theta_t = \tfrac{t-1}{t},\ \eta_t = \tfrac{3t}{4\eta},\ \textit{and}\ \tau_t = \tfrac{t}{\eta}, \tag{4.4.86}$$

where

$$\eta = 2L_{\hat{f}}+2\|A\|(N-1)+\tfrac{N\sqrt{N-1}\sigma}{\tilde{D}}\ \textit{for some}\ \tilde{D}>0,\ \sigma=\sqrt{\tfrac{9}{4}\sigma_x^2+\sigma_y^2}, \tag{4.4.87}$$

then there exists v_N that satisfies (4.4.82) *with*

$$\varepsilon_N \le \frac{36L_{\hat{f}}D^2}{N(N-1)} + \frac{36\|A\|D^2}{N} + \frac{\sigma D\big(18D/\tilde{D}+6\tilde{D}/D\big)}{\sqrt{N-1}}, \tag{4.4.88}$$

$$\mathbb{E}[\|v_N\|] \le \frac{50L_{\hat{f}}D}{N(N-1)} + \frac{\|A\|D(55+4\tilde{D}/D)}{N} + \frac{\sigma(9+25D/\tilde{D})}{\sqrt{N-1}}, \tag{4.4.89}$$

where D is defined in (4.4.51).

Proof. For the parameters in (4.4.86), it is clear that (4.4.24) and (4.4.41) hold. Furthermore, let $p = 1/4$, $q = 3/4$, then for all $t = 1,\ldots,N-1$, we have

$$\frac{q}{\eta_t} - \frac{L_{\hat{f}}}{\beta_t} - \frac{\|A\|^2\tau_t}{p} = \frac{\eta}{t} - \frac{2L_{\hat{f}}}{t+1} - \frac{4\|A\|^2 t}{\eta} \ge \frac{2L_{\hat{f}}+2\|A\|(N-1)}{t} - \frac{2L_{\hat{f}}}{t} - \frac{2\|A\|^2 t}{\|A\|(N-1)} \ge 0,$$

thus (4.4.58) holds. By Theorem 4.9, we get (4.4.82) and (4.4.83). Note that $\eta_t/\tau_t = 3/4$, and

$$\frac{1}{\beta_{N-1}\eta_{N-1}}\|\hat{x}-x_1\| \le \frac{1}{\beta_{N-1}\eta_{N-1}}D, \ \frac{1}{\beta_{N-1}\tau_{N-1}}\|\hat{y}-y_1\|$$
$$\le \frac{1}{\beta_{N-1}\eta_{N-1}}\cdot\frac{\eta_{N-1}}{\tau_{N-1}}\cdot\sqrt{\tfrac{4}{3}}D = \frac{\sqrt{3/4}D}{\beta_{N-1}\eta_{N-1}},$$

so in (4.4.82) and (4.4.83) we have

$$\varepsilon_N \le \frac{1}{\beta_{N-1}\eta_{N-1}}\left(\tfrac{20}{3}D^2 + \tfrac{17}{3}C^2\right), \tag{4.4.90}$$

$$\mathbb{E}[\|v_N\|] \le \frac{(2+\sqrt{3})D}{\beta_{N-1}\eta_{N-1}} + \frac{\sqrt{2D^2+2C^2}\left(3+\sqrt{3/4}\right)}{\beta_{N-1}\eta_{N-1}} + \frac{2\|A\|\sqrt{2D^2+2C^2}}{\beta_{N-1}}. \tag{4.4.91}$$

By (4.4.84) and the fact that $\sum_{i=1}^{N-1} i^2 \le N^2(N-1)/3$, we have

$$C = \sqrt{\textstyle\sum_{i=1}^{N-1}\frac{9\sigma_x^2 i^2}{4\eta^2} + \sum_{i=1}^{N-1}\frac{\sigma_y^2 i^2}{\eta^2}} \le \sqrt{\tfrac{1}{3\eta^2}N^2(N-1)\left(\tfrac{9\sigma_x^2}{4}+\sigma_y^2\right)} = \frac{\sigma N\sqrt{N-1}}{\sqrt{3}\eta}$$

Applying the above bound to (4.4.90) and (4.4.91), and using (4.4.87) and the fact that $\sqrt{2D^2+C^2} \le \sqrt{2}D + C$, we obtain

$$\begin{aligned}
\varepsilon_N &\le \frac{8\eta}{3N(N-1)}\left(\frac{20}{3}D^2 + \frac{17\sigma^2N^2(N-1)}{9\eta^2}\right) = \frac{8}{3N(N-1)}\left(\frac{20}{3}\eta D^2 + \frac{17\sigma^2N^2(N-1)}{9\eta}\right)\\
&\le \frac{320L_{\hat{f}}D^2}{9N(N-1)} + \frac{320\|A\|(N-1)D^2}{9N(N-1)} + \frac{160N\sqrt{N-1}\sigma D^2/\tilde{D}}{9N(N-1)} + \frac{136\sigma^2N^2(N-1)}{27N^2(N-1)^{3/2}\sigma/\tilde{D}}\\
&\le \frac{36L_{\hat{f}}D^2}{N(N-1)} + \frac{36\|A\|D^2}{N} + \frac{\sigma D\big(18D/\tilde{D}+6\tilde{D}/D\big)}{\sqrt{N-1}},\\
\mathbb{E}[\|v_N\|] &\le \frac{1}{\beta_{N-1}\eta_{N-1}}\left(2D+\sqrt{3}D+3\sqrt{2}D+\sqrt{6}D/2+3\sqrt{2}C+\sqrt{6}C/2\right)\\
&\quad + \frac{2\sqrt{2}\|A\|D}{\beta_{N-1}} + \frac{2\sqrt{2}\|A\|C}{\beta_{N-1}}\\
&\le \frac{16L_{\hat{f}}+16\|A\|(N-1)+8N\sqrt{N-1}\sigma/\tilde{D}}{3N(N-1)}\left(2+\sqrt{3}+3\sqrt{2}+\sqrt{6}/2\right)D\\
&\quad + \frac{8\sigma}{3\sqrt{N-1}}\left(\sqrt{6}+\sqrt{2}/2\right) + \frac{4\sqrt{2}\|A\|D}{N} + \frac{4\sqrt{2}\|A\|\sigma N\sqrt{N-1}}{N\sqrt{3}N\sqrt{N-1}\sigma/\tilde{D}}\\
&\le \frac{50L_{\hat{f}}D}{N(N-1)} + \frac{\|A\|D(55+4\tilde{D}/D)}{N} + \frac{\sigma(9+25D/\tilde{D})}{\sqrt{N-1}}.
\end{aligned}$$

■

Observe that the parameter settings in (4.4.86)–(4.4.87) are more complicated than the ones in (4.4.53) for the deterministic unbounded case. In particular, for the stochastic unbounded case, we need to choose a parameter $\tilde{D}$ which is not required for the deterministic case. Clearly, the optimal selection for $\tilde{D}$ minimizing the RHS of (4.4.88) is given by $\sqrt{6}D$. Note, however, that the value of D will be very difficult to estimate for the unbounded case and hence one often has to resort to a suboptimal selection for $\tilde{D}$. For example, if $\tilde{D} = 1$, then the RHS of (4.4.88) and (4.4.89) will become $\mathcal{O}(L_{\hat{f}}D^2/N^2 + \|A\|D^2/N + \sigma D^2/\sqrt{N})$ and $\mathcal{O}(L_{\hat{f}}D/N^2 + \|A\|D/N + \sigma D/\sqrt{N})$, respectively.

4.5 Stochastic Accelerated Mirror-Prox Method

Let $\mathbb{R}^n$ denote a finite dimensional vector space with inner product $\langle \cdot, \cdot \rangle$ and norm $\|\cdot\|$ (not necessarily induced by $\langle \cdot, \cdot \rangle$), and Z be a nonempty closed convex set in $\mathbb{R}^n$. Our problem of interest is to find a $u^* \in Z$ that solves the following monotone stochastic variational inequality (SVI) problem:

$$\langle \mathbb{E}_{\xi,\zeta}[\mathsf{F}(u;\xi,\zeta)], u^* - u \rangle \le 0, \forall u \in Z. \tag{4.5.1}$$

Here, the expectation is taken with respect to the random vectors ξ and ζ whose distributions are supported on $\Xi \subseteq \mathbb{R}^d$ and $\Xi' \subseteq \mathbb{R}^{d'}$, respectively, and F is given by the summation of three components with different structural properties, i.e.,

$$\mathsf{F}(u;\xi,\zeta) = \mathsf{G}(u;\xi) + \mathsf{H}(u;\zeta) + J'(u), \ \forall u \in Z. \tag{4.5.2}$$

In particular, we assume that $J'(u) \in \partial J(u)$ is a subgradient of a relatively simple and convex function J (see (4.5.9) below), $\mathsf{H}(u;\zeta)$ is an unbiased estimator of a monotone and Lipschitz continuous operator H such that $\mathbb{E}_{\zeta}[\mathsf{H}(u;\zeta)] = H(u)$,

$$\langle H(w) - H(v), w - v \rangle \ge 0, \text{ and } \|H(w) - H(v)\|_* \le M\|w - v\|, \ \forall w, v \in Z, \tag{4.5.3}$$

where $\|\cdot\|_*$ denotes the conjugate norm of $\|\cdot\|$. Moreover, we assume that $\mathsf{G}(u;\xi)$ is an unbiased estimator of the gradient for a convex and continuously differentiable function G such that $\mathbb{E}_{\xi}[\mathsf{G}(u;\xi)] = \nabla G(u)$ and

$$0 \le G(w) - G(v) - \langle \nabla G(v), w - v \rangle \le \tfrac{L}{2}\|w - v\|^2, \forall w, v \in Z. \tag{4.5.4}$$

Observe that u^* given by (4.5.1) is often called a weak solution for SVI. Recall that a related notion is a strong SVI solution (see Sect. 3.8). More specifically, letting

$$F(u) := \mathbb{E}_{\xi,\zeta}[\mathsf{F}(u;\xi,\zeta)] = \nabla G(u) + H(u) + J'(u), \tag{4.5.5}$$

we say that u^* is a strong SVI solution if it satisfies

$$\langle F(u^*), u^* - u\rangle \le 0, \forall u \in Z. \tag{4.5.6}$$

It should be noted that the operator F above might not be continuous. Problems (4.5.1) and (4.5.6) are also known as the Minty variational inequality and the Stampacchia variational inequality respectively. For any monotone operator F, it is well known that strong solutions defined in (4.5.6) are also weak solutions in (4.5.1), and the reverse is also true under mild assumptions (e.g., when F is continuous). For example, for F in (4.5.5), if $J = 0$, then the weak and strong solutions in (4.5.1) and (4.5.6) are equivalent. For the sake of notational convenience, we use $SVI(Z;G,H,J)$ or simply $SVI(Z;F)$ to denote problem (4.5.1).

In this section, we assume that there exist *stochastic first-order oracles* SFO_G and SFO_H that provide random samples of $\mathsf{G}(u;\xi)$ and $\mathsf{H}(u;\xi)$ for any test point $u \in Z$.

Assumption 10 *At the i-th call of* SFO_G *and* SFO_H *with input* $z \in Z$, *the oracles* SFO_G *and* SFO_H *output stochastic information* $\mathsf{G}(z;\xi_i)$ *and* $\mathsf{H}(z;\zeta_i)$ *respectively, such that*

$$\mathbb{E}\left[\|\mathsf{G}(u;\xi_i) - \nabla G(u)\|_*^2\right] \le \sigma_G^2 \text{ and } \mathbb{E}\left[\|\mathsf{H}(u;\zeta_i) - H(u)\|_*^2\right] \le \sigma_H^2,$$

for some $\sigma_G, \sigma_H \ge 0$, *where* ξ_i *and* ζ_i *are independently distributed random samples.*

For the sake of notational convenience, throughout this section we also denote

$$\sigma := \sqrt{\sigma_G^2 + \sigma_H^2}. \tag{4.5.7}$$

Assumption 10 basically implies that the variance associated with $\mathsf{G}(u,\xi_i)$ and $\mathsf{H}(u,\zeta_i)$ is bounded. It should be noted that deterministic VIs, denoted by $VI(Z;G,H,J)$, are special cases of SVIs with $\sigma_G = \sigma_H = 0$. The above setting covers as a special case of the regular SVIs whose operators $G(u)$ or $H(u)$ are given in the form of expectation as shown in (4.5.1). Moreover, it provides a framework to study randomized algorithms for solving deterministic VI or saddle point problems.

4.5.1 Algorithmic Framework

By examining the structural properties (e.g., gradient field G and Lipschitz continuity of H) of the SVI problems in (4.5.1), we can see that the total number of gradient and operator evaluations for solving SVI cannot be smaller than

$$\mathcal{O}\left(\sqrt{\tfrac{L}{\varepsilon}} + \tfrac{M}{\varepsilon} + \tfrac{\sigma^2}{\varepsilon^2}\right). \tag{4.5.8}$$

This is a lower complexity bound derived based on the following three observations:

(a) If $H = 0$ and $\sigma = 0$, $SVI(Z;G,0,0)$ is equivalent to a smooth optimization problem $\min_{u\in Z} G(u)$, and the complexity for minimizing $G(u)$ cannot be better than $\mathscr{O}(\sqrt{L/\varepsilon})$.
(b) If $G = 0$ and $\sigma = 0$, the complexity for solving $SVI(Z;0,H,0)$ cannot be better than $\mathscr{O}(M/\varepsilon)$.
(c) If $H = 0$, $SVI(Z;G,0,0)$ is equivalent to a stochastic smooth optimization problem, and the complexity cannot be better than $\mathscr{O}(\sigma^2/\varepsilon^2)$.

The lower complexity bound in (4.5.8) and the three observations stated above provide some important guidelines to the design of efficient algorithms to solve the SVI problem with the operator given in (4.5.5). It might seem natural to consider the more general problem (4.5.6) by combining $\nabla G(u)$ and $H(u)$ in (4.5.5) together as a single monotone operator, instead of separating them apart. Such consideration is reasonable from a generalization point of view, by noting that the convexity of function $G(u)$ is equivalent to the monotonicity of $\nabla G(u)$, and the Lipschitz conditions (4.5.3) and (4.5.4) are equivalent to a Lipschitz condition of $F(u)$ in (4.5.6) with $\|F(w) - F(v)\|_* \le (L+M)\|w - v\|$. However, from the algorithmic point of view, a special treatment of ∇G separately from H is crucial for the design of accelerated algorithms. By observations (b) and (c) above, if we consider $F := \nabla G + H$ as a single monotone operator, the complexity for solving $SVI(Z;0;F;0)$ cannot be smaller than

$$\mathscr{O}\left(\tfrac{L+M}{\varepsilon} + \tfrac{\sigma^2}{\varepsilon^2}\right),$$

which is worse than (4.5.8) in terms of the dependence on L.

In order to achieve the complexity bound in (4.5.8) for SVIs, we incorporate a multi-step acceleration scheme into the mirror-prox method in Sect. 3.8, and introduce a stochastic accelerated mirror-prox (SAMP) method that can exploit the structural properties of (4.5.1). Specifically, we assume that the following subproblem can be solved efficiently:

$$\operatorname{argmin}_{u\in Z} \langle \eta, u - z\rangle + V(z,u) + J(u). \tag{4.5.9}$$

Here $V(\cdot,\cdot)$ is the prox-function associated with a distance generating function $\omega : Z \to \mathbb{R}$ given by

$$V(z,u) := \omega(u) - \omega(z) - \langle \nabla\omega(z), u - z\rangle,\ \forall u,z \in Z. \tag{4.5.10}$$

Using the aforementioned definition of the prox-mapping, we describe the SAMP method in Algorithm 4.4.

Observe that in the SAMP algorithm we introduced two sequences, i.e., $\{\underline{w}_t\}$ and $\{\bar{w}_t\}$, that are convex combinations of iterations $\{w_t\}$ and $\{r_t\}$ as long as $\alpha_t \in [0,1]$. If $\alpha_t \equiv 1$, $G = 0$, and $J = 0$, then Algorithm 4.4 for solving $SVI(Z;0,H,0)$ is equivalent to the stochastic version of the mirror-prox method in Sect. 3.8. Moreover, if the distance generating function $\omega(\cdot) = \|\cdot\|_2^2/2$, then iterations (4.5.12) and (4.5.13) become

Algorithm 4.4 The stochastic accelerated mirror-prox (SAMP) method

Choose $r_1 \in Z$. Set $w_1 = r_1$, $\bar{w}_1 = r_1$.
For $t = 1, 2, \ldots, N-1$, calculate

$$\underline{w}_t = (1-\alpha_t)\bar{w}_t + \alpha_t r_t, \tag{4.5.11}$$

$$w_{t+1} = \text{argmin}_{u \in Z} \gamma_t [\langle \mathsf{H}(r_t;\zeta_{2t-1}) + \mathsf{G}(\underline{w}_t;\xi_t), u - r_t\rangle + J(u)] + V(r_t, u), \tag{4.5.12}$$

$$r_{t+1} = \text{argmin}_{u \in Z} \gamma_t [\langle \mathsf{H}(w_{t+1};\zeta_{2t}) + \mathsf{G}(\underline{w}_t;\xi_t), u - r_t\rangle + J(u)] + V(r_t, u), \tag{4.5.13}$$

$$\bar{w}_{t+1} = (1-\alpha_t)\bar{w}_t + \alpha_t w_{t+1}. \tag{4.5.14}$$

Output w_N^{ag}.

$$w_{t+1} = \text{argmin}_{u \in Z} \langle \gamma_t H(r_t), u - r_t\rangle + \tfrac{1}{2}\|u - r_t\|_2^2,$$
$$r_{t+1} = \text{argmin}_{u \in Z} \langle \gamma_t H(w_{t+1}), u - r_t\rangle + \tfrac{1}{2}\|u - r_t\|_2^2,$$

which are exactly the iterates of the extragradient method. On the other hand, if $H = 0$, then (4.5.12) and (4.5.13) produce the same optimizer $w_{t+1} = r_{t+1}$, and Algorithm 4.4 is equivalent to the stochastic accelerated gradient descent method in Sect. 4.2. Therefore, Algorithm 4.4 can be viewed as a hybrid algorithm of the stochastic mirror-prox method and the stochastic accelerated gradient descent method, which gives its name stochastic accelerated mirror-prox method. It is interesting to note that for any t, there are two calls of SFO_H but just one call of SFO_G.

4.5.2 Convergence Analysis

In order to analyze the convergence of Algorithm 4.4, we introduce a notion to characterize the weak solutions of $SVI(Z;G,H,J)$. For all $\tilde{u}, u \in Z$, we define

$$Q(\tilde{u}, u) := G(\tilde{u}) - G(u) + \langle H(u), \tilde{u} - u\rangle + J(\tilde{u}) - J(u). \tag{4.5.15}$$

Clearly, for F defined in (4.5.5), we have $\langle F(u), \tilde{u} - u\rangle \le Q(\tilde{u}, u)$. Therefore, if $Q(\tilde{u}, u) \le 0$ for all $u \in Z$, then $\tilde{u}$ is a weak solution of $SVI(Z;G,H,J)$. Hence when Z is bounded, it is natural to use the gap function

$$g(\tilde{u}) := \sup_{u \in Z} Q(\tilde{u}, u) \tag{4.5.16}$$

to evaluate the accuracy of a feasible solution $\tilde{u} \in Z$. However, if Z is unbounded, then $g(\tilde{z})$ may not be well defined, even when $\tilde{z} \in Z$ is a nearly optimal solution. Therefore, we need to employ a slightly modified gap function in order to measure the accuracy of candidate solutions when Z is unbounded. In the sequel, we will consider the cases of bounded and unbounded Z separately. For both cases we establish the rate of convergence of the gap functions in terms of their expectation, i.e., the "average" rate of convergence over many runs of the algorithm. Furthermore, we demonstrate that if Z is bounded, then we can also refine the rate of convergence of $g(\cdot)$ in the probability sense, under the following "light-tail" assumption:

Assumption 11 *For any i-th call on oracles* SFO_H *and* SFO_H *with any input* $u \in Z$,

$$\mathbb{E}[\exp\{\|\nabla G(u) - \mathsf{G}(u;\xi_i)\|_*^2/\sigma_G^2\}] \le \exp\{1\},$$

and

$$\mathbb{E}[\exp\{\|H(u) - \mathsf{H}(u;\zeta_i)\|_*^2/\sigma_H^2\}] \le \exp\{1\}.$$

It should be noted that Assumption 11 implies Assumption 10 by Jensen's inequality.

We start with establishing some convergence properties of Algorithm 4.4 when Z is bounded. It should be noted that the following quantity will be used throughout the convergence analysis of this algorithm:

$$\Gamma_t = \begin{cases} 1, & \text{when } t = 1 \\ (1-\alpha_t)\Gamma_{t-1}, & \text{when } t > 1. \end{cases} \tag{4.5.17}$$

To prove the convergence of the stochastic AMP algorithm, we first present some technical results. Lemma 4.12 describes some important properties of the projection (or prox-mapping) used in (4.5.12) and (4.5.13) of Algorithm 4.4. Lemma 4.13 provides a recursion related to the function $Q(\cdot,\cdot)$ defined in (4.5.15). With the help of Lemmas 4.12 and 4.13, we estimate a bound on $Q(\cdot,\cdot)$ in Lemma 4.14.

Lemma 4.12. *Given* $r, w, y \in Z$ *and* $\eta, \vartheta \in \mathbb{R}^n$ *that satisfy*

$$w = \mathrm{argmin}_{u\in Z}\langle \eta, u - r\rangle + V(r,u) + J(u), \tag{4.5.18}$$

$$y = \mathrm{argmin}_{u\in Z}\langle \vartheta, u - r\rangle + V(r,u) + J(u), \tag{4.5.19}$$

and

$$\|\vartheta - \eta\|_*^2 \le L^2\|w - r\|^2 + M^2. \tag{4.5.20}$$

Then, for all $u \in Z$,

$$\langle \vartheta, w - u\rangle + J(w) - J(u) \le V(r,u) - V(y,u) - \left(\tfrac{1}{2} - \tfrac{L^2}{2}\right)\|r - w\|^2 + \tfrac{M^2}{2}, \tag{4.5.21}$$

and

$$V(y,w) \le L^2 V(r,w) + \tfrac{M^2}{2}. \tag{4.5.22}$$

Proof. Applying Lemma 3.5 to (4.5.18) and (4.5.19), for all $u \in Z$ we have

$$\langle \eta, w-u\rangle + J(w) - J(u) \le V(r,u) - V(r,w) - V(w,u), \tag{4.5.23}$$
$$\langle \vartheta, y-u\rangle + J(y) - J(u) \le V(r,u) - V(r,y) - V(y,u). \tag{4.5.24}$$

In particular, letting $u = y$ in (4.5.23) we have

$$\langle \eta, w-y\rangle + J(w) - J(y) \le V(r,y) - V(r,w) - V(w,y). \tag{4.5.25}$$

Adding inequalities (4.5.24) and (4.5.25), then

$$\langle \vartheta, y-u\rangle + \langle \eta, w-y\rangle + J(w) - J(u) \le V(r,u) - V(y,u) - V(r,w) - V(w,y),$$

which is equivalent to

$$\langle \vartheta, w-u\rangle + J(w) - J(u) \le \langle \vartheta - \eta, w-y\rangle + V(r,u) - V(y,u) - V(r,w) - V(w,y).$$

Applying the Cauchy-Schwartz inequality and (3.1.6) to the above inequality, and using the fact that

$$\tfrac{1}{2}\|z-u\|^2 \le V(u,z), \forall u,z \in Z, \tag{4.5.26}$$

due to the strong convexity of $\omega(\cdot)$ in (4.5.10), we obtain

$$\begin{aligned}
&\langle \vartheta, w-u\rangle + J(w) - J(u)\\
&\le \|\vartheta - \eta\|_* \|w-y\| + V(r,u) - V(y,u) - V(r,w) - \tfrac{1}{2}\|w-y\|^2\\
&\le \tfrac{1}{2}\|\vartheta - \eta\|_*^2 + \tfrac{1}{2}\|w-y\|^2 + V(r,u) - V(y,u) - V(r,w) - \tfrac{1}{2}\|w-y\|^2\\
&= \tfrac{1}{2}\|\vartheta - \eta\|_*^2 + V(r,u) - V(y,u) - V(r,w).
\end{aligned} \tag{4.5.27}$$

The result in (4.5.21) then follows immediately from above relation, (4.5.20) and (4.5.26).

Moreover, observe that by setting $u = w$ and $u = y$ in (4.5.24) and (4.5.27), respectively, we have

$$\langle \vartheta, y-w\rangle + J(y) - J(w) \le V(r,w) - V(r,y) - V(y,w),$$
$$\langle \vartheta, w-y\rangle + J(w) - J(y) \le \tfrac{1}{2}\|\vartheta - \eta\|_*^2 + V(r,y) - V(r,w).$$

Adding the above two inequalities, and using (4.5.20) and (4.5.26), we have

$$0 \le \tfrac{1}{2}\|\vartheta - \eta\|_*^2 - V(y,w) \le \tfrac{L^2}{2}\|r-w\|^2 + \tfrac{M^2}{2} - V(y,w) \le L^2 V(r,w) + \tfrac{M^2}{2} - V(y,w),$$

and thus (4.5.22) holds. ■

Lemma 4.13. *For any sequences $\{r_t\}_{t\ge1}$ and $\{w_t\}_{t\ge1} \subset Z$, if the sequences $\{\bar{w}_t\}$ and $\{\underline{w}_t\}$ are generated by* (4.5.11) *and* (4.5.14), *then for all $u \in Z$,*

$$Q(\bar{w}_{t+1},u)-(1-\alpha_t)Q(\bar{w}_t,u)$$
$$\leq \alpha_t\langle \nabla G(\underline{w}_t)+H(w_{t+1}),w_{t+1}-u\rangle+\tfrac{L\alpha_t^2}{2}\|w_{t+1}-r_t\|^2+\alpha_t J(w_{t+1})-\alpha_t J(u). \tag{4.5.28}$$

Proof. Observe from (4.5.11) and (4.5.14) that $\bar{w}_{t+1}-\underline{w}_t=\alpha_t(w_{t+1}-r_t)$. This observation together with the convexity of $G(\cdot)$ imply that for all $u\in Z$,

$$\begin{aligned}
G(\bar{w}_{t+1}) &\leq G(\underline{w}_t)+\langle \nabla G(\underline{w}_t),\bar{w}_{t+1}-\underline{w}_t\rangle+\tfrac{L}{2}\|\bar{w}_{t+1}-\underline{w}_t\|^2\\
&=(1-\alpha_t)\left[G(\underline{w}_t)+\langle \nabla G(\underline{w}_t),\bar{w}_t-\underline{w}_t\rangle\right]\\
&\quad+\alpha_t\left[G(\underline{w}_t)+\langle \nabla G(\underline{w}_t),u-\underline{w}_t\rangle\right]\\
&\quad+\alpha_t\langle \nabla G(\underline{w}_t),w_{t+1}-u\rangle+\tfrac{L\alpha_t^2}{2}\|w_{t+1}-r_t\|^2\\
&\leq(1-\alpha_t)G(\bar{w}_t)+\alpha_t G(u)+\alpha_t\langle \nabla G(\underline{w}_t),w_{t+1}-u\rangle+\tfrac{L\alpha_t^2}{2}\|w_{t+1}-r_t\|^2.
\end{aligned}$$

Using the above inequality, (4.5.14), (4.5.15), and the monotonicity of $H(\cdot)$, we have

$$\begin{aligned}
&Q(\bar{w}_{t+1},u)-(1-\alpha_t)Q(\bar{w}_t,u)\\
=\;&G(\bar{w}_{t+1})-(1-\alpha_t)G(\bar{w}_t)-\alpha_t G(u)\\
&+\langle H(u),\bar{w}_{t+1}-u\rangle-(1-\alpha_t)\langle H(u),\bar{w}_t-u\rangle\\
&+J(\bar{w}_{t+1})-(1-\alpha_t)J(\bar{w}_t)-\alpha_t J(u)\\
\leq\;&G(\bar{w}_{t+1})-(1-\alpha_t)G(\bar{w}_t)-\alpha_t G(u)+\alpha_t\langle H(u),w_{t+1}-u\rangle\\
&+\alpha_t J(w_{t+1})-\alpha_t J(u)\\
\leq\;&\alpha_t\langle \nabla G(\underline{w}_t),w_{t+1}-u\rangle+\tfrac{L\alpha_t^2}{2}\|w_{t+1}-r_t\|^2+\alpha_t\langle H(w_{t+1}),w_{t+1}-u\rangle\\
&+\alpha_t J(w_{t+1})-\alpha_t J(u).
\end{aligned}$$

■

In the sequel, we will use the following notations to describe the inexactness of the first order information from SFO_H and SFO_G. At the t-th iteration, letting $\mathsf{H}(r_t;\zeta_{2t-1})$, $\mathsf{H}(w_{t+1};\zeta_{2t})$ and $\mathsf{G}(\underline{w}_t;\xi_t)$ be the output of the stochastic oracles, we denote

$$\begin{aligned}
\Delta_H^{2t-1}&:=\mathsf{H}(r_t;\zeta_{2t-1})-H(r_t),\\
\Delta_H^{2t}&:=\mathsf{H}(w_{t+1};\zeta_{2t})-H(w_{t+1}),\\
\Delta_G^{t}&:=\mathsf{G}(\underline{w}_t;\xi_t)-\nabla G(\underline{w}_t).
\end{aligned} \tag{4.5.29}$$

Lemma 4.14 below provides a bound on $Q(\bar{w}_{t+1},u)$ for all $u\in Z$.

Lemma 4.14. *Suppose that the parameters $\{\alpha_t\}$ in Algorithm 4.4 satisfy $\alpha_1=1$ and $0\leq\alpha_t<1$ for all $t>1$. Then the iterates $\{r_t\}$, $\{w_t\}$ and $\{\bar{w}_t\}$ satisfy*

$$\frac{1}{\Gamma_t}Q(\bar{w}_{t+1},u)$$
$$\leq \mathrm{B}_t(u,r_{[t]}) - \Sigma_{i=1}^t \frac{\alpha_i}{2\Gamma_i\gamma_i}\left(q - L\alpha_i\gamma_i - 3M^2\gamma_i^2\right)\|r_i - w_{i+1}\|^2 + \Sigma_{i=1}^t \Lambda_i(u),\ \forall u \in Z, \tag{4.5.30}$$

where Γ_t is defined in (4.5.17),

$$\mathrm{B}_t(u,r_{[t]}) := \Sigma_{i=1}^t \frac{\alpha_i}{\Gamma_i\gamma_i}(V(r_i,u) - V(r_{i+1},u)), \tag{4.5.31}$$

and

$$\begin{aligned}\Lambda_i(u) := &\frac{3\alpha_i\gamma_i}{2\Gamma_i}\left(\|\Delta_H^{2i}\|_*^2 + \|\Delta_H^{2i-1}\|_*^2\right) - \frac{(1-q)\alpha_i}{2\Gamma_i\gamma_i}\|r_i - w_{i+1}\|^2\\ &- \frac{\alpha_i}{\Gamma_i}\langle \Delta_H^{2i} + \Delta_G^i, w_{i+1} - u\rangle.\end{aligned} \tag{4.5.32}$$

Proof. Observe from (4.5.29) that

$$\begin{aligned}&\|\mathsf{H}(w_{t+1};\zeta_{2t}) - \mathsf{H}(r_t;\zeta_{2t-1})\|_*^2\\ &\leq \left(\|H(w_{t+1}) - H(r_t)\|_* + \|\Delta_H^{2t}\|_* + \|\Delta_H^{2t-1}\|_*\right)^2\\ &\leq 3\left(\|H(w_{t+1}) - H(r_t)\|_*^2 + \|\Delta_H^{2t}\|_*^2 + \|\Delta_H^{2t-1}\|_*^2\right)\\ &\leq 3\left(M^2\|w_{t+1} - r_t\|^2 + \|\Delta_H^{2t}\|_*^2 + \|\Delta_H^{2t-1}\|_*^2\right).\end{aligned} \tag{4.5.33}$$

Applying Lemma 4.12 to (4.5.12) and (4.5.13) (with $r = r_t, w = w_{t+1}, y = r_{t+1}, \eta = \gamma_t\mathsf{H}(r_t;\zeta_{2t-1}) + \gamma_t\mathsf{G}(\underline{w}_t;\xi_t)$, $\vartheta = \gamma_t\mathsf{H}(w_{t+1};\zeta_{2t}) + \gamma_t\mathsf{G}(\underline{w}_t;\xi_t), J = \gamma_t J$, $L^2 = 3M^2\gamma_t^2$, and $M^2 = 3\gamma_t^2(\|\Delta_H^{2t}\|_*^2 + \|\Delta_H^{2t-1}\|_*^2)$), and using (4.5.33), we have for any $u \in Z$,

$$\begin{aligned}&\gamma_t\langle \mathsf{H}(w_{t+1};\zeta_{2t}) + \mathsf{G}(\underline{w}_t;\xi_t), w_{t+1} - u\rangle + \gamma_t J(w_{t+1}) - \gamma_t J(u)\\ &\leq V(r_t,u) - V(r_{t+1},u) - \left(\tfrac{1}{2} - \tfrac{3M^2\gamma_t^2}{2}\right)\|r_t - w_{t+1}\|^2 + \tfrac{3\gamma_t^2}{2}(\|\Delta_H^{2t}\|_*^2 + \|\Delta_H^{2t-1}\|_*^2).\end{aligned}$$

Applying (4.5.29) and the above inequality to (4.5.28), we have

$$\begin{aligned}&Q(\bar{w}_{t+1},u) - (1-\alpha_t)Q(\bar{w}_t,u)\\ &\leq \alpha_t\langle \mathsf{H}(w_{t+1};\zeta_{2t}) + \mathsf{G}(\underline{w}_t;\xi_t), w_{t+1} - u\rangle + \alpha_t J(w_{t+1}) - \alpha_t J(u)\\ &\quad + \tfrac{L\alpha_t^2}{2}\|w_{t+1} - r_t\|^2 - \alpha_t\langle \Delta_H^{2t} + \Delta_G^t, w_{t+1} - u\rangle\\ &\leq \tfrac{\alpha_t}{\gamma_t}\left(V(r_t,u) - V(r_{t+1},u)\right) - \tfrac{\alpha_t}{2\gamma_t}\left(1 - L\alpha_t\gamma_t - 3M^2\gamma_t^2\right)\|r_t - w_{t+1}\|^2\\ &\quad + \tfrac{3\alpha_t\gamma_t}{2}\left(\|\Delta_H^{2t}\|_*^2 + \|\Delta_H^{2t-1}\|_*^2\right) - \alpha_t\langle \Delta_H^{2t} + \Delta_G^t, w_{t+1} - u\rangle.\end{aligned}$$

Dividing the above inequality by Γ_t and using the definition of $\Lambda_t(u)$ in (4.5.32), we obtain

$$\begin{aligned}&\tfrac{1}{\Gamma_t}Q(\bar{w}_{t+1},u) - \tfrac{1-\alpha_t}{\Gamma_t}Q(\bar{w}_t,u)\\ &\leq \tfrac{\alpha_t}{\Gamma_t\gamma_t}\left(V(r_t,u) - V(r_{t+1},u)\right)\\ &\quad - \tfrac{\alpha_t}{2\Gamma_t\gamma_t}\left(q - L\alpha_t\gamma_t - 3M^2\gamma_t^2\right)\|r_t - w_{t+1}\|^2 + \Lambda_t(u).\end{aligned}$$

Noting the fact that $\alpha_1 = 1$ and $(1-\alpha_t)/\Gamma_t = 1/\Gamma_{t-1}, t > 1$, due to (4.5.17), applying the above inequality recursively, and using the definition of $\mathrm{B}_t(\cdot,\cdot)$ in (4.5.31), we conclude (4.5.30). ■

We still need the following technical result that helps to provide a bound on the last stochastic term in (4.5.30) before proving Theorems 4.10 and 4.11.

Lemma 4.15. *Let $\theta_t, \gamma_t > 0$, $t = 1, 2, \ldots$, be given. For any $w_1 \in Z$ and any sequence $\{\Delta^t\} \subset \mathbb{R}^n$, if we define $w_1^v = w_1$ and*

$$w_{i+1}^v = \mathrm{argmin}_{u\in Z} - \gamma_i \langle \Delta^i, u\rangle + V(w_i^v, u), \ \forall i > 1, \tag{4.5.34}$$

then

$$\Sigma_{i=1}^t \theta_i \langle -\Delta^i, w_i^v - u\rangle \le \Sigma_{i=1}^t \tfrac{\theta_i}{\gamma_i}(V(w_i^v,u) - V(w_{i+1}^v,u)) + \Sigma_{i=1}^t \tfrac{\theta_i\gamma_i}{2}\|\Delta_i\|_*^2, \ \forall u \in Z. \tag{4.5.35}$$

Proof. Applying Lemma 3.5 to (4.5.34) (with $r = w_i^v$, $w = w_{i+1}^v$, $\zeta = -\gamma_i\Delta^i$, and $J = 0$), we have

$$-\gamma_i\langle \Delta^i, w_{i+1}^v - u\rangle \le V(w_i^v,u) - V(w_i^v, w_{i+1}^v) - V(w_{i+1}^v, u), \ \forall u \in Z.$$

Moreover, by Cauchy-Schwartz inequality, (3.1.6), and (4.5.26) we have

$$\begin{aligned} &-\gamma_i\langle \Delta^i, w_i^v - w_{i+1}^v\rangle \\ &\le \gamma_i\|\Delta^i\|_*\|w_i^v - w_{i+1}^v\| \le \tfrac{\gamma_i^2}{2}\|\Delta_i\|_*^2 + \tfrac{1}{2}\|w_i^v - w_{i+1}^v\|^2 \le \tfrac{\gamma_i^2}{2}\|\Delta_i\|_*^2 + V(w_i^v, w_{i+1}^v). \end{aligned}$$

Adding the above two inequalities and multiplying the resulting inequality by θ_i/γ_i, we obtain

$$-\theta_i\langle \Delta^i, w_i^v - u\rangle \le \tfrac{\theta_i\gamma_i}{2}\|\Delta_i\|_*^2 + \tfrac{\theta_i}{\gamma_i}(V(w_i^v,u) - V(w_{i+1}^v,u)).$$

Summing the above inequalities from $i = 1$ to t, we conclude (4.5.35). ■

With the help of Lemmas 4.14 and 4.15, we are now ready to prove Theorem 4.10, which provides an estimate of the gap function of SAMP in both expectation and probability.

Theorem 4.10. *Suppose that*

$$\sup_{z_1,z_2\in Z} V(z_1,z_2) \le D_Z^2. \tag{4.5.36}$$

Also assume that the parameters $\{\alpha_t\}$ and $\{\gamma_t\}$ in Algorithm 4.4 satisfy $\alpha_1 = 1$,

$$q - L\alpha_t\gamma_t - 3M^2\gamma_t^2 \ge 0 \text{ for some } q \in (0,1), \text{ and } \tfrac{\alpha_t}{\Gamma_t\gamma_t} \le \tfrac{\alpha_{t+1}}{\Gamma_{t+1}\gamma_{t+1}}, \ \forall t \ge 1, \tag{4.5.37}$$

where Γ_t is defined in (4.5.17). Then,

(a) Under Assumption 10, for all $t \geq 1$,

$$\mathbb{E}[g(\bar{w}_{t+1})] \leq \mathrm{Q}_0(t) := \tfrac{2\alpha_t}{\gamma_t} D_Z^2 + \left[4\sigma_H^2 + \left(1 + \tfrac{1}{2(1-q)}\right)\sigma_G^2\right] \Gamma_t \Sigma_{i=1}^t \tfrac{\alpha_i \gamma_i}{\Gamma_i}. \quad (4.5.38)$$

(b) Under Assumption 11, for all $\lambda > 0$ and $t \geq 1$,

$$\mathrm{Prob}\{g(\bar{w}_{t+1}) > \mathrm{Q}_0(t) + \lambda \mathrm{Q}_1(t)\} \leq 2\exp\{-\lambda^2/3\} + 3\exp\{-\lambda\}, \quad (4.5.39)$$

where

$$\begin{aligned} \mathrm{Q}_1(t) := {} & \Gamma_t(\sigma_G + \sigma_H) D_Z \sqrt{2\Sigma_{i=1}^t \left(\tfrac{\alpha_i}{\Gamma_i}\right)^2} \\ & + \left[4\sigma_H^2 + \left(1 + \tfrac{1}{2(1-q)}\right)\sigma_G^2\right] \Gamma_t \Sigma_{i=1}^t \tfrac{\alpha_i \gamma_i}{\Gamma_i}. \end{aligned} \quad (4.5.40)$$

Proof. We first provide a bound on $\mathrm{B}_t(u, r_{[t]})$. Since the sequence $\{r_i\}_{i=1}^{t+1}$ is in the bounded set Z, applying (4.5.36) and (4.5.37) to (4.5.31) we have

$$\begin{aligned} & \mathrm{B}_t(u, r_{[t]}) \\ & = \tfrac{\alpha_1}{\Gamma_1 \gamma_1} V(r_1, u) - \Sigma_{i=1}^{t-1}\left[\tfrac{\alpha_i}{\Gamma_i \gamma_i} - \tfrac{\alpha_{i+1}}{\Gamma_{i+1}\gamma_{i+1}}\right] V(r_{t+1}[i], u) - \tfrac{\alpha_t}{\Gamma_t \gamma_t} V(r_{t+1}, u) \\ & \leq \tfrac{\alpha_1}{\Gamma_1 \gamma_1} D_Z^2 - \Sigma_{i=1}^{t-1}\left[\tfrac{\alpha_i}{\Gamma_i \gamma_i} - \tfrac{\alpha_{i+1}}{\Gamma_{i+1}\gamma_{i+1}}\right] D_Z^2 = \tfrac{\alpha_t}{\Gamma_t \gamma_t} D_Z^2, \; \forall u \in Z, \end{aligned} \quad (4.5.41)$$

Applying (4.5.37) and the above inequality to (4.5.30) in Lemma 4.14, we have

$$\tfrac{1}{\Gamma_t} Q(\bar{w}_{t+1}, u) \leq \tfrac{\alpha_t}{\Gamma_t \gamma_t} D_Z^2 + \Sigma_{i=1}^t \Lambda_i(u), \; \forall u \in Z. \quad (4.5.42)$$

Letting $w_1^v = w_1$, defining w_{i+1}^v as in (4.5.34) with $\Delta^i = \Delta_H^{2i} + \Delta_G^i$ for all $i > 1$, we conclude from (4.5.31) and Lemma 4.15 (with $\theta_i = \alpha_i / \Gamma_i$) that

$$-\Sigma_{i=1}^t \tfrac{\alpha_i}{\Gamma_i} \langle \Delta_H^{2i} + \Delta_G^i, w_i^v - u \rangle \leq \mathrm{B}_t(u, w_{[t]}^v) + \Sigma_{i=1}^t \tfrac{\alpha_i \gamma_i}{2\Gamma_i} \|\Delta_H^{2i} + \Delta_G^i\|_*^2, \; \forall u \in Z. \quad (4.5.43)$$

The above inequality together with (4.5.32) and the Young's inequality yield

$$\begin{aligned} \Sigma_{i=1}^t \Lambda_i(u) = {} & -\Sigma_{i=1}^t \tfrac{\alpha_i}{\Gamma_i} \langle \Delta_H^{2i} + \Delta_G^i, w_i^v - u \rangle + \Sigma_{i=1}^t \tfrac{3\alpha_i \gamma_i}{2\Gamma_i} \left(\|\Delta_H^{2i}\|_*^2 + \|\Delta_H^{2i-1}\|_*^2\right) \\ & + \Sigma_{i=1}^t \tfrac{\alpha_i}{\Gamma_i}\left[-\tfrac{1-q}{2\gamma_i}\|r_i - w_{i+1}\|^2 - \langle \Delta_G^i, w_{i+1} - r_i \rangle\right] \\ & - \Sigma_{i=1}^t \tfrac{\alpha_i}{\Gamma_i} \langle \Delta_G^i, r_i - w_i^v \rangle - \Sigma_{i=1}^t \tfrac{\alpha_i}{\Gamma_i} \langle \Delta_H^{2i}, w_{i+1} - w_i^v \rangle \\ \leq {} & \mathrm{B}_t(u, w_{[t]}^v) + U_t, \end{aligned} \quad (4.5.44)$$

where

$$\begin{aligned}U_t := & \textstyle\sum_{i=1}^t \frac{\alpha_i\gamma_i}{2\Gamma_i}\|\Delta_H^{2i}+\Delta_G^i\|_*^2 + \sum_{i=1}^t \frac{\alpha_i\gamma_i}{2(1-q)\Gamma_i}\|\Delta_G^i\|_*^2 \\ & +\textstyle\sum_{i=1}^t \frac{3\alpha_i\gamma_i}{2\Gamma_i}\left(\|\Delta_H^{2i}\|_*^2+\|\Delta_H^{2i-1}\|_*^2\right) \\ & -\textstyle\sum_{i=1}^t \frac{\alpha_i}{\Gamma_i}\langle\Delta_G^i, r_i-w_i^v\rangle - \sum_{i=1}^t \frac{\alpha_i}{\Gamma_i}\langle\Delta_H^{2i}, w_{i+1}-w_i^v\rangle.\end{aligned} \tag{4.5.45}$$

Applying (4.5.41) and (4.5.44) to (4.5.42), we have

$$\tfrac{1}{\Gamma_t}Q(\bar{w}_{t+1},u) \le \tfrac{2\alpha_t}{\gamma_t\Gamma_t}D_Z^2 + U_t,\ \forall u\in Z,$$

or equivalently,

$$g(\bar{w}_{t+1}) \le \tfrac{2\alpha_t}{\gamma_t}D_Z^2 + \Gamma_t U_t. \tag{4.5.46}$$

Now it suffices to bound U_t, in both expectation and probability.

We prove part (a) first. By our assumptions on SO_G and SO_H and in view of (4.5.12), (4.5.13), and (4.5.34), during the i-th iteration of Algorithm 4.4, the random noise Δ_H^{2i} is independent of w_{i+1} and w_i^v, and Δ_G^i is independent of r_i and w_i^v, hence $\mathbb{E}[\langle\Delta_G^i, r_i-w_i^v\rangle] = \mathbb{E}[\langle\Delta_H^{2i}, w_{i+1}-w_i^v\rangle] = 0$. In addition, Assumption 10 implies that $\mathbb{E}[\|\Delta_G^i\|_*^2] \le \sigma_G^2$, $\mathbb{E}[\|\Delta_H^{2i-1}\|_*^2] \le \sigma_H^2$ and $\mathbb{E}[\|\Delta_H^{2i}\|_*^2] \le \sigma_H^2$, where Δ_G^i, Δ_H^{2i-1}, and Δ_H^{2i} are independent. Therefore, taking expectation on (4.5.45) we have

$$\begin{aligned}\mathbb{E}[U_t] \le & \ \mathbb{E}\Big[\textstyle\sum_{i=1}^t \frac{\alpha_i\gamma_i}{\Gamma_i}\left(\|\Delta_H^{2i}\|^2+\|\Delta_G^i\|_*^2\right) + \sum_{i=1}^t \frac{\alpha_i\gamma_i}{2(1-q)\Gamma_i}\|\Delta_G^i\|_*^2 \\ & +\textstyle\sum_{i=1}^t \frac{3\alpha_i\gamma_i}{2\Gamma_i}\left(\|\Delta_H^{2i}\|_*^2+\|\Delta_H^{2i-1}\|_*^2\right)\Big] \\ = & \textstyle\sum_{i=1}^t \frac{\alpha_i\gamma_i}{\Gamma_i}\left[4\sigma_H^2 + \left(1+\frac{1}{2(1-q)}\right)\sigma_G^2\right].\end{aligned} \tag{4.5.47}$$

Taking expectation on both sides of (4.5.46), and using (4.5.47), we obtain (4.5.38).

Next we prove part (b). Observe that the sequence $\{\langle\Delta_G^i, r_i-w_i^v\rangle\}_{i\ge1}$ is a martingale difference and hence satisfies the large-deviation theorem (see Lemma 4.1). Therefore using Assumption 11 and the fact that

$$\begin{aligned}& \mathbb{E}\left[\exp\left\{\frac{(\alpha_i\Gamma_i^{-1}\langle\Delta_G^i, r_i-w_i^v\rangle)^2}{2(\sigma_G\alpha_i\Gamma_i^{-1}D_Z)^2}\right\}\right] \\ \le & \ \mathbb{E}\left[\exp\left\{\frac{\|\Delta_G^i\|_*^2\|r_i-w_i^v\|^2}{2\sigma_G^2D_Z^2}\right\}\right] \le \mathbb{E}\left[\exp\left\{\|\Delta_G^i\|_*^2/\sigma_G^2\right\}\right] \le \exp\{1\},\end{aligned}$$

we conclude from the large-deviation theorem that

$$\mathrm{Prob}\left\{-\textstyle\sum_{i=1}^t \frac{\alpha_i}{\Gamma_i}\langle\Delta_G^i, r_i-w_i^v\rangle > \lambda\sigma_G D_Z\sqrt{2\sum_{i=1}^t\left(\frac{\alpha_i}{\Gamma_i}\right)^2}\right\} \le \exp\{-\lambda^2/3\}. \tag{4.5.48}$$

By using a similar argument we have

$$\text{Prob}\left\{-\sum_{i=1}^{t}\tfrac{\alpha_i}{\Gamma_i}\langle\Delta_H^{2i}, w_{i+1}-w_i^v\rangle > \lambda\sigma_H D_Z\sqrt{2\sum_{i=1}^{t}\left(\tfrac{\alpha_i}{\Gamma_i}\right)^2}\right\} \le \exp\{-\lambda^2/3\}. \tag{4.5.49}$$

In addition, letting $S_i = \alpha_i\gamma_i/(\Gamma_i)$ and $S = \sum_{i=1}^{t} S_i$, by Assumption 11 and the convexity of exponential functions, we have

$$\mathbb{E}\left[\exp\left\{\tfrac{1}{S}\sum_{i=1}^{t} S_i\|\Delta_G^i\|_*^2/\sigma_G^2\right\}\right] \le \mathbb{E}\left[\tfrac{1}{S}\sum_{i=1}^{t} S_i\exp\left\{\|\Delta_G^i\|_*^2/\sigma_G^2\right\}\right] \le \exp\{1\}.$$

Noting by Markov's inequality that $P(X > a) \le \mathbb{E}[X]/a$ for all nonnegative random variables X and constants $a > 0$, the above inequality implies that

$$\begin{aligned}
&\text{Prob}\left[\sum_{i=1}^{t} S_i\|\Delta_G^i\|_*^2 > (1+\lambda)\sigma_G^2 S\right]\\
=&\text{Prob}\left[\exp\left\{\tfrac{1}{S}\sum_{i=1}^{t} S_i\|\Delta_G^i\|_*^2/\sigma_G^2\right\} > \exp\{1+\lambda\}\right]\\
\le&\mathbb{E}\left[\exp\left\{\tfrac{1}{S}\sum_{i=1}^{t} S_i\|\Delta_G^i\|_*^2/\sigma_G^2\right\}\right]/\exp\{1+\lambda\}\\
\le&\exp\{-\lambda\}.
\end{aligned}$$

Recalling that $S_i = \alpha_i\gamma_i/(\Gamma_i)$ and $S = \sum_{i=1}^{t} S_i$, the above relation is equivalent to

$$\text{Prob}\left\{\left(1+\tfrac{1}{2(1-q)}\right)\sum_{i=1}^{t}\tfrac{\alpha_i\gamma_i}{\Gamma_i}\|\Delta_G^i\|_*^2 > (1+\lambda)\sigma_G^2\left(1+\tfrac{1}{2(1-q)}\right)\sum_{i=1}^{t}\tfrac{\alpha_i\gamma_i}{\Gamma_i}\right\} \le \exp\{-\lambda\}. \tag{4.5.50}$$

Using similar arguments, we also have

$$\text{Prob}\left\{\sum_{i=1}^{t}\tfrac{3\alpha_i\gamma_i}{2\Gamma_i}\|\Delta_H^{2i-1}\|_*^2 > (1+\lambda)\tfrac{3\sigma_H^2}{2}\sum_{i=1}^{t}\tfrac{\alpha_i\gamma_i}{\Gamma_i}\right\} \le \exp\{-\lambda\}, \tag{4.5.51}$$

$$\text{Prob}\left\{\sum_{i=1}^{t}\tfrac{5\alpha_i\gamma_i}{2\Gamma_i}\|\Delta_H^{2i}\|_*^2 > (1+\lambda)\tfrac{5\sigma_H^2}{2}\sum_{i=1}^{t}\tfrac{\alpha_i\gamma_i}{\Gamma_i}\right\} \le \exp\{-\lambda\}. \tag{4.5.52}$$

Using the fact that $\|\Delta_H^{2i} + \Delta_G^{2i-1}\|_*^2 \le 2\|\Delta_H^{2i}\|_*^2 + 2\|\Delta_G^{2i-1}\|_*^2$, we conclude from (4.5.46) to (4.5.52) that (4.5.39) holds. ■

There are various options for choosing the parameters $\{\alpha_t\}$ and $\{\gamma_t\}$ that satisfy (4.5.37). In the following corollary, we give one example of such parameter settings.

Corollary 4.5. *Suppose that* (4.5.36) *holds. If the stepsizes* $\{\alpha_t\}$ *and* $\{\gamma_t\}$ *in Algorithm 4.4 are set to:*

$$\alpha_t = \tfrac{2}{t+1} \text{ and } \gamma_t = \tfrac{t}{4L+3Mt+\beta(t+1)\sqrt{t}}, \tag{4.5.53}$$

where $\beta > 0$ *is a parameter. Then under Assumption 10,*

$$\mathbb{E}\left[g(\bar{w}_{t+1})\right] \le \tfrac{16LD_Z^2}{t(t+1)} + \tfrac{12MD_Z^2}{t+1} + \tfrac{\sigma D_Z}{\sqrt{t-1}}\left(\tfrac{4\beta D_Z}{\sigma} + \tfrac{16\sigma}{3\beta D_Z}\right) =: \mathcal{C}_0(t), \tag{4.5.54}$$

where σ and D_Z are defined in (4.5.7) *and* (4.5.36), *respectively. Furthermore, under Assumption 11,*

$$\text{Prob}\{g(\bar{w}_{t+1}) > \mathcal{C}_0(t) + \lambda \mathcal{C}_1(t)\} \le 2\exp\{-\lambda^2/3\} + 3\exp\{-\lambda\}, \; \forall \lambda > 0,$$

where

$$\mathcal{C}_1(t) := \tfrac{\sigma D_Z}{\sqrt{t-1}} \left(\tfrac{4\sqrt{3}}{3} + \tfrac{16\sigma}{3\beta D_Z} \right). \tag{4.5.55}$$

Proof. It is easy to check that

$$\Gamma_t = \tfrac{2}{t(t+1)} \text{ and } \tfrac{\alpha_t}{\Gamma_t \gamma_t} \le \tfrac{\alpha_{t+1}}{\Gamma_{t+1}\gamma_{t+1}}.$$

In addition, in view of (4.5.53), we have $\gamma_t \le t/(4L)$ and $\gamma_t^2 \le 1/(9M^2)$, which implies

$$\tfrac{5}{6} - L\alpha_t\gamma_t - 3M^2\gamma_t^2 \ge \tfrac{5}{6} - \tfrac{t}{4} \cdot \tfrac{2}{t+1} - \tfrac{1}{3} \ge 0.$$

Therefore the first relation in (4.5.37) holds with constant $q = 5/6$. In view of Theorem 4.10, it now suffices to show that $\mathcal{Q}_0(t) \le \mathcal{C}_0(t)$ and $\mathcal{Q}_1(t) \le \mathcal{C}_1(t)$. Observing that $\alpha_t/\Gamma_t = t$ and $\gamma_t \le 1/(\beta\sqrt{t})$, we obtain

$$\textstyle\sum_{i=1}^t \frac{\alpha_i\gamma_i}{\Gamma_i} \le \frac{1}{\beta}\sum_{i=1}^t \sqrt{i} \le \frac{1}{\beta}\int_0^{t+1} \sqrt{t}dt = \frac{1}{\beta} \cdot \frac{2(t+1)^{3/2}}{3} = \frac{2(t+1)^{3/2}}{3\beta}.$$

Using the above relation, (4.5.36), (4.5.38), (4.5.40), (4.5.53), and the fact that $\sqrt{t+1}/t \le 1/\sqrt{t-1}$ and $\sum_{i=1}^t i^2 \le t(t+1)^2/3$, we have

$$\begin{aligned}
\mathcal{Q}_0(t) &= \tfrac{4D_Z^2}{t(t+1)}\left(4L + 3Mt + \beta(t+1)\sqrt{t}\right) + \tfrac{8\sigma^2}{t(t+1)}\textstyle\sum_{i=1}^t \frac{\alpha_i\gamma_i}{\Gamma_i} \\
&\le \tfrac{16LD_Z^2}{t(t+1)} + \tfrac{12MD_Z^2}{t+1} + \tfrac{4\beta D_Z^2}{\sqrt{t}} + \tfrac{16\sigma^2\sqrt{t+1}}{3\beta t} \\
&\le \mathcal{C}_0(t),
\end{aligned}$$

and

$$\begin{aligned}
\mathcal{Q}_1(t) &= \tfrac{2(\sigma_G+\sigma_H)}{t(t+1)} D_Z \sqrt{2\textstyle\sum_{i=1}^t i^2} + \tfrac{8\sigma^2}{t(t+1)}\textstyle\sum_{i=1}^t \frac{\alpha_i\gamma_i}{\Gamma_i} \\
&\le \tfrac{2\sqrt{2}(\sigma_G+\sigma_H)D_Z}{\sqrt{3t}} + \tfrac{16\sigma^2\sqrt{t+1}}{3\beta t} \\
&\le \mathcal{C}_1(t).
\end{aligned}$$

■

We now add a few remarks about the results obtained in Corollary 4.5. First, in view of (4.5.8), (4.5.54), and (4.5.55), we can clearly see that the SAMP method is robust with respect to the estimates of σ and D_Z. Indeed, the SAMP method achieves the optimal iteration complexity for solving the SVI problem as long as

$\beta = \mathcal{O}(\sigma/D_Z)$. In particular, in this case, the number of iterations performed by the SAMP method to find an ε-solution of (4.5.1), i.e., a point $\bar{w} \in Z$ s.t. $\mathbb{E}[g(\bar{w})] \leq \varepsilon$, can be bounded by

$$\mathcal{O}\left(\sqrt{\tfrac{L}{\varepsilon}} + \tfrac{M}{\varepsilon} + \tfrac{\sigma^2}{\varepsilon^2}\right), \tag{4.5.56}$$

which implies that this algorithm allows L to be as large as $\mathcal{O}(\varepsilon^{-3/2})$ and M to be as large as $\mathcal{O}(\varepsilon^{-1})$ without significantly affecting its convergence properties. Second, for the deterministic case when $\sigma = 0$, the complexity bound in (4.5.56) achieves the best-known so-far complexity for solving problem (4.5.1) (see (4.5.8)) in terms of their dependence on the Lipschitz constant L.

In the following theorem, we demonstrate some convergence properties of Algorithm 4.4 for solving the stochastic problem $SVI(Z;G,H,J)$ when Z is unbounded. To study the convergence properties of SAMP in this case, we use a perturbation-based termination criterion based on the enlargement of a maximal monotone operator. More specifically, we say that the pair $(\tilde{v}, \tilde{u}) \in \mathbb{R}^n \times Z$ is a (ρ, ε)-approximate solution of $SVI(Z;G,H,J)$ if $\|\tilde{v}\| \leq \rho$ and $\tilde{g}(\tilde{u}, \tilde{v}) \leq \varepsilon$, where the gap function $\tilde{g}(\cdot,\cdot)$ is defined by

$$\tilde{g}(\tilde{u}, \tilde{v}) := \sup_{u \in Z} Q(\tilde{u}, u) - \langle \tilde{v}, \tilde{u} - u \rangle. \tag{4.5.57}$$

We call $\tilde{v}$ the perturbation vector associated with $\tilde{u}$. One advantage of employing this termination criterion is that the convergence analysis does not depend on the boundedness of Z.

Theorem 4.11 below describes the convergence properties of SAMP for solving SVIs with unbounded feasible sets, under the assumption that a strong solution of (4.5.6) exists.

Theorem 4.11. *Suppose that $V(r,z) := \|z - r\|^2/2$ for any $r \in Z$ and $z \in Z$. If the parameters $\{\alpha_t\}$ and $\{\gamma_t\}$ in Algorithm 4.4 are chosen such that $\alpha_1 = 1$, and for all $t > 1$,*

$$0 \leq \alpha_t < 1,\ L\alpha_t\gamma_t + 3M^2\gamma_t^2 \leq c^2 < q \text{ for some } c, q \in (0,1), \text{ and } \tfrac{\alpha_t}{\Gamma_t\gamma_t} = \tfrac{\alpha_{t+1}}{\Gamma_{t+1}\gamma_{t+1}}, \tag{4.5.58}$$

where Γ_t is defined in (4.5.17). *Then for all $t \geq 1$ there exists a perturbation vector v_{t+1} and a residual $\varepsilon_{t+1} \geq 0$ such that $\tilde{g}(\bar{w}_{t+1}, v_{t+1}) \leq \varepsilon_{t+1}$. Moreover, for all $t \geq 1$, we have*

$$\mathbb{E}[\|v_{t+1}\|] \leq \tfrac{\alpha_t}{\gamma_t}\left(2D + 2\sqrt{D^2 + C_t^2}\right), \tag{4.5.59}$$

$$\mathbb{E}[\varepsilon_{t+1}] \leq \tfrac{\alpha_t}{\gamma_t}\left[(3+6\theta)D^2 + (1+6\theta)C_t^2\right] + \tfrac{18\alpha_t^2\sigma_H^2}{\gamma_t^2}\textstyle\sum_{i=1}^t \gamma_i^3, \tag{4.5.60}$$

where

$$D := \|r_1 - u^*\|, \tag{4.5.61}$$

u^* is a strong solution of $SVI(Z;G,H,J)$,

$$\theta = \max\left\{1, \tfrac{c^2}{q-c^2}\right\} \text{ and } C_t = \sqrt{\left[4\sigma_H^2 + \left(1+\tfrac{1}{2(1-q)}\right)\sigma_G^2\right]\textstyle\sum_{i=1}^t \gamma_i^2}. \tag{4.5.62}$$

Proof. Let U_t be defined in (4.5.45). First, applying (4.5.58) and (4.5.44) to (4.5.30) in Lemma 4.14, we have

$$\begin{aligned}&\tfrac{1}{\Gamma_t} Q(\bar{w}_{t+1}, u)\\ &\le \mathrm{B}_t(u, r_{[t]}) - \tfrac{\alpha_t}{2\Gamma_t\gamma_t}\textstyle\sum_{i=1}^t \left(q-c^2\right)\|r_i - w_{i+1}\|^2 + \mathrm{B}_t(u, w^v_{[t]}) + U_t,\ \forall u\in Z.\end{aligned} \tag{4.5.63}$$

In addition, applying (4.5.58) to the definition of $\mathrm{B}_t(\cdot,\cdot)$ in (4.5.31), we obtain

$$\mathrm{B}_t(u, r_{[t]}) = \tfrac{\alpha_t}{2\Gamma_t\gamma_t}(\|r_1 - u\|^2 - \|r_{t+1} - u\|^2) \tag{4.5.64}$$

$$= \tfrac{\alpha_t}{2\Gamma_t\gamma_t}(\|r_1 - \bar{w}_{t+1}\|^2 - \|r_{t+1} - \bar{w}_{t+1}\|^2 + 2\langle r_1 - r_{t+1}, \bar{w}_{t+1} - u\rangle). \tag{4.5.65}$$

By using a similar argument and the fact that $w_1^v = w_1 = r_1$, we have

$$\mathrm{B}_t(u, w^v_{[t]}) = \tfrac{\alpha_t}{2\Gamma_t\gamma_t}(\|r_1 - u\|^2 - \|w^v_{t+1} - u\|^2) \tag{4.5.66}$$

$$= \tfrac{\alpha_t}{2\Gamma_t\gamma_t}(\|r_1 - \bar{w}_{t+1}\|^2 - \|w^v_{t+1} - \bar{w}_{t+1}\|^2 + 2\langle r_1 - w^v_{t+1}, \bar{w}_{t+1} - u\rangle). \tag{4.5.67}$$

We then conclude from (4.5.63), (4.5.65), and (4.5.67) that

$$Q(\bar{w}_{t+1}, u) - \langle v_{t+1}, \bar{w}_{t+1} - u\rangle \le \varepsilon_{t+1},\ \forall u \in Z, \tag{4.5.68}$$

where

$$v_{t+1} := \tfrac{\alpha_t}{\gamma_t}(2r_1 - r_{t+1} - w^v_{t+1}) \tag{4.5.69}$$

and

$$\begin{aligned}\varepsilon_{t+1} := \tfrac{\alpha_t}{2\gamma_t}\,&\left(2\|r_1 - \bar{w}_{t+1}\|^2 - \|r_{t+1} - \bar{w}_{t+1}\|^2 - \|w^v_{t+1} - \bar{w}_{t+1}\|^2\right.\\ &\left.- \textstyle\sum_{i=1}^t \left(q-c^2\right)\|r_i - w_{i+1}\|^2\right) + \Gamma_t U_t.\end{aligned} \tag{4.5.70}$$

It is easy to see that the residual ε_{t+1} is positive by setting $u = \bar{w}_{t+1}$ in (4.5.68). Hence $\tilde{g}(\bar{w}_{t+1}, v_{t+1}) \le \varepsilon_{t+1}$. To finish the proof, it suffices to estimate the bounds for $\mathbb{E}[\|v_{t+1}\|]$ and $\mathbb{E}[\varepsilon_{t+1}]$. Observe that by (4.5.2), (4.5.6), (4.5.15), and the convexity of G and J, we have

$$Q(\bar{w}_{t+1}, u^*) \ge \langle F(u^*), \bar{w}_{t+1} - u^*\rangle \ge 0, \tag{4.5.71}$$

where the last inequality follows from the assumption that u^* is a strong solution of *$SVI(Z;G,H,J)$*. Using the above inequality and letting $u = u^*$ in (4.5.63), we conclude from (4.5.64) and (4.5.66) that

$$2\|r_1 - u^*\|^2 - \|r_{t+1} - u^*\|^2 - \|w^v_{t+1} - u^*\|^2 - \Sigma_{i=1}^t \left(q - c^2\right) \|r_i - w_{i+1}\|^2 + \tfrac{2\Gamma_t\gamma_t}{\alpha_t} U_t$$
$$\geq \tfrac{2\gamma_t}{\alpha_t} Q(\bar{w}_{t+1}, u^*) \geq 0.$$

By the above inequality and the definition of D in (4.5.61), we have

$$\|r_{t+1} - u^*\|^2 + \|w^v_{t+1} - u^*\|^2 + \Sigma_{i=1}^t \left(q - c^2\right) \|r_i - w_{i+1}\|^2 \leq 2D^2 + \tfrac{2\Gamma_t\gamma_t}{\alpha_t} U_t. \tag{4.5.72}$$

In addition, applying (4.5.58) and the definition of C_t in (4.5.62) to (4.5.47), we have

$$\mathbb{E}[U_t] \leq \Sigma_{i=1}^t \tfrac{\alpha_t \gamma_i^2}{\Gamma_t \gamma_t} \left[4\sigma_H^2 + \left(1 + \tfrac{1}{2(1-q)}\right) \sigma_G^2\right] = \tfrac{\alpha_t}{\Gamma_t \gamma_t} C_t^2. \tag{4.5.73}$$

Combining (4.5.72) and (4.5.73), we have

$$\mathbb{E}[\|r_{t+1} - u^*\|^2] + \mathbb{E}[\|w^v_{t+1} - u^*\|^2] + \Sigma_{i=1}^t \left(q - c^2\right) \mathbb{E}[\|r_i - w_{i+1}\|^2] \leq 2D^2 + 2C_t^2. \tag{4.5.74}$$

We are now ready to prove (4.5.59). Observe from the definition of v_{t+1} in (4.5.69) and the definition of D in (4.5.61) that $\|v_{t+1}\| \leq \alpha_t(2D + \|w^v_{t+1} - u^*\| + \|r_{t+1} - u^*\|)/\gamma_t$, using the previous inequality, Jensen's inequality, and (4.5.74), we obtain

$$\mathbb{E}[\|v_{t+1}\|] \leq \tfrac{\alpha_t}{\gamma_t}(2D + \sqrt{\mathbb{E}[(\|r_{t+1} - u^*\| + \|w^v_{t+1} - u^*\|)^2]})$$
$$\leq \tfrac{\alpha_t}{\gamma_t}(2D + \sqrt{2\mathbb{E}[\|r_{t+1} - u^*\|^2 + \|w^v_{t+1} - u^*\|^2]}) \leq \tfrac{\alpha_t}{\gamma_t}(2D + 2\sqrt{D^2 + C_t^2}).$$

Our remaining goal is to prove (4.5.60). By (4.5.14) and (4.5.17), we have

$$\tfrac{1}{\Gamma_t}\bar{w}_{t+1} = \tfrac{1}{\Gamma_{t-1}}\bar{w}_t + \tfrac{\alpha_t}{\Gamma_t} w_{t+1}, \ \forall t > 1.$$

Using the assumption that $\bar{w}_1 = w_1$, we obtain

$$\bar{w}_{t+1} = \Gamma_t \Sigma_{i=1}^t \tfrac{\alpha_i}{\Gamma_i} w_{i+1}, \tag{4.5.75}$$

where by (4.5.17) we have

$$\Gamma_t \Sigma_{i=1}^t \tfrac{\alpha_i}{\Gamma_i} = 1. \tag{4.5.76}$$

Therefore, $\bar{w}_{t+1}$ is a convex combination of iterates $w_2, \ldots, w_{t+1}$. Also, by a similar argument in the proof of Lemma 4.14, applying Lemma 4.12 to (4.5.12) and (4.5.13) (with $r = r_t, w = w_{t+1}, y = r_{t+1}, \eta = \gamma_t \mathsf{H}(r_t; \zeta_{2t-1}) + \gamma_t \mathsf{G}(\underline{w}_t; \xi_t), \vartheta = \gamma_t \mathsf{H}(w_{t+1}; \zeta_{2t}) + \gamma_t \mathsf{G}(\underline{w}_t; \xi_t), J = \gamma_t J,\ L = 3M^2\gamma_t^2$ and $M^2 = 3\gamma_t^2(\|\Delta_H^{2t}\|_*^2 + \|\Delta_H^{2t-1}\|_*^2)$), and using (4.5.22) and (4.5.33), we have

$$\begin{aligned}\tfrac{1}{2}\|r_{t+1}-w_{t+1}\|^2 &\le \tfrac{3M^2\gamma_t^2}{2}\|r_t-w_{t+1}\|^2+\tfrac{3\gamma_t^2}{2}(\|\Delta_H^{2t}\|_*^2+\|\Delta_H^{2t-1}\|_*^2)\\ &\le \tfrac{c^2}{2}\|r_t-w_{t+1}\|^2+\tfrac{3\gamma_t^2}{2}(\|\Delta_H^{2t}\|_*^2+\|\Delta_H^{2t-1}\|_*^2),\end{aligned}$$

where the last inequality follows from (4.5.58).

Now using (4.5.70), (4.5.75), (4.5.76), the above inequality, and applying Jensen's inequality, we have

$$\begin{aligned}\varepsilon_{t+1}-\Gamma_t U_t &\le \tfrac{\alpha_t}{\gamma_t}\|r_1-\bar{w}_{t+1}\|^2\\ &= \tfrac{\alpha_t}{\gamma_t}\Big\|r_1-u^*+\Gamma_t\textstyle\sum_{i=1}^t\tfrac{\alpha_i}{\Gamma_i}(u^*-r_{t+1}[i])\\ &\qquad +\Gamma_t\textstyle\sum_{i=1}^t\tfrac{\alpha_i}{\Gamma_i}(r_{t+1}[i]-w_{t+1}[i])\Big\|^2\\ &\le \tfrac{3\alpha_t}{\gamma_t}\left[D^2+\Gamma_t\textstyle\sum_{i=1}^t\tfrac{\alpha_i}{\Gamma_i}\left(\|r_{i+1}-u^*\|^2+\|w_{i+1}-r_{i+1}\|^2\right)\right] \qquad (4.5.77)\\ &\le \tfrac{3\alpha_t}{\gamma_t}\Big[D^2+\Gamma_t\textstyle\sum_{i=1}^t\tfrac{\alpha_i}{\Gamma_i}\Big(\|r_{i+1}-u^*\|^2+c^2\|w_{i+1}-r_i\|^2\\ &\qquad +3\gamma_i^2(\|\Delta_H^{2i}\|_*^2+\|\Delta_H^{2i-1}\|_*^2)\|\Big)\Big].\end{aligned}$$

Noting that by (4.5.62) and (4.5.72),

$$\begin{aligned}&\Gamma_t\textstyle\sum_{i=1}^t\tfrac{\alpha_i}{\Gamma_i}(\|r_{i+1}-u^*\|^2+c^2\|w_{i+1}-r_i\|^2)\\ &\le \Gamma_t\textstyle\sum_{i=1}^t\tfrac{\alpha_i\theta}{\Gamma_i}(\|r_{i+1}-u^*\|^2+(q-c^2)\|w_{i+1}-r_i\|^2)\\ &\le \Gamma_t\textstyle\sum_{i=1}^t\tfrac{\alpha_i\theta}{\Gamma_i}(2D^2+\tfrac{2\Gamma_i\gamma_i}{\alpha_i}U_i)=2\theta D^2+2\theta\Gamma_t\textstyle\sum_{i=1}^t\gamma_i U_i,\end{aligned}$$

and that by (4.5.58),

$$\begin{aligned}&\Gamma_t\textstyle\sum_{i=1}^t\tfrac{3\alpha_i\gamma_i^2}{\Gamma_i}(\|\Delta_H^{2i}\|_*^2+\|\Delta_H^{2i-1}\|_*^2)\\ &=\Gamma_t\textstyle\sum_{i=1}^t\tfrac{3\alpha_t\gamma_i^3}{\Gamma_t\gamma_t}(\|\Delta_H^{2i}\|_*^2+\|\Delta_H^{2i-1}\|_*^2)=\tfrac{3\alpha_t}{\gamma_t}\textstyle\sum_{i=1}^t\gamma_i^3(\|\Delta_H^{2i}\|_*^2+\|\Delta_H^{2i-1}\|_*^2),\end{aligned}$$

we conclude from (4.5.73), (4.5.77), and Assumption 10 that

$$\begin{aligned}\mathbb{E}[\varepsilon_{t+1}] &\le \Gamma_t\mathbb{E}[U_t]+\tfrac{3\alpha_t}{\gamma_t}\left[D^2+2\theta D^2+2\theta\Gamma_t\textstyle\sum_{i=1}^t\gamma_i\mathbb{E}[U_i]+\tfrac{6\alpha_t\sigma_H^2}{\gamma_t}\textstyle\sum_{i=1}^t\gamma_i^3\right]\\ &\le \tfrac{\alpha_t}{\gamma_t}C_t^2+\tfrac{3\alpha_t}{\gamma_t}\left[(1+2\theta)D^2+2\theta\Gamma_t\textstyle\sum_{i=1}^t\tfrac{\alpha_i}{\Gamma_i}C_i^2+\tfrac{6\alpha_t\sigma_H^2}{\gamma_t}\textstyle\sum_{i=1}^t\gamma_i^3\right].\end{aligned}$$

Finally, observing from (4.5.62) and (4.5.76) that

$$\Gamma_t\textstyle\sum_{i=1}^t\tfrac{\alpha_i}{\Gamma_i}C_i^2\le C_t^2\Gamma_t\textstyle\sum_{i=1}^t\tfrac{\alpha_i}{\Gamma_i}=C_t^2,$$

we conclude (4.5.60) from the above inequality.

■

Below we give an example of parameters α_t and γ_t that satisfies (4.5.58).

Corollary 4.6. *Suppose that there exists a strong solution of* (4.5.1). *If the maximum number of iterations N is given, and the stepsizes $\{\alpha_t\}$ and $\{\gamma_t\}$ in Algorithm 4.4 are set to*

$$\alpha_t = \tfrac{2}{t+1} \text{ and } \gamma_t = \tfrac{t}{5L+3MN+\beta N\sqrt{N-1}}, \tag{4.5.78}$$

where σ is defined in Corollary 4.5, then there exists $v_N \in \mathbb{R}^n$ and $\varepsilon_N > 0$, such that $\tilde{g}(\bar{w}_N, v_N) \le \varepsilon_N$,

$$\mathbb{E}[\|v_N\|] \le \tfrac{40LD}{N(N-1)} + \tfrac{24MD}{N-1} + \tfrac{\sigma}{\sqrt{N-1}}\left(\tfrac{8\beta D}{\sigma} + 5\right), \tag{4.5.79}$$

and

$$\mathbb{E}[\varepsilon_N] \le \tfrac{90LD^2}{N(N-1)} + \tfrac{54MD^2}{N-1} + \tfrac{\sigma D}{\sqrt{N-1}}\left(\tfrac{18\beta D}{\sigma} + \tfrac{56\sigma}{3\beta D} + \tfrac{18\sigma}{\beta DN}\right). \tag{4.5.80}$$

Proof. Clearly, we have $\Gamma_t = 2/[t(t+1)]$, and hence (4.5.17) is satisfied. Moreover, in view of (4.5.78), we have

$$\begin{aligned} L\alpha_t\gamma_t + 3M^2\gamma_t^2 &\le \tfrac{2L}{5L+3MN} + \tfrac{3M^2N^2}{(5L+3MN)^2} \\ &= \tfrac{10L^2+6LMN+3M^2N^2}{(5L+3MN)^2} < \tfrac{5}{12} < \tfrac{5}{6}, \end{aligned}$$

which implies that (4.5.58) is satisfied with $c^2 = 5/12$ and $q = 5/6$. Observing from (4.5.78) that $\gamma_t = t\gamma_1$, setting $t = N-1$ in (4.5.62) and (4.5.78), we obtain

$$\tfrac{\alpha_{N-1}}{\gamma_{N-1}} = \tfrac{2}{\gamma_1 N(N-1)} \text{ and } C_{N-1}^2 = 4\sigma^2 \textstyle\sum_{i=1}^{N-1} \gamma_1^2 i^2 \le \tfrac{4\sigma^2\gamma_1^2 N^2(N-1)}{3}, \tag{4.5.81}$$

where C_{N-1} is defined in (4.5.62). Applying (4.5.81) to (4.5.59) we have

$$\begin{aligned} \mathbb{E}[\|v_N\|] &\le \tfrac{2}{\gamma_1 N(N-1)}(4D + 2C_{N-1}) \le \tfrac{8D}{\gamma_1 N(N-1)} + \tfrac{8\sigma}{\sqrt{3(N-1)}} \\ &\le \tfrac{40LD}{N(N-1)} + \tfrac{24MD}{N-1} + \tfrac{\sigma}{\sqrt{N-1}}\left(\tfrac{8\beta D}{\sigma} + 5\right). \end{aligned}$$

In addition, using (4.5.60), (4.5.81), and the facts that $\theta = 1$ in (4.5.62) and

$$\textstyle\sum_{i=1}^{N-1} \gamma_i^3 = \gamma_1^3 N^2(N-1)^2/4,$$

we have

$$\begin{aligned} \mathbb{E}[\varepsilon_{N-1}] &\le \tfrac{2}{\gamma_1 N(N-1)}(9D^2 + 7C_{N-1}^2) + \tfrac{72\sigma_H^2}{\gamma_1^2 N^2(N-1)^2} \cdot \tfrac{\gamma_1^3 N^2(N-1)^2}{4} \\ &\le \tfrac{18D^2}{\gamma_1 N(N-1)} + \tfrac{56\sigma^2\gamma_1 N}{3} + 18\sigma_H^2\gamma_1 \end{aligned}$$

$$\le \frac{90LD^2}{N(N-1)} + \frac{54MD^2}{N-1} + \frac{18\beta D^2}{\sqrt{N-1}} + \frac{56\sigma^2}{3\beta\sqrt{N-1}} + \frac{18\sigma_H^2}{\beta N\sqrt{N-1}}$$
$$\le \frac{90LD^2}{N(N-1)} + \frac{54MD^2}{N-1} + \frac{\sigma D}{\sqrt{N-1}}\left(\frac{18\beta D}{\sigma} + \frac{56\sigma}{3\beta D} + \frac{18\sigma}{\beta DN}\right).$$

■

Several remarks are in place for the results obtained in Theorem 4.11 and Corollary 4.6. First, similar to the bounded case (see the remark after Corollary 4.5), one may want to choose β in a way such that the right-hand side of (4.5.79) or (4.5.80) is minimized, e.g., $\beta = \mathcal{O}(\sigma/D)$. However, since the value of D will be very difficult to estimate for the unbounded case, one often has to resort to a suboptimal selection for β. For example, if $\beta = \sigma$, then the RHS of (4.5.79) and (4.5.80) will become $\mathcal{O}(LD/N^2 + MD/N + \sigma D/\sqrt{N})$ and $\mathcal{O}(LD^2/N^2 + MD^2/N + \sigma D^2/\sqrt{N})$, respectively. Second, both residuals $\|v_N\|$ and ε_N in (4.5.79) and (4.5.80) converge to 0 at the same rate (up to a constant factor). Finally, it is only for simplicity that we assume that $V(r,z) = \|z-r\|^2/2$; similar results can be achieved under assumptions that $\nabla\omega$ is Lipschitz continuous.

4.6 Stochastic Block Mirror Descent Method

In this section we consider the stochastic programming problem given by

$$f^* := \min_{x\in X}\{f(x) := \mathbb{E}[F(x,\xi)]\}. \tag{4.6.1}$$

Here $X \subseteq \mathbb{R}^n$ is a closed convex set, ξ is a random variable with support $\Xi \subseteq \mathbb{R}^d$, and $F(\cdot,\xi): X \to \mathbb{R}$ is continuous for every $\xi \in \Xi$. In addition, we assume that X has a block structure, i.e.,

$$X = X_1 \times X_2 \times \cdots \times X_b, \tag{4.6.2}$$

where $X_i \subseteq \mathbb{R}^{n_i}$, $i = 1,\ldots,b$, are closed convex sets with $n_1 + n_2 + \ldots + n_b = n$.

The block coordinate descent (BCD) method is a natural method for solving problems with X given in the form of (4.6.2). In comparison with regular first-order methods, each iteration of these methods updates only one block of variables. In particular, if each block consists of only one variable (i.e., $n_i = 1, i = 1,\ldots,b$), then the BCD method becomes the classical coordinate descent (CD) method.

Most BCD methods were designed for solving deterministic optimization problems. One possible approach for solving problem (4.6.1), based on these methods and the sample average approximation (SAA), can be described as follows. For a given set of i.i.d. samples (dataset) $\xi_k, k = 1,\ldots,N$, of ξ, we first approximate $f(\cdot)$ in (4.6.1) by $\tilde{f}(x) := \frac{1}{N}\sum_{k=1}^N F(x,\xi_k)$ and then apply the BCD methods to $\min_{x\in X}\tilde{f}(x)$. Since ξ_k, $k = 1,\ldots,N$, are fixed a priori, by recursively updating the (sub)gradient of $\tilde{f}$, the iteration cost of the BCD method can be considerably smaller than that of the gradient descent methods. However, the above SAA approach is also known for the following drawbacks: (a) the high memory requirement to store ξ_k, $k = 1,\ldots,N$; (b) the high dependence (at least linear) of the iteration cost on the sample size N, which can be expensive when dealing with large datasets; and (c) the difficulty to apply the approach to the on-line setting where one needs to update the solution whenever a new piece of data ξ_k is collected.

A different approach to solve problem (4.6.1) is to apply the stochastic gradient descent (SGD) type methods as introduced in the previous few sections. Note that all these algorithms only need to access one single ξ_k at each iteration, and hence does not require much memory. In addition, their iteration cost is independent of the sample size N. However, since these algorithms need to update the whole vector x at each iteration, their iteration cost can strongly depend on n unless the problem is very sparse.

Our main goal in this section is to present a new class of stochastic optimization methods, referred to as the stochastic block mirror descent (SBMD) methods, by incorporating the aforementioned block-coordinate decomposition into the classic stochastic mirror descent method. As a motivating example, consider an important class of SP problems with $F(x,\xi) = \psi(Bx - q,\xi)$, where B is a certain linear operator and ψ is a relatively simple function. These problems arise from many machine learning applications, where ψ is a loss function and $B \in \mathbb{R}^{m\times n}$ denotes a certain basis (or dictionary) obtained by, e.g., metric learning. Each iteration of existing SGD methods would require $\mathcal{O}(mn)$ arithmetic operations to compute Bx and becomes prohibitive if mn exceeds 10^{12} . On the other hand, by using block-coordinate decomposition with $n_i = 1$, the iteration cost of the SBMD algorithms can be significantly reduced to $\mathcal{O}(m)$, which can be further reduced if B and ξ_k are sparse (see Sect. 4.6.1.1 for more discussions). Our development has also been motivated by the situation when the bottleneck of the mirror descent method exists in the projection (or prox-mapping) subproblems, in the sense that X is decomposable and the computation of the projection over X is more expensive than that of gradient. In this case, we can also significantly reduce the iteration cost by using the block-coordinate decomposition, since each iteration of the SBMD method requires only one projection over X_i for some $1 \le i \le b$, while the mirror descent method needs to perform the projections over X_i for all $1 \le i \le b$. It should be noted, however, that our algorithm does not apply to the situation when a decomposition of X is not available.

To fix notation in this section, we use $\mathbb{R}^{n_i}$, $i = 1,\ldots,b$, to denote Euclidean spaces equipped with inner product $\langle\cdot,\cdot\rangle$ and norm $\|\cdot\|_i$ ($\|\cdot\|_{i,*}$ be the conjugate) such that $\sum_{i=1}^b n_i = n$. Let I_n be the identity matrix in $\mathbb{R}^n$ and $U_i \in \mathbb{R}^{n\times n_i}, i = 1,2,\ldots,b$, be the set of matrices satisfying $(U_1,U_2,\ldots,U_b) = I_n$. For a given $x \in \mathbb{R}^n$, we denote its i-th block by $x^{(i)} = U_i^T x$, $i = 1,\ldots,b$. Note that $x = U_1 x^{(1)} + \ldots + U_b x^{(b)}$. Moreover, we define $\|x\|^2 = \|x^{(1)}\|_1^2 + \ldots + \|x^{(b)}\|_b^2$, and denote its conjugate by $\|y\|_*^2 = \|y^{(1)}\|_{1,*}^2 + \ldots + \|y^{(b)}\|_{b,*}^2$.

4.6.1 Nonsmooth Convex Optimization

In this subsection we assume that the objective function f of problem (4.6.1) is convex but not necessarily differentiable. The SBMD method incorporates random block decomposition into the classic mirror descent method. More specifically, each iteration of this algorithm updates one block of the search point along a stochastic (sub)gradient direction given by $G_{i_k}(x_k, \xi_k) \equiv U_{i_k}^T G(x, \xi)$. Here, the index i_k is randomly chosen and $G(x, \xi)$ is an unbiased estimator of the subgradient of $f(\cdot)$, i.e.,

$$\mathbb{E}[G(x,\xi)] = g(x) \in \partial f(x), \ \ \forall x \in X. \tag{4.6.3}$$

Moreover, we assume that

$$\mathbb{E}[\|G_i(x,\xi)\|_{i,*}^2] \le M_i^2, \ i = 1,2,\ldots,b. \tag{4.6.4}$$

Clearly, by (4.6.3) and (4.6.4), we have

$$\|g_i(x)\|_{i,*}^2 = \|\mathbb{E}[G_i(x,\xi)]\|_{i,*}^2 \le \mathbb{E}[\|G_i(x,\xi)\|_{i,*}^2] \le M_i^2, \ i = 1,2,\ldots,b, \tag{4.6.5}$$

and

$$\|g(x)\|_*^2 = \textstyle\sum_{i=1}^b \|g_i(x)\|_{i,*}^2 \le \sum_{i=1}^b M_i^2. \tag{4.6.6}$$

4.6.1.1 The SBMD Algorithm for Nonsmooth Problems

We present a general scheme of the SBMD algorithm, based on Bregman's divergence, to solve stochastic convex optimization problems.

Recall that a function $\nu_i : X_i \to R$ is a distance generating function with modulus α_i with respect to $\|\cdot\|_i$, if ν_i is continuously differentiable and strongly convex with parameter α_i with respect to $\|\cdot\|_i$. Without loss of generality, we assume throughout the section that $\alpha_i = 1$ for any $i = 1,\ldots,b$ because we can always rescale $\nu_i(x)$ to $\bar{\nu}_i(x) = \nu_i(x)/\alpha_i$ in case $\alpha_i \ne 1$. Therefore, we have

$$\langle x - z, \nabla \nu_i(x) - \nabla \nu_i(z)\rangle \ge \|x - z\|_i^2 \ \ \forall x, z \in X_i.$$

The prox-function (or Bregman distance) associated with ν_i is given by

$$V_i(z,x) = \nu_i(x) - [\nu_i(z) + \langle \nabla \nu_i(z), x - z\rangle] \ \ \forall x, z \in X_i. \tag{4.6.7}$$

Suppose that the set X_i is bounded, the distance generating function ν_i also gives rise to the diameter of X_i that will be used frequently in our convergence analysis:

$$\mathscr{D}_{\nu_i, X_i} := \max_{x \in X_i} \nu_i(x) - \min_{x \in X_i} \nu_i(x). \tag{4.6.8}$$

For the sake of notational convenience, sometimes we simply denote $\mathscr{D}_{\nu_i, X_i}$ by D_i. Note that the definition of $\mathscr{D}_{\nu_i, X_i}$ slightly differs from the diameter D_{X_i} in (3.2.4). Sometimes it is a little easier to compute $\mathscr{D}_{\nu_i, X_i}$ than D_{X_i}. However, most convergence results we discuss in this section also hold by using D_{X_i} in place of $\mathscr{D}_{v_i,X_i}$ after slightly modifying some constant factors.

Letting $x_1^{(i)} = \mathrm{argmin}_{x \in X_i} v_i(x)$, $i = 1,\ldots,b$, we can easily see that for any $x \in X$,

$$V_i(x_1^{(i)}, x^{(i)}) = v_i(x^{(i)}) - v_i(x_1^{(i)}) - \langle \nabla v_i(x_1^{(i)}), x^{(i)} - x_1^{(i)} \rangle \le v_i(x^{(i)}) - v_i(x_1^{(i)}) \le D_i, \tag{4.6.9}$$

which, in view of the strong convexity of v_i, also implies that $\|x_1^{(i)} - x^{(i)}\|_i^2/2 \le D_i$. Therefore, for any $x, y \in X$, we have

$$\|x^{(i)} - y^{(i)}\|_i \le \|x^{(i)} - x_1^{(i)}\|_i + \|x_1^{(i)} - y^{(i)}\|_i \le 2\sqrt{2D_i}, \tag{4.6.10}$$

$$\|x - y\| = \sqrt{\textstyle\sum_{i=1}^b \|x^{(i)} - y^{(i)}\|_i^2} \le 2\sqrt{2\textstyle\sum_{i=1}^b D_i}. \tag{4.6.11}$$

With the above definition of the prox-mapping, we can formally describe the stochastic block mirror descent (SBMD) method as in Algorithm 4.5.

Algorithm 4.5 The stochastic block mirror descent (SBMD) algorithm

Let $x_1 \in X$, positive stepsizes $\{\gamma_k\}_{k\ge1}$, nonnegative weights $\{\theta_k\}_{k\ge1}$, and probabilities $p_i \in [0,1]$, $i = 1,\ldots,b$, s.t. $\sum_{i=1}^b p_i = 1$ be given. Set $s_1 = 0$, and $u_i = 1$ for $i = 1,\ldots,b$.

for $k = 1,\ldots,N$ **do**

1. Generate a random variable i_k according to

$$\mathrm{Prob}\{i_k = i\} = p_i, \quad i = 1,\ldots,b. \tag{4.6.12}$$

2. Update $s_k^{(i)}$, $i = 1,\ldots,b$, by

$$s_{k+1}^{(i)} = \begin{cases} s_k^{(i)} + x_k^{(i)} \sum_{j=u_{i_k}}^{k} \theta_j & i = i_k, \\ s_k^{(i)} & i \ne i_k, \end{cases} \tag{4.6.13}$$

and then set $u_{i_k} = k+1$.

3. Update $x_k^{(i)}$, $i = 1,\ldots,b$, by

$$x_{k+1}^{(i)} = \begin{cases} \arg\min_{u \in X_i} \langle G_{i_k}(x_k, \xi_k), u\rangle + \frac{1}{\gamma_k} V_i(x_k^{(i)}, u) & i = i_k, \\ x_k^{(i)} & i \ne i_k. \end{cases} \tag{4.6.14}$$

end for

Output: Set $s_{N+1}^{(i)} = s_{N+1}^{(i)} + x_N^{(i)} \sum_{j=u_i}^N \theta_j$, $i = 1,\ldots,b$, $i \ne i_N$ and $\bar{x}_N = s_{N+1} / \sum_{k=1}^N \theta_k$.

We now add a few remarks about the SBMD algorithm stated above. First, note that in this algorithm, the random variables ξ_k and i_k, $k = 1,\ldots,$ are assumed to be independent of each other. Second, each iteration of the SBMD method recursively updates the search point x_k based on $G_{i_k}(x_k, \xi_k)$, the i_k-th block of the stochastic subgradient. In addition, rather than taking the average of $\{x_k\}$ in the end of algorithm as in the mirror-descent method, we introduce an incremental block averaging scheme to compute the output of the algorithm. More specifically, we use a summation vector s_k to denote the weighted sum of x_k's and the index variables u_i, $i = 1, \ldots, b$, to record the latest iteration when the i-th block of s_k is updated. Then in (4.6.13), we add up the i_k-th block of s_k with $x_k \sum_{j=i_k}^k \theta_j$, where $\sum_{j=i_k}^k \theta_j$ is often given by explicit formula and hence easy to compute. It can be checked that by using this averaging scheme, we have

$$\bar{x}_N = \left(\sum_{k=1}^N \theta_k\right)^{-1} \sum_{k=1}^N (\theta_k x_k). \tag{4.6.15}$$

Third, observe that in addition to (4.6.13) and (4.6.14), each iteration of the SBMD method involves the computation of G_{i_k}. Whenever possible, we should update G_{i_k} recursively in order to reduce the iteration cost of the SBMD algorithm. Consider the SP problems with the objective function

$$f(x) = \mathbb{E}[\psi(Bx - q, \xi)] + \chi(x),$$

where $\psi(\cdot)$ and $\chi(\cdot)$ are relatively simple functions, $q \in \mathbb{R}^n$, and $B \in \mathbb{R}^{m \times n}$. For the sake of simplicity, let us also assume that $n_1 = \ldots = n_b = 1$. For example, in the well-known support vector machine problem (Sect. 1.4), we have $\psi(y) = \max\{\langle y, \xi \rangle, 0\}$ and $\chi(x) = \|x\|_2^2/2$. In order to compute the full vector $G(x_k, \xi_k)$, we need $\mathcal{O}(mn)$ arithmetic operations to compute the vector $Bx_k - q$, which majorizes other arithmetic operations if ψ and χ are simple. On the other hand, by recursively updating the vector $y_k = Bx_k$ in the SBMD method, we can significantly reduce the iteration cost from $\mathcal{O}(mn)$ to $\mathcal{O}(m)$. This bound can be further reduced if both ξ_k and B are sparse (i.e., the vector ξ_k and each row vector of B contain just a few nonzeros). The above example can be generalized to the case when B has $r \times b$ blocks denoted by $B_{i,j} \in \mathbb{R}^{m_i \times n_j}$, $1 \le i \le r$ and $1 \le j \le b$, and each block row $B_i = (B_{i,1}, \ldots, B_{i,b})$, $i = 1, \ldots, r$, is block-sparse.

Finally, observe that the above SBMD method is conceptual only because we have not yet specified the selection of the stepsizes $\{\gamma_k\}$, the weights $\{\theta_k\}$, and the probabilities $\{p_i\}$. We will specify these parameters after establishing some basic convergence properties of this method.

4.6.1.2 Convergence Properties of SBMD for Nonsmooth Problems

In this subsection, we discuss the main convergence properties of the SBMD method for solving general nonsmooth convex problems.

Theorem 4.12. *Let $\bar{x}_N$ be the output of the SBMD algorithm and suppose that*

$$\theta_k = \gamma_k, \quad k = 1, \ldots, N. \tag{4.6.16}$$

Then we have, for any $N \ge 1$,

$$\mathbb{E}[f(\bar{x}_N) - f(x_*)] \leq \left(\sum_{k=1}^N \gamma_k\right)^{-1} \left[\sum_{i=1}^b p_i^{-1} V_i(x_1^{(i)}, x_*^{(i)}) + \tfrac{1}{2} \sum_{k=1}^N \gamma_k^2 \sum_{i=1}^b M_i^2\right], \tag{4.6.17}$$

where x_* is an arbitrary solution of (4.6.1) and the expectation is taken with respect to (w.r.t.) $\{i_k\}$ and $\{\xi_k\}$.

Proof. For simplicity, let us denote $V_i(z,x) \equiv V_i(z^{(i)}, x^{(i)})$, $g_{i_k} \equiv g^{(i_k)}(x_k)$ (cf. (4.6.3)) and $V(z,x) = \sum_{i=1}^b p_i^{-1} V_i(z,x)$. Also let us denote $\zeta_k = (i_k, \xi_k)$ and $\zeta_{[k]} = (\zeta_1, \ldots, \zeta_k)$. By Lemma 4.3 and the definition of $x_k^{(i)}$ in (4.6.14), we have

$$V_{i_k}(x_{k+1}, x) \leq V_{i_k}(x_k, x) + \gamma_k \langle G_{i_k}(x_k, \xi_k), U_{i_k}^T(x - x_k) \rangle + \tfrac{1}{2} \gamma_k^2 \|G_{i_k}(x_k, \xi_k)\|_{i_k,*}^2.$$

Using this observation, we have, for any $k \geq 1$ and $x \in X$,

$$\begin{aligned}
&V(x_{k+1}, x) = \textstyle\sum_{i \neq i_k} p_i^{-1} V_i(x_k, x) + p_{i_k}^{-1} V_{i_k}(x_{k+1}, x) \\
&\leq \textstyle\sum_{i \neq i_k} p_i^{-1} V_i(x_k, x) + \\
&p_{i_k}^{-1} \left[V_{i_k}(x_k, x) + \gamma_k \left\langle G_{i_k}(x_k, \xi_k), U_{i_k}^T(x - x_k) \right\rangle + \tfrac{1}{2} \gamma_k^2 \|G_{i_k}(x_k, \xi_k)\|_{i_k,*}^2\right] \\
&= V(x_k, x) + \gamma_k p_{i_k}^{-1} \langle U_{i_k} G_{i_k}(x_k, \xi_k), x - x_k \rangle + \tfrac{1}{2} \gamma_k^2 p_{i_k}^{-1} \|G_{i_k}(x_k, \xi_k)\|_{i_k,*}^2 \\
&= V(x_k, x) + \gamma_k \langle g(x_k), x - x_k \rangle + \gamma_k \delta_k + \tfrac{1}{2} \gamma_k^2 \bar{\delta}_k,
\end{aligned} \tag{4.6.18}$$

where

$$\delta_k := \langle p_{i_k}^{-1} U_{i_k} G_{i_k}(x_k, \xi_k) - g(x_k), x - x_k \rangle \text{ and } \bar{\delta}_k := p_{i_k}^{-1} \|G_{i_k}(x_k, \xi_k)\|_{i_k,*}^2. \tag{4.6.19}$$

It then follows from (4.6.18) and the convexity of $f(\cdot)$ that, for any $k \geq 1$ and $x \in X$,

$$\gamma_k [f(x_k) - f(x)] \leq \gamma_k \langle g(x_k), x_k - x \rangle \leq V(x_k, x) - V(x_{k+1}, x) + \gamma_k \delta_k + \tfrac{1}{2} \gamma_k^2 \bar{\delta}_k.$$

By using the above inequalities, the convexity of $f(\cdot)$, and the fact that $\bar{x}_N = \sum_{k=1}^N (\gamma_k x_k) / \sum_{k=1}^N \gamma_k$ due to (4.6.15) and (4.6.16), we conclude that for any $N \geq 1$ and $x \in X$,

$$\begin{aligned}
f(\bar{x}_N) - f(x) &\leq \left(\textstyle\sum_{k=1}^N \gamma_k\right)^{-1} \textstyle\sum_{k=1}^N \gamma_k [f(x_k) - f(x)] \\
&\leq \left(\textstyle\sum_{k=1}^N \gamma_k\right)^{-1} \left[V(x_1, x) + \textstyle\sum_{k=1}^N \left(\gamma_k \delta_k + \tfrac{1}{2} \gamma_k^2 \bar{\delta}_k\right)\right].
\end{aligned} \tag{4.6.20}$$

Now, observe that by (4.6.3) and (4.6.12),

$$\begin{aligned}
\mathbb{E}_{\zeta_k} \left[p_{i_k}^{-1} \langle U_{i_k} G_{i_k}, x - x_k \rangle | \zeta_{[k-1]}\right] &= \textstyle\sum_{i=1}^b \mathbb{E}_{\xi_k} \left[\langle U_i G_i(x_k, \xi_k), x - x_k \rangle | \zeta_{[k-1]}\right] \\
&= \textstyle\sum_{i=1}^b \langle U_i g_i(x_k), x - x_k \rangle = \langle g(x_k), x - x_k \rangle,
\end{aligned}$$

and hence that using the independence between i_k and ξ_k, we have

$$\mathbb{E}[\delta_k | \zeta_{k-1}] = 0. \tag{4.6.21}$$

Also, by (4.6.4) and (4.6.12),

$$\mathbb{E}\left[p_{i_k}^{-1}\|G_{i_k}(x_k,\xi_k)\|_{i_k,*}^2\right] = \sum_{i=1}^b p_i p_i^{-1}\|G_i(x_k,\xi_k)\|_{i,*}^2 \le \sum_{i=1}^b M_i^2. \tag{4.6.22}$$

Our result in (4.6.17) then immediately follows by taking expectation on both sides of (4.6.20), replacing x by x_*, and using the previous observations in (4.6.21) and (4.6.22). ■

Below we provide a few specialized convergence results for the SBMD algorithm after properly selecting $\{p_i\}$, $\{\gamma_k\}$, and $\{\theta_k\}$.

Corollary 4.7. *Suppose that $\{\theta_k\}$ in Algorithm 4.5 are set to (4.6.16) and x_* is an arbitrary solution of (4.6.1).*

(a) If X is bounded, and $\{p_i\}$ and $\{\gamma_k\}$ are set to

$$p_i = \frac{\sqrt{D_i}}{\sum_{i=1}^b \sqrt{D_i}},\ i=1,\ldots,b,\ \text{and}\ \gamma_k = \gamma \equiv \frac{\sqrt{2}\sum_{i=1}^b \sqrt{D_i}}{\sqrt{N\sum_{i=1}^b M_i^2}},\ k=1,\ldots,N, \tag{4.6.23}$$

then

$$\mathbb{E}[f(\bar{x}_N) - f(x_*)] \le \sqrt{\tfrac{2}{N}}\sum_{i=1}^b \sqrt{D_i}\sqrt{\sum_{i=1}^b M_i^2}. \tag{4.6.24}$$

(b) If $\{p_i\}$ and $\{\gamma_k\}$ are set to

$$p_i = \tfrac{1}{b},\ i=1,\ldots,b,\ \text{and}\ \gamma_k = \gamma \equiv \frac{\sqrt{2b\tilde{D}}}{\sqrt{N\sum_{i=1}^b M_i^2}},\ k=1,\ldots,N, \tag{4.6.25}$$

for some $\tilde{D} > 0$, then

$$\mathbb{E}[f(\bar{x}_N) - f(x_*)] \le \sqrt{\sum_{i=1}^b M_i^2}\left(\frac{\sum_{i=1}^b V_i(x_1^{(i)},x_*^{(i)})}{\tilde{D}} + \tilde{D}\right)\frac{\sqrt{b}}{\sqrt{2N}}. \tag{4.6.26}$$

Proof. We show part (a) only, since part (b) can be proved similarly. Note that by (4.6.9) and (4.6.23), we have

$$\sum_{i=1}^b p_i^{-1} V_i(x_1,x_*) \le \sum_{i=1}^b p_i^{-1} D_i = \left(\sum_{i=1}^b \sqrt{D_i}\right)^2.$$

Using this observation, (4.6.17), and (4.6.23), we have

$$\mathbb{E}[f(\bar{x}_N) - f(x_*)] \le (N\gamma)^{-1}\left[\left(\sum_{i=1}^b \sqrt{D_i}\right)^2 + \tfrac{N\gamma^2}{2}\sum_{i=1}^b M_i^2\right] = \sqrt{\tfrac{2}{N}}\sum_{i=1}^b \sqrt{D_i}\sqrt{\sum_{i=1}^b M_i^2}.$$

■

A few remarks about the results obtained in Theorem 4.12 and Corollary 4.7 are in place. First, the parameter setting in (4.6.23) only works for the case when X is bounded, while the one in (4.6.25) also applies to the case when X is unbounded or when the bounds D_i, $i=1,\ldots,b$, are not available. It can be easily seen that the optimal choice of $\tilde{D}$ in (4.6.26) would be $\sqrt{\sum_{i=1}^b V_i(x_1,x_*)}$. In this case, (4.6.26) reduces to

$$\mathbb{E}[f(\bar{x}_N) - f(x_*)] \le \sqrt{2\sum_{i=1}^b M_i^2}\sqrt{\sum_{i=1}^b V_i(x_1, x_*)}\tfrac{\sqrt{b}}{\sqrt{N}}$$
$$\le \sqrt{2\sum_{i=1}^b M_i^2}\sqrt{\sum_{i=1}^b D_i}\tfrac{\sqrt{b}}{\sqrt{N}}, \qquad (4.6.27)$$

where the second inequality follows from (4.6.9). It is interesting to note the difference between the above bound and (4.6.24). Specifically, the bound obtained in (4.6.24) by using a nonuniform distribution $\{p_i\}$ always minorizes the one in (4.6.27) by the Cauchy–Schwartz inequality.

Second, observe that in view of (4.6.24), the total number of iterations required by the SBMD method to find an ε-solution of (4.6.1) can be bounded by

$$2\left(\sum_{i=1}^b \sqrt{D_i}\right)^2 \left(\sum_{i=1}^b M_i^2\right) \tfrac{1}{\varepsilon^2}. \qquad (4.6.28)$$

Also note that the iteration complexity of the stochastic mirror descent algorithm employed with the same $v_i(\cdot)$, $i = 1, \ldots, b$, is given by

$$2\sum_{i=1}^b D_i \left(\sum_{i=1}^b M_i^2\right) \tfrac{1}{\varepsilon^2}. \qquad (4.6.29)$$

Clearly, the bound in (4.6.28) can be larger, up to a factor of b, than the one in (4.6.29). Therefore, the total arithmetic cost of the SBMD algorithm will be comparable to or smaller than that of the mirror descent SA, if its iteration cost is smaller than that of the latter algorithm by a factor of $\mathcal{O}(b)$.

Third, in Corollary 4.7 we have used a constant stepsize policy where $\gamma_1 = \ldots = \gamma_N$. However, it should be noted that variable stepsize policies can also be used in the SBMD method.

4.6.1.3 Nonsmooth Strongly Convex Problems

In this subsection, we assume that the objective function $f(\cdot)$ in (4.6.1) is strongly convex, i.e., $\exists\ \mu > 0$ s.t.

$$f(y) \ge f(x) + \langle g(x), y - x\rangle + \mu \sum_{i=1}^b V_i(x^{(i)}, y^{(i)})\ \ \forall x, y \in X. \qquad (4.6.30)$$

In addition, for the sake of simplicity, we assume that the probability distribution of i_k is uniform, i.e.,

$$p_1 = p_2 = \ldots = p_b = \tfrac{1}{b}. \qquad (4.6.31)$$

It should be noted, however, that our analysis can be easily adapted to the case when i_k is nonuniform.

We are now ready to describe the main convergence properties of the SBMD algorithm for solving nonsmooth strongly convex problems.

Theorem 4.13. *Suppose that (4.6.30) and (4.6.31) hold. If*

$$\gamma_k \le \tfrac{b}{\mu} \qquad (4.6.32)$$

and

$$\theta_k = \tfrac{\gamma_k}{\Gamma_k} \quad \textit{with} \quad \Gamma_k = \begin{cases} 1 & k = 1 \\ \Gamma_{k-1}(1 - \tfrac{\gamma_k \mu}{b}) & k \geq 2, \end{cases} \tag{4.6.33}$$

then, for any $N \geq 1$, we have

$$\mathbb{E}[f(\bar{x}_N) - f(x_*)] \leq \left(\Sigma_{k=1}^N \theta_k\right)^{-1} \Big[(b - \gamma_1 \mu)\Sigma_{i=1}^b V_i(x_1^{(i)}, x_*^{(i)}) + \tfrac{1}{2}\Sigma_{k=1}^N \gamma_k \theta_k \Sigma_{i=1}^b M_i^2\Big], \tag{4.6.34}$$

where x_* is the optimal solution of (4.6.1).

Proof. For simplicity, let us denote $V_i(z,x) \equiv V_i(z^{(i)}, x^{(i)})$, $g_{i_k} \equiv g^{(i_k)}(x_k)$, and $V(z,x) = \Sigma_{i=1}^b p_i^{-1} V_i(z,x)$. Also let us denote $\zeta_k = (i_k, \xi_k)$ and $\zeta_{[k]} = (\zeta_1, \ldots, \zeta_k)$, and let δ_k and $\bar{\delta}_k$ be defined in (4.6.19). By (4.6.31), we have

$$V(z,x) = b\Sigma_{i=1}^b V_i(z^{(i)}, x^{(i)}). \tag{4.6.35}$$

Using this observation, (4.6.18), and (4.6.30), we obtain

$$\begin{aligned} V(x_{k+1}, x) &\leq V(x_k, x) + \gamma_k \langle g(x_k), x - x_k \rangle + \gamma_k \delta_k + \tfrac{1}{2}\gamma_k^2 \bar{\delta}_k \\ &\leq V(x_k, x) + \gamma_k \left[f(x) - f(x_k) - \tfrac{\mu}{2}\|x - x_k\|^2\right] + \gamma_k \delta_k + \tfrac{1}{2}\gamma_k^2 \bar{\delta}_k \\ &\leq \left(1 - \tfrac{\gamma_k \mu}{b}\right) V(x_k, x) + \gamma_k \left[f(x) - f(x_k)\right] + \gamma_k \delta_k + \tfrac{1}{2}\gamma_k^2 \bar{\delta}_k, \end{aligned}$$

which, in view of Lemma 3.17, then implies that

$$\tfrac{1}{\Gamma_N} V(x_{N+1}, x) \leq \left(1 - \tfrac{\gamma_1 \mu}{b}\right) V(x_1, x) + \Sigma_{k=1}^N \Gamma_k^{-1} \gamma_k \left[f(x) - f(x_k) + \delta_k + \tfrac{1}{2}\gamma_k^2 \bar{\delta}_k\right]. \tag{4.6.36}$$

Using the fact that $V(x_{N+1}, x) \geq 0$ and (4.6.33), we conclude from the above relation that

$$\Sigma_{k=1}^N \theta_k [f(x_k) - f(x)] \leq \left(1 - \tfrac{\gamma_1 \mu}{b}\right) V(x_1, x) + \Sigma_{k=1}^N \theta_k \delta_k + \tfrac{1}{2}\Sigma_{k=1}^N \gamma_k \theta_k \bar{\delta}_k. \tag{4.6.37}$$

Taking expectation on both sides of the above inequality, and using relations (4.6.21) and (4.6.22), we obtain

$$\Sigma_{k=1}^N \theta_k \mathbb{E}[f(x_k) - f(x)] \leq \left(1 - \tfrac{\gamma_1 \mu}{b}\right) V(x_1, x) + \tfrac{1}{2}\Sigma_{k=1}^N \gamma_k \theta_k \Sigma_{i=1}^b M_i^2,$$

which, in view of (4.6.15), (4.6.31), and the convexity of $f(\cdot)$, then clearly implies (4.6.34). ■

Below we provide a specialized convergence result for the SBMD method to solve nonsmooth strongly convex problems after properly selecting $\{\gamma_k\}$.

Corollary 4.8. *Suppose that (4.6.30) and (4.6.31) hold. If $\{\theta_k\}$ are set to (4.6.33) and $\{\gamma_k\}$ are set to*

$$\gamma_k = \tfrac{2b}{\mu(k+1)}, \quad k = 1, \ldots, N, \tag{4.6.38}$$

then, for any $N \geq 1$, we have

$$\mathbb{E}[f(\bar{x}_N) - f(x_*)] \le \tfrac{2b}{\mu(N+1)} \Sigma_{i=1}^b M_i^2, \tag{4.6.39}$$

where x_* is the optimal solution of (4.6.1).

Proof. It can be easily seen from (4.6.33) and (4.6.38) that

$$\Gamma_k = \tfrac{2}{k(k+1)}, \;\; \theta_k = \tfrac{\gamma_k}{\Gamma_k} = \tfrac{bk}{\mu}, \;\; b - \gamma_1 \mu = 0, \tag{4.6.40}$$

$$\Sigma_{k=1}^N \theta_k = \tfrac{bN(N+1)}{2\mu}, \;\; \Sigma_{k=1}^N \gamma_k \theta_k \le \tfrac{2b^2 N}{\mu^2}, \tag{4.6.41}$$

and

$$\Sigma_{k=1}^N \theta_k^2 = \tfrac{b^2}{\mu^2} \tfrac{N(N+1)(2N+1)}{6} \le \tfrac{b^2}{\mu^2} \tfrac{N(N+1)^2}{3}. \tag{4.6.42}$$

Hence, by (4.6.34),

$$\mathbb{E}[f(\bar{x}_N) - f(x_*)] \le \tfrac{1}{2} \left(\Sigma_{k=1}^N \theta_k\right)^{-1} \Sigma_{k=1}^N \gamma_k \theta_k \Sigma_{i=1}^b M_i^2 \le \tfrac{2b}{\mu(N+1)} \Sigma_{i=1}^b M_i^2.$$

■

In view of (4.6.39), the number of iterations performed by the SBMD method to find an ε-solution for nonsmooth strongly convex problems can be bound by $\frac{2b}{\mu\varepsilon} \Sigma_{i=1}^b M_i^2$.

4.6.1.4 Large-Deviation Properties for Nonsmooth Problems

Our goal in this subsection is to establish the large-deviation results associated with the SBMD algorithm under the following "light-tail" assumption about the random variable ξ:

$$\mathbb{E}\left\{\exp\left[\|G_i(x,\xi)\|_{i,*}^2 / M_i^2\right]\right\} \le \exp(1), \; i = 1,2,\ldots,b. \tag{4.6.43}$$

It can be easily seen that (4.6.43) implies (4.6.4) by Jensen's inequality. It should be pointed out that the above "light-tail" assumption is always satisfied for deterministic problems with bounded subgradients.

For the sake of simplicity, we only consider the case when the random variables $\{i_k\}$ in the SBMD algorithm are uniformly distributed, i.e., relation (4.6.31) holds. The following result states the large-deviation properties of the SBMD algorithm for solving general nonsmooth problems.

Theorem 4.14. *Suppose that Assumptions (4.6.43) and (4.6.31) hold. Also assume that X is bounded.*

(a) For solving general nonsmooth CP problems (i.e., (4.6.16) holds), we have

$$\begin{aligned}\mathrm{Prob}\Big\{ f(\bar{x}_N) - f(x_*) \ge b\left(\Sigma_{k=1}^N \gamma_k\right)^{-1} \Big[\Sigma_{i=1}^b V_i(x_1^{(i)}, x_*^{(i)}) + \tfrac{1}{2}\bar{M}^2 \Sigma_{k=1}^N \gamma_k^2 \\ + \lambda \bar{M}\Big(\tfrac{1}{2}\bar{M}\Sigma_{k=1}^N \gamma_k^2 + 4\sqrt{2}\sqrt{\Sigma_{i=1}^b D_i}\sqrt{\Sigma_{k=1}^N \gamma_k^2}\Big)\Big]\Big\} \le \exp\left(-\lambda^2/3\right) + \exp(-\lambda),\end{aligned} \tag{4.6.44}$$

for any $N \geq 1$ *and* $\lambda > 0$*, where* $\bar{M} = \max_{i=1,\ldots,b} M_i$ *and* x_* *is an arbitrary solution of (4.6.1).*

(b) *For solving strongly convex problems (i.e., (4.6.30), (4.6.32), and (4.6.33) hold), we have*

$$\begin{aligned}\text{Prob}\Big\{ f(\bar{x}_N) - f(x_*) \geq b\left(\Sigma_{k=1}^N \theta_k\right)^{-1} \Big[(1 - \tfrac{\gamma_1 \mu}{b})\Sigma_{i=1}^b V_i(x_1^{(i)}, x_*^{(i)}) \\ + \tfrac{1}{2}\bar{M}^2 \Sigma_{k=1}^N \gamma_k \theta_k + \lambda \bar{M}\left(\tfrac{1}{2}\bar{M}\Sigma_{k=1}^N \gamma_k \theta_k + 4\sqrt{2}\sqrt{\Sigma_{i=1}^b D_i}\sqrt{\Sigma_{k=1}^N \theta_k^2}\right)\Big]\Big\} \\ \leq \exp\left(-\lambda^2/3\right) + \exp(-\lambda),\end{aligned} \tag{4.6.45}$$

for any $N \geq 1$ *where* x_* *is the optimal solution of (4.6.1).*

Proof. We first show part (a). Note that by (4.6.43), the concavity of $\phi(t) = \sqrt{t}$ for $t \geq 0$ and the Jensen's inequality, we have, for any $i = 1, 2, \ldots, b$,

$$\mathbb{E}\left\{\exp\left[\|G_i(x,\xi)\|_{i,*}^2/(2M_i^2)\right]\right\} \leq \sqrt{\mathbb{E}\left\{\exp\left[\|G_i(x,\xi)\|_{i,*}^2/M_i^2\right]\right\}} \leq \exp(1/2). \tag{4.6.46}$$

Also note that by (4.6.21), δ_k, $k = 1, \ldots, N$, is the martingale-difference. In addition, denoting $\mathrm{M}^2 \equiv 32\, b^2 \bar{M}^2 \Sigma_{i=1}^b D_i$, we have

$$\begin{aligned}\mathbb{E}[\exp\left(\mathrm{M}^{-2}\delta_k^2\right)] &\leq \Sigma_{i=1}^b \tfrac{1}{b}\mathbb{E}\left[\exp\left(\mathrm{M}^{-2}\|x - x_k\|^2 \|bU_i^T G_i - g(x_k)\|_*^2\right)\right] \qquad (4.6.47)\\ &\leq \Sigma_{i=1}^b \tfrac{1}{b}\mathbb{E}\left\{\exp\left[2\mathrm{M}^{-2}\|x - x_k\|^2 \left(b^2\|G_i\|_{i,*}^2 + \|g(x_k)\|_*^2\right)\right]\right\}\\ &\leq \Sigma_{i=1}^b \tfrac{1}{b}\mathbb{E}\left\{\exp\left[16\mathrm{M}^{-2}\left(\Sigma_{i=1}^b D_i\right)\left(b^2\|G_i\|_{i,*}^2 + \Sigma_{i=1}^b M_i^2\right)\right]\right\}\\ &\leq \Sigma_{i=1}^b \tfrac{1}{b}\mathbb{E}\left\{\exp\left[\frac{b^2\|G_i\|_{i,*}^2 + \Sigma_{i=1}^b M_i^2}{2b^2\bar{M}^2}\right]\right\}\\ &\leq \Sigma_{i=1}^b \tfrac{1}{b}\mathbb{E}\left\{\exp\left[\frac{\|G_i\|_{i,*}^2}{2M_i^2} + \frac{1}{2}\right]\right\} \leq \exp(1),\end{aligned}$$

where the first five inequalities follow from (4.6.12) and (4.6.19), (4.6.31), (4.6.6) and (4.6.11), the definition of M, and (4.6.46), respectively. Therefore, by the large-deviation theorem on the Martingale-difference (see Lemma 4.1), we have

$$\text{Prob}\left\{\Sigma_{k=1}^N \gamma_k \delta_k \geq \lambda \mathrm{M}\sqrt{\Sigma_{k=1}^N \gamma_k^2}\right\} \leq \exp(-\lambda^2/3). \tag{4.6.48}$$

Also observe that under Assumption (4.6.43), (4.6.12), (4.6.19), and (4.6.31))

$$\begin{aligned}\mathbb{E}\left[\exp\left(\bar{\delta}_k/(b\bar{M}^2)\right)\right] &\leq \Sigma_{i=1}^b \tfrac{1}{b}\mathbb{E}\left[\exp\left(\|G_i(x_k,\xi_k)\|_{i,*}^2/\bar{M}^2\right)\right]\\ &\leq \Sigma_{i=1}^b \tfrac{1}{b}\mathbb{E}\left[\exp\left(\|G_i(x_k,\xi_k)\|_{i,*}^2/M_i^2\right)\right]\\ &\leq \Sigma_{i=1}^b \tfrac{1}{b}\exp(1) = \exp(1),\end{aligned} \tag{4.6.49}$$

where the second inequality follows from the definition of $\bar{M}$ and the third one follows from (4.6.4). Setting $\psi_k = \gamma_k^2/\Sigma_{k=1}^N \gamma_k^2$, we have $\exp\left\{\Sigma_{k=1}^N \psi_k \bar{\delta}_k/(b\bar{M}^2)\right\} \leq$

$\sum_{k=1}^{N} \psi_k \exp\{\bar{\delta}_k/(b\bar{M}^2)\}$. Using these previous two inequalities, we have

$$\mathbb{E}\left[\exp\left\{\sum_{k=1}^{N} \gamma_k^2 \bar{\delta}_k/(b\bar{M}^2 \sum_{k=1}^{N} \gamma_k^2)\right\}\right] \le \exp\{1\}.$$

It then follows from Markov's inequality that $\forall \lambda \ge 0$,

$$\begin{aligned}
&\text{Prob}\left\{\sum_{k=1}^{N} \gamma_k^2 \bar{\delta}_k > (1+\lambda)(b\bar{M}^2)\sum_{k=1}^{N} \gamma_k^2\right\} \\
&= \text{Prob}\left\{\exp\left\{\sum_{k=1}^{N} \gamma_k^2 \bar{\delta}_k/(b\bar{M}^2 \sum_{k=1}^{N} \gamma_k^2)\right\} > \exp\{(1+\lambda)\}\right\} \\
&\le \tfrac{\mathbb{E}\left[\exp\left\{\sum_{k=1}^{N} \gamma_k^2 \bar{\delta}_k/(b\bar{M}^2 \sum_{k=1}^{N} \gamma_k^2)\right\}\right]}{\exp\{1+\lambda\}} \le \tfrac{\exp\{1\}}{\exp\{1+\lambda\}} = \exp\{-\lambda\}. \qquad (4.6.50)
\end{aligned}$$

Combining (4.6.20), (4.6.48), and (4.6.50), we obtain (4.6.44).

The probabilistic bound in (4.6.45) follows from (4.6.37) and an argument similar to the one used in the proof of (4.6.44), and hence the details are skipped. ■

We now provide some specialized large-deviation results for the SBMD algorithm with different selections of $\{\gamma_k\}$ and $\{\theta_k\}$.

Corollary 4.9. *Suppose that (4.6.43) and (4.6.31) hold. Also assume that X is bounded.*

(a) If $\{\theta_k\}$ and $\{\gamma_k\}$ are set to (4.6.16), (4.6.25), and $\tilde{D}=\sqrt{\sum_{i=1}^{b} V_i(x_1^{(i)}, x_*^{(i)})}$ for general nonsmooth problems, then we have

$$\begin{aligned}
&\text{Prob}\left\{ f(\bar{x}_N) - f(x_*) \ge \tfrac{\sqrt{b\sum_{i=1}^{b} M_i^2 \sum_{i=1}^{b} D_i}}{\sqrt{2N}} \left[1+(1+\lambda)\tfrac{\bar{M}^2}{\bar{m}^2} + 8\lambda \tfrac{\bar{M}}{\bar{m}}\right]\right\} \\
&\le \exp\left(-\lambda^2/3\right) + \exp(-\lambda)
\end{aligned} \qquad (4.6.51)$$

for any $\lambda > 0$, where $\bar{m} = \min_{i=1,\ldots,b} M_i$ and x_* is an arbitrary solution of (4.6.1).

(b) If $\{\theta_k\}$ and $\{\gamma_k\}$ are set to (4.6.33) and (4.6.38) for strongly convex problems, then we have

$$\begin{aligned}
&\text{Prob}\left\{ f(\bar{x}_N) - f(x_*) \ge \tfrac{2(1+\lambda)b\sum_{i=1}^{b} M_i^2}{(N+1)\mu} \tfrac{\bar{M}^2}{\bar{m}^2} + \tfrac{8\sqrt{2}\lambda \bar{M}\sqrt{b\sum_{i=1}^{b} M_i^2 \sum_{i=1}^{b} D_i}}{\bar{m}\sqrt{3N}} \right\} \\
&\le \exp\left(-\lambda^2/3\right) + \exp(-\lambda)
\end{aligned} \qquad (4.6.52)$$

for any $\lambda > 0$, where x_* is the optimal solution of (4.6.1).

Proof. Note that by (4.6.9), we have $\sum_{i=1}^{b} V_i(x_1^{(i)}, x_*^{(i)}) \le \sum_{i=1}^{b} D_i$. Also by (4.6.25), we have

$$\sum_{k=1}^{N} \gamma_k = \left(\tfrac{2Nb\tilde{D}^2}{\sum_{i=1}^{b} M_i^2}\right)^{1/2} \quad \text{and} \quad \sum_{k=1}^{N} \gamma_k^2 = \tfrac{2b\tilde{D}^2}{\sum_{i=1}^{b} M_i^2}.$$

Using these identities and (4.6.44), we conclude that

$$\text{Prob}\left\{f(\bar{x}_N)-f(x_*)\geq b\left(\frac{\sum_{i=1}^b M_i^2}{2Nb\tilde{D}^2}\right)^{1/2}\left[\sum_{i=1}^b V_i(x_1^{(i)},x_*^{(i)})+b\tilde{D}^2\bar{M}^2\left(\sum_{i=1}^b M_i^2\right)^{-1}\right.\right.$$
$$\left.\left.+\lambda\bar{M}\left(2b\bar{M}\tilde{D}^2\left(\sum_{i=1}^b M_i^2\right)^{-1}+8b^{1/2}\tilde{D}\sqrt{\sum_{i=1}^b D_i}\left(\sum_{i=1}^b M_i^2\right)^{-\frac{1}{2}}\right)\right]\right\}$$
$$\leq \exp\left(-\lambda^2/3\right)+\exp(-\lambda).$$

Using the fact that $\sum_{i=1}^b M_i^2 \geq b\bar{m}^2$ and $\tilde{D} = \sqrt{\sum_{i=1}^b V_i(x_1^{(i)},x_*^{(i)})}$ to simplify the above relation, we obtain (4.6.51). Similarly, relation (4.6.52) follows directly from (4.6.45) and a few bounds in (6.4.80), (6.4.81), and (4.6.42). ∎

We now add a few remarks about the results obtained in Theorem 4.14 and Corollary 4.9. First, observe that by (4.6.51), the number of iterations required by the SBMD method to find an (ε,Λ)-solution of (4.6.1), i.e., a point $\bar{x}\in X$ s.t. $\text{Prob}\{f(\bar{x})-f^*\geq\varepsilon\}\leq\Lambda$ can be bounded by

$$\mathcal{O}\left(\frac{b\log^2(1/\Lambda)}{\varepsilon^2}\right)$$

after disregarding a few constant factors.

Second, it follows from (4.6.52) that the number of iterations performed by the SBMD method to find an (ε,Λ)-solution for nonsmooth strongly convex problems, after disregarding a few constant factors, can be bounded by $\mathcal{O}\left(b\log^2(1/\Lambda)/\varepsilon^2\right)$, which, in case $b=1$, is about the same as the one obtained for solving nonsmooth problems without assuming convexity. It should be noted, however, that this bound can be improved to $\mathcal{O}\left(b\log(1/\Lambda)/\varepsilon\right)$, for example, by incorporating a domain shrinking procedure (see Sect. 4.2).

4.6.2 Convex Composite Optimization

In this section, we consider a special class of convex stochastic composite optimization problems given by

$$\phi^* := \min_{x\in X}\left\{\phi(x) := f(x)+\chi(x)\right\}. \tag{4.6.53}$$

Here $\chi(\cdot)$ is a relatively simple convex function and $f(\cdot)$ defined in (4.6.1) is a smooth convex function with Lipschitz-continuous gradients $g(\cdot)$. Our goal is to present a variant of the SBMD algorithm which can make use of the smoothness properties of the objective function. More specifically, we consider convex composite optimization problems given in the form of (4.6.53), where $f(\cdot)$ is smooth and its gradients $g(\cdot)$ satisfy

$$\|g_i(x+U_i\rho_i)-g_i(x)\|_{i,*}\leq L_i\|\rho_i\|_i \;\; \forall\, \rho_i\in\mathbb{R}^{n_i},\; i=1,2,\ldots,b. \tag{4.6.54}$$

It then follows that

$$f(x+U_i\rho_i) \le f(x) + \langle g_i(x), \rho_i \rangle + \tfrac{L_i}{2}\|\rho_i\|_i^2 \;\; \forall \rho_i \in \mathbb{R}^{n_i}, x \in X. \tag{4.6.55}$$

The following assumption is made throughout this section.

Assumption 12 *The function $\chi(\cdot)$ is block separable, i.e., $\chi(\cdot)$ can be decomposed as*

$$\chi(x) = \sum_{i=1}^{b} \chi_i(x^{(i)}) \;\; \forall x \in X. \tag{4.6.56}$$

where $\chi_i : \mathbb{R}^{n_i} \to \mathbb{R}$ are closed and convex.

We are now ready to describe a variant of the SBMD algorithm for solving smooth and composite problems.

Algorithm 4.6 A variant of SBMD for convex stochastic composite optimization

Let $x_1 \in X$, positive stepsizes $\{\gamma_k\}_{k\ge1}$, nonnegative weights $\{\theta_k\}_{k\ge1}$, and probabilities $p_i \in [0,1]$, $i = 1,\ldots,b$, s.t. $\sum_{i=1}^{b} p_i = 1$ be given. Set $s_1 = 0$, $u_i = 1$ for $i = 1,\ldots,b$, and $\theta_1 = 0$.
for $k = 1,\ldots,N$ **do**

1. Generate a random variable i_k according to (4.6.12).
2. Update $s_k^{(i)}$, $i = 1,\ldots,b$, by (4.6.13) and then set $u_{i_k} = k+1$.
3. Update $x_k^{(i)}$, $i = 1,\ldots,b$, by

$$x_{k+1}^{(i)} = \begin{cases} \operatorname{argmin}_{z\in X_i} \left\langle G_i(x_k,\xi_k), z - x_k^{(i)} \right\rangle + \frac{1}{\gamma_k} V_i(x_k^{(i)}, z) + \chi_i(x) & i = i_k, \\ x_k^{(i)} & i \ne i_k. \end{cases} \tag{4.6.57}$$

end for
Output: Set $s_{N+1}^{(i)} = s_{N+1}^{(i)} + x_{N+1}^{(i)} \sum_{j=u_i}^{N+1} \theta_j$, $i = 1,\ldots,b$, and $\bar{x}_N = s_{N+1} / \sum_{k=1}^{N+1} \theta_k$.

A few remarks about the above variant of SBMD algorithm for composite convex problem in place. First, similar to Algorithm 4.5, $G(x_k, \xi_k)$ is an unbiased estimator of $g(x_k)$ (i.e., (4.6.3) holds). Moreover, in order to know exactly the effect of stochastic noises in $G(x_k, \xi_k)$, we assume that for some $\sigma_i \ge 0$,

$$E[\|G_i(x,\xi) - g_i(x)\|_{i,*}^2] \le \sigma_i^2, \; i = 1,\ldots,b. \tag{4.6.58}$$

Clearly, if $\sigma_i = 0$, $i = 1,\ldots,b$, then the problem is deterministic. For notational convenience, we also denote

$$\sigma := \left(\sum_{i=1}^{b} \sigma_i^2\right)^{1/2}. \tag{4.6.59}$$

Second, observe that the way we compute the output $\bar{x}_N$ in Algorithm 4.6 is slightly different from Algorithm 4.5. In particular, we set $\theta_1 = 0$ and compute $\bar{x}_N$ of Algorithm 4.6 as a weighted average of the search points $x_2,\ldots,x_{N+1}$, i.e.,

$$\bar{x}_N = \left(\sum_{k=2}^{N+1} \theta_k\right)^{-1} s_{N+1} = \left(\sum_{k=2}^{N+1} \theta_k\right)^{-1} \sum_{k=2}^{N+1} (\theta_k x_k), \tag{4.6.60}$$

while the output of Algorithm 4.5 is taken as a weighted average of $x_1,\ldots,x_N$.

Third, it can be easily seen from (4.6.9), (6.2), and (4.6.58) that if X is bounded, then

$$\begin{aligned}
\mathbb{E}[\|G_i(x,\xi)\|_{i,*}^2] &\le 2\|g_i(x)\|_{i,*}^2 + 2\mathbb{E}\|G_i(x,\xi)-g_i(x)\|_{i,*}^2] \le 2\|g_i(x)\|_{i,*}^2 + 2\sigma_i^2\\
&\le 2\left[2\|g_i(x)-g_i(x_1)\|_{i,*}^2 + 2\|g_i(x_1)\|_{i,*}^2\right] + 2\sigma_i^2\\
&\le 2\left[2\|g(x)-g(x_1)\|_{*}^2 + 2\|g_i(x_1)\|_{i,*}^2\right] + 2\sigma_i^2\\
&\le 4\left(\textstyle\sum_{i=1}^b L_i\right)^2 \|x-x_1\|^2 + 4\|g_i(x_1)\|_{i,*}^2 + 2\sigma_i^2\\
&\le 8b^2\bar{L}^2\textstyle\sum_{i=1}^b D_i + 4\|g_i(x_1)\|_{i,*}^2 + 2\sigma_i^2, \quad i=1,\ldots,b, \qquad (4.6.61)
\end{aligned}$$

where $\bar{L} := \max_{i=1,\ldots,b} L_i$ and the fourth inequality follows from the fact that g is Lipschitz continuous with constant $\sum_{i=1}^b L_i$. Hence, we can directly apply Algorithm 4.5 in the previous section to problem (4.6.53), and its rate of convergence is readily given by Theorems 4.12 and 4.13. However, in this section we will show that by properly selecting $\{\theta_k\}$, $\{\gamma_k\}$, and $\{p_i\}$ in the above variant of the SBMD algorithm, we can significantly improve the dependence of the rate of convergence of the SBMD algorithm on the Lipschitz constants L_i, $i=1,\ldots,b$.

We first discuss the main convergence properties of Algorithm 4.6 for convex stochastic composite optimization without assuming strong convexity.

Theorem 4.15. *Suppose that $\{i_k\}$ in Algorithm 4.6 are uniformly distributed, i.e.,* (4.6.31) *holds. Also assume that $\{\gamma_k\}$ and $\{\theta_k\}$ are chosen such that for any $k \ge 1$,*

$$\gamma_k \le \tfrac{1}{2\bar{L}}, \qquad (4.6.62)$$

$$\theta_{k+1} = b\gamma_k - (b-1)\gamma_{k+1}. \qquad (4.6.63)$$

Then, under assumptions (4.6.3) *and* (4.6.58), *we have, for any $N \ge 2$,*

$$\begin{aligned}
&E[\phi(\bar{x}_N) - \phi(x_*)]\\
&\le \left(\textstyle\sum_{k=2}^{N+1}\theta_k\right)^{-1}\left[(b-1)\gamma_1[\phi(x_1)-\phi(x_*)] + b\textstyle\sum_{i=1}^b V_i(x_1^{(i)},x_*^{(i)}) + \sigma^2\textstyle\sum_{k=1}^N \gamma_k^2\right], \qquad (4.6.64)
\end{aligned}$$

where x_* is an arbitrary solution of problem (4.6.53) and σ is defined in (4.6.59).

Proof. For simplicity, let us denote $V_i(z,x) \equiv V_i(z^{(i)},x^{(i)})$, $g_{i_k} \equiv g^{(i_k)}(x_k)$, and $V(z,x) = \sum_{i=1}^b p_i^{-1} V_i(z,x)$. Also denote $\zeta_k = (i_k,\xi_k)$ and $\zeta_{[k]} = (\zeta_1,\ldots,\zeta_k)$, and let $\delta_{i_k} = G_{i_k}(x_k,\xi_k) - g_{i_k}(x_k)$ and $\rho_{i_k} = U_{i_k}^T(x_{k+1}-x_k)$. By the definition of $\phi(\cdot)$ in (4.6.53) and (4.6.55), we have

$$\begin{aligned}
\phi(x_{k+1}) &\le f(x_k) + \langle g_{i_k}(x_k), \rho_{i_k}\rangle + \tfrac{L_{i_k}}{2}\left\|\rho_{i_k}\right\|_{i_k}^2 + \chi(x_{k+1})\\
&= f(x_k) + \langle G_{i_k}(x_k,\xi_k), \rho_{i_k}\rangle + \tfrac{L_{i_k}}{2}\left\|\rho_{i_k}\right\|_{i_k}^2 + \chi(x_{k+1}) - \langle \delta_{i_k}, \rho_{i_k}\rangle. \qquad (4.6.65)
\end{aligned}$$

Moreover, it follows from Lemma 3.5 and relation (4.6.57) that, for any $x \in X$,

$$\langle G_{i_k}(x_k,\xi_k),\rho_{i_k}\rangle + \chi_{i_k}(x_{k+1}^{(i_k)}) \le \langle G_{i_k}(x_k,\xi_k), x^{(i_k)} - x_k^{(i_k)}\rangle + \chi_{i_k}(x^{(i_k)}) \\ + \tfrac{1}{\gamma_k}\left[V_{i_k}(x_k,x) - V_{i_k}(x_{k+1},x) - V_{i_k}(x_{k+1},x_k)\right].$$

Combining the above two inequalities and using (4.6.56), we obtain

$$\begin{aligned}\phi(x_{k+1}) \le{}& f(x_k) + \left\langle G_{i_k}(x_k,\xi_k), x^{(i_k)} - x_k^{(i_k)}\right\rangle + \chi_{i_k}(x^{(i_k)}) \\ &+ \tfrac{1}{\gamma_k}\left[V_{i_k}(x_k,x) - V_{i_k}(x_{k+1},x) - V_{i_k}(x_{k+1},x_k)\right] \\ &+ \tfrac{L_{i_k}}{2}\left\|\rho_{i_k}\right\|_{i_k}^2 + \Sigma_{i\neq i_k}\chi_i(x_{k+1}^{(i)}) - \langle \delta_{i_k},\rho_{i_k}\rangle. \end{aligned} \quad (4.6.66)$$

By the strong convexity of $v_i(\cdot)$ and (4.6.62), using the simple inequality that $bu - \frac{au^2}{2} \le \frac{b^2}{2a}, \forall a > 0$, we have

$$\begin{aligned}-\tfrac{1}{\gamma_k}V_{i_k}(x_{k+1},x_k) + \tfrac{L_{i_k}}{2}\left\|\rho_{i_k}\right\|_{i_k}^2 - \langle \delta_{i_k},\rho_{i_k}\rangle &\le -\left(\tfrac{1}{2\gamma_k} - \tfrac{L_{i_k}}{2}\right)\left\|\rho_{i_k}\right\|_{i_k}^2 - \langle \delta_{i_k},\rho_{i_k}\rangle \\ &\le \tfrac{\gamma_k\|\delta_{i_k}\|_*^2}{2(1-\gamma_k L_{i_k})} \le \tfrac{\gamma_k\|\delta_{i_k}\|_*^2}{2(1-\gamma_k \bar{L})} \le \gamma_k\left\|\delta_{i_k}\right\|_*^2. \end{aligned}$$

Also observe that by the definition of x_{k+1} in (4.6.57), (4.6.14), and the definition of $V(\cdot,\cdot)$, we have $\Sigma_{i\neq i_k}\chi_i(x_{k+1}^{(i)}) = \Sigma_{i\neq i_k}\chi_i(x_k^{(i)})$ and $V_{i_k}(x_k,x) - V_{i_k}(x_{k+1},x) = [V(x_k,x) - V(x_{k+1},x)]/b$. Using these observations, we conclude from (4.6.66) that

$$\begin{aligned}\phi(x_{k+1}) \le{}& f(x_k) + \left\langle G_{i_k}(x_k,\xi_k), x^{(i_k)} - x_k^{(i_k)}\right\rangle + \tfrac{1}{b\gamma_k}\left[V(x_k,x) - V(x_{k+1},x)\right] \\ &+ \gamma_k\left\|\delta_{i_k}\right\|_*^2 + \Sigma_{i\neq i_k}\chi_i(x_k^{(i)}) + \chi_{i_k}(x^{(i_k)}). \end{aligned} \quad (4.6.67)$$

Now noting that

$$\begin{aligned}\mathbb{E}_{\zeta_k}\left[\left\langle G_{i_k}(x_k,\xi_k), x^{(i_k)} - x_k^{(i_k)}\right\rangle|\zeta_{[k-1]}\right] &= \tfrac{1}{b}\Sigma_{i=1}^b \mathbb{E}_{\xi_k}\left[\left\langle G_i(x_k,\xi_k), x^{(i)} - x_k^{(i)}\right\rangle|\zeta_{[k-1]}\right] \\ &= \tfrac{1}{b}\langle g(x_k), x - x_k\rangle \le \tfrac{1}{b}[f(x) - f(x_k)], \quad (4.6.68)\\ \mathbb{E}_{\zeta_k}\left[\|\delta_{i_k}\|_*^2|\zeta_{[k-1]}\right] &= \tfrac{1}{b}\Sigma_{i=1}^b \mathbb{E}_{\xi_k}\left[\|G_i(x_k,\xi_k) - g_i(x_k)\|_{i,*}^2|\zeta_{[k-1]}\right] \\ &\le \tfrac{1}{b}\Sigma_{i=1}^b\sigma_i^2 = \tfrac{\sigma^2}{b}, \quad (4.6.69)\\ \mathbb{E}_{\zeta_k}\left[\Sigma_{i\neq i_k}\chi_i(x_k^{(i)})|\zeta_{[k-1]}\right] &= \tfrac{1}{b}\Sigma_{j=1}^b\Sigma_{i\neq j}\chi_i(x_k^{(i)}) = \tfrac{b-1}{b}\chi(x_k), \quad (4.6.70)\\ \mathbb{E}_{\zeta_k}\left[\chi_{i_k}(x^{(i_k)})|\zeta_{[k-1]}\right] &= \tfrac{1}{b}\Sigma_{i=1}^b\chi_i(x^{(i)}) = \tfrac{1}{b}\chi(x), \quad (4.6.71)\end{aligned}$$

we conclude from (4.6.67) that

$$\mathbb{E}_{\zeta_k}\left[\phi(x_{k+1}) + \tfrac{V(x_{k+1},x)}{b\gamma_k}|\zeta_{[k-1]}\right] \le f(x_k) + \tfrac{1}{b}[f(x) - f(x_k)] + \tfrac{1}{b}\chi(x) \\ + \tfrac{V(x_k,x)}{b\gamma_k} + \tfrac{\gamma_k}{b}\sigma^2 + \tfrac{b-1}{b}\chi(x_k) = \tfrac{b-1}{b}\phi(x_k) + \tfrac{1}{b}\phi(x) + \tfrac{V(x_k,x)}{b\gamma_k} + \tfrac{\gamma_k}{b}\sigma^2,$$

which implies that

$$b\gamma_k\mathbb{E}[\phi(x_{k+1})-\phi(x)]+\mathbb{E}[V(x_{k+1},x)]\le(b-1)\gamma_k\mathbb{E}[\phi(x_k)-\phi(x)] \\ +\mathbb{E}[V(x_k,x)]+\gamma_k^2\sigma^2. \tag{4.6.72}$$

Now, summing up the above inequalities (with $x=x_*$) for $k=1,\ldots,N$, and noting that $\theta_{k+1}=b\gamma_k-(b-1)\gamma_{k+1}$, we obtain

$$\begin{aligned}&\textstyle\sum_{k=2}^N\theta_k\mathbb{E}[\phi(x_k)-\phi(x_*)]+b\gamma_N\mathbb{E}[\phi(x_{N+1})-\phi(x_*)]+\mathbb{E}[V(x_{N+1},x_*)]\\&\le(b-1)\gamma_1[\phi(x_1)-\phi(x_*)]+V(x_1,x_*)+\sigma^2\textstyle\sum_{k=1}^N\gamma_k^2,\end{aligned}$$

Using the above inequality and the facts that $V(\cdot,\cdot)\ge 0$ and $\phi(x_{N+1})\ge\phi(x_*)$, we conclude

$$\textstyle\sum_{k=2}^{N+1}\theta_k\mathbb{E}[\phi(x_k)-\phi(x_*)]\le(b-1)\gamma_1[\phi(x_1)-\phi(x_*)]+V(x_1,x_*)+\sigma^2\sum_{k=1}^N\gamma_k^2,$$

which, in view of (4.6.59), (4.6.60), and the convexity of $\phi(\cdot)$, clearly implies (4.6.64). ■

The following corollary describes a specialized convergence result of Algorithm 4.6 for solving convex stochastic composite optimization problems after properly selecting $\{\gamma_k\}$.

Corollary 4.10. *Suppose that $\{p_i\}$ in Algorithm 4.6 are set to* (4.6.31). *Also assume that $\{\gamma_k\}$ are set to*

$$\gamma_k=\gamma=\min\left\{\tfrac{1}{2\bar{L}},\tfrac{\tilde{D}}{\sigma}\sqrt{\tfrac{b}{N}}\right\} \tag{4.6.73}$$

for some $\tilde{D}>0$, and $\{\theta_k\}$ are set to (4.6.63). Then, under Assumptions (4.6.3) *and* (4.6.58), *we have*

$$\begin{aligned}\mathbb{E}\left[\phi(\bar{x}_N)-\phi(x_*)\right]\le{}&\tfrac{(b-1)[\phi(x_1)-\phi(x_*)]}{N}+\tfrac{2b\bar{L}\sum_{i=1}^bV_i(x_1^{(i)},x_*^{(i)})}{N}\\&+\tfrac{\sigma\sqrt{b}}{\sqrt{N}}\left[\tfrac{\sum_{i=1}^bV_i(x_1^{(i)},x_*^{(i)})}{\tilde{D}}+\tilde{D}\right].\end{aligned} \tag{4.6.74}$$

where x_* is an arbitrary solution of problem (4.6.53).

Proof. It follows from (4.6.63) and (4.6.73) that $\theta_k=\gamma_k=\gamma$, $k=1,\ldots,N$. Using this observation and Theorem 4.15, we obtain

$$\mathbb{E}\left[\phi(\bar{x}_N)-\phi(x_*)\right]\le\tfrac{(b-1)[\phi(x_1)-\phi(x_*)]}{N}+\tfrac{b\sum_{i=1}^bV_i(x_1^{(i)},x_*^{(i)})}{N\gamma}+\gamma\sigma^2,$$

which, in view of (4.6.73), then implies (4.6.74). ■

We now add a few remarks about the results obtained in Corollary 4.10. First, in view of (4.6.74), an optimal selection of $\tilde{D}$ would be $\sqrt{\sum_{i=1}^bV_i(x_1^{(i)},x_*^{(i)})}$. In this case, (4.6.74) reduces to

$$\mathbb{E}\left[\phi(\bar{x}_N)-\phi(x_*)\right] \le \tfrac{(b-1)[\phi(x_1)-\phi(x_*)]}{N} + \tfrac{2b\bar{L}\sum_{i=1}^b V_i(x_1^{(i)},x_*^{(i)})}{N} + \tfrac{2\sigma\sqrt{b}\sqrt{\sum_{i=1}^b D_i}}{\sqrt{N}}$$
$$\le \tfrac{(b-1)[\phi(x_1)-\phi(x_*)]}{N} + \tfrac{2b\bar{L}\sum_{i=1}^b D_i}{N} + \tfrac{2\sigma\sqrt{b}\sqrt{\sum_{i=1}^b D_i}}{\sqrt{N}}. \tag{4.6.75}$$

Second, if we directly apply Algorithm 4.5 to problem (4.6.53), then, in view of (4.6.27) and (4.6.61), we have

$$\mathbb{E}[\phi(\bar{x}_N)-\phi(x_*)] \le 2\sqrt{\textstyle\sum_{i=1}^b\left[4b^2\bar{L}^2(\sum_{i=1}^b D_i)+2\|g_i(x_1)\|_{i,*}^2+\sigma_i^2\right]}\,\tfrac{\sqrt{b}\sqrt{\sum_{i=1}^b D_i}}{\sqrt{N}}$$
$$\le \tfrac{4b^2\bar{L}\sum_{i=1}^b D_i}{\sqrt{N}} + 2\sqrt{\textstyle\sum_{i=1}^b\left(2\|g_i(x_1)\|_{i,*}^2+\sigma_i^2\right)}\,\tfrac{\sqrt{b}\sqrt{\sum_{i=1}^b D_i}}{\sqrt{N}}. \tag{4.6.76}$$

Clearly, the bound in (4.6.75) has a much weaker dependence on the Lipschitz constant $\bar{L}$ than the one in (4.6.76). In particular, we can see that $\bar{L}$ can be as large as $\mathcal{O}(\sqrt{N})$ without affecting the bound in (4.6.75), after disregarding some other constant factors. Moreover, the bound in (4.6.75) also has a much weaker dependence on the number of blocks b than the one in (4.6.76).

In the remaining part of this section, we consider the case when the objective function is strongly convex, i.e., the function $f(\cdot)$ in (4.6.53) satisfies (4.6.30). The following theorem describes some convergence properties of the SBMD algorithm for solving strongly convex composite problems.

Theorem 4.16. *Suppose that (4.6.30) and (4.6.31) hold. Also assume that the parameters $\{\gamma_k\}$ and $\{\theta_k\}$ are chosen such that for any $k \ge 1$,*

$$\gamma_k \le \min\left\{\tfrac{1}{2\bar{L}}, \tfrac{b}{\mu}\right\}, \tag{4.6.77}$$

$$\theta_{k+1} = \tfrac{b\gamma_k}{\Gamma_k} - \tfrac{(b-1)\gamma_{k+1}}{\Gamma_{k+1}} \quad \textit{with} \quad \Gamma_k = \begin{cases} 1 & k=1 \\ \Gamma_{k-1}(1-\tfrac{\gamma_k\mu}{b}) & k \ge 2. \end{cases} \tag{4.6.78}$$

Then, for any $N \ge 2$, we have

$$\mathbb{E}[\phi(\bar{x}_N)-\phi(x_*)] \le \left[\textstyle\sum_{k=2}^{N+1}\theta_k\right]^{-1}\Big[\left(b-\mu\gamma_1\right)\textstyle\sum_{i=1}^b V_i(x_1^{(i)},x_*^{(i)})$$
$$+(b-1)\gamma_1[\phi(x_1)-\phi(x_*)] + \textstyle\sum_{k=1}^N \tfrac{\gamma_k^2}{\Gamma_k}\sigma^2\Big], \tag{4.6.79}$$

where x_* is the optimal solution of problem (4.6.53).

Proof. Observe that by the strong convexity of $f(\cdot)$, for any $x \in X$, the relation in (4.6.68) can be strengthened to

$$\mathbb{E}_{\zeta_k}\left[\left\langle G_{i_k}(x_k,\xi_k),x^{(i_k)}-x_k^{(i_k)}\right\rangle|\zeta_{[k-1]}\right]$$
$$=\tfrac{1}{b}\langle g(x_k),x-x_k\rangle \le \tfrac{1}{b}[f(x)-f(x_k)-\tfrac{\mu}{2}\|x-x_k\|^2].$$

Using this observation, (4.6.69), (4.6.70), and (4.6.71), we conclude from (4.6.67) that

$$\mathbb{E}_{\zeta_k}\left[\phi(x_{k+1})+\tfrac{1}{b\gamma_k}V(x_{k+1},x)|\zeta_{[k-1]}\right] \le f(x_k)+\tfrac{1}{b}\left[f(x)-f(x_k)-\tfrac{\mu}{2}\|x-x_k\|^2\right]$$
$$+\tfrac{1}{b\gamma_k}V(x_k,x)+\tfrac{\gamma_k}{b}\sigma^2+\tfrac{b-1}{b}\chi(x_k)+\tfrac{1}{b}\chi(x)$$
$$\le \tfrac{b-1}{b}\phi(x_k)+\tfrac{1}{b}\phi(x)+\left(\tfrac{1}{b\gamma_k}-\tfrac{\mu}{b^2}\right)V(x_k,x)+\tfrac{\gamma_k}{b}\sigma^2,$$

where the last inequality follows from (4.6.35). By taking expectation w.r.t. $\xi_{[k-1]}$ on both sides of the above inequality and replacing x by x_*, we conclude that, for any $k \ge 1$,

$$\mathbb{E}[V(x_{k+1},x_*)] \le \left(1-\tfrac{\mu\gamma_k}{b}\right)\mathbb{E}[V(x_k,x_*)]+(b-1)\gamma_k\mathbb{E}[\phi(x_k)-\phi(x_*)]-$$
$$b\gamma_k\mathbb{E}\left[\phi(x_{k+1})-\phi(x_*)\right]+\gamma_k^2\sigma^2,$$

which, in view of Lemma 3.17 (with $a_k=\gamma_k\mu/(b)$, $A_k=\Gamma_k$, and $B_k=(b-1)\gamma_k[\phi(x_k)-\phi(x_*)]-b\gamma_k\mathbb{E}\left[\phi(x_{k+1})-\phi(x_*)\right]+\gamma_k^2\sigma^2$), then implies that

$$\tfrac{1}{\Gamma_N}[V(x_{k+1},x_*)] \le (1-\tfrac{\mu\gamma_1}{b})V(x_1,x_*)+(b-1)\textstyle\sum_{k=1}^N\tfrac{\gamma_k}{\Gamma_k}[\phi(x_k)-\phi(x_*)]$$
$$-b\textstyle\sum_{k=1}^N\tfrac{\gamma_k}{\Gamma_k}[\phi(x_{k+1})-\phi(x_*)]+\sum_{k=1}^N\tfrac{\gamma_k^2}{\Gamma_k}\sigma^2$$
$$\le (1-\tfrac{\mu\gamma_1}{b})V(x_1,x_*)+(b-1)\gamma_1[\phi(x_1)-\phi(x_*)]$$
$$-\textstyle\sum_{k=2}^{N+1}\theta_k[\phi(x_k)-\phi(x_*)]+\sum_{k=1}^N\tfrac{\gamma_k^2}{\Gamma_k}\sigma^2,$$

where the last inequality follows from (4.6.78) and the fact that $\phi(x_{N+1})-\phi(x_*)\ge 0$. Noting that $V(x_{N+1},x_*)\ge 0$, we conclude from the above inequality that

$$\textstyle\sum_{k=2}^{N+1}\theta_k\mathbb{E}[\phi(x_k)-\phi(x_*)] \le (1-\tfrac{\mu\gamma_1}{b})V(x_1,x_*)+(b-1)\gamma_1[\phi(x_1)-\phi(x_*)]$$
$$+\textstyle\sum_{k=1}^N\tfrac{\gamma_k^2}{\Gamma_k}\sigma^2.$$

Our result immediately follows from the above inequality, the convexity of $\phi(\cdot)$, and (4.6.60). ■

Below we specialize the rate of convergence of the SBMD method for solving strongly convex composite problems with a proper selection of $\{\gamma_k\}$.

Corollary 4.11. *Suppose that (4.6.30) and (4.6.31) hold. Also assume that* $\{\theta_k\}$ *are set to (4.6.78) and*

$$\gamma_k = 2b/(\mu(k+k_0)) \;\; \forall k \ge 1, \tag{4.6.80}$$

where

$$k_0 := \left\lfloor \frac{4b\bar{L}}{\mu} \right\rfloor.$$

Then, for any $N \geq 2$, we have

$$\mathbb{E}[\phi(\bar{x}_N) - \phi(x_*)] \leq \frac{\mu k_0^2}{N(N+1)} \Sigma_{i=1}^b V_i(x_1, x_*) + \frac{2(b-1)k_0}{N(N+1)} [\phi(x_1) - \phi(x_*)] + \frac{4b\sigma^2}{\mu(N+1)}, \tag{4.6.81}$$

where x_* is the optimal solution of problem (4.6.53).

Proof. We can check that

$$\gamma_k = \frac{2b}{\mu(k+\lfloor 4b\bar{L}/\mu \rfloor)} \leq \frac{1}{2\bar{L}}.$$

It can also be easily seen from the definition of γ_k and (4.6.78) that

$$\Gamma_k = \frac{k_0(k_0+1)}{(k+k_0)(k+k_0-1)}, \quad 1 - \frac{\gamma_1 \mu}{b} = \frac{k_0-1}{k_0+1}, \quad \forall k \geq 1, \tag{4.6.82}$$

$$\theta_k = \frac{b\gamma_k}{\Gamma_k} - \frac{(b-1)\gamma_{k+1}}{\Gamma_{k+1}} = \frac{2bk+2b(k_0-b)}{\mu k_0(k_0+1)} \geq \frac{2bk}{\mu k_0(k_0+1)}, \tag{4.6.83}$$

where the relation $k_0 \geq b$ follows from the definition of k_0 and the fact that $\bar{L} \geq \mu$. Hence,

$$\Sigma_{k=2}^{N+1} \theta_k \geq \frac{bN(N+1)}{\mu k_0(k_0+1)}, \quad \Sigma_{k=1}^N \frac{\gamma_k^2}{\Gamma_k} = \frac{4b^2}{\mu^2 k_0(k_0+1)} \Sigma_{k=1}^N \frac{k+k_0-1}{k+k_0} \leq \frac{4Nb^2}{\mu^2 k_0(k_0+1)}. \tag{4.6.84}$$

By using the above observations and (4.6.79), we have

$$\begin{aligned}
&\mathbb{E}[\phi(\bar{x}_N) - \phi(x_*)] \\
&\leq \left(\Sigma_{k=2}^{N+1} \theta_k\right)^{-1} \left[\left(1 - \frac{\mu\gamma_1}{b}\right) V(x_1, x_*) + (b-1)\gamma_1[\phi(x_1) - \phi(x_*)] + \Sigma_{k=1}^N \frac{\gamma_k^2}{\Gamma_k} \sigma^2\right]. \\
&\leq \frac{\mu k_0(k_0+1)}{bN(N+1)} \left[\frac{k_0-1}{k_0+1} V(x_1, x_*) + \frac{2b(b-1)}{\mu(k_0+1)} [\phi(x_1) - \phi(x_*)] + \frac{4Nb^2\sigma^2}{\mu^2 k_0(k_0+1)}\right] \\
&\leq \frac{\mu k_0^2}{bN(N+1)} V(x_1, x_*) + \frac{2(b-1)k_0}{N(N+1)} [\phi(x_1) - \phi(x_*)] + \frac{4b\sigma^2}{\mu(N+1)},
\end{aligned}$$

where the second inequality follows (4.6.82), (4.6.83), and (4.6.84). ∎

It is interesting to observe that, in view of (4.6.81) and the definition of k_0, the Lipschitz constant $\bar{L}$ can be as large as $\mathcal{O}(\sqrt{N})$ without affecting the rate of convergence of the SBMD algorithm, after disregarding other constant factors, for solving strongly convex stochastic composite optimization problems.

4.7 Exercises and Notes

1. Establish the rate of convergence for the stochastic mirror descent method applied to problem (4.1.1) under the following situation.

   (a) Nonsmooth and strongly convex for which (4.1.7) holds and

$$f(y) - f(x) - \langle f'(x), y - x \rangle \geq \mu V(x,y), \forall x, y \in X. \quad (4.7.85)$$

(b) Smooth and strongly convex for which both (4.1.20) and (4.7.85) hold.

2. Establish the lower complexity bound stated in (4.2.3) and (4.2.4) for stochastic optimization methods applied to problem 4.2.1.
3. Establish the rate of convergence of the accelerated stochastic gradient descent method if the objective function f in (4.2.1) is differentiable and its gradients are Hölder continuous, i.e.,

$$\|\nabla f(x) - \nabla f(y)\|_* \leq L\|x - y\|^\nu, \forall x, y \in X$$

for some $\nu \in [0,1]$.
4. Consider the matrix game problem, that is, problem (4.3.1) with

$$\phi(x,y) := y^T Ax + b^T x + c^T y,$$

where $A \in \mathbb{R}^{m\times n}$, X is the standard Euclidean ball, i.e.,

$$X := \{x \in \mathbb{R}^n : \Sigma_{j=1}^n x_j^2 \leq 1\},$$

and Y is the standard simplex, i.e.,

$$Y := \{y \in \mathbb{R}^m : y \geq 0, \Sigma_{i=1}^m y_i = 1\}.$$

Try to derive a randomized oracle for the stochastic mirror descent method discussed in Sect. 4.3 for solving this problem.
5. Consider the linearized primal–dual method in Algorithm 4.1.

   (a) Establish the rate of convergence for solving deterministic saddle point problems in (4.4.1).
   (b) Develop a stochastic version of this method and establish its rate of convergence for solving problem (4.4.1) under the stochastic first-order oracle as described in Sect. 4.4.

6. The stochastic accelerated mirror-prox method with $\alpha_t = 1$, $G = 0$, and $J = 0$ will be equivalent to a stochastic version of the mirror-prox method.

   (a) Establish the rate of convergence of this algorithm applied to monotone variational inequalities (i.e., H is monotone).
   (b) Establish the rate of convergence of this algorithm applied to generalized monotone variational inequalities, i.e., there exists an $\bar{z} \in Z$ such that

$$\langle H(z), z - \bar{z} \rangle \geq 0, \forall z \in Z.$$

Notes. Stochastic gradient descent (a.k.a, stochastic approximation) goes back to Robbins and Monro [117]. Nemirovksi and Yudin [94] first introduced the stochastic mirror descent method in 1983 and further improvement was made in

[95, 98, 108, 109]. In particular, Nemirovski et al. [95] presented a comprehensive treatment for the stochastic mirror descent method, which includes the complexity analysis for general nonsmooth, strongly convex, and convex–concave saddle point problems, the derivation of large-deviation results, and extensive numerical experimentation. Moreover, Lan, Nemirovski, and Shapiro [73] present the validation analysis, i.e., accuracy certificates for this method. Lan [62] first presented the accelerated stochastic gradient descent (a.k.a., accelerated stochastic approximation or SGD with momentum) in 2008 and the paper was formally published in [63]. A comprehensive study of this method, including the generalization for solving strongly convex composite optimization problems, multi-epoch (or multi-stage) variants, shrinking procedures, and accuracy certificates, was presented by Ghadimi and Lan in [38, 39]. Chen, Lan, and Ouyang presented the stochastic accelerated primal–dual method for solving stochastic saddle point problems with a bilinear structure in [23], and the stochastic accelerated mirror-prox method for solving a class of composite variational inequalities in [22]. Note that an earlier stochastic mirror-prox method without the acceleration steps was presented in [58]. Dang and Lan first introduced the randomized block decomposition into nonsmooth and stochastic optimization in [28]. Note that here exists a long history for block coordinate descent methods for solving deterministic optimization problems (see, e.g., [77, 80, 101, 116, 131]).

Chapter 5
Convex Finite-Sum and Distributed Optimization

In this chapter, we will study an important class of convex optimization problems whose objective function is given by the summation of many components. With wide applications in machine learning and distributed optimization, these problems can be viewed either as deterministic optimization problems with a special finite-sum structure or stochastic optimization problems with a discrete distribution. Accordingly, we will study two typical classes of randomized algorithms for solving them. While the first class incorporates random block decomposition into the dual space of the primal–dual methods for deterministic convex optimization, the second one employs variance reduction techniques into stochastic gradient descent methods for stochastic optimization.

5.1 Random Primal–Dual Gradient Method

In this section we are interested in the convex programming problem given by

$$\Psi^* := \min_{x \in X} \left\{ \Psi(x) := \tfrac{1}{m} \textstyle\sum_{i=1}^m f_i(x) + h(x) + \mu\, \nu(x) \right\}. \tag{5.1.1}$$

Here, $X \subseteq \mathbb{R}^n$ is a closed convex set, h is a relatively simple convex function, $f_i : \mathbb{R}^n \to \mathbb{R}$, $i = 1, \ldots, m$, are smooth convex functions with Lipschitz continuous gradient, i.e., $\exists L_i \geq 0$ such that

$$\|\nabla f_i(x_1) - \nabla f_i(x_2)\|_* \leq L_i \|x_1 - x_2\|, \ \forall x_1, x_2 \in \mathbb{R}^n, \tag{5.1.2}$$

$\nu : X \to \mathbb{R}$ is a strongly convex function with modulus 1 w.r.t. an arbitrary norm $\|\cdot\|$, i.e.,

$$\langle \nu'(x_1) - \nu'(x_2), x_1 - x_2 \rangle \geq \tfrac{1}{2} \|x_1 - x_2\|^2, \ \forall x_1, x_2 \in X, \tag{5.1.3}$$

and $\mu \geq 0$ is a given constant. Hence, the objective function Ψ is strongly convex whenever $\mu > 0$. For notational convenience, we also denote $f(x) \equiv \frac{1}{m} \sum_{i=1}^m f_i(x)$ and $L \equiv \frac{1}{m} \sum_{i=1}^m L_i$. It is easy to see that for some $L_f \geq 0$,

G. Lan, *First-Order and Stochastic Optimization Methods for Machine Learning*, Springer Series in the Data Sciences, https://doi.org/10.1007/978-3-030-39568-1_5

$$\|\nabla f(x_1) - \nabla f(x_2)\|_* \le L_f \|x_1 - x_2\| \le L\|x_1 - x_2\|, \ \forall x_1, x_2 \in \mathbb{R}^n. \tag{5.1.4}$$

Throughout this section, we assume that subproblems of the form

$$\mathrm{argmin}_{x\in X} \langle g, x\rangle + h(x) + \mu\, v(x) \tag{5.1.5}$$

are easy to solve for any $g \in \mathbb{R}^n$ and $\mu \ge 0$. We point out below a few examples where such an assumption is satisfied.

- If X is relatively simple, e.g., Euclidean ball, simplex or l_1 ball, and $h(x) = 0$, and $w(\cdot)$ is some properly choosing distance generating function, we can obtain closed form solutions of problem (5.1.5). This is the standard setting used in the regular first-order methods.
- If the problem is unconstrained, i.e., $X = \mathbb{R}^n$, and $h(x)$ is relatively simple, we can derive closed form solutions of (5.1.5) for some interesting cases. For example, if $h(x) = \|x\|_1$ and $w(x) = \|x\|_2^2$, then an explicit solution of (5.1.5) is readily given by its first-order optimality condition. A similar example is given by $h(x) = \sum_{i=1}^d \sigma_i(x)$ and $w(x) = \mathrm{tr}(x^T x)/2$, where $\sigma_i(x)$, $i = 1,\ldots,d$, denote the singular values of $x \in \mathbb{R}^{d\times d}$.
- If X is relatively simple and $h(x)$ is nontrivial, we can still compute closed form solutions of (5.1.5) for some interesting special cases, e.g., when X is the standard simplex, $w(x) = \sum_{i=1}^d x_i \log x_i$ and $h(x) = \sum_{i=1}^d x_i$.

The deterministic finite-sum problem (5.1.1) can model the empirical risk minimization in machine learning and statistical inferences, and hence has become the subject of intensive studies during the past few years. Our study on the finite-sum problems (5.1.1) has also been motivated by the emerging need for distributed optimization and machine learning. One typical example of the aforementioned distributed problems is *Federated Learning* (see Fig. 5.1). Under such settings, each component function f_i is associated with an agent i, $i = 1,\ldots,m$, which are connected through a distributed network. While different topologies can be considered for distributed optimization, in this section we focus on the star network where m agents are connected to one central server, and all agents only communicate with the server (see Fig. 5.1). These types of distributed optimization problems have several unique features. First, they allow for data privacy, since no local data is stored in the server. Second, network agents behave independently and they may not be responsive at the same time. Third, the communication between the server and agent can be expensive and has high latency. Under the distributed setting, methods requiring full gradient computation may incur extra communication and synchronization costs. As a consequence, methods which require fewer full gradient computations seem to be more advantageous in this regard. As a particular example, in the ℓ_2-regularized logistic regression problem, we have

$$\begin{aligned} &f_i(x) = l_i(x) := \tfrac{1}{N_i}\textstyle\sum_{j=1}^{N_i} \log(1+\exp(-b_j^i {a_j^i}^T x)),\ i = 1,\ldots,m,\\ &v(x) = R(x) := \tfrac{1}{2}\|x\|_2^2, \end{aligned} \tag{5.1.6}$$

provided that f_i is the loss function of agent i with training data $\{a_j^i, b_j^i\}_{j=1}^{N_i} \in \mathbb{R}^n \times \{-1,1\}$, and $\mu := \lambda$ is the penalty parameter. Note that another type of topol-

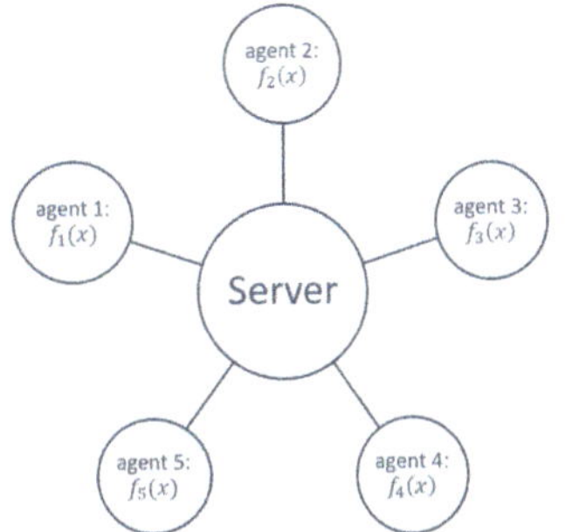

Fig. 5.1 A distributed network with five agents and one server

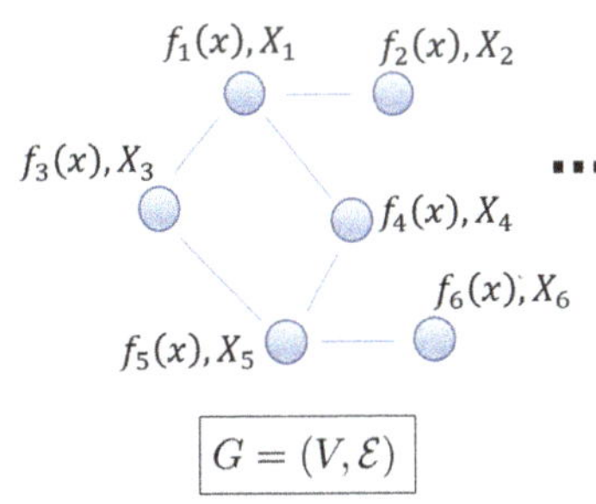

Fig. 5.2 An example of the decentralized network

ogy for distributed optimization is the multi-agent network without a central server, namely the decentralized setting, as shown in Fig. 5.2, where the agents can only communicate with their neighbors to update information (see Sect. 8.3 for discussions about decentralized algorithms).

Stochastic (sub)gradient descent (SGD) type methods, as discussed in Chap. 4, have been proven useful to solve problems given in the form of (5.1.1). Recall that SGD was originally designed to solve stochastic optimization problems given by

$$\min_{x \in X} \mathbb{E}_\xi [F(x, \xi)], \tag{5.1.7}$$

where ξ is a random variable with support $\Xi \subseteq \mathbb{R}^d$. Problem (5.1.1) can be viewed as a special case of (5.1.7) by setting ξ to be a discrete random variable supported on $\{1, \ldots, m\}$ with $\mathrm{Prob}\{\xi = i\} = p_i$ and

$$F(x,i) = (mp_i)^{-1} f_i(x) + h(x) + \mu \nu(x), i = 1, \ldots, m.$$

Observe that the subgradients of h and ν are not required due to the assumption in (5.1.5). Hence, each iteration of SGDs needs to compute the (sub)gradient of only one randomly selected f_i. As a result, their iteration cost is significantly smaller than that for deterministic first-order methods (FOM), which involves the computation of first-order information of f and thus all the m (sub)gradients of f_i's. Moreover, when f_i's are general convex Lipschitz continuous functions constants M_i, by properly specifying the probabilities $p_i = M_i / \Sigma_{i=1}^m M_i$, $i = 1, \ldots, m$, it can be shown (see Section 4.1) that the iteration complexities for both SGDs and FOMs are in the same order of magnitude. Consequently, the total number of subgradients required by SGDs can be m times smaller than those by FOMs.

Note however, that there is a significant gap on the complexity bounds between SGDs and deterministic FOMs if f_i's are smooth convex functions. For the sake of simplicity, let us focus on the strongly convex case when $\mu > 0$ and let x^* be the optimal solution of (5.1.1). In order to find a solution $\bar{x} \in X$ s.t. $\|\bar{x} - x^*\|^2 \le \varepsilon$, the total number of gradient evaluations for f_i's performed by optimal FOMs (see Sect. 3.3) can be bounded by

$$\mathcal{O}\left\{m\sqrt{\tfrac{L_f}{\mu}}\log\tfrac{1}{\varepsilon}\right\}. \tag{5.1.8}$$

On the other hand, a direct application of optimal SGDs (see Sect. 4.2) to the aforementioned stochastic optimization reformulation of (5.1.1) would yield an

$$\mathcal{O}\left\{\sqrt{\tfrac{L}{\mu}}\log\tfrac{1}{\varepsilon} + \tfrac{\sigma^2}{\mu\varepsilon}\right\} \tag{5.1.9}$$

iteration complexity bound on the number of gradient evaluations for f_i's. Here $\sigma > 0$ denotes variance of the stochastic gradients, i.e., $\mathbb{E}[\|G(x,\xi) - \nabla f(x)\|_*^2] \le \sigma^2$, where $G(x,\xi)$ is an unbiased estimator for the gradient $\nabla f(x)$. Clearly, the latter bound is significantly better than the one in (5.1.8) in terms of its dependence on m, but much worse in terms of its dependence on accuracy ε and a few other problem parameters (e.g., L and μ). It should be noted that the optimality of (5.1.9) for general stochastic programming (5.1.7) does not preclude the existence of more efficient algorithms for solving (5.1.1), because (5.1.1) is a special case of (5.1.7) with finite support Ξ.

Following the constructions in Sect. 3.3, we develop a randomized primal–dual gradient (RPDG) method, which is an incremental gradient method using only one randomly selected component ∇f_i at each iteration. This method was developed by using the following ideas: (1) a proper reformulation of (5.1.1) as a primal–dual saddle point problem with multiple dual players; and (2) the incorporation of a new non-differentiable prox-function (or Bregman distance) based on the conjugate functions of f in the dual space. Different from the game interpretation of the accelerated gradient descent method, the RPDG method incorporates an additional dual prediction step before performing the primal descent step (with a properly defined primal prox-function). We prove that the number of iterations (and hence the number of gradients) required by RPDG is bounded by

$$\mathcal{O}\left(\left(m + \sqrt{\tfrac{mL}{\mu}}\right)\log\tfrac{1}{\varepsilon}\right), \tag{5.1.10}$$

both in expectation and with high probability. The complexity bounds of the RPDG method are established in terms of not only the distance from the iterate x^k to the optimal solution, but also the primal optimality gap based on the ergodic mean of the iterates.

Moreover, we show that the number of gradient evaluations required by any randomized incremental gradient methods to find an ε-solution of (5.1.1), i.e., a point

$\bar{x} \in X$ s.t. $\mathbb{E}[\|\bar{x} - x^*\|_2^2] \le \varepsilon$, cannot be smaller than

$$\Omega\left(\left(m + \sqrt{\tfrac{mL}{\mu}}\right)\log\tfrac{1}{\varepsilon}\right), \tag{5.1.11}$$

whenever the dimension

$$n \ge (k + m/2)/\log(1/q).$$

This bound is obtained by carefully constructing a special class of separable quadratic programming problems and tightly bounding the expected distance to the optimal solution for any arbitrary distribution used to choose f_i at each iteration. Note that we assume that the distribution is chosen prior to the execution of the algorithms and is independent of the iteration number for the sake of simplicity. However, the construction can be extended for more general randomized methods. Comparing (5.1.10) with (5.1.11), we conclude that the complexity of the RPDG method is optimal if n is large enough. As a byproduct, we also derived a lower complexity bound for randomized block coordinate descent methods by utilizing the separable structure of the aforementioned worst-case instances.

Finally, we generalize RPDG for problems which are not necessarily strongly convex (i.e., $\mu = 0$) and/or involve structured nonsmooth terms f_i. We show that for all these cases, the RPDG can save $\mathcal{O}(\sqrt{m})$ times gradient computations (up to certain logarithmic factors) in comparison with the corresponding optimal deterministic FOMs at the cost of making $\mathcal{O}(\sqrt{m})$ times more calls to the prox-oracle. In particular, we show that when both the primal and dual of (5.1.1) are not strongly convex, the total number of iterations performed by the RPDG method can be bounded by $\mathcal{O}(\sqrt{m}/\varepsilon)$ (up to some logarithmic factors), which is $\mathcal{O}(\sqrt{m})$ times better, in terms of the total number of dual subproblems to be solved, than deterministic methods for solving bilinear saddle point problems (see Sect. 3.6).

5.1.1 Multi-Dual-Player Game Reformulation

We start by introducing a new saddle point reformulation of (5.1.1). Let $J_i : \mathscr{Y}_i \to \mathbb{R}$ be the conjugate functions of f_i/m and $\mathscr{Y}_i$, $i = 1,\ldots,m$, denote the dual spaces where the gradients of f_i/m reside. For the sake of notational convenience, let us denote $J(y) := \sum_{i=1}^m J_i(y_i)$, $\mathscr{Y} := \mathscr{Y}_1 \times \mathscr{Y}_2 \times \ldots \times \mathscr{Y}_m$, and $y = (y_1; y_2; \ldots; y_m)$ for any $y_i \in \mathscr{Y}_i$, $i = 1,\ldots,m$. Clearly, we can reformulate problem (5.1.1) equivalently as a saddle point problem:

$$\Psi^* := \min_{x \in X}\left\{h(x) + \mu\, \nu(x) + \max_{y \in \mathscr{Y}} \langle x, Uy\rangle - J(y)\right\}, \tag{5.1.12}$$

where $U \in \mathbb{R}^{n \times nm}$ is given by

$$U := [I, I, \ldots, I]. \tag{5.1.13}$$

Here I is the identity matrix in $\mathbb{R}^n$. Given a pair of feasible solutions $\bar{z} = (\bar{x}, \bar{y})$ and $z = (x, y)$ of (5.1.12), we define the primal–dual gap function $Q(\bar{z}, z)$ by

$$Q(\bar{z},z) := [h(\bar{x}) + \mu \nu(\bar{x}) + \langle \bar{x}, Uy\rangle - J(y)] - [h(x) + \mu \nu(x) + \langle x, U\bar{y}\rangle - J(\bar{y})]. \tag{5.1.14}$$

It is known that $\bar{z} \in Z \equiv X \times \mathscr{Y}$ is an optimal solution of (5.1.12) if and only if $Q(\bar{z}, z) \le 0$ for all $z \in Z$.

We now discuss both primal and dual prox-functions (proximity control functions) in the primal and dual spaces, respectively, associated with problem 5.1.12.

Recall that the function $\nu : X \to \mathbb{R}$ in (5.1.1) is strongly convex with modulus 1 with respect to $\|\cdot\|$. We can define a primal *prox-function* associated with ν as

$$V(x^0, x) \equiv V_\nu(x^0, x) := \nu(x) - [\nu(x^0) + \langle \nu'(x^0), x - x^0\rangle], \tag{5.1.15}$$

where $\nu'(x^0) \in \partial \nu(x^0)$ is an arbitrary subgradient of ν at x^0. Clearly, by the strong convexity of ν, we have

$$V(x^0, x) \ge \tfrac{1}{2}\|x - x^0\|^2, \quad \forall x, x^0 \in X. \tag{5.1.16}$$

Note that the prox-function $V(\cdot,\cdot)$ described above generalizes the Bregman's distance in the sense that ν is not necessarily differentiable (see Sect. 3.2). Throughout this section, we assume that the prox-mapping associated with X, ν, and h, given by

$$\arg\min_{x \in X} \left\{ \langle g, x\rangle + h(x) + \mu\, \nu(x) + \eta V(x^0, x) \right\}, \tag{5.1.17}$$

is easily computable for any $x^0 \in X, g \in \mathbb{R}^n$, $\mu \ge 0$, and $\eta > 0$. Clearly this is equivalent to the assumption that (5.1.5) is easy to solve. Whenever ν is non-differentiable, we need to specify a particular selection of the subgradient ν' before performing the prox-mapping. We assume throughout this section that such a selection of ν' is defined recursively as follows. Denote

$$x^1 = \arg\min_{x \in X} \left\{ \langle g, x\rangle + h(x) + \mu\, \nu(x) + \eta V(x^0, x) \right\}.$$

By the optimality condition, we have

$$g + h'(x^1) + (\mu + \eta)\nu'(x^1) - \eta \nu'(x^0) \in \mathscr{N}_X(x^1),$$

where $\mathscr{N}_X(x^1)$ denotes the normal cone of X at x^1 given by $\mathscr{N}_X(\bar{x}) := \{v \in \mathbb{R}^n : v^T(x - \bar{x}) \le 0, \forall x \in X\}$. Once such a $\nu'(x^1)$ satisfying the above relation is identified, we will use it as a subgradient when defining $V(x^1, x)$ in the next iteration. Note that such a subgradient can be identified without additional computational cost as long as x^1 is obtained, since one needs it to check the optimality condition of (5.1.17) when finding x^1.

Since $J_i, i = 1, \ldots, m$, are strongly convex with modulus $\sigma_i = m/L_i$ w.r.t. $\|\cdot\|_*$, we can define their associated dual prox-functions and dual prox-mappings as

$$W_i(y_i^0, y_i) := J_i(y_i) - [J_i(y_i^0) + \langle J_i'(y_i^0), y_i - y_i^0 \rangle], \tag{5.1.18}$$

$$\arg\min_{y_i \in \mathcal{Y}_i} \left\{ \langle -\tilde{x}, y_i \rangle + J_i(y_i) + \tau W_i(y_i^0, y_i) \right\}, \tag{5.1.19}$$

for any $y_i^0, y_i \in \mathcal{Y}_i$. Accordingly, we define

$$W(\tilde{y}, y) := \Sigma_{i=1}^m W_i(\tilde{y}_i, y_i). \tag{5.1.20}$$

Again, W_i may not be uniquely defined since J_i are not necessarily differentiable. However, we will discuss how to specify the particular selection of $J_i' \in \partial J_i$ later in this section.

5.1.2 Randomization on Gradient Computation

The basic idea of the randomized primal–dual gradient method (see Algorithm 5.1) is to incorporate the random block decomposition into the primal–dual method discussed in Sect. 3.6 for solving problem 5.1.12. In view of our discussions in Sect. 3.4, the computation of dual prox-mapping is equivalent to the computation of gradient, hence, the randomization of the computation of dual prox-mapping reduces to the randomized computation of gradients.

Algorithm 5.1 A randomized primal–dual gradient (RPDG) method

---

Let $x^0 = x^{-1} \in X$, and the nonnegative parameters $\{\tau_t\}$, $\{\eta_t\}$, and $\{\alpha_t\}$ be given.
Set $y_i^0 = \frac{1}{m}\nabla f_i(x^0)$, $i = 1, \ldots, m$.
for $t = 1, \ldots, k$ **do**
  Choose i_t according to $\text{Prob}\{i_t = i\} = p_i$, $i = 1, \ldots, m$.
  Update $z^t = (x^t, y^t)$ according to

$$\tilde{x}^t = \alpha_t(x^{t-1} - x^{t-2}) + x^{t-1}. \tag{5.1.21}$$

$$y_i^t = \begin{cases} \arg\min_{y_i \in \mathcal{Y}_i} \left\{ \langle -\tilde{x}^t, y_i \rangle + J_i(y_i) + \tau W_i(y_i^{t-1}, y_i) \right\}, & i = i_t, \\ y_i^{t-1}, & i \neq i_t. \end{cases} \tag{5.1.22}$$

$$\tilde{y}_i^t = \begin{cases} p_i^{-1}(y_i^t - y_i^{t-1}) + y_i^{t-1}, & i = i_t, \\ y_i^{t-1}, & i \neq i_t. \end{cases} \tag{5.1.23}$$

$$x^t = \arg\min_{x \in X} \left\{ \langle \Sigma_{i=1}^m \tilde{y}_i^t, x \rangle + h(x) + \mu\, \nu(x) + \eta_t V(x^{t-1}, x) \right\}. \tag{5.1.24}$$

end for

We now add some remarks about the randomized primal–dual gradient method. First, we only compute a randomly selected dual prox-mapping in this method in (5.1.22). Second, in addition to the primal prediction step (5.1.21), we add a new dual prediction step (5.1.23), and then use the predicted dual variable $\tilde{y}^t$ for the computation of the new search point x^t in (5.1.24). It can be easily seen that the

RPDG method reduces to the primal–dual method, as well as the accelerated gradient method, whenever the number of blocks $m = 1$ (Sect. 3.4).

The RPDG method can be viewed as a game iteratively performed by a buyer and m suppliers for finding the solutions (order quantities and product prices) of the saddle point problem in (5.1.12). In this game, both the buyer and suppliers have access to their local cost $h(x) + \mu \nu(x)$ and $J_i(y_i)$, respectively, as well as their interactive cost (or revenue) represented by a bilinear function $\langle x, y_i \rangle$. Also, the buyer has to purchase the same amount of products from each supplier (e.g., for fairness). Although there are m suppliers, in each iteration only a randomly chosen supplier can make price changes according to (5.1.22) using the predicted demand $\tilde{x}^t$. In order to understand the buyer's decision in (5.1.24), let us first denote

$$\hat{y}_i^t := \arg\min_{y_i \in \mathscr{Y}_i} \left\{ \langle -\tilde{x}^t, y_i \rangle + J_i(y_i) + \tau W_i(y_i^{t-1}, y_i) \right\}, \quad i = 1, \ldots, m; t = 1, \ldots, k. \tag{5.1.25}$$

In other words, $\hat{y}_i^t$, $i = 1, \ldots, m$, denote the prices that all the suppliers can possibly set up at iteration t. Then we can see that

$$\mathbb{E}_t[\tilde{y}_i^t] = \hat{y}_i^t. \tag{5.1.26}$$

Indeed, we have

$$y_i^t = \begin{cases} \hat{y}_i^t, & i = i_t, \\ y_i^{t-1}, & i \neq i_t. \end{cases} \tag{5.1.27}$$

Hence $\mathbb{E}_t[y_i^t] = p_i \hat{y}_i^t + (1 - p_i) y_i^{t-1}$, $i = 1, \ldots, m$. Using this identity in the definition of $\tilde{y}^t$ in (5.1.23), we obtain (5.1.26). Instead of using $\sum_{i=1}^m \hat{y}_i^t$ in determining this order in (5.1.24), the buyer notices that only one supplier has made a change in the price, and thus uses $\sum_{i=1}^m \tilde{y}_i^t$ to predict the case when all the dual players would modify the prices simultaneously.

In order to implement the above RPDG method, we shall explicitly specify the selection of the subgradient J'_{i_t} in the definition of the dual prox-mapping in (5.1.22). Denoting $\underline{x}_i^0 = x^0$, $i = 1, \ldots, m$, we can easily see from $y_i^0 = \frac{1}{m} \nabla f_i(x^0)$ that $\underline{x}_i^0 \in \partial J_i(y_i^0)$, $i = 1, \ldots, m$. Using this relation and letting $J'_i(y_i^{t-1}) = \underline{x}_i^{t-1}$ in the definition of $W_i(y_i^{t-1}, y_i)$ in (5.1.22) (see (5.1.18)), we then conclude from Lemma 3.6 and (5.1.22) that for any $t \geq 1$,

$$\begin{aligned} \underline{x}_{i_t}^t &= (\tilde{x}^t + \tau_t \underline{x}_{i_t}^{t-1})/(1 + \tau_t), \quad \underline{x}_i^t = \underline{x}_i^{t-1}, \ \forall i \neq i_t; \\ y_{i_t}^t &= \tfrac{1}{m} \nabla f_{i_t}(\underline{x}_{i_t}^t), \quad y_i^t = y_i^{t-1}, \ \forall i \neq i_t. \end{aligned}$$

Moreover, observe that the computation of x^t in (5.1.24) requires an involved computation of $\sum_{i=1}^m \tilde{y}_i^t$. In order to save computational time, we suggest to compute this quantity in a recursive manner as follows. Let us denote $g^t \equiv \sum_{i=1}^m y_i^t$. Clearly, in view of the fact that $y_i^t = y_i^{t-1}$, $\forall i \neq i_t$, we have

$$g^t = g^{t-1} + (y_{i_t}^t - y_{i_t}^{t-1}).$$

Also, by the definition of g^t and (5.1.23), we have

$$\begin{aligned}\Sigma_{i=1}^m \tilde{y}_i^t &= \Sigma_{i\neq i_t} y_i^{t-1} + p_{i_t}^{-1}(y_{i_t}^t - y_{i_t}^{t-1}) + y_{i_t}^{t-1}\\ &= \Sigma_{i=1}^m y_i^{t-1} + p_{i_t}^{-1}(y_{i_t}^t - y_{i_t}^{t-1})\\ &= g^{t-1} + p_{i_t}^{-1}(y_{i_t}^t - y_{i_t}^{t-1}).\end{aligned}$$

Incorporating these two ideas mentioned above, we present an efficient implementation of the RPDG method in Algorithm 5.2.

Algorithm 5.2 An efficient implementation of the RPDG method

Let $x^0 = x^{-1} \in X$, and nonnegative parameters $\{\alpha_t\}$, $\{\tau_t\}$, and $\{\eta_t\}$ be given.
Set $\underline{x}_i^0 = x^0$, $y_i^0 = \frac{1}{m}\nabla f_i(x^0)$, $i = 1,\ldots,m$, and $g^0 = \Sigma_{i=1}^m y_i^0$.
for $t = 1,\ldots,k$ **do**
Choose i_t according to $\text{Prob}\{i_t = i\} = p_i$, $i = 1,\ldots,m$.
Update $z^t := (x^t, y^t)$ by

$$\tilde{x}^t = \alpha_t(x^{t-1} - x^{t-2}) + x^{t-1}. \tag{5.1.28}$$

$$\underline{x}_i^t = \begin{cases}(1+\tau_t)^{-1}\left(\tilde{x}^t + \tau_t \underline{x}_i^{t-1}\right), & i = i_t,\\ \underline{x}_i^{t-1}, & i \neq i_t.\end{cases} \tag{5.1.29}$$

$$y_i^t = \begin{cases}\frac{1}{m}\nabla f_i(\underline{x}_i^t), & i = i_t,\\ y_i^{t-1}, & i \neq i_t.\end{cases} \tag{5.1.30}$$

$$x^t = \arg\min_{x\in X}\left\{\langle g^{t-1} + p_{i_t}^{-1}(y_{i_t}^t - y_{i_t}^{t-1}), x\rangle + h(x) + \mu\, \nu(x) + \eta_t V(x^{t-1}, x)\right\}. \tag{5.1.31}$$

$$g^t = g^{t-1} + y_{i_t}^t - y_{i_t}^{t-1}. \tag{5.1.32}$$

end for

Clearly, the RPDG method is an incremental gradient type method since each iteration of this algorithm involves the computation of the gradient ∇f_{i_t} of only one component function. As shown in the following subsection, such a randomization scheme can lead to significant savings on the total number of gradient evaluations, at the expense of more primal prox-mappings.

It should also be noted that due to the randomness in the RPDG method, we cannot guarantee that $\underline{x}_i^t \in X$ for all $i = 1,\ldots,m$, and $t \geq 1$ in general, even though we do have all the iterates $x^t \in X$. That is why we need to make the assumption that f_i's are differentiable over $\mathbb{R}^n$ for the RPDG method. We will address this issue later in Sect. 5.2 by developing a different randomized algorithm.

5.1.3 Convergence for Strongly Convex Problems

Our goal in this subsection is to describe the convergence properties of the RPDG method for the strongly convex case when $\mu > 0$. Generalization of the RPDG method for the non-strongly convex case will be discussed in Sect. 5.1.5.

We first introduce some basic tools and general results about the RPDG method in Sects. 5.1.3.1 and 5.1.3.2, respectively. Then we describe the main convergence properties in Sect. 5.1.3.3. Moreover, in Sect. 5.1.4, we derive the lower complexity bound for randomized algorithms for solving finite-sum optimization problems.

5.1.3.1 Some Basic Tools

The following result provides a few different bounds on the diameter of the dual feasible sets $\mathscr{G}$ and $\mathscr{Y}$ in (5.2.14) and (5.1.12).

Lemma 5.1. *Let $x^0 \in X$ be given, and $y_i^0 = \frac{1}{m}\nabla f_i(x^0)$, $i = 1,\ldots,m$. Assume that $J_i'(y_i^0) = x^0$ in the definition of $W(y^0,y)$ in (5.1.18).*

(a) For any $x \in X$ and $y_i = \frac{1}{m}\nabla f_i(x)$, $i = 1,\ldots,m$, we have

$$W(y^0,y) \le \frac{L_f}{2}\|x^0 - x\|^2 \le L_f V(x^0,x). \tag{5.1.33}$$

(b) If $x^* \in X$ is an optimal solution of (5.1.1) and $y_i^* = \frac{1}{m}\nabla f_i(x^*)$, $i = 1,\ldots,m$, then

$$W(y^0,y^*) \le \Psi(x^0) - \Psi(x^*). \tag{5.1.34}$$

Proof. We first show part (a). It follows from the definitions of $W(y^0,y)$ and J_i that

$$\begin{aligned}
W(y^0,y) &= J(y) - J(y^0) - \Sigma_{i=1}^m \langle J_i'(y_i^0), y_i - y_i^0\rangle \\
&= \langle x, Uy\rangle - f(x) + f(x^0) - \langle x^0, Uy^0\rangle - \langle x^0, U(y - y^0)\rangle \\
&= f(x^0) - f(x) - \langle Uy, x^0 - x\rangle \\
&\le \frac{L_f}{2}\|x^0 - x\|^2 \le L_f V(x^0,x),
\end{aligned}$$

where the last inequality follows from (5.1.16). We now show part (b). By the above relation, the convexity of h and ν, and the optimality of (x^*,y^*), we have

$$\begin{aligned}
&W(y^0,y^*) = f(x^0) - f(x^*) - \langle Uy^*, x^0 - x^*\rangle \\
&= f(x^0) - f(x^*) + \langle h'(x^*) + \mu\nu'(x^*), x^0 - x^*\rangle - \langle Uy^* + h'(x^*) + \mu\nu'(x^*), x^0 - x^*\rangle \\
&\le f(x^0) - f(x^*) + \langle h'(x^*) + \mu\nu'(x^*), x^0 - x^*\rangle \le \Psi(x^0) - \Psi(x^*).
\end{aligned}$$

∎

The following lemma gives an important bound for the primal optimality gap $\Psi(\bar{x}) - \Psi(x^*)$ for some $\bar{x} \in X$.

Lemma 5.2. *Let $(\bar{x}, \bar{y}) \in Z$ be a given pair of feasible solutions of (5.1.12), and $z^* = (x^*, y^*)$ be a pair of optimal solutions of (5.1.12). Then, we have*

$$\Psi(\bar{x}) - \Psi(x^*) \leq Q((\bar{x}, \bar{y}), z^*) + \frac{L_f}{2}\|\bar{x} - x^*\|^2. \tag{5.1.35}$$

Proof. Let $\bar{y}_* = (\frac{1}{m}\nabla f_1(\bar{x}); \frac{1}{m}\nabla f_2(\bar{x}); \ldots; \frac{1}{m}\nabla f_m(\bar{x}))$, and by the definition of $Q(\cdot,\cdot)$ in (5.1.14), we have

$$\begin{aligned}
Q((\bar{x}, \bar{y}), z^*) &= [h(\bar{x}) + \mu\nu(\bar{x}) + \langle \bar{x}, U y^* \rangle - J(y^*)] - [h(x^*) + \mu\nu(x^*) + \langle x^*, U\bar{y} \rangle - J(\bar{y})] \\
&\geq [h(\bar{x}) + \mu\nu(\bar{x}) + \langle \bar{x}, U\bar{y}_* \rangle - J(\bar{y}_*)] + \langle \bar{x}, U(y^* - \bar{y}_*) \rangle - J(y^*) + J(\bar{y}_*) \\
&\quad - \left[h(x^*) + \mu\nu(x^*) + \max_{y \in \mathcal{Y}} \{\langle x^*, Uy \rangle - J(y)\} \right] \\
&= \Psi(\bar{x}) - \Psi(x^*) + \langle \bar{x}, U(y^* - \bar{y}_*) \rangle - \langle x^*, Uy^* \rangle + f(x^*) + \langle \bar{x}, U\bar{y}_* \rangle - f(\bar{x}) \\
&= \Psi(\bar{x}) - \Psi(x^*) + f(x^*) - f(\bar{x}) + \langle \bar{x} - x^*, \nabla f(x^*) \rangle \\
&\geq \Psi(\bar{x}) - \Psi(x^*) - \frac{L_f}{2}\|\bar{x} - x^*\|^2,
\end{aligned}$$

where the second equality follows from the fact that $J_i, i = 1, \ldots, m$, are the conjugate functions of f_i. ∎

5.1.3.2 General Results for RPDG

We will establish some general convergence results in Proposition 5.1. Before showing Proposition 5.1 we will develop a few technical results. Lemma 5.3 below characterizes the solutions of the prox-mapping in (5.1.17) and (5.1.19). This result slightly generalizes Lemma 3.5.

Lemma 5.3. *Let U be a closed convex set and a point $\tilde{u} \in U$ be given. Also let $w : U \to \mathbb{R}$ be a convex function and*

$$W(\tilde{u}, u) = w(u) - w(\tilde{u}) - \langle w'(\tilde{u}), u - \tilde{u} \rangle, \tag{5.1.36}$$

for some $w'(\tilde{u}) \in \partial w(\tilde{u})$. Assume that the function $q : U \to \mathbb{R}$ satisfies

$$q(u_1) - q(u_2) - \langle q'(u_2), u_1 - u_2 \rangle \geq \mu_0 W(u_2, u_1), \ \forall u_1, u_2 \in U \tag{5.1.37}$$

for some $\mu_0 \geq 0$. Also assume that the scalars μ_1 and μ_2 are chosen such that $\mu_0 + \mu_1 + \mu_2 \geq 0$. If

$$u^* \in \mathrm{Argmin}\{q(u) + \mu_1 w(u) + \mu_2 W(\tilde{u}, u) : u \in U\}, \tag{5.1.38}$$

then for any $u \in U$, we have

$$q(u^*)+\mu_1 w(u^*)+\mu_2 W(\tilde{u},u^*)+(\mu_0+\mu_1+\mu_2)W(u^*,u) \le q(u)+\mu_1 w(u)+\mu_2 W(\tilde{u},u).$$

Proof. Let $\phi(u) := q(u)+\mu_1 w(u)+\mu_2 W(\tilde{u},u)$. It can be easily checked that for any $u_1,u_2 \in U$,

$$W(\tilde{u},u_1) = W(\tilde{u},u_2)+\langle W'(\tilde{u},u_2),u_1-u_2\rangle + W(u_2,u_1),$$
$$w(u_1) = w(u_2)+\langle w'(u_2),u_1-u_2\rangle + W(u_2,u_1).$$

Using these relations and (5.1.37), we conclude that

$$\phi(u_1)-\phi(u_2)-\langle \phi'(u_2),u_1-u_2\rangle \ge (\mu_0+\mu_1+\mu_2)W(u_2,u_1) \tag{5.1.39}$$

for any $u_1,u_2 \in Y$, which together with the fact that $\mu_0+\mu_1+\mu_2 \ge 0$ then imply that ϕ is convex. Since u^* is an optimal solution of (5.1.38), we have $\langle \phi'(u^*),u-u^*\rangle \ge 0$. Combining this inequality with (5.1.39), we conclude that

$$\phi(u)-\phi(u^*) \ge (\mu_0+\mu_1+\mu_2)W(u^*,u),$$

from which the result immediately follows. ■

The following simple result provides a few identities related to y^t and $\tilde{y}^t$ that will be useful for the analysis of the RPDG algorithm.

Lemma 5.4. *Let y^t, $\tilde{y}^t$, and $\hat{y}^t$ be defined in (5.1.22), (5.1.23), and (5.1.25), respectively. Then we have, for any $i = 1,\ldots,m$ and $t = 1,\ldots,k$,*

$$\mathbb{E}_t[W_i(y_i^{t-1},y_i^t)] = p_i W_i(y_i^{t-1},\hat{y}_i^t), \tag{5.1.40}$$
$$\mathbb{E}_t[W_i(y_i^t,y_i)] = p_i W_i(\hat{y}_i^t,y_i)+(1-p_i)W_i(y_i^{t-1},y_i), \tag{5.1.41}$$

for any $y \in \mathscr{Y}$, where $\mathbb{E}_t$ denotes the conditional expectation w.r.t. i_t given $i_1,\ldots,i_{t-1}$.

Proof. (5.1.40) follows immediately from the facts that $\text{Prob}_t\{y_i^t = \hat{y}_i^t\} = \text{Prob}_t\{i_t = i\} = p_i$ and $\text{Prob}_t\{y_i^t = y_i^{t-1}\} = 1-p_i$. Here Prob_t denotes the conditional probability w.r.t. i_t given $i_1,\ldots,i_{t-1}$. Similarly, we can show (5.1.41). ■

We now prove an important recursion about the RPDG method.

Lemma 5.5. *Let the gap function Q be defined in (5.1.14). Also let x^t and $\hat{y}^t$ be defined in (5.1.24) and (5.1.25), respectively. Then for any $t \ge 1$, we have*

$$\begin{aligned}\mathbb{E}[Q((x^t,\hat{y}^t),z)] \le\ & \mathbb{E}\left[\eta_t V(x^{t-1},x)-(\mu+\eta_t)V(x^t,x)-\eta_t V(x^{t-1},x^t)\right]\\ &+\textstyle\sum_{i=1}^m \mathbb{E}\left[\left(p_i^{-1}(1+\tau_t)-1\right)W_i(y_i^{t-1},y_i)-p_i^{-1}(1+\tau_t)W_i(y_i^t,y_i)\right]\\ &+\mathbb{E}\left[\langle \tilde{x}^t-x^t,U(\hat{y}^t-y)\rangle-\tau_t p_{i_t}^{-1}W_{i_t}(y_{i_t}^{t-1},y_{i_t}^t)\right],\quad \forall z \in Z. \end{aligned}\tag{5.1.42}$$

Proof. It follows from Lemma 5.3 applied to (5.1.24) that $\forall x \in X$,

$$\begin{aligned}&\langle x^t - x, U\tilde{y}^t\rangle + h(x^t) + \mu\nu(x^t) - h(x) - \mu\nu(x)\\ &\le \eta_t V(x^{t-1}, x) - (\mu + \eta_t)V(x^t, x) - \eta_t V(x^{t-1}, x^t).\end{aligned} \tag{5.1.43}$$

Moreover, by Lemma 5.3 applied to (5.1.25), we have, for any $i = 1, \ldots, m$ and $t = 1, \ldots, k$,

$$\langle -\tilde{x}^t, \hat{y}_i^t - y_i\rangle + J_i(\hat{y}_i^t) - J_i(y_i) \le \tau_t W_i(y_i^{t-1}, y_i) - (1+\tau_t)W_i(\hat{y}_i^t, y_i) - \tau_t W_i(y_i^{t-1}, \hat{y}_i^t).$$

Summing up these inequalities over $i = 1, \ldots, m$, we have, $\forall y \in \mathscr{Y}$,

$$\begin{aligned}&\langle -\tilde{x}^t, U(\hat{y}^t - y)\rangle + J(\hat{y}^t) - J(y)\\ &\le \textstyle\sum_{i=1}^m \left[\tau_t W_i(y_i^{t-1}, y_i) - (1+\tau_t)W_i(\hat{y}_i^t, y_i) - \tau_t W_i(y_i^{t-1}, \hat{y}_i^t)\right].\end{aligned} \tag{5.1.44}$$

Using the definition of Q in (5.1.14), (5.1.43), and (5.1.44), we have

$$\begin{aligned}Q((x^t, \hat{y}^t), z) \le\ & \eta_t V(x^{t-1}, x) - (\mu + \eta_t)V(x^t, x) - \eta_t V(x^{t-1}, x^t)\\ &+ \textstyle\sum_{i=1}^m \left[\tau_t W_i(y_i^{t-1}, y_i) - (1+\tau_t)W_i(\hat{y}_i^t, y_i) - \tau_t W_i(y_i^{t-1}, \hat{y}_i^t)\right]\\ &+ \langle \tilde{x}^t, U(\hat{y}^t - y)\rangle - \langle x^t, U(\tilde{y}^t - y)\rangle + \langle x, U(\tilde{y}^t - \hat{y}^t)\rangle.\end{aligned} \tag{5.1.45}$$

Also observe that by (5.1.22), (5.1.26), (5.1.40), and (5.1.41),

$$\begin{aligned}W_i(y_i^{t-1}, y_i^t) &= 0, \ \forall i \ne i_t,\\ \mathbb{E}[\langle x, U(\tilde{y}^t - \hat{y}^t)\rangle] &= 0,\\ \mathbb{E}[\langle \tilde{x}^t, U\hat{y}^t\rangle] &= \mathbb{E}[\langle \tilde{x}^t, U\tilde{y}^t\rangle],\\ \mathbb{E}[W_i(y_i^{t-1}, \hat{y}_i^t)] &= \mathbb{E}[p_i^{-1} W_i(y_i^{t-1}, y_i^t)]\\ \mathbb{E}[W_i(\hat{y}_i^t, y_i)] &= p_i^{-1}\mathbb{E}[W_i(y_i^t, y_i)] - (p_i^{-1} - 1)\mathbb{E}[W_i(y_i^{t-1}, y_i)].\end{aligned}$$

Taking expectation on both sides of (5.1.45) and using the above observations, we obtain (5.1.42). ■

We are now ready to establish a general convergence result for RPDG.

Proposition 5.1. *Suppose that* $\{\tau_t\}$, $\{\eta_t\}$, *and* $\{\alpha_t\}$ *in the RPDG method satisfy*

$$\theta_t\left(p_i^{-1}(1+\tau_t) - 1\right) \le p_i^{-1}\theta_{t-1}(1+\tau_{t-1}), i = 1, \ldots, m; t = 2, \ldots, k, \tag{5.1.46}$$

$$\theta_t\eta_t \le \theta_{t-1}(\mu + \eta_{t-1}), t = 2, \ldots, k, \tag{5.1.47}$$

$$\tfrac{\eta_k}{4} \ge \tfrac{L_i(1-p_i)^2}{m\tau_k p_i}, i = 1, \ldots, m, \tag{5.1.48}$$

$$\tfrac{\eta_{t-1}}{2} \ge \tfrac{L_i\alpha_t}{m\tau_t p_i} + \tfrac{(1-p_j)^2 L_j}{m\tau_{t-1}p_j}, \ i, j \in \{1, \ldots, m\}; t = 2, \ldots, k, \tag{5.1.49}$$

$$\tfrac{\eta_k}{2} \ge \tfrac{\sum_{i=1}^m (p_i L_i)}{m(1+\tau_k)}, \tag{5.1.50}$$

$$\alpha_t\theta_t = \theta_{t-1}, t = 2, \ldots, k, \tag{5.1.51}$$

for some $\theta_t \geq 0$, $t = 1,\ldots,k$. *Then, for any* $k \geq 1$ *and any given* $z \in Z$, *we have*

$$\begin{aligned}\Sigma_{t=1}^k \theta_t \mathbb{E}[Q((x^t,\hat{y}^t),z)] &\leq \eta_1\theta_1 V(x^0,x) - (\mu+\eta_k)\theta_k\mathbb{E}[V(x^k,x)]\\ &\quad + \Sigma_{i=1}^m \theta_1\left(p_i^{-1}(1+\tau_1)-1\right)W_i(y_i^0,y_i).\end{aligned} \tag{5.1.52}$$

Proof. Multiplying both sides of (5.1.42) by θ_t and summing up the resulting inequalities, we have

$$\begin{aligned}\mathbb{E}[\Sigma_{t=1}^k \theta_t Q((x^t,\hat{y}^t),z)] &\leq \mathbb{E}\left[\Sigma_{t=1}^k \theta_t\left(\eta_t V(x^{t-1},x) - (\mu+\eta_t)V(x^t,x) - \eta_t V(x^{t-1},x^t)\right)\right]\\ &+ \Sigma_{i=1}^m \mathbb{E}\left\{\Sigma_{t=1}^k \theta_t\left[\left(p_i^{-1}(1+\tau_t)-1\right)W_i(y_i^{t-1},y_i) - p_i^{-1}(1+\tau_t)W_i(y_i^t,y_i)\right]\right\}\\ &+ \mathbb{E}\left[\Sigma_{t=1}^k \theta_t\left(\langle \tilde{x}^t - x^t, U(\tilde{y}^t - y)\rangle - \tau_t p_{i_t}^{-1} W_{i_t}(y_{i_t}^{t-1},y_{i_t}^t)\right)\right],\end{aligned}$$

which, in view of the assumptions in (5.1.47) and (5.1.46), then implies that

$$\begin{aligned}\mathbb{E}[\Sigma_{t=1}^k \theta_t Q((x^t,\hat{y}^t),z)] &\leq \eta_1\theta_1 V(x^0,x) - (\mu+\eta_k)\theta_k\mathbb{E}[V(x^k,x)]\\ &+ \Sigma_{i=1}^m \mathbb{E}\left[\theta_1\left(p_i^{-1}(1+\tau_1)-1\right)W_i(y_i^0,y_i) - p_i^{-1}\theta_k(1+\tau_k)W_i(y_i^k,y_i)\right]\\ &- \mathbb{E}\left[\Sigma_{t=1}^k \theta_t \Delta_t\right],\end{aligned} \tag{5.1.53}$$

where

$$\Delta_t := \eta_t V(x^{t-1},x^t) - \langle \tilde{x}^t - x^t, U(\tilde{y}^t - y)\rangle + \tau_t p_{i_t}^{-1} W_{i_t}(y_{i_t}^{t-1},y_{i_t}^t). \tag{5.1.54}$$

We now provide a bound on $\Sigma_{t=1}^k \theta_t \Delta_t$ in (5.1.53). Note that by (5.1.21), we have

$$\begin{aligned}\langle \tilde{x}^t - x^t, U(\tilde{y}^t - y)\rangle &= \langle x^{t-1} - x^t, U(\tilde{y}^t - y)\rangle - \alpha_t\langle x^{t-2} - x^{t-1}, U(\tilde{y}^t - y)\rangle\\ &= \langle x^{t-1} - x^t, U(\tilde{y}^t - y)\rangle - \alpha_t\langle x^{t-2} - x^{t-1}, U(\tilde{y}^{t-1} - y)\rangle\\ &\quad - \alpha_t\langle x^{t-2} - x^{t-1}, U(\tilde{y}^t - \tilde{y}^{t-1})\rangle\\ &= \langle x^{t-1} - x^t, U(\tilde{y}^t - y)\rangle - \alpha_t\langle x^{t-2} - x^{t-1}, U(\tilde{y}^{t-1} - y)\rangle\\ &\quad - \alpha_t p_{i_t}^{-1}\langle x^{t-2} - x^{t-1}, y_{i_t}^t - y_{i_t}^{t-1}\rangle\\ &\quad - \alpha_t(p_{i_{t-1}}^{-1} - 1)\langle x^{t-2} - x^{t-1}, y_{i_{t-1}}^{t-2} - y_{i_{t-1}}^{t-1}\rangle,\end{aligned} \tag{5.1.55}$$

where the last identity follows from the observation that by (5.1.22) and (5.1.23),

$$\begin{aligned}U(\tilde{y}^t - \tilde{y}^{t-1}) &= \Sigma_{i=1}^m\left\{\left[p_i^{-1}(y_i^t - y_i^{t-1}) + y_i^{t-1}\right] - \left[p_i^{-1}(y_i^{t-1} - y_i^{t-2}) + y_i^{t-2}\right]\right\}\\ &= \Sigma_{i=1}^m\left\{\left[p_i^{-1}y_i^t - (p_i^{-1}-1)y_i^{t-1}\right] - \left[p_i^{-1}y_i^{t-1} - (p_i^{-1}-1)y_i^{t-2}\right]\right\}\\ &= \Sigma_{i=1}^m\left[p_i^{-1}(y_i^t - y_i^{t-1}) + (p_i^{-1}-1)(y_i^{t-2} - y_i^{t-1})\right]\\ &= p_{i_t}^{-1}(y_{i_t}^t - y_{i_t}^{t-1}) + (p_{i_{t-1}}^{-1} - 1)(y_{i_{t-1}}^{t-2} - y_{i_{t-1}}^{t-1}).\end{aligned}$$

Using relation (5.1.55) in the definition of Δ_t in (5.1.54), we have

$$\begin{aligned}\Sigma_{t=1}^k \theta_t \Delta_t = \Sigma_{t=1}^k \theta_t \big[&\eta_t V(x^{t-1}, x^t) \\ &- \langle x^{t-1} - x^t, U(\tilde{y}^t - y)\rangle + \alpha_t \langle x^{t-2} - x^{t-1}, U(\tilde{y}^{t-1} - y)\rangle \\ &+ \alpha_t p_{i_t}^{-1} \langle x^{t-2} - x^{t-1}, y_{i_t}^t - y_{i_t}^{t-1}\rangle \\ &\quad + \alpha_t (p_{i_{t-1}}^{-1} - 1)\langle x^{t-2} - x^{t-1}, y_{i_{t-1}}^{t-2} - y_{i_{t-1}}^{t-1}\rangle \\ &+ p_{i_t}^{-1} \tau_t W_{i_t}(y_{i_t}^{t-1}, y_{i_t}^t)\big]. \end{aligned} \tag{5.1.56}$$

Observe that by (5.1.51) and the fact that $x^{-1} = x^0$,

$$\begin{aligned}&\Sigma_{t=1}^k \theta_t \left[\langle x^{t-1} - x^t, U(\tilde{y}^t - y)\rangle - \alpha_t \langle x^{t-2} - x^{t-1}, U(\tilde{y}^{t-1} - y)\rangle\right] \\ &= \theta_k \langle x^{k-1} - x^k, U(\tilde{y}^k - y)\rangle \\ &= \theta_k \langle x^{k-1} - x^k, U(y^k - y)\rangle + \theta_k \langle x^{k-1} - x^k, U(\tilde{y}^k - y^k)\rangle \\ &= \theta_k \langle x^{k-1} - x^k, U(y^k - y)\rangle + \theta_k (p_{i_k}^{-1} - 1)\langle x^{k-1} - x^k, y_{i_k}^k - y_{i_k}^{k-1}\rangle, \end{aligned}$$

where the last identity follows from the definitions of y^k and $\tilde{y}^k$ in (5.1.22) and (5.1.23), respectively. Also, by the strong convexity of V and W_i, we have

$$V(x^{t-1}, x^t) \ge \tfrac{1}{2}\|x^{t-1} - x^t\|^2 \quad \text{and} \quad W_{i_t}(y_{i_t}^{t-1}, y_{i_t}^t) \ge \tfrac{m}{2L_{i_t}}\|y_{i_t}^{t-1} - y_{i_t}^t\|^2.$$

Using the previous three relations in (5.1.56), we have

$$\begin{aligned}\Sigma_{t=1}^k \theta_t \Delta_t \ge \Sigma_{t=1}^k \theta_t \big[&\tfrac{\eta_t}{2}\|x^{t-1} - x^t\|^2 + \alpha_t p_{i_t}^{-1} \langle x^{t-2} - x^{t-1}, y_{i_t}^t - y_{i_t}^{t-1}\rangle \\ &+ \alpha_t (p_{i_{t-1}}^{-1} - 1)\langle x^{t-2} - x^{t-1}, y_{i_{t-1}}^{t-2} - y_{i_{t-1}}^{t-1}\rangle + \tfrac{m\tau_t}{2L_{i_t} p_{i_t}}\|y_{i_t}^{t-1} - y_{i_t}^t\|^2\big] \\ &- \theta_k \langle x^{k-1} - x^k, U(y^k - y)\rangle - \theta_k (p_{i_k}^{-1} - 1)\langle x^{k-1} - x^k, y_{i_k}^k - y_{i_k}^{k-1}\rangle. \end{aligned}$$

Regrouping the terms in the above relation, and using the fact that $x^{-1} = x^0$, we obtain

$$\begin{aligned}\Sigma_{t=1}^k \theta_t \Delta_t \ge \theta_k &\left[\tfrac{\eta_k}{4}\|x^{k-1} - x^k\|^2 - \langle x^{k-1} - x^k, U(y^k - y)\rangle\right] \\ &+ \theta_k \left[\tfrac{\eta_k}{4}\|x^{k-1} - x^k\|^2 - (p_{i_k}^{-1} - 1)\langle x^{k-1} - x^k, y_{i_k}^k - y_{i_k}^{k-1}\rangle \right.\\ &\qquad \left. + \tfrac{m\tau_k}{4L_{i_k} p_{i_k}}\|y_{i_k}^{k-1} - y_{i_k}^k\|^2\right] \\ &+ \Sigma_{t=2}^k \theta_t \left[\tfrac{\alpha_t}{p_{i_t}}\langle x^{t-2} - x^{t-1}, y_{i_t}^t - y_{i_t}^{t-1}\rangle + \tfrac{m\tau_t}{4L_{i_t} p_{i_t}}\|y_{i_t}^{t-1} - y_{i_t}^t\|^2\right] \\ &+ \Sigma_{t=2}^k \left[\alpha_t \theta_t (p_{i_{t-1}}^{-1} - 1)\langle x^{t-2} - x^{t-1}, y_{i_{t-1}}^{t-2} - y_{i_{t-1}}^{t-1}\rangle \right.\\ &\qquad \left. + \tfrac{m\tau_{t-1}\theta_{t-1}}{4L_{i_{t-1}} p_{i_{t-1}}}\|y_{i_{t-1}}^{t-2} - y_{i_{t-1}}^{t-1}\|^2\right] \\ &+ \Sigma_{t=2}^k \tfrac{\theta_{t-1}\eta_{t-1}}{2}\|x^{t-2} - x^{t-1}\|^2 \end{aligned}$$

$$\begin{aligned}
&\geq \theta_k \left[\tfrac{\eta_k}{4}\|x^{k-1}-x^k\|^2 - \langle x^{k-1}-x^k, U(y^k-y)\rangle\right] \\
&\quad + \theta_k\left(\tfrac{\eta_k}{4} - \tfrac{L_{i_k}(1-p_{i_k})^2}{m\tau_k p_{i_k}}\right)\|x^{k-1}-x^k\|^2 \\
&\quad + \textstyle\sum_{t=2}^k \left[\tfrac{\theta_{t-1}\eta_{t-1}}{2} - \tfrac{L_{i_t}\alpha_t^2\theta_t}{m\tau_t p_{i_t}} - \tfrac{\alpha_t^2\theta_t^2(1-p_{i_{t-1}})^2 L_{i_{t-1}}}{m\tau_{t-1}\theta_{t-1}p_{i_{t-1}}}\right]\|x^{t-2}-x^{t-1}\|^2 \\
&= \theta_k \left[\tfrac{\eta_k}{4}\|x^{k-1}-x^k\|^2 - \langle x^{k-1}-x^k, U(y^k-y)\rangle\right] \\
&\quad + \theta_k\left(\tfrac{\eta_k}{4} - \tfrac{L_{i_k}(1-p_{i_k})^2}{m\tau_k p_{i_k}}\right)\|x^{k-1}-x^k\|^2 \\
&\quad + \textstyle\sum_{t=2}^k \theta_{t-1}\left(\tfrac{\eta_{t-1}}{2} - \tfrac{L_{i_t}\alpha_t}{m\tau_t p_{i_t}} - \tfrac{(1-p_{i_{t-1}})^2 L_{i_{t-1}}}{m\tau_{t-1}p_{i_{t-1}}}\right)\|x^{t-2}-x^{t-1}\|^2 \\
&\geq \theta_k \left[\tfrac{\eta_k}{4}\|x^{k-1}-x^k\|^2 - \langle x^{k-1}-x^k, U(y^k-y)\rangle\right],
\end{aligned} \tag{5.1.57}$$

where the second inequality follows from

$$b\langle u,v\rangle + a\|v\|^2/2 \geq -b^2\|u\|^2/(2a), \forall a>0 \tag{5.1.58}$$

due to the Cauchy-Schwarz inequality and (3.1.6), and the last inequality follows from (5.1.48) and (5.1.49). Plugging the bound (5.1.57) into (5.1.53), we have

$$\begin{aligned}
&\textstyle\sum_{t=1}^k \theta_t \mathbb{E}[Q((x^t,\hat{y}^t),z)] \leq \theta_1\eta_1 V(x^0,x) - \theta_k(\mu+\eta_k)\mathbb{E}[V(x^k,x)] \\
&\quad + \textstyle\sum_{i=1}^m \theta_1\left(p_i^{-1}(1+\tau_1)-1\right) W_i(y_i^0,y_i) \\
&\quad - \theta_k \mathbb{E}\left[\tfrac{\eta_k}{4}\|x^{k-1}-x^k\|^2 - \langle x^{k-1}-x^k, U(y^k-y)\rangle + \textstyle\sum_{i=1}^m p_i^{-1}(1+\tau_k)W_i(y_i^k,y_i)\right].
\end{aligned}$$

Also observe that by (5.1.50) and (5.1.58),

$$\begin{aligned}
&\tfrac{\eta_k}{4}\|x^{k-1}-x^k\|^2 - \langle x^{k-1}-x^k, U(y^k-y)\rangle + \textstyle\sum_{i=1}^m p_i^{-1}(1+\tau_k)W_i(y_i^k,y_i) \\
&\geq \tfrac{\eta_k}{4}\|x^{k-1}-x^k\|^2 + \textstyle\sum_{i=1}^m\left[-\langle x^{k-1}-x^k, y_i^k - y_i\rangle + \tfrac{m(1+\tau_k)}{2L_i p_i}\|y_i^k-y_i\|^2\right] \\
&\geq \left(\tfrac{\eta_k}{4} - \tfrac{\sum_{i=1}^m (p_i L_i)}{2m(1+\tau_k)}\right)\|x^{k-1}-x^k\|^2 \geq 0.
\end{aligned}$$

The result then immediately follows by combining the above two conclusions. ■

5.1.3.3 Main Convergence Results

We are now ready to establish the main convergence properties for the RPDG method applied to strongly convex problems with $\mu > 0$.

Theorem 5.1. *Suppose that* $\{\tau_t\}$, $\{\eta_t\}$, *and* $\{\alpha_t\}$ *in the RPDG method are set to*

$$\tau_t = \tau, \quad \eta_t = \eta, \quad \textit{and} \quad \alpha_t = \alpha, \tag{5.1.59}$$

for any $t \geq 1$ such that

$$(1-\alpha)(1+\tau) \leq p_i, i = 1,\ldots,m, \tag{5.1.60}$$
$$\eta \leq \alpha(\mu+\eta), \tag{5.1.61}$$
$$\eta \tau p_i \geq 4L_i/m, i = 1,\ldots,m, \tag{5.1.62}$$

for some $\alpha \in (0,1)$. Then, for any $k \geq 1$, we have

$$\mathbb{E}[V(x^k,x^*)] \leq \left(1+\tfrac{L_f\alpha}{(1-\alpha)\eta}\right)\alpha^k V(x^0,x^*), \tag{5.1.63}$$
$$\mathbb{E}[\Psi(\bar{x}^k)-\Psi(x^*)] \leq \alpha^{k/2}\left(\alpha^{-1}\eta + \tfrac{3-2\alpha}{1-\alpha}L_f + \tfrac{2L_f^2\alpha}{(1-\alpha)\eta}\right)V(x^0,x^*), \tag{5.1.64}$$

where $\bar{x}^k = (\sum_{t=1}^k \theta_t)^{-1}\sum_{t=1}^k(\theta_t x^t)$ with

$$\theta_t = \tfrac{1}{\alpha^t},\ \forall t = 1,\ldots,k, \tag{5.1.65}$$

and x^* denotes the optimal solution of problem (5.1.1), and the expectation is taken w.r.t. $i_1,\ldots,i_k$.

Proof. It can be easily checked that the conditions in (5.1.46)–(5.1.51) are satisfied with our requirements (5.1.59)–(5.1.62) of $\{\tau_t\}$, $\{\eta_t\}$, $\{\alpha_t\}$, and $\{\theta_t\}$. Using the fact that $Q((x^t,\hat{y}^t),z^*) \geq 0$, we then conclude from (5.1.52) (with $x = x^*$ and $y = y^*$) that, for any $k \geq 1$,

$$\mathbb{E}[V(x^k,x^*)] \leq \tfrac{1}{\theta_k(\mu+\eta)}\left[\theta_1\eta V(x^0,x^*) + \tfrac{\theta_1\alpha}{1-\alpha}W(y^0,y^*)\right] \leq \left(1+\tfrac{L_f\alpha}{(1-\alpha)\eta}\right)\alpha^k V(x^0,x^*),$$

where the first inequality follows from (5.1.59) and (5.1.60), and the second inequality follows from (5.1.61) and (5.1.33).

Let us denote $\bar{y}^k \equiv (\sum_{t=1}^k\theta_t)^{-1}\sum_{t=1}^k(\theta_t\hat{y}^t)$, $\bar{z}^k = (\bar{x}^k,\bar{y}^k)$. In view of (5.2.60), the convexity of $\|\cdot\|$, and (5.1.16), we have

$$\begin{aligned}\mathbb{E}[\Psi(\bar{x}^k)-\Psi(x^*)] &\leq \mathbb{E}[Q(\bar{z}^k,z^*)] + \tfrac{L_f}{2}(\textstyle\sum_{t=1}^k\theta_t)^{-1}\mathbb{E}[\sum_{t=1}^k\theta_t\|x^t-x^*\|^2]\\ &\leq \mathbb{E}[Q(\bar{z}^k,z^*)] + L_f(\textstyle\sum_{t=1}^k\theta_t)^{-1}\mathbb{E}[\sum_{t=1}^k\theta_t V(x^t,x^*)].\end{aligned} \tag{5.1.66}$$

Using (5.1.52) (with $x = x^*$ and $y = y^*$), the fact that $V(x^k,x^*) \geq 0$, and (5.1.65) (or (5.1.65)), we obtain

$$\begin{aligned}\mathbb{E}[Q(\bar{z}^k,z^*)] &\leq \left(\textstyle\sum_{t=1}^k\theta_t\right)^{-1}\sum_{t=1}^k\theta_t\mathbb{E}[Q((x^t,\hat{y}^t),z^*)]\\ &\leq \alpha^k\left(\alpha^{-1}\eta + \tfrac{L_f}{1-\alpha}\right)V(x^0,x^*).\end{aligned}$$

We conclude from (5.1.63) and the definition of $\{\theta_t\}$ that

$$(\textstyle\sum_{t=1}^k\theta_t)^{-1}\mathbb{E}[\sum_{t=1}^k\theta_t V(x^t,x^*)]$$

$$
\begin{aligned}
&= (\textstyle\sum_{t=1}^k \alpha^{-t})^{-1} \sum_{t=1}^k \alpha^{-t}(1 + \frac{L_f \alpha}{(1-\alpha)\eta})\alpha^t V(x^0, x^*)\\
&\le \tfrac{1-\alpha}{\alpha^{-k}-1} \textstyle\sum_{t=1}^k \frac{\alpha^t}{\alpha^{3t/2}}(1 + \frac{L_f \alpha}{(1-\alpha)\eta})V(x^0, x^*)\\
&= \tfrac{1-\alpha}{\alpha^{-k}-1} \tfrac{\alpha^{-k/2}-1}{1-\alpha^{1/2}}(1 + \tfrac{L_f \alpha}{(1-\alpha)\eta})V(x^0, x^*)\\
&= \tfrac{1+\alpha^{1/2}}{1+\alpha^{-k/2}}(1 + \tfrac{L_f \alpha}{(1-\alpha)\eta})V(x^0, x^*)\\
&\le 2\alpha^{k/2}(1 + \tfrac{L_f \alpha}{(1-\alpha)\eta})V(x^0, x^*).
\end{aligned}
$$

Using the above two relations, and (5.1.66), we obtain

$$
\begin{aligned}
\mathbb{E}[\Psi(\bar{x}^k) - \Psi(x^*)] &\le \alpha^k \left(\alpha^{-1}\eta + \tfrac{L_f}{1-\alpha}\right)V(x^0, x^*) + L_f 2\alpha^{k/2}\left(1 + \tfrac{L_f \alpha}{(1-\alpha)\eta}\right)V(x^0, x^*)\\
&\le \alpha^{k/2}\left(\alpha^{-1}\eta + \tfrac{3-2\alpha}{1-\alpha}L_f + \tfrac{2L_f^2 \alpha}{(1-\alpha)\eta}\right)V(x^0, x^*).
\end{aligned}
$$

■

We now provide a few specific selections of p_i, τ, η, and α satisfying (5.1.60)–(5.1.62) and establish the complexity of the RPDG method for computing a stochastic ε-solution of problem (5.1.1), i.e., a point $\bar{x} \in X$ s.t. $\mathbb{E}[V(\bar{x}, x^*)] \le \varepsilon$, as well as a stochastic (ε, λ)-solution of problem (5.1.1), i.e., a point $\bar{x} \in X$ s.t. $\text{Prob}\{V(\bar{x}, x^*) \le \varepsilon\} \ge 1 - \lambda$ for some $\lambda \in (0, 1)$. Moreover, in view of (5.1.64), similar complexity bounds of the RPDG method can be established in terms of the primal optimality gap, i.e., $\mathbb{E}[\Psi(\bar{x}) - \Psi^*]$.

The following corollary shows the convergence of RPDG under a nonuniform distribution for the random variables i_t, $t = 1, \ldots, k$.

Corollary 5.1. *Suppose that $\{i_t\}$ in the RPDG method are distributed over $\{1, \ldots, m\}$ according to*

$$
p_i = \text{Prob}\{i_t = i\} = \tfrac{1}{2m} + \tfrac{L_i}{2mL}, i = 1, \ldots, m. \tag{5.1.67}
$$

Also assume that $\{\tau_t\}$, $\{\eta_t\}$, and $\{\alpha_t\}$ are set to (5.1.59) with

$$
\begin{aligned}
\tau &= \tfrac{\sqrt{(m-1)^2 + 4mC} - (m-1)}{2m},\\
\eta &= \tfrac{\mu\sqrt{(m-1)^2 + 4mC} + \mu(m-1)}{2},\\
\alpha &= 1 - \tfrac{1}{(m+1) + \sqrt{(m-1)^2 + 4mC}},
\end{aligned} \tag{5.1.68}
$$

where

$$
C = \tfrac{8L}{\mu}. \tag{5.1.69}
$$

Then for any $k \ge 1$, we have

$$
\mathbb{E}[V(x^k, x^*)] \le (1 + \tfrac{3L_f}{\mu})\alpha^k V(x^0, x^*), \tag{5.1.70}
$$

$$
\mathbb{E}[\Psi(\bar{x}^k) - \Psi^*] \le \alpha^{k/2}(1-\alpha)^{-1}\left[\mu + 2L_f + \tfrac{L_f^2}{\mu}\right]V(x^0, x^*). \tag{5.1.71}
$$

As a consequence, the number of iterations performed by the RPDG method to find a stochastic ε-solution and a stochastic (ε,λ)-solution of (5.1.1), in terms of the distance to the optimal solution, i.e., $\mathbb{E}[V(x^k,x^*)]$, can be bounded by $K(\varepsilon,C)$ and $K(\lambda\varepsilon,C)$, respectively, where

$$K(\varepsilon,C) := \left[(m+1)+\sqrt{(m-1)^2+4mC}\right] \log\left[(1+\tfrac{3L_f}{\mu})\tfrac{V(x^0,x^*)}{\varepsilon}\right]. \tag{5.1.72}$$

Similarly, the total number of iterations performed by the RPDG method to find a stochastic ε-solution and a stochastic (ε,λ)-solution of (5.1.1), in terms of the primal optimality gap, i.e., $\mathbb{E}[\Psi(\bar{x}^k)-\Psi^*]$, can be bounded by $\tilde{K}(\varepsilon,C)$ and $\tilde{K}(\lambda\varepsilon,C)$, respectively, where

$$\tilde{K}(\varepsilon,C) := 2\left[(m+1)+\sqrt{(m-1)^2+4mC}\right] \log\left[2(\mu+2L_f+\tfrac{L_f^2}{\mu})(m+\sqrt{mC})\tfrac{V(x^0,x^*)}{\varepsilon}\right]. \tag{5.1.73}$$

Proof. It follows from (5.1.68) that

$$(1-\alpha)(1+\tau)=1/(2m)\le p_i,\ \ (1-\alpha)\eta=(\alpha-1/2)\mu\le\alpha\mu, \quad \text{and } \eta\tau p_i=\mu C p_i\ge 4L_i/m,$$

and hence that the conditions in (5.1.60)–(5.1.62) are satisfied. Notice that by the fact that $\alpha\ge 3/4,\ \forall m\ge 1$ and (5.1.68), we have

$$1+\tfrac{L_f\alpha}{(1-\alpha)\eta}=1+L_f\tfrac{\alpha}{(\alpha-1/2)\mu}\le 1+\tfrac{3L_f}{\mu}.$$

Using the above bound in (5.1.63), we obtain (5.1.70). It follows from the facts $(1-\alpha)\eta\le\alpha\mu$, $1/2\le\alpha\le 1,\forall m\ge 1$, and $\eta\ge\mu\sqrt{C}>2\mu$ that

$$\alpha^{-1}\eta+\tfrac{3-2\alpha}{1-\alpha}L_f+\tfrac{2L_f^2\alpha}{(1-\alpha)\eta}\le(1-\alpha)^{-1}(\mu+2L_f+\tfrac{L_f^2}{\mu}).$$

Using the above bound in (5.1.64), we obtain (5.1.71). Denoting $D\equiv(1+\tfrac{3L_f}{\mu})V(x^0,x^*)$, we conclude from (5.1.70) and the fact that $\log x\le x-1$ for any $x\in(0,1)$ that

$$\mathbb{E}[V(x^{K(\varepsilon,C)},x^*)]\le D\alpha^{\frac{\log(D/\varepsilon)}{1-\alpha}}\le D\alpha^{\frac{\log(D/\varepsilon)}{-\log\alpha}}\le D\alpha^{\frac{\log(\varepsilon/D)}{\log\alpha}}=\varepsilon.$$

Moreover, by Markov's inequality, (5.1.70), and the fact that $\log x\le x-1$ for any $x\in(0,1)$, we have

$$\begin{aligned}\text{Prob}\{V(x^{K(\lambda\varepsilon,C)},x^*)>\varepsilon\}&\le\tfrac{1}{\varepsilon}\mathbb{E}[V(x^{K(\lambda\varepsilon,C)},x^*)]\le\tfrac{D}{\varepsilon}\alpha^{\frac{\log(D/(\lambda\varepsilon))}{1-\alpha}}\\&\le\tfrac{D}{\varepsilon}\alpha^{\frac{\log(\lambda\varepsilon/D)}{\log\alpha}}=\lambda.\end{aligned}$$

The proofs for the complexity bounds in terms of the primal optimality gap is similar and hence the details are skipped. ■

The nonuniform distribution in (5.1.67) requires the estimation of the Lipschitz constants L_i, $i = 1,\ldots,m$. In case such information is not available, we can use a uniform distribution for i_t, and as a result, the complexity bounds will depend on a larger condition number given by

$$\max_{i=1,\ldots,m} L_i/\mu.$$

However, if we do have $L_1 = L_2 = \cdots = L_m$, then the results obtained by using a uniform distribution is slightly sharper than the one by using a nonuniform distribution in Corollary 5.1.

Corollary 5.2. *Suppose that $\{i_t\}$ in the RPDG method are uniformly distributed over $\{1,\ldots,m\}$ according to*

$$p_i = \text{Prob}\{i_t = i\} = \tfrac{1}{m}, i = 1,\ldots,m. \tag{5.1.74}$$

Also assume that $\{\tau_t\}$, $\{\eta_t\}$, and $\{\alpha_t\}$ are set to (5.1.59) with

$$\begin{aligned} \tau &= \tfrac{\sqrt{(m-1)^2+4m\bar{C}}-(m-1)}{2m}, \\ \eta &= \tfrac{\mu\sqrt{(m-1)^2+4m\bar{C}}+\mu(m-1)}{2}, \\ \alpha &= 1 - \tfrac{2}{(m+1)+\sqrt{(m-1)^2+4m\bar{C}}}, \end{aligned} \tag{5.1.75}$$

where

$$\bar{C} := \tfrac{4}{\mu} \max_{i=1,\ldots,m} L_i. \tag{5.1.76}$$

Then we have

$$\mathbb{E}[V(x^k,x^*)] \le (1 + \tfrac{L_f}{\mu})\alpha^k V(x^0,x^*), \tag{5.1.77}$$

$$\mathbb{E}[\Psi(\bar{x}^k) - \Psi^*] \le \alpha^{k/2}(1-\alpha)^{-1}\left(\mu + 2L_f + \tfrac{L_f^2}{\mu}\right) V(x^0,x^*). \tag{5.1.78}$$

for any $k \ge 1$. As a consequence, the number of iterations performed by the RPDG method to find a stochastic ε-solution and a stochastic (ε,λ)-solution of (5.1.1), in terms of the distance to the optimal solution, i.e., $\mathbb{E}[V(x^k,x^*)]$, can be bounded by $K_u(\varepsilon,\bar{C})$ and $K_u(\lambda\varepsilon,\bar{C})$, respectively, where

$$K_u(\varepsilon,\bar{C}) := \tfrac{(m+1)+\sqrt{(m-1)^2+4m\bar{C}}}{2} \log\left[(1+\tfrac{L_f}{\mu})\tfrac{V(x^0,x^*)}{\varepsilon}\right].$$

Similarly, the total number of iterations performed by the RPDG method to find a stochastic ε-solution and a stochastic (ε,λ)-solution of (5.1.1), in terms of the primal optimality gap, i.e., $\mathbb{E}[\Psi(\bar{x}^k) - \Psi^*]$, can be bounded by $\tilde{K}(\varepsilon,\bar{C})/2$ and $\tilde{K}(\lambda\varepsilon,\bar{C})/2$, respectively, where $\tilde{K}(\varepsilon,\bar{C})$ is defined in (5.1.73).

Proof. It follows from (5.1.75) that

$$(1-\alpha)(1+\tau)=1/m=p_i,\ \ (1-\alpha)\eta-\alpha\mu=0,\ \text{ and }\ \eta\tau=\mu\bar{C}\geq 4L_i,$$

and hence that the conditions in (5.1.60)–(5.1.62) are satisfied. By the identity $(1-\alpha)\eta=\alpha\mu$, we have

$$1+\tfrac{L_f\alpha}{(1-\alpha)\eta}=1+\tfrac{L_f}{\mu}.$$

Using the above bound in (5.1.63), we obtain (5.1.77). Moreover, note that $\eta\geq\mu\sqrt{\bar{C}}\geq 2\mu$ and $2/3\leq\alpha\leq 1,\forall m\geq 1$ we have

$$\alpha^{-1}\eta+\tfrac{3-2\alpha}{1-\alpha}L_f+\tfrac{2L_f^2\alpha}{(1-\alpha)\eta}\leq(1-\alpha)^{-1}(\mu+2L_f+\tfrac{L_f^2}{\mu}).$$

Using the above bound in (5.1.64), we obtain (5.1.78). The proofs for the complexity bounds are similar to those in Corollary 5.1 and hence the details are skipped. ■

Comparing the complexity bounds obtained from Corollaries 5.1 and 5.2 with those of any optimal deterministic first-order method, they differ in a factor of $\mathcal{O}(\sqrt{mL_f/L})$, whenever $\sqrt{mC}\log(1/\varepsilon)$ is dominating in (5.1.72). Clearly, when L_f and L are in the same order of magnitude, RPDG can save up to $\mathcal{O}(\sqrt{m})$ gradient evaluations for the component function f_i than the deterministic first-order methods. However, it should be pointed out that L_f can be much smaller than L. In particular, when $L_i=L_j,\forall i,j\in\{1,\ldots,m\},L_f=L/m$. In the next subsection, we will construct examples in such extreme cases to obtain the lower complexity bound for general randomized incremental gradient methods.

5.1.4 Lower Complexity Bound for Randomized Methods

Our goal in this subsection is to demonstrate that the complexity bounds obtained in Theorem 5.1, and Corollaries 5.1 and 5.2 for the RPDG method are essentially not improvable. Observe that although there exist rich lower complexity bounds in the literature for deterministic first-order methods, the study on lower complexity bounds for randomized methods are still quite limited.

To derive the performance limit of the incremental gradient methods, we consider a special class of unconstrained and separable strongly convex optimization problems given in the form of

$$\min_{x_i\in\mathbb{R}^{\tilde{n}},i=1,\ldots,m}\left\{\Psi(x):=\textstyle\sum_{i=1}^m\frac{1}{m}\left[f_i(x_i)+\frac{\mu}{2}\|x_i\|_2^2\right]\right\}. \tag{5.1.79}$$

Here $\tilde{n}\equiv n/m\in\{1,2,\ldots\}$ and $\|\cdot\|_2$ denotes standard Euclidean norm. To fix the notation, we also denote $x=(x_1,\ldots,x_m)$. Moreover, we assume that f_i's are quadratic functions given by

$$f_i(x_i) = \tfrac{\mu m(\mathcal{Q}-1)}{4}\left[\tfrac{1}{2}\langle Ax_i, x_i\rangle - \langle e_1, x_i\rangle\right], \tag{5.1.80}$$

where $e_1 := (1,0,\ldots,0)$ and A is a symmetric matrix in $\mathbb{R}^{\tilde{n}\times\tilde{n}}$ given by

$$A = \begin{pmatrix} 2 & -1 & 0 & 0 & \cdots & 0 & 0 & 0 \\ -1 & 2 & -1 & 0 & \cdots & 0 & 0 & 0 \\ 0 & -1 & 2 & -1 & \cdots & 0 & 0 & 0 \\ \cdots & \cdots & \cdots & \cdots & \cdots & \cdots & \cdots & \\ 0 & 0 & 0 & 0 & \cdots & -1 & 2 & -1 \\ 0 & 0 & 0 & 0 & \cdots & 0 & -1 & \kappa \end{pmatrix} \quad \text{with} \quad \kappa = \tfrac{\sqrt{\mathcal{Q}}+3}{\sqrt{\mathcal{Q}}+1}. \tag{5.1.81}$$

Note that the tridiagonal matrix A above consists of a different diagonal element κ. It can be easily checked that $A \succeq 0$ and its maximum eigenvalue does not exceed 4. Indeed, for any $s \equiv (s_1, \ldots, s_{\tilde{n}}) \in \mathbb{R}^{\tilde{n}}$, we have

$$\begin{aligned} \langle As, s\rangle &= s_1^2 + \textstyle\sum_{i=1}^{\tilde{n}-1}(s_i - s_{i+1})^2 + (\kappa-1)s_{\tilde{n}}^2 \geq 0 \\ \langle As, s\rangle &\leq s_1^2 + \textstyle\sum_{i=1}^{\tilde{n}-1} 2(s_i^2 + s_{i+1}^2) + (\kappa-1)s_{\tilde{n}}^2 \\ &= 3s_1^2 + 4\textstyle\sum_{i=2}^{\tilde{n}-1} s_i^2 + (\kappa+1)s_{\tilde{n}}^2 \leq 4\|s\|_2^2, \end{aligned}$$

where the last inequality follows from the fact that $\kappa \leq 3$. Therefore, for any $\mathcal{Q} > 1$, the component functions f_i in (5.1.80) are convex and their gradients are Lipschitz continuous with constant bounded by $L_i = \mu m(\mathcal{Q}-1)$, $i = 1, \ldots, m$.

The following result provides an explicit expression for the optimal solution of (5.1.79).

Lemma 5.6. *Let q be defined in (5.1.91), $x_{i,j}^*$ is the j-th element of x_i, and define*

$$x_{i,j}^* = q^j, i = 1, \ldots, m; j = 1, \ldots, \tilde{n}. \tag{5.1.82}$$

Then x^* is the unique optimal solution of (5.1.79).

Proof. It can be easily seen that q is the smallest root of the equation

$$q^2 - 2\tfrac{\mathcal{Q}+1}{\mathcal{Q}-1}q + 1 = 0. \tag{5.1.83}$$

Note that x^* satisfies the optimality condition of (5.1.79), i.e.,

$$\left(A + \tfrac{4}{\mathcal{Q}-1}I\right)x_i^* = e_1, \quad i = 1, \ldots, m. \tag{5.1.84}$$

Indeed, we can write the coordinate form of (5.1.84) as

$$2\tfrac{\mathcal{Q}+1}{\mathcal{Q}-1}x_{i,1}^* - x_{i,2}^* = 1, \tag{5.1.85}$$

$$x_{i,j+1}^* - 2\tfrac{\mathcal{Q}+1}{\mathcal{Q}-1}x_{i,j}^* + x_{i,j-1}^* = 0, \; j = 2, 3, \ldots, \tilde{n}-1, \tag{5.1.86}$$

$$-(\kappa + \tfrac{4}{\mathcal{Q}-1})x_{i,\tilde{n}}^* + x_{i,\tilde{n}-1}^* = 0, \tag{5.1.87}$$

where the first two equations follow directly from the definition of x^* and relation (5.1.83), and the last equation is implied by the definitions of κ and x^* in (5.1.81) and (5.1.82), respectively. ∎

We consider a general class of randomized incremental gradient methods which sequentially acquire the gradient of a randomly selected component function f_{i_t} at iteration t. More specifically, we assume that the independent random variables i_t, $t = 1,2,\ldots$, satisfy

$$\mathrm{Prob}\{i_t = i\} = p_i \quad \text{and} \quad \textstyle\sum_{i=1}^m p_i = 1, \;\; p_i \geq 0, i = 1,\ldots,m. \tag{5.1.88}$$

Moreover, we assume that these methods generate a sequence of test points $\{x^k\}$ such that

$$x^k \in x^0 + \mathrm{Lin}\{\nabla f_{i_1}(x^0),\ldots,\nabla f_{i_k}(x^{k-1})\}, \tag{5.1.89}$$

where Lin denotes the linear span.

Theorem 5.2 below describes the performance limit of the above randomized incremental gradient methods for solving (5.1.79). We also need a few technical results to establish the lower complexity bounds.

Lemma 5.7.

(a) For any $x > 1$, we have

$$\log(1 - \tfrac{1}{x}) \geq -\tfrac{1}{x-1}. \tag{5.1.90}$$

(b) Let $\rho, q, \bar{q} \in (0,1)$ be given. If we have

$$\tilde{n} \geq \tfrac{t \log \bar{q} + \log(1-\rho)}{2 \log q},$$

for any $t \geq 0$, then

$$\bar{q}^t - q^{2\tilde{n}} \geq \rho \bar{q}^t (1 - q^{2\tilde{n}}).$$

Proof. We first show part (a). Denote $\phi(x) = \log(1 - \frac{1}{x}) + \frac{1}{x-1}$. It can be easily seen that $\lim_{x \to +\infty} \phi(x) = 0$. Moreover, for any $x > 1$, we have

$$\phi'(x) = \tfrac{1}{x(x-1)} - \tfrac{1}{(x-1)^2} = \tfrac{1}{x-1}\left(\tfrac{1}{x} - \tfrac{1}{x-1}\right) < 0,$$

which implies that ϕ is a strictly decreasing function for $x > 1$. Hence, we must have $\phi(x) > 0$ for any $x > 1$. Part (b) follows from the following simple calculation.

$$\bar{q}^t - q^{2\tilde{n}} - \rho \bar{q}^t (1 - q^{2\tilde{n}}) = (1-\rho)\bar{q}^t - q^{2\tilde{n}} + \rho \bar{q}^t q^{2\tilde{n}} \geq (1-\rho)\bar{q}^t - q^{2\tilde{n}} \geq 0.$$

∎

We are now ready to describe our main results regarding the lower complexity bound.

Theorem 5.2. *Let x^* be the optimal solution of problem (5.1.79) and denote*

$$q := \tfrac{\sqrt{Q}-1}{\sqrt{Q}+1}. \tag{5.1.91}$$

Then the iterates $\{x^k\}$ generated by any randomized incremental gradient method must satisfy

$$\frac{\mathbb{E}[\|x^k-x^*\|_2^2]}{\|x^0-x^*\|_2^2} \geq \frac{1}{2}\exp\left(-\frac{4k\sqrt{Q}}{m(\sqrt{Q}+1)^2-4\sqrt{Q}}\right) \tag{5.1.92}$$

for any

$$n \geq \underline{n}(m,k) \equiv \frac{m\log\left[\left(1-(1-q^2)/m\right)^k/2\right]}{2\log q}. \tag{5.1.93}$$

Proof. Without loss of generality, we may assume that the initial point $x_i^0 = 0$, $i = 1,\ldots,m$. Indeed, the incremental gradient methods described in (5.1.89) are invariant with respect to a simultaneous shift of the decision variables. In other words, the sequence of iterates $\{x^k\}$, which is generated by such a method for minimizing the function $\Psi(x)$ starting from x^0, is just a shift of the sequence generated for minimizing $\bar{\Psi}(x) = \Psi(x+x^0)$ starting from the origin.

Now let k_i, $i = 1,\ldots,m$, denote the number of times that the gradients of the component function f_i are computed from iteration 1 to k. Clearly k_i's are binomial random variables supported on $\{0,1,\ldots,k\}$ such that $\sum_{i=1}^m k_i = k$. Also observe that we must have $x_{i,j}^k = 0$ for any $k \geq 0$ and $k_i + 1 \leq j \leq \tilde{n}$, because each time the gradient ∇f_i is computed, the incremental gradient methods add at most one more nonzero entry to the i-th component of x^k due to the structure of the gradient ∇f_i. Therefore, we have

$$\frac{\|x^k-x^*\|_2^2}{\|x^0-x^*\|_2^2} = \frac{\sum_{i=1}^m \|x_i^k-x_i^*\|_2^2}{\sum_{i=1}^m \|x_i^*\|^2} \geq \frac{\sum_{i=1}^m \sum_{j=k_i+1}^{\tilde{n}} (x_{i,j}^*)^2}{\sum_{i=1}^m \sum_{j=1}^{\tilde{n}} (x_{i,j}^*)^2} = \frac{\sum_{i=1}^m (q^{2k_i}-q^{2\tilde{n}})}{m(1-q^{2\tilde{n}})}. \tag{5.1.94}$$

Observing that for any $i = 1,\ldots,m$,

$$\mathbb{E}[q^{2k_i}] = \sum_{t=0}^k \left[q^{2t}\binom{k}{t} p_i^t (1-p_i)^{k-t}\right] = [1-(1-q^2)p_i]^k,$$

we then conclude from (5.1.94) that

$$\frac{\mathbb{E}[\|x^k-x^*\|_2^2]}{\|x^0-x^*\|_2^2} \geq \frac{\sum_{i=1}^m [1-(1-q^2)p_i]^k - mq^{2\tilde{n}}}{m(1-q^{2\tilde{n}})}.$$

Noting that $[1-(1-q^2)p_i]^k$ is convex w.r.t. p_i for any $p_i \in [0,1]$ and $k \geq 1$, by minimizing the RHS of the above bound w.r.t. p_i, $i = 1,\ldots,m$, subject to $\sum_{i=1}^m p_i = 1$ and $p_i \geq 0$, we conclude that

$$\frac{\mathbb{E}[\|x^k-x^*\|_2^2]}{\|x^0-x^*\|_2^2} \geq \frac{[1-(1-q^2)/m]^k - q^{2\tilde{n}}}{1-q^{2\tilde{n}}} \geq \frac{1}{2}[1-(1-q^2)/m]^k, \tag{5.1.95}$$

for any $n \geq \underline{n}(m,k)$ (see (5.1.93)) and possible selection of p_i, $i = 1,\ldots,m$ satisfying (5.1.88), where the last inequality follows from Lemma 5.7(b). Noting that

$$1-(1-q^2)/m = 1-\left[1-\left(\tfrac{\sqrt{Q}-1}{\sqrt{Q}+1}\right)^2\right]\tfrac{1}{m} = 1-\tfrac{1}{m}+\tfrac{1}{m}\left(1-\tfrac{2}{\sqrt{Q}+1}\right)^2$$
$$= 1-\tfrac{4}{m(\sqrt{Q}+1)}+\tfrac{4}{m(\sqrt{Q}+1)^2} = 1-\tfrac{4\sqrt{Q}}{m(\sqrt{Q}+1)^2},$$

we then conclude from (5.1.95) and Lemma 5.7(a) that

$$\tfrac{\mathbb{E}[\|x^k-x^*\|_2^2]}{\|x^0-x^*\|_2^2} \geq \tfrac{1}{2}\left[1-\tfrac{4\sqrt{Q}}{m(\sqrt{Q}+1)^2}\right]^k = \tfrac{1}{2}\exp\left(k\log\left(1-\tfrac{4\sqrt{Q}}{m(\sqrt{Q}+1)^2}\right)\right)$$
$$\geq \tfrac{1}{2}\exp\left(-\tfrac{4k\sqrt{Q}}{m(\sqrt{Q}+1)^2-4\sqrt{Q}}\right).$$

■

As an immediate consequence of Theorem 5.2, we obtain a lower complexity bound for randomized incremental gradient methods.

Corollary 5.3. *The number of gradient evaluations performed by any randomized incremental gradient methods for finding a solution $\bar{x} \in X$ of problem (5.1.1) such that $\mathbb{E}[\|\bar{x}-x^*\|_2^2] \leq \varepsilon$ cannot be smaller than*

$$\Omega\left\{\left(\sqrt{m\mathscr{C}}+m\right)\log\tfrac{\|x^0-x^*\|_2^2}{\varepsilon}\right\}$$

if n is sufficiently large, where $\mathscr{C} = L/\mu$ and $L = \frac{1}{m}\sum_{i=1}^m L_i$.

Proof. It follows from (5.1.92) that the number of iterations k required by any randomized incremental gradient methods to find an approximate solution $\bar{x}$ must satisfy

$$k \geq \left(\tfrac{m(\sqrt{Q}+1)^2}{4\sqrt{Q}}-1\right)\log\tfrac{\|x^0-x^*\|_2^2}{2\varepsilon} \geq \left[\tfrac{m}{2}\left(\tfrac{\sqrt{Q}}{2}+1\right)-1\right]\log\tfrac{\|x^0-x^*\|_2^2}{2\varepsilon}. \quad (5.1.96)$$

Noting that for the worst-case instance in (5.1.79), we have $L_i = \mu m(Q-1)$, $i = 1,\ldots,m$, and hence that $L = \frac{1}{m}\sum_{i=1}^m L_i = m\mu(Q-1)$. Using this relation, we conclude that

$$k \geq \left[\tfrac{1}{2}\left(\tfrac{\sqrt{m\mathscr{C}+m^2}}{2}+m\right)-1\right]\log\tfrac{\|x^0-x^*\|_2^2}{2\varepsilon} =: \underline{k}.$$

The above bound holds when $n \geq \underline{n}(m,\underline{k})$. ■

In view of Theorem 5.2, we can also derive a lower complexity bound for randomized block coordinate descent methods, which update one randomly selected block of variables at each iteration for $\min_{x\in X}\Psi(x)$. Here Ψ is smooth and strongly convex such that

$$\tfrac{\mu_\Psi}{2}\|x-y\|_2^2 \leq \Psi(x)-\Psi(y)-\langle\nabla\Psi(y),x-y\rangle \leq \tfrac{L_\Psi}{2}\|x-y\|_2^2, \forall x,y \in X.$$

Corollary 5.4. *The number of iterations performed by any randomized block coordinate descent methods for finding a solution $\bar{x} \in X$ of $\min_{x\in X}\Psi(x)$ such that $\mathbb{E}[\|\bar{x}-x^*\|_2^2] \leq \varepsilon$ cannot be smaller than*

$$\Omega\left\{\left(m\sqrt{Q_\Psi}\right)\log\frac{\|x^0-x^*\|_2^2}{\varepsilon}\right\}$$

if n is sufficiently large, where $Q_\Psi = L_\Psi/\mu_\Psi$ *denotes the condition number of* Ψ.

Proof. The worst-case instances in (5.1.79) have a block separable structure. Therefore, any randomized incremental gradient methods are equivalent to randomized block coordinate descent methods. The result then immediately follows from (5.1.96). ■

5.1.5 Generalization to Problems Without Strong Convexity

In this section, we generalize the RPDG method for solving a few different types of convex optimization problems which are not necessarily smooth and strongly convex.

5.1.5.1 Smooth Problems with Bounded Feasible Sets

Our goal in this subsection is to generalize RPDG for solving smooth problems without strong convexity (i.e., $\mu = 0$). Different from the accelerated gradient descent method, it is difficult to develop a simple stepsize policy for $\{\tau_t\}$, $\{\eta_t\}$, and $\{\alpha_t\}$ which can guarantee the convergence of this method unless a weaker termination criterion is used. In order to obtain stronger convergence results, we will discuss a different approach obtained by applying the RPDG method to a slightly perturbed problem of (5.1.1).

In order to apply this perturbation approach, we will assume that X is bounded (see Sect. 5.1.5.3 for possible extensions), i.e., given $x_0 \in X$, $\exists D_X \geq 0$ s.t.

$$\max_{x\in X} V(x_0,x) \leq D_X^2. \tag{5.1.97}$$

Now we define the perturbation problem as

$$\Psi_\delta^* := \min_{x\in X}\{\Psi_\delta(x) := f(x) + h(x) + \delta V(x_0,x)\}, \tag{5.1.98}$$

for some fixed $\delta > 0$. It is well known that an approximate solution of (5.1.98) will also be an approximate solution of (5.1.1) if δ is sufficiently small. More specifically, it is easy to verify that

$$\Psi^* \leq \Psi_\delta^* \leq \Psi^* + \delta D_X^2, \tag{5.1.99}$$

$$\Psi(x) \leq \Psi_\delta(x) \leq \Psi(x) + \delta D_X^2, \ \ \forall x \in X. \tag{5.1.100}$$

The following result describes the complexity associated with this perturbation approach for solving smooth problems without strong convexity (i.e., $\mu = 0$).

Proposition 5.2. *Let us apply the RPDG method with the parameter settings in Corollary 5.1 to the perturbation problem (5.1.98) with*

$$\delta = \frac{\varepsilon}{2D_X^2}, \tag{5.1.101}$$

for some $\varepsilon > 0$. Then we can find a solution $\bar{x} \in X$ s.t. $\mathbb{E}[\Psi(\bar{x}) - \Psi^*] \le \varepsilon$ in at most

$$\mathscr{O}\left\{\left(m + \sqrt{\frac{mLD_X^2}{\varepsilon}}\right) \log \frac{mL_f D_X}{\varepsilon}\right\} \tag{5.1.102}$$

iterations. Moreover, we can find a solution $\bar{x} \in X$ s.t. $\mathrm{Prob}\{\Psi(\bar{x}) - \Psi^* > \varepsilon\} \le \lambda$ for any $\lambda \in (0,1)$ in at most

$$\mathscr{O}\left\{\left(m + \sqrt{\frac{mLD_X^2}{\varepsilon}}\right) \log \frac{mL_f D_X}{\lambda\varepsilon}\right\} \tag{5.1.103}$$

iterations.

Proof. Let x_δ^* be the optimal solution of (5.1.98). Denote $C := 16LD_X^2/\varepsilon$ and

$$K := 2\left[(m+1) + \sqrt{(m-1)^2 + 4mC}\right] \log\left[(m + \sqrt{mC})(\delta + 2L_f + \tfrac{L_f^2}{\delta})\tfrac{4D_X^2}{\varepsilon}\right].$$

It can be easily seen that

$$\Psi(\bar{x}^K) - \Psi^* \le \Psi_\delta(\bar{x}^K) - \Psi_\delta^* + \delta D_X^2 = \Psi_\delta(\bar{x}^K) - \Psi_\delta^* + \tfrac{\varepsilon}{2}.$$

Note that problem (5.1.98) is given in the form of (5.1.1) with the strongly convex modulus $\mu = \delta$, and $h(x) = h(x) - \delta\langle \nu'(x_0), x\rangle$. Hence by applying Corollary 5.1, we have

$$\mathbb{E}[\Psi_\delta(\bar{x}^K) - \Psi_\delta^*] \le \tfrac{\varepsilon}{2}.$$

Combining these two inequalities, we have $\mathbb{E}[\Psi(\bar{x}^K) - \Psi^*] \le \varepsilon$, which implies the bound in (5.1.102). The bound in (5.1.103) can be shown similarly and hence the details are skipped. ■

Observe that if we apply a deterministic optimal first-order method (e.g., Nesterov's method or the PDG method), the total number of gradient evaluations for ∇f_i, $i = 1, \ldots, m$, would be given by

$$m\sqrt{\frac{L_f D_X^2}{\varepsilon}}.$$

Comparing this bound with (5.1.102), we can see that the number of gradient evaluations performed by the RPDG method can be $\mathscr{O}\left(\sqrt{m}\log^{-1}(mL_f D_X/\varepsilon)\right)$ times smaller than these deterministic methods when L and L_f are in the same order of magnitude.

5.1.5.2 Structured Nonsmooth Problems

In this subsection, we assume that the component functions f_i are nonsmooth but can be approximated closely by smooth ones. More specifically, we assume that

$$f_i(x) := \max_{y_i \in Y_i} \langle A_i x, y_i \rangle - q_i(y_i). \tag{5.1.104}$$

One can approximate $f_i(x)$ and f, respectively, by

$$\tilde{f}_i(x,\delta) := \max_{y_i \in Y_i} \langle A_i x, y_i \rangle - q_i(y_i) - \delta w_i(y_i) \text{ and } \tilde{f}(x,\delta) = \tfrac{1}{m}\textstyle\sum_{i=1}^m \tilde{f}_i(x,\delta), \tag{5.1.105}$$

where $w_i(y_i)$ is a strongly convex function with modulus 1 such that

$$0 \le w_i(y_i) \le D_{Y_i}^2, \quad \forall y_i \in Y_i. \tag{5.1.106}$$

In particular, we can easily show that

$$\tilde{f}_i(x,\delta) \le f_i(x) \le \tilde{f}_i(x,\delta) + \delta D_{Y_i}^2 \text{ and } \tilde{f}(x,\delta) \le f(x) \le \tilde{f}(x,\delta) + \delta D_Y^2, \tag{5.1.107}$$

for any $x \in X$, where $D_Y^2 = \frac{1}{m}\sum_{i=1}^m D_{Y_i}^2$. Moreover, $f_i(\cdot,\delta)$ and $f(\cdot,\delta)$ are continuously differentiable and their gradients are Lipschitz continuous with constants given by

$$\tilde{L}_i = \frac{\|A_i\|^2}{\delta} \text{ and } \tilde{L} = \frac{\sum_{i=1}^m \|A_i\|^2}{m\delta} = \frac{\|A\|^2}{m\delta}, \tag{5.1.108}$$

respectively. As a consequence, we can apply the RPDG method to solve the approximation problem

$$\tilde{\Psi}_\delta^* := \min_{x\in X} \{\tilde{\Psi}_\delta(x) := \tilde{f}(x,\delta) + h(x) + \mu\nu(x)\}. \tag{5.1.109}$$

The following result provides complexity bounds of the RPDG method for solving the above structured nonsmooth problems for the case when $\mu > 0$.

Proposition 5.3. *Let us apply the RPDG method with the parameter settings in Corollary 5.1 to the approximation problem (5.1.109) with*

$$\delta = \frac{\varepsilon}{2D_Y^2}, \tag{5.1.110}$$

for some $\varepsilon > 0$. Then we can find a solution $\bar{x} \in X$ s.t. $\mathbb{E}[\Psi(\bar{x}) - \Psi^*] \le \varepsilon$ in at most

$$\mathscr{O}\left\{\|A\| D_Y \sqrt{\frac{m}{\mu\varepsilon}} \log \frac{\|A\| D_X D_Y}{m\mu\varepsilon}\right\} \tag{5.1.111}$$

iterations. Moreover, we can find a solution $\bar{x} \in X$ s.t. $\text{Prob}\{\Psi(\bar{x}) - \Psi^* > \varepsilon\} \le \lambda$ for any $\lambda \in (0,1)$ in at most

$$\mathscr{O}\left\{\|A\| D_Y \sqrt{\frac{m}{\mu\varepsilon}} \log \frac{\|A\| D_X D_Y}{\lambda m\mu\varepsilon}\right\} \tag{5.1.112}$$

iterations.

Proof. It follows from (5.1.107) and (5.1.109) that

$$\Psi(\bar{x}^k) - \Psi^* \le \tilde{\Psi}_\delta(\bar{x}^k) - \tilde{\Psi}_\delta^* + \delta D_Y^2 = \tilde{\Psi}_\delta(\bar{x}^k) - \tilde{\Psi}_\delta^* + \tfrac{\varepsilon}{2}. \tag{5.1.113}$$

Using relation (5.1.108) and Corollary 5.1, we conclude that a solution $\bar{x}^k \in X$ satisfying $\mathbb{E}[\tilde{\Psi}_\delta(\bar{x}^k) - \tilde{\Psi}_\delta^*] \le \varepsilon/2$ can be found in

$$\mathcal{O}\left\{\|A\|D_Y\sqrt{\tfrac{m}{\mu\varepsilon}}\log\left[(m+\sqrt{\tfrac{m\tilde{L}}{\mu}})\left(\mu+2\tilde{L}+\tfrac{\tilde{L}^2}{\mu}\right)\tfrac{D_X^2}{\varepsilon}\right]\right\}$$

iterations. This observation together with (5.1.113) and the definition of $\tilde{L}$ in (5.1.108) then imply the bound in (5.1.111). The bound in (5.1.112) follows similarly from (5.1.113) and Corollary 5.1, and hence the details are skipped. ■

The following result holds for the RPDG method applied to the above structured nonsmooth problems when $\mu = 0$.

Proposition 5.4. *Let us apply the RPDG method with the parameter settings in Corollary 5.1 to the approximation problem (5.1.109) with δ in (5.1.110) for some $\varepsilon > 0$. Then we can find a solution $\bar{x} \in X$ s.t. $\mathbb{E}[\Psi(\bar{x}) - \Psi^*] \le \varepsilon$ in at most*

$$\mathcal{O}\left\{\tfrac{\sqrt{m}\|A\|D_X D_Y}{\varepsilon}\log\tfrac{\|A\|D_X D_Y}{m\varepsilon}\right\}$$

iterations. Moreover, we can find a solution $\bar{x} \in X$ s.t. $\text{Prob}\{\Psi(\bar{x}) - \Psi^* > \varepsilon\} \le \lambda$ *for any $\lambda \in (0,1)$ in at most*

$$\mathcal{O}\left\{\tfrac{\sqrt{m}\|A\|D_X D_Y}{\varepsilon}\log\tfrac{m\|A\|D_X D_Y}{\lambda m\varepsilon}\right\}$$

iterations.

Proof. Similar to the arguments used in the proof of Proposition 5.3, our results follow from (5.1.113), and an application of Proposition 5.2 to problem (5.1.109). ■

By Propositions 5.3 and 5.4, the total number of gradient computations for $\tilde{f}(\cdot,\delta)$ performed by the RPDG method, after disregarding the logarithmic factors, can be $\mathcal{O}(\sqrt{m})$ times smaller than those required by deterministic first-order methods.

5.1.5.3 Unconstrained Smooth Problems

In this subsection, we set $X = \mathbb{R}^n$, $h(x) = 0$, and $\mu = 0$ in (5.1.1) and consider the basic convex programming problem of

$$f^* := \min_{x\in\mathbb{R}^n}\left\{f(x) := \tfrac{1}{m}\Sigma_{i=1}^m f_i(x)\right\}. \tag{5.1.114}$$

We assume that the set of optimal solutions X^* of this problem is nonempty.

We will still use the perturbation-based approach as described in Sect. 5.1.5.1 by solving the perturbation problem given by

$$f_\delta^* := \min_{x \in \mathbb{R}^n} \left\{ f_\delta(x) := f(x) + \tfrac{\delta}{2}\|x - x^0\|_2^2, \right\} \tag{5.1.115}$$

for some $x^0 \in X, \delta > 0$, where $\|\cdot\|_2$ denotes the Euclidean norm. Also let L_δ denote the Lipschitz constant for $f_\delta(x)$. Clearly, $L_\delta = L + \delta$. Since the problem is unconstrained and the information on the size of the optimal solution is unavailable, it is hard to estimate the total number of iterations by using the absolute accuracy in terms of $\mathbb{E}[f(\bar{x}) - f^*]$. Instead, we define the relative accuracy associated with a given $\bar{x} \in X$ by

$$R_{ac}(\bar{x}, x^0, f^*) := \tfrac{2[f(\bar{x}) - f^*]}{L(1 + \min_{u \in X^*} \|x^0 - u\|_2^2)}. \tag{5.1.116}$$

We are now ready to establish the complexity of the RPDG method applied to (5.1.114) in terms of $R_{ac}(\bar{x}, x^0, f^*)$.

Proposition 5.5. *Let us apply the RPDG method with the parameter settings in Corollary 5.1 to the perturbation problem (5.1.115) with*

$$\delta = \tfrac{L\varepsilon}{2}, \tag{5.1.117}$$

for some $\varepsilon > 0$. Then we can find a solution $\bar{x} \in X$ s.t. $\mathbb{E}[R_{ac}(\bar{x}, x^0, f^*)] \le \varepsilon$ in at most

$$\mathcal{O}\left\{\sqrt{\tfrac{m}{\varepsilon}} \log \tfrac{m}{\varepsilon}\right\} \tag{5.1.118}$$

iterations. Moreover, we can find a solution $\bar{x} \in X$ s.t. $\mathrm{Prob}\{R_{ac}(\bar{x}, x^0, f^*) > \varepsilon\} \le \lambda$ *for any $\lambda \in (0,1)$ in at most*

$$\mathcal{O}\left\{\sqrt{\tfrac{m}{\varepsilon}} \log \tfrac{m}{\lambda\varepsilon}\right\} \tag{5.1.119}$$

iterations.

Proof. Let x_δ^* be the optimal solution of (5.1.115). Also let x^* be the optimal solution of (5.1.114) that is closest to x^0, i.e., $x^* = \mathrm{argmin}_{u \in X^*} \|x^0 - u\|_2$. It then follows from the strong convexity of f_δ that

$$\begin{aligned}
\tfrac{\delta}{2}\|x_\delta^* - x^*\|_2^2 &\le f_\delta(x^*) - f_\delta(x_\delta^*) \\
&= f(x^*) + \tfrac{\delta}{2}\|x^* - x^0\|_2^2 - f_\delta(x_\delta^*) \\
&\le \tfrac{\delta}{2}\|x^* - x^0\|_2^2,
\end{aligned}$$

which implies that

$$\|x_\delta^* - x^*\|_2 \le \|x^* - x^0\|_2. \tag{5.1.120}$$

Moreover, using the definition of f_δ and the fact that x^* is feasible to (5.1.115), we have

$$f^* \le f_\delta^* \le f^* + \tfrac{\delta}{2}\|x^* - x^0\|_2^2,$$

which implies that

$$
\begin{aligned}
f(\bar{x}^K) - f^* &\le f_\delta(\bar{x}^K) - f_\delta^* + f_\delta^* - f^* \\
&\le f_\delta(\bar{x}^K) - f_\delta^* + \tfrac{\delta}{2}\|x^* - x^0\|_2^2.
\end{aligned}
$$

Now suppose that we run the RPDG method applied to (5.1.115) for K iterations. Then by Corollary 5.1, we have

$$
\begin{aligned}
\mathbb{E}[f_\delta(\bar{x}^K) - f_\delta^*] &\le \alpha^{K/2}(1-\alpha)^{-1}\left(\delta + 2L_\delta + \tfrac{L_\delta^2}{\delta}\right)\|x^0 - x_\delta^*\|_2^2 \\
&\le \alpha^{K/2}(1-\alpha)^{-1}\left(\delta + 2L_\delta + \tfrac{L_\delta^2}{\delta}\right)[\|x^0 - x^*\|_2^2 + \|x^* - x_\delta^*\|_2^2] \\
&= 2\alpha^{K/2}(1-\alpha)^{-1}\left(3\delta + 2L + \tfrac{(L+\delta)^2}{\delta}\right)\|x^0 - x^*\|_2^2,
\end{aligned}
$$

where the last inequality follows from (5.1.120) and α is defined in (5.1.68) with $C = 8L_\delta/\delta = \frac{8(L+\delta)}{\delta} = 8(2/\varepsilon + 1)$. Combining the above two relations, we have

$$
\mathbb{E}[f(\bar{x}^K) - f^*] \le \left[2\alpha^{K/2}(1-\alpha)^{-1}\left(3\delta + 2L + \tfrac{(L+\delta)^2}{\delta}\right) + \tfrac{\delta}{2}\right][\|x^0 - x^*\|_2^2.
$$

Dividing both sides of the above inequality by $L(1 + \|x^0 - x^*\|_2^2)/2$, we obtain

$$
\begin{aligned}
\mathbb{E}[R_{ac}(\bar{x}^K, x^0, f^*)] &\le \tfrac{2}{L}\left[2\alpha^{K/2}(1-\alpha)^{-1}\left(3\delta + 2L + \tfrac{(L+\delta)^2}{\delta}\right) + \tfrac{\delta}{2}\right] \\
&\le 4\left(m + 2\sqrt{2m(\tfrac{2}{\varepsilon}+1)}\right)\left(3\varepsilon + 4 + (2+\varepsilon)(\tfrac{2}{\varepsilon}+1)\right)\alpha^{K/2} + \tfrac{\varepsilon}{2},
\end{aligned}
$$

which clearly implies the bound in (5.1.118). The bound in (5.1.119) also follows from the above inequality and the Markov's inequality. ■

By Proposition 5.5, the total number of gradient evaluations for the component functions f_i required by the RPDG method can be $\mathcal{O}(\sqrt{m}\log^{-1}(m/\varepsilon))$ times smaller than those performed by deterministic optimal first-order methods.

5.2 Random Gradient Extrapolation Method

In the last section, we have introduced a randomized primal–dual gradient (RPDG) method, which can be viewed as a randomized version of the accelerated gradient methods in Sect. 3.3, for solving finite-sum and distributed optimization problems. As discussed earlier, one potential problem associated with RPDG is that it requires a restrictive assumption that each f_i has to be differentiable and has Lipschitz continuous gradients over the whole $\mathbb{R}^n$ due to its primal extrapolation step. Moreover, RPDG has a complicated algorithmic scheme, which contains a primal extrapolation step and a gradient (dual) prediction step in addition to solving a primal proximal subproblem, and thus leads to an intricate primal–dual convergence analysis. Our goal in this section is to address these issues by presenting a novel randomized first-order method, namely randomized gradient extrapolation method (RGEM). Before discussing RGEM, we first need to introduce the gradient extrapolation method, a new optimal first-order method inspired by the game interpretation of accelerated gradient descent method.

More specifically, we consider the finite-sum convex programming problem given in the form of

$$\psi^* := \min_{x \in X} \left\{ \psi(x) := \tfrac{1}{m} \Sigma_{i=1}^m f_i(x) + \mu \nu(x) \right\}. \tag{5.2.1}$$

Here, $X \subseteq \mathbb{R}^n$ is a closed convex set, $f_i : X \to \mathbb{R}, \ \ i = 1, \ldots, m$, are smooth convex functions with Lipschitz continuous gradients over X, i.e., $\exists L_i \geq 0$ such that

$$\|\nabla f_i(x_1) - \nabla f_i(x_2)\|_* \leq L_i \|x_1 - x_2\|, \ \ \forall x_1, x_2 \in X, \tag{5.2.2}$$

$\nu : X \to \mathbb{R}$ is a strongly convex function with modulus 1 w.r.t. a norm $\|\cdot\|$, i.e.,

$$\nu(x_1) - \nu(x_2) - \langle \nu'(x_2), x_1 - x_2 \rangle \geq \tfrac{1}{2}\|x_1 - x_2\|^2, \ \ \forall x_1, x_2 \in X, \tag{5.2.3}$$

where $\nu'(\cdot)$ denotes any subgradient (or gradient) of $\nu(\cdot)$ and $\mu \geq 0$ is a given constant. Hence, the objective function ψ is strongly convex whenever $\mu > 0$. For notational convenience, we also denote $f(x) \equiv \frac{1}{m}\Sigma_{i=1}^m f_i(x)$, $L \equiv \frac{1}{m}\Sigma_{i=1}^m L_i$, and $\hat{L} = \max_{i=1,\ldots,m} L_i$. It is easy to see that for some $L_f \geq 0$,

$$\|\nabla f(x_1) - \nabla f(x_2)\|_* \leq L_f \|x_1 - x_2\| \leq L\|x_1 - x_2\|, \ \ \forall x_1, x_2 \in X. \tag{5.2.4}$$

Observe that problem (5.2.1) is slightly simpler than the finite-sum optimization problem (5.1.1) in the previous section since the convex term h does not appear in (5.2.1). However, it is relatively easy to extend our discussions here in this section to solve the more general finite-sum optimization problem in (5.1.1).

We also consider a class of stochastic finite-sum optimization problems given by

$$\psi^* := \min_{x \in X} \left\{ \psi(x) := \tfrac{1}{m}\Sigma_{i=1}^m \mathbb{E}_{\xi_i}[F_i(x, \xi_i)] + \mu \nu(x) \right\}, \tag{5.2.5}$$

where ξ_i's are random variables with support $\Xi_i \subseteq \mathbb{R}^d$. It can be easily seen that (5.2.5) is a special case of (5.2.1) with $f_i(x) = \mathbb{E}_{\xi_i}[F_i(x, \xi_i)]$, $i = 1, \ldots, m$. However, different from deterministic finite-sum optimization problems, only noisy gradient information of each component function f_i can be accessed for the stochastic finite-sum optimization problem in (5.2.5). By considering the stochastic finite-sum optimization problem, we are interested in not only the deterministic empirical risk minimization, but also the generalization risk for distributed machine learning which allows the private data for each agent to be collected in an online (steaming) fashion. In the distributed learning example given in (5.1.6), for minimization of the generalized risk, f_i's are given in the form of expectation, i.e.,

$$f_i(x) = l_i(x) := \mathbb{E}_{\xi_i}[\log(1 + \exp(-\xi_i^T x))], \ i = 1, \ldots, m,$$

where the random variable ξ_i models the underlying distribution for training dataset of agent i. It should be noted, however, that little attention in the study of randomized first-order methods has been paid to the stochastic finite-sum problem in (5.2.5). We present in this section a new stochastic optimization algorithm which requires only a small number of communication rounds (i.e., $\mathcal{O}(\log 1/\varepsilon)$), but can achieve the optimal $\mathcal{O}(1/\varepsilon)$ sampling complexity for solving the stochastic finite-sum problem in (5.2.5).

5.2.1 Gradient Extrapolation Method

Our goal in this section is to introduce a new algorithmic framework, referred to as the gradient extrapolation method (GEM), for solving the convex optimization problem given by

$$\psi^* := \min_{x\in X} \{\psi(x) := f(x) + \mu\nu(x)\}. \tag{5.2.6}$$

We show that GEM can be viewed as a dual of the accelerated gradient descent method although these two algorithms appear to be quite different. Moreover, GEM possesses some nice properties which enable us to develop and analyze the random gradient extrapolation method for distributed and stochastic optimization.

5.2.1.1 The Algorithm

Similar to Sect. 5.1, we define a *prox-function* associated with ν as

$$V(x^0,x) \equiv V_\nu(x^0,x) := \nu(x) - \left[\nu(x^0) + \langle \nu'(x^0), x - x^0\rangle\right], \tag{5.2.7}$$

where $\nu'(x^0) \in \partial\nu(x^0)$ is an arbitrary subgradient of ν at x^0. By the strong convexity of ν, we have

$$V(x^0,x) \geq \frac{1}{2}\|x - x^0\|^2, \quad \forall x, x^0 \in X. \tag{5.2.8}$$

It should be pointed out that the prox-function $V(\cdot,\cdot)$ described above is a generalized Bregman distance in the sense that ν is not necessarily differentiable. Throughout this section, we assume that the prox-mapping associated with X and ν, given by

$$\text{argmin}_{x\in X}\left\{\langle g, x\rangle + \mu\nu(x) + \eta V(x^0,x)\right\}, \tag{5.2.9}$$

is easily computable for any $x^0 \in X, g \in \mathbb{R}^n, \mu \geq 0, \eta > 0$. Note that whenever ν is non-differentiable, we need to specify a particular selection of the subgradient ν' before performing the prox-mapping. We assume throughout this section that such a selection of ν' is defined recursively in a way similar to Sect. 5.1 with $h = 0$.

We are now ready to describe the gradient extrapolation method (GEM). As shown in Algorithm 5.3, GEM starts with a gradient extrapolation step (5.2.10) to compute $\tilde{g}^t$ from the two previous gradients g^{t-1} and g^{t-2}. Based on $\tilde{g}^t$, it performs a proximal gradient descent step in (5.2.11) and updates the output solution $\underline{x}^t$. Finally, the gradient at $\underline{x}^t$ is computed for gradient extrapolation in the next iteration.

Algorithm 5.3 An optimal gradient extrapolation method (GEM)

Input: Let $x^0 \in X$, and the nonnegative parameters $\{\alpha_t\}$, $\{\eta_t\}$, and $\{\tau_t\}$ be given.
Set $\underline{x}^0 = x^0$ and $g^{-1} = g^0 = \nabla f(x^0)$.
for $t = 1, 2, \ldots, k$ **do**

$$\tilde{g}^t = \alpha_t(g^{t-1} - g^{t-2}) + g^{t-1}. \tag{5.2.10}$$

$$x^t = \operatorname{argmin}_{x \in X}\{\langle \tilde{g}^t, x\rangle + \mu\nu(x) + \eta_t V(x^{t-1}, x)\}. \tag{5.2.11}$$

$$\underline{x}^t = (x^t + \tau_t \underline{x}^{t-1})/(1+\tau_t). \tag{5.2.12}$$

$$g^t = \nabla f(\underline{x}^t). \tag{5.2.13}$$

end for
Output: $\underline{x}^k$.

We now show that GEM can be viewed as the dual of the accelerated gradient descent (AGD) method. To see such a relationship, we will first rewrite GEM in a primal–dual form. Let us consider the dual space $\mathcal{G}$, where the gradients of f reside, and equip it with the conjugate norm $\|\cdot\|_*$. Let $J_f : \mathcal{G} \to \mathbb{R}$ be the conjugate function of f such that $f(x) := \max_{g \in \mathcal{G}}\{\langle x, g\rangle - J_f(g)\}$. We can reformulate the original problem in (5.2.6) as the following saddle point problem:

$$\psi^* := \min_{x \in X}\left\{\max_{g \in \mathcal{G}}\{\langle x, g\rangle - J_f(g)\} + \mu\,\nu(x)\right\}. \tag{5.2.14}$$

It is clear that J_f is strongly convex with modulus $1/L_f$ w.r.t. $\|\cdot\|_*$. Therefore, we can define its associated dual generalized Bregman distance and dual prox-mappings as

$$W_f(g^0, g) := J_f(g) - [J_f(g^0) + \langle J_f'(g^0), g - g^0\rangle], \tag{5.2.15}$$

$$\operatorname{argmin}_{g \in \mathcal{G}}\{\langle -\tilde{x}, g\rangle + J_f(g) + \tau W_f(g^0, g)\}, \tag{5.2.16}$$

for any $g^0, g \in \mathcal{G}$.

Lemma 3.6 in Sect. 5.1 shows that the computation of the dual prox-mapping associated with W_f is equivalent to the computation of ∇f. Using this result, we can see that the GEM iteration can be written in a primal–dual form. Given $(x^0, g^{-1}, g^0) \in X \times \mathcal{G} \times \mathcal{G}$, it updates (x^t, g^t) by

$$\tilde{g}^t = \alpha_t(g^{t-1} - g^{t-2}) + g^{t-1}, \tag{5.2.17}$$

$$x^t = \operatorname{argmin}_{x \in X}\{\langle \tilde{g}^t, x\rangle + \mu\nu(x) + \eta_t V(x^{t-1}, x)\}, \tag{5.2.18}$$

$$g^t = \operatorname{argmin}_{g \in \mathcal{G}}\{\langle -\tilde{x}^t, g\rangle + J_f(g) + \tau_t W_f(g^{t-1}, g)\}, \tag{5.2.19}$$

with a specific selection of $J_f'(g^{t-1}) = \underline{x}^{t-1}$ in $W_f(g^{t-1}, g)$. Indeed, by denoting $\underline{x}^0 = x^0$, we can easily see from $g^0 = \nabla f(\underline{x}^0)$ that $\underline{x}^0 \in \partial J_f(g^0)$. Now assume that $g^{t-1} = \nabla f(\underline{x}^{t-1})$ and hence that $\underline{x}^{t-1} \in \partial J_f(g^{t-1})$. By the definition of g^t in (5.2.19) and Lemma 3.6, we conclude that $g^t = \nabla f(\underline{x}^t)$ with $\underline{x}^t = (x^t + \tau_t \underline{x}^{t-1})/(1+\tau_t)$, which are exactly the definitions given in (5.2.12) and (5.2.13).

Recall that in the AGD method given $(x^{t-1}, \bar{x}^{t-1}) \in X \times X$, it updates $(x^t, \bar{x}^t)$ by

$$\underline{x}^t = (1-\lambda_t)\bar{x}^{t-1} + \lambda_t x^{t-1}, \tag{5.2.20}$$

$$g^t = \nabla f(\underline{x}^t), \tag{5.2.21}$$

$$x^t = \operatorname{argmin}_{x \in X} \left\{ \langle g^t, x \rangle + \mu \nu(x) + \eta_t V(x^{t-1}, x) \right\}, \tag{5.2.22}$$

$$\bar{x}^t = (1-\lambda_t)\bar{x}^{t-1} + \lambda_t x^t, \tag{5.2.23}$$

for some $\lambda_t \in [0,1]$. Moreover, in view of the discussions in Sect. 3.4, we can show that (5.2.20)–(5.2.23) can be viewed as a specific instantiation of the following primal–dual updates:

$$\tilde{x}^t = \alpha_t (x^{t-1} - x^{t-2}) + x^{t-1}, \tag{5.2.24}$$

$$g^t = \operatorname{argmin}_{g \in \mathcal{G}} \left\{ \langle -\tilde{x}^t, g \rangle + J_f(g) + \tau_t W_f(g^{t-1}, g) \right\}, \tag{5.2.25}$$

$$x^t = \operatorname{argmin}_{x \in X} \left\{ \langle g^t, x \rangle + \mu \nu(x) + \eta_t V(x^{t-1}, x) \right\}. \tag{5.2.26}$$

Comparing (5.2.17)–(5.2.19) with (5.2.24)–(6.6.13), we can clearly see that GEM is a dual version of AGD, obtained by switching the primal and dual variables in each equation of (5.2.24)–(6.6.13). The major difference exists in that the extrapolation step in GEM is performed in the dual space while the one in AGD is performed in the primal space. In fact, extrapolation in the dual space will help us to greatly simplify and further enhance the randomized primal–dual gradient methods in Sect. 5.1. Another interesting fact is that in GEM, the gradients are computed for the output solutions $\{x^t\}$. On the other hand, the output solutions in the AGD method are given by $\{\bar{x}^t\}$ while the gradients are computed for the extrapolation sequence $\{\underline{x}^t\}$.

5.2.1.2 Convergence of GEM

We set out to establish the convergence properties of the GEM method for solving (5.2.6). Observe that our analysis is carried out completely in the primal space and does not rely on the primal–dual interpretation described in the previous section. This type of analysis technique also differs significantly from that of AGD.

We will need to establish some basic properties for a smooth convex function.

Lemma 5.8. *If $f : X \to \mathbb{R}$ has Lipschitz continuous gradients with constant L, then*

$$\tfrac{1}{2L}\|\nabla f(x) - \nabla f(z)\|_*^2 \le f(x) - f(z) - \langle \nabla f(z), x - z \rangle \ \ \forall x, z \in X.$$

Proof. Denote $\phi(x) = f(x) - f(z) - \langle \nabla f(z), x - z \rangle$. Clearly ϕ also has L-Lipschitz continuous gradients. It is easy to check that $\nabla\phi(z) = 0$, and hence that $\min_x \phi(x) = \phi(z) = 0$, which implies

$$\begin{aligned}
\phi(z) &\le \phi(x - \tfrac{1}{L}\nabla\phi(x)) \\
&= \phi(x) + \int_0^1 \langle \nabla\phi\left(x - \tfrac{\tau}{L}\nabla\phi(x)\right), -\tfrac{1}{L}\nabla\phi(x)\rangle d\tau \\
&= \phi(x) + \langle \nabla\phi(x), -\tfrac{1}{L}\nabla\phi(x)\rangle + \int_0^1 \langle \nabla\phi\left(x - \tfrac{\tau}{L}\nabla\phi(x)\right) - \nabla\phi(x), -\tfrac{1}{L}\nabla\phi(x)\rangle d\tau \\
&\le \phi(x) - \tfrac{1}{L}\|\nabla\phi(x)\|_*^2 + \int_0^1 L\|\tfrac{\tau}{L}\nabla\phi(x)\|_* \, \|\tfrac{1}{L}\nabla\phi(x)\|_* d\tau \\
&= \phi(x) - \tfrac{1}{2L}\|\nabla\phi(x)\|_*^2.
\end{aligned}$$

Therefore, we have $\frac{1}{2L}\|\nabla\phi(x)\|_*^2 \le \phi(x) - \phi(z) = \phi(x)$, and the result follows immediately from this relation. ■

We first establish some general convergence properties for GEM for both smooth convex ($\mu = 0$) and strongly convex cases ($\mu > 0$).

Theorem 5.3. *Suppose that $\{\eta_t\}$, $\{\tau_t\}$, and $\{\alpha_t\}$ in GEM satisfy*

$$\theta_{t-1} = \alpha_t \theta_t, \;\; t = 2, \ldots, k, \tag{5.2.27}$$
$$\theta_t \eta_t \le \theta_{t-1}(\mu + \eta_{t-1}), \;\; t = 2, \ldots, k, \tag{5.2.28}$$
$$\theta_t \tau_t = \theta_{t-1}(1 + \tau_{t-1}), \; t = 2, \ldots, k, \tag{5.2.29}$$
$$\alpha_t L_f \le \tau_{t-1}\eta_t, \;\; t = 2, \ldots, k, \tag{5.2.30}$$
$$2L_f \le \tau_k(\mu + \eta_k), \tag{5.2.31}$$

for some $\theta_t \ge 0$, $t = 1, \ldots, k$. Then, for any $k \ge 1$ and any given $x \in X$, we have

$$\begin{aligned}
&\theta_k(1 + \tau_k)[\psi(\underline{x}^k) - \psi(x)] + \tfrac{\theta_k(\mu + \eta_k)}{2} V(x^k, x) \\
&\quad \le \theta_1 \tau_1 [\psi(x^0) - \psi(x)] + \theta_1 \eta_1 V(x^0, x).
\end{aligned} \tag{5.2.32}$$

Proof. Applying Lemma 3.5 to (5.2.11), we obtain

$$\begin{aligned}
&\langle x^t - x, \alpha_t(g^{t-1} - g^{t-2}) + g^{t-1}\rangle + \mu\nu(x^t) - \mu\nu(x) \\
&\quad \le \eta_t V(x^{t-1}, x) - (\mu + \eta_t)V(x^t, x) - \eta_t V(x^{t-1}, x^t).
\end{aligned} \tag{5.2.33}$$

Moreover, using the definition of ψ, the convexity of f, and the fact that $g^t = \nabla f(\underline{x}^t)$, we have

$$\begin{aligned}
&(1 + \tau_t) f(\underline{x}^t) + \mu\nu(x^t) - \psi(x) \\
&\le (1 + \tau_t) f(\underline{x}^t) + \mu\nu(x^t) - \mu\nu(x) - [f(\underline{x}^t) + \langle g^t, x - \underline{x}^t\rangle] \\
&= \tau_t [f(\underline{x}^t) - \langle g^t, \underline{x}^t - \underline{x}^{t-1}\rangle] - \langle g^t, x - x^t\rangle + \mu\nu(x^t) - \mu\nu(x) \\
&\le -\tfrac{\tau_t}{2L_f}\|g^t - g^{t-1}\|_*^2 + \tau_t f(\underline{x}^{t-1}) - \langle g^t, x - x^t\rangle + \mu\nu(x^t) - \mu\nu(x),
\end{aligned}$$

where the first equality follows from the definition of $\underline{x}^t$ in (5.2.12), and the last inequality follows from Lemma 5.8. In view of (5.2.33), we then have

$$\begin{aligned}&(1+\tau_t)f(\underline{x}^t)+\mu\nu(x^t)-\psi(x)\\&\le -\tfrac{\tau_t}{2L_f}\|g^t-g^{t-1}\|_*^2+\tau_t f(\underline{x}^{t-1})+\langle x^t-x,g^t-g^{t-1}-\alpha_t(g^{t-1}-g^{t-2})\rangle\\&\quad+\eta_t V(x^{t-1},x)-(\mu+\eta_t)V(x^t,x)-\eta_t V(x^{t-1},x^t).\end{aligned}$$

Multiplying both sides of the above inequality by θ_t, and summing up the resulting inequalities from $t=1$ to k, we obtain

$$\begin{aligned}&\textstyle\sum_{t=1}^k\theta_t(1+\tau_t)f(\underline{x}^t)+\sum_{t=1}^k\theta_t[\mu\nu(x^t)-\psi(x)]\\&\le -\textstyle\sum_{t=1}^k\frac{\theta_t\tau_t}{2L_f}\|g^t-g^{t-1}\|_*^2+\sum_{t=1}^k\theta_t\tau_t f(\underline{x}^{t-1})\\&\quad+\textstyle\sum_{t=1}^k\theta_t\langle x^t-x,g^t-g^{t-1}-\alpha_t(g^{t-1}-g^{t-2})\rangle\\&\quad+\textstyle\sum_{t=1}^k\theta_t[\eta_t V(x^{t-1},x)-(\mu+\eta_t)V(x^t,x)-\eta_t V(x^{t-1},x^t)].\end{aligned}\tag{5.2.34}$$

Now by (5.2.27) and the fact that $g^{-1}=g^0$, we have

$$\begin{aligned}&\textstyle\sum_{t=1}^k\theta_t\langle x^t-x,g^t-g^{t-1}-\alpha_t(g^{t-1}-g^{t-2})\rangle\\&=\textstyle\sum_{t=1}^k\theta_t[\langle x^t-x,g^t-g^{t-1}\rangle-\alpha_t\langle x^{t-1}-x,g^{t-1}-g^{t-2}\rangle]\\&\quad-\textstyle\sum_{t=2}^k\theta_t\alpha_t\langle x^t-x^{t-1},g^{t-1}-g^{t-2}\rangle\\&=\theta_k\langle x^k-x,g^k-g^{k-1}\rangle-\textstyle\sum_{t=2}^k\theta_t\alpha_t\langle x^t-x^{t-1},g^{t-1}-g^{t-2}\rangle.\end{aligned}\tag{5.2.35}$$

Moreover, in view of (5.2.28), (5.2.29), and the definition of $\underline{x}^t$ (5.2.12), we obtain

$$\begin{aligned}&\textstyle\sum_{t=1}^k\theta_t[\eta_t V(x^{t-1},x)-(\mu+\eta_t)V(x^t,x)]\\&\overset{(5.2.28)}{\le}\theta_1\eta_1 V(x^0,x)-\theta_k(\mu+\eta_k)V(x^k,x),\end{aligned}\tag{5.2.36}$$

$$\begin{aligned}&\textstyle\sum_{t=1}^k\theta_t[(1+\tau_t)f(\underline{x}^t)-\tau_t f(\underline{x}^{t-1})]\\&\overset{(5.2.29)}{=}\theta_k(1+\tau_k)f(\underline{x}^k)-\theta_1\tau_1 f(\underline{x}^0),\end{aligned}\tag{5.2.37}$$

$$\textstyle\sum_{t=1}^k\theta_t\overset{(5.2.29)}{=}\sum_{t=2}^k[\theta_t\tau_t-\theta_{t-1}\tau_{t-1}]+\theta_k=\theta_k(1+\tau_k)-\theta_1\tau_1,\tag{5.2.38}$$

$$\begin{aligned}&\theta_k(1+\tau_k)\underline{x}^k\\&\overset{(5.2.12)}{=}\theta_k(x^k+\tfrac{\tau_k}{1+\tau_{k-1}}x^{k-1}+\cdots+\textstyle\prod_{t=2}^k\frac{\tau_t}{1+\tau_{t-1}}x^1+\prod_{t=2}^k\frac{\tau_t}{1+\tau_{t-1}}\tau_1x^0)\\&\overset{(5.2.29)}{=}\textstyle\sum_{t=1}^k\theta_t x^t+\theta_1\tau_1x^0.\end{aligned}\tag{5.2.39}$$

The last two relations (cf. (5.2.38) and (5.2.39)), in view of the convexity of $\nu(\cdot)$, also imply that

$$\theta_k(1+\tau_k)\mu\nu(\underline{x}^k)\le\textstyle\sum_{t=1}^k\theta_t\mu\nu(x^t)+\theta_1\tau_1\mu\nu(x^0).$$

Therefore, by (5.2.34)–(5.2.39), and the definition of ψ, we conclude that

$$\begin{aligned}
&\theta_k(1+\tau_k)[\psi(\underline{x}^k)-\psi(x)]\\
&\le \textstyle\sum_{t=2}^k\left[-\frac{\theta_{t-1}\tau_{t-1}}{2L_f}\|g^{t-1}-g^{t-2}\|_*^2-\theta_t\alpha_t\langle x^t-x^{t-1},g^{t-1}-g^{t-2}\rangle-\theta_t\eta_t V(x^{t-1},x^t)\right]\\
&\quad-\theta_k\left[\frac{\tau_k}{2L_f}\|g^k-g^{k-1}\|_*^2-\langle x^k-x,g^k-g^{k-1}\rangle+(\mu+\eta_k)V(x^k,x)\right]+\theta_1\eta_1V(x^0,x)\\
&\quad+\theta_1\tau_1[\psi(x^0)-\psi(x)]-\theta_1\eta_1V(x^0,x^1).
\end{aligned} \tag{5.2.40}$$

By the strong convexity of $V(\cdot,\cdot)$ in (5.2.8), the simple relation in (5.1.58), and the conditions in (5.2.30) and (5.2.31), we have

$$\begin{aligned}
&-\textstyle\sum_{t=2}^k\left[\frac{\theta_{t-1}\tau_{t-1}}{2L_f}\|g^{t-1}-g^{t-2}\|_*^2+\theta_t\alpha_t\langle x^t-x^{t-1},g^{t-1}-g^{t-2}\rangle+\theta_t\eta_t V(x^{t-1},x^t)\right]\\
&\quad\le \textstyle\sum_{t=2}^k\frac{\theta_t}{2}\left(\frac{\alpha_t L_f}{\tau_{t-1}}-\eta_t\right)\|x^{t-1}-x^t\|^2\le 0\\
&-\theta_k\left[\frac{\tau_k}{2L_f}\|g^k-g^{k-1}\|_*^2-\langle x^k-x,g^k-g^{k-1}\rangle+\frac{(\mu+\eta_k)}{2}V(x^k,x)\right]\\
&\quad\le\frac{\theta_k}{2}\left(\frac{L_f}{\tau_k}-\frac{\mu+\eta_k}{2}\right)\|x^k-x\|^2\le 0.
\end{aligned}$$

Using the above relations in (5.2.40), we obtain (8.2). ■

We are now ready to establish the optimal convergence behavior of GEM as a consequence of Theorem 5.3. We first provide a constant step-size policy which guarantees an optimal linear rate of convergence for the strongly convex case ($\mu>0$).

Corollary 5.5. *Let x^* be an optimal solution of (5.2.1), x^k, and $\underline{x}^k$ be defined in (5.2.11) and (5.2.12), respectively. Suppose that $\mu>0$, and that $\{\tau_t\}$, $\{\eta_t\}$, and $\{\alpha_t\}$ are set to*

$$\tau_t\equiv\tau=\sqrt{\tfrac{2L_f}{\mu}},\quad \eta_t\equiv\eta=\sqrt{2L_f\mu},\quad \text{and}\quad \alpha_t\equiv\alpha=\tfrac{\sqrt{2L_f/\mu}}{1+\sqrt{2L_f/\mu}},\ \forall t=1,\ldots,k. \tag{5.2.41}$$

Then,

$$V(x^k,x^*)\le 2\alpha^k[V(x^0,x^*)+\tfrac{1}{\mu}(\psi(x^0)-\psi^*)], \tag{5.2.42}$$

$$\psi(\underline{x}^k)-\psi^*\le\alpha^k\left[\mu V(x^0,x^*)+\psi(x^0)-\psi^*\right]. \tag{5.2.43}$$

Proof. Let us set $\theta_t=\alpha^{-t}$, $t=1,\ldots,k$. It is easy to check that the selection of $\{\tau_t\},\{\eta_t\}$ and $\{\alpha_t\}$ in (5.2.41) satisfies conditions (5.2.27)–(5.2.31). In view of Theorem 5.3 and (5.2.41), we have

$$\begin{aligned}
\psi(\underline{x}^k)-\psi(x^*)+\tfrac{\mu+\eta}{2(1+\tau)}V(x^k,x^*)&\le\tfrac{\theta_1\tau}{\theta_k(1+\tau)}[\psi(x^0)-\psi(x^*)]+\tfrac{\theta_1\eta}{\theta_k(1+\tau)}V(x^0,x^*)\\
&=\alpha^k[\psi(x^0)-\psi(x^*)+\mu V(x^0,x^*)].
\end{aligned}$$

It also follows from the above relation, the fact $\psi(\underline{x}^k) - \psi(x^*) \geq 0$, and (5.2.41) that

$$\begin{aligned} V(x^k, x^*) &\leq \tfrac{2(1+\tau)\alpha^k}{\mu+\eta}[\mu V(x^0, x^*) + \psi(x^0) - \psi(x^*)] \\ &= 2\alpha^k[V(x^0, x^*) + \tfrac{1}{\mu}(\psi(x^0) - \psi(x^*))]. \end{aligned}$$

■

We now provide a stepsize policy which guarantees the optimal rate of convergence for the smooth case ($\mu = 0$). Observe that in smooth case we can estimate the solution quality for the sequence $\{\underline{x}^k\}$ only.

Corollary 5.6. *Let x^* be an optimal solution of* (5.2.1), *and $\underline{x}^k$ be defined in* (5.2.12). *Suppose that $\mu = 0$, and that $\{\tau_t\}$, $\{\eta_t\}$, and $\{\alpha_t\}$ are set to*

$$\tau_t = \tfrac{t}{2}, \quad \eta_t = \tfrac{4L_f}{t}, \quad \text{and} \quad \alpha_t = \tfrac{t}{t+1}, \ \forall t = 1, \ldots, k. \tag{5.2.44}$$

Then,

$$\psi(\underline{x}^k) - \psi(x^*) = f(\underline{x}^k) - f(x^*) \leq \tfrac{2}{(k+1)(k+2)}[f(x^0) - f(x^*) + 8L_f V(x^0, x^*)]. \tag{5.2.45}$$

Proof. Let us set $\theta_t = t + 1$, $t = 1, \ldots, k$. It is easy to check that the parameters in (5.2.44) satisfy conditions (5.2.30)–(5.2.31). In view of (8.2) and (5.2.44), we conclude that

$$\psi(\underline{x}^k) - \psi(x^*) \leq \tfrac{2}{(k+1)(k+2)}[\psi(x^0) - \psi(x^*) + 8L_f V(x^0, x^*)].$$

■

In Corollary 5.7, we improve the above complexity result in terms of the dependence on $f(x^0) - f(x^*)$ by using a different step-size policy and a slightly more involved analysis for the smooth case ($\mu = 0$).

Corollary 5.7. *Let x^* be an optimal solution of* (5.2.1), *x^k, and $\underline{x}^k$ be defined in* (5.2.11) *and* (5.2.12), *respectively. Suppose that $\mu = 0$, and that $\{\tau_t\}$, $\{\eta_t\}$, and $\{\alpha_t\}$ are set to*

$$\tau_t = \tfrac{t-1}{2}, \quad \eta_t = \tfrac{6L_f}{t}, \quad \text{and} \quad \alpha_t = \tfrac{t-1}{t}, \ \forall t = 1, \ldots, k. \tag{5.2.46}$$

Then, for any $k \geq 1$,

$$\psi(\underline{x}^k) - \psi(x^*) = f(\underline{x}^k) - f(x^*) \leq \tfrac{12L_f}{k(k+1)} V(x^0, x^*). \tag{5.2.47}$$

Proof. If we set $\theta_t = t$, $t = 1, \ldots, k$. It is easy to check that the parameters in (5.2.46) satisfy conditions (5.2.27)–(5.2.29) and (5.2.31). However, condition (5.2.30) only holds for $t = 3, \ldots, k$, i.e.,

$$\alpha_t L_f \leq \tau_{t-1}\eta_t, \ t = 3, \ldots, k. \tag{5.2.48}$$

In view of (5.2.40) and the fact that $\tau_1 = 0$, we have

$$
\begin{aligned}
&\theta_k(1+\tau_k)[\psi(\underline{x}^k) - \psi(x)] \\
&\le -\theta_2[\alpha_2\langle x^2 - x^1, g^1 - g^0\rangle + \eta_2 V(x^1,x^2)] - \theta_1\eta_1 V(x^0,x^1) \\
&\quad - \textstyle\sum_{t=3}^k \left[\frac{\theta_{t-1}\tau_{t-1}}{2L_f}\|g^{t-1}-g^{t-2}\|_*^2 + \theta_t\alpha_t\langle x^t - x^{t-1}, g^{t-1}-g^{t-2}\rangle + \theta_t\eta_t V(x^{t-1},x^t)\right] \\
&\quad - \theta_k\left[\frac{\tau_k}{2L_f}\|g^k - g^{k-1}\|_*^2 - \langle x^k - x, g^k - g^{k-1}\rangle + (\mu+\eta_k)V(x^k,x)\right] + \theta_1\eta_1 V(x^0,x) \\
&\le \tfrac{\theta_1\alpha_2}{2\eta_2}\|g^1 - g^0\|_*^2 - \tfrac{\theta_1\eta_1}{2}\|x^1 - x^0\|^2 + \textstyle\sum_{t=3}^k \frac{\theta_t}{2}\left(\frac{\alpha_t L_f}{\tau_{t-1}} - \eta_t\right)\|x^{t-1} - x^t\|^2 \\
&\quad + \tfrac{\theta_k}{2}\left(\tfrac{L_f}{\tau_k} - \tfrac{\eta_k}{2}\right)\|x^k - x\|^2 + \theta_1\eta_1 V(x^0,x) - \tfrac{\theta_k\eta_k}{2}V(x^k,x) \\
&\le \tfrac{\theta_1\alpha_2 L_f^2}{2\eta_2}\|\underline{x}^1 - \underline{x}^0\|^2 - \tfrac{\theta_1\eta_1}{2}\|x^1 - x^0\|^2 + \theta_1\eta_1 V(x^0,x) - \tfrac{\theta_k\eta_k}{2}V(x^k,x) \\
&\le \theta_1\left(\frac{\alpha_2 L_f^2}{2\eta_2} - \eta_1\right)\|x^1 - x^0\|^2 + \theta_1\eta_1 V(x^0,x) - \tfrac{\theta_k\eta_k}{2}V(x^k,x),
\end{aligned}
$$

where the second inequality follows from the simple relation in (5.1.58) and (5.2.8), the third inequality follows from (5.2.48), (5.2.31), the definition of g^t in (5.2.13) and (5.2.4), and the last inequality follows from the facts that $\underline{x}^0 = x^0$ and $\underline{x}^1 = x^1$ (due to $\tau_1 = 0$). Therefore, by plugging the parameter setting in (5.2.46) into the above inequality, we conclude that

$$
\begin{aligned}
\psi(\underline{x}^k) - \psi^* = f(\underline{x}^k) - f(x^*) &\le [\theta_k(1+\tau_k)]^{-1}[\theta_1\eta_1 V(x^0,x^*) - \tfrac{\theta_k\eta_k}{2}V(x^k,x)] \\
&\le \tfrac{12L_f}{k(k+1)}V(x^0,x^*).
\end{aligned}
$$

■

In view of the results obtained in the above three corollaries, GEM exhibits optimal rates of convergence for both strongly convex and smooth cases. Different from the classical AGD method, GEM performs extrapolation on the gradients, rather than the iterates. This fact will help us to develop an enhanced randomized incremental gradient method than RPDG, i.e., the random gradient extrapolation method, with a much simpler analysis.

5.2.2 *Deterministic Finite-Sum Problems*

We present in this subsection a randomized version of GEM and discuss its convergence properties for solving the deterministic finite-sum problem in (5.2.1).

5.2.2.1 Algorithmic Framework

The basic scheme of RGEM is formally stated in Algorithm 5.4. This algorithm simply initializes the gradient as $y_i^{-1} = y_i^0 = \mathbf{0}$, $i = 1, \ldots, m$. At each iteration, RGEM requires the new gradient information of only one randomly selected component function f_i, but maintains m pairs of search points and gradients $(\underline{x}_i^t, y_i^t)$, $i = 1, \ldots, m$, which are stored, possibly by their corresponding agents in the distributed network. More specifically, it first performs a gradient extrapolation step in (5.2.49) and the primal proximal mapping in (5.2.50). Then a randomly selected block $\underline{x}_{i_t}^t$ is updated in (5.2.51) and the corresponding component gradient ∇f_{i_t} is computed in (5.2.52). As can be seen from Algorithm 5.4, RGEM does not require any exact gradient evaluations.

Note that the computation of x^t in (5.2.50) requires an involved computation of $\frac{1}{m}\sum_{i=1}^m \tilde{y}_i^t$. In order to save computational time when implementing this algorithm, we suggest to compute this quantity in a recursive manner as follows. Let us denote $g^t \equiv \frac{1}{m}\sum_{i=1}^m y_i^t$, $t = 1, \ldots, k$. Clearly, in view of the fact that $y_i^t = y_i^{t-1}$, $\forall i \neq i_t$, we have

$$g^t = g^{t-1} + \tfrac{1}{m}(y_{i_t}^t - y_{i_t}^{t-1}). \tag{5.2.54}$$

Also, by the definition of g^t and (5.2.49), we have

$$\tfrac{1}{m}\sum_{i=1}^m \tilde{y}_i^t = \tfrac{1}{m}\sum_{i=1}^m y_i^{t-1} + \tfrac{\alpha_t}{m}(y_{i_{t-1}}^{t-1} - y_{i_{t-1}}^{t-2}) = g^{t-1} + \tfrac{\alpha_t}{m}(y_{i_{t-1}}^{t-1} - y_{i_{t-1}}^{t-2}). \tag{5.2.55}$$

Algorithm 5.4 A random gradient extrapolation method (RGEM)

Input: Let $x^0 \in X$, and the nonnegative parameters $\{\alpha_t\}$, $\{\eta_t\}$, and $\{\tau_t\}$ be given.
Initialization:
Set $\underline{x}_i^0 = x^0$, $y_i^{-1} = y_i^0 = \mathbf{0}$, $i = 1, \ldots, m$.
for $t = 1, \ldots, k$ **do**
Choose i_t according to $\text{Prob}\{i_t = i\} = \frac{1}{m}, i = 1, \ldots, m$.

$$\tilde{y}_i^t = y_i^{t-1} + \alpha_t(y_i^{t-1} - y_i^{t-2}), \forall i, \tag{5.2.49}$$

$$x^t = \text{argmin}_{x \in X}\left\{\langle \tfrac{1}{m}\sum_{i=1}^m \tilde{y}_i^t, x\rangle + \mu\nu(x) + \eta V(x^{t-1}, x)\right\}, \tag{5.2.50}$$

$$\underline{x}_i^t = \begin{cases} (1+\tau_t)^{-1}(x^t + \tau_t \underline{x}_i^{t-1}), & i = i_t, \\ \underline{x}_i^{t-1}, & i \neq i_t. \end{cases} \tag{5.2.51}$$

$$y_i^t = \begin{cases} \nabla f_i(\underline{x}_i^t), & i = i_t, \\ y_i^{t-1}, & i \neq i_t. \end{cases} \tag{5.2.52}$$

end for
Output: For some $\theta_t > 0$, $t = 1, \ldots, k$, set

$$\underline{x}^k := (\textstyle\sum_{t=1}^k \theta_t)^{-1} \sum_{t=1}^k \theta_t x^t. \tag{5.2.53}$$

Using these two ideas mentioned above, we can compute $\frac{1}{m}\sum_{i=1}^{m}\tilde{y}_i^t$ in two steps: (i) initialize $g^0 = \mathbf{0}$, and update g^t as in (5.2.54) after the gradient evaluation step (5.2.52); (ii) replace (5.2.49) by (5.2.55) to compute $\frac{1}{m}\sum_{i=1}^{m}\tilde{y}_i^t$. Also note that the difference $y_{i_t}^t - y_{i_t}^{t-1}$ can be saved as it is used in both (5.2.54) and (5.2.55) for the next iteration. These enhancements will be incorporated into the distributed setting in Sect. 5.2.4 to possibly save communication costs.

It is also interesting to observe the differences between RGEM and RPDG (Sect. 5.1). RGEM has only one extrapolation step (5.2.49) which combines two types of predictions. One is to predict future gradients using historic data, and the other is to obtain an estimator of the current exact gradient of f from the randomly updated gradient information of f_i. However, RPDG method needs two extrapolation steps in both the primal and dual spaces. Due to the existence of the primal extrapolation step, RPDG cannot guarantee the search points where it performs gradient evaluations to fall within the feasible set X. Hence, it requires the assumption that f_i's are differentiable with Lipschitz continuous gradients over $\mathbb{R}^n$. Such a strong assumption is not required by RGEM, since all the primal iterates generated by RGEM stay within the feasible region X. As a result, RGEM can deal with a much wider class of problems than RPDG. Moreover, in RGEM we do not need to compute the exact gradients at the initial point $\underline{x}_i^0$, but simply set them as $y_i^0 = 0$. It can be seen that under the L-smooth assumption on gradients (cf. (5.2.4)), there exists $0 \le \sigma_0 < +\infty$ such that

$$\tfrac{1}{m}\textstyle\sum_{i=1}^{m}\|\nabla f_i(x^0)\|_*^2 = \sigma_0^2. \tag{5.2.56}$$

5.2.2.2 Convergence Analysis

Our main goal in this subsection is to establish the convergence of RGEM for solving problem (5.2.1).

Comparing RGEM in Algorithm 5.4 with GEM in Algorithm 5.3, we can see that RGEM is a direct randomization of GEM. Therefore, inheriting from GEM, its convergence analysis is carried out completely in the primal space. However, the analysis for RGEM is more challenging especially because we need to (1) build up the relationship between $\frac{1}{m}\sum_{i=1}^{m}f_i(\underline{x}_i^k)$ and $f(\underline{x}^k)$, for which we exploit the function Q defined in (6.6.26) as an intermediate tool; (2) bound the error caused by inexact gradients at the initial point; and (3) analyze the accumulated error caused by randomization and noisy stochastic gradients.

Before establishing the convergence of RGEM, we first provide some important technical results. Let $\hat{\underline{x}}_i^t$ and $\hat{y}_i^t, i = 1,\ldots,m, t \ge 1$ be defined as

$$\hat{\underline{x}}_i^t = (1+\tau_t)^{-1}(x^t + \tau_t \underline{x}_i^{t-1}), \tag{5.2.57}$$

$$\hat{y}_i^t = \nabla f_i(\hat{\underline{x}}_i^t). \tag{5.2.58}$$

The following simple result demonstrates a few identities related to $\underline{x}_i^t$ (cf. (5.2.51)) and y_i^t (cf. (5.2.52)).

Lemma 5.9. *Let x^t and y_i^t be defined in (5.2.50) and (5.2.52), respectively, and $\hat{\underline{x}}_i^t$ and $\hat{y}_i^t$ be defined as in (5.2.57) and (5.2.58), respectively. Then we have, for any $i = 1,\ldots,m$ and $t = 1,\ldots,k$,*

$$
\begin{aligned}
\mathbb{E}_t[y_i^t] &= \tfrac{1}{m}\hat{y}_i^t + (1-\tfrac{1}{m})y_i^{t-1},\\
\mathbb{E}_t[\underline{x}_i^t] &= \tfrac{1}{m}\hat{\underline{x}}_i^t + (1-\tfrac{1}{m})\underline{x}_i^{t-1},\\
\mathbb{E}_t[f_i(\underline{x}_i^t)] &= \tfrac{1}{m}f_i(\hat{\underline{x}}_i^t) + (1-\tfrac{1}{m})f_i(\underline{x}_i^{t-1}),\\
\mathbb{E}_t[\|\nabla f_i(\underline{x}_i^t) - \nabla f_i(\underline{x}_i^{t-1})\|_*^2] &= \tfrac{1}{m}\|\nabla f_i(\hat{\underline{x}}_i^t) - \nabla f_i(\underline{x}_i^{t-1})\|_*^2,
\end{aligned}
$$

where $\mathbb{E}_t$ denotes the conditional expectation w.r.t. i_t given $i_1,\ldots,i_{t-1}$ when y_i^t is defined in (5.2.52).

Proof. This first equality follows immediately from the facts that $\mathrm{Prob}_t\{y_i^t = \hat{y}_i^t\} = \mathrm{Prob}_t\{i_t = i\} = \frac{1}{m}$ and $\mathrm{Prob}_t\{y_i^t = y_i^{t-1}\} = 1 - \frac{1}{m}$. Here Prob_t denotes the conditional probability w.r.t. i_t given $i_1,\ldots,i_{t-1}$. Similarly, we can prove the rest equalities. ■

We define the following function Q to help us analyze the convergence properties of RGEM. Let $\underline{x}, x \in X$ be two feasible solutions of (5.2.1) (or (5.2.5)), we define the corresponding $Q(\underline{x},x)$ by

$$
Q(\underline{x},x) := \langle \nabla f(x), \underline{x} - x\rangle + \mu\nu(\underline{x}) - \mu\nu(x). \tag{5.2.59}
$$

It is obvious that if we fix $x = x^*$, an optimal solution of (5.2.1) (or (5.2.5)), by the convexity of ν and the optimality condition of x^*, for any feasible solution $\underline{x}$, we can conclude that

$$
Q(\underline{x},x^*) \ge \langle \nabla f(x^*) + \mu\nu'(x^*), \underline{x} - x^*\rangle \ge 0.
$$

Moreover, observing that f is smooth, we conclude that

$$
\begin{aligned}
Q(\underline{x},x^*) &= f(x^*) + \langle \nabla f(x^*), \underline{x} - x^*\rangle + \mu\nu(\underline{x}) - \psi(x^*)\\
&\ge -\tfrac{L_f}{2}\|\underline{x} - x^*\|^2 + \psi(\underline{x}) - \psi(x^*).
\end{aligned} \tag{5.2.60}
$$

The following lemma establishes an important relationship regarding Q.

Lemma 5.10. *Let x^t be defined in* (5.2.50), *and $x \in X$ be any feasible solution of* (5.2.1) *or* (5.2.5). *Suppose that τ_t in RGEM satisfy*

$$
\theta_t(m(1+\tau_t) - 1) = \theta_{t-1}m(1+\tau_{t-1}),\ \ t = 2,\ldots,k, \tag{5.2.61}
$$

for some $\theta_t \ge 0$, $t = 1,\ldots,k$. Then, we have

$$
\begin{aligned}
\textstyle\sum_{t=1}^k \theta_t \mathbb{E}[Q(x^t,x)] \le{}& \theta_k(1+\tau_k)\textstyle\sum_{i=1}^m \mathbb{E}[f_i(\underline{x}_i^k)] + \textstyle\sum_{t=1}^k \theta_t \mathbb{E}[\mu\nu(x^t) - \psi(x)]\\
&- \theta_1(m(1+\tau_1) - 1)[\langle x^0 - x, \nabla f(x)\rangle + f(x)].
\end{aligned} \tag{5.2.62}
$$

Proof. In view of the definition of Q in (6.6.26), we have

$$
\begin{aligned}
Q(x^t,x) &= \tfrac{1}{m}\Sigma_{i=1}^m \langle \nabla f_i(x), x^t - x\rangle + \mu\nu(x^t) - \mu\nu(x)\\
&\overset{(5.2.57)}{=} \tfrac{1}{m}\Sigma_{i=1}^m [(1+\tau_t)\langle \underline{x}_i^t - x, \nabla f_i(x)\rangle - \tau_t\langle \underline{x}_i^{t-1} - x, \nabla f_i(x)\rangle]\\
&\quad + \mu\nu(x^t) - \mu\nu(x).
\end{aligned}
$$

Taking expectation on both sides of the above relation over $\{i_1,\ldots,i_k\}$, and using Lemma 5.11, we obtain

$$
\begin{aligned}
\mathbb{E}[Q(x^t,x)] = \Sigma_{i=1}^m \mathbb{E}[(1+\tau_t)\langle \underline{x}_i^t - x, \nabla f_i(x)\rangle\\
- ((1+\tau_t) - \tfrac{1}{m})\langle \underline{x}_i^{t-1} - x, \nabla f_i(x)\rangle] + \mathbb{E}[\mu\nu(x^t) - \mu\nu(x)].
\end{aligned}
$$

Multiplying both sides of the above inequality by θ_t, and summing up the resulting inequalities from $t=1$ to k, we conclude that

$$
\begin{aligned}
&\Sigma_{t=1}^k \theta_t \mathbb{E}[Q(x^t,x)]\\
&= \Sigma_{i=1}^m \Sigma_{t=1}^k \mathbb{E}[\theta_t(1+\tau_t)\langle \underline{x}_i^t - x, \nabla f_i(x)\rangle - \theta_t((1+\tau_t) - \tfrac{1}{m})\langle \underline{x}_i^{t-1} - x, \nabla f_i(x)\rangle]\\
&\quad + \Sigma_{t=1}^k \theta_t \mathbb{E}[\mu\nu(x^t) - \mu\nu(x)].
\end{aligned}
$$

Note that by (5.2.61) and the fact that $\underline{x}_i^0 = x^0$, $i = 1,\ldots,m$, we have

$$
\begin{aligned}
\Sigma_{t=1}^k \theta_t &= \Sigma_{t=2}^k [\theta_t m(1+\tau_t) - \theta_{t-1} m(1+\tau_{t-1})] + \theta_1\\
&= \theta_k m(1+\tau_k) - \theta_1(m(1+\tau_1) - 1), \qquad (5.2.63)\\
\Sigma_{t=1}^k [\theta_t(1+\tau_t)&\langle \underline{x}_i^t - x, \nabla f_i(x)\rangle - \theta_t((1+\tau_t) - \tfrac{1}{m})\langle \underline{x}_i^{t-1} - x, \nabla f_i(x)\rangle]\\
&= \theta_k(1+\tau_k)\langle \underline{x}_i^k - x, \nabla f_i(x)\rangle - \theta_1((1+\tau_1) - \tfrac{1}{m})\langle x^0 - x, \nabla f_i(x)\rangle,
\end{aligned}
$$

for $i = 1,\ldots,m$. Combining the above three relations and using the convexity of f_i, we obtain

$$
\begin{aligned}
&\Sigma_{t=1}^k \theta_t \mathbb{E}[Q(x^t,x)]\\
&\le \theta_k(1+\tau_k)\Sigma_{i=1}^m \mathbb{E}[f_i(\underline{x}_i^k) - f_i(x)] - \theta_1(m(1+\tau_1) - 1)\langle x^0 - x, \nabla f(x)\rangle\\
&\quad + \Sigma_{t=1}^k \theta_t \mathbb{E}[\mu\nu(x^t) - \mu\nu(x)],
\end{aligned}
$$

which in view of (5.2.63) implies (5.2.62). ∎

We now prove the main convergence properties for RGEM to solve (5.2.1). Observe that RGEM starts with $y_i^0 = \mathbf{0}$, $i = 1,\ldots,m$, and only updates the corresponding i_t-block of $(\underline{x}_i^t, y_i^t)$, $i = 1,\ldots,m$, according to (5.2.51) and (5.2.52), respectively. Therefore, for y_i^t generated by RGEM, we have

$$
y_i^t = \begin{cases} \mathbf{0}, & \text{if the } i\text{-th block has never been updated for the first } t \text{ iterations},\\ \nabla f_i(\underline{x}_i^t), & \text{o.w.} \end{cases} \qquad (5.2.64)
$$

Throughout this subsection, we assume that there exists $\sigma_0 \geq 0$ which is the upper bound of the initial gradients, i.e., (5.2.56) holds. Proposition 5.6 below establishes some general convergence properties of RGEM for solving strongly convex problems.

Proposition 5.6. *Let x^t and $\underline{x}^k$ be defined as in* (5.2.50) *and* (5.2.53)*, respectively, and x^* be an optimal solution of* (5.2.1)*. Let σ_0 be defined in* (5.2.56)*, and suppose that $\{\eta_t\}$, $\{\tau_t\}$, and $\{\alpha_t\}$ in RGEM satisfy* (5.2.61) *and*

$$m\theta_{t-1} = \alpha_t\theta_t, \;\; t \geq 2, \tag{5.2.65}$$

$$\theta_t\eta_t \leq \theta_{t-1}(\mu + \eta_{t-1}), \;\; t \geq 2, \tag{5.2.66}$$

$$2\alpha_t L_i \leq m\tau_{t-1}\eta_t, \;\; i = 1,\ldots,m; \; t \geq 2, \tag{5.2.67}$$

$$4L_i \leq \tau_k(\mu + \eta_k), \;\; i = 1,\ldots,m, \tag{5.2.68}$$

for some $\theta_t \geq 0$, $t = 1,\ldots,k$. Then, for any $k \geq 1$, we have

$$\begin{aligned} \mathbb{E}[Q(\underline{x}^k, x^*)] &\leq (\textstyle\sum_{t=1}^k \theta_t)^{-1}\tilde{\Delta}_{0,\sigma_0}, \\ \mathbb{E}[V(x^k, x^*)] &\leq \tfrac{2\tilde{\Delta}_{0,\sigma_0}}{\theta_k(\mu+\eta_k)}, \end{aligned} \tag{5.2.69}$$

where

$$\begin{aligned} \tilde{\Delta}_{0,\sigma_0} := \theta_1(m(1+\tau_1)-1)(\psi(x^0)-\psi^*) + \theta_1\eta_1 V(x^0,x^*) \\ + \textstyle\sum_{t=1}^k (\tfrac{m-1}{m})^{t-1}\tfrac{2\theta_t\alpha_{t+1}}{m\eta_{t+1}}\sigma_0^2. \end{aligned} \tag{5.2.70}$$

Proof. In view of the definition of x^t in (5.2.50) and Lemma 3.5, we have

$$\begin{aligned} &\langle x^t - x, \tfrac{1}{m}\textstyle\sum_{i=1}^m \tilde{y}_i^t\rangle + \mu\nu(x^t) - \mu\nu(x) \\ &\leq \eta_t V(x^{t-1},x) - (\mu+\eta_t)V(x^t,x) - \eta_t V(x^{t-1},x^t). \end{aligned} \tag{5.2.71}$$

Moreover, using the definition of ψ in (5.2.1), the convexity of f_i, and the fact that $\hat{y}_i^t = \nabla f_i(\hat{\underline{x}}_i^t)$ (see (5.2.58) with y_i^t defined in (5.2.52)), we obtain

$$\begin{aligned} &\tfrac{1+\tau_t}{m}\textstyle\sum_{i=1}^m f_i(\hat{\underline{x}}_i^t) + \mu\nu(x^t) - \psi(x) \\ &\leq \tfrac{1+\tau_t}{m}\textstyle\sum_{i=1}^m f_i(\hat{\underline{x}}_i^t) + \mu\nu(x^t) - \mu\nu(x) - \tfrac{1}{m}\textstyle\sum_{i=1}^m [f_i(\hat{\underline{x}}_i^t) + \langle \hat{y}_i^t, x - \hat{\underline{x}}_i^t\rangle] \\ &= \tfrac{\tau_t}{m}\textstyle\sum_{i=1}^m [f_i(\hat{\underline{x}}_i^t) + \langle \hat{y}_i^t, \underline{x}_i^{t-1} - \hat{\underline{x}}_i^t\rangle] + \mu\nu(x^t) - \mu\nu(x) - \tfrac{1}{m}\textstyle\sum_{i=1}^m \langle \hat{y}_i^t, x - x^t\rangle \\ &\leq -\tfrac{\tau_t}{2m}\textstyle\sum_{i=1}^m \tfrac{1}{L_i}\|\nabla f_i(\hat{\underline{x}}_i^t) - \nabla f_i(\underline{x}_i^{t-1})\|_*^2 + \tfrac{\tau_t}{m}\textstyle\sum_{i=1}^m f_i(\underline{x}_i^{t-1}) \\ &\quad + \mu\nu(x^t) - \mu\nu(x) - \tfrac{1}{m}\textstyle\sum_{i=1}^m \langle \hat{y}_i^t, x - x^t\rangle, \end{aligned} \tag{5.2.72}$$

where the first equality follows from the definition of $\hat{\underline{x}}_i^t$ in (5.2.57), and the last inequality follows from the smoothness of f_i (see Lemma 5.8) and (5.2.58). It then follows from (5.2.71) and the definition of $\tilde{y}_i^t$ in (5.2.49) that

$$\begin{aligned}
&\tfrac{1+\tau_t}{m}\textstyle\sum_{i=1}^m f_i(\hat{\underline{x}}_i^t) + \mu \nu(x^t) - \psi(x) \\
&\le -\tfrac{\tau_t}{2m}\textstyle\sum_{i=1}^m \tfrac{1}{L_i}\|\nabla f_i(\hat{\underline{x}}_i^t) - \nabla f_i(\underline{x}_i^{t-1})\|_*^2 + \tfrac{\tau_t}{m}\textstyle\sum_{i=1}^m f_i(\underline{x}_i^{t-1}) \\
&\quad + \langle x^t - x, \tfrac{1}{m}\textstyle\sum_{i=1}^m [\hat{y}_i^t - y_i^{t-1} - \alpha_t(y_i^{t-1} - y_i^{t-2})]\rangle \\
&\quad + \eta_t V(x^{t-1},x) - (\mu+\eta_t)V(x^t,x) - \eta_t V(x^{t-1},x^t).
\end{aligned}$$

Therefore, taking expectation on both sides of the above relation over $\{i_1,\ldots,i_k\}$, and using Lemma 5.11, we have

$$\begin{aligned}
&\mathbb{E}[(1+\tau_t)\textstyle\sum_{i=1}^m f_i(\underline{x}_i^t) + \mu \nu(x^t) - \psi(x)] \\
&\le \mathbb{E}[-\tfrac{\tau_t}{2L_{i_t}}\|\nabla f_{i_t}(\underline{x}_{i_t}^t) - \nabla f_{i_t}(\underline{x}_{i_t}^{t-1})\|_*^2 + \tfrac{1}{m}\textstyle\sum_{i=1}^m (m(1+\tau_t)-1) f_i(\underline{x}_i^{t-1})] \\
&\quad + \mathbb{E}\{\langle x^t - x, \tfrac{1}{m}\textstyle\sum_{i=1}^m [m(y_i^t - y_i^{t-1}) - \alpha_t(y_i^{t-1} - y_i^{t-2})]\rangle\} \\
&\quad + \mathbb{E}[\eta_t V(x^{t-1},x) - (\mu+\eta_t)V(x^t,x) - \eta_t V(x^{t-1},x^t)].
\end{aligned}$$

Multiplying both sides of the above inequality by θ_t, and summing up the resulting inequalities from $t=1$ to k, we obtain

$$\begin{aligned}
&\textstyle\sum_{t=1}^k\sum_{i=1}^m \mathbb{E}[\theta_t(1+\tau_t) f_i(\underline{x}_i^t)] + \textstyle\sum_{t=1}^k \theta_t \mathbb{E}[\mu \nu(x^t) - \psi(x)] \\
&\le \textstyle\sum_{t=1}^k \theta_t \mathbb{E}\left[-\tfrac{\tau_t}{2L_{i_t}}\|\nabla f_{i_t}(\underline{x}_{i_t}^t) - \nabla f_{i_t}(\underline{x}_{i_t}^{t-1})\|_*^2 + \textstyle\sum_{i=1}^m ((1+\tau_t) - \tfrac{1}{m}) f_i(\underline{x}_i^{t-1})\right] \\
&\quad + \textstyle\sum_{t=1}^k\sum_{i=1}^m \theta_t \mathbb{E}[\langle x^t - x, y_i^t - y_i^{t-1} - \tfrac{\alpha_t}{m}(y_i^{t-1} - y_i^{t-2})\rangle] \\
&\quad + \textstyle\sum_{t=1}^k \theta_t \mathbb{E}[\eta_t V(x^{t-1},x) - (\mu+\eta_t)V(x^t,x) - \eta_t V(x^{t-1},x^t)]. \qquad (5.2.73)
\end{aligned}$$

Now by (5.2.65), and the facts that $y_i^{-1} = y_i^0$, $i=1,\ldots,m$, and that we only update $y_{i_t}^t$ (see (5.2.52)), we have

$$\begin{aligned}
&\textstyle\sum_{t=1}^k\sum_{i=1}^m \theta_t \langle x^t - x, y_i^t - y_i^{t-1} - \tfrac{\alpha_t}{m}(y_i^{t-1} - y_i^{t-2})\rangle \\
&= \textstyle\sum_{t=1}^k \theta_t \langle x^t - x, y_{i_t}^t - y_{i_t}^{t-1}\rangle - \tfrac{\theta_t\alpha_t}{m}\langle x^{t-1} - x, y_{i_{t-1}}^{t-1} - y_{i_{t-1}}^{t-2}\rangle \\
&\quad - \textstyle\sum_{t=2}^k \tfrac{\theta_t\alpha_t}{m}\langle x^t - x^{t-1}, y_{i_{t-1}}^{t-1} - y_{i_{t-1}}^{t-2}\rangle \\
&\overset{(5.2.65)}{=} \theta_k \langle x^k - x, y_{i_k}^k - y_{i_k}^{k-1}\rangle - \textstyle\sum_{t=2}^k \tfrac{\theta_t\alpha_t}{m}\langle x^t - x^{t-1}, y_{i_{t-1}}^{t-1} - y_{i_{t-1}}^{t-2}\rangle.
\end{aligned}$$

Moreover, in view of (5.2.66), (5.2.61), and the fact that $\underline{x}_i^0 = x^0$, $i=1,\ldots,m$, we obtain

$$\begin{aligned}
&\textstyle\sum_{t=1}^k \theta_t[\eta_t V(x^{t-1},x) - (\mu+\eta_t)V(x^t,x)] \\
&\quad \overset{(5.2.66)}{\le} \theta_1\eta_1 V(x^0,x) - \theta_k(\mu+\eta_k)V(x^k,x), \\
&\textstyle\sum_{t=1}^k\sum_{i=1}^m \theta_t(1+\tau_t) f_i(\underline{x}_i^t) - \theta_t((1+\tau_t) - \tfrac{1}{m}) f_i(\underline{x}_i^{t-1}) \\
&\quad \overset{(5.2.61)}{=} \textstyle\sum_{i=1}^m \theta_k(1+\tau_k) f_i(\underline{x}_i^k) - \theta_1(m(1+\tau_1)-1) f(x^0),
\end{aligned}$$

which together with (5.2.73), (5.2.64), and the fact that $\theta_1\eta_1 V(x^0,x^1)\geq 0$ imply that

$$\begin{aligned}
&\theta_k(1+\tau_k)\Sigma_{i=1}^m \mathbb{E}[f_i(\underline{x}_i^k)]+\Sigma_{t=1}^k\theta_t\mathbb{E}[\mu\nu(x^t)-\psi(x)]+\tfrac{\theta_k(\mu+\eta_k)}{2}\mathbb{E}[V(x^k,x)]\\
&\leq \theta_1(m(1+\tau_1)-1)f(x^0)+\theta_1\eta_1 V(x^0,x)\\
&\quad+\Sigma_{t=2}^k\mathbb{E}\Big[-\tfrac{\theta_t\alpha_t}{m}\langle x^t-x^{t-1},y_{i_{t-1}}^{t-1}-y_{i_{t-1}}^{t-2}\rangle\\
&\quad-\theta_t\eta_t V(x^{t-1},x^t)-\tfrac{\theta_{t-1}\tau_{t-1}}{2L_{i_{t-1}}}\|y_{i_{t-1}}^{t-1}-\nabla f_{i_{t-1}}(\underline{x}_{i_{t-1}}^{t-2})\|_*^2\Big]\\
&\quad+\theta_k\mathbb{E}\Big[\langle x^k-x,y_{i_k}^k-y_{i_k}^{k-1}\rangle-\tfrac{(\mu+\eta_k)}{2}V(x^k,x)-\tfrac{\tau_k}{2L_{i_k}}\|y_{i_k}^k-\nabla f_{i_k}(\underline{x}_{i_k}^{k-1})\|_*^2\Big].
\end{aligned}\tag{5.2.74}$$

By the strong convexity of $V(\cdot,\cdot)$ in (5.2.8), the simple relations that $b\langle u,v\rangle-a\|v\|^2/2\leq b^2\|u\|^2/(2a),\forall a>0$ and $\|a+b\|^2\leq 2\|a\|^2+2\|b\|^2$, we have

$$\begin{aligned}
&\Sigma_{t=2}^k\Big[-\tfrac{\theta_t\alpha_t}{m}\langle x^t-x^{t-1},y_{i_{t-1}}^{t-1}-y_{i_{t-1}}^{t-2}\rangle-\theta_t\eta_t V(x^{t-1},x^t)-\tfrac{\theta_{t-1}\tau_{t-1}}{2L_{i_{t-1}}}\|y_{i_{t-1}}^{t-1}-\nabla f_{i_{t-1}}(\underline{x}_{i_{t-1}}^{t-2})\|_*^2\Big]\\
&\overset{(5.2.8)}{\leq}\Sigma_{t=2}^k\Big[-\tfrac{\theta_t\alpha_t}{m}\langle x^t-x^{t-1},y_{i_{t-1}}^{t-1}-y_{i_{t-1}}^{t-2}\rangle-\tfrac{\theta_t\eta_t}{2}\|x^{t-1}-x^t\|^2\\
&\quad-\tfrac{\theta_{t-1}\tau_{t-1}}{2L_{i_{t-1}}}\|y_{i_{t-1}}^{t-1}-\nabla f_{i_{t-1}}(\underline{x}_{i_{t-1}}^{t-2})\|_*^2\Big]\\
&\leq\Sigma_{t=2}^k\Big[\tfrac{\theta_{t-1}\alpha_t}{2m\eta_t}\|y_{i_{t-1}}^{t-1}-y_{i_{t-1}}^{t-2}\|_*^2-\tfrac{\theta_{t-1}\tau_{t-1}}{2L_{i_{t-1}}}\|y_{i_{t-1}}^{t-1}-\nabla f_{i_{t-1}}(\underline{x}_{i_{t-1}}^{t-2})\|_*^2\Big]\\
&\leq\Sigma_{t=2}^k\Big[\Big(\tfrac{\theta_{t-1}\alpha_t}{m\eta_t}-\tfrac{\theta_{t-1}\tau_{t-1}}{2L_{i_{t-1}}}\Big)\|y_{i_{t-1}}^{t-1}-\nabla f_{i_{t-1}}(\underline{x}_{i_{t-1}}^{t-2})\|_*^2+\tfrac{\theta_{t-1}\alpha_t}{m\eta_t}\|\nabla f_{i_{t-1}}(\underline{x}_{i_{t-1}}^{t-2})-y_{i_{t-1}}^{t-2}\|_*^2\Big]
\end{aligned}$$

which in view of conditions in (5.2.67) implies that

$$\begin{aligned}
&\Sigma_{t=2}^k\Big[-\tfrac{\theta_t\alpha_t}{m}\langle x^t-x^{t-1},y_{i_{t-1}}^{t-1}-y_{i_{t-1}}^{t-2}\rangle-\theta_t\eta_t V(x^{t-1},x^t)\\
&\qquad-\tfrac{\theta_{t-1}\tau_{t-1}}{2L_{i_{t-1}}}\|y_{i_{t-1}}^{t-1}-\nabla f_{i_{t-1}}(\underline{x}_{i_{t-1}}^{t-2})\|_*^2\Big]\\
&\overset{(5.2.67)}{\leq}\Sigma_{t=2}^k\tfrac{\theta_{t-1}\alpha_t}{m\eta_t}\Big[\|\nabla f_{i_{t-1}}(\underline{x}_{i_{t-1}}^{t-2})-y_{i_{t-1}}^{t-2}\|_*^2\Big].
\end{aligned}$$

Similarly, in view of (5.2.68), we obtain

$$\begin{aligned}
&\theta_k\Big[\langle x^k-x,y_{i_k}^k-y_{i_k}^{k-1}\rangle-\tfrac{(\mu+\eta_k)}{2}V(x^k,x)-\tfrac{\tau_k}{2L_{i_k}}\|y_{i_k}^k-\nabla f_{i_k}(\underline{x}_{i_k}^{k-1})\|_*^2\Big]\\
&\leq\tfrac{2\theta_k}{\mu+\eta_k}\Big[\|\nabla f_{i_k}(\underline{x}_{i_k}^{k-1})-y_{i_k}^{k-1}\|_*^2\Big]\leq\tfrac{2\theta_k\alpha_{k+1}}{m\eta_{k+1}}\Big[\|\nabla f_{i_k}(\underline{x}_{i_k}^{k-1})-y_{i_k}^{k-1}\|_*^2\Big],
\end{aligned}$$

where the last inequality follows from the fact that $m\eta_{k+1}\leq\alpha_{k+1}(\mu+\eta_k)$ (induced from (5.2.65) and (5.2.66)). Therefore, combining the above three relations, we conclude that

$$\begin{aligned}
&\theta_k(1+\tau_k)\Sigma_{i=1}^m\mathbb{E}[f_i(\underline{x}_i^k)]+\Sigma_{t=1}^k\theta_t\mathbb{E}[\mu\nu(x^t)-\psi(x)]+\tfrac{\theta_k(\mu+\eta_k)}{2}\mathbb{E}[V(x^k,x)]\\
&\quad\leq\theta_1(m(1+\tau_1)-1)f(x^0)+\theta_1\eta_1 V(x^0,x)\\
&\quad+\Sigma_{t=1}^k\tfrac{2\theta_t\alpha_{t+1}}{m\eta_{t+1}}\mathbb{E}[\|\nabla f_{i_t}(\underline{x}_{i_t}^{t-1})-y_{i_t}^{t-1}\|_*^2].
\end{aligned}\tag{5.2.75}$$

We now provide a bound on $\mathbb{E}[\|\nabla f_{i_t}(\underline{x}_{i_t}^{t-1}) - y_{i_t}^{t-1}\|_*^2]$. In view of (5.2.64), we have

$$\|\nabla f_{i_t}(\underline{x}_{i_t}^{t-1}) - y_{i_t}^{t-1}\|_*^2 = \begin{cases} \|\nabla f_{i_t}(\underline{x}_{i_t}^{t-1})\|_*^2, & \text{no update on the } i_t\text{-th block until iteration } t; \\ 0, & \text{o.w.} \end{cases}$$

Let us denote event $\mathscr{B}_{i_t} :=$ {the i_t-th block has never been updated until iteration t}, for all $t = 1,\ldots,k$, we have

$$\mathbb{E}[\|\nabla f_{i_t}(\underline{x}_{i_t}^{t-1}) - y_{i_t}^{t-1}\|_*^2] = \mathbb{E}[\|\nabla f_{i_t}(\underline{x}_{i_t}^{t-1})\|_*^2 | \mathscr{B}_{i_t}]\text{Prob}\{\mathscr{B}_{i_t}\} \le \left(\tfrac{m-1}{m}\right)^{t-1}\sigma_0^2,$$

where the last inequality follows from the definitions of $\mathscr{B}_{i_t}$, $\underline{x}_i^t$ in (5.2.51), and σ_0^2 in (5.2.56). Fixing $x = x^*$, and using the above result in (5.2.75), we then conclude from (5.2.75) and Lemma 5.10 that

$$\begin{aligned} 0 &\le \textstyle\sum_{t=1}^k \theta_t \mathbb{E}[Q(x^t, x^*)] \\ &\le \theta_1(m(1+\tau_1)-1)[f(x^0) - \langle x^0 - x^*, \nabla f(x^*)\rangle - f(x^*)] \\ &\quad + \theta_1\eta_1 V(x^0, x^*) + \textstyle\sum_{t=1}^k \left(\tfrac{m-1}{m}\right)^{t-1} \tfrac{2\theta_t\alpha_{t+1}}{m\eta_{t+1}}\sigma_0^2 - \tfrac{\theta_k(\mu+\eta_k)}{2}\mathbb{E}[V(x^k, x^*)], \end{aligned}$$

which, in view of the relation $-\langle x^0 - x^*, \nabla f(x^*)\rangle \le \langle x^0 - x^*, \mu\nu'(x^*)\rangle \le \mu\nu(x^0) - \mu\nu(x^*)$ and the convexity of $Q(\cdot, x^*)$, implies the first result in (5.2.69). Moreover, we can also conclude from the above inequality that

$$\begin{aligned} &\tfrac{\theta_k(\mu+\eta_k)}{2}\mathbb{E}[V(x^k, x^*)] \\ &\le \theta_1(m(1+\tau_1)-1)[\psi(x^0) - \psi(x^*)] + \theta_1\eta_1 V(x^0, x^*) + \textstyle\sum_{t=1}^k \left(\tfrac{m-1}{m}\right)^{t-1} \tfrac{2\theta_t\alpha_{t+1}}{m\eta_{t+1}}\sigma_0^2, \end{aligned}$$

from which the second result in (5.2.69) follows. ■

With the help of Proposition 5.6, we are now ready to establish the convergence properties of RGEM.

Theorem 5.4. *Let x^* be an optimal solution of (5.2.1), x^k and $\underline{x}^k$ be defined in (5.2.50) and (5.2.53), respectively, and $\hat{L} = \max_{i=1,\ldots,m} L_i$. Also let $\{\tau_t\}$, $\{\eta_t\}$, and $\{\alpha_t\}$ be set to*

$$\tau_t \equiv \tau = \tfrac{1}{m(1-\alpha)} - 1, \quad \eta_t \equiv \eta = \tfrac{\alpha}{1-\alpha}\mu, \quad \text{and} \quad \alpha_t \equiv m\alpha. \tag{5.2.76}$$

If (5.2.56) holds and α is set as

$$\alpha = 1 - \frac{1}{m + \sqrt{m^2 + 16m\hat{L}/\mu}}, \tag{5.2.77}$$

then

$$\mathbb{E}[V(x^k, x^*)] \le \frac{2\Delta_{0,\sigma_0}\alpha^k}{\mu}, \tag{5.2.78}$$

$$\mathbb{E}[\psi(\underline{x}^k) - \psi(x^*)] \le 16 \max\left\{m, \tfrac{\hat{L}}{\mu}\right\} \Delta_{0,\sigma_0} \alpha^{k/2}, \tag{5.2.79}$$

where

$$\Delta_{0,\sigma_0} := \mu V(x^0, x^*) + \psi(x^0) - \psi^* + \tfrac{\sigma_0^2}{m\mu}. \tag{5.2.80}$$

Proof. Letting $\theta_t = \alpha^{-t}$, $t = 1, \ldots, k$, we can easily check that parameter setting in (5.2.76) with α defined in (5.2.77) satisfies conditions (5.2.61) and (5.2.65)–(5.2.68) stated in Proposition 5.6. It then follows from (5.2.76) and (5.2.69) that

$$\mathbb{E}[Q(\underline{x}^k, x^*)] \le \tfrac{\alpha^k}{1-\alpha^k}\left[\mu V(x^0, x^*) + \psi(x^0) - \psi^* + \tfrac{2m(1-\alpha)^2\sigma_0^2}{(m-1)\mu}\textstyle\sum_{t=1}^k \left(\tfrac{m-1}{m\alpha}\right)^t\right],$$
$$\mathbb{E}[V(x^k, x^*)] \le 2\alpha^k\left[V(x^0, x^*) + \tfrac{\psi(x^0)-\psi^*}{\mu} + \tfrac{2m(1-\alpha)^2\sigma_0^2}{(m-1)\mu^2}\textstyle\sum_{t=1}^k \left(\tfrac{m-1}{m\alpha}\right)^t\right], \forall k \ge 1.$$

Also observe that $\alpha \ge \frac{2m-1}{2m}$, we then have

$$\textstyle\sum_{t=1}^k \left(\tfrac{m-1}{m\alpha}\right)^t \le \sum_{t=1}^k \left(\tfrac{2(m-1)}{2m-1}\right)^t \le 2(m-1).$$

Combining the above three relations and the fact that $m(1-\alpha) \le 1/2$, we have

$$\mathbb{E}[Q(\underline{x}^k, x^*)] \le \tfrac{\alpha^k}{1-\alpha^k}\Delta_{0,\sigma_0},$$
$$\mathbb{E}[V(x^k, x^*)] \le 2\alpha^k \Delta_{0,\sigma_0}/\mu, \ \forall k \ge 1, \tag{5.2.81}$$

where Δ_{0,σ_0} is defined in (5.2.80). The second relation immediately implies our bound in (5.2.78). Moreover, by the strong convexity of $V(\cdot,\cdot)$ in (5.2.8) and (5.2.78), we have

$$\begin{aligned}
\tfrac{L_f}{2}\mathbb{E}[\|\underline{x}^k - x^*\|^2] &\le \tfrac{L_f}{2}(\textstyle\sum_{t=1}^k \theta_t)^{-1}\sum_{t=1}^k \theta_t \mathbb{E}[\|x^t - x^*\|^2] \\
&\overset{(5.2.8)}{\le} L_f \tfrac{(1-\alpha)\alpha^k}{1-\alpha^k}\textstyle\sum_{t=1}^k \alpha^{-t}\mathbb{E}[V(x^t, x^*)] \\
&\overset{(5.2.78)}{\le} \tfrac{L_f(1-\alpha)\alpha^k}{1-\alpha^k}\textstyle\sum_{t=1}^k \tfrac{2\Delta_{0,\sigma_0}}{\mu} = \tfrac{2L_f(1-\alpha)\Delta_{0,\sigma_0}k\alpha^k}{\mu(1-\alpha^k)}.
\end{aligned}$$

Combining the above relation with the first inequality in (5.2.81) and (5.2.60), we obtain

$$\begin{aligned}
\mathbb{E}[\psi(\underline{x}^k) - \psi(x^*)] &\overset{(5.2.60)}{\le} \mathbb{E}[Q(\underline{x}^k, x^*)] + \tfrac{L_f}{2}\mathbb{E}[\|\underline{x}^k - x^*\|^2] \\
&\le \left(1 + \tfrac{2L_f(1-\alpha)}{\mu}k\right)\tfrac{\Delta_{0,\sigma_0}\alpha^k}{1-\alpha^k} = \left(\tfrac{1}{1-\alpha} + \tfrac{2L_f}{\mu}k\right)\tfrac{\Delta_{0,\sigma_0}\alpha^k(1-\alpha)}{1-\alpha^k}.
\end{aligned}$$

Observing that

$$\tfrac{1}{1-\alpha} \le \tfrac{16}{3}\max\{m, \hat{L}/\mu\},$$
$$\tfrac{2L_f}{\mu} \le \tfrac{16}{3}\max\{m, \hat{L}/\mu\},$$

$$
\begin{aligned}
(k+1)\tfrac{\alpha^k(1-\alpha)}{1-\alpha^k} &= \left(\textstyle\sum_{t=1}^k \tfrac{\alpha^t}{\alpha^t}+1\right)\tfrac{\alpha^k(1-\alpha)}{1-\alpha^k} \le \left(\textstyle\sum_{t=1}^k \tfrac{\alpha^t}{\alpha^{3t/2}}+1\right)\tfrac{\alpha^k(1-\alpha)}{1-\alpha^k}\\
&\le \tfrac{1-\alpha^{k/2}}{\alpha^{k/2}(1-\alpha^{1/2})}\tfrac{\alpha^k(1-\alpha)}{1-\alpha^k} + \alpha^k \le 2\alpha^{k/2}+\alpha^k \le 3\alpha^{k/2},
\end{aligned}
$$

we have

$$
\mathbb{E}[\psi(\underline{x}^k)-\psi(x^*)] \le \tfrac{16}{3}\max\left\{m,\tfrac{\hat{L}}{\mu}\right\}\tfrac{(k+1)\alpha^k(1-\alpha)\Delta_{0,\sigma_0}}{1-\alpha^k} \le 16\max\left\{m,\tfrac{\hat{L}}{\mu}\right\}\Delta_{0,\sigma_0}\alpha^{k/2}.
$$

■

In view of Theorem 5.4, we can provide bounds on the total number of gradient evaluations performed by RGEM to find a stochastic ε-solution of problem (5.2.1), i.e., a point $\bar{x}\in X$ s.t. $\mathbb{E}[\psi(\bar{x})-\psi^*]\le\varepsilon$. Theorem 5.4 implies the number of gradient evaluations of f_i performed by RGEM to find a stochastic ε-solution of (5.2.1) can be bounded by

$$
\begin{aligned}
K(\varepsilon,C,\sigma_0^2) &= 2\left(m+\sqrt{m^2+16mC}\right)\log\tfrac{16\max\{m,C\}\Delta_{0,\sigma_0}}{\varepsilon}\\
&= \mathcal{O}\left\{\left(m+\sqrt{\tfrac{m\hat{L}}{\mu}}\right)\log\tfrac{1}{\varepsilon}\right\}. \qquad (5.2.82)
\end{aligned}
$$

Here $C=\hat{L}/\mu$. Therefore, whenever $\sqrt{mC}\log(1/\varepsilon)$ is dominating, and L_f and $\hat{L}$ are in the same order of magnitude, RGEM can save up to $\mathcal{O}(\sqrt{m})$ gradient evaluations of the component function f_i than the optimal deterministic first-order methods. More specifically, RGEM does not require any exact gradient computation and its iteration cost is similar to pure stochastic gradient descent. It should be pointed out that while the rates of convergence of RGEM obtained in Theorem 5.4 is stated in terms of expectation, we can develop large-deviation results for these rates of convergence using similar techniques in Sect. 5.1 for solving strongly convex problems.

Furthermore, if a one-time exact gradient evaluation is available at the initial point, i.e., $y_i^{-1}=y_i^0=\nabla f_i(x^0)$, $i=1,\ldots,m$, we can employ a more aggressive stepsize policy with

$$
\alpha = 1-\tfrac{2}{m+\sqrt{m^2+8m\hat{L}/\mu}}.
$$

Similarly, we can demonstrate that the number of gradient evaluations of f_i performed by RGEM with this initialization method to find a stochastic ε-solution can be bounded by

$$
\left(m+\sqrt{m^2+8mC}\right)\log\left(\tfrac{6\max\{m,C\}\Delta_{0,0}}{\varepsilon}\right)+m = \mathcal{O}\left\{\left(m+\sqrt{\tfrac{m\hat{L}}{\mu}}\right)\log\tfrac{1}{\varepsilon}\right\}.
$$

It is worth noting that according to the parameter setting in (5.2.76), we have

$$
\eta = \left(\tfrac{1}{1-\alpha}-1\right)\mu = \left(m+\sqrt{m^2+16m\hat{L}/\mu}\right)\mu-\mu = \Omega(m\mu+\sqrt{mL\mu}).
$$

In some statistical learning applications with l_2 regularization (i.e., $\omega(x) = \|x\|_2^2/2$), one usually chooses $\mu = \Omega(1/m)$. For these applications, the stepsize of RGEM is in the order of $1/\sqrt{L}$, which is larger than $1/L$ for those un-accelerated methods.

5.2.3 *Stochastic Finite-Sum Problems*

We discuss in this subsection the stochastic finite-sum optimization and online learning problems, where only noisy gradient information of f_i can be accessed via a stochastic first-order (SFO) oracle. In particular, for any given point $\underline{x}_i^t \in X$, the SFO oracle outputs a vector $G_i(\underline{x}_i^t, \xi_i^t)$ s.t.

$$\mathbb{E}_\xi[G_i(\underline{x}_i^t, \xi_i^t)] = \nabla f_i(\underline{x}_i^t),\ i = 1,\ldots,m, \tag{5.2.83}$$

$$\mathbb{E}_\xi[\|G_i(\underline{x}_i^t, \xi_i^t) - \nabla f_i(\underline{x}_i^t)\|_*^2] \le \sigma^2,\ i = 1,\ldots,m. \tag{5.2.84}$$

We also assume that throughout this subsection that the $\|\cdot\|$ is associated with the inner product $\langle\cdot,\cdot\rangle$.

As shown in Algorithm 5.5, the RGEM method for stochastic finite-sum optimization is naturally obtained by replacing the gradient evaluation of f_i in Algorithm 5.4 (see (5.2.52)) with a stochastic gradient estimator of f_i given in (5.2.85). In particular, at each iteration, we collect B_t number of stochastic gradients of only one randomly selected component f_i and take their average as the stochastic estimator of ∇f_i. Moreover, it needs to be mentioned that the way RGEM initializes gradients, i.e., $y^{-1} = y^0 = \mathbf{0}$, is very important for stochastic optimization, since it is usually impossible to compute exact gradient for expectation functions even at the initial point.

Algorithm 5.5 RGEM for stochastic finite-sum optimization

This algorithm is the same as Algorithm 5.4 except that (5.2.52) is replaced by

$$y_i^t = \begin{cases} \frac{1}{B_t}\sum_{j=1}^{B_t} G_i(\underline{x}_i^t, \xi_{i,j}^t), & i = i_t, \\ y_i^{t-1}, & i \neq i_t. \end{cases} \tag{5.2.85}$$

Here, $G_i(\underline{x}_i^t, \xi_{i,j}^t)$, $j = 1,\ldots,B_t$, are stochastic gradients of f_i computed by the SFO oracle at $\underline{x}_i^t$.

Under the standard assumptions in (5.2.83) and (5.2.84) for stochastic optimization, and with proper choices of algorithmic parameters, we can show that RGEM can achieve the optimal $\mathcal{O}\{\sigma^2/(\mu^2\varepsilon)\}$ rate of convergence (up to a certain logarithmic factor) for solving strongly convex problems given in the form of (5.2.5) in terms of the number of stochastic gradients of f_i.

Before establishing the convergence of RGEM, we first provide some important technical results. Let $\hat{\underline{x}}_i^t$ be defined in (5.2.57) and

$$\hat{y}_i^t = \tfrac{1}{B_t}\textstyle\sum_{j=1}^{B_t} G_i(\hat{\underline{x}}_i^t, \xi_{i,j}^t). \tag{5.2.86}$$

Note the above definition of $\hat{y}_i^t t$ is different from the one in (5.2.85) or (5.2.85).

The following simple result demonstrates a few identities related to $\underline{x}_i^t$ (cf. (5.2.51)) and y_i^t (cf. (5.2.52) or (5.2.85)). For notation convenience, we use $\mathbb{E}_{[i_k]}$ for taking expectation over $\{i_1,\ldots,i_k\}$, $\mathbb{E}_\xi$ for expectations over $\{\xi^1,\ldots,\xi^k\}$, and $\mathbb{E}$ for expectations over all random variables.

Lemma 5.11. *Let x^t and y_i^t be defined in (5.2.50) and* (5.2.85), *respectively, and $\hat{\underline{x}}_i^t$ and $\hat{y}_i^t$ be defined as in* (5.2.57) *and* (5.2.86), *respectively. Then we have, for any $i=1,\ldots,m$ and $t=1,\ldots,k$,*

$$\begin{aligned}
\mathbb{E}_t[y_i^t] &= \tfrac{1}{m}\hat{y}_i^t + (1-\tfrac{1}{m})y_i^{t-1},\\
\mathbb{E}_t[\underline{x}_i^t] &= \tfrac{1}{m}\hat{\underline{x}}_i^t + (1-\tfrac{1}{m})\underline{x}_i^{t-1},\\
\mathbb{E}_t[f_i(\underline{x}_i^t)] &= \tfrac{1}{m}f_i(\hat{\underline{x}}_i^t) + (1-\tfrac{1}{m})f_i(\underline{x}_i^{t-1}),\\
\mathbb{E}_t[\|\nabla f_i(\underline{x}_i^t) - \nabla f_i(\underline{x}_i^{t-1})\|_*^2] &= \tfrac{1}{m}\|\nabla f_i(\hat{\underline{x}}_i^t) - \nabla f_i(\underline{x}_i^{t-1})\|_*^2,
\end{aligned}$$

where $\mathbb{E}_t$ denotes the conditional expectation w.r.t. i_t given $i_1,\ldots,i_{t-1}$ when y_i^t is defined in (5.2.52), *and w.r.t. i_t given $i_1,\ldots,i_{t-1},\xi_1^t,\ldots,\xi_m^t$ when y_i^t is defined in* (5.2.85), *respectively.*

Proof. This first equality follows immediately from the facts that $\text{Prob}_t\{y_i^t = \hat{y}_i^t\} = \text{Prob}_t\{i_t = i\} = \frac{1}{m}$ and $\text{Prob}_t\{y_i^t = y_i^{t-1}\} = 1 - \frac{1}{m}$. Here Prob_t denotes the conditional probability w.r.t i_t given $i_1,\ldots,i_{t-1},\xi_1^t,\ldots,\xi_m^t$. Similarly, we can prove the rest equalities. ■

Note that the parameter $\{B_t\}$ in Algorithm 5.5 denotes the batch size used to compute $y_{i_t}^t$ in (5.2.85). Since we now assume that $\|\cdot\|$ is associated with a certain inner product, it can be easily seen from (5.2.85), and the two assumptions we have for the stochastic gradients computed by the SFO oracle, i.e., (5.2.83) and (5.2.84), that

$$\mathbb{E}_\xi[y_{i_t}^t] = \nabla f_{i_t}(\underline{x}_{i_t}^t) \text{ and } \mathbb{E}_\xi[\|y_{i_t}^t - \nabla f_{i_t}(\underline{x}_{i_t}^t)\|_*^2] \le \tfrac{\sigma^2}{B_t},\ \forall i_t, t = 1,\ldots,k, \tag{5.2.87}$$

and hence $y_{i_t}^t$ is an unbiased estimator of $\nabla f_{i_t}(\underline{x}_{i_t}^t)$. Moreover, for y_i^t generated by Algorithm 5.5, we can see that

$$y_i^t = \begin{cases} \mathbf{0}, & \text{no update on the } i\text{-th block for the first } t \text{ iterations;}\\ \frac{1}{B_l}\sum_{j=1}^{B_l} G_i(\underline{x}_i^l, \xi_{i,j}^l), & \text{latest update happened at } l\text{-th iteration, for } 1 \le l \le t. \end{cases} \tag{5.2.88}$$

We now establish some general convergence properties for Algorithm 5.5.

Proposition 5.7. *Let x^t and $\underline{x}^k$ be defined as in* (5.2.50) *and* (5.2.53)*, respectively, and x^* be an optimal solution of* (5.2.5)*. Suppose that σ_0 and σ are defined in* (5.2.56) *and* (5.2.84)*, respectively, and $\{\eta_t\}$, $\{\tau_t\}$, and $\{\alpha_t\}$ in Algorithm 5.5 satisfy* (5.2.61), (5.2.65), (5.2.66), *and* (5.2.68) *for some $\theta_t \geq 0$, $t = 1, \ldots, k$. Moreover, if*

$$3\alpha_t L_i \leq m\tau_{t-1}\eta_t,\ i = 1, \ldots, m; t \geq 2, \tag{5.2.89}$$

then for any $k \geq 1$, we have

$$\begin{aligned} \mathbb{E}[Q(\underline{x}^k, x^*)] &\leq (\textstyle\sum_{t=1}^k \theta_t)^{-1} \tilde{\Delta}_{0,\sigma_0,\sigma}, \\ \mathbb{E}[V(x^k, x^*)] &\leq \tfrac{2\tilde{\Delta}_{0,\sigma_0,\sigma}}{\theta_k(\mu+\eta_k)}, \end{aligned} \tag{5.2.90}$$

where

$$\tilde{\Delta}_{0,\sigma_0,\sigma} := \tilde{\Delta}_{0,\sigma_0} + \textstyle\sum_{t=2}^k \tfrac{3\theta_{t-1}\alpha_t\sigma^2}{2m\eta_t B_{t-1}} + \sum_{t=1}^k \tfrac{2\theta_t\alpha_{t+1}}{m^2\eta_{t+1}} \sum_{l=1}^{t-1} (\tfrac{m-1}{m})^{t-1-l} \tfrac{\sigma^2}{B_l}, \tag{5.2.91}$$

with $\tilde{\Delta}_{0,\sigma_0}$ defined in (5.2.70).

Proof. Observe that in Algorithm 5.5 y_i^t is updated as in (5.2.85). Therefore, according to (5.2.86), we have

$$\hat{y}_i^t = \tfrac{1}{B_t} \textstyle\sum_{j=1}^{B_t} G_i(\hat{\underline{x}}_i^t, \xi_{i,j}^t),\ i = 1, \ldots, m,\ t \geq 1,$$

which together with the first relation in (5.2.87) imply that $\mathbb{E}_\xi[\langle \hat{y}_i^t, x - \hat{\underline{x}}_i^t \rangle] = \mathbb{E}_\xi[\langle \nabla f_i(\hat{\underline{x}}_i^t), x - \hat{\underline{x}}_i^t \rangle]$. Hence, we can rewrite (5.2.72) as

$$\begin{aligned} &\mathbb{E}_\xi[\tfrac{1+\tau_t}{m} \textstyle\sum_{i=1}^m f_i(\hat{\underline{x}}_i^t) + \mu\nu(x^t) - \psi(x)] \\ &\leq \mathbb{E}_\xi \left[\tfrac{1+\tau_t}{m} \textstyle\sum_{i=1}^m f_i(\hat{\underline{x}}_i^t) + \mu\nu(x^t) - \mu\nu(x) - \tfrac{1}{m} \sum_{i=1}^m [f_i(\hat{\underline{x}}_i^t) + \langle \nabla f_i(\hat{\underline{x}}_i^t), x - \hat{\underline{x}}_i^t \rangle] \right] \\ &= \mathbb{E}_\xi \left[\tfrac{1+\tau_t}{m} \textstyle\sum_{i=1}^m f_i(\hat{\underline{x}}_i^t) + \mu\nu(x^t) - \mu\nu(x) - \tfrac{1}{m} \sum_{i=1}^m [f_i(\hat{\underline{x}}_i^t) + \langle \hat{y}_i^t, x - \hat{\underline{x}}_i^t \rangle] \right] \\ &\leq \mathbb{E}_\xi \big[-\tfrac{\tau_t}{2m} \textstyle\sum_{i=1}^m \tfrac{1}{L_i} \|\nabla f_i(\hat{\underline{x}}_i^t) - \nabla f_i(\underline{x}_i^{t-1})\|_*^2 + \tfrac{\tau_t}{m} \sum_{i=1}^m f_i(\underline{x}_i^{t-1}) \\ &\quad + \langle x^t - x, \tfrac{1}{m} \textstyle\sum_{i=1}^m [\hat{y}_i^t - y_i^{t-1} - \alpha_t(y_i^{t-1} - y_i^{t-2})] \rangle \\ &\quad + \eta_t V(x^{t-1}, x) - (\mu + \eta_t) V(x^t, x) - \eta_t V(x^{t-1}, x^t) \big], \end{aligned}$$

where the last inequality follows from (5.2.71). Following the same procedure as in the proof of Proposition 5.6, we obtain the following similar relation (cf. (5.2.74))

$$\begin{aligned} &\theta_k(1+\tau_k) \textstyle\sum_{i=1}^m \mathbb{E}[f_i(\underline{x}_i^k)] + \sum_{t=1}^k \theta_t \mathbb{E}[\mu\nu(x^t) - \psi(x)] + \tfrac{\theta_k(\mu+\eta_k)}{2} \mathbb{E}[V(x^k, x)] \\ &\quad \leq \theta_1(m(1+\tau_1) - 1) f(x^0) + \theta_1\eta_1 V(x^0, x) \\ &\quad + \textstyle\sum_{t=2}^k \mathbb{E} \Big[-\tfrac{\theta_t\alpha_t}{m} \langle x^t - x^{t-1}, y_{i_{t-1}}^{t-1} - y_{i_{t-1}}^{t-2} \rangle \\ &\qquad - \theta_t\eta_t V(x^{t-1}, x^t) - \tfrac{\theta_{t-1}\tau_{t-1}}{2L_{i_{t-1}}} \|\nabla f_{i_{t-1}}(\underline{x}_{i_{t-1}}^{t-1}) - \nabla f_{i_{t-1}}(\underline{x}_{i_{t-1}}^{t-2})\|_*^2 \Big] \end{aligned}$$

$$+\theta_k\mathbb{E}\left[\langle x^k-x, y_{i_k}^k-y_{i_k}^{k-1}\rangle-\tfrac{(\mu+\eta_k)}{2}V(x^k,x)\right.$$
$$\left.-\tfrac{\tau_k}{2L_{i_k}}\|\nabla f_{i_k}(\underline{x}_{i_k}^k)-\nabla f_{i_k}(\underline{x}_{i_k}^{k-1})\|_*^2\right].$$

By the strong convexity of $V(\cdot,\cdot)$ in (5.2.8), and the simple relation in (5.1.58), we have, for $t=2,\ldots,k$,

$$\begin{aligned}
&\mathbb{E}[-\tfrac{\theta_t\alpha_t}{m}\langle x^t-x^{t-1}, y_{i_{t-1}}^{t-1}-y_{i_{t-1}}^{t-2}\rangle-\theta_t\eta_t V(x^{t-1},x^t)\\
&\quad-\tfrac{\theta_{t-1}\tau_{t-1}}{2L_{i_{t-1}}}\|\nabla f_{i_{t-1}}(\underline{x}_{i_{t-1}}^{t-1})-\nabla f_{i_{t-1}}(\underline{x}_{i_{t-1}}^{t-2})\|_*^2]\\
&\overset{(5.2.8)}{\le}\mathbb{E}[-\tfrac{\theta_t\alpha_t}{m}\langle x^t-x^{t-1}, y_{i_{t-1}}^{t-1}-\nabla f_{i_{t-1}}(\underline{x}_{i_{t-1}}^{t-1})+\nabla f_{i_{t-1}}(\underline{x}_{i_{t-1}}^{t-1})-\nabla f_{i_{t-1}}(\underline{x}_{i_{t-1}}^{t-2})\\
&\quad+\nabla f_{i_{t-1}}(\underline{x}_{i_{t-1}}^{t-2})-y_{i_{t-1}}^{t-2}\rangle]\\
&\quad-\mathbb{E}\left[\tfrac{\theta_t\eta_t}{2}\|x^{t-1}-x^t\|^2+\tfrac{\theta_{t-1}\tau_{t-1}}{2L_{i_{t-1}}}\|\nabla f_{i_{t-1}}(\underline{x}_{i_{t-1}}^{t-1})-\nabla f_{i_{t-1}}(\underline{x}_{i_{t-1}}^{t-2})\|_*^2\right]\\
&\le\mathbb{E}\left[\left(\tfrac{3\theta_{t-1}\alpha_t}{2m\eta_t}-\tfrac{\theta_{t-1}\tau_{t-1}}{2L_{i_{t-1}}}\right)\|\nabla f_{i_{t-1}}(\underline{x}_{i_{t-1}}^{t-1})-\nabla f_{i_{t-1}}(\underline{x}_{i_{t-1}}^{t-2})\|_*^2\right]\\
&\quad+\tfrac{3\theta_{t-1}\alpha_t}{2m\eta_t}\mathbb{E}\left[\|y_{i_{t-1}}^{t-1}-\nabla f_{i_{t-1}}(\underline{x}_{i_{t-1}}^{t-1})\|_*^2+\|\nabla f_{i_{t-1}}(\underline{x}_{i_{t-1}}^{t-2})-y_{i_{t-1}}^{t-2}\|_*^2\right]\\
&\overset{(5.2.89)}{\le}\tfrac{3\theta_{t-1}\alpha_t}{2m\eta_t}\mathbb{E}\left[\|y_{i_{t-1}}^{t-1}-\nabla f_{i_{t-1}}(\underline{x}_{i_{t-1}}^{t-1})\|_*^2+\|\nabla f_{i_{t-1}}(\underline{x}_{i_{t-1}}^{t-2})-y_{i_{t-1}}^{t-2}\|_*^2\right].
\end{aligned}$$

Similarly, we can also obtain

$$\begin{aligned}
&\mathbb{E}\left[\langle x^k-x, y_{i_k}^k-y_{i_k}^{k-1}\rangle-\tfrac{(\mu+\eta_k)}{2}V(x^k,x)-\tfrac{\tau_k}{2L_{i_k}}\|f_{i_k}(\underline{x}_{i_k}^k)-\nabla f_{i_k}(\underline{x}_{i_k}^{k-1})\|_*^2\right]\\
&\overset{(5.2.87),(5.2.8)}{\le}\mathbb{E}\left[\langle x^k-x, \nabla f_{i_k}(\underline{x}_{i_k}^k)-\nabla f_{i_k}(\underline{x}_{i_k}^{k-1})+\nabla f_{i_k}(\underline{x}_{i_k}^{k-1})-y_{i_k}^{k-1}\rangle\right]\\
&\quad-\mathbb{E}\left[\tfrac{(\mu+\eta_k)}{4}\|x^k-x\|^2+\tfrac{\tau_k}{2L_{i_k}}\|f_{i_k}(\underline{x}_{i_k}^k)-\nabla f_{i_k}(\underline{x}_{i_k}^{k-1})\|_*^2\right]\\
&\le\mathbb{E}\left[\left(\tfrac{2}{\mu+\eta_k}-\tfrac{\tau_k}{2L_{i_k}}\right)\|\nabla f_{i_k}(\underline{x}_{i_k}^k)-\nabla f_{i_k}(\underline{x}_{i_k}^{k-1})\|_*^2+\tfrac{2}{\mu+\eta_k}\|\nabla f_{i_k}(\underline{x}_{i_k}^{k-1})-y_{i_k}^{k-1}\|_*^2\right]\\
&\overset{(5.2.68)}{\le}\mathbb{E}\left[\tfrac{2}{\mu+\eta_k}\|\nabla f_{i_k}(\underline{x}_{i_k}^{k-1})-y_{i_k}^{k-1}\|_*^2\right].
\end{aligned}$$

Combining the above three relations, and using the fact that $m\eta_{k+1}\le\alpha_{k+1}(\mu+\eta_k)$ (induced from (5.2.65) and (5.2.66)), we have

$$\begin{aligned}
&\theta_k(1+\tau_k)\textstyle\sum_{i=1}^m\mathbb{E}[f_i(\underline{x}_i^k)]+\sum_{t=1}^k\theta_t\mathbb{E}[\mu\nu(x^t)-\psi(x)]+\tfrac{\theta_k(\mu+\eta_k)}{2}\mathbb{E}[V(x^k,x)]\\
&\quad\le\theta_1(m(1+\tau_1)-1)f(x^0)+\theta_1\eta_1V(x^0,x)\\
&\qquad+\textstyle\sum_{t=2}^k\tfrac{3\theta_{t-1}\alpha_t}{2m\eta_t}\mathbb{E}[\|y_{i_{t-1}}^{t-1}-\nabla f_{i_{t-1}}(\underline{x}_{i_{t-1}}^{t-1})\|_*^2]+\sum_{t=1}^k\tfrac{2\theta_t\alpha_{t+1}}{m\eta_{t+1}}\mathbb{E}[\|\nabla f_{i_t}(\underline{x}_{i_t}^{t-1})-y_{i_t}^{t-1}\|_*^2].
\end{aligned}$$

Moreover, in view of the second relation in (5.2.87), we have

$$\mathbb{E}[\|y_{i_{t-1}}^{t-1}-\nabla f_{i_{t-1}}(\underline{x}_{i_{t-1}}^{t-1})\|_*^2]\le\tfrac{\sigma^2}{B_{t-1}},\ \forall t\ge2.$$

Let us denote $\mathscr{E}_{i_t,t}:=\max\{l: i_l=i_t, l<t\}$ with $\mathscr{E}_{i_t,t}=0$ denoting the event that the i_t-th block has never been updated until iteration t, we can also conclude that for any $t\ge1$

$$\mathbb{E}[\|\nabla f_{i_t}(\underline{x}_{i_t}^{t-1}) - y_{i_t}^{t-1}\|_*^2] = \sum_{l=0}^{t-1}\mathbb{E}\left[\|\nabla f_{i_l}(\underline{x}_{i_l}^{l}) - y_{i_l}^{l}\|_*^2|\{\mathscr{E}_{i_t,t} = l\}\right] \text{Prob}\{\mathscr{E}_{i_t,t} = l\}$$
$$\leq (\tfrac{m-1}{m})^{t-1}\sigma_0^2 + \sum_{l=1}^{t-1}\tfrac{1}{m}(\tfrac{m-1}{m})^{t-1-l}\tfrac{\sigma^2}{B_l},$$

where the first term in the inequality corresponds to the case when the i_t-block has never been updated for the first $t-1$ iterations, and the second term represents that its latest update for the first $t-1$ iterations happened at the l-th iteration. Hence, using Lemma 5.10 and following the same argument as in the proof of Proposition 5.6, we obtain our results in (5.2.90). ■

We are now ready to prove Theorem 5.5, which establishes an optimal complexity bound (up to a logarithmic factor) on the number of calls to the SFO oracle and a linear rate of convergence in terms of the communication complexity for solving problem (5.2.5).

Theorem 5.5. *Let x^* be an optimal solution of (5.2.5), x^k and $\underline{x}^k$ be generated by Algorithm 5.5, and $\hat{L} = \max_{i=1,\ldots,m} L_i$. Suppose that σ_0 and σ are defined in* (5.2.56) *and* (5.2.84), *respectively. Given the iteration limit k, let $\{\tau_t\}$, $\{\eta_t\}$, and $\{\alpha_t\}$ be set to* (5.2.76) *with α being set as* (5.2.77), *and we also set*

$$B_t = \lceil k(1-\alpha)^2\alpha^{-t}\rceil,\ t = 1,\ldots,k, \tag{5.2.92}$$

then

$$\mathbb{E}[V(x^k,x^*)] \leq \frac{2\alpha^k\Delta_{0,\sigma_0,\sigma}}{\mu}, \tag{5.2.93}$$

$$\mathbb{E}[\psi(\underline{x}^k) - \psi(x^*)] \leq 16\max\left\{m, \tfrac{\hat{L}}{\mu}\right\}\Delta_{0,\sigma_0,\sigma}\alpha^{k/2}, \tag{5.2.94}$$

where the expectation is taken w.r.t. $\{i_t\}$ and $\{\xi_i^t\}$ and

$$\Delta_{0,\sigma_0,\sigma} := \mu V(x^0,x^*) + \psi(x^0) - \psi(x^*) + \frac{\sigma_0^2/m + 5\sigma^2}{\mu}. \tag{5.2.95}$$

Proof. Let us set $\theta_t = \alpha^{-t}$, $t = 1,\ldots,k$. It is easy to check that the parameter setting in (5.2.76) with α defined in (5.2.77) satisfies conditions (5.2.61), (5.2.65), (5.2.66), (5.2.68), and (5.2.89) as required by Proposition 5.7. By (5.2.76), the definition of B_t in (5.2.92), and the fact that $\alpha \geq \frac{2m-1}{2m} > (m-1)/m$, we have

$$\sum_{t=2}^k \frac{3\theta_{t-1}\alpha_t\sigma^2}{2m\eta_t B_{t-1}} \leq \sum_{t=2}^k \frac{3\sigma^2}{2\mu(1-\alpha)k} \leq \frac{3\sigma^2}{2\mu(1-\alpha)},$$

$$\sum_{t=1}^k \frac{2\theta_t\alpha_{t+1}}{m^2\eta_{t+1}}\sum_{l=1}^{t-1}(\tfrac{m-1}{m})^{t-1-l}\tfrac{\sigma^2}{B_l} \leq \frac{2\sigma^2}{\alpha\mu m(1-\alpha)k}\sum_{t=1}^k(\tfrac{m-1}{m\alpha})^{t-1}\sum_{l=1}^{t-1}(\tfrac{m\alpha}{m-1})^l$$
$$\leq \frac{2\sigma^2}{\mu(1-\alpha)m\alpha k}\sum_{t=1}^k(\tfrac{m-1}{m\alpha})^{t-1}(\tfrac{m\alpha}{m-1})^{t-1}$$
$$\frac{1}{1-(m-1)/(m\alpha)}$$
$$\leq \frac{2\sigma^2}{\mu(1-\alpha)}\frac{1}{m\alpha-(m-1)} \leq \frac{4\sigma^2}{\mu(1-\alpha)}.$$

Hence, similar to the proof of Theorem 5.4, using the above relations and (5.2.76) in (5.2.90), we obtain

$$\mathbb{E}[Q(\underline{x}^k, x^*)] \le \tfrac{\alpha^k}{1-\alpha^k}\left[\Delta_{0,\sigma_0} + \tfrac{5\sigma^2}{\mu}\right],$$
$$\mathbb{E}[V(x^k, x^*)] \le 2\alpha^k\left[\Delta_{0,\sigma_0} + \tfrac{5\sigma^2}{\mu^2}\right],$$

where Δ_{0,σ_0} is defined in (5.2.80). The second relation implies our results in (5.2.93). Moreover, (5.2.94) follows from the same argument as we used in proving Theorem 5.4. ■

In view of (5.2.94), the number of iterations performed by RGEM to find a stochastic ε-solution of (5.2.5) can be bounded by

$$\hat{K}(\varepsilon, C, \sigma_0^2, \sigma^2) := 2\left(m + \sqrt{m^2 + 16mC}\right)\log\frac{16\max\{m,C\}\Delta_{0,\sigma_0,\sigma}}{\varepsilon}. \tag{5.2.96}$$

Furthermore, in view of (5.2.93) this iteration complexity bound can be improved to

$$\bar{K}(\varepsilon, \alpha, \sigma_0^2, \sigma^2) := \log_{1/\alpha}\tfrac{2\tilde{\Delta}_{0,\sigma_0,\sigma}}{\mu\varepsilon}, \tag{5.2.97}$$

in terms of finding a point $\bar{x} \in X$ s.t. $\mathbb{E}[V(\bar{x}, x^*)] \le \varepsilon$. Therefore, the corresponding number of stochastic gradient evaluations performed by RGEM for solving problem (5.2.5) can be bounded by

$$\textstyle\sum_{t=1}^k B_t \le k\sum_{t=1}^k (1-\alpha)^2\alpha^{-t} + k = \mathcal{O}\left\{\left(\tfrac{\Delta_{0,\sigma_0,\sigma}}{\mu\varepsilon} + m + \sqrt{mC}\right)\log\tfrac{\Delta_{0,\sigma_0,\sigma}}{\mu\varepsilon}\right\},$$

which together with (5.2.95) imply that the total number of required stochastic gradients or samples of the random variables ξ_i, $i = 1, \ldots, m$, can be bounded by

$$\tilde{\mathcal{O}}\left\{\tfrac{\sigma_0^2/m+\sigma^2}{\mu^2\varepsilon} + \tfrac{\mu V(x^0,x^*)+\psi(x^0)-\psi^*}{\mu\varepsilon} + m + \sqrt{\tfrac{m\hat{L}}{\mu}}\right\}.$$

Observe that this bound does not depend on the number of terms m for small enough ε. This complexity bound in fact is in the same order of magnitude (up to a logarithmic factor) as the complexity bound achieved by the stochastic accelerated gradient descent method (Sect. 4.2), which uniformly sample all the random variables ξ_i, $i = 1, \ldots, m$. However, this latter approach will thus involve much higher communication costs in the distributed setting (see Sect. 5.2.4 for more discussions).

5.2.4 Distributed Implementation

This subsection is devoted to RGEMs (see Algorithm 5.4 and Algorithm 5.5) from two different perspectives, i.e., the server and the activated agent under a distributed setting. We also discuss the communication costs incurred by RGEM under this setting.

Both the server and agents in the distributed network start with the same global initial point x^0, i.e., $\underline{x}_i^0 = x^0$, $i = 1,\ldots,m$, and the server also sets $\Delta y = \mathbf{0}$ and $g^0 = \mathbf{0}$. During the process of RGEM, the server updates iterate x^t and calculates the output solution $\underline{x}^k$ (cf. (5.2.53)) which is given by $\mathrm{sum}x/\mathrm{sum}\theta$. Each agent only stores its local variable $\underline{x}_i^t$ and updates it according to information received from the server (i.e., x^t) when activated. The activated agent also needs to upload the changes of gradient Δy_i to the server. Note that line 5 of RGEM from the i_t-th agent's perspective is optional if the agent saves historic gradient information from the last update.

RGEM The server's perspective

1: **while** $t \le k$ **do**
2: $\quad x^t \leftarrow \mathrm{argmin}_{x\in X}\left\{\langle g^{t-1} + \frac{\alpha_t}{m}\Delta y, x\rangle + \mu \nu(x) + \eta_t V(x^{t-1}, x)\right\}$
3: $\quad \mathrm{sum}x \leftarrow \mathrm{sum}x + \theta_t x^t$
4: $\quad \mathrm{sum}\theta \leftarrow \mathrm{sum}\theta + \theta_t$
5: $\quad$ Send signal to the i_t-th agent where i_t is selected uniformly from $\{1,\ldots,m\}$
6: $\quad$ **if** i_t-th agent is responsive **then**
7: $\quad\quad$ Send current iterate x^t to i_t-th agent
8: $\quad\quad$ **if** Receive feedback Δy **then**
9: $\quad\quad\quad g^t \leftarrow g^{t-1} + \Delta y$
10: $\quad\quad\quad t \leftarrow t+1$
11: $\quad\quad$ **else goto** *Line 5*
12: $\quad\quad$ **end if**
13: $\quad$ **else goto** *Line 5*
14: $\quad$ **end if**
15: **end while**

RGEM The activated i_t-th agent's perspective

1: Download the current iterate x^t from the server
2: **if** $t = 1$ **then**
3: $\quad y_i^{t-1} \leftarrow \mathbf{0}$
4: **else**
5: $\quad y_i^{t-1} \leftarrow \nabla f_i(\underline{x}_i^{t-1})$ ▷ Optional
6: **end if**
7: $\underline{x}_i^t \leftarrow (1+\tau_t)^{-1}(x^t + \tau_t \underline{x}_i^{t-1})$
8: $y_i^t \leftarrow \nabla f_i(\underline{x}_i^t)$
9: Upload the local changes to the server, i.e., $\Delta y_i = y_i^t - y_i^{t-1}$

We now add some remarks about the potential benefits of RGEM for distributed optimization and machine learning. First, since RGEM does not require any exact gradient evaluation of f, it does not need to wait for the responses from all agents in order to compute an exact gradient. Each iteration of RGEM only involves communication between the server and the activated i_t-th agent. In fact, RGEM will move to the next iteration in case no response is received from the i_t-th agent. This scheme works under the assumption that the probability for any agent being responsive or available at a certain point of time is equal. Second, since each iteration of RGEM involves only a constant number of communication rounds between the server and one selected agent, the communication complexity for RGEM under distributed setting can be bounded by

$$\mathcal{O}\left\{\left(m+\sqrt{\tfrac{m\hat{L}}{\mu}}\right)\log\tfrac{1}{\varepsilon}\right\}.$$

Therefore, it can save up to $\mathcal{O}\{\sqrt{m}\}$ rounds of communication than the optimal deterministic first-order methods.

For solving distributed stochastic finite-sum optimization problems (5.2.5), RGEM from the i_t-th agent's perspective will be slightly modified as follows.

RGEM The activated i_t-th agent's perspective for solving (5.2.5)

1: Download the current iterate x^t from the server
2: **if** $t = 1$ **then**
3: $\quad y_i^{t-1} \leftarrow \mathbf{0}$ $\quad\triangleright$ Assuming RGEM saves y_i^{t-1} for $t \geq 2$ at the latest update
4: **end if**
5: $\underline{x}_i^t \leftarrow (1+\tau_t)^{-1}(x^t + \tau_t \underline{x}_i^{t-1})$
6: $y_i^t \leftarrow \frac{1}{B_t}\sum_{j=1}^{B_t} G_i(\underline{x}_i^t, \xi_{i,j}^t)$ $\quad\triangleright$ B_t is the batch size, and G_i's are the stochastic gradients given by SFO
7: Upload the local changes to the server, i.e., $\Delta y_i = y_i^t - y_i^{t-1}$

Similar to the case for the deterministic finite-sum optimization, the total number of communication rounds performed by the above RGEM can be bounded by

$$\mathcal{O}\left\{\left(m+\sqrt{\tfrac{m\hat{L}}{\mu}}\right)\log\tfrac{1}{\varepsilon}\right\},$$

for solving (5.2.5). Each round of communication only involves the server and a randomly selected agent. This communication complexity seems to be optimal, since it matches the lower complexity bound (5.1.11) established in Sect. 5.1.4. Moreover, the sampling complexity, i.e., the total number of samples to be collected by all the agents, is also nearly optimal and comparable to the case when all these samples are collected in a centralized location and processed by the stochastic accelerated gradient descent method.

5.3 Variance-Reduced Mirror Descent

In the previous two sections, we have derived a few stochastic algorithms for finite-sum optimization by introducing randomization into accelerated gradient descent (or primal–dual) methods in Chap. 3. In this section, we will study these finite-sum problems from a different perspective by viewing them as some special stochastic optimization problems with finite support. Our goal is to improve the stochastic optimization methods in Chap. 4 by introducing novel gradient estimators with reduced variance.

More specifically, we consider the problem of

$$\min_{x\in X}\{\Psi(x) := f(x)+h(x)\}, \tag{5.3.1}$$

where $X \subseteq \mathbb{R}^m$ is a closed convex set, f is the average of m smooth convex component functions f_i, i.e., $f(x) = \Sigma_{i=1}^m f_i(x)/m$, and h is a simple but possibly non-differentiable convex function. We assume that for $\forall$ i=1,2,...,m, $\exists L_i > 0$, s.t.

$$\|\nabla f_i(x) - \nabla f_i(y)\|_* \le L_i\|x-y\|, \ \forall x,y \in X.$$

Clearly, f has Lipschitz continuous gradients with constant

$$L_f \le L \equiv \tfrac{1}{m}\Sigma_{i=1}^m L_i.$$

Moreover, we assume that the objective function f is possibly strongly convex, i.e., $\exists\ \mu \ge 0$ s.t.

$$f(y) \ge f(x) + \langle\nabla f(x), y-x\rangle + \mu V(x,y), \forall x,y \in X. \tag{5.3.2}$$

It is worth noting that in comparison with problem (5.1.1) and (5.2.1), we do not need to explicitly put a strongly convex term into the objective function of (5.3.1), because the stochastic algorithm we will introduce in this section does not rely on the strong convexity.

In the basic stochastic gradient (mirror) descent method, we can use $\nabla f_{i_t}(x)$ for a randomly selected component $i_t \in \{1,\ldots,m\}$ as an unbiased estimator of the exact gradient $\nabla f(x)$. The variance of this gradient estimator will remain as a constant throughout the algorithm. In contrast to this basic algorithm, variance-reduced mirror descent intends to find an unbiased gradient estimator whose variance will vanish as the algorithm converges. The basic scheme of the variance-reduced mirror descent method is described in Algorithm 5.6.

The variance-reduced mirror descent method is a multi-epoch algorithm. Each epoch of this method contains T iterations and requires the computation of a full gradient at the point $\tilde{x}$. The gradient $\nabla f(\tilde{x})$ will then be used to define a gradient estimator G_t for $\nabla f(x^{t-1})$ at each iteration t. We will show that G_t has smaller variance than the aforementioned estimator $\nabla f_{i_t}(x^{t-1})$, notwithstanding both are unbiased estimators for $\nabla f(x^{t-1})$.

Algorithm 5.6 A variance-reduced mirror descent method

Input: $x^0, \gamma, T, \{\theta_t\}$.
$\tilde{x}^0 = x^0$.
for $s = 1, 2, \ldots$ **do**
  Set $\tilde{x} = \tilde{x}^{s-1}$ and $\tilde{g} = \nabla f(\tilde{x})$.
  Set $x_1 = x^{s-1}$ and $T = T_s$.
  probability $Q = \{q_1, \ldots, q_m\}$ on $\{1, \ldots, m\}$.
  for $t = 1, 2, \ldots, T$ **do**
    Pick $i_t \in \{1, \ldots, m\}$ randomly according to Q.
    $G_t = (\nabla f_{i_t}(x_t) - \nabla f_{i_t}(\tilde{x}))/(q_{i_t} m) + \tilde{g}$.
    $x_{t+1} = \mathrm{argmin}_{x \in X} \{\gamma[\langle G_t, x\rangle + h(x)] + V(x_t, x)\}$.
  end for
  set $x^s = x_{T+1}$ and $\tilde{x}^s = \sum_{t=2}^T (\theta_t x_t)/\sum_{t=2}^T \theta_t$.
end for

Different from the randomized methods discussed in the previous two sections, variance-reduced mirror descent needs to compute full gradients from time to time, which may incur extra delays caused by synchronization under distributed settings for optimization and machine learning. However, this method does not require the computation of gradients at m search points $\underline{x}_i$, $i = 1, \ldots, m$. Moreover, one does not need to save the gradients $\nabla f_i(\tilde{x})$ by computation two gradient components $\nabla f_{i_t}(x_t)$ and $\nabla f_{i_t}(\tilde{x})$ in each iteration. Therefore, this algorithm does not require much memory. As we will show later in the next section, an accelerated version of variance-reduced mirror descent method can achieve comparable or even slightly better rate of convergence than RPGD and RGEM. In this section, we focus on the convergence analysis of the basic scheme of variance-reduced mirror descent method without acceleration.

We now discuss a few results that will be used to establish the convergence of variance-reduced stochastic mirror descent method.

The following result follows as a consequence of Lemma 5.8.

Lemma 5.12. *Let x^* be an optimal solution of* (5.3.1). *Then we have*

$$\tfrac{1}{m}\textstyle\sum_{i=1}^m \tfrac{1}{mq_i}\|\nabla f_i(x) - \nabla f_i(x^*)\|_*^2 \le 2L_Q\left[\Psi(x) - \Psi(x^*)\right], \ \forall x \in X, \tag{5.3.3}$$

where

$$L_Q := \tfrac{1}{m} \max_{i=1,\ldots,m} \tfrac{L_i}{q_i}. \tag{5.3.4}$$

Proof. By Lemma 5.8 (with $f = f_i$), we have

$$\|\nabla f_i(x) - \nabla f_i(x^*)\|_*^2 \le 2L_i\left[f_i(x) - f_i(x^*) - \langle \nabla f_i(x^*), x - x^*\rangle\right].$$

Dividing this inequality by $1/(m^2 q_i)$, and summing over $i = 1, \ldots, m$, we obtain

$$\tfrac{1}{m}\textstyle\sum_{i=1}^m \tfrac{1}{mq_i}\|\nabla f_i(x) - \nabla f_i(x^*)\|_*^2 \le 2L_Q\left[f(x) - f(x^*) - \langle \nabla f(x^*), x - x^*\rangle\right]. \tag{5.3.5}$$

By the optimality of x^*, we have $\langle \nabla f(x^*) + h'(x^*), x - x^* \rangle \geq 0$ for any $x \in X$, which in view of the convexity of h implies that $\langle \nabla f(x^*), x - x^* \rangle \geq h(x^*) - h(x)$ for any $x \in X$. The result then follows by combining the previous two conclusions. ■

In the sequel, we still denote $\delta_t := G_t - g(x_t)$, where $g(x_t) = \nabla f(x_t)$. Lemma 5.13 below shows that if the algorithm converges, then the variance of δ_t will become smaller and smaller.

Lemma 5.13. *Conditionally on* $x_1, \ldots, x_t$,

$$\mathbb{E}[\delta_t] = 0, \tag{5.3.6}$$

$$\mathbb{E}[\|\delta_t\|_*^2] \leq 2L_Q[f(\tilde{x}) - f(x_t) - \langle \nabla f(x_t), \tilde{x} - x_t],\tag{5.3.7}$$

$$\mathbb{E}[\|\delta_t\|_*^2] \leq 4L_Q[\Psi(x_t) - \Psi(x^*) + \Psi(\tilde{x}) - \Psi(x^*)]. \tag{5.3.8}$$

Proof. Taking expectation with respect to i_t conditionally on $x_1, \ldots, x_t$, we obtain

$$\mathbb{E}\left[\tfrac{1}{mq_{i_t}} \nabla f_{i_t}(x_t)\right] = \textstyle\sum_{i=1}^m \tfrac{q_i}{mq_i} \nabla f_i(x_t) = \sum_{i=1}^m \tfrac{1}{m} \nabla f_i(x_t) = \nabla f(x_t).$$

Similarly we have $\mathbb{E}\left[\tfrac{1}{mq_{i_t}} \nabla f_{i_t}(\tilde{x})\right] = \nabla f(\tilde{x})$. Therefore,

$$\mathbb{E}[G_t] = \mathbb{E}\left[\tfrac{1}{mq_{i_t}} (\nabla f_{i_t}(x_t) - \nabla f_{i_t}(\tilde{x})) + \nabla f(\tilde{x})\right] = \nabla f(x_t). \tag{5.3.9}$$

To bound the variance, we have

$$\begin{aligned}
\mathbb{E}[\|\delta_t\|_*^2] =& \mathbb{E}[\|\tfrac{1}{mq_{i_t}} (\nabla f_{i_t}(x_t) - \nabla f_{i_t}(\tilde{x})) + \nabla f(\tilde{x}) - \nabla f(x_t)\|_*^2] \\
=& \mathbb{E}[\tfrac{1}{(mq_{i_t})^2} \|\nabla f_{i_t}(x_t) - \nabla f_{i_t}(\tilde{x})\|_*^2] - \|\nabla f(x_t) - \nabla f(\tilde{x})\|_*^2 \\
\leq& \mathbb{E}[\tfrac{1}{(mq_{i_t})^2} \|\nabla f_{i_t}(x_t) - \nabla f_{i_t}(\tilde{x})\|_*^2].
\end{aligned}$$

The above relation, in view of relation 5.3.5 (with x and x^* replaced by $\tilde{x}$ and x_t), then implies (5.3.7). Moreover,

$$\begin{aligned}
\mathbb{E}[\tfrac{1}{(mq_{i_t})^2} \|\nabla f_{i_t}(x_t) - \nabla f_{i_t}(\tilde{x})\|_*^2] &= \mathbb{E}[\tfrac{1}{(mq_{i_t})^2} \|\nabla f_{i_t}(x_t) - \nabla f_i(x_*) + \nabla f_i(x_*) \\
&\quad - \nabla f_{i_t}(\tilde{x})\|_*^2] \\
\leq \mathbb{E}[\tfrac{2}{(mq_{i_t})^2} \|\nabla f_{i_t}(x_t) - \nabla f_{i_t}(x^*)\|_*^2] &+ \mathbb{E}[\tfrac{2}{(mq_{i_t})^2} \|\nabla f_{i_t}(\tilde{x}) - \nabla f_{i_t}(x^*)\|_*^2] \\
= \tfrac{2}{m} \textstyle\sum_{i=1}^m \tfrac{1}{mq_i} \|\nabla f_i(x_t) - \nabla f_i(x^*)\|_*^2 &+ \tfrac{2}{m} \textstyle\sum_{i=1}^m \tfrac{1}{mq_i} \|\nabla f_i(\tilde{x}) - \nabla f_i(x^*)\|_*^2,
\end{aligned}$$

which together with Lemma 5.12 then imply (5.3.8). ■

We now show the possible progress made by each iteration of the variance-reduced mirror descent method. This result resembles Lemma 4.2 for the original stochastic mirror descent method.

Lemma 5.14. *If the stepsize γ satisfies $L\gamma \le 1/2$, then for any $x \in X$,*

$$\gamma[\Psi(x_{t+1}) - \Psi(x)] + V(x_{t+1},x) \le (1-\gamma\mu)V(x_t,x) + \gamma\langle\delta_t, x - x_t\rangle + \gamma^2\|\delta_t\|_*^2. \tag{5.3.10}$$

Moreover, conditionally on $i_1,\ldots,i_{t-1}$

$$\begin{aligned}&\gamma\mathbb{E}[\Psi(x_{t+1}) - \Psi(x^*)] + \mathbb{E}[V(x_{t+1},x^*)]\\&\le (1-\gamma\mu)V(x_t,x^*) + 4L_Q\gamma^2[\Psi(x_t) - \Psi(x^*) + \Psi(\tilde{x}) - \Psi(x^*)].\end{aligned} \tag{5.3.11}$$

Proof. Similar to (4.1.26), we have

$$\begin{aligned}\gamma f(x_{t+1}) &\le \gamma[f(x_t) + \langle G_t, x_{t+1} - x_t\rangle] + V(x_t,x_{t+1}) + \tfrac{\gamma^2\|\delta_t\|_*^2}{2(1-L\gamma)}\\&\le \gamma[f(x_t) + \langle G_t, x_{t+1} - x_t\rangle] + V(x_t,x_{t+1}) + \gamma^2\|\delta_t\|_*^2,\end{aligned}$$

where the last inequality follows from the assumption that $L\gamma \le 1/2$. Moreover, it follows from Lemma 3.5 that

$$\begin{aligned}&\gamma[f(x_t) + \langle G_t, x_{t+1} - x_t\rangle + h(x_{t+1})] + V(x_t,x_{t+1})\\&\le \gamma[f(x_t) + \langle G_t, x - x_t\rangle + h(x)] + V(x_t,x) - V(x_{t+1},x)\\&= \gamma[f(x_t) + \langle g(x_t), x - x_t\rangle + h(x)] + \gamma\langle\delta_t, x - x_t\rangle + V(x_t,x) - V(x_{t+1},x)\\&\le \gamma[\Psi(x) - \mu V(x_t,x)] + \gamma\langle\delta_t, x - x_t\rangle + V(x_t,x) - V(x_{t+1},x),\end{aligned}$$

where the last inequality follows from the convexity of $f(\cdot)$. Combining the above two conclusions and rearranging the terms, we obtain (5.3.10). Using Lemma 5.13 and taking expectation on both sides of (5.3.10) with respect to i_t, we obtain (5.3.11). ■

With the help of Lemma 5.14, we will be able to establish the rate of convergence for the stochastic variance-reduced mirror descent method for solving a few different classes of finite-sum optimization problems.

5.3.1 Smooth Problems Without Strong Convexity

In this subsection, we assume that f is not necessarily strongly convex, i.e., $\mu = 0$ in (5.3.2). We will present a general convergence result first and then discuss the selection of some algorithmic parameters (e.g., γ and T_s).

Theorem 5.6. *Suppose that the algorithmic parameters for the variance-reduced stochastic mirror descent method satisfy*

$$\theta_t = 1, t \ge 1, \tag{5.3.12}$$

$$4L_Q\gamma \le 1, \tag{5.3.13}$$

$$w_s := (1 - 4L_Q\gamma)(T_{s-1} - 1) - 4L_Q\gamma T_s > 0, s \ge 2. \tag{5.3.14}$$

Then we have

$$\mathbb{E}[\Psi(\bar{x}^S) - \Psi(x^*)] \le \frac{\gamma(1+4L_Q\gamma T_1)[\Psi(x^0)-\Psi(x^*)]+V(x^0,x^*)}{\gamma\sum_{s=1}^S w_s}, \tag{5.3.15}$$

where

$$\bar{x}^S = \frac{\sum_{s=1}^S (w_s \tilde{x}^s)}{\sum_{s=1}^S w_s}. \tag{5.3.16}$$

Proof. Summing up (5.3.11) (with $\mu = 0$) over $t = 1,\ldots,T$ and taking expectation with respect to random variables $i_1,\ldots,i_T$, we obtain

$$\begin{aligned}&\gamma\mathbb{E}[\Psi(x_{T+1}) - \Psi(x^*)] + (1-4L_Q\gamma)\gamma\textstyle\sum_{t=2}^T \mathbb{E}[\Psi(x_t) - \Psi(x^*)] + \mathbb{E}[V(x_{T+1},x^*)]\\ &\le 4L_Q\gamma^2[\Psi(x_1) - \Psi(x^*)] + 4L_Q\gamma^2 T[\Psi(\tilde{x}) - \Psi(x^*)] + V(x_1,x^*).\end{aligned}$$

Now consider a fixed epoch s with iteration limit $T = T_s$, input $x_1 = x^{s-1}$ and $\tilde{x} = \tilde{x}^{s-1}$, and output $x^s = x_{T+1}$ and $\tilde{x}^s = \sum_{t=2}^T x^t/(T-1)$ due to $\theta_t = 1$. By the above inequality and the convexity of Ψ, we have

$$\begin{aligned}&\gamma\mathbb{E}[\Psi(x^s) - \Psi(x^*)] + (1-4L_Q\gamma)\gamma(T_s-1)\mathbb{E}[\Psi(\tilde{x}^s) - \Psi(x^*)] + \mathbb{E}[V(x^s,x^*)]\\ &\le 4L_Q\gamma^2[\Psi(x^{s-1}) - \Psi(x^*)] + 4L_Q\gamma^2 T_s[\Psi(\tilde{x}^{s-1}) - \Psi(x^*)] + V(x^{s-1},x^*)\\ &\le \gamma[\Psi(x^{s-1}) - \Psi(x^*)] + 4L_Q\gamma^2 T_s[\Psi(\tilde{x}^{s-1}) - \Psi(x^*)] + V(x^{s-1},x^*),\end{aligned}$$

where the last inequality follows from the assumption that $4L_Q\gamma \le 1$. Summing up the above inequalities over $s = 1,\ldots,S$, and taking full expectation over all random variables, we have

$$\begin{aligned}&\gamma\mathbb{E}[\Psi(x^S) - \Psi(x^*)] + (1-4L_Q\gamma)\gamma(T_S-1)\mathbb{E}[\Psi(\tilde{x}^S) - \Psi(x^*)]\\ &+ \gamma\textstyle\sum_{s=1}^{S-1}[(1-4L_Q\gamma)(T_{s-1}-1) - 4L_Q\gamma T_s]\mathbb{E}[\Psi(\tilde{x}^s) - \Psi(x^*)] + \mathbb{E}[V(x^S,x^*)]\\ &\le \gamma[\Psi(x^0) - \Psi(x^*)] + 4L_Q\gamma^2 T_1[\Psi(\tilde{x}^0) - \Psi(x^*)] + V(x^0,x^*)\\ &= \gamma(1+4L_Q\gamma T_1)[\Psi(x^0) - \Psi(x^*)] + V(x^0,x^*),\end{aligned}$$

which, in view of (5.3.16) and the convexity of Ψ, then implies the result in (5.3.15). ■

In light of Theorem 5.6, we now provide some specific rules to select γ and T_s and establish the complexity of the resulting algorithm. Certainly it is possible to develop many other rules to specify these parameters.

Corollary 5.8. *Suppose that $\theta = 1$, $\gamma = 1/(16L_Q)$, and*

$$T_s = 2T_{s-1}, s = 2,3,\ldots, \tag{5.3.17}$$

with $T_1 = 7$. Then for any $S \ge 1$, we have

$$\mathbb{E}[\Psi(\bar{x}^S) - \Psi(x^*)] \le \tfrac{8}{2^S-1}\left[\tfrac{11}{4}(\Psi(x^0) - \Psi(x^*)) + 16L_Q V(x^0,x^*)\right]. \tag{5.3.18}$$

Moreover, the total number of gradient computations required to find an ε-solution of (5.3.1), i.e., a point $\bar{x} \in X$ s.t. $\Psi(\bar{x}) - \Psi(x^*) \le \varepsilon$, can be bounded by

$$\mathcal{O}\left\{m \log \tfrac{\Psi(x^0)-\Psi(x^*)+L_Q V(x^0,x^*)}{\varepsilon} + \tfrac{\Psi(x^0)-\Psi(x^*)+L_Q V(x^0,x^*)}{\varepsilon}\right\}. \tag{5.3.19}$$

Proof. Note that by (5.3.17), we have

$$\begin{aligned} w_s &= \tfrac{3}{4}(T_{s-1}-1) - \tfrac{1}{4}T_s = \tfrac{1}{8}T_s - \tfrac{3}{4} \\ &\ge \tfrac{1}{8}T_s - \tfrac{3}{28}T_s = \tfrac{1}{56}T_s, \end{aligned} \tag{5.3.20}$$

where the inequality follows from the fact that $T_s \ge 7$. Using this observation and (5.3.17), we have

$$\textstyle\sum_{s=1}^S w_s \ge \tfrac{1}{8}(2^S - 1).$$

Using these relations in (5.3.15), we obtain (5.3.18). In view of (5.3.18), the total number of epochs S to find an ε-solution of (5.3.1) can be bounded by $\bar{S} \equiv \log(\Psi(x^0) - \Psi(x^*) + L_Q V(x^0,x^*))/\varepsilon$. Hence the total number of gradient computations can be bounded $m\bar{S} + \sum_{s=1}^{\bar{S}} T_s$, which is bounded by (5.3.19). ■

Recall that directly applying the mirror descent method would require $\mathcal{O}(m/\varepsilon)$ gradient computations for each component functions f_i to find an ε-solution of (5.3.1). It is interesting to see that the total number of gradient computations required by the variance-reduced mirror descent method improves this bound significantly in terms of its dependence on m. Moreover, the bound in (5.3.19) is also smaller than a direct application of the stochastic mirror descent method to (5.3.1) as long as

$$m \le \mathcal{O}\left\{\tfrac{1}{\log(1/\varepsilon)}\left(\tfrac{1}{\varepsilon^2} - \tfrac{1}{\varepsilon}\right)\right\}$$

after disregarding some constant factors.

5.3.2 Smooth and Strongly Convex Problems

In this subsection, we assume that f is strongly convex, i.e., $\mu > 0$ in (5.3.2).

The following result shows the progress made by each epoch of the variance-reduced mirror descent method.

Theorem 5.7. *Suppose that the algorithmic parameters for the variance-reduced mirror descent method satisfy*

$$\theta_t = (1-\gamma\mu)^{-t}, t \ge 0, \tag{5.3.21}$$

$$1 \ge 2L\gamma, \tag{5.3.22}$$

$$1 \ge \gamma\mu + 4L_Q\gamma. \tag{5.3.23}$$

Then we have $\Delta_s \le \rho \Delta_{s-1}$ for any $s \ge 1$, where

$$\begin{aligned}\Delta_s &:= \gamma \mathbb{E}[\Psi(x^s) - \Psi(x^*)] + \mathbb{E}[V(x^s, x^*)] \\ &\quad + (1 - \gamma\mu - 4L_Q\gamma)\mu^{-1}(\theta_T - \theta_1)\mathbb{E}[\Psi(\tilde{x}^s) - \Psi(x^*)], \qquad (5.3.24)\\ \rho &:= \max\left\{\tfrac{1}{\theta_T}, \tfrac{4L_Q\gamma(\theta_T - 1)}{(1-\gamma\mu-4L_Q\gamma)(\theta_T - \theta_1)}\right\}. \qquad (5.3.25)\end{aligned}$$

Proof. Multiplying both sides of (5.3.11) by θ_t in (5.3.21), we have

$$\begin{aligned}&\gamma\theta_t \mathbb{E}[\Psi(x_{t+1}) - \Psi(x^*)] + \theta_t \mathbb{E}[V(x_{t+1}, x^*)] \\ &\le 4L_Q\gamma^2\theta_t[\Psi(x_t) - \Psi(x^*) + \Psi(\tilde{x}) - \Psi(x^*)] + \theta_{t-1}V(x_t, x^*).\end{aligned}$$

Summing up the previous inequality over $t = 1, \ldots, T$ and taking expectation with respect to the history of random variables $i_1, \ldots, i_T$, we obtain

$$\begin{aligned}&\gamma\theta_T \mathbb{E}[\Psi(x_{T+1}) - \Psi(x^*)] \\ &+ (1 - \gamma\mu - 4L_Q\gamma)\gamma\Sigma_{t=2}^T \theta_t \mathbb{E}[\Psi(x_t) - \Psi(x^*)] + \theta_T \mathbb{E}[V(x_{T+1}, x^*)] \\ &\le 4L_Q\gamma^2[\Psi(x_1) - \Psi(x^*)] + 4L_Q\gamma^2\Sigma_{t=1}^T \theta_t[\Psi(\tilde{x}) - \Psi(x^*)] + V(x_1, x^*).\end{aligned}$$

Now consider a fixed epoch s with input $x_1 = x^{s-1}$ and $\tilde{x} = \tilde{x}^{s-1}$, and output $x^s = x_{T+1}$ and $\tilde{x}^s = \Sigma_{t=2}^T \theta_t x^t / \Sigma_{t=2}^T \theta_t$. By the above inequality, the convexity of Ψ, and the fact that

$$\Sigma_{t=s}^T \theta_t = (\gamma\mu)^{-1}\Sigma_{t=s}^T(\theta_t - \theta_{t-1}) = (\gamma\mu)^{-1}(\theta_T - \theta_{s-1}), \qquad (5.3.26)$$

we have

$$\begin{aligned}&\gamma\theta_T \mathbb{E}[\Psi(x^s) - \Psi(x^*)] \\ &+ (1 - \gamma\mu - 4L_Q\gamma)\mu^{-1}(\theta_T - \theta_1)\mathbb{E}[\Psi(\tilde{x}^s) - \Psi(x^*)] + \theta_T \mathbb{E}[V(x^s, x^*)] \\ &\le 4L_Q\gamma^2[\Psi(x^{s-1}) - \Psi(x^*)] + 4L_Q\gamma\mu^{-1}(\theta_T - 1)[\Psi(\tilde{x}^{s-1}) - \Psi(x^*)] + V(x^{s-1}, x^*) \\ &\le \gamma[\Psi(x^{s-1}) - \Psi(x^*)] + 4L_Q\gamma\mu^{-1}(\theta_T - 1)[\Psi(\tilde{x}^{s-1}) - \Psi(x^*)] + V(x^{s-1}, x^*),\end{aligned}$$

where the last inequality follows from the fact $4L_Q\gamma \le 1$ due to (5.3.23). The result then immediately follows the above inequality. ■

We are now ready to establish the convergence of the variance-reduced mirror descent method.

Corollary 5.9. *Suppose that the algorithmic parameters for the variance-reduced mirror descent method are set to*

$$\gamma = \tfrac{1}{21L}, \qquad (5.3.27)$$

$$q_i = \tfrac{L_i}{\Sigma_{i=1}^m L_i}, \qquad (5.3.28)$$

$$T \ge 2, \qquad (5.3.29)$$

$$\theta_t = (1 - \gamma\mu)^{-t}, t \ge 0, \qquad (5.3.30)$$

Then for any $s \geq 1$, we have

$$\Delta_s \leq \left\{\max\left[(1-\tfrac{\mu}{21L})^T, \tfrac{1}{2}\right]\right\}^s \Delta_0, \tag{5.3.31}$$

where Δ_s is defined in (5.3.24). In particular, if T is set to a constant factor of m, denoted by $o\{m\}$, then the total number of gradients required to find an ε-solution of problem (5.3.1) can be bounded by $\mathcal{O}\{(\frac{L}{\mu}+m)\log\frac{\Delta_0}{\varepsilon}\}$.

Proof. It is easy to check that $L_Q = \frac{1}{m}\Sigma_{i=1}^m L_i = L$. Moreover, whenever $T \geq 2$, we have $\frac{\theta_T - 1}{\theta_T - \theta_1} \leq 2$. Indeed,

$$\frac{\theta_T-1}{\theta_T-\theta_1} - 2 = \frac{1-(1-\gamma\mu)^T}{1-(1-\gamma\mu)^{T-1}} - 2 = \frac{(1-\gamma\mu)^{T-1}(1+\gamma\mu)-1}{1-(1-\gamma\mu)^{T-1}} \leq 0. \tag{5.3.32}$$

Hence,

$$\frac{4L_Q\gamma(\theta_T-1)}{(1-\gamma\mu-4L_Q\gamma)(\theta_T-\theta_1)} \leq \frac{8L_Q\gamma}{1-\gamma\mu-4L_Q\gamma} \leq \frac{8L\gamma}{1-5L\gamma} \leq \frac{1}{2}. \tag{5.3.33}$$

Using these relations in Proposition 5.7, we obtain (5.3.31). Now assume that $T = o\{m\}$. We consider two cases. If $(1-\frac{\mu}{21L})^T \leq \frac{1}{2}$, then by (5.3.31) the total number of epochs can be bounded by $\mathcal{O}\{\log(\Delta_0/\varepsilon)\}$, which implies that the total number of gradient computations $(s \times (m+T))$ can be bounded by $\mathcal{O}\{m\log(\Delta_0/\varepsilon)\}$. Now if $(1-\frac{\mu}{21L})^T \geq \frac{1}{2}$, then it follows from (5.3.31) that both the total number of iterations $(s \times T)$ and the total number of gradient computations $(s \times (m+T))$ are bounded by $\mathcal{O}\{\frac{L}{\mu}\log(\Delta_0/\varepsilon)$. The result follows by combining these two cases. ■

The complexity bound obtained in Corollary 5.9 significantly improves the $\mathcal{O}(1/(\mu\varepsilon))$ complexity bound for a direction application of the stochastic mirror descent method since the former one only depends on $\log(1/\varepsilon)$. However, the stochastic mirror descent and stochastic accelerated gradient descent methods still have their own merits. First, they can be used to handle different classes of problems such as general nonsmooth problems, as well as stochastic optimization problems with continuous (rather than discrete) random variables. Second, suppose that one applies the variance-reduced mirror descent method to solve the sample average approximation problem of $\min_{x\in X}\mathbb{E}[F(x,\xi)]$. In other words, we will collect and save a sample $\{\xi_1,\ldots,\xi_m\}$ of size $m = \mathcal{O}(1/\varepsilon)$ and use the variance reduction methods to solve the deterministic counterpart given by $\min_{x\in X}\frac{1}{m}\Sigma_{i=1}^m F(x,\xi_i)$. The total number of gradient computations performed by this approach will be worse than those by stochastic mirror descent method and stochastic accelerated gradient descent method up to a factor of $\mathcal{O}(\log(1/\varepsilon))$. Third, the stochastic mirror descent and stochastic accelerated gradient descent methods do not require us to save the sample $\{\xi_1,\ldots,\xi_m\}$ and can be used to process online learning and optimization problems.

5.4 Variance-Reduced Accelerated Gradient Descent

In this section, we introduce a novel randomized incremental gradient method, namely, the variance-reduced accelerated gradient algorithm, for solving the finite-sum optimization problem in (5.3.1) whose objective function consists of the average of m smooth components together with a simple convex term. We demonstrate that the variance-reduced accelerated gradient method exhibits a unified optimal rates of convergence for solving both convex and strongly convex problems. In particular, for solving smooth convex problems which are not necessarily strongly convex, the variance-reduced accelerated gradient algorithm does not require to add any strongly convex perturbation into the objective, but can directly achieve the optimal $\mathcal{O}(m\log m + \sqrt{mL/\varepsilon})$ complexity bound. This bound can be improved to $\mathcal{O}(m\log(1/\varepsilon))$ under certain lower accuracy regime. Here L and ε denote the Lipschitz constant of the gradients and target accuracy respectively. Moreover, for solving strongly convex problems with modulus μ, the variance-reduced accelerated gradient method equipped with a unified step-size policy can adjust itself according to the value of the conditional number (L/μ), and achieve the optimal linear rate of convergence when the conditional number is relatively small, and the best-known sublinear rate of convergence that is independent of the condition number otherwise. In addition, we show that the variance-reduced accelerated gradient method exhibits an accelerated linear rate of convergence for solving a wide class of weakly strongly convex problems, which only satisfy a certain error bound condition rather than strong convexity.

The basic scheme of the variance-reduced accelerated gradient method is formally described in Algorithm 5.7. This algorithm consists of multiple epochs. In each epoch (or outer loop), it first computes the full gradient $\nabla f(\tilde{x})$ at the point $\tilde{x}$, which will then be repeatedly used to define a gradient estimator G_t at each iteration of the inner loop. This is the same as the variance reduction technique introduced in Sect. 5.3. The inner loop has a similar algorithmic scheme to the stochastic accelerated gradient method in Sect. 4.2 with a constant step-size policy. Indeed, the parameters used in the inner loop, i.e., $\{\gamma_s\}, \{\alpha_s\}$, and $\{p_s\}$, only depend on the index of epoch s. Each iteration of the inner loop requires the gradient information of only one randomly selected component function f_{i_t}, and maintains three primal sequences.

Algorithm 5.7 Variance-reduced accelerated gradient method

Input: $x^0, \gamma, T, \{\theta_t\}$.
$\tilde{x}^0 = x^0$.
for $s = 1, 2, \ldots$ **do**
  Set $\tilde{x} = \tilde{x}^{s-1}$ and $\tilde{g} = \nabla F(\tilde{x})$.
  Set $x_0 = x^{s-1}$, $\bar{x}_0 = \tilde{x}$ and $T = T_s$.
  for $t = 1, 2, \ldots, T$ **do**
    Pick $i_t \in \{1, \ldots, m\}$ randomly according to $Q = \{q_1, \ldots, q_m\}$ on $\{1, \ldots, m\}$.
    $\underline{x}_t = [(1+\mu\gamma_s)(1-\alpha_s-p_s)\bar{x}_{t-1} + \alpha_s x_{t-1} + (1+\mu\gamma_s)p_s\tilde{x}]/[1+\mu\gamma_s(1-\alpha_s)]$.
    $G_t = (\nabla f_{i_t}(\underline{x}_t) - \nabla f_{i_t}(\tilde{x}))/(q_{i_t} n) + \tilde{g}$.
    $x_t = \arg\min_{x \in X} \{\gamma_s [\langle G_t, x\rangle + h(x) + \mu V(\underline{x}_t, x)] + V(x_{t-1}, x)\}$.
    $\bar{x}_t = (1-\alpha_s-p_s)\bar{x}_{t-1} + \alpha_s x_t + p_s\tilde{x}$.
  end for
  set $x^s = x_T$ and $\tilde{x}^s = \sum_{t=1}^{T}(\theta_t \bar{x}_t)/\sum_{t=1}^{T}\theta_t$.
end for

The variance-reduced accelerated gradient method reduces to the variance-reduced mirror descent method if $\alpha_s = 1$ and $p_s = 0$. In this case, the algorithm only maintains one primal sequence $\{x_t\}$ and exhibits a non-accelerated rate of convergence with complexity bounded by $\mathcal{O}\{(m + L/\mu)\log(1/\varepsilon\}$ for solving (5.3.1).

In the sequel, we define

$$l_f(z,x) := f(z) + \langle \nabla f(z), x - z\rangle, \tag{5.4.1}$$

$$\delta_t := G_t - \nabla f(\underline{x}_t), \tag{5.4.2}$$

$$x_{t-1}^+ := \tfrac{1}{1+\mu\gamma_s}(x_{t-1} + \mu\gamma_s \underline{x}_t). \tag{5.4.3}$$

Similar to the results (i.e., (5.3.6) and (5.3.7)) in Lemma 5.13, we have, conditionally on $x_1, \ldots, x_t$,

$$\mathbb{E}[\delta_t] = 0, \tag{5.4.4}$$

$$\mathbb{E}[\|\delta_t\|_*^2] \le 2L_Q[f(\tilde{x}) - f(x_t) - \langle \nabla f(x_t), \tilde{x} - x_t\rangle]. \tag{5.4.5}$$

Also observe that the auxiliary point x_{t-1}^+ has been used in the original accelerated gradient method and its stochastic counterpart. Using the above definition of x_{t-1}^+, and the definitions of $\underline{x}_t$ and $\bar{x}_t$ in Algorithm 5.7, we have

$$\begin{aligned}\bar{x}_t - \underline{x}_t &= (1-\alpha_s-p_s)\bar{x}_{t-1} + \alpha_s x_t + p_s\tilde{x} - \underline{x}_t \\ &= \alpha_s x_t + \tfrac{1}{1+\mu\gamma_s}\{[1+\mu\gamma_s(1-\alpha_s)]\underline{x}_t - \alpha_s x_{t-1}\} - \underline{x}_t \\ &= \alpha_s(x_t - x_{t-1}^+).\end{aligned} \tag{5.4.6}$$

The following result examines the optimality conditions associated with the definition of x_t in Algorithm 5.7.

Lemma 5.15. *For any $x \in X$, we have*

$$\gamma_s[l_f(\underline{x}_t, x_t) - l_f(\underline{x}_t, x) + h(x_t) - h(x)] \le \gamma_s \mu V(\underline{x}_t, x) + V(x_{t-1}, x) \\ - (1 + \mu\gamma_s)V(x_t, x) - \tfrac{1+\mu\gamma_s}{2}\|x_t - x_{t-1}^+\|^2 - \gamma_s\langle\delta_t, x_t - x\rangle.$$

Proof. It follows from Lemma 3.5 and the definition of x_t in Algorithm 5.7 that

$$\gamma_s[\langle G_t, x_t - x\rangle + h(x_t) - h(x) + \mu V(\underline{x}_t, x_t)] + V(x_{t-1}, x_t) \\ \le \gamma_s \mu V(\underline{x}_t, x) + V(x_{t-1}, x) - (1 + \mu\gamma_s)V(x_t, x).$$

Also observe that

$$\langle G_t, x_t - x\rangle = \langle \nabla f(\underline{x}_t), x_t - x\rangle + \langle\delta_t, x_t - x\rangle \\ = l_f(\underline{x}_t, x_t) - l_f(\underline{x}_t, x) + \langle\delta_t, x_t - x\rangle$$

and

$$\gamma_s \mu V(\underline{x}_t, x_t) + V(x_{t-1}, x_t) \ge \tfrac{1}{2}\left(\mu\gamma_s\|x_t - \underline{x}_t\|^2 + \|x_t - x_{t-1}\|^2\right) \\ \ge \tfrac{1+\mu\gamma_s}{2}\|x_t - x_{t-1}^+\|^2,$$

where the last inequality follows from the definition of x_{t-1}^+ in (5.4.3) and the convexity of $\|\cdot\|$. The result then follows by combining the above three relations. ■

We now show the possible progress made by each inner iteration of the variance-reduced accelerated gradient method.

Lemma 5.16. *Assume that $\alpha_s \in [0,1]$, $p_s \in [0,1]$, and $\gamma_s > 0$ satisfy*

$$1 + \mu\gamma_s - L\alpha_s\gamma_s > 0, \tag{5.4.7}$$

$$p_s - \tfrac{L_Q\alpha_s\gamma_s}{1+\mu\gamma_s - L\alpha_s\gamma_s} \ge 0. \tag{5.4.8}$$

Then, conditional on $x_1, \ldots, x_{t-1}$, we have

$$\tfrac{\gamma_s}{\alpha_s}\mathbb{E}[\psi(\bar{x}_t) - \psi(x)] + (1+\mu\gamma_s)\mathbb{E}[V(x_t, x)] \le \tfrac{\gamma_s}{\alpha_s}(1-\alpha_s-p_s)\mathbb{E}[\psi(\bar{x}_{t-1}) - \psi(x)] \\ + \tfrac{\gamma_s p_s}{\alpha_s}\mathbb{E}[\psi(\tilde{x}) - \psi(x)] + \mathbb{E}[V(x_{t-1}, x)] \tag{5.4.9}$$

for any $x \in X$.

Proof. Note that by the smoothness of f and the definition of $\bar{x}_t$, we have

$$f(\bar{x}_t) \le l_f(\underline{x}_t, \bar{x}_t) + \tfrac{L}{2}\|\bar{x}_t - \underline{x}_t\|^2 \\ = (1-\alpha_s - p_s)l_f(\underline{x}_t, \bar{x}_{t-1}) + \alpha_s l_f(\underline{x}_t, x_t) + p_s l_f(\underline{x}_t, \tilde{x}) + \tfrac{L\alpha_s^2}{2}\|x_t - x_{t-1}^+\|^2.$$

The above inequality, in view of Lemma 5.15 and the (strong) convexity of f, then implies that

$$
\begin{aligned}
f(\bar{x}_t) &\le (1-\alpha_s-p_s)l_f(\underline{x}_t,\bar{x}_{t-1}) \\
&\quad + \alpha_s\left[l_f(\underline{x}_t,x)+h(x)-h(x_t)+\mu V(\underline{x}_t,x)+\tfrac{1}{\gamma_s}V(x_{t-1},x)-\tfrac{1+\mu\gamma_s}{\gamma_s}V(x_t,x)\right] \\
&\quad + p_s l_f(\underline{x}_t,\tilde{x}) - \tfrac{\alpha_s}{2\gamma_s}(1+\mu\alpha_s-L\alpha_s\gamma_s)\|x_t-x_{t-1}^+\|^2 - \alpha_s\langle\delta_t,x_t-x\rangle \\
&\le (1-\alpha_s-p_s)f(\bar{x}_{t-1}) \\
&\quad + \alpha_s\left[\Psi(x)-h(x_t)+\tfrac{1}{\gamma_s}V(x_{t-1},x)-\tfrac{1+\mu\gamma_s}{\gamma_s}V(x_t,x)\right] \\
&\quad + p_s l_f(\underline{x}_t,\tilde{x}) - \tfrac{\alpha_s}{2\gamma_s}(1+\mu\alpha_s-L\alpha_s\gamma_s)\|x_t-x_{t-1}^+\|^2 \\
&\quad - \alpha_s\langle\delta_t,x_t-x_{t-1}^+\rangle - \alpha_s\langle\delta_t,x_{t-1}^+-x\rangle \\
&\le (1-\alpha_s-p_s)f(\bar{x}_{t-1}) \\
&\quad + \alpha_s\left[\Psi(x)-h(x_t)+\tfrac{1}{\gamma_s}V(x_{t-1},x)-\tfrac{1+\mu\gamma_s}{\gamma_s}V(x_t,x)\right] \\
&\quad + p_s l_f(\underline{x}_t,\tilde{x}) + \tfrac{\alpha_s\gamma_s\|\delta_t\|_*^2}{2(1+\mu\gamma_s-L\alpha_s\gamma_s)} + \alpha_s\langle\delta_t,x_{t-1}^+-x\rangle.
\end{aligned}
$$

Note that by (5.4.4), (5.4.5), (5.4.8), and the convexity of f, we have, conditional on $x_1,\ldots,x_{t-1}$,

$$
\begin{aligned}
&p_s l_f(\underline{x}_t,\tilde{x}) + \tfrac{\alpha_s\gamma_s\mathbb{E}[\|\delta_t\|_*^2]}{2(1+\mu\gamma_s-L\alpha_s\gamma_s)} + \alpha_s\mathbb{E}[\langle\delta_t,x_{t-1}^+-x\rangle] \\
&\le p_s l_f(\underline{x}_t,\tilde{x}) + \tfrac{L_Q\alpha_s\gamma_s}{1+\mu\gamma_s-L\alpha_s\gamma_s}[f(\tilde{x})-l_f(\underline{x}_t,\tilde{x})] \\
&\le \left(p_s - \tfrac{L_Q\alpha_s\gamma_s}{1+\mu\gamma_s-L\alpha_s\gamma_s}\right)l_f(\underline{x}_t,\tilde{x}) + \tfrac{L_Q\alpha_s\gamma_s}{1+\mu\gamma_s-L\alpha_s\gamma_s}f(\tilde{x}) \\
&\le p_s f(\tilde{x}).
\end{aligned}
$$

Moreover, by the convexity of h, we have $h(\bar{x}_t) \le (1-\alpha_s-p_s)h(\bar{x}_{t-1})+\alpha_s h(x_t)+p_s h(\tilde{x})$. Summing up the previous three conclusions we obtain

$$
\begin{aligned}
\mathbb{E}[\Psi(\bar{x}_t)] &\le (1-\alpha_s-p_s)\Psi(\bar{x}_{t-1})+p_s\Psi(\tilde{x})+\alpha_s\Psi(x) && (5.4.10)\\
&\quad + \tfrac{\alpha_s}{\gamma_s}\left[V(x_{t-1},x)-(1+\mu\gamma_s)V(x_t,x)\right]. && (5.4.11)
\end{aligned}
$$

The result then follows by subtracting $\Psi(x)$ from both sides of the above inequality. ■

Utilizing this result in Lemma 5.16, we will discuss the convergence properties of the variance-reduced accelerated gradient method for solving smooth problems, smooth and strongly convex problems, and smooth problems satisfying a certain error bound condition in the following few subsections.

5.4.1 Smooth Problems Without Strong Convexity

In this subsection, we consider the case when f is not necessarily strongly convex, i.e., $\mu = 0$ in (5.3.2).

The following result shows possible decrease of function values in each epoch of the variance-reduced accelerated gradient method for solving smooth finite-sum convex optimization problems.

Lemma 5.17. *Assume that for each epoch s, $s \geq 1$, α_s, γ_s, p_s, and T_s are chosen such that (5.4.7)–(5.4.8) hold. Also, let us set*

$$\theta_t = \begin{cases} \frac{\gamma_s}{\alpha_s}(\alpha_s + p_s) & t = 1, \ldots, T_s - 1 \\ \frac{\gamma_s}{\alpha_s} & t = T_s. \end{cases} \tag{5.4.12}$$

Moreover, let us denote

$$\mathcal{L}_s := \tfrac{\gamma_s}{\alpha_s} + (T_s - 1)\tfrac{\gamma_s(\alpha_s+p_s)}{\alpha_s}, \ \mathcal{R}_s := \tfrac{\gamma_s}{\alpha_s}(1 - \alpha_s) + (T_s - 1)\tfrac{\gamma_s p_s}{\alpha_s}, \tag{5.4.13}$$

and assume that

$$w_s := \mathcal{L}_s - \mathcal{R}_{s+1} \geq 0, \forall s \geq 1. \tag{5.4.14}$$

Then we have

$$\begin{aligned} &\mathcal{L}_s \mathbb{E}[\Psi(\tilde{x}^s) - \Psi(x)] + (\textstyle\sum_{j=1}^{s-1} w_j)\mathbb{E}[\Psi(\bar{x}^s) - \Psi(x)] \\ &\quad \leq \mathcal{R}_1 \mathbb{E}[\Psi(\tilde{x}^0) - \Psi(x)] + V(x^0, x) - V(x^s, x), \end{aligned} \tag{5.4.15}$$

where

$$\bar{x}^s := (\textstyle\sum_{j=1}^{s-1} w_j)\sum_{j=1}^{s-1}(w_j \tilde{x}^j). \tag{5.4.16}$$

Proof. Using our assumptions on α_s, γ_s and p_s, the fact that $\mu = 0$, we have

$$\begin{aligned} \tfrac{\gamma_s}{\alpha_s}\mathbb{E}[\Psi(\bar{x}_t) - \Psi(x)] &\leq \tfrac{\gamma_s}{\alpha_s}(1 - \alpha_s - p_s)\mathbb{E}[\Psi(\bar{x}_{t-1}) - \Psi(x)] \\ &\quad + \tfrac{\gamma_s p_s}{\alpha_s}\mathbb{E}[\Psi(\tilde{x}) - \Psi(x)] + V(x_{t-1}, x) - V(x_t, x). \end{aligned}$$

Summing up these inequalities for $t = 1, \ldots, T_s$, using the definition of θ_t in (5.4.12) and the fact that $\bar{x}_0 = \tilde{x}$, and rearranging the terms, we have

$$\begin{aligned} \textstyle\sum_{t=1}^{T_s} \theta_t \mathbb{E}[\Psi(\bar{x}_t) - \Psi(x)] &\leq \left[\tfrac{\gamma_s}{\alpha_s}(1 - \alpha_s) + (T_s - 1)\tfrac{\gamma_s p_s}{\alpha_s}\right]\mathbb{E}[\Psi(\tilde{x}) - \Psi(x)] \\ &\quad + V(x_0, x) - V(x_T, x). \end{aligned}$$

Now using the facts that $x^s = x_T$, $\tilde{x}^s = \sum_{t=1}^{T}(\theta_t x_t)/\sum_{t=1}^{T_s}\theta_t$, $\tilde{x} = \tilde{x}^{s-1}$, and the convexity of Ψ, we have

$$\begin{aligned} \textstyle\sum_{t=1}^{T_s} \theta_t \mathbb{E}[\Psi(\tilde{x}^s) - \Psi(x)] &\leq \left[\tfrac{\gamma_s}{\alpha_s}(1 - \alpha_s) + (T_s - 1)\tfrac{\gamma_s p_s}{\alpha_s}\right]\mathbb{E}[\Psi(\tilde{x}^{s-1}) - \Psi(x)] \\ &\quad + V(x^{s-1}, x) - V(x^s, x), \end{aligned}$$

which, in view of the fact that $\sum_{t=1}^{T_s}\theta_t = \frac{\gamma_s}{\alpha_s} + (T_s - 1)\frac{\gamma_s(\alpha_s+p_s)}{\alpha_s}$, then implies that

$$\mathcal{L}_s \mathbb{E}[\Psi(\tilde{x}^s) - \Psi(x)] \leq \mathcal{R}_s \mathbb{E}[\Psi(\tilde{x}^{s-1}) - \Psi(x)] + V(x^{s-1}, x) - V(x^s, x).$$

Summing over the above relations, using the convexity of Ψ, and rearranging the terms, we then obtain (5.4.15). ■

With the help of Lemma 5.17 we are now ready to establish the main convergence properties of the variance-reduced accelerated gradient method for the case when $\mu = 0$ in (5.3.2).

Theorem 5.8. *Assume that θ_t are defined in (5.4.12) and the probabilities q_i are set to $L_i/\sum_{i=1}^m L_i$ for $i = 1,\ldots,m$. Moreover, let us denote $s_0 := \lfloor \log_2 m \rfloor + 1$ and assume that*

$$T_s = \begin{cases} 2^{s-1}, & s \le s_0 \\ T_{s_0}, & s > s_0 \end{cases}, \quad \gamma_s = \tfrac{1}{3L\alpha_s}, \quad \text{and } p_s = \tfrac{1}{2}, \tag{5.4.17}$$

with

$$\alpha_s = \begin{cases} \tfrac{1}{2}, & s \le s_0 \\ \tfrac{2}{s-s_0+4}, & s > s_0 \end{cases}. \tag{5.4.18}$$

Then for any $x \in X$,

$$\mathbb{E}[\Psi(\tilde{x}^s) - \Psi(x)] \le \begin{cases} 2^{-(s+1)} D_0, & 1 \le s \le s_0, \\ \tfrac{8LD_0}{(s-s_0+4)^2(m+1)}, & s > s_0, \end{cases} \tag{5.4.19}$$

where

$$D_0 := 2[\psi(x^0) - \psi(x^*)] + 3LV(x^0, x^*). \tag{5.4.20}$$

Proof. Note that by the definition of L_Q in (5.3.4) and the selection of q_i, we have $L_Q = L$. Observe that both conditions in (5.4.7) and (5.4.8) are satisfied since

$$1 + \mu\gamma_s - L\alpha_s\gamma_s = 1 - L\alpha_s\gamma_s = \tfrac{2}{3}$$

and

$$p_s - \tfrac{L_Q\alpha_s\gamma_s}{1+\mu\gamma_s - L\alpha_s\gamma_s} = p_s - \tfrac{1}{2} = 0.$$

Next, letting $\mathscr{L}_s$ and $\mathscr{R}_s$ be defined in (5.4.13), we will show that $\mathscr{L}_s \ge \mathscr{R}_{s+1}$ for any $s \ge 1$. Indeed, if $1 \le s < s_0$, we have $\alpha_{s+1} = \alpha_s$, $\gamma_{s+1} = \gamma_s$, $T_{s+1} = 2T_s$, and hence

$$\begin{aligned} w_s = \mathscr{L}_s - \mathscr{R}_{s+1} &= \tfrac{\gamma_s}{\alpha_s}[1 + (T_s - 1)(\alpha_s + p_s) - (1-\alpha_s) - (2T_s - 1)p_s] \\ &= \tfrac{\gamma_s}{\alpha_s}[T_s(\alpha_s - p_s)] = 0. \end{aligned}$$

Moreover, if $s \ge s_0$, we have

$$\begin{aligned} w_s = \mathscr{L}_s - \mathscr{R}_{s+1} &= \tfrac{\gamma_s}{\alpha_s} - \tfrac{\gamma_{s+1}}{\alpha_{s+1}}(1-\alpha_{s+1}) + (T_{s_0} - 1)\left[\tfrac{\gamma_s(\alpha_s+p_s)}{\alpha_s} - \tfrac{\gamma_{s+1}p_{s+1}}{\alpha_{s+1}}\right] \\ &= \tfrac{1}{12L} + \tfrac{(T_{s_0}-1)[2(s-s_0+4)-1]}{24L} \ge 0. \end{aligned}$$

Using these observations in (5.4.15) iteratively, we then conclude that

$$\begin{aligned}\mathscr{L}_s\mathbb{E}[\Psi(\tilde{x}^s)-\Psi(x)] &\le \mathscr{R}_1\mathbb{E}[\Psi(\tilde{x}^0)-\Psi(x)]+V(x^0,x)-V(x^s,x)\\ &\le \tfrac{2}{3L}[\Psi(x^0)-\Psi(x)]+V(x^0,x)\end{aligned}$$

for any $s \ge 1$, where the last identity follows from the fact that $\mathscr{R}_1 = \frac{2}{3L}$. It can be easily seen that if $s \le s_0$, $\mathscr{L}_s = \frac{2^{s+1}}{3L}$. In addition, if $s \ge s_0$, we have

$$\begin{aligned}\mathscr{L}_s &= \tfrac{1}{3L\alpha_s^2}\left[1+(T_s-1)(\alpha_s+\tfrac{1}{2})\right]\\ &= \tfrac{(s-s_0+4)(T_{s_0}-1)}{6L}+\tfrac{(s-s_0+4)^2(T_{s_0}+1)}{24L}\\ &\ge \tfrac{(s-s_0+4)^2(m+1)}{24L}.\end{aligned} \tag{5.4.21}$$

The result then follows immediately by combining the previous three inequalities. ∎

In view of Theorem 5.8, we can bound the total number of gradient computations of f_i as follows.

Corollary 5.10. *The total number of gradient evaluation performed by the variance-reduced accelerated gradient method can be bounded by*

$$\bar{N} := \begin{cases} \mathscr{O}\left\{m\log\frac{D_0}{\varepsilon}\right\}, & m\varepsilon \ge D_0,\\ \mathscr{O}\left\{\sqrt{\frac{mD_0}{\varepsilon}}+m\log m\right\}, & o.w.,\end{cases} \tag{5.4.22}$$

where D_0 is defined in (5.4.20).

Proof. First let us consider the regime of lower accuracy and/or large number of components, i.e., when $m\varepsilon \ge D_0$. In this case we need to run the algorithm for at most s_0 epochs because we can easily check that

$$\tfrac{D_0}{2^{s_0-1}-1} \le \varepsilon.$$

More precisely, the number of epochs can be bounded by

$$S_l := \min\left\{\left\lceil 1+\log_2\left(\tfrac{D_0}{\varepsilon}\right)\right\rceil, s_0\right\}.$$

Hence the total number of gradient evaluations can be bounded by

$$\begin{aligned}mS_l+\Sigma_{s=1}^{S_l}T_s &= mS_l+\Sigma_{s=1}^{S_l}2^{s-1} \le mS_l+2^{S_l}\\ &= \mathscr{O}\left\{\min\left(m\log\tfrac{D_0}{\varepsilon}, m\log m\right)\right\} = \mathscr{O}\left\{m\log\tfrac{D_0}{\varepsilon}\right\},\end{aligned} \tag{5.4.23}$$

where the last identity follows from the assumption that $m\varepsilon \ge D_0$. Now let us consider the regime for high accuracy and/or smaller number of components, i.e., when

$m\varepsilon < D_0$. In this case, we may need to run the algorithm for more than s_0 epochs. More precisely, the total number of epochs can be bounded by

$$S_h := \left\lceil \sqrt{\tfrac{16D_0}{(m+1)\varepsilon}} + s_0 - 4 \right\rceil.$$

Note that the total number of gradient evaluations needed for the first s_0 epochs can be bounded by $ms_0 + \sum_{s=1}^{s_0} T_s$ while the total number of gradient evaluations for the remaining epochs can be bounded by $(T_{s_0} + m)(S_h - s_0)$. As a consequence, the total number of gradient evaluations can be bounded by

$$\begin{aligned} ms_0 + \textstyle\sum_{s=1}^{s_0} T_s + (T_{s_0} + m)(S_h - s_0) &= \textstyle\sum_{s=1}^{s_0} T_s + (T_{s_0} + m)S_h \\ &= \mathcal{O}\left\{\sqrt{\tfrac{mD_0}{\varepsilon}} + m\log m\right\}. \end{aligned} \tag{5.4.24}$$

■

We now make a few observations regarding the results obtained in Corollary 5.10. First, whenever the required accuracy ε is low and/or the number of components m is large, the variance-reduced accelerated gradient method can achieve a fast linear rate of convergence even if the objective function is not strongly convex. Otherwise, it exhibits an optimal sublinear rate of convergence with complexity bounded by $\mathcal{O}\{\sqrt{mD_0/\varepsilon} + m\log m\}$. Second, whenever $\sqrt{mD_0/\varepsilon}$ is dominating in the second case of (5.4.22), the variance-reduced accelerated gradient method can save up to $\mathcal{O}(\sqrt{m})$ gradient evaluations of the component function f_i than the optimal deterministic first-order methods for solving (5.3.1).

5.4.2 Smooth and Strongly Convex Problems

In this subsection we consider the case when f is possibly strongly convex, including the situation when the problem is almost non-strongly convex, i.e., $\mu \approx 0$. Our goal is to provide a unified step-size policy which allows the variance-reduced accelerated gradient method to achieve an optimal rate of convergence for finite-sum optimization in (5.3.1) regardless of its strong convexity.

We first discuss how to specify the algorithmic parameters used in the variance-reduced accelerated gradient method. In fact, our selection of $\{T_s\}$, $\{\gamma_s\}$, and $\{p_s\}$ will be exactly the same as those in (5.4.17) for the smooth convex case, but with α_s given by

$$\alpha_s = \begin{cases} \frac{1}{2}, & s \le s_0, \\ \max\left\{\frac{2}{s-s_0+4}, \min\{\sqrt{\frac{m\mu}{3L}}, \frac{1}{2}\}\right\}, & s > s_0. \end{cases} \tag{5.4.25}$$

Such a selection of α_s enables us to consider different classes of problems with different problem parameters such as L/μ and m. However, the selection of $\{\theta_t\}$, i.e., the weights to take the average of the iterates for each epoch, will be more complicated. Specifically, denoting $s_0 := \lfloor \log m \rfloor + 1$, we assume that the weights $\{\theta_t\}$ are set to (5.4.12) if $1 \le s \le s_0$ or $s_0 < s \le s_0 + \sqrt{\frac{12L}{m\mu}} - 4$, $m < \frac{3L}{4\mu}$. Otherwise, we set them to

$$\theta_t = \begin{cases} \Gamma_{t-1} - (1 - \alpha_s - p_s)\Gamma_t, & 1 \le t \le T_s - 1, \\ \Gamma_{t-1}, & t = T_s, \end{cases} \tag{5.4.26}$$

where $\Gamma_t = (1 + \mu\gamma_s)^t$. The selection of these weights come from the convergence analysis of the algorithms as we will see later.

Below we consider four different cases and establish the convergence properties of the variance-reduced accelerated gradient method in each case.

Lemma 5.18. *If $s \le s_0$, then for any $x \in X$,*

$$\mathbb{E}[\psi(\tilde{x}^s) - \psi(x)] \le 2^{-(s+1)} D_0, \quad 1 \le s \le s_0,$$

where D_0 is defined in (5.4.20).

Proof. In this case, we have $\alpha_s = p_s = \frac{1}{2}$, $\gamma_s = \frac{2}{3L}$, and $T_s = 2^{s-1}$. It then follows from (5.4.9) that

$$\tfrac{\gamma_s}{\alpha_s}\mathbb{E}[\psi(\bar{x}_t) - \psi(x)] + (1 + \mu\gamma_s)\mathbb{E}[V(x_t, x)] \le \tfrac{\gamma_s}{2\alpha_s}\mathbb{E}[\psi(\tilde{x}) - \psi(x)] + \mathbb{E}[V(x_{t-1}, x)].$$

Summing up the above relation from $t = 1$ to T_s, we have

$$\begin{aligned} &\tfrac{\gamma_s}{\alpha_s}\textstyle\sum_{t=1}^{T_s}\mathbb{E}[\psi(\bar{x}_t) - \psi(x)] + \mathbb{E}[V(x_{T_s}, x)] + \mu\gamma_s\textstyle\sum_{t=1}^{T_s}\mathbb{E}[V(x_t, x)] \\ &\qquad \le \tfrac{\gamma_s T_s}{2\alpha_s}\mathbb{E}[\psi(\tilde{x}) - \psi(x)] + \mathbb{E}[V(x_0, x)]. \end{aligned}$$

Note that in this case θ_t are chosen as in (5.4.12), i.e., $\theta_t = \frac{\gamma_s}{\alpha_s}$, $t = 1, \ldots, T_s$ in the definition of $\tilde{x}^s$, we then have

$$\begin{aligned} \tfrac{4T_s}{3L}\mathbb{E}[\psi(\tilde{x}^s) - \psi(x)] + \mathbb{E}[V(x^s, x)] &\le \tfrac{4T_s}{6L}\mathbb{E}[\psi(\tilde{x}^{s-1}) - \psi(x)] + \mathbb{E}[V(x^{s-1}, x)] \\ &= \tfrac{4T_{s-1}}{3L}\mathbb{E}[\psi(\tilde{x}^{s-1}) - \psi(x)] + \mathbb{E}[V(x^{s-1}, x)], \end{aligned}$$

where we use the facts that $\tilde{x} = \tilde{x}^{s-1}$, $x_0 = x^{s-1}$, and $x^s = x_{T_s}$ in the epoch s and the parameter settings in (5.4.17). Applying this inequality recursively, we then have

$$\begin{aligned} \tfrac{4T_s}{3L}\mathbb{E}[\psi(\tilde{x}^s) - \psi(x)] + \mathbb{E}[V(x^s, x)] &\le \tfrac{2}{3L}\mathbb{E}[\psi(\tilde{x}^0) - \psi(x)] + V(x^0, x) \\ &= \tfrac{2}{3L}\mathbb{E}[\psi(x^0) - \psi(x)] + V(x^0, x). \end{aligned} \tag{5.4.27}$$

By plugging $T_s = 2^{s-1}$ into the above inequality, we obtain the result. ∎

Lemma 5.19. *If $s \ge s_0$ and $m \ge \frac{3L}{4\mu}$,*

$$\mathbb{E}[\psi(\tilde{x}^s) - \psi(x^*)] \le \left(\tfrac{4}{5}\right)^s D_0,$$

where x^* is an optimal solution of (5.3.1).

Proof. In this case, we have $\alpha_s = p_s = \frac{1}{2}$, $\gamma_s = \gamma = \frac{2}{3L}$, and $T_s \equiv T_{s_0} = 2^{s_0-1}, s \geq s_0$. It then follows from (5.4.9) that

$$\tfrac{4}{3L}\mathbb{E}[\psi(\bar{x}_t) - \psi(x)] + (1+\tfrac{2\mu}{3L})\mathbb{E}[V(x_t,x)] \leq \tfrac{2}{3L}\mathbb{E}[\psi(\tilde{x}) - \psi(x)] + \mathbb{E}[V(x_{t-1},x)].$$

Multiplying both sides of the above inequality by $\Gamma_{t-1} = (1+\frac{2\mu}{3L})^{t-1}$, we obtain

$$\begin{aligned}\tfrac{4}{3L}\Gamma_{t-1}\mathbb{E}[\psi(\bar{x}_t) - \psi(x)] + \Gamma_t\mathbb{E}[V(x_t,x)] &\leq \tfrac{2}{3L}\Gamma_{t-1}\mathbb{E}[\psi(\tilde{x}) - \psi(x)] \\ &\quad + \Gamma_{t-1}\mathbb{E}[V(x_{t-1},x)].\end{aligned}$$

Note that θ_t are chosen as in (5.4.26) when $s \geq s_0$, i.e., $\theta_t = \Gamma_{t-1} = (1+\frac{2\mu}{3L})^{t-1}$, $t = 1,\ldots,T_s$, $s \geq s_0$. Summing up the above inequality for $t = 1,\ldots,T_s$ we have

$$\begin{aligned}&\tfrac{4}{3L}\textstyle\sum_{t=1}^{T_s}\theta_t\mathbb{E}[\psi(\bar{x}_t) - \psi(x)] + \Gamma_{T_s}\mathbb{E}[V(x_{T_s},x)] \\ &\quad \leq \tfrac{2}{3L}\textstyle\sum_{t=1}^{T_s}\theta_t\mathbb{E}[\psi(\tilde{x}) - \psi(x)] + \mathbb{E}[V(x_0,x)],\ s \geq s_0.\end{aligned}$$

Observe that for $s \geq s_0$, $m \geq T_s \equiv T_{s_0} = 2^{\lfloor \log_2 m \rfloor} \geq m/2$, and hence that

$$\Gamma_{T_s} = (1+\tfrac{2\mu}{3L})^{T_s} = (1+\tfrac{2\mu}{3L})^{T_{s_0}} \geq 1 + \tfrac{2\mu T_{s_0}}{3L} \geq 1 + \tfrac{T_{s_0}}{2m} \geq \tfrac{5}{4},\ \forall s \geq s_0, \tag{5.4.28}$$

and using the facts that $\tilde{x}^s = \sum_{t=1}^{T_s}(\theta_t\bar{x}_t)/\sum_{t=1}^{T_s}\theta_t$, $\tilde{x} = \tilde{x}^{s-1}$, $x_0 = x^{s-1}$, and $x_{T_s} = x^s$ in the s epoch, and $\psi(\tilde{x}^s) - \psi(x^*) \geq 0$, we conclude from the above inequalities that

$$\begin{aligned}&\tfrac{5}{4}\left\{\tfrac{2}{3L}\mathbb{E}[\psi(\tilde{x}^s) - \psi(x^*)] + (\textstyle\sum_{t=1}^{T_s}\theta_t)^{-1}\mathbb{E}[V(x^s,x^*)]\right\} \\ &\quad \leq \tfrac{2}{3L}\mathbb{E}[\psi(\tilde{x}^{s-1}) - \psi(x^*)] + (\textstyle\sum_{t=1}^{T_s}\theta_t)^{-1}\mathbb{E}[V(x^{s-1},x^*)], s \geq s_0.\end{aligned}$$

Applying this relation recursively for $s \geq s_0$, we then obtain

$$\begin{aligned}&\tfrac{2}{3L}\mathbb{E}[\psi(\tilde{x}^s) - \psi(x^*)] + (\textstyle\sum_{t=1}^{T_s}\theta_t)^{-1}\mathbb{E}[V(x^s,x^*)] \\ &\leq \left(\tfrac{4}{5}\right)^{s-s_0}\left\{\tfrac{2}{3L}\mathbb{E}[\psi(\tilde{x}^{s_0}) - \psi(x^*)] + (\textstyle\sum_{t=1}^{T_s}\theta_t)^{-1}\mathbb{E}[V(x^{s_0},x^*)]\right\} \\ &\leq \left(\tfrac{4}{5}\right)^{s-s_0}\left\{\tfrac{2}{3L}\mathbb{E}[\psi(\tilde{x}^{s_0}) - \psi(x^*)] + \tfrac{1}{T_{s_0}}\mathbb{E}[V(x^{s_0},x^*)]\right\},\end{aligned}$$

where the last inequality follows from $\sum_{t=1}^{T_s}\theta_t \geq T_s = T_{s_0}$. Plugging (5.4.27) into the above inequality, we have

$$\mathbb{E}[\psi(\tilde{x}^s) - \psi(x^*)] \leq \left(\tfrac{4}{5}\right)^{s-s_0}\tfrac{D_0}{2T_{s_0}} = \left(\tfrac{4}{5}\right)^{s-s_0}\tfrac{D_0}{2^{s_0}} \leq \left(\tfrac{4}{5}\right)^{s}D_0,\ s \geq s_0.$$

■

Lemma 5.20. *If* $s_0 < s \leq s_0 + \sqrt{\frac{12L}{m\mu}} - 4$ *and* $m < \frac{3L}{4\mu}$, *then for any* $x \in X$,

$$\mathbb{E}[\psi(\tilde{x}^s) - \psi(x)] \leq \tfrac{16D_0}{(s-s_0+4)^2 m}.$$

Proof. In this case, $\frac{1}{2} \geq \frac{2}{s-s_0+4} \geq \sqrt{\frac{m\mu}{3L}}$. Therefore, we set θ_t as in (5.4.12), $\alpha_s = \frac{2}{s-s_0+4}$, $p_s = \frac{1}{2}$, $\gamma_s = \frac{1}{3L\alpha_s}$, and $T_s \equiv T_{s_0}$. Observe that the parameter setting in this case is the same as the smooth case in Theorem 5.8. Hence, by following the same procedure as in the proof of Theorem 5.8, we can obtain

$$\begin{aligned} \mathscr{L}_s\mathbb{E}[\psi(\tilde{x}^s) - \psi(x)] + \mathbb{E}[V(x^s,x)] &\leq \mathscr{R}_{s_0+1}\mathbb{E}[\psi(\tilde{x}^{s_0}) - \psi(x)] + \mathbb{E}[V(x^{s_0},x)] \\ &\leq \mathscr{L}_{s_0}\mathbb{E}[\psi(\tilde{x}^{s_0}) - \psi(x)] + \mathbb{E}[V(x^{s_0},x)] \\ &\leq \frac{D_0}{3L}, \end{aligned} \tag{5.4.29}$$

where the last inequality follows from the fact that $\mathscr{L}_{s_0} \geq \frac{2T_{s_0}}{3L}$ and the relation in (5.4.27). The result then follows by noting that $\mathscr{L}_s \geq \frac{(s-s_0+4)^2 m}{48L}$ (see (5.4.21)). ■

Lemma 5.21. *If $s > \bar{s}_0 := s_0 + \sqrt{\frac{12L}{m\mu}} - 4$ and $m < \frac{3L}{4\mu}$, then*

$$\mathbb{E}[\psi(\tilde{x}^s) - \psi(x^*)] \leq \left(1 + \sqrt{\frac{\mu}{3mL}}\right)^{\frac{-m(s-\bar{s}_0)}{2}} \frac{D_0}{3L/4\mu}, \tag{5.4.30}$$

where x^* is an optimal solution of (5.3.1).

Proof. In this case, $\frac{1}{2} \geq \sqrt{\frac{m\mu}{3L}} \geq \frac{2}{s-s_0+4}$. Therefore, we use constant step-size policy that $\alpha_s \equiv \sqrt{\frac{m\mu}{3L}}$, $p_s \equiv \frac{1}{2}$, $\gamma_s \equiv \frac{1}{3L\alpha_s} = \frac{1}{\sqrt{3mL\mu}}$, and $T_s \equiv T_{s_0}$. Also note that in this case θ_t are chosen as in (5.4.26). Multiplying both sides of (5.4.9) by $\Gamma_{t-1} = (1+\mu\gamma_s)^{t-1}$, we obtain

$$\begin{aligned} \tfrac{\gamma_s}{\alpha_s}\Gamma_{t-1}\mathbb{E}[\psi(\bar{x}_t) - \psi(x)] + \Gamma_t\mathbb{E}[V(x_t,x)] &\leq \tfrac{\Gamma_{t-1}\gamma_s}{\alpha_s}(1-\alpha_s-p_s)\mathbb{E}[\psi(\bar{x}_{t-1}) - \psi(x)] \\ &+ \tfrac{\Gamma_{t-1}\gamma_s p_s}{\alpha_s}\mathbb{E}[\psi(\tilde{x}) - \psi(x)] + \Gamma_{t-1}\mathbb{E}[V(x_{t-1},x)]. \end{aligned}$$

Summing up the above inequality from $t = 1,\ldots,T_s$ and using the fact that $\bar{x}_0 = \tilde{x}$, we arrive at

$$\begin{aligned} &\tfrac{\gamma_s}{\alpha_s}\textstyle\sum_{t=1}^{T_s}\theta_t\mathbb{E}[\psi(\bar{x}_t) - \psi(x)] + \Gamma_{T_s}\mathbb{E}[V(x_{T_s},x)] \\ &\leq \tfrac{\gamma_s}{\alpha_s}\left[1-\alpha_s-p_s+p_s\textstyle\sum_{t=1}^{T_s}\Gamma_{t-1}\right]\mathbb{E}[\psi(\tilde{x}) - \psi(x)] + \mathbb{E}[V(x_0,x)]. \end{aligned}$$

Now using the facts that $x^s = x_{T_s}$, $x_0 = x^{s-1}$, $\tilde{x}^s = \sum_{t=1}^{T_s}(\theta_t\bar{x}_t)/\sum_{t=1}^{T_s}\theta_t$, $\tilde{x} = \tilde{x}^{s-1}$, $T_s = T_{s_0}$, and the convexity of ψ, we obtain

$$\begin{aligned} &\tfrac{\gamma_s}{\alpha_s}\textstyle\sum_{t=1}^{T_{s_0}}\theta_t\mathbb{E}[\psi(\tilde{x}^s) - \psi(x)] + \Gamma_{T_{s_0}}\mathbb{E}[V(x^s,x)] \\ &\leq \tfrac{\gamma_s}{\alpha_s}\left[1-\alpha_s-p_s+p_s\textstyle\sum_{t=1}^{T_{s_0}}\Gamma_{t-1}\right]\mathbb{E}[\psi(\tilde{x}^{s-1}) - \psi(x)] + \mathbb{E}[V(x^{s-1},x)] \end{aligned} \tag{5.4.31}$$

for any $s > \bar{s}_0$. Moreover, we have

$$\begin{aligned}\sum_{t=1}^{T_{s_0}} \theta_t &= \Gamma_{T_{s_0}-1} + \sum_{t=1}^{T_{s_0}-1} (\Gamma_{t-1} - (1-\alpha_s - p_s)\Gamma_t) \\ &= \Gamma_{T_{s_0}}(1-\alpha_s - p_s) + \sum_{t=1}^{T_{s_0}} (\Gamma_{t-1} - (1-\alpha_s - p_s)\Gamma_t) \\ &= \Gamma_{T_{s_0}}(1-\alpha_s - p_s) + [1-(1-\alpha_s - p_s)(1+\mu\gamma_s)]\sum_{t=1}^{T_{s_0}} \Gamma_{t-1}.\end{aligned}$$

Observe that for any $T > 1$ and $0 \le \delta T \le 1$, $(1+\delta)^T \le 1 + 2T\delta$, $\alpha_s = \sqrt{\frac{m\mu}{3L}} \ge \sqrt{\frac{T_{s_0}\mu}{3L}}$ and hence that

$$\begin{aligned}1-(1-\alpha_s - p_s)(1+\mu\gamma_s) &\ge (1+\mu\gamma_s)(\alpha_s - \mu\gamma_s + p_s) \\ &\ge (1+\mu\gamma_s)(T_{s_0}\mu\gamma_s - \mu\gamma_s + p_s) \\ &= p_s(1+\mu\gamma_s)[2(T_{s_0}-1)\mu\gamma_s + 1] \\ &\ge p_s(1+\mu\gamma_s)^{T_{s_0}} = p_s\Gamma_{T_{s_0}}.\end{aligned}$$

Then we conclude that $\sum_{t=1}^{T_{s_0}} \theta_t \ge \Gamma_{T_{s_0}}\left[1-\alpha_s - p_s + p_s\sum_{t=1}^{T_{s_0}}\Gamma_{t-1}\right]$. Together with (5.4.31) and the fact that $\psi(\tilde{x}^s) - \psi(x^*) \ge 0$, we have

$$\begin{aligned}&\Gamma_{T_{s_0}}\left\{\tfrac{\gamma_s}{\alpha_s}\left[1-\alpha_s - p_s + p_s\sum_{t=1}^{T_{s_0}}\Gamma_{t-1}\right]\mathbb{E}[\psi(\tilde{x}^s) - \psi(x^*)] + \mathbb{E}[V(x^s,x^*)]\right\} \\ &\quad\le \tfrac{\gamma_s}{\alpha_s}\left[1-\alpha_s - p_s + p_s\sum_{t=1}^{T_{s_0}}\Gamma_{t-1}\right]\mathbb{E}[\psi(\tilde{x}^{s-1}) - \psi(x^*)] + \mathbb{E}[V(x^{s-1},x^*)].\end{aligned}$$

Applying the above relation recursively for $s > \bar{s}_0 = s_0 + \sqrt{\frac{12L}{m\mu}} - 4$, and also noting that $\Gamma_t = (1+\mu\gamma_s)^t$ and the constant step-size policy in this case, we obtain

$$\begin{aligned}&\tfrac{\gamma_s}{\alpha_s}\left[1-\alpha_s - p_s + p_s\sum_{t=1}^{T_{s_0}}\Gamma_{t-1}\right]\mathbb{E}[\psi(\tilde{x}^s) - \psi(x^*)] + \mathbb{E}[V(x^s,x^*)] \\ &\le (1+\mu\gamma_s)^{-T_{s_0}(s-\bar{s}_0)}\left\{\tfrac{\gamma_s}{\alpha_s}\left[1-\alpha_s - p_s + p_s\sum_{t=1}^{T_{s_0}}\Gamma_{t-1}\right]\right. \\ &\quad\left.\mathbb{E}[\psi(\tilde{x}^{\bar{s}_0}) - \psi(x^*)] + \mathbb{E}[V(x^{\bar{s}_0},x^*)]\right\}.\end{aligned}$$

According to the parameter settings in this case, i.e., $\alpha_s \equiv \sqrt{\frac{m\mu}{3L}}, p_s \equiv \frac{1}{2}, \gamma_s \equiv \frac{1}{3L\alpha_s} = \frac{1}{\sqrt{3mL\mu}}$, and $\bar{s}_0 = s_0 + \sqrt{\frac{12L}{m\mu}} - 4$, we have $\frac{\gamma_s}{\alpha_s}\left[1-\alpha_s - p_s + p_s\sum_{t=1}^{T_{s_0}}\Gamma_{t-1}\right] \ge \frac{\gamma_s p_s T_{s_0}}{\alpha_s} = \frac{T_{s_0}}{2m\mu} = \frac{(\bar{s}_0 - s_0 + 4)^2 T_{s_0}}{24L}$. Using this observation in the above inequality, we then conclude that

$$\begin{aligned}\mathbb{E}[\psi(\tilde{x}^s) - \psi(x^*)] &\le (1+\mu\gamma_s)^{-T_{s_0}(s-\bar{s}_0)}\left[\mathbb{E}[\psi(\tilde{x}^{\bar{s}_0}) - \psi(x^*)]\right. \\ &\quad\left.+\tfrac{24L}{(\bar{s}_0 - s_0 + 4)^2 T_{s_0}}\mathbb{E}[V(x^{\bar{s}_0},x^*)]\right] \\ &\le (1+\mu\gamma_s)^{-T_{s_0}(s-\bar{s}_0)}\tfrac{24L}{(\bar{s}_0 - s_0 + 4)^2 T_{s_0}}\left[\mathcal{L}_{\bar{s}_0}\mathbb{E}[\psi(\tilde{x}^{\bar{s}_0}) - \psi(x^*)] + \mathbb{E}[V(x^{\bar{s}_0},x^*)]\right]\end{aligned}$$

$$
\begin{aligned}
&\leq (1+\mu\gamma_s)^{-T_{s_0}(s-\bar{s}_0)} \frac{24L}{(\bar{s}_0-s_0+4)^2 T_{s_0}} \frac{D_0}{3L} \\
&\leq (1+\mu\gamma_s)^{-T_{s_0}(s-\bar{s}_0)} \frac{16D_0}{(\bar{s}_0-s_0+4)^2 m} \\
&= (1+\mu\gamma_s)^{-T_{s_0}(s-\bar{s}_0)} \frac{D_0}{3L/4\mu},
\end{aligned}
$$

where the second inequality follows from the fact that $\mathscr{L}_{\bar{s}_0} \geq \frac{(\bar{s}_0-s_0+4)^2 T_{s_0}}{24L} = \frac{T_{s_0}}{2m\mu}$ due to (5.4.21), the third inequality follows from (5.4.29) in Case 3, and last inequality follows from $T_{s_0} = 2^{\lfloor \log_2 m \rfloor} \geq m/2$. ■

Putting the above four technical results together, we obtain the following main result of this subsection.

Theorem 5.9. *Suppose that the probabilities q_i's are set to $L_i/\sum_{i=1}^m L_i$ for $i = 1,\ldots,m$. Moreover, let us denote $s_0 := \lfloor \log m \rfloor + 1$ and assume that the weights $\{\theta_t\}$ are set to* (5.4.12) *if $1 \leq s \leq s_0$ or $s_0 < s \leq s_0 + \sqrt{\frac{12L}{m\mu}} - 4$, $m < \frac{3L}{4\mu}$. Otherwise, they are set to (5.4.26). If the parameters $\{T_s\}$, $\{\gamma_s\}$, and $\{p_s\}$ set to* (5.4.17) *with $\{\alpha_s\}$ given by (5.4.25), then we have*

$$
\mathbb{E}[\psi(\tilde{x}^s) - \psi(x^*)] \leq \begin{cases}
2^{-(s+1)} D_0, & 1 \leq s \leq s_0, \\
\left(\frac{4}{5}\right)^s D_0, & s > s_0 \text{ and } m \geq \frac{3L}{4\mu}, \\
\frac{16D_0}{(s-s_0+4)^2 m}, & s_0 < s \leq s_0 + \sqrt{\frac{12L}{m\mu}} - 4 \\
& \text{and } m < \frac{3L}{4\mu}, \\
\left(1+\sqrt{\frac{\mu}{3mL}}\right)^{\frac{-m(s-\bar{s}_0)}{2}} \frac{D_0}{3L/4\mu}, & s_0 + \sqrt{\frac{12L}{m\mu}} - 4 = \bar{s}_0 < s \\
& \text{and } m < \frac{3L}{4\mu},
\end{cases} \tag{5.4.32}
$$

where x^* is an optimal solution of (5.3.1) and D_0 is defined as in (5.4.20).

We are now ready to derive the complexity bound in terms of the total number of gradient evaluations.

Corollary 5.11. *The total number of gradient evaluations of f_i performed by Algorithm 5.7 to find a stochastic ε-solution of* (2.0.1) *can be bounded by*

$$
\bar{N} := \begin{cases}
\mathscr{O}\left\{ m \log \frac{D_0}{\varepsilon} \right\}, & m \geq \frac{D_0}{\varepsilon} \text{ or } m \geq \frac{3L}{4\mu}, \\
\mathscr{O}\left\{ m \log m + \sqrt{\frac{mD_0}{\varepsilon}} \right\}, & m < \frac{D_0}{\varepsilon} \leq \frac{3L}{4\mu}, \\
\mathscr{O}\left\{ m \log m + \sqrt{\frac{mL}{\mu}} \log \frac{D_0/\varepsilon}{3L/4\mu} \right\}, & m < \frac{3L}{4\mu} \leq \frac{D_0}{\varepsilon}.
\end{cases} \tag{5.4.33}
$$

where D_0 is defined as in (5.4.20).

Proof. First, it is clear that the first case and the third case correspond to the results of the smooth case discussed in Corollary 5.10. As a consequence, the total number of gradient evaluations can also be bounded by (5.4.23) and (5.4.24), respectively. Second, for the second case of (5.4.32), it is easy to check that the variance-reduced accelerated gradient method needs to run at most $S := \mathcal{O}\{\log D_0/\varepsilon\}$ epochs, and hence the total number of gradient evaluations can be bounded by

$$mS + \sum_{s=1}^{S} T_s \le 2mS = \mathcal{O}\left\{m \log \tfrac{D_0}{\varepsilon}\right\}. \tag{5.4.34}$$

Finally, for the last case of (5.4.32), since the variance-reduced accelerated gradient method only needs to run at most $S' = \bar{s}_0 + 2\sqrt{\frac{3L}{m\mu}} \log \frac{D_0/\varepsilon}{3L/4\mu}$ epochs, the total number of gradient evaluations can be bounded by

$$\begin{aligned}
\sum_{s=1}^{S'}(m+T_s) &= \sum_{s=1}^{s_0}(m+T_s) + \sum_{s=s_0+1}^{\bar{s}_0}(m+T_{s_0}) + (m+T_{s_0})(S'-\bar{s}_0) \\
&\le 2m\log m + 2m(\sqrt{\tfrac{12L}{m\mu}} - 4) + 4m\sqrt{\tfrac{3L}{m\mu}} \log \tfrac{D_0/\varepsilon}{3L/4\mu} \\
&= \mathcal{O}\left\{m\log m + \sqrt{\tfrac{mL}{\mu}} \log \tfrac{D_0/\varepsilon}{3L/4\mu}\right\}.
\end{aligned}$$

■

Observe that the complexity bound (5.4.33) is a unified convergence result for the variance-reduced accelerated gradient method to solve deterministic smooth finite-sum optimization problems (2.0.1). When the strongly convex modulus μ of the objective function is large enough, i.e., $3L/\mu < D_0/\varepsilon$, the variance-reduced accelerated gradient method exhibits an optimal linear rate of convergence as can be seen from the third case of (5.4.33). If μ is relatively small, this algorithm treats the finite-sum problem (5.3.1) as a smooth problem without strongly convexity, which leads to the same complexity bounds as in Corollary 5.10.

5.4.3 Problems Satisfying an Error-Bound Condition

In this subsection, we establish the convergence properties of the variance-reduced accelerated gradient method to solve some more general classes of finite-sum optimization problems. In particular, we investigate a class of weakly strongly convex problems, i.e., the objective function $\phi(x)$ satisfies the error bound condition given by

$$V(x, X^*) \le \tfrac{1}{\bar{\mu}}(\psi(x) - \psi^*), \ \forall x \in X, \tag{5.4.35}$$

where X^* denotes the set of optimal solutions of (5.3.1).

Many optimization problems satisfy the above error bound condition, including linear systems, quadratic programs, linear matrix inequalities, and composite problems with strongly convex outer function and polyhedral inner functions. Even though these problems are not strongly convex, by properly restarting the variance-reduced accelerated gradient method we can solve them with an accelerated linear rate of convergence as shown in the following result.

Theorem 5.10. *Assume that the probabilities q_i's are set to $L_i/\sum_{i=1}^m L_i$ for $i = 1,\ldots,m$, and θ_t are defined as* (5.4.12). *Moreover, let us set parameters $\{\gamma_s\}$, $\{p_s\}$, and $\{\alpha_s\}$ as in* (5.4.17) *and* (5.4.18) *with*

$$T_s = \begin{cases} T_1 2^{s-1}, & s \le s_0 \\ T_{s_0}, & s > s_0 \end{cases}, \tag{5.4.36}$$

where $s_0 = 4, s = s_0 + 4\sqrt{\frac{L}{\bar{\mu}m}}$ and $T_1 = \min\{m, \frac{L}{\bar{\mu}}\}$. Then under (5.4.35), *for any $x^* \in X^*$,*

$$\mathbb{E}[\psi(\tilde{x}^s) - \psi(x^*)] \le \tfrac{5}{16}[\psi(x^0) - \psi(x^*)]. \tag{5.4.37}$$

Moreover, if we restart Algorithm 5.7 $k = \log\frac{\psi(x^0)-\psi(x^*)}{\varepsilon}$ times and each time has s iterations, then

$$\mathbb{E}[\psi(\tilde{x}^{sk}) - \psi(x^*)] \le \left(\tfrac{5}{16}\right)^k [\psi(x^0) - \psi(x^*)] \le \varepsilon,$$

and the total number of gradient evaluations of f_i to find a stochastic ε-solution of (2.0.1) *can be bounded by*

$$\bar{N} := k(\textstyle\sum_s (m + T_s)) = \mathscr{O}\left(m + \sqrt{\tfrac{mL}{\bar{\mu}}}\right)\log\frac{\psi(x^0)-\psi(x^*)}{\varepsilon}. \tag{5.4.38}$$

Proof. Similar to the smooth case, according to (5.4.15), for any $x \in X$, we have

$$\begin{aligned}\mathscr{L}_s\mathbb{E}[\psi(\tilde{x}^s) - \psi(x)] &\le \mathscr{R}_1\mathbb{E}[\psi(\tilde{x}^0) - \psi(x)] + \mathbb{E}[V(x^0,x) - V(x^s,x)] \\ &\le \mathscr{R}_1[\psi(x^0) - \psi(x)] + V(x^0,x).\end{aligned}$$

Then we use x^* to replace x and use the relation of (5.4.35) to obtain

$$\mathscr{L}_s\mathbb{E}[\psi(\tilde{x}^s) - \psi(x^*)] \le \mathscr{R}_1[\psi(x^0) - \psi(x^*)] + \tfrac{1}{u}[\psi(x) - \psi(x^*)].$$

Now, we compute $\mathscr{L}_s$ and $\mathscr{R}_1$. According to (5.4.21), we have $\mathscr{L}_s \ge \frac{(s-s_0+4)^2(T_{s_0}+1)}{24L}$. We have $\mathscr{R}_1 = \frac{2T_1}{3L}$ by plugging the parameters γ_1, p_1, α_1, and T_1 into (5.4.13). Thus, we prove (5.4.37) as follows (recall that $s_0 = 4$ and $s = s_0 + 4\sqrt{\frac{L}{\bar{\mu}m}}$):

$$\begin{aligned}\mathbb{E}[\psi(\tilde{x}^s) - \psi(x^*)] &\le \frac{16T_1 + 24L/\bar{\mu}}{(s-s_0+4)^2 T_1 2^{s_0-1}}[\psi(x^0) - \psi(x^*)] \\ &\le \frac{16 + 24L/(\bar{\mu}T_1)}{(s-s_0+4)^2 2^{s_0-1}}[\psi(x^0) - \psi(x^*)] \\ &\le \frac{5}{16}\frac{L/(\bar{\mu}T_1)}{1+L/(\bar{\mu}m)}[\psi(x^0) - \psi(x^*)] \\ &\le \tfrac{5}{16}[\psi(x^0) - \psi(x^*)],\end{aligned}$$

where the last inequality follows from $T_1 = \min\{m, \frac{L}{\bar{\mu}}\}$. Finally, we plug $k = \log \frac{\psi(x^0)-\psi(x^*)}{\varepsilon}$, $s_0 = 4, s = s_0 + 4\sqrt{\frac{L}{\bar{\mu}m}}$ and $T_1 = \min\{m, \frac{L}{\bar{\mu}}\}$ to prove (5.4.38):

$$\bar{N} := k(\Sigma_s(m+T_s)) \le k(ms + T_1 2^{s_0}(s - s_0 + 1)) = \mathscr{O}\left(m + \sqrt{\tfrac{mL}{\bar{\mu}}}\right) \log \tfrac{\psi(x^0)-\psi(x^*)}{\varepsilon}.$$

■

5.5 Exercises and Notes

1. Figure out one way to use the function values f_i and its gradients ∇f_i, rather than the conjugate functions J_i, in the analysis of the primal–dual gradient method.
2. It is possible to specialize the mirror-prox method for solving the saddle point problem in (5.2.14). Please state this algorithm in such a way that only gradient computation rather than dual prox-mapping is involved.
3. Consider the finite-sum problem in (5.3.1). Also assume that (5.3.2) holds with $\mu > 0$. Show the rate of convergence of the following variance-reduced gradient method applied to solve this problem. Given the value of x^{k-1} and of each $f_i'(\phi_i^{k-1})$ at the end of iteration $k-1$, the updates for iteration k are as follows:

   (a) Pick a j uniformly at random.
   (b) Take $\phi_j^k = x^{k-1}$, and store $f_j'(\phi_j^k)$ in the table. All other entries in the table remain unchanged. The quantity ϕ_j^k is not explicitly stored.
   (c) Update x using $f_j'(\phi_j^k)$, $f_j'(\phi_j^{k-1})$ and the table average:

$$w^k = x^{k-1} - \gamma\left[f_j'(\phi_j^k) - f_j'(\phi_j^{k-1}) + \tfrac{1}{n}\Sigma_{i=1}^n f_i'(\phi_i^{k-1})\right], \tag{5.5.39}$$

$$x^k = \operatorname{argmin}_{x\in X}\left\{\gamma[\langle w^k, x\rangle + h(x)] + V(x_t, x)\right\}. \tag{5.5.40}$$

4. Try to establish the convergence of the variance-reduced mirror descent for solving stochastic finite-sum problems in (5.2.5).
5. Try to establish the convergence of the variance-reduced accelerated gradient method for solving stochastic finite-sum problems in (5.2.5).

Notes. The randomized primal–dual gradient (RPDG) method was first presented by Lan and Zhou [71] in 2015. They also established a lower complexity bound for solving a class of finite-sum optimization problems. The gradient extrapolation methods for stochastic and distributed optimization were first presented by Lan and Zhou in [70]. RPDG and the catalyst scheme by Lin et al. [78] are the first two incremental gradient methods that can achieve the lower complexity bound in [71]. Allen-zhu [2] presented the Katyusha method which combines a special class of accelerated SGD with variance reduction technique. Lan et al. [74] introduced a new variance-reduced accelerated gradient method called Varag, which is uniformly optimal for convex and strongly convex problems, as well as problems satisfying an error-bound condition. Our introduction to variance-reduced accelerated gradient method follows [74]. Earlier important progresses on random incremental gradient method were made in [15, 29, 56, 121] among many others. In particular, Schimidt et al. [121] presented a stochastic average gradient (SAG) method, which uses randomized sampling of f_i to update the gradients, and can achieve a linear rate of convergence, i.e., an $\mathcal{O}\{m+(mL/\mu)\log(1/\varepsilon)\}$ complexity bound, to solve unconstrained finite-sum problems (5.2.1). Johnson and Zhang later in [56] presented a stochastic variance-reduced gradient (SVRG) method, which computes an estimator of ∇f by iteratively updating the gradient of one randomly selected f_i of the current exact gradient information and reevaluating the exact gradient from time to time. [140] later extended SVRG to solve proximal finite-sum problems (5.2.1). All these methods exhibit an $\mathcal{O}\{(m+L/\mu)\log(1/\varepsilon)\}$ complexity bound, and [29] also presented an improved SAG method, called SAGA, that can achieve such a complexity result. Related stochastic dual methods (e.g., [122, 123, 141]) may involve the solution of a more complicated subproblem. In spite of these developments, variance reduction was not incorporated into the more general mirror descent method until Sect. 5.3 of this monograph to the best of our knowledge.

Chapter 6
Nonconvex Optimization

In the last few chapters, we have discussed a few stochastic gradient descent type methods and established their convergence rates for solving different convex optimization problems. Note however that convexity has played an important role in our analysis. In this chapter, we focus on stochastic optimization problems which are not necessarily convex. We first introduce some new stochastic optimization algorithms, including the randomized stochastic gradient and stochastic accelerated gradient descent methods, for solving these nonconvex problems. We establish the complexity of these methods for computing an approximate stationary point of a nonlinear programming problem. We will also discuss variants of these methods for solving a class of simulation-based optimization problems in which only stochastic zeroth-order information is available. In addition, we study variance-reduced mirror descent methods for solving nonconvex finite-sum problems, as well as indirect acceleration schemes through proximal point methods for solving nonconvex finite-sum and multi-block problems.

6.1 Unconstrained Nonconvex Stochastic Optimization

In this section, we study the classical unconstrained nonlinear programming (NLP) problem given in the form of

$$f^* := \inf_{x\in\mathbb{R}^n} f(x), \tag{6.1.1}$$

where $f : \mathbb{R}^n \to \mathbb{R}$ is a differentiable (not necessarily convex), bounded from below, and its gradient $\nabla f(\cdot)$ satisfies

$$\|\nabla f(y) - \nabla f(x)\| \le L\|y-x\|, \quad \forall x,y \in \mathbb{R}^n.$$

G. Lan, *First-Order and Stochastic Optimization Methods for Machine Learning*,
Springer Series in the Data Sciences, https://doi.org/10.1007/978-3-030-39568-1_6

We say that $f \in \mathscr{C}_L^{1,1}(\mathbb{R}^n)$ if it is differentiable and the above assumption is satisfied. Clearly, we have

$$|f(y) - f(x) - \langle \nabla f(x), y - x \rangle| \leq \tfrac{L}{2}\|y - x\|^2, \quad \forall x, y \in \mathbb{R}^n. \tag{6.1.2}$$

If, in addition, $f(\cdot)$ is convex, then

$$f(y) - f(x) - \langle \nabla f(x), y - x \rangle \geq \tfrac{1}{2L}\|\nabla f(y) - \nabla f(x)\|^2, \tag{6.1.3}$$

and

$$\langle \nabla f(y) - \nabla f(x), y - x \rangle \geq \tfrac{1}{L}\|\nabla f(y) - \nabla f(x)\|^2, \quad \forall x, y \in \mathbb{R}^n. \tag{6.1.4}$$

However, different from the standard NLP, we assume throughout this section that we only have access to noisy function values or gradients about the objective function f in (6.1.1). In particular, in the basic setting, we assume that problem (6.1.1) is to be solved by iterative algorithms which acquire the gradients of f via subsequent calls to a stochastic first-order oracle (SFO). At iteration k of the algorithm, x_k being the input, the SFO outputs a *stochastic gradient* $G(x_k, \xi_k)$, where ξ_k, $k \geq 1$, are random variables whose distributions P_k are supported on $\Xi_k \subseteq \mathbb{R}^d$. The following assumptions are made for the Borel functions $G(x_k, \xi_k)$.

Assumption 13 *For any $k \geq 1$, we have*

$$a)\ \mathbb{E}[G(x_k, \xi_k)] = \nabla f(x_k), \tag{6.1.5}$$

$$b)\ \mathbb{E}\left[\|G(x_k, \xi_k) - \nabla f(x_k)\|^2\right] \leq \sigma^2, \tag{6.1.6}$$

for some parameter $\sigma \geq 0$.

Observe that, by (6.1.5), $G(x_k, \xi_k)$ is an unbiased estimator of $\nabla f(x_k)$ and, by (6.1.6), the variance of the random variable $\|G(x_k, \xi_k) - \nabla f(x_k)\|$ is bounded. It is worth noting that in the standard setting for stochatic pogramming (SP), the random vectors ξ_k, $k = 1, 2, \ldots$, are independent of each other (and also of x_k). Our assumption here is slightly weaker since we do not need to assume ξ_k, $k = 1, 2, \ldots$, to be independent.

Our study on the aforementioned SP problems has been motivated by a few interesting applications which are briefly outlined as follows.

- In many *machine learning* problems, we intend to minimize a regularized loss function $f(\cdot)$ given by

$$f(x) = \int_{\Xi} L(x, \xi) dP(\xi) + r(x), \tag{6.1.7}$$

  where either the loss function $L(x, \xi)$ or the regularization $r(x)$ is nonconvex.
- Another important class of problems originate from the so-called *endogenous uncertainty* in SP. More specifically, the objective functions for these SP problems are given in the form of

$$f(x) = \int_{\Xi(x)} F(x,\xi)dP_x(\xi), \tag{6.1.8}$$

where the support $\Xi(x)$ and the distribution function P_x of the random vector ξ depend on x. The function f in (6.1.8) is usually nonconvex even if $F(x,\xi)$ is convex with respect to x. For example, if the support Ξ does not depend on x, it is often possible to represent $dP_x = H(x)dP$ for some fixed distribution P. Typically this transformation results in a nonconvex integrand function. Other techniques have also been developed to compute unbiased estimators for the gradient of $f(\cdot)$ in (6.1.8).

- Finally, in *simulation-based optimization*, the objective function is given by $f(x) = \mathbb{E}_\xi[F(x,\xi)]$, where $F(\cdot,\xi)$ is not given explicitly, but through a black-box simulation procedure. Therefore, we do not know if the function f is convex or not. Moreover, in these cases, we usually only have access to stochastic zeroth-order information about the function values of $f(\cdot)$ rather than its gradients.

The complexity of the gradient descent method for solving problem (6.1.1) has been well understood under the deterministic setting (i.e., $\sigma = 0$ in (6.1.6)). In particular, it is well known that after running the method for at most $N = \mathcal{O}(1/\varepsilon)$ steps, we have $\min_{k=1,\ldots,N} \|\nabla f(x_k)\|^2 \le \varepsilon$. Note, however, such an analysis is not applicable to the stochastic setting (i.e., $\sigma > 0$ in (6.1.6)). Moreover, even if we have $\min_{k=1,\ldots,N} \|\nabla f(x_k)\|^2 \le \varepsilon$, to find the best solution from $\{x_1,\ldots,x_N\}$ is still difficult since $\|\nabla f(x_k)\|$ is not known exactly. This section proceeds as follows. First, to solve the aforementioned nonconvex SP problem, we present a randomized stochastic gradient descent (RSGD) method by introducing the following modifications to the classical stochastic gradient descent. Instead of taking average of the iterates as in the stochastic mirror descent for convex SP, we randomly select a solution $\bar{x}$ from $\{x_1,\ldots,x_N\}$ according to a certain probability distribution as the output. We show that such a solution satisfies $\mathbb{E}[\|\nabla f(\bar{x})\|^2] \le \varepsilon$ after running the method for at most $N = \mathcal{O}(1/\varepsilon^2)$ iterations. Moreover, if $f(\cdot)$ is convex, we show that the relation $\mathbb{E}[f(\bar{x}) - f^*] \le \varepsilon$ always holds. We demonstrate that such a complexity result is nearly optimal for solving convex SP problems (see the discussions after Corollary 6.1). It should not be too surprising to see that the complexity for the stochastic case is much worse than that for the deterministic case. For example, in the convex case, it is known from the previous chapter that the complexity for finding an solution $\bar{x}$ satisfying $f(\bar{x}) - f^* \le \varepsilon$ will be substantially increased from $\mathcal{O}(1/\sqrt{\varepsilon})$ to $\mathcal{O}(1/\varepsilon^2)$ as one moves from the deterministic to stochastic setting.

Second, in order to improve the large deviation properties and hence the reliability of the RSGD method, we present a two-phase randomized stochastic gradient descent (2-RSGD) method by introducing a post-optimization phase to evaluate a short list of solutions generated by several independent runs of the RSGD method. We show that the complexity of the 2-RSGD method for computing an (ε,Λ)-*solution* of problem (6.1.1), i.e., a point $\bar{x}$ such that $\text{Prob}\{\|\nabla f(\bar{x})\|^2 \le \varepsilon\} \ge 1-\Lambda$ for some $\varepsilon > 0$ and $\Lambda \in (0,1)$, can be bounded by

$$\mathcal{O}\left\{\frac{\log(1/\Lambda)\sigma^2}{\varepsilon}\left[\frac{1}{\varepsilon} + \frac{\log(1/\Lambda)}{\Lambda}\right]\right\}.$$

We further show that, under certain light-tail assumption about the SFO, the above complexity bound can be reduced to

$$\mathcal{O}\left\{\frac{\log(1/\Lambda)\sigma^2}{\varepsilon}\left(\frac{1}{\varepsilon}+\log\frac{1}{\Lambda}\right)\right\}.$$

Third, we specialize the RSGD method for the case where only stochastic zeroth-order information is available. There exists a somewhat long history for the development of zeroth-order (or derivative-free) methods in nonlinear programming. However, only few complexity results are available for these types of methods, mostly for convex programming and deterministic nonconvex programming problems. By incorporating a Gaussian smoothing technique into the aforementioned RSGD method, we present a randomized stochastic gradient free (RSGF) method for solving a class of simulation-based optimization problems and demonstrate that its iteration complexity for finding the aforementioned ε-solution (i.e., $\mathbb{E}[\|\nabla f(\bar{x})\|^2] \le \varepsilon$) can be bounded by $\mathcal{O}(n/\varepsilon^2)$. Moreover, the same RSGF algorithm possesses an $\mathcal{O}(n/\varepsilon^2)$ complexity bound, in terms of $\mathbb{E}[f(\bar{x}) - f^*] \le \varepsilon$, for solving smooth convex SP problems.

6.1.1 Stochastic First-Order Methods

Our goal in this section is to present and analyze a new class of SGD algorithms for solving general smooth nonlinear (possibly nonconvex) SP problems. More specifically, we present the RSGD method and establish its convergence properties in Sect. 6.1.1.1, and then introduce the 2-RSGD method which can significantly improve the large-deviation properties of the RSGD method in Sect. 6.1.1.2.

We assume throughout this section that Assumption 13 holds. In some cases, Assumption 13 is augmented by the following "light-tail" assumption.

Assumption 14 *For any $x \in \mathbb{R}^n$ and $k \ge 1$, we have*

$$\mathbb{E}\left[\exp\{\|G(x,\xi_k) - \nabla f(x)\|^2/\sigma^2\}\right] \le \exp\{1\}. \tag{6.1.9}$$

It can be easily seen that Assumption 14 implies Assumption 13(b) by Jensen's inequality.

6.1.1.1 The Randomized Stochastic Gradient Method

The convergence of existing SGD methods requires $f(\cdot)$ to be convex. Moreover, in order to guarantee the convexity of $f(\cdot)$, one often need to assume that the random variables ξ_k, $k \ge 1$, to be independent of the search sequence $\{x_k\}$. Below we present a new SGD-type algorithm that can deal with both convex and nonconvex SP problems, and allow random noises to be dependent on the search sequence. This algorithm is obtained by incorporating a certain randomization scheme into the classical SGD method.

A Randomized Stochastic Gradient (RSGD) Method

Input: Initial point x_1, iteration limit N, stepsizes $\{\gamma_k\}_{k\geq 1}$ and probability mass function $P_R(\cdot)$ supported on $\{1,\ldots,N\}$.
Step 0. Let R be a random variable with probability mass function P_R.
Step $k = 1,\ldots,R$. Call the stochastic first-order oracle for computing $G(x_k,\xi_k)$ and set

$$x_{k+1} = x_k - \gamma_k G(x_k,\xi_k). \tag{6.1.10}$$

Output x_R.

A few remarks about the above RSGD method are in order. First, in comparison with the classical SGD, we have used a random iteration count, R, to terminate the execution of the RSGD algorithm. Equivalently, one can view such a randomization scheme from a slightly different perspective described as follows. Instead of terminating the algorithm at the R-th step, one can also run the RSGD algorithm for N iterations but randomly choose a search point x_R (according to P_R) from its trajectory as the output of the algorithm. Clearly, using the latter scheme, we just need to run the algorithm for the first R iterations and the remaining $N-R$ iterations are surpluses. Note, however, that the primary goal of introducing the random iteration count R is to derive new complexity results for nonconvex SP, rather than saving the computational efforts in the last $N-R$ iterations of the algorithm. Indeed, if R is uniformly distributed, the computational gain from such a randomization scheme is simply a factor of 2. Second, the RSGD algorithm described above is conceptual only because we have not specified the selection of the stepsizes $\{\gamma_k\}$ and the probability mass function P_R yet. We will address this issue after establishing some basic convergence properties of the RSGD method.

The following result describes some convergence properties of the RSGD method.

Theorem 6.1. *Suppose that the stepsizes $\{\gamma_k\}$ and the probability mass function $P_R(\cdot)$ in the RSGD method are chosen such that $\gamma_k < 2/L$ and*

$$P_R(k) := \text{Prob}\{R = k\} = \frac{2\gamma_k - L\gamma_k^2}{\sum_{k=1}^N (2\gamma_k - L\gamma_k^2)}, \quad k = 1,\ldots,N. \tag{6.1.11}$$

Then, under Assumption 13,

(a) for any $N \geq 1$, we have

$$\tfrac{1}{L}\mathbb{E}[\|\nabla f(x_R)\|^2] \leq \frac{D_f^2 + \sigma^2 \sum_{k=1}^N \gamma_k^2}{\sum_{k=1}^N (2\gamma_k - L\gamma_k^2)}, \tag{6.1.12}$$

where the expectation is taken with respect to R and $\xi_{[N]} := (\xi_1,\ldots,\xi_N)$,

$$D_f := \left[\frac{2(f(x_1) - f^*)}{L}\right]^{\frac{1}{2}}, \tag{6.1.13}$$

and f^* denotes the optimal value of problem (6.1.1);

(b) if, in addition, problem (6.1.1) is convex with an optimal solution x^*, then, for any $N \geq 1$,

$$\mathbb{E}[f(x_R) - f^*] \leq \frac{D_X^2 + \sigma^2 \sum_{k=1}^N \gamma_k^2}{\sum_{k=1}^N (2\gamma_k - L\gamma_k^2)}, \tag{6.1.14}$$

where the expectation is taken with respect to R and $\xi_{[N]}$, and

$$D_X := \|x_1 - x^*\|. \tag{6.1.15}$$

Proof. Denote $\delta_k \equiv G(x_k, \xi_k) - \nabla f(x_k)$, $k \geq 1$. We first show part (a). Using the assumption that $f \in \mathscr{C}_L^{1,1}(\mathbb{R}^n)$, (6.1.2) and (6.1.10), we have, for any $k = 1, \ldots, N$,

$$\begin{aligned}
f(x_{k+1}) &\leq f(x_k) + \langle \nabla f(x_k), x_{k+1} - x_k \rangle + \tfrac{L}{2}\gamma_k^2 \|G(x_k, \xi_k)\|^2 \\
&= f(x_k) - \gamma_k \langle \nabla f(x_k), G(x_k, \xi_k) \rangle + \tfrac{L}{2}\gamma_k^2 \|G(x_k, \xi_k)\|^2 \\
&= f(x_k) - \gamma_k \|\nabla f(x_k)\|^2 - \gamma_k \langle \nabla f(x_k), \delta_k \rangle \\
&\quad + \tfrac{L}{2}\gamma_k^2 \left[\|\nabla f(x_k)\|^2 + 2\langle \nabla f(x_k), \delta_k \rangle + \|\delta_k\|^2\right] \\
&= f(x_k) - \left(\gamma_k - \tfrac{L}{2}\gamma_k^2\right) \|\nabla f(x_k)\|^2 \\
&\quad - \left(\gamma_k - L\gamma_k^2\right) \langle \nabla f(x_k), \delta_k \rangle + \tfrac{L}{2}\gamma_k^2 \|\delta_k\|^2.
\end{aligned} \tag{6.1.16}$$

Summing up the above inequalities and rearranging the terms, we obtain

$$\begin{aligned}
\textstyle\sum_{k=1}^N \left(\gamma_k - \tfrac{L}{2}\gamma_k^2\right) \|\nabla f(x_k)\|^2 &\leq f(x_1) - f(x_{N+1}) - \textstyle\sum_{k=1}^N \left(\gamma_k - L\gamma_k^2\right) \langle \nabla f(x_k), \\
&\quad \delta_k \rangle + \tfrac{L}{2} \textstyle\sum_{k=1}^N \gamma_k^2 \|\delta_k\|^2 \\
&\leq f(x_1) - f^* - \textstyle\sum_{k=1}^N \left(\gamma_k - L\gamma_k^2\right) \langle \nabla f(x_k), \delta_k \rangle \\
&\quad + \tfrac{L}{2} \textstyle\sum_{k=1}^N \gamma_k^2 \|\delta_k\|^2,
\end{aligned} \tag{6.1.17}$$

where the last inequality follows from the fact that $f(x_{N+1}) \geq f^*$. Note that the search point x_k is a function of the history $\xi_{[k-1]}$ of the generated random process and hence is random. Taking expectations (with respect to $\xi_{[N]}$) on both sides of (6.1.17) and noting that under Assumption 13, $\mathbb{E}[\|\delta_k\|^2] \leq \sigma^2$, and

$$\mathbb{E}[\langle \nabla f(x_k), \delta_k \rangle | \xi_{[k-1]}] = 0, \tag{6.1.18}$$

we obtain

$$\textstyle\sum_{k=1}^N \left(\gamma_k - \tfrac{L}{2}\gamma_k^2\right) \mathbb{E}_{\xi_{[N]}} \|\nabla f(x_k)\|^2 \leq f(x_1) - f^* + \tfrac{L\sigma^2}{2} \sum_{k=1}^N \gamma_k^2. \tag{6.1.19}$$

Dividing both sides of the above inequality by $L\sum_{k=1}^N \left(\gamma_k - L\gamma_k^2/2\right)$ and noting that

$$\mathbb{E}[\|\nabla f(x_R)\|^2] = \mathbb{E}_{R,\xi_{[N]}}[\|\nabla f(x_R)\|^2] = \frac{\sum_{k=1}^N (2\gamma_k - L\gamma_k^2) \mathbb{E}_{\xi_{[N]}} \|\nabla f(x_k)\|^2}{\sum_{k=1}^N (2\gamma_k - L\gamma_k^2)},$$

we conclude

$$\tfrac{1}{L}\mathbb{E}[\|\nabla f(x_R)\|^2] \leq \frac{1}{\sum_{k=1}^N (2\gamma_k - L\gamma_k^2)} \left[\frac{2(f(x_1) - f^*)}{L} + \sigma^2 \textstyle\sum_{k=1}^N \gamma_k^2\right],$$

which, in view of (6.1.13), clearly implies (6.1.12).

We now show that part (b) holds. Denote $v_k \equiv \|x_k - x^*\|$. First observe that, for any $k = 1,\ldots,N$,

$$
\begin{aligned}
v_{k+1}^2 &= \|x_k - \gamma_k G(x_k,\xi_k) - x^*\|^2\\
&= v_k^2 - 2\gamma_k\langle G(x_k,\xi_k), x_k - x^*\rangle + \gamma_k^2\|G(x_k,\xi_k)\|^2\\
&= v_k^2 - 2\gamma_k\langle \nabla f(x_k)+\delta_k, x_k - x^*\rangle + \gamma_k^2\left(\|\nabla f(x_k)\|^2 + 2\langle \nabla f(x_k),\delta_k\rangle + \|\delta_k\|^2\right).
\end{aligned}
$$

Moreover, in view of (6.1.4) and the fact that $\nabla f(x^*) = 0$, we have

$$
\tfrac{1}{L}\|\nabla f(x_k)\|^2 \le \langle \nabla f(x_k), x_k - x^*\rangle. \tag{6.1.20}
$$

Combining the above two relations, we obtain, for any $k = 1,\ldots,N$,

$$
\begin{aligned}
v_{k+1}^2 &\le v_k^2 - (2\gamma_k - L\gamma_k^2)\langle \nabla f(x_k), x_k - x^*\rangle - 2\gamma_k\langle x_k - \gamma_k\nabla f(x_k) - x^*, \delta_k\rangle + \gamma_k^2\|\delta_k\|^2\\
&\le v_k^2 - (2\gamma_k - L\gamma_k^2)[f(x_k) - f^*] - 2\gamma_k\langle x_k - \gamma_k\nabla f(x_k) - x^*, \delta_k\rangle + \gamma_k^2\|\delta_k\|^2,
\end{aligned}
$$

where the last inequality follows from the convexity of $f(\cdot)$ and the fact that $\gamma_k \le 2/L$. Summing up the above inequalities and rearranging the terms, we have

$$
\begin{aligned}
\textstyle\sum_{k=1}^N (2\gamma_k - L\gamma_k^2)[f(x_k) - f^*] &\le v_1^2 - v_{N+1}^2 - 2\textstyle\sum_{k=1}^N \gamma_k\langle x_k - \gamma_k\nabla f(x_k) - x^*, \delta_k\rangle\\
&\quad + \textstyle\sum_{k=1}^N \gamma_k^2\|\delta_k\|^2\\
&\le D_X^2 - 2\textstyle\sum_{k=1}^N \gamma_k\langle x_k - \gamma_k\nabla f(x_k) - x^*, \delta_k\rangle\\
&\quad + \textstyle\sum_{k=1}^N \gamma_k^2\|\delta_k\|^2,
\end{aligned}
$$

where the last inequality follows from (6.1.15) and the fact that $v_{N+1} \ge 0$. The rest of the proof is similar to that of part (a) and hence the details are skipped. ∎

We now describe a possible strategy for the selection of the stepsizes $\{\gamma_k\}$ in the RSGD method. For the sake of simplicity, let us assume that a constant stepsize policy is used, i.e., $\gamma_k = \gamma$, $k = 1,\ldots,N$, for some $\gamma \in (0, 2/L)$. Note that the assumption of constant stepsizes does not hurt the efficiency estimate of the RSGD method. The following corollary of Theorem 6.1 is obtained by appropriately choosing the parameter γ.

Corollary 6.1. *Suppose that the stepsizes $\{\gamma_k\}$ are set to*

$$
\gamma_k = \min\left\{\tfrac{1}{L}, \tfrac{\tilde{D}}{\sigma\sqrt{N}}\right\}, k = 1,\ldots,N, \tag{6.1.21}
$$

for some $\tilde{D} > 0$. Also assume that the probability mass function $P_R(\cdot)$ is set to (6.1.11). Then, under Assumption 13, we have

$$
\tfrac{1}{L}\mathbb{E}[\|\nabla f(x_R)\|^2] \le \mathscr{B}_N := \tfrac{LD_f^2}{N} + \left(\tilde{D} + \tfrac{D_f^2}{\tilde{D}}\right)\tfrac{\sigma}{\sqrt{N}}, \tag{6.1.22}
$$

where D_f is defined in (6.1.13). If, in addition, problem (6.1.1) is convex with an optimal solution x^*, then

$$\mathbb{E}[f(x_R) - f^*] \le \tfrac{LD_X^2}{N} + \left(\tilde{D} + \tfrac{D_X^2}{\tilde{D}}\right)\tfrac{\sigma}{\sqrt{N}}, \tag{6.1.23}$$

where D_X is defined in (6.1.15).

Proof. Noting that by (6.1.21), we have

$$\begin{aligned}
\tfrac{D_f^2+\sigma^2\sum_{k=1}^N\gamma_k^2}{\sum_{k=1}^N(2\gamma_k-L\gamma_k^2)} &= \tfrac{D_f^2+N\sigma^2\gamma_1^2}{N\gamma_1(2-L\gamma_1)} \le \tfrac{D_f^2+N\sigma^2\gamma_1^2}{N\gamma_1} = \tfrac{D_f^2}{N\gamma_1} + \sigma^2\gamma_1 \\
&\le \tfrac{D_f^2}{N}\max\left\{L, \tfrac{\sigma\sqrt{N}}{\tilde{D}}\right\} + \sigma^2\tfrac{\tilde{D}}{\sigma\sqrt{N}} \\
&\le \tfrac{LD_f^2}{N} + \left(\tilde{D} + \tfrac{D_f^2}{\tilde{D}}\right)\tfrac{\sigma}{\sqrt{N}},
\end{aligned}$$

which together with (6.1.12) then imply (6.1.22). Relation (6.1.23) follows similarly from the above inequality (with D_f replaced by D_X) and (6.1.14). ■

We now make a few remarks about the results obtained in Theorem 6.1 and Corollary 6.1. First, as can be seen from (6.1.19), instead of randomly selecting a solution x_R from $\{x_1, \ldots, x_N\}$, another possibility would be to output the solution $\hat{x}_N$ such that

$$\|\nabla f(\hat{x}_N)\| = \min_{k=1,\ldots,N}\|\nabla f(x_k)\|. \tag{6.1.24}$$

We can show that $\mathbb{E}\|\nabla f(\hat{x}_N)\|$ goes to zero with similar rates of convergence as in (6.1.12) and (6.1.22). However, to use this strategy would require some extra computational effort to compute $\|\nabla f(x_k)\|$ for all $k = 1, \ldots, N$. Since $\|\nabla f(x_k)\|$ cannot be computed exactly, to estimate them by using Monte-Carlo simulation would incur additional approximation errors and raise some reliability issues. On the other hand, the above RSGD method does not require any extra computational effort for estimating the gradients $\|\nabla f(x_k)\|$, $k = 1, \ldots, N$.

Second, observe that in the stepsize policy (6.1.21), we need to specify a parameter $\tilde{D}$. While the RSGD method converges for any arbitrary $\tilde{D} > 0$, it can be easily seen from (6.1.22) and (6.1.23) that an optimal selection of $\tilde{D}$ would be D_f and D_X, respectively, for solving nonconvex and convex SP problems. With such selections, the bounds in (6.1.22) and (6.1.23), respectively, reduce to

$$\tfrac{1}{L}\mathbb{E}[\|\nabla f(x_R)\|^2] \le \tfrac{LD_f^2}{N} + \tfrac{2D_f\sigma}{\sqrt{N}}. \tag{6.1.25}$$

and

$$\mathbb{E}[f(x_R) - f^*] \le \tfrac{LD_X^2}{N} + \tfrac{2D_X\sigma}{\sqrt{N}}. \tag{6.1.26}$$

Note however, that the exact values of D_f or D_X are rarely known a priori and one often needs to set $\tilde{D}$ to a suboptimal value, e.g., certain upper bounds on D_f or D_X.

Third, a possible drawback for the above RSGD method is that one needs to estimate L to obtain an upper bound on γ_k (see, e.g., (6.1.21)), which will also possibly affect the selection of P_R (see (6.1.11)). Note that similar requirements also exist for some deterministic first-order methods (e.g., gradient descent and accelerated gradient descent methods). While under the deterministic setting one can somehow relax such requirements by using certain line-search procedures to enhance the practical performance of these methods, it is more difficult to devise similar line-search procedures for the stochastic setting, since the exact values of $f(x_k)$ and $\nabla f(x_k)$ are not available. It should be noted, however, that we do not need very accurate estimate for L in the RSGD method. Indeed, it can be easily checked that the RSGD method exhibits an $\mathcal{O}(1/\sqrt{N})$ rate of convergence if the stepsizes $\{\gamma_k\}$ are set to

$$\min\left\{\tfrac{1}{qL}, \tfrac{\tilde{D}}{\sigma\sqrt{N}}\right\}, \quad k = 1,\ldots,N$$

for any $q \in [1, \sqrt{N}]$. In other words, we can overestimate the value of L by a factor up to $\sqrt{N}$ and the resulting RSGD method still exhibits similar rate of convergence. A common practice in stochastic optimization is to estimate L by using the stochastic gradients computed at a small number of trial points. It is also worth noting that, although in general the selection of P_R will depend on γ_k and hence on L, such a dependence is not necessary in some special cases. In particular, if the stepsizes $\{\gamma_k\}$ are chosen according to a constant stepsize policy (e.g., (6.1.21)), then R is uniformly distributed on $\{1,\ldots,N\}$. It should be stressed that the persisting dependency of the stepsize on the Lipschitz constant seems essentially impossible to overcome at present and it is a potentially challenging direction of future research for stochastic methods.

Fourth, it is interesting to note that the RSGD method allows us to have a unified treatment for both nonconvex and convex SP problems in view of the specification of $\{\gamma_k\}$ and $P_R(\cdot)$ (cf., (6.1.11) and (6.1.21)). Recall from Chap. 4 that the optimal rate of convergence for solving smooth convex SP problems is given by

$$\mathcal{O}\left(\tfrac{LD_X^2}{N^2} + \tfrac{D_X\sigma}{\sqrt{N}}\right).$$

Comparing (6.1.26) with the above bound, the RSGD method possesses a nearly optimal rate of convergence, since the second term in (6.1.26) is unimprovable while the first term in (6.1.26) can be much improved.

Finally, observe that we can use different stepsize policy other than the constant one in (6.1.21). In particular, it can be shown that the RSGD method with the following two stepsize policies will exhibit similar rates of convergence as those in Corollary 6.1.

- *Increasing stepsize policy*:

$$\gamma_k = \min\left\{\tfrac{1}{L}, \tfrac{\tilde{D}\sqrt{k}}{\sigma N}\right\}, k = 1,\ldots,N.$$

- *Decreasing stepsize policy*:

$$\gamma_k = \min\left\{\tfrac{1}{L}, \tfrac{\tilde{D}}{\sigma(kN)^{1/4}}\right\}, k = 1,\ldots,N.$$

Intuitively speaking, one may want to choose decreasing stepsizes which, according to the definition of $P_R(\cdot)$ in (6.1.11), can stop the algorithm earlier. On the other hand, as the algorithm moves forward and local information about the gradient gets better, choosing increasing stepsizes might be a better option. We expect that the practical performance of these stepsize policies will depend on each problem instance to be solved.

While Theorem 6.1 and Corollary 6.1 establish the expected convergence performance over many runs of the RSGD method, we are also interested in the large-deviation properties for a single run of this method. In particular, we are interested in establishing its complexity for computing an (ε, Λ)-*solution* of problem (6.1.1), i.e., a point $\bar{x}$ satisfying $\mathrm{Prob}\{\|\nabla f(\bar{x})\|^2 \le \varepsilon\} \ge 1 - \Lambda$ for some $\varepsilon > 0$ and $\Lambda \in (0,1)$. By using (6.1.22) and Markov's inequality, we have

$$\mathrm{Prob}\left\{\|\nabla f(x_R)\|^2 \ge \lambda L \mathscr{B}_N\right\} \le \tfrac{1}{\lambda}, \ \forall \lambda > 0. \tag{6.1.27}$$

It then follows that the number of calls to SFO performed by the RSGD method for finding an (ε, Λ)-solution, after disregarding a few constant factors, can be bounded by

$$\mathscr{O}\left\{\tfrac{1}{\Lambda\varepsilon} + \tfrac{\sigma^2}{\Lambda^2\varepsilon^2}\right\}. \tag{6.1.28}$$

The above complexity bound is rather pessimistic in terms of its dependence on Λ. We will investigate one possible way to significantly improve it in next subsection.

6.1.1.2 A Two-Phase Randomized Stochastic Gradient Method

In this subsection, we describe a variant of the RSGD method which can considerably improve the complexity bound in (6.1.28). This procedure consists of two phases: an optimization phase used to generate a list of candidate solutions via a few independent runs of the RSGD method and a post-optimization phase in which a solution is selected from this candidate list.

A Two-Phase RSGD (2-RSGD) Method

Input: Initial point x_1, number of runs S, iteration limit N, and sample size T.
Optimization phase:
For $s = 1,\ldots,S$

Call the RSGD method with input x_1, iteration limit N, stepsizes $\{\gamma_k\}$ in (6.1.21) and probability mass function P_R in (6.1.11). Let $\bar{x}_s$ be the output of this procedure.

Post-optimization phase:
Choose a solution $\bar{x}^*$ from the candidate list $\{\bar{x}_1,\ldots,\bar{x}_S\}$ such that

$$\|g(\bar{x}^*)\| = \min_{s=1,\ldots,S} \|g(\bar{x}_s)\|, \;\; g(\bar{x}_s) := \tfrac{1}{T}\textstyle\sum_{k=1}^T G(\bar{x}_s, \xi_k), \tag{6.1.29}$$

where $G(x, \xi_k)$, $k = 1, \ldots, T$, are the stochastic gradients returned by the SFO.

Observe that in (6.1.29), we define the best solution $\bar{x}^*$ as the one with the smallest value of $\|g(\bar{x}_s)\|$, $s = 1, \ldots, S$. Alternatively, one can choose $\bar{x}^*$ from $\{\bar{x}_1, \ldots, \bar{x}_S\}$ such that

$$\tilde{f}(\bar{x}^*) = \min_{1,\ldots,S} \tilde{f}(\bar{x}_s), \;\; \tilde{f}(\bar{x}_s) = \tfrac{1}{T}\textstyle\sum_{k=1}^T F(\bar{x}_s, \xi_k). \tag{6.1.30}$$

In the 2-RSGD method described above, the number of calls to the SFO are given by $S \times N$ and $S \times T$, respectively, for the optimization phase and post-optimization phase. Also note that we can possibly recycle the same sequence $\{\xi_k\}$ across all gradient estimations in the post-optimization phase of 2-RSGD method. We will provide in Theorem 6.2 below certain bounds on S, N, and T to compute an (ε, Λ)-solution of problem (6.1.1).

We need the following results regarding the large deviations of vector valued martingales (see, e.g., Theorem 2.1 of [57]).

Lemma 6.1. *Assume that we are given a polish space with Borel probability measure μ and a sequence of $\mathscr{F}_0 = \{\emptyset, \Omega\} \subseteq \mathscr{F}_1 \subseteq \mathscr{F}_2 \subseteq \ldots$ of σ-sub-algebras of Borel σ-algebra of Ω. Let $\zeta_i \in \mathbb{R}^n$, $i = 1, \ldots, \infty$, be a martingale-difference sequence of Borel functions on Ω such that ζ_i is $\mathscr{F}_i$ measurable and $\mathbb{E}[\zeta_i|i-1] = 0$, where $\mathbb{E}[\cdot|i]$, $i = 1, 2, \ldots$, denotes the conditional expectation w.r.t. $\mathscr{F}_i$ and $\mathbb{E} \equiv \mathbb{E}[\cdot|0]$ is the expectation w.r.t. μ.*

(a) If $\mathbb{E}[\|\zeta_i\|^2] \le \sigma_i^2$ for any $i \ge 1$, then $\mathbb{E}[\|\sum_{i=1}^N \zeta_i\|^2] \le \sum_{i=1}^N \sigma_i^2$. As a consequence, we have

$$\forall N \ge 1, \lambda \ge 0 : \mathrm{Prob}\left\{\|\textstyle\sum_{i=1}^N \zeta_i\|^2 \ge \lambda \sum_{i=1}^N \sigma_i^2\right\} \le \tfrac{1}{\lambda};$$

(b) If $\mathbb{E}\left\{\exp\left(\|\zeta_i\|^2/\sigma_i^2\right)|i-1\right\} \le \exp(1)$ almost surely for any $i \ge 1$, then

$$\forall N \ge 1, \lambda \ge 0 : \mathrm{Prob}\left\{\|\textstyle\sum_{i=1}^N \zeta_i\| \ge \sqrt{2}(1+\lambda)\sqrt{\sum_{i=1}^N \sigma_i^2}\right\} \le \exp(-\lambda^2/3).$$

We are now ready to describe the main convergence properties of the 2-RSGD method. More specifically, Theorem 6.2(a) below shows the convergence rate of this algorithm for a given set of parameters (S, N, T), while Theorem 6.2(b) establishes the complexity of the 2-RSGD method for computing an (ε, Λ)-solution of problem (6.1.1).

Theorem 6.2. *Under Assumption 13, the following statements hold for the 2-RSGD method applied to problem (6.1.1).*

(a) Let $\mathscr{B}_N$ be defined in (6.1.22). We have

$$\mathrm{Prob}\left\{\|\nabla f(\bar{x}^*)\|^2 \ge 2\left(4L\mathscr{B}_N + \tfrac{3\lambda\sigma^2}{T}\right)\right\} \le \tfrac{S+1}{\lambda} + 2^{-S}, \;\; \forall \lambda > 0; \tag{6.1.31}$$

(b) Let $\varepsilon > 0$ and $\Lambda \in (0,1)$ be given. If the parameters (S,N,T) are set to

$$S = S(\Lambda) := \lceil \log(2/\Lambda) \rceil, \tag{6.1.32}$$

$$N = N(\varepsilon) := \left\lceil \max\left\{ \tfrac{32L^2 D_f^2}{\varepsilon}, \left[32L\left(\tilde{D} + \tfrac{D_f^2}{\tilde{D}} \right) \tfrac{\sigma}{\varepsilon} \right]^2 \right\} \right\rceil, \tag{6.1.33}$$

$$T = T(\varepsilon,\Lambda) := \left\lceil \tfrac{24(S+1)\sigma^2}{\Lambda\varepsilon} \right\rceil, \tag{6.1.34}$$

then the 2-RSGD method can compute an (ε,Λ)-solution of problem (6.1.1) after taking at most

$$S(\Lambda)\,[N(\varepsilon) + T(\varepsilon,\Lambda)] \tag{6.1.35}$$

calls to the stochastic first-order oracle.

Proof. We first show part (a). Observe that by the definition of $\bar{x}^*$ in (6.1.29), we have

$$\begin{aligned}
\|g(\bar{x}^*)\|^2 &= \min_{s=1,\ldots,S} \|g(\bar{x}_s)\|^2 = \min_{s=1,\ldots,S} \|\nabla f(\bar{x}_s) + g(\bar{x}_s) - \nabla f(\bar{x}_s)\|^2 \\
&\le \min_{s=1,\ldots,S} \{2\|\nabla f(\bar{x}_s)\|^2 + 2\|g(\bar{x}_s) - \nabla f(\bar{x}_s)\|^2\} \\
&\le 2 \min_{s=1,\ldots,S} \|\nabla f(\bar{x}_s)\|^2 + 2 \max_{s=1,\ldots,S} \|g(\bar{x}_s) - \nabla f(\bar{x}_s)\|^2,
\end{aligned}$$

which implies that

$$\begin{aligned}
\|\nabla f(\bar{x}^*)\|^2 \le 2\|g(\bar{x}^*)\|^2 + 2\|\nabla f(\bar{x}^*) - g(\bar{x}^*)\|^2 \le 4 \min_{s=1,\ldots,S} \|\nabla f(\bar{x}_s)\|^2 \\
+4 \max_{s=1,\ldots,S} \|g(\bar{x}_s) - \nabla f(\bar{x}_s)\|^2 + 2\|\nabla f(\bar{x}^*) - g(\bar{x}^*)\|^2.
\end{aligned} \tag{6.1.36}$$

We now provide certain probabilistic upper bounds to the three terms in the right-hand side of the above inequality. First, using the fact that $\bar{x}_s$, $1 \le s \le S$, are independent and relation (6.1.27) (with $\lambda = 2$), we have

$$\mathrm{Prob}\left\{ \min_{s=1,\ldots,S} \|\nabla f(\bar{x}_s)\|^2 \ge 2L\mathscr{B}_N \right\} = \prod_{s=1}^{S} \mathrm{Prob}\left\{ \|\nabla f(\bar{x}_s)\|^2 \ge 2L\mathscr{B}_N \right\} \le 2^{-S}. \tag{6.1.37}$$

Moreover, denoting $\delta_{s,k} = G(\bar{x}_s, \xi_k) - \nabla f(\bar{x}_s)$, $k = 1,\ldots,T$, we have $g(\bar{x}_s) - \nabla f(\bar{x}_s) = \sum_{k=1}^{T} \delta_{s,k}/T$. Using this observation, Assumption 13 and Lemma 6.1(a), we conclude that, for any $s = 1,\ldots,S$,

$$\mathrm{Prob}\left\{ \|g(\bar{x}_s) - \nabla f(\bar{x}_s)\|^2 \ge \tfrac{\lambda\sigma^2}{T} \right\} = \mathrm{Prob}\left\{ \|\textstyle\sum_{k=1}^{T} \delta_{s,k}\|^2 \ge \lambda T\sigma^2 \right\} \le \tfrac{1}{\lambda}, \ \forall \lambda > 0,$$

which implies that

$$\mathrm{Prob}\left\{ \max_{s=1,\ldots,S} \|g(\bar{x}_s) - \nabla f(\bar{x}_s)\|^2 \ge \tfrac{\lambda\sigma^2}{T} \right\} \le \tfrac{S}{\lambda}, \ \forall \lambda > 0, \tag{6.1.38}$$

and that

$$\text{Prob}\left\{\|g(\bar{x}^*)-\nabla f(\bar{x}^*)\|^2 \ge \tfrac{\lambda\sigma^2}{T}\right\} \le \tfrac{1}{\lambda},\ \forall \lambda > 0. \tag{6.1.39}$$

The result then follows by combining relations (6.1.36), (6.1.37), (6.1.38), and (6.1.39).

We now show that part (b) holds. Since the 2-RSGD method needs to call the RSGD method S times with iteration limit $N(\varepsilon)$ in the optimization phase, and estimate the gradients $g(\bar{x}_s)$, $s=1,\ldots,S$ with sample size $T(\varepsilon)$ in the post-optimization phase, the total number of calls to the stochastic first-order oracle is bounded by $S[N(\varepsilon)+T(\varepsilon)]$. It remains to show that $\bar{x}^*$ is an (ε,Λ)-solution of problem (6.1.1). Noting that by the definitions of θ_1 and $N(\varepsilon)$, respectively, in (6.1.22) and (6.1.33), we have

$$\mathscr{B}_{N(\varepsilon)} = \tfrac{LD_f^2}{N(\varepsilon)} + \left(\tilde{D}+\tfrac{D_f^2}{\tilde{D}}\right)\tfrac{\sigma}{\sqrt{N(\varepsilon)}} \le \tfrac{\varepsilon}{32L} + \tfrac{\varepsilon}{32L} = \tfrac{\varepsilon}{16L}.$$

Using the above observation, (6.1.34) and setting $\lambda = [2(S+1)]/\Lambda$ in (6.1.31), we have

$$4LB_{N(\varepsilon)} + \tfrac{3\lambda\sigma^2}{T(\varepsilon)} = \tfrac{\varepsilon}{4} + \tfrac{\lambda\Lambda\varepsilon}{8(S+1)} = \tfrac{\varepsilon}{2},$$

which, together with relations (6.1.31) and (6.1.32), and the selection of λ, then imply that

$$\text{Prob}\left\{\|\nabla f(\bar{x}^*)\|^2 \ge \varepsilon\right\} \le \tfrac{\Lambda}{2} + 2^{-S} \le \Lambda.$$

■

It is interesting to compare the complexity bound in (6.1.35) with the one in (6.1.28). In view of (6.1.32), (6.1.33), and (6.1.34), the complexity bound in (6.1.35), after disregarding a few constant factors, is equivalent to

$$\mathscr{O}\left\{\tfrac{\log(1/\Lambda)}{\varepsilon} + \tfrac{\sigma^2}{\varepsilon^2}\log\tfrac{1}{\Lambda} + \tfrac{\log^2(1/\Lambda)\sigma^2}{\Lambda\varepsilon}\right\}. \tag{6.1.40}$$

The above bound can be considerably smaller than the one in (6.1.28) up to a factor of $1/\left[\Lambda^2\log(1/\Lambda)\right]$, when the second terms are the dominating ones in both bounds.

The following result shows that the bound (6.1.35) obtained in Theorem 6.2 can be further improved under certain light-tail assumption of SFO.

Corollary 6.2. *Under Assumptions 13 and 14, the following statements hold for the 2-RSGD method applied to problem (6.1.1).*

(a) Let $\mathscr{B}_N$ is defined in (6.1.22). We have, $\forall \lambda > 0$,

$$\text{Prob}\left\{\|\nabla f(\bar{x}^*)\|^2 \ge 4\left[2L\mathscr{B}_N + 3(1+\lambda)^2\tfrac{\sigma^2}{T}\right]\right\} \le (S+1)\exp(-\lambda^2/3)+2^{-S}; \tag{6.1.41}$$

(b) Let $\varepsilon > 0$ and $\Lambda \in (0,1)$ be given. If S and N are set to $S(\Lambda)$ and $N(\varepsilon)$ as in (6.1.32) and (6.1.33), respectively, and the sample size T is set to

$$T = T'(\varepsilon,\Lambda) := \tfrac{24\sigma^2}{\varepsilon}\left[1+\left(3\ln\tfrac{2(S+1)}{\Lambda}\right)^{\frac{1}{2}}\right]^2, \tag{6.1.42}$$

then the 2-RSGD method can compute an (ε,Λ)-solution of problem (6.1.1) in at most

$$S(\Lambda)\left[N(\varepsilon)+T'(\varepsilon,\Lambda)\right] \tag{6.1.43}$$

calls to the stochastic first-order oracle.

Proof. We provide the proof of part (a) only, since part (b) follows immediately from part (a) and an argument similar to the one used in the proof of Theorem 6.2(b). Denoting $\delta_{s,k} = G(\bar{x}_s,\xi_k) - \nabla f(\bar{x}_s)$, $k = 1,\ldots,T$, we have $g(\bar{x}_s) - \nabla f(\bar{x}_s) = \sum_{k=1}^T \delta_{s,k}/T$. Using this observation, Assumption 14, and Lemma 6.1(b), we conclude that, for any $s = 1,\ldots,S$ and $\lambda > 0$,

$$\begin{aligned}&\mathrm{Prob}\left\{\|g(\bar{x}_s) - \nabla f(\bar{x}_s)\|^2 \geq 2(1+\lambda)^2\tfrac{\sigma^2}{T}\right\}\\ &= \mathrm{Prob}\left\{\|\textstyle\sum_{k=1}^T \delta_{s,k}\| \geq \sqrt{2T}(1+\lambda)\sigma\right\} \leq \exp(-\lambda^2/3),\end{aligned}$$

which implies that

$$\mathrm{Prob}\left\{\max_{s=1,\ldots,S}\|g(\bar{x}_s) - \nabla f(\bar{x}_s)\|^2 \geq 2(1+\lambda)^2\tfrac{\sigma^2}{T}\right\} \leq S\exp(-\lambda^2/3),\ \forall\lambda > 0, \tag{6.1.44}$$

and that

$$\mathrm{Prob}\left\{\|g(\bar{x}^*) - \nabla f(\bar{x}^*)\|^2 \geq 2(1+\lambda)^2\tfrac{\sigma^2}{T}\right\} \leq \exp(-\lambda^2/3),\ \forall\lambda > 0. \tag{6.1.45}$$

The result in part (a) then follows by combining relations (6.1.36), (6.1.37), (6.1.44), and (6.1.45). ■

In view of (6.1.32), (6.1.33), and (6.1.42), the bound in (6.1.43), after disregarding a few constant factors, is equivalent to

$$\mathcal{O}\left\{\tfrac{\log(1/\Lambda)}{\varepsilon} + \tfrac{\sigma^2}{\varepsilon^2}\log\tfrac{1}{\Lambda} + \tfrac{\log^2(1/\Lambda)\sigma^2}{\varepsilon}\right\}. \tag{6.1.46}$$

Clearly, the third term of the above bound is significantly smaller than the corresponding one in (6.1.40) by a factor of $1/\Lambda$.

6.1.2 Stochastic Zeroth-Order Methods

Our problem of interest in this section is problem (6.1.1) with f given in the form of expectation, i.e.,

$$f^* := \inf_{x\in\mathbb{R}^n}\left\{f(x) := \int_{\Xi} F(x,\xi)dP(\xi)\right\}. \tag{6.1.47}$$

Moreover, we assume that $F(x,\xi) \in \mathscr{C}_L^{1,1}(\mathbb{R}^n)$ almost surely, which clearly implies $f(x) \in \mathscr{C}_L^{1,1}(\mathbb{R}^n)$. Our goal in this subsection is to specialize the RSGD and 2-RSGD methods, respectively, in Sects. 6.1.2.1 and 6.1.2.2, to deal with the situation when only stochastic zeroth-order information of f is available.

6.1.2.1 The Randomized Stochastic Gradient Free Method

Throughout this section, we assume that f is represented by a *stochastic zeroth-order oracle* (SZO). More specifically, at the k-th iteration, x_k and ξ_k being the input, the SZO outputs the quantity $F(x_k, \xi_k)$ such that the following assumption holds:

Assumption 15 *For any $k \geq 1$, we have*

$$\mathbb{E}[F(x_k, \xi_k)] = f(x_k). \tag{6.1.48}$$

To exploit zeroth-order information, we consider a smooth approximation of the objective function f. It is well known that the convolution of f with any nonnegative, measurable, and bounded function $\psi : \mathbb{R}^n \to \mathbb{R}$ satisfying $\int_{\mathbb{R}^n} \psi(u)\, du = 1$ is an approximation of f which is at least as smooth as f. One of the most important examples of the function ψ is the probability density function. Here, we use the Gaussian distribution in the convolution. Let u be n-dimensional standard Gaussian random vector and $\mu > 0$ be the smoothing parameter. Then, a smooth approximation of f is defined as

$$f_\mu(x) = \tfrac{1}{(2\pi)^{n/2}} \int f(x+\mu u) e^{-\frac{1}{2}\|u\|^2} du = \mathbb{E}_u[f(x+\mu u)]. \tag{6.1.49}$$

The following result describes some properties of $f_\mu(\cdot)$ (see [102]).

Lemma 6.2. *If $f \in \mathscr{C}_L^{1,1}(\mathbb{R}^n)$, then*

(a) f_μ is also Lipschitz continuously differentiable with gradient Lipschitz constant $L_\mu \leq L$ and

$$\nabla f_\mu(x) = \frac{1}{(2\pi)^{\frac{n}{2}}} \int \tfrac{f(x+\mu v) - f(x)}{\mu} v e^{-\frac{1}{2}\|v\|^2} dv. \tag{6.1.50}$$

(b) For any $x \in \mathbb{R}^n$, we have

$$|f_\mu(x) - f(x)| \leq \tfrac{\mu^2}{2} Ln, \tag{6.1.51}$$

$$\|\nabla f_\mu(x) - \nabla f(x)\| \leq \tfrac{\mu}{2} L(n+3)^{\frac{3}{2}}, \tag{6.1.52}$$

$$\mathbb{E}_v\left[\left\|\tfrac{f(x+\mu v) - f(x)}{\mu} v\right\|^2\right] \leq 2(n+4)\|\nabla f(x)\|^2 + \tfrac{\mu^2}{2} L^2 (n+6)^3. \tag{6.1.53}$$

(c) f_μ is also convex provided f is convex.

It immediately follows from (6.1.52) that

$$\|\nabla f_\mu(x)\|^2 \le 2\|\nabla f(x)\|^2 + \tfrac{\mu^2}{2}L^2(n+3)^3, \tag{6.1.54}$$

$$\|\nabla f(x)\|^2 \le 2\|\nabla f_\mu(x)\|^2 + \tfrac{\mu^2}{2}L^2(n+3)^3. \tag{6.1.55}$$

Moreover, denoting

$$f_\mu^* := \min_{x\in\mathbb{R}^n} f_\mu(x), \tag{6.1.56}$$

we conclude from (6.1.57) that $|f_\mu^* - f^*| \le \mu^2 Ln/2$ and hence that

$$-\mu^2 Ln \le [f_\mu(x) - f_\mu^*] - [f(x) - f^*] \le \mu^2 Ln. \tag{6.1.57}$$

Below we modify the RSGD method in subsection (6.1.1.1) to use stochastic zeroth-order rather than first-order information for solving problem (6.1.47).

A Randomized Stochastic Gradient Free (RSGF) Method

Input: Initial point x_1, iteration limit N, stepsizes $\{\gamma_k\}_{k\ge1}$, probability mass function $P_R(\cdot)$ supported on $\{1,\ldots,N\}$.
Step 0. Let R be a random variable with probability mass function P_R.
Step $k = 1,\ldots,R$. Generate u_k by Gaussian random vector generator and call the stochastic zeroth-order oracle for computing $G_\mu(x_k,\xi_k,u_k)$ given by

$$G_\mu(x_k,\xi_k,u_k) = \tfrac{F(x_k+\mu u_k,\xi_k)-F(x_k,\xi_k)}{\mu}u_k. \tag{6.1.58}$$

Set

$$x_{k+1} = x_k - \gamma_k G_\mu(x_k,\xi_k,u_k). \tag{6.1.59}$$

Output x_R.

Note that $G_\mu(x_k,\xi_k,u_k)$ is an unbiased estimator of $\nabla f_\mu(x_k)$. Indeed, by (6.1.50) and Assumption 15, we have

$$\mathbb{E}_{\xi,u}[G_\mu(x,\xi,u)] = \mathbb{E}_u\left[\mathbb{E}_\xi[G_\mu(x,\xi,u)|u]\right] = \nabla f_\mu(x). \tag{6.1.60}$$

Hence, if the variance $\tilde\sigma^2 \equiv \mathbb{E}_{\xi,u}[\|G_\mu(x,\xi,u) - \nabla f_\mu(x)\|^2]$ is bounded, we can directly apply the convergence results in Theorem 6.1 to the above RSGF method. However, there still exist a few problems in this approach. First, we do not know an explicit expression of the bound $\tilde\sigma^2$. Second, this approach does not provide any information regarding how to appropriately specify the smoothing parameter μ. The latter issue is critical for the implementation of the RSGF method.

By applying the approximation results in Lemma 6.2 to the functions $F(\cdot,\xi_k)$, $k = 1,\ldots,N$, and using a slightly different convergence analysis than the one in Theorem 6.1, we are able to obtain much refined convergence results for the above RSGF method.

Theorem 6.3. *Suppose that the stepsizes $\{\gamma_k\}$ and the probability mass function $P_R(\cdot)$ in the RSGF method are chosen such that $\gamma_k < 1/[2(n+4)L]$ and*

$$P_R(k) := \text{Prob}\{R=k\} = \frac{\gamma_k - 2L(n+4)\gamma_k^2}{\sum_{k=1}^N [\gamma_k - 2L(n+4)\gamma_k^2]}, \quad k=1,\ldots,N. \tag{6.1.61}$$

Then, under Assumptions 13 and 15,

(a) for any $N \geq 1$, we have

$$\frac{1}{L}\mathbb{E}[\|\nabla f(x_R)\|^2] \leq \frac{1}{\sum_{k=1}^N [\gamma_k - 2L(n+4)\gamma_k^2]} \Big[D_f^2 + 2\mu^2(n+4) \\ (1+L(n+4)^2 \textstyle\sum_{k=1}^N (\frac{\gamma_k}{4} + L\gamma_k^2)) + 2(n+4)\sigma^2 \sum_{k=1}^N \gamma_k^2\Big], \tag{6.1.62}$$

where the expectation is taken with respect to R, $\xi_{[N]}$, and $u_{[N]}$, and D_f is defined in (6.1.13);

(b) if, in addition, problem (6.1.47) is convex with an optimal solution x^*, then, for any $N \geq 1$,

$$\mathbb{E}[f(x_R) - f^*] \leq \frac{1}{2\sum_{k=1}^N [\gamma_k - 2(n+4)L\gamma_k^2]} \Big[D_X^2 + 2\mu^2 L(n+4) \\ \textstyle\sum_{k=1}^N \left[\gamma_k + L(n+4)^2\gamma_k^2\right] + 2(n+4)\sigma^2 \sum_{k=1}^N \gamma_k^2\Big], \tag{6.1.63}$$

where the expectation is taken with respect to R, $\xi_{[N]}$, and $u_{[N]}$, and D_X is defined in (6.1.15).

Proof. Let $\zeta_k \equiv (\xi_k, u_k)$, $k \geq 1$, $\zeta_{[N]} := (\zeta_1, \ldots, \zeta_N)$, and $\mathbb{E}_{\zeta[N]}$ denote the expectation w.r.t. $\zeta_{[N]}$. Also denote $\Delta_k \equiv G_\mu(x_k, \xi_k, u_k) - \nabla f_\mu(x_k) \equiv G_\mu(x_k, \zeta_k) - \nabla f_\mu(x_k)$, $\quad k \geq 1$. Using the fact that $f \in \mathscr{C}_L^{1,1}(\mathbb{R}^n)$, Lemma 6.2(a), (6.1.2) and (6.1.59), we have, for any $k = 1, \ldots, N$,

$$\begin{aligned} f_\mu(x_{k+1}) &\leq f_\mu(x_k) - \gamma_k \langle \nabla f_\mu(x_k), G_\mu(x_k, \zeta_k)\rangle + \tfrac{L}{2}\gamma_k^2 \|G_\mu(x_k, \zeta_k)\|^2 \\ &= f_\mu(x_k) - \gamma_k \|\nabla f_\mu(x_k)\|^2 - \gamma_k \langle \nabla f_\mu(x_k), \Delta_k\rangle \\ &\quad + \tfrac{L}{2}\gamma_k^2 \|G_\mu(x_k, \zeta_k)\|^2. \end{aligned} \tag{6.1.64}$$

Summing up these inequalities, rearranging the terms, and noting that $f_\mu^* \leq f_\mu(x_{N+1})$, we obtain

$$\begin{aligned} \textstyle\sum_{k=1}^N \gamma_k \|\nabla f_\mu(x_k)\|^2 \leq f_\mu(x_1) - f_\mu^* - \sum_{k=1}^N \gamma_k \langle \nabla f_\mu(x_k), \Delta_k\rangle \\ + \tfrac{L}{2}\textstyle\sum_{k=1}^N \gamma_k^2 \|G_\mu(x_k, \zeta_k)\|^2. \end{aligned} \tag{6.1.65}$$

Now, observe that by (6.1.60),

$$\mathbb{E}[\langle \nabla f_\mu(x_k), \Delta_k\rangle | \zeta_{[k-1]}] = 0, \tag{6.1.66}$$

and that by the assumption $F(\cdot, \xi_k) \in \mathscr{C}_L^{1,1}(\mathbb{R}^n)$, (6.1.53) (with $f = F(\cdot, \xi_k)$), and (6.1.58),

$$\begin{aligned}\mathbb{E}[\|G_\mu(x_k,\zeta_k)\|^2|\zeta_{[k-1]}] &\le 2(n+4)\mathbb{E}[\|G(x_k,\xi_k)\|^2|\zeta_{[k-1]}] + \tfrac{\mu^2}{2}L^2(n+6)^3\\ &\le 2(n+4)\left[\mathbb{E}[\|\nabla f(x_k)\|^2|\zeta_{[k-1]}]+\sigma^2\right]\\ &\quad +\tfrac{\mu^2}{2}L^2(n+6)^3,\end{aligned} \tag{6.1.67}$$

where the second inequality follows from Assumption 13. Taking expectations with respect to $\zeta_{[N]}$ on both sides of (6.1.65) and using the above two observations, we obtain

$$\begin{aligned}&\sum_{k=1}^N \gamma_k \mathbb{E}_{\zeta[N]}\left[\|\nabla f_\mu(x_k)\|^2\right] \le f_\mu(x_1) - f_\mu^*\\ &\quad + \tfrac{L}{2}\sum_{k=1}^N \gamma_k^2 \left\{2(n+4)\left[\mathbb{E}_{\zeta[N]}[\|\nabla f(x_k)\|^2]+\sigma^2\right] + \tfrac{\mu^2}{2}L^2(n+6)^3\right\}.\end{aligned}$$

The above conclusion together with (6.1.54) and (6.1.57) then imply that

$$\begin{aligned}&\sum_{k=1}^N \gamma_k \left[\mathbb{E}_{\zeta[N]}[\|\nabla f(x_k)\|^2] - \tfrac{\mu^2}{2}L^2(n+3)^3\right] \le 2\left[f(x_1)-f^*\right] + 2\mu^2 Ln\\ &+2L(n+4)\sum_{k=1}^N \gamma_k^2 \mathbb{E}_{\zeta[N]}[\|\nabla f(x_k)\|^2] + \left[2L(n+4)\sigma^2 + \tfrac{\mu^2}{2}L^3(n+6)^3\right]\sum_{k=1}^N \gamma_k^2.\end{aligned} \tag{6.1.68}$$

By rearranging the terms and simplifying the constants, we have

$$\begin{aligned}&\sum_{k=1}^N \left\{\left[\gamma_k - 2L(n+4)\gamma_k^2\right]\mathbb{E}_{\zeta[N]}[\|\nabla f(x_k)\|^2]\right\}\\ &\le 2\left[f(x_1)-f^*\right] + 2L(n+4)\sigma^2\sum_{k=1}^N\gamma_k^2 + 2\mu^2 Ln\\ &\quad + \tfrac{\mu^2}{2}L^2\sum_{k=1}^N\left[(n+3)^3\gamma_k + L(n+6)^3\gamma_k^2\right]\\ &\le 2\left[f(x_1)-f^*\right] + 2L(n+4)\sigma^2\sum_{k=1}^N\gamma_k^2\\ &\quad +2\mu^2 L(n+4)\left[1 + L(n+4)^2\sum_{k=1}^N\left(\tfrac{\gamma_k}{4}+L\gamma_k^2\right)\right].\end{aligned} \tag{6.1.69}$$

Dividing both sides of the above inequality by $\sum_{k=1}^N\left[\gamma_k - 2L(n+4)\gamma_k^2\right]$ and noting that

$$\mathbb{E}[\|\nabla f(x_R)\|^2] = \mathbb{E}_{R,\zeta[N]}[\|\nabla f(x_R)\|^2] = \frac{\sum_{k=1}^N\left\{\left[\gamma_k - 2L(n+4)\gamma_k^2\right]\mathbb{E}_{\zeta[N]}\|\nabla f(x_k)\|^2\right\}}{\sum_{k=1}^N\left[\gamma_k - 2L(n+4)\gamma_k^2\right]},$$

we obtain (6.1.62).

We now show part (b). Denote $v_k \equiv \|x_k - x^*\|$. First observe that, for any $k = 1,\ldots,N$,

$$\begin{aligned}v_{k+1}^2 &= \|x_k - \gamma_k G_\mu(x_k,\zeta_k) - x^*\|^2\\ &= v_k^2 - 2\gamma_k\langle\nabla f_\mu(x_k)+\Delta_k, x_k - x^*\rangle + \gamma_k^2\|G_\mu(x_k,\zeta_k)\|^2.\end{aligned}$$

and hence that

$$\begin{aligned}v_{N+1}^2 =& v_1^2 - 2\sum_{k=1}^N\gamma_k\langle\nabla f_\mu(x_k), x_k - x^*\rangle - 2\sum_{k=1}^N\gamma_k\langle\Delta_k, x_k - x^*\rangle\\ &+\sum_{k=1}^N\gamma_k^2\|G_\mu(x_k,\zeta_k)\|^2.\end{aligned}$$

Taking expectation w.r.t. $\zeta_{\zeta[N]}$ on both sides of the above equality, using relation (6.1.67) and noting that by (6.1.60), $\mathbb{E}[\langle\Delta_k, x_k - x^*\rangle|\zeta_{[k-1]}] = 0$, we obtain

$$\begin{aligned}
\mathbb{E}_{\zeta[N]}[v_{N+1}^2] &\le v_1^2 - 2\textstyle\sum_{k=1}^N \gamma_k \mathbb{E}_{\zeta[N]}[\langle \nabla f_\mu(x_k), x_k - x^*\rangle] \\
&\quad + 2(n+4)\textstyle\sum_{k=1}^N \gamma_k^2 \mathbb{E}_{\zeta[N]}[\|\nabla f(x_k)\|^2] \\
&\quad + \left[2(n+4)\sigma^2 + \tfrac{\mu^2}{2}L^2(n+6)^3\right] \textstyle\sum_{k=1}^N \gamma_k^2 \\
&\le v_1^2 - 2\textstyle\sum_{k=1}^N \gamma_k \mathbb{E}_{\zeta[N]}[f_\mu(x_k) - f_\mu(x^*)] \\
&\quad + 2(n+4)L\textstyle\sum_{k=1}^N \gamma_k^2 \mathbb{E}_{\zeta[N]}[f(x_k) - f^*] \\
&\quad + \left[2(n+4)\sigma^2 + \tfrac{\mu^2}{2}L^2(n+6)^3\right] \textstyle\sum_{k=1}^N \gamma_k^2 \\
&\le v_1^2 - 2\textstyle\sum_{k=1}^N \gamma_k \mathbb{E}_{\zeta[N]}\left[f(x_k) - f^* - \mu^2 L n\right] \\
&\quad + 2(n+4)L\textstyle\sum_{k=1}^N \gamma_k^2 \mathbb{E}_{\zeta[N]}[f(x_k) - f^*] \\
&\quad + \left[2(n+4)\sigma^2 + \tfrac{\mu^2}{2}L^2(n+6)^3\right] \textstyle\sum_{k=1}^N \gamma_k^2,
\end{aligned}$$

where the second inequality follows from (6.1.20) and the convexity of f_μ, and the last inequality follows from (6.1.57). Rearranging the terms in the above inequality, using the facts that $v_{N+1}^2 \ge 0$ and $f(x_k) \ge f^*$, and simplifying the constants, we have

$$\begin{aligned}
&2\textstyle\sum_{k=1}^N \left[\gamma_k - 2(n+4)L\gamma_k^2)\right] \mathbb{E}_{\zeta[N]}[f(x_k) - f^*] \\
&\le 2\textstyle\sum_{k=1}^N \left[\gamma_k - (n+4)L\gamma_k^2)\right] \mathbb{E}_{\zeta[N]}[f(x_k) - f^*] \\
&\le v_1^2 + 2\mu^2 L(n+4)\textstyle\sum_{k=1}^N \gamma_k + 2(n+4)\left[L^2\mu^2(n+4)^2 + \sigma^2\right] \textstyle\sum_{k=1}^N \gamma_k^2.
\end{aligned}$$

The rest of proof is similar to part (a) and hence the details are skipped. ∎

Similar to the RSGD method, we can specialize the convergence results in Theorem 6.3 for the RSGF method with a constant stepsize policy.

Corollary 6.3. *Suppose that the stepsizes $\{\gamma_k\}$ are set to*

$$\gamma_k = \tfrac{1}{\sqrt{n+4}} \min\left\{\tfrac{1}{4L\sqrt{n+4}}, \tfrac{\tilde{D}}{\sigma\sqrt{N}}\right\}, \quad k = 1, \ldots, N, \tag{6.1.70}$$

for some $\tilde{D} > 0$. Also assume that the probability mass function $P_R(\cdot)$ is set to (6.1.61) and μ is chosen such that

$$\mu \le \tfrac{D_f}{(n+4)\sqrt{2N}} \tag{6.1.71}$$

where D_f and D_X are defined in (6.1.13) and (6.1.15), respectively. Then, under Assumptions 13 and 15, we have

$$\tfrac{1}{L}\mathbb{E}[\|\nabla f(x_R)\|^2] \le \bar{\mathscr{B}}_N := \tfrac{12(n+4)LD_f^2}{N} + \tfrac{4\sigma\sqrt{n+4}}{\sqrt{N}}\left(\tilde{D} + \tfrac{D_f^2}{\tilde{D}}\right). \tag{6.1.72}$$

If, in addition, problem (6.1.47) is convex with an optimal solution x^* and μ is chosen such that

$$\mu \le \tfrac{D_X}{\sqrt{(n+4)}},$$

then,

$$\mathbb{E}[f(x_R)-f^*] \le \frac{5L(n+4)D_X^2}{N} + \frac{2\sigma\sqrt{n+4}}{\sqrt{N}}\left(\tilde{D}+\frac{D_X^2}{\tilde{D}}\right). \qquad (6.1.73)$$

Proof. We prove (6.1.72) only since relation (6.1.73) can be shown by using similar arguments. First note that by (6.1.70), we have

$$\gamma_k \le \frac{1}{4(n+4)L}, \; k=1,\ldots,N, \qquad (6.1.74)$$

$$\sum_{k=1}^N \left[\gamma_k - 2L(n+4)\gamma_k^2\right] = N\gamma_1\left[1-2L(n+4)\gamma_1\right] \ge \frac{N\gamma_1}{2}. \qquad (6.1.75)$$

Therefore, using the above inequalities and (6.1.62), we obtain

$$\begin{aligned}\tfrac{1}{L}\mathbb{E}[\|\nabla f(x_R)\|^2] &\le \frac{2D_f^2+4\mu^2(n+4)}{N\gamma_1} + \mu^2 L(n+4)^3 + 4(n+4)\left[\mu^2L^2(n+4)^2+\sigma^2\right]\gamma_1 \\ &\le \frac{2D_f^2+4\mu^2(n+4)}{N}\max\left\{4L(n+4), \frac{\sigma\sqrt{(n+4)N}}{\tilde{D}}\right\} \\ &\quad + \mu^2L(n+4)^2\left[(n+4)+1\right] + \frac{4\sqrt{n+4}\tilde{D}\sigma}{\sqrt{N}},\end{aligned}$$

which, in view of (6.1.71), then implies that

$$\begin{aligned}\tfrac{1}{L}\mathbb{E}[\|\nabla f(x_R)\|^2] &\le \frac{2D_f^2}{N}\left[1+\frac{1}{(n+4)N}\right]\left[4L(n+4)+\frac{\sigma\sqrt{(n+4)N}}{\tilde{D}}\right] \\ &\quad + \frac{LD_f^2}{2N}\left[(n+4)+1\right] + \frac{4\sqrt{n+4}\tilde{D}\sigma}{\sqrt{N}} \\ &= \frac{LD_f^2}{N}\left[\frac{17(n+4)}{2}+\frac{8}{N}+\frac{1}{2}\right] + \frac{2\sigma\sqrt{n+4}}{\sqrt{N}}\left[\frac{D_f^2}{\tilde{D}}\left(1+\frac{1}{(n+4)N}\right)+2\tilde{D}\right] \\ &\le \frac{12L(n+4)D_f^2}{N} + \frac{4\sigma\sqrt{n+4}}{\sqrt{N}}\left(\tilde{D}+\frac{D_f^2}{\tilde{D}}\right).\end{aligned}$$

■

A few remarks about the results obtained in Corollary 6.1 are in order. First, similar to the RSGD method, we use the same selection of stepsizes $\{\gamma_k\}$ and probability mass function $P_R(\cdot)$ in RSGF method for both convex and nonconvex SP problems. In particular, in view of (6.1.72), the iteration complexity of the RSGF method for finding an ε-solution of problem (6.1.47) can be bounded by $\mathcal{O}(n/\varepsilon^2)$. Moreover, in view of (6.1.73), if the problem is convex, a solution $\bar{x}$ satisfying $\mathbb{E}[f(\bar{x})-f^*] \le \varepsilon$ can also be found in $\mathcal{O}(n/\varepsilon^2)$ iterations.

Second, we need to specify $\tilde{D}$ for the stepsize policy in (6.1.70). According to (6.1.72) and (6.1.73), an optimal selection of $\tilde{D}$ would be D_f and D_X, respectively, for the nonconvex and convex case. With such selections, the bounds in (6.1.72) and (6.1.73), respectively, reduce to

$$\tfrac{1}{L}\mathbb{E}[\|\nabla f(x_R)\|^2] \le \frac{12(n+4)LD_f^2}{N} + \frac{8\sqrt{n+4}D_f\sigma}{\sqrt{N}}, \qquad (6.1.76)$$

$$\mathbb{E}[f(x_R)-f^*] \le \frac{5L(n+4)D_X^2}{N} + \frac{4\sqrt{n+4}D_X\sigma}{\sqrt{N}}. \qquad (6.1.77)$$

Similar to the RSGD method, we can establish the complexity of the RSGF method for finding an (ε,Λ)-solution of problem (6.1.47) for some $\varepsilon > 0$ and $\Lambda \in (0,1)$. More specifically, by using (6.1.72) and Markov's inequality, we have

$$\text{Prob}\left\{\|\nabla f(x_R)\|^2 \geq \lambda L\bar{\mathscr{B}}_N\right\} \leq \tfrac{1}{\lambda}, \ \forall \lambda > 0, \tag{6.1.78}$$

which implies that the total number of calls to the SZO performed by the RSGF method for finding an (ε,Λ)-solution of (6.1.47) can be bounded by

$$\mathscr{O}\left\{\frac{nL^2D_f^2}{\Lambda\varepsilon} + \frac{nL^2}{\Lambda^2}\left(\tilde{D} + \frac{D_f^2}{\tilde{D}}\right)^2 \frac{\sigma^2}{\varepsilon^2}\right\}. \tag{6.1.79}$$

We will investigate a possible approach to improve the above complexity bound in next subsection.

6.1.2.2 A Two-Phase Randomized Stochastic Gradient Free Method

In this section, we modify the 2-RSGD method to improve the complexity bound in (6.1.79) for finding an (ε,Λ)-solution of problem (6.1.47).

A Two-Phase RSGF (2-RSGF) Method

Input: Initial point x_1, number of runs S, iteration limit N, and sample size T.
Optimization phase:
For $s = 1,\ldots,S$

Call the RSGF method with input x_1, iteration limit N, stepsizes $\{\gamma_k\}$ in (6.1.70), probability mass function P_R in (6.1.61), and the smoothing parameter μ satisfying (6.1.71). Let $\bar{x}_s$ be the output of this procedure.

Post-optimization phase:
Choose a solution $\bar{x}^*$ from the candidate list $\{\bar{x}_1,\ldots,\bar{x}_S\}$ such that

$$\|g_\mu(\bar{x}^*)\| = \min_{s=1,\ldots,S} \|g_\mu(\bar{x}_s)\|, \ g_\mu(\bar{x}_s) := \tfrac{1}{T}\textstyle\sum_{k=1}^T G_\mu(\bar{x}_s,\xi_k,u_k), \tag{6.1.80}$$

where $G_\mu(x,\xi,u)$ is defined in (6.1.58).

The main convergence properties of the 2-RSGF method are summarized in Theorem 6.4. More specifically, Theorem 6.4(a) establishes the rate of convergence of the 2-RSGF method with a given set of parameters (S,N,T), while Theorem 6.4(b) shows the complexity of this method for finding an (ε,Λ)-solution of problem (6.1.47).

Theorem 6.4. *Under Assumptions 13 and 15, the following statements hold for the 2-RSGF method applied to problem (6.1.47).*

(a) Let $\bar{\mathscr{B}}_N$ be defined in (6.1.72). We have

$$\text{Prob}\left\{\|\nabla f(\bar{x}^*)\|^2 \geq 8L\bar{\mathscr{B}}_N+\tfrac{3(n+4)L^2D_f^2}{2N}+\tfrac{24(n+4)\lambda}{T}\left[L\bar{\mathscr{B}}_N+\tfrac{(n+4)L^2D_f^2}{N}+\sigma^2\right]\right\}$$
$$\leq \tfrac{S+1}{\lambda}+2^{-S},\ \forall\lambda>0; \tag{6.1.81}$$

(b) Let $\varepsilon>0$ and $\Lambda\in(0,1)$ be given. If S is set to $S(\Lambda)$ as in (6.1.32), and the iteration limit N and sample size T, respectively, are set to

$$N=\hat{N}(\varepsilon):=\max\left\{\tfrac{12(n+4)(6LD_f)^2}{\varepsilon},\left[72L\sqrt{n+4}\left(\tilde{D}+\tfrac{D_f^2}{\tilde{D}}\right)\tfrac{\sigma}{\varepsilon}\right]^2\right\}, \tag{6.1.82}$$

$$T=\hat{T}(\varepsilon,\Lambda):=\tfrac{24(n+4)(S+1)}{\Lambda}\max\left\{1,\tfrac{6\sigma^2}{\varepsilon}\right\}, \tag{6.1.83}$$

then the 2-RSGF method can compute an (ε,Λ)-solution of problem (6.1.47) after taking at most

$$2S(\Lambda)\left[\hat{N}(\varepsilon)+\hat{T}(\varepsilon,\Lambda)\right] \tag{6.1.84}$$

calls to the SZO.

Proof. First, observe that by (6.1.52), (6.1.71), and (6.1.72), we have

$$\|\nabla f_\mu(x)-\nabla f(x)\|^2 \leq \tfrac{\mu^2}{4}L^2(n+3)^3 \leq \tfrac{(n+4)L^2D_f^2}{8N}. \tag{6.1.85}$$

Using this observation and the definition of $\bar{x}^*$ in (6.1.80), we obtain

$$\begin{aligned}
\|g_\mu(\bar{x}^*)\|^2 &= \min_{s=1,\ldots,S}\|g_\mu(\bar{x}_s)\|^2 = \min_{s=1,\ldots,S}\|\nabla f(\bar{x}_s)+g_\mu(\bar{x}_s)-\nabla f(\bar{x}_s)\|^2\\
&\leq \min_{s=1,\ldots,S}\{2\left[\|\nabla f(\bar{x}_s)\|^2+\|g_\mu(\bar{x}_s)-\nabla f(\bar{x}_s)\|^2\right]\}\\
&\leq \min_{s=1,\ldots,S}\{2\left[\|\nabla f(\bar{x}_s)\|^2+2\|g_\mu(\bar{x}_s)-\nabla f_\mu(\bar{x}_s)\|^2+2\|\nabla f_\mu(\bar{x}_s)-\nabla f(\bar{x}_s)\|^2\right]\}\\
&\leq 2\min_{s=1,\ldots,S}\|\nabla f(\bar{x}_s)\|^2+4\max_{s=1,\ldots,S}\|g_\mu(\bar{x}_s)-\nabla f(\bar{x}_s)\|^2+\tfrac{(n+4)L^2D_f^2}{2N},
\end{aligned}$$

which implies that

$$\begin{aligned}
\|\nabla f(\bar{x}^*)\|^2 &\leq 2\|g_\mu(\bar{x}^*)\|^2+2\|\nabla f(\bar{x}^*)-g_\mu(\bar{x}^*)\|^2\\
&\leq 2\|g_\mu(\bar{x}^*)\|^2+4\|\nabla f_\mu(\bar{x}^*)-g_\mu(\bar{x}^*)\|^2+4\|\nabla f(\bar{x}^*)-\nabla f_\mu(\bar{x}^*)\|^2\\
&\leq 4\min_{s=1,\ldots,S}\|\nabla f(\bar{x}_s)\|^2+8\max_{s=1,\ldots,S}\|g_\mu(\bar{x}_s)-\nabla f(\bar{x}_s)\|^2+\tfrac{(n+4)L^2D_f^2}{N}\\
&\quad+4\|\nabla f_\mu(\bar{x}^*)-g_\mu(\bar{x}^*)\|^2+4\|\nabla f(\bar{x}^*)-\nabla f_\mu(\bar{x}^*)\|^2\\
&\leq 4\min_{s=1,\ldots,S}\|\nabla f(\bar{x}_s)\|^2+8\max_{s=1,\ldots,S}\|g_\mu(\bar{x}_s)-\nabla f(\bar{x}_s)\|^2\\
&\quad+4\|\nabla f_\mu(\bar{x}^*)-g_\mu(\bar{x}^*)\|^2+\tfrac{3(n+4)L^2D_f^2}{2N},
\end{aligned} \tag{6.1.86}$$

where the last inequality also follows from (6.1.85). We now provide certain probabilistic bounds on the individual terms in the right-hand side of the above inequality. Using (6.1.78) (with $\lambda = 2$), we obtain

$$\text{Prob}\left\{\min_{s=1,\ldots,S} \|\nabla f(\bar{x}_s)\|^2 \geq 2L\bar{\mathscr{B}}_N\right\} = \prod_{s=1}^{S} \text{Prob}\left\{\|\nabla f(\bar{x}_s)\|^2 \geq 2L\bar{\mathscr{B}}_N\right\} \leq 2^{-S}. \tag{6.1.87}$$

Moreover, denote $\Delta_{s,k} = G_\mu(\bar{x}_s, \xi_k, u_k) - \nabla f_\mu(\bar{x}_s)$, $k = 1, \ldots, T$. Note that, similar to (6.1.67), we have

$$\begin{aligned}
\mathbb{E}[\|G_\mu(\bar{x}_s, \xi_k, u_k)\|^2] &\leq 2(n+4)[\mathbb{E}[\|G(\bar{x}_s, \xi)\|^2] + \tfrac{\mu^2}{2} L^2 (n+6)^3 \\
&\leq 2(n+4)[\mathbb{E}[\|\nabla f(\bar{x}_s)\|^2] + \sigma^2] + 2\mu^2 L^2 (n+4)^3.
\end{aligned}$$

It then follows from the previous inequality, (6.1.71), and (6.1.72) that

$$\begin{aligned}
\mathbb{E}[\|\Delta_{s,k}\|^2] &= \mathbb{E}[\|G_\mu(\bar{x}_s, \xi_k, u_k) - \nabla f_\mu(\bar{x}_s)\|^2] \leq \mathbb{E}[\|G_\mu(\bar{x}_s, \xi_k, u_k)\|^2] \\
&\leq 2(n+4)\left[L\bar{\mathscr{B}}_N + \sigma^2\right] + 2\mu^2 L^2 (n+4)^3 \\
&\leq 2(n+4)\left[L\bar{\mathscr{B}}_N + \sigma^2 + \tfrac{L^2 D_f^2}{2N}\right] =: \mathscr{D}_N.
\end{aligned} \tag{6.1.88}$$

Noting that $g_\mu(\bar{x}_s) - \nabla f_\mu(\bar{x}_s) = \sum_{k=1}^{T} \Delta_{s,k}/T$, we conclude from (6.1.88), Assumption 13, and Lemma 6.1(a) that, for any $s = 1, \ldots, S$,

$$\begin{aligned}
&\text{Prob}\left\{\|g_\mu(\bar{x}_s) - \nabla f_\mu(\bar{x}_s)\|^2 \geq \tfrac{\lambda \mathscr{D}_N}{T}\right\} = \text{Prob}\left\{\|\textstyle\sum_{k=1}^{T} \Delta_{s,k}\|^2 \geq \lambda T \mathscr{D}_N\right\} \\
&\leq \tfrac{1}{\lambda}, \ \forall \lambda > 0,
\end{aligned}$$

which implies that

$$\text{Prob}\left\{\max_{s=1,\ldots,S} \|g_\mu(\bar{x}_s) - \nabla f_\mu(\bar{x}_s)\|^2 \geq \tfrac{\lambda \mathscr{D}_N}{T}\right\} \leq \tfrac{S}{\lambda}, \ \forall \lambda > 0, \tag{6.1.89}$$

and that

$$\text{Prob}\left\{\|g_\mu(\bar{x}^*) - \nabla f_\mu(\bar{x}^*)\|^2 \geq \tfrac{\lambda \mathscr{D}_N}{T}\right\} \leq \tfrac{1}{\lambda}, \ \forall \lambda > 0. \tag{6.1.90}$$

The result then follows by combining relations (6.1.86), (6.1.87), (6.1.88), (6.1.89), and (6.1.90).

We now show part (b) holds. Clearly, the total number of calls to SZO in the 2-RSGF method is bounded by $2S[\hat{N}(\varepsilon) + \hat{T}(\varepsilon)]$. It then suffices to show that $\bar{x}^*$ is an (ε, Λ)-solution of problem (6.1.47). Noting that by the definitions of $\bar{\mathscr{B}}(N)$ and $\hat{N}(\varepsilon)$, respectively, in (6.1.72) and (6.1.82), we have

$$\bar{\mathscr{B}}_{\hat{N}(\varepsilon)} = \frac{12(n+4)LD_f^2}{\hat{N}(\varepsilon)} + \frac{4\sigma\sqrt{n+4}}{\sqrt{\hat{N}(\varepsilon)}}\left(\tilde{D} + \tfrac{D_f^2}{\tilde{D}}\right) \leq \tfrac{\varepsilon}{36L} + \tfrac{\varepsilon}{18L} = \tfrac{\varepsilon}{12L}.$$

Hence, we have

$$8L\bar{B}_{\hat{N}(\varepsilon)} + \frac{3(n+4)L^2D_f^2}{2\hat{N}(\varepsilon)} \le \frac{2\varepsilon}{3} + \frac{\varepsilon}{288} \le \frac{17\varepsilon}{24}.$$

Moreover, by setting $\lambda = [2(S+1)]/\Lambda$ and using (6.1.82) and (6.1.83), we obtain

$$\begin{aligned}\frac{24(n+4)\lambda}{T}\left[L\bar{\mathscr{B}}_{\hat{N}(\varepsilon)} + \frac{(n+4)L^2D_f^2}{\hat{N}(\varepsilon)} + \sigma^2\right] &\le \frac{24(n+4)\lambda}{T}\left(\frac{\varepsilon}{12} + \frac{\varepsilon}{432} + \sigma^2\right) \\ &\le \frac{\varepsilon}{12} + \frac{\varepsilon}{432} + \frac{\varepsilon}{6} \le \frac{7\varepsilon}{24}.\end{aligned}$$

Using these two observations and relation (6.1.81) with $\lambda = [2(S+1)]/\Lambda$, we conclude that

$$\begin{aligned}\text{Prob}\left\{\nabla f(\bar{x}^*)\|^2 \ge \varepsilon\right\} \le \text{Prob}&\left\{\|\nabla f(\bar{x}^*)\|^2 \ge 8L\bar{\mathscr{B}}_{\hat{N}(\varepsilon)} + \frac{3(n+4)L^2D_f^2}{2\hat{N}(\varepsilon)}\right. \\ &\left. + \frac{24(n+4)\lambda}{T}\left[L\bar{\mathscr{B}}_{\hat{N}(\varepsilon)} + \frac{(n+4)L^2D_f^2}{\hat{N}(\varepsilon)} + \sigma^2\right]\right\} \\ \le &\frac{S+1}{\lambda} + 2^{-S} = \Lambda.\end{aligned}$$

■

Observe that in the view of (6.1.32), (6.1.82), and (6.1.83), the total number of calls to SZO performed by the 2-RSGF method can be bounded by

$$\mathscr{O}\left\{\frac{nL^2D_f^2\log(1/\Lambda)}{\varepsilon} + nL^2\left(\tilde{D} + \frac{D_f^2}{\tilde{D}}\right)^2\frac{\sigma^2}{\varepsilon^2}\log\frac{1}{\Lambda} + \frac{n\log^2(1/\Lambda)}{\Lambda}\left(1 + \frac{\sigma^2}{\varepsilon}\right)\right\}. \tag{6.1.91}$$

The above bound is considerably smaller than the one in (6.1.79), up to a factor of $\mathscr{O}\left(1/[\Lambda^2\log(1/\Lambda)]\right)$, when the second terms are the dominating ones in both bounds.

6.2 Nonconvex Stochastic Composite Optimization

In this section, we consider the following problem

$$\Psi^* := \min_{x\in X}\{\Psi(x) := f(x) + h(x)\}, \tag{6.2.1}$$

where X is a closed convex set in Euclidean space $\mathbb{R}^n$, $f : X \to \mathbb{R}$ is continuously differentiable, but possibly nonconvex, and h is a simple convex function with known structure, but possibly nonsmooth (e.g., $h(x) = \|x\|_1$ or $h(x) \equiv 0$). We also assume that the gradient of f is L-Lipschitz continuous for some $L > 0$, i.e.,

$$\|\nabla f(y) - \nabla f(x)\| \le L\|y - x\|, \quad \text{for any } x, y \in X, \tag{6.2.2}$$

or equivalently,

$$|f(y) - f(x) - \langle \nabla f(x), y - x \rangle| \leq \tfrac{L}{2}\|y - x\|^2, \qquad \text{for any } x, y \in X, \tag{6.2.3}$$

and Ψ is bounded below over X, i.e., Ψ^* is finite. Although f is Lipschitz continuously differentiable, we assume that only the noisy gradient of f is available via subsequent calls to a *stochastic first-order oracle* (SFO) as discussed in the previous section. Specifically, at the k-th call, $k \geq 1$, for the input $x_k \in X$, SFO would output a *stochastic gradient* $G(x_k, \xi_k)$, where ξ_k is a random variable whose distribution is supported on $\Xi_k \subseteq \mathbb{R}^d$. Throughout the section, we assume the same Assumption 13 as in the previous section about the Borel functions $G(x_k, \xi_k)$.

In last section, we presented a randomized stochastic gradient (RSGD) method, for solving the unconstrained nonconvex SP problem, i.e., problem (6.2.1) with $h \equiv 0$ and $X = \mathbb{R}^n$. While the RSGD algorithm and its variants can handle the unconstrained nonconvex SP problems, their convergence cannot be guaranteed for stochastic composite optimization problems in (6.2.1) where $X \neq \mathbb{R}^n$ and/or $h(\cdot)$ is non-differentiable.

Our goal in this section mainly consists of developing variants of the RSGD algorithm by taking a mini-batch of samples at each iteration of our algorithm to deal with the constrained composite problems while preserving the complexity results. More specifically, we first modify the scheme of the RSGD algorithm to propose a randomized stochastic mirror descent (RSMD) algorithm to solve constrained nonconvex stochastic composite problems. Unlike the RSGD algorithm, at each iteration of the RSMD algorithm, we take multiple samples such that the total number of calls to the SFO to find a solution $\bar{x} \in X$ such that $\mathbb{E}[\|g_X(\bar{x})\|^2] \leq \varepsilon$ is still $\mathcal{O}(\sigma^2/\varepsilon^2)$, where $g_X(\bar{x})$ is a generalized projected gradient of Ψ at $\bar{x}$ over X. In addition, our RSMD algorithm is in a more general setting depending on a general distance function rather than Euclidean distance. This would be particularly useful for special structured constrained set (e.g., X being a standard simplex). Second, we present a two-phase randomized stochastic mirror descent (2-RSMD) algorithm, the RSMD algorithm with a post-optimization phase, to improve the large-deviation results of the RSMD algorithm. We show that the complexity of this approach can be further improved under a light-tail assumption about the SFO. Third, under the assumption that the gradient of f is also bounded on X, we specialize the RSMD algorithm to give a randomized stochastic gradient free mirror descent (RSMDF) algorithm, which only uses the stochastic zeroth-order information.

The remaining part of this section is organized as follows. We first describe some properties of the projection based on a general distance function in Sect. 6.2.1. In Sect. 6.2.2, a deterministic first-order method for problem (6.2.1) is proposed, which mainly provides a basis for our stochastic algorithms developed in later sections. Then, by incorporating a randomized scheme, we present the RSMD and 2-RSMD algorithms for solving the SP problem (6.2.1) in Sect. 6.2.3. In Sect. 6.2.4, we discuss how to generalize the RSMD algorithm to the case when only zeroth-order information is available.

6.2.1 Some Properties of Prox-Mapping

As shown in the previous chapters for the convex setting, a general distance generating function, instead of the usual Euclidean distance function, helps us to design algorithms that can adjust to the geometry of the feasible set. Moreover, sometimes non-Euclidean prox-mapping can be easier to compute. Our goal in this section is to generalize such constructions for the nonconvex setting.

Recall a function $v : X \to \mathbb{R}$ is said to be a *distance generating function* with modulus 1 with respect to $\|\cdot\|$, if v is continuously differentiable and strongly convex satisfying

$$\langle x - z, \nabla v(x) - \nabla v(z) \rangle \geq \|x - z\|^2, \; \forall x, z \in X. \tag{6.2.4}$$

Then, the *prox-function* associated with v is defined as

$$V(z,x) = v(x) - [v(z) + \langle \nabla v(z), x - z \rangle]. \tag{6.2.5}$$

In this section, we assume that the prox-function V is chosen such that the generalized projection problem given by

$$x^+ = \text{argmin}_{u \in X} \left\{ \langle g, u \rangle + \tfrac{1}{\gamma} V(x,u) + h(u) \right\} \tag{6.2.6}$$

is easily solvable for any $\gamma > 0$, $g \in \mathbb{R}^n$ and $x \in X$. Apparently, different choices of v can be used in the definition of prox-function.

In order to discuss some important properties of the generalized projection defined in (6.2.6), let us first define

$$P_X(x, g, \gamma) = \tfrac{1}{\gamma}(x - x^+), \tag{6.2.7}$$

where x^+ is given in (6.2.6). We can see that $P_X(x, \nabla f(x), \gamma)$ (see also (3.8.18)) can be viewed as a generalized projected gradient (or gradient mapping) of Ψ at x. Indeed, if $X = \mathbb{R}^n$ and h vanishes, we would have $P_X(x, \nabla f(x), \gamma) = \nabla f(x) = \nabla \Psi(x)$. For more general h, the following result shows that as the size of $P_X(x, \nabla f(x), \gamma)$ vanishes, x^+ approaches to a stationary point of problem (6.2.1).

Lemma 6.3. *Let $x \in \mathbb{R}^n$ be given and denote $g \equiv \nabla f(x)$. Assume that the distance generating function v has L_v-Lipschitz gradients. If $\|P_X(x, g, \gamma)\| \leq \varepsilon$ for some $\gamma > 0$, then*

$$-\nabla f(x^+) \in \partial h(x^+) + N_X(x^+) + B(\varepsilon(\gamma L + L_v)),$$

where $\partial h(\cdot)$ denotes the subdifferential of $h(\cdot)$, N_X denotes the normal cone given by

$$N_X(\bar{x}) := \{d \in \mathbb{R}^n : \langle d, x - \bar{x} \leq 0 \forall x \in X\}, \tag{6.2.8}$$

and $B(r) := \{x \in \mathbb{R}^n : \|x\| \leq r\}$.

Proof. By the optimality condition of (6.4.42), we have $-\nabla f(x) - \frac{1}{\gamma}(\nabla v(x^+) - \nabla v(x)) \in \partial h(x^+) + N_X(x^+)$, which implies that

$$-\nabla f(x^+) + \left[\nabla f(x^+) - \nabla f(x) - \tfrac{1}{\gamma}(\nabla v(x^+) - \nabla v(x))\right] \in \partial h(x^+) + N_X(x^+). \tag{6.2.9}$$

Our conclusion immediately follows from the above relation and the simple fact that

$$\begin{aligned}
\|\nabla f(x^+) - \nabla f(x) - \tfrac{1}{\gamma}(\nabla v(x^+) - \nabla v(x))\| &\le L\|x^+ - x\| + \tfrac{L_v}{\gamma}\|x^+ - x\| \\
&\le L\|x^+ - x\| + \tfrac{L_v}{\gamma}\|x^+ - x\| \\
&= (\gamma L + L_v)\|P_X(x,g,\gamma)\|.
\end{aligned}$$

■

The following lemma provides a bound for the size of $P_X(x,g,\gamma)$.

Lemma 6.4. *Let x^+ be given in (6.2.6). Then, for any $x \in X$, $g \in \mathbb{R}^n$, and $\gamma > 0$, we have*

$$\langle g, P_X(x,g,\gamma)\rangle \ge \|P_X(x,g,\gamma)\|^2 + \tfrac{1}{\gamma}\left[h(x^+) - h(x)\right]. \tag{6.2.10}$$

Proof. By the optimality condition of (6.2.6) and the definition of prox-function in (6.2.5), there exists a $p \in \partial h(x^+)$ such that

$$\langle g + \tfrac{1}{\gamma}\left[\nabla v(x^+) - \nabla v(x)\right] + p, u - x^+\rangle \ge 0, \qquad \text{for any } u \in X.$$

Letting $u = x$ in the above inequality, by the convexity of h and (6.2.4), we obtain

$$\begin{aligned}
\langle g, x - x^+\rangle &\ge \tfrac{1}{\gamma}\langle \nabla v(x^+) - \nabla v(x), x^+ - x\rangle + \langle p, x^+ - x\rangle \\
&\ge \tfrac{1}{\gamma}\|x^+ - x\|^2 + \left[h(x^+) - h(x)\right],
\end{aligned}$$

which in the view of (6.2.7) and $\gamma > 0$ clearly imply (6.2.10). ■

It is well known that the Euclidean projection is Lipschitz continuous. Below, we show that this property also holds for the general prox-mapping.

Lemma 6.5. *Let x_1^+ and x_2^+ be given in (6.2.6) with g replaced by g_1 and g_2 respectively. Then,*

$$\|x_2^+ - x_1^+\| \le \gamma\|g_2 - g_1\|. \tag{6.2.11}$$

Proof. By the optimality condition of (6.2.6), for any $u \in X$, there exist $p_1 \in \partial h(x_1^+)$ and $p_2 \in \partial h(x_2^+)$ such that

$$\langle g_1 + \tfrac{1}{\gamma}\left[\nabla v(x_1^+) - \nabla v(x)\right] + p_1, u - x_1^+\rangle \ge 0, \tag{6.2.12}$$

and

$$\langle g_2 + \tfrac{1}{\gamma}\left[\nabla v(x_2^+) - \nabla v(x)\right] + p_2, u - x_2^+\rangle \ge 0. \tag{6.2.13}$$

Letting $u = x_2^+$ in (6.2.12), by the convexity of h, we have

$$\begin{aligned}
\langle g_1, x_2^+ - x_1^+\rangle &\ge \tfrac{1}{\gamma}\langle \nabla v(x) - \nabla v(x_1^+), x_2^+ - x_1^+\rangle + \langle p_1, x_1^+ - x_2^+\rangle \\
&\ge \tfrac{1}{\gamma}\langle \nabla v(x_2^+)
\end{aligned}$$

$$-\nabla v(x_1^+), x_2^+ - x_1^+\rangle + \tfrac{1}{\gamma}\langle \nabla v(x) - \nabla v(x_2^+), x_2^+ - x_1^+\rangle$$
$$+h(x_1^+) - h(x_2^+). \qquad (6.2.14)$$

Similarly, letting $u = x_1^+$ in (6.2.13), we have

$$\begin{aligned}\langle g_2, x_1^+ - x_2^+\rangle &\geq \tfrac{1}{\gamma}\langle \nabla v(x) - \nabla v(x_2^+), x_1^+ - x_2^+\rangle + \langle p_2, x_2^+ - x_1^+\rangle \\ &\geq \tfrac{1}{\gamma}\langle \nabla v(x) - \nabla v(x_2^+), x_1^+ - x_2^+\rangle + h(x_2^+) - h(x_1^+). \qquad (6.2.15)\end{aligned}$$

Summing up (6.2.14) and (6.2.15), by the strong convexity (6.2.4) of v, we obtain

$$\|g_1 - g_2\|\|x_2^+ - x_1^+\| \geq \langle g_1 - g_2, x_2^+ - x_1^+\rangle \geq \tfrac{1}{\gamma}\|x_2^+ - x_1^+\|^2,$$

which gives (6.2.11). ■

As a consequence of the above lemma, we have $P_X(x,\cdot,\gamma)$ is Lipschitz continuous.

Proposition 6.1. *Let $P_X(x,g,\gamma)$ be defined in (6.2.7). Then, for any g_1 and g_2 in $\mathbb{R}^n$, we have*

$$\|P_X(x,g_1,\gamma) - P_X(x,g_2,\gamma)\| \leq \|g_1 - g_2\|. \qquad (6.2.16)$$

Proof. Noticing (6.2.7), (6.2.12), and (6.2.13), we have

$$\|P_X(x,g_1,\gamma) - P_X(x,g_2,\gamma)\| = \|\tfrac{1}{\gamma}(x - x_1^+) - \tfrac{1}{\gamma}(x - x_2^+)\| = \tfrac{1}{\gamma}\|x_2^+ - x_1^+\| \leq \|g_1 - g_2\|,$$

where the last inequality follows from (6.2.11). ■

The following lemma characterizes the solution of the generalized projection and its proof follows as a special case of Lemma 3.5.

Lemma 6.6. *Let x^+ be given in (6.2.6). Then, for any $u \in X$, we have*

$$\langle g, x^+\rangle + h(x^+) + \tfrac{1}{\gamma}V(x,x^+) \leq \langle g, u\rangle + h(u) + \tfrac{1}{\gamma}[V(x,u) - V(x^+,u)]. \qquad (6.2.17)$$

6.2.2 Nonconvex Mirror Descent Methods

In this subsection, we consider the problem (6.2.1) with $f \in \mathscr{C}_L^{1,1}(X)$, and for each input $x_k \in X$, we assume that the exact gradient $\nabla f(x_k)$ is available. Using the exact gradient information, we give a deterministic nonconvex mirror descent (MD) algorithm for solving (6.2.1), which provides a basis for us to develop the stochastic first-order algorithms in the next subsection.

A Nonconvex Mirror Descent (MD) Algorithm

Input: initial point $x_1 \in X$, total number of iterations N, and the stepsizes $\{\gamma_k\}$ with $\gamma_k > 0$, $k \geq 1$.

Step $k = 1,\ldots,N$. Compute

$$x_{k+1} = \text{argmin}_{u\in X}\left\{\langle \nabla f(x_k), u\rangle + \tfrac{1}{\gamma_k}V(x_k,u) + h(u)\right\}. \tag{6.2.18}$$

Output: $x_R \in \{x_k,\ldots,x_N\}$ such that

$$R = \text{argmin}_{k\in\{1,\ldots,N\}}\|g_{X,k}\|, \tag{6.2.19}$$

where the $g_{X,k}$ is given by

$$g_{X,k} = P_X(x_k, \nabla f(x_k), \gamma_k). \tag{6.2.20}$$

We can see that the above algorithm outputs the iterate with the minimum norm of the generalized projected gradients. In practice, one may choose the solution with the minimum function value as the output of the algorithm. However, since f may not be a convex function, we cannot provide theoretical performance guarantee for such a selection of the output solution. In the above algorithm, we have not specified the selection of the stepsizes $\{\gamma_k\}$. We will return to this issue after establishing the following convergence results.

Theorem 6.5. *Suppose that the stepsizes $\{\gamma_k\}$ in the nonconvex MD algorithm are chosen such that $0 < \gamma_k \le 2/L$ with $\gamma_k < 2/L$ for at least one k. Then, we have*

$$\|g_{X,R}\|^2 \le \frac{LD_\Psi^2}{\sum_{k=1}^N(\gamma_k - L\gamma_k^2/2)}, \tag{6.2.21}$$

where

$$g_{X,R} = P_X(x_R, \nabla f(x_R), \gamma_R) \quad \textit{and} \quad D_\Psi := \left[\tfrac{(\Psi(x_1)-\Psi^*)}{L}\right]^{\frac{1}{2}}. \tag{6.2.22}$$

Proof. Since $f \in \mathscr{C}_L^{1,1}(X)$, it follows from (6.2.3), (6.2.7), (6.2.18), and (6.2.20) that for any $k = 1,\ldots,N$, we have

$$\begin{aligned} f(x_{k+1}) &\le f(x_k) + \langle \nabla f(x_k), x_{k+1} - x_k\rangle + \tfrac{L}{2}\|x_{k+1} - x_k\|^2 \\ &= f(x_k) - \gamma_k\langle \nabla f(x_k), g_{X,k}\rangle + \tfrac{L}{2}\gamma_k^2\|g_{X,k}\|^2. \end{aligned} \tag{6.2.23}$$

Then, by Lemma 6.4 with $x = x_k$, $\gamma = \gamma_k$, and $g = \nabla f(x_k)$, we obtain

$$f(x_{k+1}) \le f(x_k) - \left[\gamma_k\|g_{X,k}\|^2 + h(x_{k+1}) - h(x_k)\right] + \tfrac{L}{2}\gamma_k^2\|g_{X,k}\|^2,$$

which implies

$$\Psi(x_{k+1}) \le \Psi(x_k) - \left(\gamma_k - \tfrac{L}{2}\gamma_k^2\right)\|g_{X,k}\|^2. \tag{6.2.24}$$

Summing up the above inequalities for $k = 1,\ldots,N$, by (6.2.19) and $\gamma_k \le 2/L$, we have

$$\begin{aligned} \|g_{X,R}\|^2\textstyle\sum_{k=1}^N\left(\gamma_k - \tfrac{L}{2}\gamma_k^2\right) &\le \textstyle\sum_{k=1}^N\left(\gamma_k - \tfrac{L}{2}\gamma_k^2\right)\|g_{X,k}\|^2 \\ &\le \Psi(x_1) - \Psi(x_{k+1}) \le \Psi(x_1) - \Psi^*. \end{aligned} \tag{6.2.25}$$

By our assumption, $\sum_{k=1}^{N}\left(\gamma_k - L\gamma_k^2/2\right) > 0$. Hence, dividing both sides of the above inequality by $\sum_{k=1}^{N}\left(\gamma_k - L\gamma_k^2/2\right)$, we obtain (6.2.21). ∎

The following corollary shows a specialized complexity result for the nonconvex MD algorithm with one proper constant stepsize policy.

Corollary 6.4. *Suppose that in the nonconvex MD algorithm the stepsizes* $\gamma_k = 1/L$ *for all* $k = 1,\ldots,N$. *Then, we have*

$$\|g_{X,R}\|^2 \leq \tfrac{2L^2 D_\Psi^2}{N}. \tag{6.2.26}$$

Proof. With the constant stepsizes $\gamma_k = 1/L$ for all $k = 1,\ldots,N$, we have

$$\frac{LD_\Psi^2}{\sum_{k=1}^{N}(\gamma_k - L\gamma_k^2/2)} = \frac{2L^2 D_\Psi^2}{N}, \tag{6.2.27}$$

which together with (6.2.21) clearly imply (6.2.26). ∎

6.2.3 Nonconvex Stochastic Mirror Descent Methods

In this subsection, we consider problem (6.2.1), but the exact gradient of f is not available. We assume that only noisy first-order information of f is available via subsequent calls to the stochastic first-order oracle SFO. In particular, given the k-th iteration $x_k \in X$ of our algorithm, the SFO will output the stochastic gradient $G(x_k, \xi_k)$, where ξ_k is a random vector whose distribution is supported on $\Xi_k \subseteq \mathbb{R}^d$. We assume the stochastic gradient $G(x_k, \xi_k)$ satisfies Assumption 13.

This subsection proceeds as follows. In Sect. 6.2.3.1, we present a stochastic variant of the nonconvex MD algorithm incorporated with a randomized stopping criterion, called the RSMD algorithm. Then, in Sect. 6.2.3.2, we describe a two phase RSMD algorithm, called the 2-RSMD algorithm, which can significantly reduce the large-deviations resulted from the RSMD algorithm. We assume throughout this section that the norm $\|\cdot\|$ is associated with the inner product $\langle\cdot,\cdot\rangle$.

6.2.3.1 A Randomized Stochastic Mirror Descent Method

Convexity of the objective function often plays an important role in establishing the convergence results for the current SGD algorithms. Similar to the RSGD method, in this subsection we give an SGD-type algorithm which does not require the convexity of the objective function. Moreover, this weaker requirement enables the algorithm to deal with the case in which the random noises $\{\xi_k\}, k \geq 1$ could depend on the iterates $\{x_k\}$.

A Randomized Stochastic Mirror Descent (RSMD) Algorithm

Input: initial point $x_1 \in X$, iteration limit N, the stepsizes $\{\gamma_k\}$ with $\gamma_k > 0$, $k \geq 1$, the batch sizes $\{m_k\}$ with $m_k > 0$, $k \geq 1$, and the probability mass function P_R supported on $\{1,\ldots,N\}$.
Step 0. Let R be a random variable with probability mass function P_R.
Step $k = 1,\ldots,R-1$. Call the SFO m_k times to obtain $G(x_k, \xi_{k,i})$, $i = 1,\ldots,m_k$, set

$$G_k = \tfrac{1}{m_k}\textstyle\sum_{i=1}^{m_k} G(x_k, \xi_{k,i}), \tag{6.2.28}$$

and compute

$$x_{k+1} = \arg\min_{u\in X}\left\{\langle G_k, u\rangle + \tfrac{1}{\gamma_k}V(x_k,u) + h(u)\right\}. \tag{6.2.29}$$

Output: x_R.

We use a randomized iteration count to terminate the RSMD algorithm. In this algorithm, we also need to specify the stepsizes $\{\gamma_k\}$, the batch sizes $\{m_k\}$, and probability mass function P_R. We will again address these issues after presenting some convergence results of the RSMD algorithm.

Theorem 6.6. *Suppose that the stepsizes $\{\gamma_k\}$ in the RSMD algorithm are chosen such that $0 < \gamma_k \leq 1/L$ with $\gamma_k < 1/L$ for at least one k, and the probability mass function P_R is chosen such that for any $k = 1,\ldots,N$,*

$$P_R(k) := \text{Prob}\{R = k\} = \tfrac{\gamma_k - L\gamma_k^2}{\sum_{k=1}^N(\gamma_k - L\gamma_k^2)}. \tag{6.2.30}$$

Then, under Assumption 13,

(a) for any $N \geq 1$, we have

$$\mathbb{E}[\|\tilde{g}_{X,R}\|^2] \leq \tfrac{LD_\Psi^2 + \sigma^2\sum_{k=1}^N(\gamma_k/m_k)}{\sum_{k=1}^N(\gamma_k - L\gamma_k^2)}, \tag{6.2.31}$$

where the expectation is taken with respect to R and $\xi_{[N]} := (\xi_1,\ldots,\xi_N)$, D_Ψ is defined in (6.2.22), and the stochastic projected gradient

$$\tilde{g}_{X,k} := P_X(x_k, G_k, \gamma_k), \tag{6.2.32}$$

with P_X defined in(6.2.7);

(b) if, in addition, f in problem (6.2.1) is convex with an optimal solution x^*, and the stepsizes $\{\gamma_k\}$ are nondecreasing, i.e.,

$$0 \leq \gamma_1 \leq \gamma_2 \leq \ldots \leq \gamma_N \leq \tfrac{1}{L}, \tag{6.2.33}$$

we have

$$\mathbb{E}\left[\Psi(x_R) - \Psi(x^*)\right] \leq \tfrac{(1-L\gamma_1)V(x_1,x^*) + (\sigma^2/2)\sum_{k=1}^N(\gamma_k^2/m_k)}{\sum_{k=1}^N(\gamma_k - L\gamma_k^2)}, \tag{6.2.34}$$

where the expectation is taken with respect to R and $\xi_{[N]}$. Similarly, if the stepsizes $\{\gamma_k\}$ are nonincreasing, i.e.,

$$\tfrac{1}{L} \geq \gamma_1 \geq \gamma_2 \geq \ldots \geq \gamma_N \geq 0, \tag{6.2.35}$$

we have

$$\mathbb{E}\left[\Psi(x_R) - \Psi(x^*)\right] \leq \frac{(1-L\gamma_N)\bar{V}(x^*)+(\sigma^2/2)\sum_{k=1}^N(\gamma_k^2/m_k)}{\sum_{k=1}^N(\gamma_k - L\gamma_k^2)}, \tag{6.2.36}$$

where $\bar{V}(x^*) := \max_{u\in X} V(u, x^*)$.

Proof. Let $\delta_k \equiv G_k - \nabla f(x_k)$, $k \geq 1$. Since f is smooth, it follows from (6.2.3), (6.2.7), (6.2.29) and (6.2.32) that, for any $k = 1,\ldots,N$, we have

$$\begin{aligned} f(x_{k+1}) &\leq f(x_k) + \langle \nabla f(x_k), x_{k+1} - x_k\rangle + \tfrac{L}{2}\|x_{k+1} - x_k\|^2 \\ &= f(x_k) - \gamma_k\langle \nabla f(x_k), \tilde{g}_{X,k}\rangle + \tfrac{L}{2}\gamma_k^2\|\tilde{g}_{X,k}\|^2 \\ &= f(x_k) - \gamma_k\langle G_k, \tilde{g}_{X,k}\rangle + \tfrac{L}{2}\gamma_k^2\|\tilde{g}_{X,k}\|^2 + \gamma_k\langle \delta_k, \tilde{g}_{X,k}\rangle. \end{aligned} \tag{6.2.37}$$

So, by Lemma 6.4 with $x = x_k$, $\gamma = \gamma_k$ and $g = G_k$, we obtain

$$\begin{aligned} f(x_{k+1}) \leq f(x_k) - \left[\gamma_k\|\tilde{g}_{X,k}\|^2 + h(x_{k+1}) - h(x_k)\right] + \tfrac{L}{2}\gamma_k^2\|\tilde{g}_{X,k}\|^2 \\ + \gamma_k\langle \delta_k, g_{X,k}\rangle + \gamma_k\langle \delta_k, \tilde{g}_{X,k} - g_{X,k}\rangle, \end{aligned}$$

where the projected gradient $g_{X,k}$ is defined in (6.2.20). Then, from the above inequality, (6.2.20) and (6.2.32), we obtain

$$\begin{aligned} \Psi(x_{k+1}) &\leq \Psi(x_k) - \left(\gamma_k - \tfrac{L}{2}\gamma_k^2\right)\|\tilde{g}_{X,k}\|^2 + \gamma_k\langle \delta_k, g_{X,k}\rangle + \gamma_k\|\delta_k\|\|\tilde{g}_{X,k} - g_{X,k}\| \\ &\leq \Psi(x_k) - \left(\gamma_k - \tfrac{L}{2}\gamma_k^2\right)\|\tilde{g}_{X,k}\|^2 + \gamma_k\langle \delta_k, g_{X,k}\rangle + \gamma_k\|\delta_k\|^2, \end{aligned} \tag{6.2.38}$$

where the last inequality follows from Proposition 6.1 with $x = x_k, \gamma = \gamma_k, g_1 = G_k$ and $g_2 = \nabla f(x_k)$. Summing up the above inequalities for $k = 1,\ldots,N$ and noticing that $\gamma_k \leq 1/L$, we obtain

$$\begin{aligned} \textstyle\sum_{k=1}^N\left(\gamma_k - L\gamma_k^2\right)\|\tilde{g}_{X,k}\|^2 &\leq \textstyle\sum_{k=1}^N\left(\gamma_k - \tfrac{L}{2}\gamma_k^2\right)\|\tilde{g}_{X,k}\|^2 \\ &\leq \Psi(x_1) - \Psi(x_{k+1}) + \textstyle\sum_{k=1}^N\left\{\gamma_k\langle \delta_k, g_{X,k}\rangle + \gamma_k\|\delta_k\|^2\right\} \\ &\leq \Psi(x_1) - \Psi^* + \textstyle\sum_{k=1}^N\left\{\gamma_k\langle \delta_k, g_{X,k}\rangle + \gamma_k\|\delta_k\|^2\right\}. \end{aligned} \tag{6.2.39}$$

Notice that the iterate x_k is a function of the history $\xi_{[k-1]}$ of the generated random process and hence is random. By part (a) of Assumption 13, we have $\mathbb{E}[\langle \delta_k, g_{X,k}\rangle | \xi_{[k-1]}] = 0$. In addition, denoting $\delta_{k,i} \equiv G(x_k, \xi_{k,i}) - \nabla f(x_k)$, $i = 1,\ldots,m_k$, $k = 1,\ldots,N$, $S_j = \sum_{i=1}^j \delta_{k,i}$, $j = 1,\ldots,m_k$, and $S_0 = 0$, and noting that $\mathbb{E}[\langle S_{i-1}, \delta_{k,i}\rangle | S_{i-1}] = 0$ for all $i = 1,\ldots,m_k$, we have

$$\mathbb{E}[\|S_{m_k}\|^2] = \mathbb{E}\left[\|S_{m_k-1}\|^2 + 2\langle S_{m_k-1}, \delta_{k,m_k}\rangle + \|\delta_{k,m_k}\|^2\right]$$
$$= \mathbb{E}[\|S_{m_k-1}\|^2] + \mathbb{E}[\|\delta_{k,m_k}\|^2] = \ldots = \textstyle\sum_{i=1}^{m_k}\mathbb{E}\|\delta_{k,i}\|^2,$$

which, in view of (6.2.28) and Assumption 13(b), then implies that

$$\mathbb{E}[\|\delta_k\|^2] = \mathbb{E}\left[\left\|\tfrac{1}{m_k}\textstyle\sum_{i=1}^{m_k}\delta_{k,i}\right\|^2\right] = \tfrac{1}{m_k^2}\mathbb{E}[\|S_{m_k}\|^2] = \tfrac{1}{m_k^2}\textstyle\sum_{i=1}^{m_k}\mathbb{E}[\|\delta_{k,i}\|^2] \le \tfrac{\sigma^2}{m_k}. \tag{6.2.40}$$

With these observations, now taking expectations with respect to $\xi_{[N]}$ on both sides of (6.2.39), we get

$$\textstyle\sum_{k=1}^N \left(\gamma_k - L\gamma_k^2\right)\mathbb{E}\|\tilde{g}_{X,k}\|^2 \le \Psi(x_1) - \Psi^* + \sigma^2\textstyle\sum_{k=1}^N(\gamma_k/m_k).$$

Then, since $\sum_{k=1}^N \left(\gamma_k - L\gamma_k^2\right) > 0$ by our assumption, dividing both sides of the above inequality by $\sum_{k=1}^N \left(\gamma_k - L\gamma_k^2\right)$ and noticing that

$$\mathbb{E}[\|\tilde{g}_{X,R}\|^2] = \frac{\sum_{k=1}^N(\gamma_k - L\gamma_k^2)\mathbb{E}\|\tilde{g}_{X,k}\|^2}{\sum_{k=1}^N(\gamma_k - L\gamma_k^2)},$$

we have (6.2.31) holds.

We now show part (b) of the theorem. By Lemma 6.6 with $x = x_k, \gamma = \gamma_k, g = G_k$, and $u = x^*$, we have

$$\langle G_k, x_{k+1}\rangle + h(x_{k+1}) + \tfrac{1}{\gamma_k}V(x_k, x_{k+1}) \le \langle G_k, x^*\rangle + h(x^*)$$
$$+ \tfrac{1}{\gamma_k}[V(x_k, x^*) - V(x_{k+1}, x^*)],$$

which together with (6.2.3) and definition of δ_k give

$$f(x_{k+1}) + \langle\nabla f(x_k) + \delta_k, x_{k+1}\rangle + h(x_{k+1}) + \tfrac{1}{\gamma_k}V(x_k, x_{k+1})$$
$$\le f(x_k) + \langle\nabla f(x_k), x_{k+1} - x_k\rangle + \tfrac{L}{2}\|x_{k+1} - x_k\|^2 + \langle\nabla f(x_k) + \delta_k, x^*\rangle + h(x^*)$$
$$+ \tfrac{1}{\gamma_k}[V(x_k, x^*) - V(x_{k+1}, x^*)].$$

Simplifying the above inequality, we have

$$\Psi(x_{k+1}) \le f(x_k) + \langle\nabla f(x_k), x^* - x_k\rangle + h(x^*) + \langle\delta_k, x^* - x_{k+1}\rangle + \tfrac{L}{2}\|x_{k+1} - x_k\|^2$$
$$- \tfrac{1}{\gamma_k}V(x_k, x_{k+1}) + \tfrac{1}{\gamma_k}[V(x_k, x^*) - V(x_{k+1}, x^*)].$$

Then, it follows from the convexity of f, (6.2.4), and (6.2.5) that

$$\Psi(x_{k+1}) \le f(x^*) + h(x^*) + \langle\delta_k, x^* - x_{k+1}\rangle + \left(\tfrac{L}{2} - \tfrac{1}{2\gamma_k}\right)\|x_{k+1} - x_k\|^2$$
$$+ \tfrac{1}{\gamma_k}[V(x_k, x^*) - V(x_{k+1}, x^*)]$$
$$= \Psi(x^*) + \langle\delta_k, x^* - x_k\rangle + \langle\delta_k, x_k - x_{k+1}\rangle + \tfrac{L\gamma_k - 1}{2\gamma_k}\|x_{k+1} - x_k\|^2$$
$$+ \tfrac{1}{\gamma_k}[V(x_k, x^*) - V(x_{k+1}, x^*)]$$
$$\le \Psi(x^*) + \langle\delta_k, x^* - x_k\rangle + \|\delta_k\|\|x_k - x_{k+1}\| - \tfrac{1 - L\gamma_k}{2\gamma_k}\|x_{k+1} - x_k\|^2$$

$$+\tfrac{1}{\gamma_k}[V(x_k,x^*)-V(x_{k+1},x^*)]$$
$$\le \Psi(x^*)+\langle\delta_k,x^*-x_k\rangle+\tfrac{\gamma_k}{2(1-L\gamma_k)}\|\delta_k\|^2+\tfrac{1}{\gamma_k}[V(x_k,x^*)-V(x_{k+1},x^*)],$$

where the last inequality follows from the fact that $ax-bx^2/2\le a^2/(2b)$. Noticing $\gamma_k\le 1/L$, multiplying both sides of the above inequality by $(\gamma_k-L\gamma_k^2)$, and summing them up for $k=1,\ldots,N$, we obtain

$$\begin{aligned}\textstyle\sum_{k=1}^N(\gamma_k-L\gamma_k^2)\,[\Psi(x_{k+1})-\Psi(x^*)]\le \textstyle\sum_{k=1}^N(\gamma_k-L\gamma_k^2)\langle\delta_k,x^*-x_k\rangle\\ +\textstyle\sum_{k=1}^N\tfrac{\gamma_k^2}{2}\|\delta_k\|^2\\ +\textstyle\sum_{k=1}^N(1-L\gamma_k)\,[V(x_k,x^*)-V(x_{k+1},x^*)].\end{aligned}\quad(6.2.41)$$

Now, if the increasing stepsize condition (6.2.33) is satisfied, we have from $V(x_{N+1},x^*)\ge 0$ that

$$\begin{aligned}&\textstyle\sum_{k=1}^N(1-L\gamma_k)\,[V(x_k,x^*)-V(x_{k+1},x^*)]\\ &=(1-L\gamma_1)V(x_1,x^*)+\textstyle\sum_{k=2}^N(1-L\gamma_k)V(x_k,x^*)-\textstyle\sum_{k=1}^N(1-L\gamma_k)V(x_{k+1},x^*)\\ &\le(1-L\gamma_1)V(x_1,x^*)+\textstyle\sum_{k=2}^N(1-L\gamma_{k-1})V(x_k,x^*)-\textstyle\sum_{k=1}^N(1-L\gamma_k)V(x_{k+1},x^*)\\ &=(1-L\gamma_1)V(x_1,x^*)-(1-L\gamma_N)V(x_{N+1},x^*)\\ &\le(1-L\gamma_1)V(x_1,x^*).\end{aligned}$$

Taking expectation on both sides of (6.2.41) with respect to $\xi_{[N]}$, again using the observations that $\mathbb{E}[\|\delta_k^2\|]\le\sigma^2/m_k$ and $\mathbb{E}[\langle\delta_k,g_{X,k}\rangle|\xi_{[k-1]}]=0$, then it follows from the above inequality that

$$\textstyle\sum_{k=1}^N(\gamma_k-L\gamma_k^2)\mathbb{E}_{\xi_{[N]}}\,[\Psi(x_{k+1})-\Psi(x^*)]\le(1-L\gamma_1)V(x_1,x^*)+\tfrac{\sigma^2}{2}\textstyle\sum_{k=1}^N(\gamma_k^2/m_k).$$

Finally, (6.2.34) follows from the above inequality and the arguments similar to the proof in part (a). Now, if the decreasing stepsize condition (6.2.35) is satisfied, we have from the definition $\bar V(x^*):=\max_{u\in X}V(u,x^*)\ge 0$ and $V(x_{N+1},x^*)\ge 0$ that

$$\begin{aligned}&\textstyle\sum_{k=1}^N(1-L\gamma_k)\,[V(x_k,x^*)-V(x_{k+1},x^*)]\\ &=(1-L\gamma_1)V(x_1,x^*)+L\textstyle\sum_{k=1}^{N-1}(\gamma_k-\gamma_{k+1})V(x_{k+1},x^*)-(1-L\gamma_N)V(x_{N+1},x^*)\\ &\le(1-L\gamma_1)\bar V(x^*)+L\textstyle\sum_{k=1}^{N-1}(\gamma_k-\gamma_{k+1})\bar V(x^*)-(1-L\gamma_N)V(x_{N+1},x^*)\\ &\le(1-L\gamma_N)\bar V(x^*),\end{aligned}$$

which together with (6.2.41) and similar arguments used above would give (6.2.36). ■

A few remarks about Theorem 6.6 are in place. First, if f is convex and the batch sizes $m_k=1$, then by properly choosing the stepsizes $\{\gamma_k\}$ (e.g., $\gamma_k=\mathcal{O}(1/\sqrt{k})$ for k large), we can still guarantee a nearly optimal rate of convergence for the RSMD algorithm (see (6.2.34) or (6.2.36)). However, if f is possibly nonconvex and $m_k=1$, then the right-hand side of (6.2.31) is bounded from below by

$$\frac{LD_{\Psi}^2+\sigma^2\sum_{k=1}^N\gamma_k}{\sum_{k=1}^N(\gamma_k-L\gamma_k^2)} \geq \sigma^2,$$

which cannot guarantee the convergence of the RSMD algorithm, no matter how the stepsizes $\{\gamma_k\}$ are specified. This is exactly the reason why we consider taking multiple samples $G(x_k,\xi_{k,i})$, $i=1,\ldots,m_k$, for some $m_k>1$ at each iteration of the RSMD method.

Second, we need to estimate L to ensure the condition on the stepsize γ_k. However, we do not need a very accurate estimation for L (see the discussion after Corollary 6.1 for more details in the similar case).

Third, from (6.2.39) in the proof of Theorem 6.6, we see that the stepsize policies can be further relaxed to get a similar result as (6.2.31). More specifically, we can have the following corollary.

Corollary 6.5. *Suppose that the stepsizes $\{\gamma_k\}$ in the RSMD algorithm are chosen such that $0<\gamma_k\leq 2/L$ with $\gamma_k<2/L$ for at least one k, and the probability mass function P_R is chosen such that for any $k=1,\ldots,N$,*

$$P_R(k) := \text{Prob}\{R=k\} = \frac{\gamma_k-L\gamma_k^2/2}{\sum_{k=1}^N(\gamma_k-L\gamma_k^2/2)}. \tag{6.2.42}$$

Then, under Assumption 13, we have

$$\mathbb{E}[\|\tilde{g}_{X,R}\|^2] \leq \frac{LD_{\Psi}^2+\sigma^2\sum_{k=1}^N(\gamma_k/m_k)}{\sum_{k=1}^N(\gamma_k-L\gamma_k^2/2)}, \tag{6.2.43}$$

where the expectation is taken with respect to R and $\xi_{[N]}:=(\xi_1,\ldots,\xi_N)$.

Based on Theorem 6.6, we can establish the following complexity results of the RSMD algorithm with proper selection of stepsizes $\{\gamma_k\}$ and batch sizes $\{m_k\}$ at each iteration.

Corollary 6.6. *Suppose that in the RSMD algorithm the stepsizes $\gamma_k=1/(2L)$ for all $k=1,\ldots,N$, and the probability mass function P_R is chosen as (6.2.30). Also assume that the batch sizes $m_k=m$, $k=1,\ldots,N$, for some $m\geq 1$. Then under Assumption 13, we have*

$$\mathbb{E}[\|g_{X,R}\|^2] \leq \frac{8L^2D_{\Psi}^2}{N}+\frac{6\sigma^2}{m} \quad \text{and} \quad \mathbb{E}[\|\tilde{g}_{X,R}\|^2] \leq \frac{4L^2D_{\Psi}^2}{N}+\frac{2\sigma^2}{m}, \tag{6.2.44}$$

where $g_{X,R}$ and $\tilde{g}_{X,R}$ are defined in (6.2.20) and (6.2.32), respectively. If, in addition, f in the problem (6.2.1) is convex with an optimal solution x^*, then

$$\mathbb{E}\left[\Psi(x_R)-\Psi(x^*)\right] \leq \frac{2LV(x_1,x^*)}{N}+\frac{\sigma^2}{2Lm}. \tag{6.2.45}$$

Proof. By (6.2.31), we have

$$\mathbb{E}[\|\tilde{g}_{X,R}\|^2] \leq \frac{LD_{\Psi}^2+\frac{\sigma^2}{m}\sum_{k=1}^N\gamma_k}{\sum_{k=1}^N(\gamma_k-L\gamma_k^2)},$$

which together with $\gamma_k = 1/(2L)$ for all $k = 1,\ldots,N$ imply that

$$\mathbb{E}[\|\tilde{g}_{X,R}\|^2] = \frac{LD_\Psi^2 + \frac{\sigma^2 N}{2mL}}{\frac{N}{4L}} = \frac{4L^2 D_\Psi^2}{N} + \frac{2\sigma^2}{m}.$$

Then, by Proposition 6.1 with $x = x_R, \gamma = \gamma_R, g_1 = \nabla f(x_R), g_2 = G_k$, we have from the above inequality and (6.2.40) that

$$\begin{aligned}\mathbb{E}[\|g_{X,R}\|^2] &\le 2\mathbb{E}[\|\tilde{g}_{X,R}\|^2] + 2\mathbb{E}[\|g_{X,R} - \tilde{g}_{X,R}\|^2] \\ &\le 2\left(\frac{4L^2 D_\Psi^2}{N} + \frac{2\sigma^2}{m}\right) + 2\mathbb{E}\left[\|G_k - \nabla f(x_R)\|^2\right] \\ &\le \frac{8L^2 D_\Psi^2}{N} + \frac{6\sigma^2}{m}.\end{aligned}$$

Moreover, since $\gamma_k = 1/(2L)$ for all $k = 1,\ldots,N$, the stepsize conditions (6.2.33) are satisfied. Hence, if the problem is convex, (6.2.45) can be derived in a similar way as (6.2.34). ∎

Note that all the bounds in the above corollary depend on m. Indeed, if m is set to some fixed positive integer constant, then the second terms in the above results will always majorize the first terms when N is sufficiently large. Hence, the appropriate choice of m should be balanced with the number of iterations N, which would eventually depend on the total computational budget given by the user. The following corollary shows an appropriate choice of m depending on the total number of calls to the SFO.

Corollary 6.7. *Suppose that all the conditions in Corollary 6.6 are satisfied. Given a fixed total number of calls $\bar{N}$ to the* SFO, *if the number of calls to the* SFO *(number of samples) at each iteration of the RSMD algorithm is*

$$m = \left\lceil \min\left\{\max\left\{1, \frac{\sigma\sqrt{6\bar{N}}}{4L\tilde{D}}\right\}, \bar{N}\right\}\right\rceil, \tag{6.2.46}$$

for some $\tilde{D} > 0$, then we have $\mathbb{E}[\|g_{X,R}\|^2]/L \le \mathscr{B}_{\bar{N}}$, where

$$\mathscr{B}_{\bar{N}} := \frac{16LD_\Psi^2}{\bar{N}} + \frac{4\sqrt{6}\sigma}{\sqrt{\bar{N}}}\left(\frac{D_\Psi^2}{\tilde{D}} + \tilde{D}\max\left\{1, \frac{\sqrt{6}\sigma}{4L\tilde{D}\sqrt{\bar{N}}}\right\}\right). \tag{6.2.47}$$

If, in addition, f in problem (6.2.1) is convex, then $\mathbb{E}[\Psi(x_R) - \Psi(x^*)] \le \mathscr{C}_{\bar{N}}$, where x^* is an optimal solution and

$$\mathscr{C}_{\bar{N}} := \frac{4LV(x_1,x^*)}{\bar{N}} + \frac{\sqrt{6}\sigma}{\sqrt{\bar{N}}}\left(\frac{V(x_1,x^*)}{\tilde{D}} + \frac{\tilde{D}}{3}\max\left\{1, \frac{\sqrt{6}\sigma}{4L\tilde{D}\sqrt{\bar{N}}}\right\}\right). \tag{6.2.48}$$

Proof. Given the total number of calls to the stochastic first-order oracle $\bar{N}$ and the number m of calls to the SFO at each iteration, the RSMD algorithm can perform at most $N = \lfloor \bar{N}/m \rfloor$ iterations. Obviously, $N \ge \bar{N}/(2m)$. With this observation and (6.2.44), we have

$$\begin{aligned}
\mathbb{E}[\|g_{X,R}\|^2] &\le \tfrac{16mL^2D_\Psi^2}{\bar N} + \tfrac{6\sigma^2}{m} \\
&\le \tfrac{16L^2D_\Psi^2}{\bar N}\left(1 + \tfrac{\sigma\sqrt{6\bar N}}{4L\tilde D}\right) + \max\left\{\tfrac{4\sqrt{6}L\tilde D\sigma}{\sqrt{\bar N}}, \tfrac{6\sigma^2}{\bar N}\right\} \\
&= \tfrac{16L^2D_\Psi^2}{\bar N} + \tfrac{4\sqrt{6}L\sigma}{\sqrt{\bar N}}\left(\tfrac{D_\Psi^2}{\tilde D} + \tilde D \max\left\{1, \tfrac{\sqrt{6}\sigma}{4L\tilde D\sqrt{\bar N}}\right\}\right), \qquad (6.2.49)
\end{aligned}$$

which gives (6.2.47). The bound (6.2.48) can be obtained in a similar way. ■

We now would like to add a few remarks about the above results in Corollary 6.7. First, although we use the constant value for $m_k = m$ at each iteration, one can also choose it adaptively during the execution of the RSMD algorithm while monitoring the convergence. For example, in practice m_k could adaptively depend on $\sigma_k^2 := \mathbb{E}\left[\|G(x_k,\xi_k) - \nabla f(x_k)\|^2\right]$. Another example is to choose growing batch sizes where one uses a smaller number of samples in the beginning of the algorithm. In particular, by setting

$$m_k = \left\lceil \min\left\{\tfrac{\sigma(k^2\bar N)^{\frac{1}{4}}}{L\tilde D}, \bar N\right\}\right\rceil,$$

we can easily see that the RSMD algorithm still achieves the same rates of convergence as those obtained by using constant batch sizes in Corollary 6.7. Second, we need to specify the parameter $\tilde D$ in (6.2.46). It can be seen from (6.2.47) and (6.2.48) that when $\bar N$ is relatively large such that

$$\max\left\{1, \sqrt{6}\sigma/(4L\tilde D\sqrt{\bar N})\right\} = 1, \quad \text{i.e.,} \quad \bar N \ge 3\sigma^2/(8L^2\tilde D^2), \qquad (6.2.50)$$

an optimal choice of $\tilde D$ would be D_Ψ and $\sqrt{3V(x_1,x^*)}$ for solving nonconvex and convex SP problems, respectively. With this selection of $\tilde D$, the bounds in (6.2.47) and (6.2.48), respectively, reduce to

$$\tfrac{1}{L}\mathbb{E}[\|g_{X,R}\|^2] \le \tfrac{16LD_\Psi^2}{\bar N} + \tfrac{8\sqrt{6}D_\Psi\sigma}{\sqrt{\bar N}} \qquad (6.2.51)$$

and

$$\mathbb{E}[\Psi(x^*) - \Psi(x_1)] \le \tfrac{4LV(x_1,x^*)}{\bar N} + \tfrac{2\sqrt{2V(x_1,x^*)}\sigma}{\sqrt{\bar N}}. \qquad (6.2.52)$$

Third, the stepsize policy in Corollary 6.6 and the probability mass function (6.2.30) together with the number of samples (6.2.46) at each iteration of the RSMD algorithm provide a unified strategy for solving both convex and nonconvex SP problems. In particular, the RSMD algorithm exhibits a nearly optimal rate of convergence for solving smooth convex SP problems, since the second term in (6.2.52) is unimprovable, while the first term in (6.2.52) can be considerably improved.

6.2.3.2 A Two-Phase Randomized Stochastic Mirror Descent Method

In the previous subsection, we present the expected complexity results over many runs of the RSMD algorithm. Indeed, we are also interested in the performance of a single run of RSMD. In particular, we intend to establish the complexity results for finding an (ε,Λ)-*solution* of the problem (6.2.1), i.e., a point $x \in X$ satisfying $\text{Prob}\{\|g_X(x)\|^2 \leq \varepsilon\} \geq 1-\Lambda$, for some $\varepsilon > 0$ and $\Lambda \in (0,1)$. Noticing that by the Markov's inequality and (6.2.47), we can directly show that

$$\text{Prob}\left\{\|g_{X,R}\|^2 \geq \gamma\lambda L\mathscr{B}_{\bar{N}}\right\} \leq \tfrac{1}{\lambda}, \qquad \text{for any } \lambda > 0. \tag{6.2.53}$$

This implies that the total number of calls to the SFO performed by the RSMD algorithm for finding an (ε,Λ)-*solution*, after disregarding a few constant factors, can be bounded by

$$\mathscr{O}\left\{\tfrac{1}{\Lambda\varepsilon} + \tfrac{\sigma^2}{\Lambda^2\varepsilon^2}\right\}. \tag{6.2.54}$$

In this subsection, we present an approach to improve the dependence of the above bound on Λ. More specifically, we propose a variant of the RSMD algorithm which has two phases: an optimization phase and a post-optimization phase. The optimization phase consists of independent single runs of the RSMD algorithm to generate a list of candidate solutions, and in the post-optimization phase, we choose a solution x^* from these candidate solutions generated by the optimization phase. For the sake of simplicity, we assume throughout this subsection that the norm $\|\cdot\|$ in $\mathbb{R}^n$ is the standard Euclidean norm.

A Two Phase RSMD (2-RSMD) Algorithm

Input: Given initial point $x_1 \in X$, number of runs S, total $\bar{N}$ of calls to the SFO in each run of the RSMD algorithm, and sample size T in the post-optimization phase.

Optimization phase:

For $s = 1,\ldots,S$

Call the RSMD algorithm with initial point x_1, iteration limit $N = \lfloor \bar{N}/m \rfloor$ with m given by (6.2.46), stepsizes $\gamma_k = 1/(2L)$ for $k = 1,\ldots,N$, batch sizes $m_k = m$, and probability mass function P_R in (6.2.30).

Let $\bar{x}_s = x_{R_s}$, $s = 1,\ldots,S$, be the outputs of this phase.

Post-optimization phase:

Choose a solution $\bar{x}^*$ from the candidate list $\{\bar{x}_1,\ldots,\bar{x}_S\}$ such that

$$\|\bar{g}_X(\bar{x}^*)\| = \min_{s=1,\ldots,S} \|\bar{g}_X(\bar{x}_s)\|, \quad \bar{g}_X(\bar{x}_s) := P_X(\bar{x}_s, \bar{G}_T(\bar{x}_s), \gamma_{R_s}), \tag{6.2.55}$$

where $\bar{G}_T(x) = \frac{1}{T}\sum_{k=1}^T G(x,\xi_k)$ and $P_X(x,g,\gamma)$ is defined in (6.2.7).

Output: $\bar{x}^*$.

In the 2-RSMD algorithm, the total number of calls of SFO in the optimization phase and post-optimization phase is bounded by $S \times \bar{N}$ and $S \times T$, respectively. In the next theorem, we provide certain bounds of S, $\bar{N}$, and T for finding an (ε, Λ)-*solution* of problem (6.2.1).

We are now ready to state the main convergence properties for the 2-RSMD algorithm.

Theorem 6.7. *Under Assumption 13, the following statements hold for the 2-RSMD algorithm applied to problem (6.2.1).*

(a) Let $\mathscr{B}_{\bar{N}}$ be defined in (6.2.47). Then, for all $\lambda > 0$

$$\text{Prob}\left\{\|g_X(\bar{x}^*)\|^2 \geq 2\left(4L\mathscr{B}_{\bar{N}} + \tfrac{3\lambda\sigma^2}{T}\right)\right\} \leq \tfrac{S}{\lambda} + 2^{-S}; \tag{6.2.56}$$

(b) Let $\varepsilon > 0$ and $\Lambda \in (0,1)$ be given. If the parameters $(S, \bar{N}, T)$ are set to

$$S(\Lambda) := \lceil \log_2(2/\Lambda) \rceil, \tag{6.2.57}$$

$$\bar{N}(\varepsilon) := \left\lceil \max\left\{ \tfrac{512L^2D_\Psi^2}{\varepsilon}, \left[\left(\tilde{D} + \tfrac{D_\Psi^2}{\tilde{D}}\right)\tfrac{128\sqrt{6}L\sigma}{\varepsilon}\right]^2, \tfrac{3\sigma^2}{8L^2\tilde{D}^2} \right\} \right\rceil, \tag{6.2.58}$$

$$T(\varepsilon, \Lambda) := \left\lceil \tfrac{24S(\Lambda)\sigma^2}{\Lambda\varepsilon} \right\rceil, \tag{6.2.59}$$

then the 2-RSMD algorithm computes an (ε, Λ)-solution of the problem (6.2.1) after taking at most

$$S(\Lambda)\,[\bar{N}(\varepsilon) + T(\varepsilon, \Lambda)] \tag{6.2.60}$$

calls of the stochastic first-order oracle.

Proof. We first show part (a). Let $g_X(\bar{x}_s) = P_X(\bar{x}_s, \nabla f(\bar{x}_s), \gamma_{R_s})$. Then, it follows from the definition of $\bar{x}^*$ in (6.2.55) that

$$\begin{aligned}
\|\bar{g}_X(\bar{x}^*)\|^2 &= \min_{s=1,\ldots,S} \|\bar{g}_X(\bar{x}_s)\|^2 = \min_{s=1,\ldots,S} \|g_X(\bar{x}_s) + \bar{g}_X(\bar{x}_s) - g_X(\bar{x}_s)\|^2 \\
&\leq \min_{s=1,\ldots,S} \{2\|g_X(\bar{x}_s)\|^2 + 2\|\bar{g}_X(\bar{x}_s) - g_X(\bar{x}_s)\|^2\} \\
&\leq 2 \min_{s=1,\ldots,S} \|g_X(\bar{x}_s)\|^2 + 2 \max_{s=1,\ldots,S} \|\bar{g}_X(\bar{x}_s) - g_X(\bar{x}_s)\|^2,
\end{aligned}$$

which implies that

$$\begin{aligned}
\|g_X(\bar{x}^*)\|^2 &\leq 2\|\bar{g}_X(\bar{x}^*)\|^2 + 2\|g_X(\bar{x}^*) - \bar{g}_X(\bar{x}^*)\|^2 \\
&\leq 4 \min_{s=1,\ldots,S} \|g_X(\bar{x}_s)\|^2 + 4 \max_{s=1,\ldots,S} \|\bar{g}_X(\bar{x}_s) - g_X(\bar{x}_s)\|^2 \\
&\quad + 2\|g_X(\bar{x}^*) - \bar{g}_X(\bar{x}^*)\|^2 \\
&\leq 4 \min_{s=1,\ldots,S} \|g_X(\bar{x}_s)\|^2 + 6 \max_{s=1,\ldots,S} \|\bar{g}_X(\bar{x}_s) - g_X(\bar{x}_s)\|^2.
\end{aligned} \tag{6.2.61}$$

We now provide certain probabilistic bounds to the two terms in the right-hand side of the above inequality. First, from the fact that $\bar{x}_s$, $1 \le s \le S$, are independent and (6.2.53) (with $\lambda = 2$), we have

$$\text{Prob}\left\{\min_{s\in\{1,2,\ldots,S\}} \|g_X(\bar{x}_s)\|^2 \ge 2L\mathscr{B}_{\bar{N}}\right\} = \prod_{s=1}^{S} \text{Prob}\left\{\|g_X(\bar{x}_s)\|^2 \ge 2L\mathscr{B}_{\bar{N}}\right\} \le 2^{-S}. \tag{6.2.62}$$

Moreover, denoting $\delta_{s,k} = G(\bar{x}_s, \xi_k) - \nabla f(\bar{x}_s)$, $k = 1,\ldots,T$, by Proposition 6.1 with $x = \bar{x}_s, \gamma = \gamma_{R_s}, g_1 = \bar{G}_T(\bar{x}_s), g_2 = \nabla f(\bar{x}_s)$, we have

$$\|\bar{g}_X(\bar{x}_s) - g_X(\bar{x}_s)\| \le \|\textstyle\sum_{k=1}^{T} \delta_{s,k}/T\|. \tag{6.2.63}$$

From the above inequality, Assumption 13, and Lemma 6.1(a), for any $\lambda > 0$ and any $s = 1,\ldots,S$, we have

$$\text{Prob}\left\{\|\bar{g}_X(\bar{x}_s) - g_X(\bar{x}_s)\|^2 \ge \tfrac{\lambda\sigma^2}{T}\right\} \le \text{Prob}\left\{\|\textstyle\sum_{k=1}^{T}\delta_{s,k}\|^2 \ge \lambda T\sigma^2\right\} \le \tfrac{1}{\lambda},$$

which implies

$$\text{Prob}\left\{\max_{s=1,\ldots,S} \|\bar{g}_X(\bar{x}_s) - g_X(\bar{x}_s)\|^2 \ge \tfrac{\lambda\sigma^2}{T}\right\} \le \tfrac{S}{\lambda}. \tag{6.2.64}$$

Then, the conclusion (6.2.56) follows from (6.2.61), (6.2.62), and (6.2.64).

We now show part (b). With the settings in part (b), it is easy to see that the total number of calls of the SFO in the 2-RSMD algorithm is bounded up by (6.2.60). Hence, we only need to show that the $\bar{x}^*$ returned by the 2-RSMD algorithm is indeed an (ε,Λ)-solution of the problem (6.2.1). With the choice of $\bar{N}(\varepsilon)$ in (6.2.58), we can see that (6.2.50) holds. So, we have from (6.2.47) and (6.2.58) that

$$\mathscr{B}_{\bar{N}(\varepsilon)} = \tfrac{16LD_\Psi^2}{\bar{N}(\varepsilon)} + \tfrac{4\sqrt{6}\sigma}{\sqrt{\bar{N}(\varepsilon)}}\left(\tilde{D} + \tfrac{D_\Psi^2}{\tilde{D}}\right) \le \tfrac{\varepsilon}{32L} + \tfrac{\varepsilon}{32L} = \tfrac{\varepsilon}{16L}.$$

By the above inequality and (6.2.59), setting $\lambda = 2S/\Lambda$ in (6.2.56), we have

$$8L\mathscr{B}_{\bar{N}(\varepsilon)} + \tfrac{6\lambda\sigma^2}{T(\varepsilon,\Lambda)} \le \tfrac{\varepsilon}{2} + \tfrac{\lambda\Lambda\varepsilon}{4S} = \varepsilon,$$

which together with (6.2.56), (6.2.57), and $\lambda = 2S/\Lambda$ imply

$$\text{Prob}\left\{\|g_X(\bar{x}^*)\|^2 \ge \varepsilon\right\} \le \tfrac{\Lambda}{2} + 2^{-S} \le \Lambda.$$

Hence,$\bar{x}^*$ is an (ε,Λ)-solution of the problem (6.2.1). ■

Now, it is interesting to compare the complexity bound in (6.2.60) with the one in (6.2.54). In view of (6.2.57), (6.2.58), and (6.2.59), the complexity bound in (6.2.60) for finding an (ε,Λ)-solution, after discarding a few constant factors, is equivalent to

$$\mathscr{O}\left\{\tfrac{1}{\varepsilon}\log_2\tfrac{1}{\Lambda} + \tfrac{\sigma^2}{\varepsilon^2}\log_2\tfrac{1}{\Lambda} + \tfrac{\sigma^2}{\Lambda\varepsilon}\log_2^2\tfrac{1}{\Lambda}\right\}. \tag{6.2.65}$$

When the second terms are the dominating terms in both bounds, the above bound (6.2.65) can be considerably smaller than the one in (6.2.54) up to a factor of $1/\left[\Lambda^2\log_2(1/\Lambda)\right]$.

The following theorem shows that under the "light-tail" Assumption 14, the bound (6.2.60) in Theorem 6.7 can be further improved.

Corollary 6.8. *Under Assumptions 13 and 14, the following statements hold for the 2-RSMD algorithm applied to problem (6.2.1).*

(a) Let $\mathscr{B}_{\bar{N}}$ is defined in (6.2.47). Then, for all $\lambda > 0$

$$\text{Prob}\left\{\|g_X(\bar{x}^*)\|^2 \geq \left[8L\mathscr{B}_{\bar{N}} + \tfrac{12(1+\lambda)^2\sigma^2}{T}\right]\right\} \leq S\exp(-\tfrac{\lambda^2}{3}) + 2^{-S}; \tag{6.2.66}$$

(b) Let $\varepsilon > 0$ and $\Lambda \in (0,1)$ be given. If S and $\bar{N}$ are set to $S(\Lambda)$ and $\bar{N}(\varepsilon)$ as in (6.2.57) and (6.2.58), respectively, and the sample size T is set to

$$T'(\varepsilon,\Lambda) := \tfrac{24\sigma^2}{\varepsilon}\left[1 + \left(3\log_2\tfrac{2S(\Lambda)}{\Lambda}\right)^{\frac{1}{2}}\right]^2, \tag{6.2.67}$$

then the 2-RSMD algorithm can compute an (ε,Λ)-solution of the problem (6.2.1) after taking at most

$$S(\Lambda)\left[\bar{N}(\varepsilon) + T'(\varepsilon,\Lambda)\right] \tag{6.2.68}$$

calls to the stochastic first-order oracle.

Proof. We only give a sketch of the proof for part (a). The proof of part (b) follows from part (a) and similar arguments for proving (b) part of Theorem 6.7. Now, denoting $\delta_{s,k} = G(\bar{x}_s, \xi_k) - \nabla f(\bar{x}_s)$, $k = 1,\ldots,T$, again by Proposition 6.1, we have (6.2.63) holds. Then, by Assumption 14 and Lemma 6.1(b), for any $\lambda > 0$ and any $s = 1,\ldots,S$, we have

$$\begin{aligned}&\text{Prob}\left\{\|\bar{g}_X(\bar{x}_s) - g_X(\bar{x}_s)\|^2 \geq (1+\lambda)^2\tfrac{2\sigma^2}{T}\right\}\\ &\leq \text{Prob}\left\{\|\Sigma_{k=1}^T \delta_{s,k}\| \geq \sqrt{2T}(1+\lambda)\sigma\right\} \leq \exp(-\tfrac{\lambda^2}{3}),\end{aligned}$$

which implies that for any $\lambda > 0$

$$\text{Prob}\left\{\max_{s=1,\ldots,S}\|\bar{g}_X(\bar{x}_s) - g_X(\bar{x}_s)\|^2 \geq (1+\lambda)^2\tfrac{2\sigma^2}{T}\right\} \leq S\exp(-\tfrac{\lambda^2}{3}), \tag{6.2.69}$$

Then, the conclusion (6.2.66) follows from (6.2.61), (6.2.62), and (6.2.69). ■

In view of (6.2.57), (6.2.58), and (6.2.67), the bound in (6.2.68), after discarding a few constant factors, is equivalent to

$$\mathscr{O}\left\{\tfrac{1}{\varepsilon}\log_2\tfrac{1}{\Lambda} + \tfrac{\sigma^2}{\varepsilon^2}\log_2\tfrac{1}{\Lambda} + \tfrac{\sigma^2}{\varepsilon}\log_2^2\tfrac{1}{\Lambda}\right\}. \tag{6.2.70}$$

Clearly, the third term of the above bound is smaller than the third term in (6.2.65) by a factor of $1/\Lambda$.

In the remaining part of this section, we briefly discuss another variant of the 2-RSMD algorithm, namely, 2-RSMD-V algorithm which can possibly improve the practical performance of the 2-RSMD algorithm. Similar to the 2-RSMD algorithm, this variant also consists of two phases. The only difference exists in that the S runs of the RSMD algorithm in the optimization phase are not independent of each other and the output of each run is used as the initial point of the next run, although the post-optimization phase of the 2-RSMD-V algorithm is the same as that of the 2-RSMD algorithm. We now formally state the optimization phase of the 2-RSMD-V algorithm as follows.

Optimization Phase of 2-RSMD-V Algorithm
For $s = 1,\ldots,S$

Call the RSMD algorithm with initial point $\bar{x}_{s-1}$ where $\bar{x}_0 = x_1$ and $\bar{x}_s = x_{R_s}$, $s = 1,\ldots,S$, are the outputs of the s-th run of the RSMD algorithm, iteration limit $N = \lfloor \bar{N}/m \rfloor$ with m given by (6.2.46), stepsizes $\gamma_k = 1/(2L)$ for $k = 1,\ldots,N$, batch sizes $m_k = m$, and probability mass function P_R in (6.2.30).

As mentioned above, in the 2-RSMD-V algorithm, unlike the 2-RSMD algorithm, the S candidate solutions are not independent and hence the analysis of Theorem 6.7 cannot be directly applied. However, by slightly modifying the proof of Theorem 3, we can show that the above 2-RSMD-V algorithm exhibits similar convergence behavior as the 2-RSMD algorithm under certain more restrictive conditions.

Corollary 6.9. *Suppose that the feasible set X is bounded and Assumption 13 holds. Then, the complexity of the 2-RSMD-V algorithm to find an (ε,Λ)-solution of problem (6.2.1) is bounded by (6.2.65). If in addition, Assumption 14 holds, then this complexity bound improves to (6.2.70).*

Proof. Denote $\bar{\Psi} = \max_{x\in X} \Psi(x)$ and let E_s be the event that $\|g_X(\bar{x}_s)\|^2 \geq 2L\hat{\mathscr{B}}_{\bar{N}}$ where

$$\hat{\mathscr{B}}_{\bar{N}} := \frac{16(\bar{\Psi}-\Psi^*)}{\bar{N}} + \frac{4\sqrt{6}\sigma}{\sqrt{\bar{N}}}\left(\frac{\bar{\Psi}-\Psi^*}{L\tilde{D}} + \tilde{D}\max\left\{1, \frac{\sqrt{6}\sigma}{4L\tilde{D}\sqrt{\bar{N}}}\right\}\right).$$

Now note that due to the boundedness of X and continuity of f, $\bar{\Psi}$ is finite and therefore the bound $\hat{\mathscr{B}}_{\bar{N}}$ is valid. Also observe that by (6.2.53) (with $\lambda = 2$) together with the fact that $\hat{\mathscr{B}}_{\bar{N}} \geq \mathscr{B}_{\bar{N}}$, we have

$$\mathrm{Prob}\left\{E_s \mid \bigcap_{j=1}^{s-1} E_j\right\} \leq \tfrac{1}{2}, \quad s = 1,2,\ldots,S,$$

which consequently implies that

$$\mathrm{Prob}\left\{\min_{s\in\{1,2,\ldots,S\}} \|g_X(\bar{x}_s)\|^2 \geq 2L\hat{\mathscr{B}}_{\bar{N}}\right\}$$

$$= \text{Prob}\left\{\bigcap_{s=1}^{S} E_s\right\} = \prod_{s=1}^{S} \text{Prob}\left\{E_s \mid \bigcap_{j=1}^{s-1} E_j\right\} \leq 2^{-S}.$$

Observing that the above inequality is similar to (6.2.62), the rest of proof is almost identical to those of Theorem 6.7 and Corollary 6.8 and hence we skip the details. ■

6.2.4 Stochastic Zeroth-Order Methods for Composite Problems

In this section, we discuss how to specialize the RSMD algorithm to deal with the situations where only noisy function values of the problem (6.2.1) are available. More specifically, we assume that we can only access the noisy zeroth-order information of f by a *stochastic zeroth-order oracle* (SZO). For any input x_k and ξ_k, the SZO would output a quantity $F(x_k, \xi_k)$, where x_k is the k-th iterate of our algorithm and ξ_k is a random variable whose distribution is supported on $\Xi \in \mathbb{R}^d$ (noting that Ξ does not depend on x_k). Throughout this section, we assume $F(x_k, \xi_k)$ is an unbiased estimator of $f(x_k)$ that satisfies Assumption 15.

We are going to apply the randomized smoothing techniques to explore the zeroth-order information of f. Hence, throughout this section, we also assume $F(\cdot, \xi_k)$ is smooth, i.e., it is differentiable and its gradients are Lipschitz continuous with constant L, almost surely with respect to $\xi_k \in \Xi$, which together with Assumption 15 imply f is smooth and its gradients are Lipschitz continuous with constant L. Also, throughout this section, we assume that $\|\cdot\|$ is the standard Euclidean norm.

Similar to Sect. 6.1.2.1, let us define the approximated stochastic gradient of f at x_k as in (6.1.58) and define $G(x_k, \xi_k) = \nabla_x F(x_k, \xi_k)$. We assume the Assumption 1 holds for $G(x_k, \xi_k)$. Then, by the Assumption 15 and Lemma 6.2(a), we directly get

$$\mathbb{E}_{v,\xi_k}[G_\mu(x_k, \xi_k, v)] = \nabla f_\mu(x_k), \tag{6.2.71}$$

where the expectation is taken with respect to v and ξ_k.

Now based on the RSMD algorithm, we state an algorithm which only uses zeroth-order information to solve problem (6.2.1).

A Randomized Stochastic Gradient Free Mirror Descent (RSMDF) Algorithm

Input: Given initial point $x_1 \in X$, iteration limit N, the stepsizes $\{\gamma_k\}$ with $\gamma_k > 0$, $k \geq 1$, the batch sizes $\{m_k\}$ with $m_k > 0$, $k \geq 1$, and the probability mass function P_R supported on $\{1, \ldots, N\}$.

Step 0. Let R be a random variable with probability mass function P_R.

Step $k = 1, \ldots, R-1$. Call the SZO m_k times to obtain $G_\mu(x_k, \xi_{k,i}, v_{k,i})$, $i = 1, \ldots, m_k$, set

$$G_{\mu,k} = \tfrac{1}{m_k} \textstyle\sum_{i=1}^{m_k} G_\mu(x_k, \xi_{k,i}, v_{k,i}) \tag{6.2.72}$$

and compute

$$x_{k+1} = \arg\min_{u \in X} \left\{\langle G_{\mu,k}, u\rangle + \tfrac{1}{\gamma_k} V(x_k, u) + h(u)\right\}. \tag{6.2.73}$$

Output: x_R.

Compared with RSMD algorithm, we can see at the k-th iteration, the RSMDF algorithm simply replaces the stochastic gradient G_k by the approximated stochastic gradient $G_{\mu,k}$. By (6.2.71), $G_{\mu,k}$ can be simply viewed as an unbiased stochastic gradient of the smoothed function f_μ. However, to apply the results developed in the previous section, we still need an estimation of the bound on the variations of the stochastic gradient $G_{\mu,k}$. In addition, the role that the smoothing parameter μ plays and the proper selection of μ in the RSMDF algorithm are still not clear now. We answer these questions in the following series of theorems and their corollaries.

Theorem 6.8. *Suppose that the stepsizes $\{\gamma_k\}$ in the RSMDF algorithm are chosen such that $0 < \gamma_k \le 1/L$ with $\gamma_k < 1/L$ for at least one k, and the probability mass function P_R is chosen as (6.2.30). If $\|\nabla f(x)\| \le M$ for all $x \in X$, then under Assumptions 13 and 15,*

(a) for any $N \ge 1$, we have

$$\mathbb{E}[\|\bar{g}_{\mu,X,R}\|^2] \le \frac{LD_\psi^2 + \mu^2 Ln + \tilde{\sigma}^2 \sum_{k=1}^N (\gamma_k/m_k)}{\sum_{k=1}^N (\gamma_k - L\gamma_k^2)}, \tag{6.2.74}$$

where the expectation is taken with respect to R, $\xi_{[N]}$ and $v_{[N]} := (v_1, \ldots, v_N)$, D_ψ is defined in (6.2.22),

$$\tilde{\sigma}^2 = 2(n+4)[M^2 + \sigma^2 + \mu^2 L^2 (n+4)^2], \tag{6.2.75}$$

and

$$\bar{g}_{\mu,X,k} = P_X(x_k, G_{\mu,k}, \gamma_k), \tag{6.2.76}$$

with P_X defined in(6.2.7);

(b) if, in addition, f in problem (6.2.1) is convex with an optimal solution x^*, and the stepsizes $\{\gamma_k\}$ are nondecreasing as (6.2.33), we have

$$\mathbb{E}\left[\Psi(x_R) - \Psi(x^*)\right] \le \frac{(1-L\gamma_1)V(x_1,x^*) + (\tilde{\sigma}^2/2)\sum_{k=1}^N (\gamma_k^2/m_k)}{\sum_{k=1}^N (\gamma_k - L\gamma_k^2)} + \mu^2 Ln, \tag{6.2.77}$$

where the expectation is taken with respect to R, $\xi_{[N]}$, and $v_{[N]}$.

Proof. By our assumption that $F(\cdot, \xi_k)$ is smooth almost surely and (6.1.53) (applying $f = F(\cdot, \xi_k)$), we have

$$\begin{aligned}
\mathbb{E}_{v_k,\xi_k}[\|G_\mu(x_k, \xi_k, v_k)\|^2] &= \mathbb{E}_{\xi_k}[\mathbb{E}_{v_k}[\|G_\mu(x_k, \xi_k, v_k)\|^2]] \\
&\le 2(n+4)[\mathbb{E}_{\xi_k}[\|G(x_k, \xi)\|^2] + \tfrac{\mu^2}{2} L^2 (n+6)^3 \\
&\le 2(n+4)[\mathbb{E}_{\xi_k}[\|\nabla f(x_k)\|^2] + \sigma^2] + 2\mu^2 L^2 (n+4)^3,
\end{aligned}$$

where the last inequality follows from Assumption 1 with $G(x_k, \xi_k) = \nabla_x F(x_k, \xi_k)$. Then, from (6.2.71), the above inequality, and $\|\nabla f(x_k)\| \le M$, we have

$$\begin{aligned}
&\mathbb{E}_{v_k,\xi_k}[\|G_\mu(x_k,\xi_k,v_k)-\nabla f_\mu(x_k)\|^2] \\
&= \mathbb{E}_{v_k,\xi_k}[\|G_\mu(x_k,\xi_k,v_k)\|^2+\|\nabla f_\mu(x_k)\|^2-2\langle G_\mu(x_k,\xi_k,v_k),\nabla f_\mu(x_k)\rangle] \\
&= \mathbb{E}_{v_k,\xi_k}[\|G_\mu(x_k,\xi_k,v_k)\|^2]+\|\nabla f_\mu(x_k)\|^2-2\langle \mathbb{E}_{v_k,\xi_k}[G_\mu(x_k,\xi_k,v_k)],\nabla f_\mu(x_k)\rangle \\
&= \mathbb{E}_{v_k,\xi_k}[\|G_\mu(x_k,\xi_k,v_k)\|^2]+\|\nabla f_\mu(x_k)\|^2-2\|\nabla f_\mu(x_k)\|^2 \\
&\le \mathbb{E}_{v_k,\xi_k}[\|G_\mu(x_k,\xi_k,v_k)\|^2] \le 2(n+4)[M^2+\sigma^2+\mu^2L^2(n+4)^2]=\tilde{\sigma}^2. \quad (6.2.78)
\end{aligned}$$

Now let $\Psi_\mu(x)=f_\mu(x)+h(x)$ and $\Psi_\mu^*=\min_{x\in X}\Psi_\mu(x)$. We have from (6.1.57) that

$$|(\Psi_\mu(x)-\Psi_\mu^*)-(\Psi(x)-\Psi^*)|\le \mu^2Ln. \quad (6.2.79)$$

By Lemma (6.2)(a), we have $L_\mu\le L$ and therefore f_μ is smooth and its gradients are Lipschitz continuous with constant L. With this observation, noticing (6.2.71) and (6.2.78), viewing $G_\mu(x_k,\xi_k,v_k)$ as a stochastic gradient of f_μ, then by part (a) of Theorem 6.6 we can directly get

$$\mathbb{E}[\|\bar{g}_{\mu,X,R}\|^2]\le \frac{LD_{\Psi_\mu}^2+\tilde{\sigma}^2\sum_{k=1}^N(\gamma_k/m_k)}{\sum_{k=1}^N(\gamma_k-L\gamma_k^2)},$$

where $D_{\Psi_\mu}=[(\Psi_\mu(x_1)-\Psi_\mu^*)/L]^{1/2}$ and the expectation is taken with respect to R, $\xi_{[N]}$, and $v_{[N]}$. Then, the conclusion (6.2.74) follows the above inequality and (6.2.79).

We now show part (b). Since f is convex, by Lemma (6.2)(c), f_μ is also convex. Again by (6.2.79), we have

$$\mathbb{E}\left[\Psi(x_R)-\Psi(x^*)\right]\le \mathbb{E}\left[\Psi_\mu(x_R)-\Psi_\mu(x^*)\right]+\mu^2Ln.$$

Then, by this inequality and the convexity of f_μ, it follows from part (b) of Theorem 6.6 and similar arguments in showing the part (a) of this theorem, the conclusion (6.2.77) holds. ∎

Using the previous Theorem 6.8, similar to the Corollary 6.6, we can give the following corollary on the RSMDF algorithm with a certain constant stepsize and batch size at each iteration.

Corollary 6.10. *Suppose that in the RSMDF algorithm the stepsizes* $\gamma_k=1/(2L)$ *for all* $k=1,\ldots,N$*, the batch sizes* $m_k=m$ *for all* $k=1,\ldots,N$*, and the probability mass function* P_R *is set to (6.2.30). Then under Assumptions 13 and 15, we have*

$$\mathbb{E}[\|\bar{g}_{\mu,X,R}\|^2]\le \frac{4L^2D_\Psi^2+4\mu^2L^2n}{N}+\frac{2\tilde{\sigma}^2}{m} \quad (6.2.80)$$

and

$$\mathbb{E}[\|g_{X,R}\|^2]\le \frac{\mu^2L^2(n+3)^2}{2}+\frac{16L^2D_\Psi^2+16\mu^2L^2n}{N}+\frac{12\tilde{\sigma}^2}{m}, \quad (6.2.81)$$

where the expectation is taken with respect to R*,* $\xi_{[N]}$*, and* $v_{[N]}$*, and* $\tilde{\sigma}$*,* $\bar{g}_{\mu,X,R}$*, and* $g_{X,R}$ *are defined in (6.2.75), (6.2.76), and (6.2.20), respectively.*

If, in addition, f in the problem (6.2.1) is convex with an optimal solution x^*, then

$$\mathbb{E}\left[\Psi(x_R) - \Psi(x^*)\right] \le \tfrac{2LV(x_1,x^*)}{N} + \tfrac{\tilde{\sigma}^2}{2Lm} + \mu^2 Ln. \tag{6.2.82}$$

Proof. Equation (6.2.80) immediately follows from (6.2.74) with $\gamma_k = 1/(2L)$ and $m_k = m$ for all $k = 1,\ldots,N$. Now let $g_{\mu,X,R} = P_X(x_R, \nabla f_\mu(x_R), \gamma_R)$, we have from (6.1.52) and Proposition 6.1 with $x = x_R$, $\gamma = \gamma_R$, $g_1 = \nabla f(x_R)$, and $g_2 = \nabla f_\mu(x_R)$ that

$$\mathbb{E}[\|g_{X,R} - g_{\mu,X,R}\|^2] \le \tfrac{\mu^2 L^2 (n+3)^2}{4}. \tag{6.2.83}$$

Similarly, by Proposition 6.1 with $x = x_R$, $\gamma = \gamma_R$, $g_1 = \bar{G}_{\mu,k}$, and $g_2 = \nabla f_\mu(x_R)$, we have

$$\mathbb{E}[\|\bar{g}_{\mu,X,R} - g_{\mu,X,R}\|^2] \le \tfrac{\tilde{\sigma}^2}{m}. \tag{6.2.84}$$

Then, it follows from (6.2.83), (6.2.84), and (6.2.80) that

$$\begin{aligned}
\mathbb{E}[\|g_{X,R}\|^2] &\le 2\mathbb{E}[\|g_{X,R} - g_{\mu,X,R}\|^2] + 2\mathbb{E}[\|g_{\mu,X,R}\|^2] \\
&\le \tfrac{\mu^2 L^2 (n+3)^2}{2} + 4\mathbb{E}[\|g_{\mu,X,R} - \bar{g}_{\mu,X,R}\|^2] + 4\mathbb{E}[\|\bar{g}_{\mu,X,R}\|^2] \\
&\le \tfrac{\mu^2 L^2 (n+3)^2}{2} + \tfrac{12\tilde{\sigma}^2}{m} + \tfrac{16L^2 D_\Psi^2 + 16\mu^2 L^2 n}{N}.
\end{aligned}$$

Moreover, if f is convex, then (6.2.82) immediately follows from (6.2.77), and the constant stepsizes $\gamma_k = 1/(2L)$ for all $k = 1,\ldots,N$. ∎

Similar to the Corollary 6.6 for the RSMD algorithm, the above results also depend on the number of samples m at each iteration. In addition, the above results depend on the smoothing parameter μ as well. The following corollary, analogous to the Corollary 6.7, shows how to choose m and μ appropriately.

Corollary 6.11. *Suppose that all the conditions in Corollary 6.10 are satisfied. Given a fixed total number of calls to the SZO $\bar{N}$, if the smoothing parameter satisfies*

$$\mu \le \tfrac{D_\Psi}{\sqrt{(n+4)\bar{N}}}, \tag{6.2.85}$$

and the number of calls to the SZO at each iteration of the RSMDF method is

$$m = \left\lceil \min\left\{ \max\left\{ \tfrac{\sqrt{(n+4)(M^2+\sigma^2)\bar{N}}}{L\tilde{D}}, n+4 \right\}, \bar{N} \right\} \right\rceil, \tag{6.2.86}$$

for some $\tilde{D} > 0$, then we have $1/L\,\mathbb{E}[\|g_{X,R}\|^2] \le \bar{\mathscr{B}}_{\bar{N}}$, where

$$\bar{\mathscr{B}}_{\bar{N}} := \tfrac{(24\theta_2+41)LD_\Psi^2(n+4)}{\bar{N}} + \tfrac{32\sqrt{(n+4)(M^2+\sigma^2)}}{\sqrt{\bar{N}}} \left(\tfrac{D_\Psi^2}{\tilde{D}} + \tilde{D}\theta_1 \right), \tag{6.2.87}$$

and

$$\theta_1 = \max\left\{1, \tfrac{\sqrt{(n+4)(M^2+\sigma^2)}}{L\tilde{D}\sqrt{\bar{N}}}\right\} \quad \textit{and} \quad \theta_2 = \max\left\{1, \tfrac{n+4}{\bar{N}}\right\}. \tag{6.2.88}$$

If, in addition, f in the problem (6.2.1) is convex and the smoothing parameter satisfies

$$\mu \leq \sqrt{\frac{V(x_1,x^*)}{(n+4)\bar{N}}}, \tag{6.2.89}$$

then $\mathbb{E}[\Psi(x_R) - \Psi(x^*)] \leq \bar{\mathscr{C}}_{\bar{N}}$, where x^* is an optimal solution and

$$\bar{\mathscr{C}}_{\bar{N}} := \frac{(5+\theta_2)LV(x_1,x^*)(n+4)}{\bar{N}} + \frac{\sqrt{(n+4)(M^2+\sigma^2)}}{\sqrt{\bar{N}}}\left(\frac{4V(x_1,x^*)}{\tilde{D}} + \tilde{D}\theta_1\right). \tag{6.2.90}$$

Proof. By the definitions of θ_1 and θ_2 in (6.2.88) and m in (6.2.86), we have

$$m = \left\lceil \max\left\{\frac{\sqrt{(n+4)(M^2+\sigma^2)\bar{N}}}{L\tilde{D}\theta_1}, \frac{n+4}{\theta_2}\right\}\right\rceil. \tag{6.2.91}$$

Given the total number of calls to the SZO $\bar{N}$ and the number m of calls to the SZO at each iteration, the RSMDF algorithm can perform at most $N = \lfloor \bar{N}/m \rfloor$ iterations. Obviously, $N \geq \bar{N}/(2m)$. With this observation $\bar{N} \geq m$, $\theta_1 \geq 1$ and $\theta_2 \geq 1$, by (6.2.81), (6.2.85), and (6.2.91), we have

$$\begin{aligned}
&\mathbb{E}[\|g_{X,R}\|^2] \\
\leq\ & \frac{L^2D_\Psi^2(n+3)}{2\bar{N}} + \frac{24(n+4)(M^2+\sigma^2)}{m} + \frac{24L^2D_\Psi^2(n+4)^2}{m\bar{N}} + \frac{32L^2D_\Psi^2 m}{\bar{N}}\left(1+\frac{1}{\bar{N}}\right) \\
\leq\ & \frac{L^2D_\Psi^2(n+4)}{2\bar{N}} + \frac{24\theta_1 L\tilde{D}\sqrt{(n+4)(M^2+\sigma^2)}}{\sqrt{\bar{N}}} + \frac{24\theta_2 L^2D_\Psi^2(n+4)}{\bar{N}} \\
&+ \frac{32L^2D_\Psi^2}{\bar{N}}\left(\frac{\sqrt{(n+4)(M^2+\sigma^2)\bar{N}}}{L\tilde{D}\theta_1} + \frac{n+4}{\theta_2}\right) + \frac{32L^2D_\Psi^2}{\bar{N}} \\
\leq\ & \frac{L^2D_\Psi^2(n+4)}{2\bar{N}} + \frac{24\theta_1 L\tilde{D}\sqrt{(n+4)(M^2+\sigma^2)}}{\sqrt{\bar{N}}} + \frac{24\theta_2 L^2D_\Psi^2(n+4)}{\bar{N}} \\
&+ \frac{32LD_\Psi^2\sqrt{(n+4)(M^2+\sigma^2)}}{\tilde{D}\sqrt{\bar{N}}} + \frac{32L^2D_\Psi^2(n+4)}{\bar{N}} + \frac{32L^2D_\Psi^2}{\bar{N}},
\end{aligned}$$

which after integrating the terms give (6.2.87). The conclusion (6.2.90) follows similarly by (6.2.89) and (6.2.82). ∎

We now would like to add a few remarks about the above results in Corollary 6.11. First, the above complexity bounds are similar to those of the first-order RSMD method in Corollary 6.7 in terms of their dependence on the total number of stochastic oracle $\bar{N}$ called by the algorithm. However, for the zeroth-order case, the complexity in Corollary 6.11 also depends on the size of the gradient M and the problem dimension n. Second, the value of $\tilde{D}$ has not been specified. It can be easily seen from (6.2.87) and (6.2.90) that when $\bar{N}$ is relatively large such that $\theta_1 = 1$ and $\theta_2 = 1$, i.e.,

$$\bar{N} \geq \max\left\{\frac{(n+4)^2(M^2+\sigma^2)}{L^2\tilde{D}^2}, n+4\right\}, \tag{6.2.92}$$

the optimal choice of $\tilde{D}$ would be D_Ψ and $2\sqrt{V(x_1,x^*)}$ for solving nonconvex and convex SP problems, respectively. With this selection of $\tilde{D}$, the bounds in (6.2.87) and (6.2.90), respectively, reduce to

$$\tfrac{1}{L}\mathbb{E}[\|g_{X,R}\|^2] \le \frac{65LD_\Psi^2(n+4)}{\bar N} + \frac{64\sqrt{(n+4)(M^2+\sigma^2)}}{\sqrt{\bar N}} \tag{6.2.93}$$

and

$$\mathbb{E}[\Psi(x_R)-\Psi(x^*)] \le \frac{6LV(x_1,x^*)(n+4)}{\bar N} + \frac{4\sqrt{V(x_1,x^*)(n+4)(M^2+\sigma^2)}}{\sqrt{\bar N}}. \tag{6.2.94}$$

Third, the complexity result in (6.2.90) implies that when Ψ is convex, if ε sufficiently small, then the number of calls to the SZO to find a solution $\bar x$ such that $\mathbb{E}[\Psi(\bar x)-\Psi^*]\le\varepsilon$ can be bounded by $\mathcal{O}(n/\varepsilon^2)$, which only linearly depends on the dimension n.

6.3 Nonconvex Stochastic Block Mirror Descent

In this section, we consider a special class of stochastic composite optimization problems given by

$$\phi^* := \min_{x\in X}\{\phi(x) := f(x)+\chi(x)\}. \tag{6.3.1}$$

Here, $f(\cdot)$ is smooth (but not necessarily convex) and its gradients $\nabla f(\cdot)$ satisfy

$$\|\nabla f_i(x+U_i\rho_i)-\nabla f_i(x)\|_{i,*} \le L_i\|\rho_i\|_i \ \forall\, \rho_i\in\mathbb{R}^{n_i},\ i=1,2,...,b. \tag{6.3.2}$$

It then follows that

$$f(x+U_i\rho_i) \le f(x)+\langle\nabla f_i(x),\rho_i\rangle+\tfrac{L_i}{2}\|\rho_i\|_i^2 \ \forall\rho_i\in\mathbb{R}^{n_i}, x\in X. \tag{6.3.3}$$

Moreover, the nonsmooth component $\chi(\cdot)$ is still convex and separable, i.e.

$$\chi(x) = \textstyle\sum_{i=1}^b \chi_i(x^{(i)}) \ \forall x\in X, \tag{6.3.4}$$

where $\chi_i:\mathbb{R}^{n_i}\to\mathbb{R}$ are closed and convex. In addition, we assume that X has a block structure, i.e.,

$$X = X_1\times X_2\times\cdots\times X_b, \tag{6.3.5}$$

where $X_i\subseteq\mathbb{R}^{n_i}$, $i=1,\ldots,b$, are closed convex sets with $n_1+n_2+\ldots+n_b=n$. Our goal is to generalize the stochastic block mirror descent method introduced in Sect. 4.6 for solving the above nonconvex stochastic composite optimization problem.

Similar to Sect. 4.6, we use $\mathbb{R}^{n_i}$, $i=1,\ldots,b$, to denote Euclidean spaces equipped with inner product $\langle\cdot,\cdot\rangle$ and norm $\|\cdot\|_i$ ($\|\cdot\|_{i,*}$ be the conjugate) such that $\sum_{i=1}^b n_i = n$. Let I_n be the identity matrix in $\mathbb{R}^n$ and $U_i\in\mathbb{R}^{n\times n_i}, i=1,2,\ldots,b$, be the set of matrices satisfying $(U_1,U_2,\ldots,U_b)=I_n$. For a given $x\in\mathbb{R}^n$, we denote its i-th block by $x^{(i)}=U_i^Tx$, $i=1,\ldots,b$. Note that $x=U_1x^{(1)}+\ldots+U_bx^{(b)}$. Moreover, we define

$\|x\|^2 = \|x^{(1)}\|_1^2 + \ldots + \|x^{(b)}\|_b^2$, and denote its conjugate by $\|y\|_*^2 = \|y^{(1)}\|_{1,*}^2 + \ldots + \|y^{(b)}\|_{b,*}^2$.

Let $v_i : X_i \to R$ be the distance generating function with modulus 1 with respect to $\|\cdot\|_i$, and V_i be the associated prox-function. For a given $x \in X_i$ and $y \in \mathbb{R}^{n_i}$, we define the prox-mapping as

$$\mathrm{argmin}_{z \in X_i} \langle y, z - x \rangle + \tfrac{1}{\gamma} V_i(x,z) + \chi_i(z). \tag{6.3.6}$$

We need to assume that the prox-functions $V_i(\cdot,\cdot)$, $i = 1,\ldots,b$, satisfy a quadratic growth condition:

$$V_i(x^{(i)}, z^{(i)}) \le \tfrac{Q}{2} \|z^{(i)} - x^{(i)}\|_i^2 \;\; \forall z^{(i)}, x^{(i)} \in X_i, \tag{6.3.7}$$

for some $Q > 0$.

In order to discuss the convergence of the SBMD algorithm for solving nonconvex composite problems, we need to first define an appropriate termination criterion. Note that if $X = \mathbb{R}^n$ and $\chi(x) = 0$, then a natural way to evaluate the quality of a candidate solution x will be $\|\nabla f(x)\|$. For more general nonconvex composite problems, we introduce the notion of composite projected gradient so as to evaluate the quality of a candidate solution. More specifically, for a given $x \in X$, $y \in \mathbb{R}^n$, and a constant $\gamma > 0$, we define $P_X(x,y,\gamma) \equiv (P_{X_1}(x,y,\gamma),\ldots,P_{X_b}(x,y,\gamma))$ by

$$P_{X_i}(x,y,\gamma) := \tfrac{1}{\gamma}[x_i - x_i^+], \;\; i = 1,\ldots,b, \tag{6.3.8}$$

where

$$x_i^+ := \mathrm{argmin}_{z \in X_i} \langle y, z - x_i \rangle + \tfrac{1}{\gamma} V_i(x_i, z) + \chi_i(z).$$

In particular, if $y = \nabla f(x)$, then we call $P_X(x, \nabla f(x), \gamma)$ the composite projected gradient of x w.r.t. γ. It can be easily seen that $P_X(x, \nabla f(x), \gamma) = \nabla f(x)$ when $X = \mathbb{R}^n$ and $\chi(x) = 0$. Proposition 6.2 below relates the composite projected gradient to the first-order optimality condition of the composite problem under a more general setting.

Proposition 6.2. *Let $x \in X$ be given and $P_X(x,y,\gamma)$ be defined as in (6.3.8) for some $\gamma > 0$. Also let us denote $x^+ := x - \gamma P_X(x, g(x), \gamma)$. Then there exists $p_i \in \partial \chi_i(U_i^T x^+)$ s.t.*

$$U_i^T g(x^+) + p_i \in -N_{X_i}(U_i^T x^+) + B_i\left((L_i \gamma + Q)\|P_X(x, g(x), \gamma)\|_i\right), \;\; i = 1,\ldots,b, \tag{6.3.9}$$

where $B_i(\varepsilon) := \{v \in \mathbb{R}^{n_i} : \|v\|_{i,*} \le \varepsilon\}$ and $N_{X_i}(U_i^T x^+)$ denotes the normal cone of X_i at $U_i^T x^+$.

Proof. By the definition of x^+, (6.3.6), and (6.3.8), we have $U_i^T x^+ = P_{X_i}(U_i^T x, U_i^T g(x), \gamma)$. Using the above relation and the optimality condition of (6.3.6), we conclude that there exists $p_i \in \partial \chi_i(U_i^T x^+)$ s.t.

$$\langle U_i^T g(x) + \tfrac{1}{\gamma}\left[\nabla v_i(U_i^T x^+) - \nabla v_i(U_i^T x)\right] + p_i, u - U_i^T x^+ \rangle \ge 0, \;\; \forall u \in X_i.$$

Now, denoting $\zeta = U_i^T[g(x) - g(x^+) + \frac{1}{\gamma}[\nabla v_i(U_i^T x^+) - \nabla v_i(U_i^T x)]$, we conclude from the above relation that $U_i^T g(x^+) + p_i + \zeta \in -N_{X_i}(U_i^T x^+)$. Also noting that by $\|U_i^T[g(x^+) - g(x)]\|_{i,*} \le L_i \|U_i^T(x^+ - x)\|_i$ and $\|\nabla v_i(U_i^T x^+) - \nabla v_i(U_i^T x)\|_{i,*} \le Q\|U_i^T(x^+ - x)\|_i$,

$$\begin{aligned}\|\zeta\|_{i,*} &\le \left(L_i + \tfrac{Q}{\gamma}\right)\|U_i^T(x^+ - x)\|_i = \left(L_i + \tfrac{Q}{\gamma}\right)\gamma\|U_i^T P_X(x, g(x), \gamma)\|_i \\ &= (L_i\gamma + Q)\|U_i^T P_X(x, g(x), \gamma)\|_i.\end{aligned}$$

Relation (6.3.9) then immediately follows from the above two relations. ■

A common practice in the gradient descent methods for solving nonconvex problems (for the simple case when $X = \mathbb{R}^n$ and $\chi(x) = 0$) is to choose the output solution $\bar{x}_N$ so that

$$\|g(\bar{x}_N)\|_* = \min_{k=1,\dots,N} \|g(x_k)\|_*, \tag{6.3.10}$$

where x_k, $k = 1, \dots, N$, is the trajectory generated by the gradient descent method. However, such a procedure requires the computation of the whole vector $g(x_k)$ at each iteration and hence can be expensive if n is large. In this section, we address this problem by introducing a randomization scheme into the SBMD algorithm as follows. Instead of taking the best solution from the trajectory as in (6.3.10), we randomly select $\bar{x}_N$ from $x_1, \dots, x_N$ according to a certain probability distribution. The basic scheme of this algorithm is described as follows.

Algorithm 6.1 The nonconvex SBMD algorithm

Let $x_1 \in X$, stepsizes $\{\gamma_k\}_{k\ge1}$ s.t. $\gamma_k < 2/L_i$, $i = 1, \dots, b$, and probabilities $p_i \in [0,1]$, $i = 1, \dots, b$, s.t. $\sum_{i=1}^b p_i = 1$ be given.

for $k = 1, \dots, N$ **do**

1. Generate a random variable i_k according to

$$\text{Prob}\{i_k = i\} = p_i, \quad i = 1, \dots, b. \tag{6.3.11}$$

2. Compute the i_k-th block of the (stochastic) gradient G_{i_k} of $f(\cdot)$ at x_k satisfying

$$\mathbb{E}[G_{i_k}] = U_{i_k}^T g(x_k) \ \text{ and } \ \mathbb{E}[\|G_{i_k} - U_{i_k}^T g(x_k)\|_{i_k,*}^2] \le \bar{\sigma}_k^2, \tag{6.3.12}$$

and update x_k by

$$x_{k+1}^{(i)} = \begin{cases} P_{X_i}(x_k^{(i)}, G_{i_k}(x_k, \xi_k), \gamma_k) & i = i_k, \\ x_k^{(i)} & i \ne i_k. \end{cases} \tag{6.3.13}$$

end for

Set $\bar{x}_N = x_R$ randomly according to

$$\text{Prob}(R = k) = \frac{\gamma_k \min_{i=1,\dots,b} p_i\left(1 - \frac{L_i}{2}\gamma_k\right)}{\sum_{k=1}^N \gamma_k \min_{i=1,\dots,b} p_i\left(1 - \frac{L_i}{2}\gamma_k\right)}, \quad k = 1, \dots, N. \tag{6.3.14}$$

We add a few remarks about the above nonconvex SBMD algorithm. First, observe that we have not yet specified how the gradient G_{i_k} is computed. If the problem is deterministic, then we can simply set $G_{i_k} = U_{i_k}^T g(x_k)$ and $\bar{\sigma}_k = 0$. However, if the problem is stochastic, then the computation of G_{i_k} is a little complicated and we cannot simply set $G_{i_k} = U_{i_k}^T \nabla F(x_k, \xi_k)$. Second, the probability of choosing x_R, $R = 1,\ldots,N$, as the output solution is given by (6.3.14). Such a randomization scheme was shown to be critical to establish the complexity for nonconvex stochastic optimization as shown earlier in this chapter.

Before establishing the convergence properties of the above nonconvex SBMD algorithm, we will first present a technical result which summarizes some important properties about the composite prox-mapping and projected gradient. Note that this result generalizes a few results in Sect. 6.2.1.

Lemma 6.7. *Let x_{k+1} be defined in (6.3.13), and denote $g_k \equiv P_X(x_k, \nabla f(x_k), \gamma_k)$ and $\tilde{g}_k \equiv P_{X_{i_k}}(x_k, U_{i_k} G_{i_k}, \gamma_k)$. We have*

$$\langle G_{i_k}, \tilde{g}_k \rangle \geq \|\tilde{g}_k\|^2 + \tfrac{1}{\gamma_k}\left[\chi(x_{k+1}) - \chi(x_k)\right], \tag{6.3.15}$$

$$\|\tilde{g}_k - U_{i_k}^T g_k\|_{i_k} \leq \|G_{i_k} - U_{i_k} \nabla f(x_k)\|_{i_k,*}. \tag{6.3.16}$$

Proof. By the optimality condition of (6.3.6) and the definition of x_{k+1} in (6.3.13), there exists $p \in \partial \chi_{i_k}(x_{k+1})$ such that

$$\langle G_{i_k} + \tfrac{1}{\gamma_k}\left[\nabla \nu_{i_k}(U_{i_k}^T x_{k+1}) - \nabla \nu_{i_k}(U_{i_k}^T x_k)\right] + p, \tfrac{1}{\gamma_k}(u - U_{i_k}^T x_{k+1})\rangle \geq 0, \;\; \forall u \in X_{i_k}. \tag{6.3.17}$$

Letting $u = U_{i_k}^T x_k$ in the above inequality and rearranging terms, we obtain

$$\begin{aligned}
\langle G_{i_k}, \tfrac{1}{\gamma_k} U_{i_k}^T (x_k - x_{k+1})\rangle &\geq \tfrac{1}{\gamma_k^2}\langle \nabla \nu_{i_k}(U_{i_k}^T x_{k+1}) - \nabla \nu_{i_k}(U_{i_k}^T x_k), U_{i_k}^T (x_{k+1} - x_k)\rangle \\
&\quad + \langle p, \tfrac{1}{\gamma_k} U_{i_k}^T (x_{k+1} - x_k)\rangle \\
&\geq \tfrac{1}{\gamma_k^2}\langle \nabla \nu_{i_k}(U_{i_k}^T x_{k+1}) - \nabla \nu_{i_k}(U_{i_k}^T x_k), U_{i_k}^T (x_{k+1} - x_k)\rangle \\
&\quad + \tfrac{1}{\gamma_k}\left[\chi_{i_k}(U_{i_k}^T x_{k+1}) - \chi_{i_k}(U_{i_k}^T x_k)\right] \\
&\geq \tfrac{1}{\gamma_k^2}\|U_{i_k}^T (x_{k+1} - x_k)\|^2 + \tfrac{1}{\gamma_k}\left[\chi_{i_k}(U_{i_k}^T x_{k+1}) - \chi_{i_k}(U_{i_k}^T x_k)\right] \\
&= \tfrac{1}{\gamma_k^2}\|U_{i_k}^T (x_{k+1} - x_k)\|^2 + \tfrac{1}{\gamma_k}[\chi(x_{k+1}) - \chi(x_k)],
\end{aligned} \tag{6.3.18}$$

where the second and third inequalities, respectively, follow from the convexity of χ_{i_k} and the strong convexity of ν, and the last identity follows from the definition of x_{k+1} and the separability assumption about χ in (6.3.4). The above inequality, in view of the fact that $\gamma_k \tilde{g}_k = U_{i_k}^T (x_k - x_{k+1})$ due to (6.3.8) and (6.3.13), then implies (6.3.15).

Now we show that (6.3.16) holds. Let us denote $x_{k+1}^+ = x_k - \gamma_k g_k$. By the optimality condition of (6.3.6) and the definition of g_k, we have, for some $q \in \partial \chi_{i_k}(x_{k+1}^+)$,

$$\langle U_{i_k}^T \nabla(x_k) + \tfrac{1}{\gamma_k}\left[\nabla v_{i_k}(U_{i_k}^T x_{k+1}^+) - \nabla v_{i_k}(U_{i_k}^T x_k)\right] + q, \tfrac{1}{\gamma_k}(u - U_{i_k}^T x_{k+1}^+)\rangle \geq 0, \;\; \forall u \in X_{i_k}. \tag{6.3.19}$$

Letting $u = U_{i_k}^T x_{k+1}^+$ in (6.3.17) and using an argument similar to (6.3.18), we have

$$\begin{aligned}\langle G_{i_k}, \tfrac{1}{\gamma_k} U_{i_k}^T (x_{k+1}^+ - x_{k+1})\rangle \geq\ & \tfrac{1}{\gamma_k^2}\langle \nabla v_{i_k}(U_{i_k}^T x_{k+1}) - \nabla v_{i_k}(U_{i_k}^T x_k), U_{i_k}^T (x_{k+1} - x_{k+1}^+)\rangle \\ & + \tfrac{1}{\gamma_k}\left[\chi_{i_k}(U_{i_k}^T x_{k+1}) - \chi_{i_k}(U_{i_k}^T x_{k+1}^+)\right].\end{aligned}$$

Similarly, letting $u = U_{i_k}^T x_{k+1}$ in (6.3.19), we have

$$\begin{aligned}&\langle U_{i_k}^T \nabla(x_k), \tfrac{1}{\gamma_k} U_{i_k}^T (x_{k+1} - x_{k+1}^+)\rangle \geq \tfrac{1}{\gamma_k^2}\langle \nabla v_{i_k}(U_{i_k}^T x_{k+1}^+) - \nabla v_{i_k}(U_{i_k}^T x_k), \\ &U_{i_k}^T (x_{k+1}^+ - x_{k+1})\rangle + \tfrac{1}{\gamma_k}\left[\chi_{i_k}(U_{i_k}^T x_{k+1}^+) - \chi_{i_k}(U_{i_k}^T x_{k+1})\right].\end{aligned}$$

Summing up the above two inequalities, we obtain

$$\begin{aligned}&\langle G_{i_k} - U_{i_k}^T \nabla(x_k), U_{i_k}^T (x_{k+1}^+ - x_{k+1})\rangle \geq \\ &\tfrac{1}{\gamma_k}\langle \nabla v_{i_k}(U_{i_k}^T x_{k+1}) - \nabla v_{i_k}(U_{i_k}^T x_{k+1}^+), U_{i_k}^T (x_{k+1} - x_{k+1}^+)\rangle \geq \tfrac{1}{\gamma_k}\|U_{i_k}^T (x_{k+1} - x_{k+1}^+)\|_{i_k}^2,\end{aligned}$$

which, in view of the Cauchy–Schwarz inequality, then implies that

$$\tfrac{1}{\gamma_k}\|U_{i_k}^T (x_{k+1} - x_{k+1}^+)\|_{i_k} \leq \|G_{i_k} - U_{i_k}^T \nabla(x_k)\|_{i_k,*}.$$

Using the above relation and (6.3.8), we have

$$\begin{aligned}\|\tilde{g}_k - U_{i_k}^T g_k\|_{i_k} &= \|\tfrac{1}{\gamma_k} U_{i_k}^T (x_k - x_{k+1}) - \tfrac{1}{\gamma_k} U_{i_k}^T (x_k - x_{k+1}^+)\|_{i_k} \\ &= \tfrac{1}{\gamma_k}\|U_{i_k}^T (x_{k+1}^+ - x_{k+1})\|_{i_k} \leq \|G_{i_k} - U_{i_k}^T \nabla(x_k)\|_{i_k,*}.\end{aligned}$$

■

We are now ready to describe the main convergence properties of the nonconvex SBMD algorithm.

Theorem 6.9. *Let $\bar{x}_N = x_R$ be the output of the nonconvex SBMD algorithm. We have*

$$\mathbb{E}[\|P_X(x_R, g(x_R), \gamma_R)\|^2] \leq \frac{\phi(x_1) - \phi^* + 2\sum_{k=1}^N \gamma_k \bar{\sigma}_k^2}{\sum_{k=1}^N \gamma_k \min_{i=1,\ldots,b} p_i \left(1 - \frac{L_i}{2}\gamma_k\right)} \tag{6.3.20}$$

for any $N \geq 1$, where the expectation is taken w.r.t. i_k, G_{i_k}, and R.

Proof. Denote $\delta_k \equiv G_{i_k} - U_{i_k}^T \nabla f(x_k)$, $g_k \equiv P_X(x_k, \nabla f(x_k), \gamma_k)$ and $\tilde{g}_k \equiv P_{X_{i_k}}(x_k, U_{i_k} G_{i_k}, \gamma_k)$ for any $k \geq 1$. Note that by (6.3.13) and (6.3.8), we have $x_{k+1} - x_k = -\gamma_k U_{i_k} \tilde{g}_k$. Using this observation and (6.3.3), we have, for any $k = 1, \ldots, N$,

$$\begin{aligned}f(x_{k+1}) &\leq f(x_k) + \langle \nabla f(x_k), x_{k+1} - x_k\rangle + \tfrac{L_{i_k}}{2}\|x_{k+1} - x_k\|^2 \\ &= f(x_k) - \gamma_k \langle \nabla f(x_k), U_{i_k} \tilde{g}_k\rangle + \tfrac{L_{i_k}}{2}\gamma_k^2 \|\tilde{g}_k\|_{i_k}^2 \\ &= f(x_k) - \gamma_k \langle G_{i_k}, \tilde{g}_k\rangle + \tfrac{L_{i_k}}{2}\gamma_k^2 \|\tilde{g}_k\|_{i_k}^2 + \gamma_k \langle \delta_k, \tilde{g}_k\rangle.\end{aligned}$$

Using the above inequality and Lemma 6.7, we obtain

$$f(x_{k+1}) \le f(x_k) - \left[\gamma_k \|\tilde{g}_k\|_{i_k}^2 + \chi(x_{k+1}) - \chi(x_k)\right] + \tfrac{L_{i_k}}{2}\gamma_k^2 \|\tilde{g}_k\|_{i_k}^2 + \gamma_k \langle \delta_k, \tilde{g}_k \rangle,$$

which, in view of the fact that $\phi(x) = f(x) + \chi(x)$, then implies that

$$\phi(x_{k+1}) \le \phi(x_k) - \gamma_k \left(1 - \tfrac{L_{i_k}}{2}\gamma_k\right) \|\tilde{g}_k\|_{i_k}^2 + \gamma_k \langle \delta_k, \tilde{g}_k \rangle. \tag{6.3.21}$$

Also observe that by (6.3.16), the definition of $\tilde{g}_k$, and the fact $U_{i_k}^T g_k = P_{X_{i_k}}(x_k, \nabla f(x_k), \gamma_k)$,

$$\|\tilde{g}_k - U_{i_k}^T g_k\|_{i_k} \le \|G_{i_k} - U_{i_k}^T \nabla f(x_k)\|_{i_k,*} = \|\delta_k\|_{i_k,*},$$

and hence that

$$\begin{aligned}
\|U_{i_k}^T g_k\|_{i_k}^2 &= \|\tilde{g}_k + U_{i_k}^T g_k - \tilde{g}_k\|_{i_k}^2 \le 2\|\tilde{g}_k\|_{i_k}^2 + 2\|U_{i_k}^T g_k - \tilde{g}_k\|_{i_k}^2 \\
&\le 2\|\tilde{g}_k\|_{i_k}^2 + 2\|\delta_k\|_{i_k,*}^2, \\
\langle \delta_k, \tilde{g}_k \rangle &= \langle \delta_k, U_{i_k}^T g_k \rangle + \langle \delta_k, \tilde{g}_k - U_{i_k}^T g_k \rangle \le \langle \delta_k, U_{i_k}^T g_k \rangle + \|\delta_k\|_{i_k,*} \|\tilde{g}_k - U_{i_k}^T g_k\|_{i_k} \\
&\le \langle \delta_k, U_{i_k}^T g_k \rangle + \|\delta_k\|_{i_k,*}^2.
\end{aligned}$$

By using the above two bounds and (6.3.21), we obtain

$$\phi(x_{k+1}) \le \phi(x_k) - \gamma_k \left(1 - \tfrac{L_{i_k}}{2}\gamma_k\right) \left(\tfrac{1}{2}\|U_{i_k}^T g_k\|_{i_k}^2 - \|\delta_k\|_{i_k,*}^2\right) + \gamma_k \langle \delta_k, U_{i_k}^T g_k \rangle + \gamma_k \|\delta_k\|_{i_k,*}^2$$

for any $k = 1, \ldots, N$. Summing up the above inequalities and rearranging the terms, we obtain

$$\begin{aligned}
\textstyle\sum_{k=1}^N \frac{\gamma_k}{2} \left(1 - \frac{L_{i_k}}{2}\gamma_k\right) \|U_{i_k}^T g_k\|_{i_k}^2 &\le \phi(x_1) - \phi(x_{k+1}) + \textstyle\sum_{k=1}^N \left[\gamma_k \langle \delta_k, U_{i_k}^T g_k \rangle + \gamma_k \|\delta_k\|_{i_k,*}^2\right] \\
&\quad + \textstyle\sum_{k=1}^N \gamma_k \left(1 - \frac{L_{i_k}}{2}\gamma_k\right) \|\delta_k\|_{i_k,*}^2 \\
&\le \phi(x_1) - \phi^* + \textstyle\sum_{k=1}^N \left[\gamma_k \langle \delta_k, U_{i_k}^T g_k \rangle + 2\gamma_k \|\delta_k\|_{i_k,*}^2\right],
\end{aligned}$$

where the last inequality follows from the facts that $\phi(x_{k+1}) \ge \phi^*$ and $L_{i_k}\gamma_k^2 \|\delta_k\|_{i_k,*}^2 \ge 0$. Now denoting $\zeta_k = G_{i_k}$, $\zeta_{[k]} = \{\zeta_1, \ldots, \zeta_k\}$, and $i_{[k]} = \{i_1, \ldots, i_k\}$, taking expectation on both sides of the above inequality w.r.t. $\zeta_{[N]}$ and $i_{[N]}$, and noting that by (6.3.11) and (6.3.12),

$$\begin{aligned}
\mathbb{E}_{\zeta_k}\left[\langle \delta_k, U_{i_k}^T g_k \rangle | i_{[k]}, \zeta_{[k-1]}\right] &= \mathbb{E}_{\zeta_k}\left[\langle G_{i_k} - U_{i_k}^T \nabla f(x_k), U_{i_k}^T g_k \rangle | i_{[k]}, \zeta_{[k-1]}\right] = 0, \\
\mathbb{E}_{\zeta_{[N]}, i_{[N]}}\left[\|\delta_k\|_{i_k,*}^2\right] &\le \bar{\sigma}_k^2, \\
\mathbb{E}_{i_k}\left[\left(1 - \tfrac{L_{i_k}}{2}\gamma_k\right) \|U_{i_k}^T g_k\|^2 | \zeta_{[k-1]}, i_{[k-1]}\right] &= \textstyle\sum_{i=1}^b p_i \left(1 - \frac{L_i}{2}\gamma_k\right) \|U_i^T g_k\|^2 \\
&\ge \left(\textstyle\sum_{i=1}^b \|U_i^T g_k\|^2\right) \min_{i=1,\ldots,b} p_i \left(1 - \tfrac{L_i}{2}\gamma_k\right) \\
&= \|g_k\|^2 \min_{i=1,\ldots,b} p_i \left(1 - \tfrac{L_i}{2}\gamma_k\right),
\end{aligned}$$

we conclude that

$$\Sigma_{k=1}^N \gamma_k \min_{i=1,\ldots,b} p_i \left(1 - \tfrac{L_i}{2}\gamma_k\right) \mathbb{E}_{\zeta_{[N]},i_{[N]}} \left[\|g_k\|^2\right] \le \phi(x_1) - \phi^* + 2\Sigma_{k=1}^N \gamma_k \bar{\sigma}_k^2.$$

Dividing both sides of the above inequality by $\Sigma_{k=1}^N \gamma_k \min_{i=1,\ldots,b} p_i \left(1 - \tfrac{L_i}{2}\gamma_k\right)$, and using the probability distribution of R given in (6.3.14), we obtain (6.3.20). ■

We now discuss some consequences for Theorem 6.9. More specifically, we discuss the rate of convergence of the nonconvex SBMD algorithm for solving deterministic and stochastic problems, respectively, in Corollaries 6.12 and 6.13.

Corollary 6.12. *Consider the deterministic case when* $\bar{\sigma}_k = 0$, $k = 1,\ldots,N$, *in (6.3.12). Suppose that the random variable* $\{i_k\}$ *is uniformly distributed, i.e.,*

$$p_1 = p_2 = \ldots = p_b = \tfrac{1}{b}. \tag{6.3.22}$$

If $\{\gamma_k\}$ *are set to*

$$\gamma_k = \tfrac{1}{\bar{L}}, k = 1,\ldots,N, \tag{6.3.23}$$

where $\bar{L} := \max_{i=1,\ldots,b} L_i$, *then we have, for any* $N \ge 1$,

$$\mathbb{E}[\|P_X(x_R, \nabla f(x_R), \gamma_R)\|^2] \le \tfrac{2b\bar{L}[\phi(x_1)-\phi^*]}{N}. \tag{6.3.24}$$

Proof. By our assumptions about p_i and (6.3.23), we have

$$\min_{i=1,\ldots,b} p_i \left(1 - \tfrac{L_i}{2}\gamma_k\right) = \tfrac{1}{b} \min_{i=1,\ldots,b} \left(1 - \tfrac{L_i}{2}\gamma_k\right) \ge \tfrac{1}{2b}, \tag{6.3.25}$$

which, in view of (6.3.20) and the fact that $\bar{\sigma}_k = 0$, then implies that, for any $N \ge 1$,

$$\mathbb{E}[\|P_X(x_R, \nabla f(x_R), \gamma_R)\|^2] \le \tfrac{2b[\phi(x_1)-\phi^*]}{N} \tfrac{1}{\bar{L}} = \tfrac{2b\bar{L}[\phi(x_1)-\phi^*]}{N}.$$

■

Now, let us consider the stochastic case when $f(\cdot)$ is given in the form of expectation (see (4.6.1)). Suppose that the norms $\|\cdot\|_i$ are inner product norms in $\mathbb{R}^{n_i}$ and that

$$\mathbb{E}[\|U_i \nabla F(x,\xi) - g_i(x)\|] \le \sigma \;\; \forall x \in X \tag{6.3.26}$$

for any $i = 1,\ldots,b$. Also assume that G_{i_k} is computed by using a mini-batch approach with size T_k, i.e.,

$$G_{i_k} = \tfrac{1}{T_k} \Sigma_{t=1}^{T_k} U_{i_k} \nabla F(x_k, \xi_{k,t}), \tag{6.3.27}$$

for some $T_k \ge 1$, where $\xi_{k,1},\ldots,\xi_{k,T_k}$ are i.i.d. samples of ξ.

Corollary 6.13. *Assume that the random variables* $\{i_k\}$ *are uniformly distributed (i.e., (6.3.22) holds). Also assume that* G_{i_k} *is computed by (6.3.27) for* $T_k = T$ *and* *that* $\{\gamma_k\}$ *are set to (6.3.23). Then we have*

$$\mathbb{E}[\|P_X(x_R, \nabla f(x_R), \gamma_R)\|^2] \leq \tfrac{2b\bar{L}[\phi(x_1)-\phi^*]}{N} + \tfrac{4b\sigma^2}{T} \tag{6.3.28}$$

for any $N \geq 1$*, where* $\bar{L} := \max_{i=1,\ldots,b} L_i$.

Proof. Denote $\delta_{k,t} \equiv U_{i_k}[\nabla F(x_k, \xi_{k,t}) - \nabla f(x_k)]$ and $S_t = \sum_{i=1}^t \delta_{k,i}$. Noting that $\mathbb{E}[\langle S_{t-1}, \delta_{k,t}\rangle | S_{t-1}] = 0$ for all $t = 1, \ldots, T_k$, we have

$$\begin{aligned}
\mathbb{E}[\|S_{T_k}\|^2] &= \mathbb{E}\left[\|S_{T_k-1}\|^2 + 2\langle S_{T_k-1}, \delta_{k,T_k}\rangle + \|\delta_{k,T_k}\|^2\right] \\
&= \mathbb{E}[\|S_{T_k-1}\|^2] + \mathbb{E}[\|\delta_{k,T_k}\|^2] = \ldots = \textstyle\sum_{t=1}^{T_k} \|\delta_{k,t}\|^2,
\end{aligned}$$

which together with (6.3.27) then imply that the conditions in (6.3.12) hold with $\bar{\sigma}_k^2 = \sigma^2 / T_k$. It then follows from the previous observation and (6.3.20) that

$$\begin{aligned}
\mathbb{E}[\|P_X(x_R, \nabla f(x_R), \gamma_R)\|^2] &\leq \tfrac{2b[\phi(x_1)-\phi^*]}{\frac{N}{\bar{L}}} + \tfrac{4b}{N} \textstyle\sum_{k=1}^N \tfrac{\sigma^2}{T_k} \\
&\leq \tfrac{2b\bar{L}[\phi(x_1)-\phi^*]}{N} + \tfrac{4b\sigma^2}{T}.
\end{aligned}$$

■

In view of Corollary 6.13, in order to find an ε solution of problem (6.3.1), we need to have

$$N = \mathcal{O}\left(\tfrac{b\bar{L}}{\varepsilon}[\phi(x_1) - \phi^*]\right) \text{ and } T = \mathcal{O}\left(\tfrac{b\sigma^2}{\varepsilon}\right), \tag{6.3.29}$$

which implies that the total number of samples of ξ required can be bounded by

$$\mathcal{O}\left(b^2 \bar{L} \sigma^2 [\phi(x_1) - \phi^*]/\varepsilon^2\right).$$

6.4 Nonconvex Stochastic Accelerated Gradient Descent

In this section we aim to generalize the accelerated gradient descent (AGD) method, originally designed for smooth convex optimization, to solve more general NLP (possibly nonconvex and stochastic) problems, and thus to present a unified treatment and analysis for convex, nonconvex, and stochastic optimization.

Our study has also been motivated by the following more practical considerations in solving nonlinear programming problems, in addition to the theoretical development of the AGD method. First, many general nonlinear objective functions are locally convex. A unified treatment for both convex and nonconvex problems will help us to make use of such local convex properties. In particular, we intend to understand whether one can apply the well-known aggressive stepsize policy in the AGD method under a more general setting to benefit from such local convexity. Second, many nonlinear objective functions arising from sparse optimization and machine learning consist of both convex and nonconvex components, corresponding to the data fidelity and sparsity regularization terms respectively. One interesting question is whether one can design more efficient algorithms for solving these nonconvex composite problems by utilizing their convexity structure. Third, the convexity of some objective functions represented by a black-box procedure is usually unknown, e.g., in simulation-based optimization. A unified treatment and analysis can thus help us to deal with such structural ambiguity. Fourth, in some cases, the objective functions are nonconvex w.r.t. a few decision variables jointly, but convex w.r.t. each one of them separately. Many machine learning/imaging processing problems are given in this form. Current practice is to first run an NLP solver to find a stationary point, and then a CP solver after one variable is fixed. A more powerful, unified treatment for both convex and nonconvex problems is desirable to better handle these types of problems.

In this section, we first consider the classic NLP problem given in the form of

$$\Psi^* = \min_{x \in \mathbb{R}^n} \Psi(x), \tag{6.4.1}$$

where $\Psi(\cdot)$ is a smooth (possibly nonconvex) function with Lipschitz continuous gradients, i.e., $\exists L_\Psi > 0$ such that

$$\|\nabla\Psi(y) - \nabla\Psi(x)\| \le L_\Psi \|y - x\| \quad \forall x, y \in \mathbb{R}^n, \tag{6.4.2}$$

In addition, we assume that $\Psi(\cdot)$ is bounded from below. We demonstrate that the AGD method, when employed with a certain stepsize policy, can find an ε-solution of (6.4.1), i.e., a point $\bar{x}$ such that $\|\nabla\Psi(\bar{x})\|^2 \le \varepsilon$, in at most $\mathcal{O}(1/\varepsilon)$ iterations, which is the best-known complexity bound possessed by first-order methods to solve general NLP problems. Note that if Ψ is convex and a more aggressive stepsize policy is applied in the AGD method, then the aforementioned complexity bound can be improved to $\mathcal{O}(1/\varepsilon^{1/3})$. In fact, by incorporating a certain regularization technique this bound can be improved to $\mathcal{O}([1/\varepsilon^{\frac{1}{4}}] \ln 1/\varepsilon)$ which is optimal, up to a logarithmic factor, for convex programming.

We then consider a class of composite problems given by

$$\min_{x \in \mathbb{R}^n} \Psi(x) + \chi(x), \quad \Psi(x) := f(x) + h(x), \tag{6.4.3}$$

where f is smooth, but possibly nonconvex, and its gradients are Lipschitz continuous with constant L_f, h is smooth, convex, and its gradients are Lipschitz continuous with constant L_h, and χ is a simple convex (possibly non-smooth) function with bounded domain (e.g., $\chi(x) = I_X(x)$ with $I_X(\cdot)$ being the indicator function of a convex compact set $X \subset \mathbb{R}^n$). Clearly, we have Ψ is smooth and its gradients are Lipschitz continuous with constant $L_\Psi = L_f + L_h$. Since χ is possibly non-differentiable, we need to employ a different termination criterion based on the generalized projected gradient (or gradient mapping) $P^n_{\mathbb{R}}(\cdot,\cdot,\cdot)$ (see (6.3.8) and more precisely (6.4.43)) below to analyze the complexity of the AGD method. We will show that the complexity bound associated with the AGD method improves the one established in Sect. 6.2.2 for the nonconvex mirror descent method applied to problem (6.4.3) in terms of their dependence on the Lipschitz constant L_h. In addition, it is significantly better than the latter bound when L_f is small enough (see Sect. 6.4.1.2 for more details).

Finally, we consider stochastic NLP problems in the form of (6.4.1) or (6.4.3), where only noisy first-order information about Ψ is available via subsequent calls to a stochastic first-order oracle (SFO). More specifically, at the k-th call, $x_k \in \mathbb{R}^n$ being the input, the SFO outputs a stochastic gradient $G(x_k, \xi_k)$, where $\{\xi_k\}_{k\geq 1}$ are random vectors whose distributions P_k are supported on $\Xi_k \subseteq \mathbb{R}^d$. The following assumptions are also made for the stochastic gradient $G(x_k, \xi_k)$.

Assumption 16 *For any $x \in \mathbb{R}^n$ and $k \geq 1$, we have*

$$(a)\ \mathbb{E}[G(x,\xi_k)] = \nabla\Psi(x), \tag{6.4.4}$$

$$(b)\ \mathbb{E}\left[\|G(x,\xi_k) - \nabla\Psi(x)\|^2\right] \leq \sigma^2. \tag{6.4.5}$$

As shown in the previous two sections, the randomized stochastic gradient (RSGD) methods can be applied to solve these problems. However, the RSGD method and its variants are only nearly optimal for solving convex SP problems. Based on the AGD method, we present a randomized stochastic AGD (RSAGD) method for solving general stochastic NLP problems and show that if $\Psi(\cdot)$ is nonconvex, then the RSAGD method can find an ε-solution of (6.4.1), i.e., a point $\bar{x}$ s.t. $\mathbb{E}[\|\nabla\Psi(\bar{x})\|^2] \leq \varepsilon$ in at most

$$\mathscr{O}\left(\tfrac{L_\Psi}{\varepsilon} + \tfrac{L_\Psi\sigma^2}{\varepsilon^2}\right) \tag{6.4.6}$$

calls to the SFO. Moreover, if $\Psi(\cdot)$ is convex, then the RSAGD exhibits an optimal rate of convergence in terms of function optimality gap, similar to the accelerated SGD method. In this case, the complexity bound in (6.4.6) in terms of the residual of gradients can be improved to

$$\mathscr{O}\left(\frac{L_\Psi^{2/3}}{\varepsilon^{1/3}} + \frac{L_\Psi^{2/3}\sigma^2}{\varepsilon^{4/3}}\right).$$

We also generalize these complexity analyses to a class of nonconvex stochastic composite optimization problems by introducing a mini-batch approach into the RSAGD method and improve a few complexity results presented in Sect. 6.2 for solving these stochastic composite optimization problems.

6.4.1 Nonconvex Accelerated Gradient Descent

Our goal in this section is to show that the AGD method, which is originally designed for smooth convex optimization, also converges for solving nonconvex optimization problems after incorporating some proper modification. More specifically, we first present an AGD method for solving a general class of nonlinear optimization problems in Sect. 6.4.1.1 and then describe the AGD method for solving a special class of nonconvex composite optimization problems in Sect. 6.4.1.2.

6.4.1.1 Minimization of Smooth Functions

In this subsection, we assume that $\Psi(\cdot)$ is a differentiable nonconvex function, bounded from below and its gradient satisfies in (6.4.2). It then follows that

$$|\Psi(y) - \Psi(x) - \langle \nabla\Psi(x), y - x\rangle| \le \tfrac{L_\Psi}{2}\|y - x\|^2 \quad \forall x, y \in \mathbb{R}^n. \tag{6.4.7}$$

While the gradient descent method converges for solving the above class of nonconvex optimization problems, it does not achieve the optimal rate of convergence, in terms of the function optimality gap, when $\Psi(\cdot)$ is convex. On the other hand, the original AGD method is optimal for solving convex optimization problems, but does not necessarily converge for solving nonconvex optimization problems. Below, we present a modified AGD method and show that by properly specifying the stepsize policy, it not only achieves the optimal rate of convergence for convex optimization, but also exhibits the best-known rate of convergence for solving general smooth NLP problems by using first-order methods.

Algorithm 6.2 The accelerated gradient descent (AGD) algorithm

Input: $x_0 \in \mathbb{R}^n$, $\{\alpha_k\}$ s.t. $\alpha_1 = 1$ and $\alpha_k \in (0,1)$ for any $k \ge 2$, $\{\beta_k > 0\}$, and $\{\lambda_k > 0\}$.
0. Set the initial points $\bar{x}_0 = x_0$ and $k = 1$.
1. Set

$$\underline{x}_k = (1-\alpha_k)\bar{x}_{k-1} + \alpha_k x_{k-1}. \tag{6.4.8}$$

2. Compute $\nabla\Psi(\underline{x}_k)$ and set

$$x_k = x_{k-1} - \lambda_k \nabla\Psi(\underline{x}_k), \tag{6.4.9}$$

$$\bar{x}_k = \underline{x}_k - \beta_k \nabla\Psi(\underline{x}_k). \tag{6.4.10}$$

3. Set $k \leftarrow k+1$ and go to step 1.

Note that, if $\beta_k = \alpha_k\lambda_k \;\; \forall k \ge 1$, then we have $\bar{x}_k = \alpha_k x_k + (1-\alpha_k)\bar{x}_{k-1}$. In this case, the above AGD method is equivalent to the accelerated gradient descent method discussed in Sect. 3.3. On the other hand, if $\beta_k = \lambda_k, \;\; k = 1, 2, \ldots$, then it can be shown by induction that $\underline{x}_k = x_{k-1}$ and $\bar{x}_k = x_k$. In this case, Algorithm 6.2 reduces to the gradient descent method. We will show in this subsection that the above AGD method actually converges for different selections of $\{\alpha_k\}$, $\{\beta_k\}$, and $\{\lambda_k\}$ in both convex and nonconvex cases.

We are now ready to describe the main convergence properties of the AGD method.

Theorem 6.10. *Let $\{\underline{x}_k, \bar{x}_k\}_{k\geq 1}$ be computed by Algorithm 6.2 and Γ_k and*

$$\Gamma_k := \begin{cases} 1, & k=1, \\ (1-\alpha_k)\Gamma_{k-1}, & k\geq 2. \end{cases} \quad (6.4.11)$$

(a) If $\{\alpha_k\}$, $\{\beta_k\}$, and $\{\lambda_k\}$ are chosen such that

$$C_k := 1 - L_\Psi \lambda_k - \tfrac{L_\Psi(\lambda_k-\beta_k)^2}{2\alpha_k\Gamma_k\lambda_k}\left(\textstyle\sum_{\tau=k}^N \Gamma_\tau\right) > 0 \quad 1\leq k\leq N, \quad (6.4.12)$$

then for any $N\geq 1$, we have

$$\min_{k=1,\ldots,N} \|\nabla\Psi(\underline{x}_k)\|^2 \leq \tfrac{\Psi(x_0)-\Psi^*}{\sum_{k=1}^N \lambda_k C_k}. \quad (6.4.13)$$

(b) Suppose that $\Psi(\cdot)$ is convex and that an optimal solution x^* exists for problem (6.4.1). If $\{\alpha_k\}$, $\{\beta_k\}$, and $\{\lambda_k\}$ are chosen such that

$$\alpha_k\lambda_k \leq \beta_k < \tfrac{1}{L_\Psi}, \quad (6.4.14)$$

$$\tfrac{\alpha_1}{\lambda_1\Gamma_1} \geq \tfrac{\alpha_2}{\lambda_2\Gamma_2} \geq \ldots, \quad (6.4.15)$$

then for any $N\geq 1$, we have

$$\min_{k=1,\ldots,N} \|\nabla\Psi(\underline{x}_k)\|^2 \leq \tfrac{\|x_0-x^*\|^2}{\lambda_1\sum_{k=1}^N \Gamma_k^{-1}\beta_k(1-L_\Psi\beta_k)}, \quad (6.4.16)$$

$$\Psi(\bar{x}_N) - \Psi(x^*) \leq \tfrac{\Gamma_N\|x_0-x^*\|^2}{2\lambda_1}. \quad (6.4.17)$$

Proof. We first show part (a). Denote $\Delta_k := \nabla\Psi(x_{k-1}) - \nabla\Psi(\underline{x}_k)$. By (6.4.2) and (6.4.8), we have

$$\|\Delta_k\| = \|\nabla\Psi(x_{k-1}) - \nabla\Psi(\underline{x}_k)\| \leq L_\Psi\|x_{k-1}-\underline{x}_k\| = L_\Psi(1-\alpha_k)\|\bar{x}_{k-1}-x_{k-1}\|. \quad (6.4.18)$$

Also by (6.4.7) and (6.4.9), we have

$$\begin{aligned}
\Psi(x_k) &\leq \Psi(x_{k-1}) + \langle\nabla\Psi(x_{k-1}), x_k - x_{k-1}\rangle + \tfrac{L_\Psi}{2}\|x_k - x_{k-1}\|^2 \\
&= \Psi(x_{k-1}) + \langle\Delta_k + \nabla\Psi(\underline{x}_k), -\lambda_k\nabla\Psi(\underline{x}_k)\rangle + \tfrac{L_\Psi\lambda_k^2}{2}\|\nabla\Psi(\underline{x}_k)\|^2 \\
&= \Psi(x_{k-1}) - \lambda_k\left(1-\tfrac{L_\Psi\lambda_k}{2}\right)\|\nabla\Psi(\underline{x}_k)\|^2 - \lambda_k\langle\Delta_k, \nabla\Psi(\underline{x}_k)\rangle \\
&\leq \Psi(x_{k-1}) - \lambda_k\left(1-\tfrac{L_\Psi\lambda_k}{2}\right)\|\nabla\Psi(\underline{x}_k)\|^2 + \lambda_k\|\Delta_k\|\cdot\|\nabla\Psi(\underline{x}_k)\|, \quad (6.4.19)
\end{aligned}$$

where the last inequality follows from the Cauchy–Schwarz inequality. Combining the previous two inequalities, we obtain

$$\begin{aligned}
\Psi(x_k) \leq \Psi(x_{k-1}) &- \lambda_k\left(1-\tfrac{L_\Psi\lambda_k}{2}\right)\|\nabla\Psi(\underline{x}_k)\|^2 \\
&+ L_\Psi(1-\alpha_k)\lambda_k\|\nabla\Psi(\underline{x}_k)\|\cdot\|\bar{x}_{k-1}-x_{k-1}\|
\end{aligned}$$

$$
\begin{aligned}
&\le \Psi(x_{k-1}) - \lambda_k \left(1 - \tfrac{L_\Psi \lambda_k}{2}\right) \|\nabla \Psi(\underline{x}_k)\|^2 \\
&\quad + \tfrac{L_\Psi \lambda_k^2}{2} \|\nabla \Psi(\underline{x}_k)\|^2 + \tfrac{L_\Psi (1-\alpha_k)^2}{2} \|\bar{x}_{k-1} - x_{k-1}\|^2 \\
&= \Psi(x_{k-1}) - \lambda_k (1 - L_\Psi \lambda_k) \|\nabla \Psi(\underline{x}_k)\|^2 + \tfrac{L_\Psi (1-\alpha_k)^2}{2} \|\bar{x}_{k-1} - x_{k-1}\|^2,
\end{aligned} \tag{6.4.20}
$$

where the second inequality follows from the fact that $ab \le (a^2+b^2)/2$. Now, by (6.4.8), (6.4.9), and (6.4.10), we have

$$
\begin{aligned}
\bar{x}_k - x_k &= (1-\alpha_k)\bar{x}_{k-1} + \alpha_k x_{k-1} - \beta_k \nabla \Psi(\underline{x}_k) - [x_{k-1} - \lambda_k \nabla \Psi(\underline{x}_k)] \\
&= (1-\alpha_k)(\bar{x}_{k-1} - x_{k-1}) + (\lambda_k - \beta_k) \nabla \Psi(\underline{x}_k).
\end{aligned}
$$

Dividing both sides of the above equality by Γ_k, summing them up, and noting (6.4.11), we obtain

$$
\bar{x}_k - x_k = \Gamma_k \textstyle\sum_{\tau=1}^k \left(\tfrac{\lambda_\tau - \beta_\tau}{\Gamma_\tau}\right) \nabla \Psi(\underline{x}_\tau).
$$

Using the above identity, the Jensen's inequality for $\|\cdot\|^2$, and the fact that

$$
\textstyle\sum_{\tau=1}^k \tfrac{\alpha_\tau}{\Gamma_\tau} = \tfrac{\alpha_1}{\Gamma_1} + \sum_{\tau=2}^k \tfrac{1}{\Gamma_\tau}\left(1 - \tfrac{\Gamma_\tau}{\Gamma_{\tau-1}}\right) = \tfrac{1}{\Gamma_1} + \sum_{\tau=2}^k \left(\tfrac{1}{\Gamma_\tau} - \tfrac{1}{\Gamma_{\tau-1}}\right) = \tfrac{1}{\Gamma_k}, \tag{6.4.21}
$$

we have

$$
\begin{aligned}
\|\bar{x}_k - x_k\|^2 &= \left\|\Gamma_k \textstyle\sum_{\tau=1}^k \left(\tfrac{\lambda_\tau - \beta_\tau}{\Gamma_\tau}\right) \nabla \Psi(\underline{x}_\tau)\right\|^2 = \left\|\Gamma_k \textstyle\sum_{\tau=1}^k \tfrac{\alpha_\tau}{\Gamma_\tau}\left[\left(\tfrac{\lambda_\tau - \beta_\tau}{\alpha_\tau}\right) \nabla \Psi(\underline{x}_\tau)\right]\right\|^2 \\
&\le \Gamma_k \textstyle\sum_{\tau=1}^k \tfrac{\alpha_\tau}{\Gamma_\tau} \left\|\left(\tfrac{\lambda_\tau - \beta_\tau}{\alpha_\tau}\right) \nabla \Psi(\underline{x}_\tau)\right\|^2 = \Gamma_k \textstyle\sum_{\tau=1}^k \tfrac{(\lambda_\tau - \beta_\tau)^2}{\Gamma_\tau \alpha_\tau} \|\nabla \Psi(\underline{x}_\tau)\|^2.
\end{aligned} \tag{6.4.22}
$$

Replacing the above bound in (6.4.20), we obtain

$$
\begin{aligned}
\Psi(x_k) &\le \Psi(x_{k-1}) - \lambda_k (1 - L_\Psi \lambda_k) \|\nabla \Psi(\underline{x}_k)\|^2 \\
&\quad + \tfrac{L_\Psi \Gamma_{k-1} (1-\alpha_k)^2}{2} \textstyle\sum_{\tau=1}^{k-1} \tfrac{(\lambda_\tau - \beta_\tau)^2}{\Gamma_\tau \alpha_\tau} \|\nabla \Psi(\underline{x}_\tau)\|^2 \\
&\le \Psi(x_{k-1}) - \lambda_k (1 - L_\Psi \lambda_k) \|\nabla \Psi(\underline{x}_k)\|^2 + \tfrac{L_\Psi \Gamma_k}{2} \textstyle\sum_{\tau=1}^k \tfrac{(\lambda_\tau - \beta_\tau)^2}{\Gamma_\tau \alpha_\tau} \|\nabla \Psi(\underline{x}_\tau)\|^2
\end{aligned} \tag{6.4.23}
$$

for any $k \ge 1$, where the last inequality follows from the definition of Γ_k in (6.4.11) and the fact that $\alpha_k \in (0,1]$ for all $k \ge 1$. Summing up the above inequalities and using the definition of C_k in (6.4.12), we have

$$
\begin{aligned}
\Psi(x_N) &\le \Psi(x_0) - \textstyle\sum_{k=1}^N \lambda_k (1 - L_\Psi \lambda_k) \|\nabla \Psi(\underline{x}_k)\|^2 \\
&\quad + \tfrac{L_\Psi}{2} \textstyle\sum_{k=1}^N \Gamma_k \sum_{\tau=1}^k \tfrac{(\lambda_\tau - \beta_\tau)^2}{\Gamma_\tau \alpha_\tau} \|\nabla \Psi(\underline{x}_\tau)\|^2 \\
&= \Psi(x_0) - \textstyle\sum_{k=1}^N \lambda_k (1 - L_\Psi \lambda_k) \|\nabla \Psi(\underline{x}_k)\|^2 \\
&\quad + \tfrac{L_\Psi}{2} \textstyle\sum_{k=1}^N \tfrac{(\lambda_k - \beta_k)^2}{\Gamma_k \alpha_k} \left(\sum_{\tau=k}^N \Gamma_\tau\right) \|\nabla \Psi(\underline{x}_k)\|^2 \\
&= \Psi(x_0) - \textstyle\sum_{k=1}^N \lambda_k C_k \|\nabla \Psi(\underline{x}_k)\|^2.
\end{aligned} \tag{6.4.24}
$$

Rearranging the terms in the above inequality and noting that $\Psi(x_N) \geq \Psi^*$, we obtain

$$\min_{k=1,\ldots,N} \|\nabla\Psi(\underline{x}_k)\|^2 \left(\sum_{k=1}^N \lambda_k C_k\right) \leq \sum_{k=1}^N \lambda_k C_k \|\nabla\Psi(\underline{x}_k)\|^2 \leq \Psi(x_0) - \Psi^*,$$

which, in view of the assumption that $C_k > 0$, clearly implies (6.4.13).

We now show part (b). First, note that by (6.4.10), we have

$$\begin{aligned}\Psi(\bar{x}_k) &\leq \Psi(\underline{x}_k) + \langle\nabla\Psi(\underline{x}_k), \bar{x}_k - \underline{x}_k\rangle + \tfrac{L_\Psi}{2}\|\bar{x}_k - \underline{x}_k\|^2 \\ &= \Psi(\underline{x}_k) - \beta_k\|\nabla\Psi(\underline{x}_k)\|^2 + \tfrac{L_\Psi\beta_k^2}{2}\|\nabla\Psi(\underline{x}_k)\|^2. \end{aligned} \tag{6.4.25}$$

Also by the convexity of $\Psi(\cdot)$ and (6.4.8),

$$\begin{aligned}&\Psi(\underline{x}_k) - [(1-\alpha_k)\Psi(\bar{x}_{k-1}) + \alpha_k\Psi(x)] \\ &\quad = \alpha_k[\Psi(\underline{x}_k) - \Psi(x)] + (1-\alpha_k)[\Psi(\underline{x}_k) - \Psi(\bar{x}_{k-1})] \\ &\quad \leq \alpha_k\langle\nabla\Psi(\underline{x}_k), \underline{x}_k - x\rangle + (1-\alpha_k)\langle\nabla\Psi(\underline{x}_k), \underline{x}_k - \bar{x}_{k-1}\rangle \\ &\quad = \langle\nabla\Psi(\underline{x}_k), \alpha_k(\underline{x}_k - x) + (1-\alpha_k)(\underline{x}_k - \bar{x}_{k-1})\rangle = \alpha_k\langle\nabla\Psi(\underline{x}_k), x_{k-1} - x\rangle. \end{aligned} \tag{6.4.26}$$

It also follows from (6.4.9) that

$$\begin{aligned}&\|x_{k-1} - x\|^2 - 2\lambda_k\langle\nabla\Psi(\underline{x}_k), x_{k-1} - x\rangle + \lambda_k^2\|\nabla\Psi(\underline{x}_k)\|^2 \\ &\quad = \|x_{k-1} - \lambda_k\nabla\Psi(\underline{x}_k) - x\|^2 = \|x_k - x\|^2,\end{aligned}$$

and hence that

$$\alpha_k\langle\nabla\Psi(\underline{x}_k), x_{k-1} - x\rangle = \tfrac{\alpha_k}{2\lambda_k}\left[\|x_{k-1} - x\|^2 - \|x_k - x\|^2\right] + \tfrac{\alpha_k\lambda_k}{2}\|\nabla\Psi(\underline{x}_k)\|^2. \tag{6.4.27}$$

Combining (6.4.25), (6.4.26), and (6.4.27), we obtain

$$\begin{aligned}\Psi(\bar{x}_k) &\leq (1-\alpha_k)\Psi(\bar{x}_{k-1}) + \alpha_k\Psi(x) + \tfrac{\alpha_k}{2\lambda_k}\left[\|x_{k-1} - x\|^2 - \|x_k - x\|^2\right] \\ &\quad - \beta_k\left(1 - \tfrac{L_\Psi\beta_k}{2} - \tfrac{\alpha_k\lambda_k}{2\beta_k}\right)\|\nabla\Psi(\underline{x}_k)\|^2 \\ &\leq (1-\alpha_k)\Psi(\bar{x}_{k-1}) + \alpha_k\Psi(x) + \tfrac{\alpha_k}{2\lambda_k}\left[\|x_{k-1} - x\|^2 - \|x_k - x\|^2\right] \\ &\quad - \tfrac{\beta_k}{2}(1 - L_\Psi\beta_k)\|\nabla\Psi(\underline{x}_k)\|^2, \end{aligned} \tag{6.4.28}$$

where the last inequality follows from the assumption in (6.4.14). Subtracting $\Psi(x)$ from both sides of the above inequality and using Lemma 3.17 and the fact that $\alpha_1 = 1$, we conclude that

$$\begin{aligned}\tfrac{\Psi(\bar{x}_N) - \Psi(x)}{\Gamma_N} &\leq \sum_{k=1}^N \tfrac{\alpha_k}{2\lambda_k\Gamma_k}\left[\|x_{k-1} - x\|^2 - \|x_k - x\|^2\right] \\ &\quad - \sum_{k=1}^N \tfrac{\beta_k}{2\Gamma_k}(1 - L_\Psi\beta_k)\|\nabla\Psi(\underline{x}_k)\|^2\end{aligned}$$

$$\leq \tfrac{\|x_0-x\|^2}{2\lambda_1} - \textstyle\sum_{k=1}^N \tfrac{\beta_k}{2\Gamma_k}(1-L_\Psi \beta_k)\|\nabla\Psi(\underline{x}_k)\|^2 \ \ \forall x \in \mathbb{R}^n, \tag{6.4.29}$$

where the second inequality follows from the simple relation that

$$\textstyle\sum_{k=1}^N \tfrac{\alpha_k}{\lambda_k \Gamma_k}\left[\|x_{k-1}-x\|^2 - \|x_k - x\|^2\right] \leq \tfrac{\alpha_1\|x_0-x\|^2}{\lambda_1\Gamma_1} = \tfrac{\|x_0-x\|^2}{\lambda_1} \tag{6.4.30}$$

due to (6.4.15) and the fact that $\alpha_1 = \Gamma_1 = 1$. Hence, (6.4.17) immediately follows from the above inequality and the assumption in (6.4.14). Moreover, fixing $x = x^*$, rearranging the terms in (6.4.29), and noting the fact that $\Psi(\bar{x}_N) \geq \Psi(x^*)$, we obtain

$$\begin{aligned}\min_{k=1,\ldots,N}\|\nabla\Psi(\underline{x}_k)\|^2 \textstyle\sum_{k=1}^N \tfrac{\beta_k}{2\Gamma_k}(1-L_\Psi\beta_k) &\leq \textstyle\sum_{k=1}^N \tfrac{\beta_k}{2\Gamma_k}(1-L_\Psi\beta_k)\|\nabla\Psi(\underline{x}_k)\|^2 \\ &\leq \tfrac{\|x^*-x_0\|^2}{2\lambda_1},\end{aligned}$$

which together with (6.4.14) clearly imply (6.4.16). ■

We add a few observations about Theorem 6.10. First, in view of (6.4.28), it is possible to use a different assumption than the one in (6.4.14) on the stepsize policies for the convex case. In particular, we only need

$$2 - L_\Psi\beta_k - \tfrac{\alpha_k\lambda_k}{\beta_k} > 0 \tag{6.4.31}$$

to show the convergence of the AGD method for minimizing smooth convex problems. However, since the condition given by (6.4.14) is required for minimizing composite problems in Sects. 6.4.1.2 and 6.4.2.2, we state this assumption for the sake of simplicity. Second, there are various options for selecting $\{\alpha_k\}$, $\{\beta_k\}$, and $\{\lambda_k\}$ to guarantee the convergence of the AGD algorithm. Below we provide some of these selections for solving both convex and nonconvex problems.

Corollary 6.14. *Suppose that $\{\alpha_k\}$ and $\{\beta_k\}$ in the AGD method are set to*

$$\alpha_k = \tfrac{2}{k+1} \ \ \textit{and} \ \ \beta_k = \tfrac{1}{2L_\Psi}. \tag{6.4.32}$$

(a) If $\{\lambda_k\}$ satisfies

$$\lambda_k \in \left[\beta_k, \left(1+\tfrac{\alpha_k}{4}\right)\beta_k\right] \ \ \forall k \geq 1, \tag{6.4.33}$$

then for any $N \geq 1$, we have

$$\min_{k=1,\ldots,N}\|\nabla\Psi(\underline{x}_k)\|^2 \leq \tfrac{6L_\Psi[\Psi(x_0)-\Psi^*]}{N}. \tag{6.4.34}$$

(b) Assume that $\Psi(\cdot)$ is convex and that an optimal solution x^* exists for problem (6.4.1). If $\{\lambda_k\}$ satisfies

$$\lambda_k = \tfrac{k\beta_k}{2} \ \ \forall k \geq 1, \tag{6.4.35}$$

then for any $N \geq 1$, we have

$$\min_{k=1,\ldots,N} \|\nabla\Psi(\underline{x}_k)\|^2 \leq \tfrac{96L_\Psi^2\|x_0-x^*\|^2}{N(N+1)(N+2)}, \tag{6.4.36}$$

$$\Psi(\bar{x}_N) - \Psi(x^*) \leq \tfrac{4L_\Psi\|x_0-x^*\|^2}{N(N+1)}. \tag{6.4.37}$$

Proof. We first show part (a). Note that by (6.4.11) and (6.4.32), we have

$$\Gamma_k = \tfrac{2}{k(k+1)}, \tag{6.4.38}$$

which implies that

$$\textstyle\sum_{\tau=k}^N \Gamma_\tau = \sum_{\tau=k}^N \tfrac{2}{\tau(\tau+1)} = 2\sum_{\tau=k}^N \left(\tfrac{1}{\tau} - \tfrac{1}{\tau+1}\right) \leq \tfrac{2}{k}. \tag{6.4.39}$$

It can also be easily seen from (6.4.33) that $0 \leq \lambda_k - \beta_k \leq \alpha_k\beta_k/4$. Using these observations, (6.4.32), and (6.4.33), we have

$$\begin{aligned}
C_k &= 1 - L_\Psi\left[\lambda_k + \tfrac{(\lambda_k-\beta_k)^2}{2\alpha_k\Gamma_k\lambda_k}\left(\textstyle\sum_{\tau=k}^N \Gamma_\tau\right)\right] \\
&\geq 1 - L_\Psi\left[\left(1+\tfrac{\alpha_k}{4}\right)\beta_k + \tfrac{\alpha_k^2\beta_k^2}{16}\tfrac{1}{k\alpha_k\Gamma_k\beta_k}\right] \\
&= 1 - \beta_k L_\Psi\left(1 + \tfrac{\alpha_k}{4} + \tfrac{1}{16}\right) \\
&\geq 1 - \beta_k L_\Psi \tfrac{21}{16} = \tfrac{11}{32}, \\
\lambda_k C_k &\geq \tfrac{11\beta_k}{32} = \tfrac{11}{64L_\Psi} \geq \tfrac{1}{6L_\Psi}.
\end{aligned} \tag{6.4.40}$$

Combining the above relation with (6.4.13), we obtain (6.4.34).

We now show part (b). Observe that by (6.4.32) and (6.4.35), we have

$$\begin{aligned}
\alpha_k\lambda_k &= \tfrac{k}{k+1}\beta_k < \beta_k, \\
\tfrac{\alpha_1}{\lambda_1\Gamma_1} &= \tfrac{\alpha_2}{\lambda_2\Gamma_2} = \ldots = 4L_\Psi,
\end{aligned}$$

which implies that conditions (6.4.14) and (6.4.15) hold. Moreover, we have

$$\textstyle\sum_{k=1}^N \Gamma_k^{-1}\beta_k(1-L_\Psi\beta_k) = \tfrac{1}{4L_\Psi}\sum_{k=1}^N \Gamma_k^{-1} = \tfrac{1}{8L_\Psi}\sum_{k=1}^N (k+k^2) = \tfrac{N(N+1)(N+2)}{24L_\Psi}. \tag{6.4.41}$$

Using (6.4.38) and the above bounds in (6.4.16) and (6.4.17), we obtain (6.4.36) and (6.4.37). ■

We now add a few remarks about the results obtained in Corollary 6.14. First, the rate of convergence in (6.4.34) for the AGD method is in the same order of magnitude as that for the gradient descent method. It is also worth noting that by choosing $\lambda_k = \beta_k$ in (6.4.33), the rate of convergence for the AGD method just changes up to a constant factor. However, in this case, the AGD method is reduced to the gradient descent method as mentioned earlier in this subsection. Second, if the problem is convex, by choosing more aggressive stepsize $\{\lambda_k\}$ in (6.4.35), the AGD method exhibits the optimal rate of convergence in (6.4.37). Moreover, with such a selection of $\{\lambda_k\}$, the AGD method can find a solution $\bar{x}$ such that $\|\nabla\Psi(\bar{x})\|^2 \le \varepsilon$ in at most $\mathcal{O}(1/\varepsilon^{\frac{1}{3}})$ iterations according to (6.4.36).

Observe that $\{\lambda_k\}$ in (6.4.33) for general nonconvex problems is in the order of $\mathcal{O}(1/L_\Psi)$, while the one in (6.4.35) for convex problems are more aggressive (in $\mathcal{O}(k/L_\Psi)$). An interesting question is whether we can apply the same stepsize policy in (6.4.35) for solving general NLP problems no matter whether they are convex or not. We will discuss such a uniform treatment for solving both convex and nonconvex composite problems in next subsection.

6.4.1.2 Minimization of Nonconvex Composite Functions

In this subsection, we consider a special class of NLP problems given in the form of (6.4.3). Our goal in this subsection is to show that we can employ a more aggressive stepsize policy in the AGD method, similar to the one used in the convex case (see Theorem 6.10(b) and Corollary 6.14(b)), to solve these composite problems, even if $\Psi(\cdot)$ is possibly nonconvex.

Throughout this subsection, we make the following assumption about the convex (possibly non-differentiable) component $\chi(\cdot)$ in (6.4.3).

Assumption 17 *There exists a constant M such that $\|x^+(y,c)\| \le M$ for any $c \in (0,+\infty)$ and $x, y \in \mathbb{R}^n$, where*

$$x^+(y,c) := \operatorname{argmin}_{u\in\mathbb{R}^n} \left\{\langle y,u\rangle + \tfrac{1}{2c}\|u-x\|^2 + \chi(u)\right\}. \tag{6.4.42}$$

Next result shows certain situations which assure that Assumption 17 is satisfied. Note that we skip its proof since it is simple.

Lemma 6.8. *Assumption 17 is satisfied if any one of the following statements holds.*

(a) $\chi(\cdot)$ is a proper closed convex function with bounded domain.
(b) There exists a constant M such that $\|x^+(y)\| \le M$ for any $x, y \in \mathbb{R}^n$, where

$$x^+(y) \equiv x^+(y,+\infty) := \operatorname{argmin}_{u\in\mathbb{R}^n} \left\{\langle y,u\rangle + \chi(u)\right\}.$$

Based on the above result, we can give the following examples. Let $X \subseteq \mathbb{R}^n$ be a given convex compact set. It can be easily seen that Assumption 17 holds if $\chi(x) = I_X(x)$. Here I_X is the indicator function of X given by

$$I_X(x) = \begin{cases} 0 & x \in X, \\ +\infty & x \notin X. \end{cases}$$

Another important example is given by $\chi(x) = I_X(x) + \|x\|_1$, where $\|\cdot\|_1$ denotes the l_1 norm.

Observe that $x^+(y,c)$ in (6.4.42) also gives rise to an important quantity that will be used frequently in our convergence analysis, i.e.,

$$P_X(x,y,c) := \tfrac{1}{c}[x - x^+(y,c)]. \tag{6.4.43}$$

In particular, if $y = \nabla\Psi(x)$, then $P_X(x,y,c)$ is called the gradient mapping at x, which has been used as a termination criterion for solving constrained or composite NLP problems. It can be easily seen that $P_X(x,\nabla\Psi(x),c) = \nabla\Psi(x)$ for any $c > 0$ when $\chi(\cdot) = 0$. For more general $\chi(\cdot)$, we can show that as the size of $P_X(x,\nabla\Psi(x),c)$ vanishes, $x^+(\nabla\Psi(x),c)$ approaches to a stationary point of problem (6.4.3). Indeed, let $x \in \mathbb{R}^n$ be given and denote $g \equiv \nabla\Psi(x)$. If $\|P_X(x,g,c)\| \le \varepsilon$ for some $c > 0$, then by Lemma 6.3

$$-\nabla\Psi(x^+(g,c)) \in \partial\chi(x^+(g,c)) + B(\varepsilon(cL_\Psi + 1)), \tag{6.4.44}$$

where $\partial\chi(\cdot)$ denotes the subdifferential of $\chi(\cdot)$ and $B(r) := \{x \in \mathbb{R}^n : \|x\| \le r\}$. Moreover, it follows from Lemma 6.1 that for any $y_1, y_2 \in \mathbb{R}^n$, we have

$$\|P_X(x,y_1,c) - P_X(x,y_2,c)\| \le \|y_1 - y_2\|. \tag{6.4.45}$$

We are now ready to describe the AGD algorithm for solving problem (6.4.3), which differs from Algorithm 6.2 only in Step 2.

Algorithm 6.3 The AGD method for composite optimization

Replace (6.4.9) and (6.4.10) in Step 2 of the Algorithm 1, respectively, by

$$x_k = \operatorname{argmin}_{u\in\mathbb{R}^n}\left\{\langle\nabla\Psi(\underline{x}_k),u\rangle + \tfrac{1}{2\lambda_k}\|u - x_{k-1}\|^2 + \chi(u)\right\}, \tag{6.4.46}$$

$$\bar{x}_k = \operatorname{argmin}_{u\in\mathbb{R}^n}\left\{\langle\nabla\Psi(\underline{x}_k),u\rangle + \tfrac{1}{2\beta_k}\|u - \underline{x}_k\|^2 + \chi(u)\right\}. \tag{6.4.47}$$

A few remarks about Algorithm 6.3 are in place. First, observe that the subproblems (6.4.46) and (6.4.47) are given in the form of (6.4.42) and hence that under Assumption 17, the search points x_k and x_k^{ag} $\forall k \ge 1$ will stay in a bounded set. Second, we need to assume that $\chi(\cdot)$ is simple enough so that the subproblems (6.4.46) and (6.4.47) are easily computable. Third, in view of (6.4.43) and (6.4.47), we have

$$P_X(\underline{x}_k,\nabla\Psi(\underline{x}_k),\beta_k) = \tfrac{1}{\beta_k}(\underline{x}_k - \bar{x}_k). \tag{6.4.48}$$

We will use $\|P_X(\underline{x}_k,\nabla\Psi(\underline{x}_k),\beta_k)\|$ as a termination criterion in the above AGD method for composite optimization.

Before establishing the convergence of the above AGD method, we first state a technical result which shows that the relation in (6.4.7) can be enhanced for composite functions.

Lemma 6.9. *Let $\Psi(\cdot)$ be defined in (6.4.3). For any $x, y \in \mathbb{R}^n$, we have*

$$-\tfrac{L_f}{2}\|y - x\|^2 \le \Psi(y) - \Psi(x) - \langle\nabla\Psi(x), y - x\rangle \le \tfrac{L_\Psi}{2}\|y - x\|^2. \tag{6.4.49}$$

Proof. We only need to show the first relation since the second one follows from (6.4.7).

$$\begin{aligned}
\Psi(y)-\Psi(x) &= \int_0^1 \langle \nabla\Psi(x+t(y-x)), y-x\rangle dt \\
&= \int_0^1 \langle \nabla f(x+t(y-x)), y-x\rangle dt + \int_0^1 \langle \nabla h(x+t(y-x)), y-x\rangle dt \\
&= \langle \nabla f(x), y-x\rangle + \int_0^1 \langle \nabla f(x+t(y-x)) - \nabla f(x), y-x\rangle dt \\
&+ \langle \nabla h(x), y-x\rangle + \int_0^1 \langle \nabla h(x+t(y-x)) - \nabla h(x), y-x\rangle dt \\
&\geq \langle \nabla f(x), y-x\rangle + \int_0^1 \langle \nabla f(x+t(y-x)) - \nabla f(x), y-x\rangle dt \\
&\quad + \langle \nabla h(x), y-x\rangle \\
&\geq \langle \nabla\Psi(x), y-x\rangle - \tfrac{L_f}{2}\|y-x\|^2 \ \ \forall x,y\in\mathbb{R}^n,
\end{aligned}$$

where the first inequality follows from the fact that $\langle \nabla h(x+t(y-x) - \nabla h(x), y-x\rangle \geq 0$ due to the convexity of h, and the last inequality follows from the fact that

$$\langle \nabla f(x+t(y-x)) - \nabla f(x), y-x\rangle \geq -\|f(x+t(y-x)) - \nabla f(x)\|\|y-x\| \geq -L_f t\|y-x\|^2.$$

■

We are now ready to describe the main convergence properties of Algorithm 2 for solving problem (6.4.3).

Theorem 6.11. *Suppose that Assumption 17 holds and that $\{\alpha_k\}$, $\{\beta_k\}$, and $\{\lambda_k\}$ in Algorithm 6.3 are chosen such that (6.4.14) and (6.4.15) hold. Also assume that an optimal solution x^* exists for problem (6.4.3). Then for any $N\geq 1$, we have*

$$\begin{aligned}
\min_{k=1,\ldots,N} \|P_X(\underline{x}_k, \nabla\Psi(\underline{x}_k), \beta_k)\|^2 \leq 2\left[\textstyle\sum_{k=1}^N \Gamma_k^{-1}\beta_k(1-L_\Psi\beta_k)\right]^{-1} \\
\left[\tfrac{\|x_0-x^*\|^2}{2\lambda_1} + \tfrac{L_f}{\Gamma_N}(\|x^*\|^2+M^2)\right],
\end{aligned} \tag{6.4.50}$$

where $P_X(\cdot,\cdot,\cdot)$ is defined in (6.4.43). If, in addition, $L_f=0$, then we have

$$\Phi(\bar{x}_N) - \Phi(x^*) \leq \tfrac{\Gamma_N\|x_0-x^*\|^2}{2\lambda_1}, \tag{6.4.51}$$

where $\Phi(x)\equiv\Psi(x)+\chi(x)$.

Proof. By the assumption that Ψ is smooth, we have

$$\Psi(\bar{x}_k) \leq \Psi(\underline{x}_k) + \langle \nabla\Psi(\underline{x}_k), \bar{x}_k - \underline{x}_k\rangle + \tfrac{L_\Psi}{2}\|\bar{x}_k - \underline{x}_k\|^2. \tag{6.4.52}$$

Also by Lemma 6.9, we have

$$\Psi(\underline{x}_k) - [(1-\alpha_k)\Psi(\bar{x}_{k-1}) + \alpha_k\Psi(x)]$$

$$
\begin{aligned}
&\le \alpha_k \left[\langle \nabla \Psi(\underline{x}_k), \underline{x}_k - x\rangle + \tfrac{L_f}{2}\|\underline{x}_k - x\|^2\right] \\
&\quad + (1-\alpha_k)\left[\langle \nabla \Psi(\underline{x}_k), \underline{x}_k - \bar{x}_{k-1}\rangle + \tfrac{L_f}{2}\|\underline{x}_k - \bar{x}_{k-1}\|^2\right] \\
&= \langle \nabla \Psi(\underline{x}_k), \underline{x}_k - \alpha_k x - (1-\alpha_k)\bar{x}_{k-1}\rangle \\
&\quad + \tfrac{L_f \alpha_k}{2}\|\underline{x}_k - x\|^2 + \tfrac{L_f(1-\alpha_k)}{2}\|\underline{x}_k - \bar{x}_{k-1}\|^2 \\
&\le \langle \nabla \Psi(\underline{x}_k), \underline{x}_k - \alpha_k x - (1-\alpha_k)\bar{x}_{k-1}\rangle \\
&\quad + \tfrac{L_f \alpha_k}{2}\|\underline{x}_k - x\|^2 + \tfrac{L_f\alpha_k^2(1-\alpha_k)}{2}\|\bar{x}_{k-1} - x_{k-1}\|^2, \qquad (6.4.53)
\end{aligned}
$$

where the last inequality follows from the fact that $\underline{x}_k - \bar{x}_{k-1} = \alpha_k(\bar{x}_{k-1} - x_{k-1})$ due to (6.4.8). Now, using the optimality condition of subproblem (6.4.46) and letting $p \in \partial \chi(x_k)$, we have, for any $x \in \mathbb{R}^n$,

$$
\begin{aligned}
\tfrac{1}{2\lambda_k}\left[\|\underline{x}_k - x\|^2 - \|\bar{x}_k - x\|^2 - \|\bar{x}_k - \underline{x}_k\|^2\right] &= \tfrac{1}{\lambda_k}\langle x - x_k, x_k - x_{k-1}\rangle \\
&\ge \langle \nabla \Psi(\underline{x}_k) + p, x_k - x\rangle,
\end{aligned}
$$

which together with the convexity of $\chi(\cdot)$ then imply that

$$
\langle \nabla \Psi(\underline{x}_k), x_k - x\rangle + \chi(x_k) \le \chi(x) + \tfrac{1}{2\lambda_k}\left[\|x_{k-1} - x\|^2 - \|x_k - x\|^2 - \|x_k - x_{k-1}\|^2\right] \qquad (6.4.54)
$$

for any $x \in \mathbb{R}^n$. Similarly, we obtain

$$
\langle \nabla \Psi(\underline{x}_k), \bar{x}_k - x\rangle + \chi(\bar{x}_k) \le \chi(x) + \tfrac{1}{2\beta_k}\left[\|\underline{x}_k - x\|^2 - \|\bar{x}_k - x\|^2 - \|\bar{x}_k - \underline{x}_k\|^2\right]. \qquad (6.4.55)
$$

Letting $x = \alpha_k x_k + (1-\alpha_k)\bar{x}_{k-1}$ in (6.4.55), we have

$$
\begin{aligned}
&\langle \nabla \Psi(\underline{x}_k), \bar{x}_k - \alpha_k x_k - (1-\alpha_k)\bar{x}_{k-1}\rangle + \chi(\bar{x}_k) \\
&\quad \le \chi(\alpha_k x_k + (1-\alpha_k)\bar{x}_{k-1}) + \tfrac{1}{2\beta_k}\left[\|\underline{x}_k - \alpha_k x_k - (1-\alpha_k)\bar{x}_{k-1}\|^2 - \|\bar{x}_k - \underline{x}_k\|^2\right] \\
&\quad \le \alpha_k \chi(x_k) + (1-\alpha_k)\chi(\bar{x}_{k-1}) + \tfrac{1}{2\beta_k}\left[\alpha_k^2\|x_k - x_{k-1}\|^2 - \|\bar{x}_k - \underline{x}_k\|^2\right],
\end{aligned}
$$

where the last inequality follows from the convexity of χ and (6.4.8). Summing up the above inequality with (6.4.54) (with both sides multiplied by α_k), we obtain

$$
\begin{aligned}
\langle \nabla \Psi(\underline{x}_k), \bar{x}_k - \alpha_k x - (1-\alpha_k)\bar{x}_{k-1}\rangle + \chi(\bar{x}_k) &\le (1-\alpha_k)\chi(\bar{x}_{k-1}) \\
&\quad + \alpha_k \chi(x) + \tfrac{\alpha_k}{2\lambda_k}\left[\|x_{k-1} - x\|^2 - \|x_k - x\|^2\right] \\
+ \tfrac{\alpha_k(\lambda_k\alpha_k - \beta_k)}{2\beta_k\lambda_k}\|x_k - x_{k-1}\|^2 - \tfrac{1}{2\beta_k}\|\bar{x}_k - \underline{x}_k\|^2 &\le (1-\alpha_k)\chi(\bar{x}_{k-1}) \\
+ \alpha_k \chi(x) + \tfrac{\alpha_k}{2\lambda_k}\left[\|x_{k-1} - x\|^2 - \|x_k - x\|^2\right] &- \tfrac{1}{2\beta_k}\|\bar{x}_k - \underline{x}_k\|^2, \qquad (6.4.56)
\end{aligned}
$$

where the last inequality follows from the assumption that $\alpha_k\lambda_k \le \beta_k$. Combining (6.4.52), (6.4.53), and (6.4.56), and using the definition $\Phi(x) \equiv \Psi(x) + \chi(x)$, we have

$$
\Phi(\bar{x}_k) \le (1-\alpha_k)\Phi(\bar{x}_{k-1}) + \alpha_k\Phi(x) - \tfrac{1}{2}\left(\tfrac{1}{\beta_k} - L_\Psi\right)\|\bar{x}_k - \underline{x}_k\|^2
$$

$$+\tfrac{\alpha_k}{2\lambda_k}\left[\|x_{k-1}-x\|^2-\|x_k-x\|^2\right]+\tfrac{L_f\alpha_k}{2}\|\underline{x}_k -x\|^2+\tfrac{L_f\alpha_k^2(1-\alpha_k)}{2}\|\bar{x}_{k-1}-x_{k-1}\|^2. \quad (6.4.57)$$

Subtracting $\Phi(x)$ from both sides of the above inequality, rearranging the terms, and using Lemma 3.17 and relation (6.4.30), we obtain

$$\tfrac{\Phi(\bar{x}_N)-\Phi(x)}{\Gamma_N}+\textstyle\sum_{k=1}^N\tfrac{1-L_\Psi\beta_k}{2\beta_k\Gamma_k}\|\bar{x}_k-\underline{x}_k\|^2 \le \tfrac{\|x_0-x\|^2}{2\lambda_1}+\tfrac{L_f}{2}\textstyle\sum_{k=1}^N\tfrac{\alpha_k}{\Gamma_k}\left[\|\underline{x}_k-x\|^2+\alpha_k(1-\alpha_k)\|\bar{x}_{k-1}-x_{k-1}\|^2\right].$$

Now letting $x=x^*$ in the above inequality, and observing that by Assumption 17 and (6.4.8),

$$\begin{aligned}
&\|\underline{x}_k-x^*\|^2+\alpha_k(1-\alpha_k)\|\bar{x}_{k-1}-x_{k-1}\|^2\\
&\le 2[\|x^*\|^2+\|\underline{x}_k\|^2+\alpha_k(1-\alpha_k)\|\bar{x}_{k-1}-x_{k-1}\|^2]\\
&=2[\|x^*\|^2+(1-\alpha_k)^2\|\bar{x}_{k-1}\|^2+\alpha_k^2\|x_{k-1}\|^2+\alpha_k(1-\alpha_k)(\|\bar{x}_{k-1}\|^2+\|x_{k-1}\|^2)]\\
&=2[\|x^*\|^2+(1-\alpha_k)\|\bar{x}_{k-1}\|^2+\alpha_k\|x_{k-1}\|^2]\le 2(\|x^*\|^2+M^2), \qquad (6.4.58)
\end{aligned}$$

we obtain

$$\begin{aligned}
\tfrac{\Phi(\bar{x}_N)-\Phi(x^*)}{\Gamma_N}+\textstyle\sum_{k=1}^N\tfrac{1-L_\Psi\beta_k}{2\beta_k\Gamma_k}\|\bar{x}_k-\underline{x}_k\|^2 &\le \tfrac{\|x_0-x\|^2}{2\lambda_1}+L_f\textstyle\sum_{k=1}^N\tfrac{\alpha_k}{\Gamma_k}(\|x^*\|^2+M^2)\\
&=\tfrac{\|x_0-x\|^2}{2\lambda_1}+\tfrac{L_f}{\Gamma_N}(\|x^*\|^2+M^2), \qquad (6.4.59)
\end{aligned}$$

where the last inequality follows from (6.4.21). The above relation, in view of (6.4.14) and the assumption $L_f=0$, then clearly implies (6.4.51). Moreover, it follows from the above relation, (6.4.48), and the fact $\Phi(\bar{x}_N)-\Phi(x^*)\ge 0$ that

$$\begin{aligned}
\textstyle\sum_{k=1}^N\tfrac{\beta_k(1-L_\Psi\beta_k)}{2\Gamma_k}\|P_X(\underline{x}_k,\nabla\Psi(\underline{x}_k),\beta_k)\|^2 &= \textstyle\sum_{k=1}^N\tfrac{1-L_\Psi\beta_k}{2\beta_k\Gamma_k}\|\bar{x}_k-\underline{x}_k\|^2\\
&\le \tfrac{\|x_0-x^*\|^2}{2\lambda_1}+\tfrac{L_f}{\Gamma_N}(\|x^*\|^2+M^2),
\end{aligned}$$

which, in view of (6.4.14), then clearly implies (6.4.50). ■

As shown in Theorem 6.11, we can have a uniform treatment for both convex and nonconvex composite problems. More specifically, we allow the same stepsize policies in Theorem 6.10(b) to be used for both convex and nonconvex composite optimizations. In the next result, we specialize the results obtained in Theorem 6.11 for a particular selection of $\{\alpha_k\}$, $\{\beta_k\}$, and $\{\lambda_k\}$.

Corollary 6.15. *Suppose that Assumption 17 holds and that* $\{\alpha_k\}$, $\{\beta_k\}$, *and* $\{\lambda_k\}$ *in Algorithm 6.3 are set to (6.4.32) and (6.4.35). Also assume that an optimal solution* x^* *exists for problem (6.4.3). Then for any* $N\ge 1$, *we have*

$$\min_{k=1,\ldots,N}\|P_X(\underline{x}_k,\nabla\Psi(\underline{x}_k),\beta_k)\|^2 \le 24L_\Psi\left[\tfrac{4L_\Psi\|x_0-x^*\|^2}{N(N+1)(N+2)}+\tfrac{L_f}{N}(\|x^*\|^2+M^2)\right]. \qquad (6.4.60)$$

If, in addition, $L_f = 0$, then we have

$$\Phi(\bar{x}_N) - \Phi(x^*) \le \tfrac{4L_\Psi \|x_0 - x^*\|^2}{N(N+1)}. \tag{6.4.61}$$

Proof. The results directly follow by plugging the value of Γ_k in (6.4.38), the value of λ_1 in (6.4.35), and the bound (6.4.41) into (6.4.50) and (6.4.51), respectively. ■

Clearly, it follows from (6.4.60) that after running the AGD method for at most $N = \mathcal{O}(L_\Psi^{\frac{2}{3}}/\varepsilon^{\frac{1}{3}} + L_\Psi L_f/\varepsilon)$ iterations, we have $-\nabla\Psi(\bar{x}_N) \in \partial\chi(\bar{x}_N) + B(\sqrt{\varepsilon})$. Using the fact that $L_\Psi = L_f + L_h$, we can easily see that if either the smooth convex term $h(\cdot)$ or the nonconvex term $f(\cdot)$ becomes zero, then the previous complexity bound reduces to $\mathcal{O}(L_f^2/\varepsilon)$ or $\mathcal{O}(L_h^2/\varepsilon^{\frac{1}{3}})$, respectively.

It is interesting to compare the rate of convergence obtained in (6.4.60) with the one obtained for the nonconvex mirror descent method applied to problem (6.4.3). More specifically, let $\{p_k\}$ and $\{\nu_k\}$, respectively, denote the iterates and stepsizes in the nonconvex mirror descent method. Also assume that the component $\chi(\cdot)$ in (6.4.3) is Lipschitz continuous with Lipschitz constant L_χ. Then, by (6.2.26), we have

$$\begin{aligned}
\min_{k=1,\ldots,N} \|P_X(p_k, \nabla\Psi(p_k), \nu_k)\|^2 &\le \tfrac{L_\Psi[\Phi(p_0) - \Phi(x^*)]}{N} \\
&\le \tfrac{L_\Psi}{N}\left(\|\nabla\Psi(x^*)\| + L_\chi\right)\left(\|x^*\| + \|p_0\|\right) + \tfrac{L_\Psi^2}{N}\left(\|x^*\|^2 + \|p_0\|^2\right),
\end{aligned} \tag{6.4.62}$$

where the last inequality follows from

$$\begin{aligned}
\Phi(p_0) - \Phi(x^*) &= \Psi(p_0) - \Psi(x^*) + \chi(p_0) - \chi(x^*) \\
&\le \langle \nabla\Psi(x^*), p_0 - x^*\rangle + \tfrac{L_\Psi}{2}\|p_0 - x^*\|^2 + L_\chi\|p_0 - x^*\| \\
&\le \left(\|\nabla\Psi(x^*)\| + L_\chi\right)\|p_0 - x^*\| + \tfrac{L_\Psi}{2}\|p_0 - x^*\|^2 \\
&\le \left(\|\nabla\Psi(x^*)\| + L_\chi\right)\left(\|x^*\| + \|p_0\|\right) + L_\Psi\left(\|x^*\|^2 + \|p_0\|^2\right).
\end{aligned}$$

Comparing (6.4.60) with (6.4.62), we can make the following observations. First, the bound in (6.4.60) does not depend on L_χ while the one in (6.4.62) may depend on L_χ. Second, if the second terms in both (6.4.60) and (6.4.62) are the dominating ones, then the rate of convergence of the AGD method is bounded by $\mathcal{O}(L_\Psi L_f/N)$, which is better than the $\mathcal{O}(L_\Psi^2/N)$ rate of convergence possessed by the projected gradient method, in terms of their dependence on the Lipschitz constant L_h. Third, consider the case when $L_f = \mathcal{O}(L_h/N^2)$. By (6.4.60), we have

$$\min_{k=1,\ldots,N} \|P_X(\underline{x}_k, \nabla\Psi(\underline{x}_k), \beta_k)\|^2 \le \frac{96L_\Psi^2\|x_0 - x^*\|^2}{N^3}\left(1 + \frac{L_f N^2(\|x^*\|^2 + M^2))}{4(L_f + L_h)\|x_0 - x^*\|^2}\right),$$

which implies that the rate of convergence of the AGD method is bounded by

$$\mathcal{O}\left(\frac{L_h^2}{N^3}\left[\|x_0 - x^*\|^2 + \|x^*\|^2 + M^2\right]\right).$$

The previous bound is significantly better than the $\mathcal{O}(L_h^2/N)$ rate of convergence possessed by the mirror descent method for this particular case. Finally, it should be noted, however, that the nonconvex mirror descent method in Sect. 6.2.2 can be used to solve more general problems as it does not require the domain of χ to be bounded. Instead, it only requires the objective function $\Phi(x)$ to be bounded from below.

6.4.2 Stochastic Accelerated Gradient Descent Method

Our goal in this section is to present a stochastic counterpart of the AGD algorithm for solving stochastic optimization problems. More specifically, we discuss the convergence of this algorithm for solving general smooth (possibly nonconvex) SP problems in Sect. 6.4.2.1, and for a special class of composite SP problems in Sect. 6.4.2.2.

6.4.2.1 Minimization of Stochastic Smooth Functions

In this subsection, we consider problem (6.4.1), where Ψ is differentiable, bounded from below, and its gradients are Lipschitz continuous with constant L_Ψ. Moreover, we assume that the first-order information of $\Psi(\cdot)$ is obtained by the SFO, which satisfies Assumption 16. It should also be mentioned that in the standard setting for SP, the random vectors ξ_k, $k = 1,2,\ldots$, are independent of each other. However, our assumption here is slightly weaker, since we do not need to require ξ_k, $k = 1,2,\ldots$, to be independent.

While Nesterov's original accelerated gradient descent method has been generalized in Sect. 4.2 to achieve the optimal rate of convergence for solving both smooth and nonsmooth convex SP problem, it is unclear whether it converges for nonconvex SP problems. On the other hand, although the randomized stochastic gradient (RSGD) method converges for nonconvex SP problems, it cannot achieve the optimal rate of convergence when applied to convex SP problems. Below, we present a new SGD-type algorithm, namely, the randomized stochastic AGD (RSAGD) method which not only converges for nonconvex SP problems, but also achieves an optimal rate of convergence when applied to convex SP problems by properly specifying the stepsize policies.

The RSAGD method is obtained by replacing the exact gradients in Algorithm 6.2 with the stochastic ones and incorporating a randomized termination criterion for nonconvex SP as in the RSGD method. This algorithm is formally described as follows.

Algorithm 6.4 The randomized stochastic AGD (RSAGD) algorithm

Input: $x_0 \in \mathbb{R}^n$, $\{\alpha_k\}$ s.t. $\alpha_1 = 1$ and $\alpha_k \in (0,1)$ for any $k \geq 2$, $\{\beta_k > 0\}$ and $\{\lambda_k > 0\}$, iteration limit $N \geq 1$, and probability mass function $P_R(\cdot)$ s.t.

$$\text{Prob}\{R = k\} = p_k, \quad k = 1,\ldots,N. \tag{6.4.63}$$

0. Set $\bar{x}_0 = x_0$ and $k = 1$. Let R be a random variable with probability mass function P_R.
1. Set $\underline{x}_k$ to (6.4.8).
2. Call the SFO for computing $G(\underline{x}_k, \xi_k)$ and set

$$x_k = x_{k-1} - \lambda_k G(\underline{x}_k, \xi_k), \tag{6.4.64}$$
$$\bar{x}_k = \underline{x}_k - \beta_k G(\underline{x}_k, \xi_k). \tag{6.4.65}$$

3. If $k = R$, **terminate** the algorithm. Otherwise, set $k = k+1$ and go to step 1.

We now add a few remarks about the above RSAGD algorithm. First, similar to our discussion in the previous section, if $\alpha_k = 1, \ \ \beta_k = \lambda_k \ \ \forall k \geq 1$, then the above algorithm reduces to the classical SGD algorithm. Moreover, if $\beta_k = \lambda_k \ \ \forall k \geq 1$, the above algorithm reduces to the stochastic accelerated gradient descent method in Sect. 4.2. Second, we have used a random number R to terminate the above RSAGD method for solving general (not necessarily convex) NLP problems. Equivalently, one can run the RSAGD method for N iterations and then randomly select the search points $(\underline{x}_R, \bar{x}_R)$ as the output of Algorithm 6.4 from the trajectory $(\underline{x}_k, \bar{x}_k)$, $k = 1,\ldots,N$. Note, however, that the remaining $N - R$ iterations will be surplus.

We are now ready to describe the main convergence properties of the RSAGD algorithm applied to problem (6.4.1) under the stochastic setting.

Theorem 6.12. *Let $\{\underline{x}_k, \bar{x}_k\}_{k\geq 1}$ be computed by Algorithm 6.4 and Γ_k be defined in (6.4.11). Also suppose that Assumption 16 holds.*

(a) If $\{\alpha_k\}$, $\{\beta_k\}$, $\{\lambda_k\}$, and $\{p_k\}$ are chosen such that (6.4.12) holds and

$$p_k = \frac{\lambda_k C_k}{\sum_{\tau=1}^N \lambda_\tau C_\tau}, \quad k = 1,\ldots,N, \tag{6.4.66}$$

where C_k is defined in (6.4.12), then for any $N \geq 1$, we have

$$\mathbb{E}[\|\nabla\Psi(\underline{x}_R)\|^2] \leq \frac{1}{\sum_{k=1}^N \lambda_k C_k}\left[\Psi(x_0) - \Psi^* + \frac{L_\Psi \sigma^2}{2}\sum_{k=1}^N \lambda_k^2\left(1 + \frac{(\lambda_k - \beta_k)^2}{\alpha_k \Gamma_k \lambda_k^2}\sum_{\tau=k}^N \Gamma_\tau\right)\right], \tag{6.4.67}$$

where the expectation is taken with respect to R and $\xi_{[N]} := (\xi_1,\ldots,\xi_N)$.

(b) Suppose that $\Psi(\cdot)$ is convex and that an optimal solution x^* exists for problem (6.4.1). If $\{\alpha_k\}$, $\{\beta_k\}$, $\{\lambda_k\}$, and $\{p_k\}$ are chosen such that (6.4.15) holds,

$$\alpha_k \lambda_k \leq L_\Psi \beta_k^2, \ \ \beta_k < 1/L_\Psi, \tag{6.4.68}$$

and

$$p_k = \frac{\Gamma_k^{-1}\beta_k(1 - L_\Psi \beta_k)}{\sum_{\tau=1}^N \Gamma_\tau^{-1}\beta_\tau(1 - L_\Psi \beta_\tau)} \tag{6.4.69}$$

for all $k = 1,...,N$, then for any $N \geq 1$, we have

$$\mathbb{E}[\|\nabla\Psi(\underline{x}_R)\|^2] \leq \frac{(2\lambda_1)^{-1}\|x_0-x^*\|^2+L_\Psi\sigma^2\sum_{k=1}^N\Gamma_k^{-1}\beta_k^2}{\sum_{k=1}^N\Gamma_k^{-1}\beta_k(1-L_\Psi\beta_k)}, \tag{6.4.70}$$

$$\mathbb{E}[\Psi(\bar{x}_R)-\Psi(x^*)] \leq \frac{\sum_{k=1}^N\beta_k(1-L_\Psi\beta_k)\left[(2\lambda_1)^{-1}\|x_0-x^*\|^2+L_\Psi\sigma^2\sum_{j=1}^k\Gamma_j^{-1}\beta_j^2\right]}{\sum_{k=1}^N\Gamma_k^{-1}\beta_k(1-L_\Psi\beta_k)}. \tag{6.4.71}$$

Proof. We first show part (a). Denote $\delta_k := G(\underline{x}_k,\xi_k)-\nabla\Psi(\underline{x}_k)$ and $\Delta_k := \nabla\Psi(x_{k-1})-\nabla\Psi(\underline{x}_k)$. By (6.4.7) and (6.4.64), we have

$$\begin{aligned}\Psi(x_k) &\leq \Psi(x_{k-1})+\langle\nabla\Psi(x_{k-1}),x_k-x_{k-1}\rangle+\tfrac{L_\Psi}{2}\|x_k-x_{k-1}\|^2\\ &= \Psi(x_{k-1})+\langle\Delta_k+\nabla\Psi(\underline{x}_k),-\lambda_k[\nabla\Psi(\underline{x}_k)+\delta_k]\rangle+\tfrac{L_\Psi\lambda_k^2}{2}\|\nabla\Psi(\underline{x}_k)+\delta_k\|^2\\ &= \Psi(x_{k-1})+\langle\Delta_k+\nabla\Psi(\underline{x}_k),-\lambda_k\nabla\Psi(\underline{x}_k)\rangle-\lambda_k\langle\nabla\Psi(x_{k-1}),\delta_k\rangle\\ &\quad+\tfrac{L_\Psi\lambda_k^2}{2}\|\nabla\Psi(\underline{x}_k)+\delta_k\|^2\\ &\leq \Psi(x_{k-1})-\lambda_k\left(1-\tfrac{L_\Psi\lambda_k}{2}\right)\|\nabla\Psi(\underline{x}_k)\|^2+\lambda_k\|\Delta_k\|\,\|\nabla\Psi(\underline{x}_k)\|+\tfrac{L_\Psi\lambda_k^2}{2}\|\delta_k\|^2\\ &\quad-\lambda_k\langle\nabla\Psi(x_{k-1})-L_\Psi\lambda_k\nabla\Psi(\underline{x}_k),\delta_k\rangle,\end{aligned}$$

which, in view of (6.4.18) and the fact that $ab \leq (a^2+b^2)/2$, then implies that

$$\begin{aligned}\Psi(x_k) &\leq \Psi(x_{k-1})-\lambda_k\left(1-\tfrac{L_\Psi\lambda_k}{2}\right)\|\nabla\Psi(\underline{x}_k)\|^2+\lambda_kL_\Psi(1-\alpha_k)\|\bar{x}_{k-1}\\ &\qquad-x_{k-1}\|\,\|\nabla\Psi(\underline{x}_k)\|\\ &\quad+\tfrac{L_\Psi\lambda_k^2}{2}\|\delta_k\|^2-\lambda_k\langle\nabla\Psi(x_{k-1})-L_\Psi\lambda_k\nabla\Psi(\underline{x}_k),\delta_k\rangle\\ &\leq \Psi(x_{k-1})-\lambda_k(1-L_\Psi\lambda_k)\|\nabla\Psi(\underline{x}_k)\|^2+\tfrac{L_\Psi(1-\alpha_k)^2}{2}\|\bar{x}_{k-1}-x_{k-1}\|^2\\ &\qquad+\tfrac{L_\Psi\lambda_k^2}{2}\|\delta_k\|^2\\ &\quad-\lambda_k\langle\nabla\Psi(x_{k-1})-L_\Psi\lambda_k\nabla\Psi(\underline{x}_k),\delta_k\rangle.\end{aligned}$$

Noting that similar to (6.4.22), we have

$$\begin{aligned}\|\bar{x}_{k-1}-x_{k-1}\|^2 &\leq \Gamma_{k-1}\textstyle\sum_{\tau=1}^{k-1}\frac{(\lambda_\tau-\beta_\tau)^2}{\Gamma_\tau\alpha_\tau}\|\nabla\Psi(\underline{x}_\tau)+\delta_k\|^2\\ &= \Gamma_{k-1}\textstyle\sum_{\tau=1}^{k-1}\frac{(\lambda_\tau-\beta_\tau)^2}{\Gamma_\tau\alpha_\tau}\left[\|\nabla\Psi(\underline{x}_\tau)\|^2+\|\delta_\tau\|^2+2\langle\nabla\Psi(\underline{x}_\tau),\delta_\tau\rangle\right].\end{aligned}$$

Combining the previous two inequalities and using the fact that $\Gamma_{k-1}(1-\alpha_k)^2 \leq \Gamma_k$, we obtain

$$\begin{aligned}\Psi(x_k) &\leq \Psi(x_{k-1})-\lambda_k(1-L_\Psi\lambda_k)\|\nabla\Psi(\underline{x}_k)\|^2+\tfrac{L_\Psi\lambda_k^2}{2}\|\delta_k\|^2-\lambda_k\langle\nabla\Psi(x_{k-1})\\ &\qquad-L_\Psi\lambda_k\nabla\Psi(\underline{x}_k),\delta_k\rangle\\ &\quad+\tfrac{L_\Psi\Gamma_k}{2}\textstyle\sum_{\tau=1}^k\frac{(\lambda_\tau-\beta_\tau)^2}{\Gamma_\tau\alpha_\tau}\left[\|\nabla\Psi(\underline{x}_\tau)\|^2+\|\delta_\tau\|^2+2\langle\nabla\Psi(\underline{x}_\tau),\delta_\tau\rangle\right].\end{aligned}$$

Summing up the above inequalities, we obtain

$$\begin{aligned}
\Psi(x_N) &\le \Psi(x_0) - \textstyle\sum_{k=1}^N \lambda_k(1 - L_\Psi \lambda_k)\|\nabla\Psi(\underline{x}_k)\|^2 - \sum_{k=1}^N \lambda_k \langle \nabla\Psi(x_{k-1}) \\
&\quad - L_\Psi \lambda_k \nabla\Psi(\underline{x}_k), \delta_k\rangle \\
&\quad + \textstyle\sum_{k=1}^N \frac{L_\Psi \lambda_k^2}{2}\|\delta_k\|^2 + \frac{L_\Psi}{2}\sum_{k=1}^N \Gamma_k \sum_{\tau=1}^k \frac{(\lambda_\tau - \beta_\tau)^2}{\Gamma_\tau \alpha_\tau}\left[\|\nabla\Psi(\underline{x}_\tau)\|^2 + \|\delta_\tau\|^2\right. \\
&\quad \left. +2\langle \nabla\Psi(\underline{x}_\tau), \delta_\tau\rangle\right] \\
&= \Psi(x_0) - \textstyle\sum_{k=1}^N \lambda_k C_k \|\nabla\Psi(\underline{x}_k)\|^2 + \frac{L_\Psi}{2}\sum_{k=1}^N \lambda_k^2 \\
&\quad \left(1 + \frac{(\lambda_k - \beta_k)^2}{\alpha_k \Gamma_k \lambda_k^2}\textstyle\sum_{\tau=k}^N \Gamma_\tau\right)\|\delta_k\|^2 - \textstyle\sum_{k=1}^N b_k,
\end{aligned}$$

where $b_k = \langle v_k, \delta_k\rangle$ and $v_k = \lambda_k \nabla\Psi(x_{k-1}) - \left[L_\Psi \lambda_k^2 + \frac{L_\Psi(\lambda_k - \beta_k)^2}{\Gamma_k \alpha_k}\left(\sum_{\tau=k}^N \Gamma_\tau\right)\right]$ $\nabla\Psi(\underline{x}_k)$. Taking expectation w.r.t. $\xi_{[N]}$ on both sides of the above inequality and noting that under Assumption 16, $\mathbb{E}[\|\delta_k\|^2] \le \sigma^2$ and $\{b_k\}$ is a martingale difference since v_k only depends on $\xi_{[k-1]}$ and hence $\mathbb{E}[b_k|\xi_{[N]}] = \mathbb{E}[b_k|\xi_{[k-1]}] = \mathbb{E}[\langle v_k, \delta_k\rangle|\xi_{[k-1]}] = \langle v_k, \mathbb{E}[\delta_k|\xi_{[k-1]}]\rangle = 0$, we have

$$\begin{aligned}
\textstyle\sum_{k=1}^N \lambda_k C_k \mathbb{E}_{\xi_{[N]}}[\|\nabla\Psi(\underline{x}_k)\|^2] &\le \Psi(x_0) - \mathbb{E}_{\xi_{[N]}}[\Psi(x_N)] + \frac{L_\Psi \sigma^2}{2}\textstyle\sum_{k=1}^N \lambda_k^2 \\
&\quad \left(1 + \frac{(\lambda_k - \beta_k)^2}{\alpha_k \Gamma_k \lambda_k^2}\textstyle\sum_{\tau=k}^N \Gamma_\tau\right).
\end{aligned}$$

Dividing both sides of the above relation by $\sum_{k=1}^N \lambda_k C_k$, and using the facts that $\Psi(x_N) \ge \Psi^*$ and

$$\mathbb{E}[\|\nabla\Psi(\underline{x}_R)\|^2] = \mathbb{E}_{R,\xi_{[N]}}[\|\nabla\Psi(\underline{x}_R)\|^2] = \frac{\sum_{k=1}^N \lambda_k C_k \mathbb{E}_{\xi_{[N]}}[\|\nabla\Psi(\underline{x}_k)\|^2]}{\sum_{k=1}^N \lambda_k C_k},$$

we obtain (6.4.67).

We now show part (b). By (6.4.7), (6.4.65), and (6.4.26), we have

$$\begin{aligned}
\Psi(\bar{x}_k) &\le \Psi(\underline{x}_k) + \langle \nabla\Psi(\underline{x}_k), \bar{x}_k - \underline{x}_k\rangle + \frac{L_\Psi}{2}\|\bar{x}_k - \underline{x}_k\|^2 \\
&= \Psi(\underline{x}_k) - \beta_k\|\nabla\Psi(\underline{x}_k)\|^2 + \beta_k\langle \nabla\Psi(\underline{x}_k), \delta_k\rangle + \frac{L_\Psi \beta_k^2}{2}\|\nabla\Psi(\underline{x}_k) + \delta_k\|^2 \\
&\le (1 - \alpha_k)\Psi(\bar{x}_{k-1}) + \alpha_k \Psi(x) + \alpha_k\langle \nabla\Psi(\underline{x}_k), x_{k-1} - x\rangle \\
&\quad -\beta_k\|\nabla\Psi(\underline{x}_k)\|^2 + \beta_k\langle \nabla\Psi(\underline{x}_k), \delta_k\rangle + \frac{L_\Psi \beta_k^2}{2}\|\nabla\Psi(\underline{x}_k) + \delta_k\|^2. \qquad (6.4.72)
\end{aligned}$$

Similar to (6.4.27), we have

$$\alpha_k\langle \nabla\Psi(\underline{x}_k) + \delta_k, x_{k-1} - x\rangle = \frac{\alpha_k}{2\lambda_k}\left[\|x_{k-1} - x\|^2 - \|x_k - x\|^2\right] + \frac{\alpha_k \lambda_k}{2}\|\nabla\Psi(\underline{x}_k) + \delta_k\|^2.$$

Combining the above two inequalities and using the fact that

$$\|\nabla\Psi(\underline{x}_k) + \delta_k\|^2 = \|\nabla\Psi(\underline{x}_k)\|^2 + \|\delta_k\|^2 + 2\langle \nabla\Psi(\underline{x}_k), \delta_k\rangle,$$

we obtain

$$\Psi(\bar{x}_k) \le (1-\alpha_k)\Psi(\bar{x}_{k-1}) + \alpha_k \Psi(x) + \tfrac{\alpha_k}{2\lambda_k}\left[\|x_{k-1}-x\|^2 - \|x_k - x\|^2\right]$$
$$- \beta_k\left(1 - \tfrac{L_\Psi \beta_k}{2} - \tfrac{\alpha_k\lambda_k}{2\beta_k}\right)\|\nabla\Psi(\underline{x}_k)\|^2 + \left(\tfrac{L_\Psi\beta_k^2+\alpha_k\lambda_k}{2}\right)\|\delta_k\|^2$$
$$+ \langle \delta_k, (\beta_k + L_\Psi\beta_k^2 + \alpha_k\lambda_k)\nabla\Psi(\underline{x}_k) + \alpha_k(x - x_{k-1})\rangle.$$

Subtracting $\Psi(x)$ from both sides of the above inequality, and using Lemma 3.17 and (6.4.30), we have

$$\tfrac{\Psi(\bar{x}_N)-\Psi(x)}{\Gamma_N} \le \tfrac{\|x_0-x\|^2}{2\lambda_1} - \textstyle\sum_{k=1}^N \tfrac{\beta_k}{2\Gamma_k}\left(2 - L_\Psi\beta_k - \tfrac{\alpha_k\lambda_k}{\beta_k}\right)\|\nabla\Psi(\underline{x}_k)\|^2$$
$$+ \textstyle\sum_{k=1}^N \left(\tfrac{L_\Psi\beta_k^2+\alpha_k\lambda_k}{2\Gamma_k}\right)\|\delta_k\|^2 + \textstyle\sum_{k=1}^N b_k' \ \ \forall x \in \mathbb{R}^n,$$

where $b_k' = \Gamma_k^{-1}\langle \delta_k, (\beta_k + L_\Psi\beta_k^2 + \alpha_k\lambda_k)\nabla\Psi(\underline{x}_k) + \alpha_k(x - x_{k-1})\rangle$. The above inequality together with the first relation in (6.4.68) then imply that

$$\tfrac{\Psi(\bar{x}_N)-\Psi(x)}{\Gamma_N} \le \tfrac{\|x_0-x\|^2}{2\lambda_1} - \textstyle\sum_{k=1}^N \tfrac{\beta_k}{\Gamma_k}(1 - L_\Psi\beta_k)\|\nabla\Psi(\underline{x}_k)\|^2$$
$$+ \textstyle\sum_{k=1}^N \tfrac{L_\Psi\beta_k^2}{\Gamma_k}\|\delta_k\|^2 + \textstyle\sum_{k=1}^N b_k' \ \ \forall x \in \mathbb{R}^n.$$

Taking expectation (with respect to $\xi_{[N]}$) on both sides of the above relation, and noting that under Assumption 16, $\mathbb{E}[\|\delta_k\|^2] \le \sigma^2$ and $\{b_k'\}$ is a martingale difference by the similar reasoning for $\{b_k\}$ in part (a), we obtain, $\forall x \in \mathbb{R}^n$,

$$\tfrac{1}{\Gamma_N}\mathbb{E}_{\xi_{[N]}}[\Psi(\bar{x}_N) - \Psi(x)] \le \tfrac{\|x_0-x\|^2}{2\lambda_1} - \textstyle\sum_{k=1}^N \tfrac{\beta_k}{\Gamma_k}(1 - L_\Psi\beta_k)\mathbb{E}_{\xi_{[N]}}[\|\nabla\Psi(\underline{x}_k)\|^2]$$
$$+\sigma^2 \textstyle\sum_{k=1}^N \tfrac{L_\Psi\beta_k^2}{\Gamma_k}. \tag{6.4.73}$$

Now, fixing $x = x^*$ and noting that $\Psi(\bar{x}_N) \ge \Psi(x^*)$, we have

$$\textstyle\sum_{k=1}^N \tfrac{\beta_k}{\Gamma_k}(1 - L_\Psi\beta_k)\mathbb{E}_{\xi_{[N]}}[\|\nabla\Psi(\underline{x}_k)\|^2] \le \tfrac{\|x_0-x^*\|^2}{2\lambda_1} + \sigma^2\textstyle\sum_{k=1}^N \tfrac{L_\Psi\beta_k^2}{\Gamma_k},$$

which, in view of the definition of $\underline{x}_R$, then implies (6.4.70). It also follows from (6.4.73) and (6.4.68) that, for any $N \ge 1$,

$$\mathbb{E}_{\xi_{[N]}}[\Psi(\bar{x}_N) - \Psi(x^*)] \le \Gamma_N\left(\tfrac{\|x_0-x\|^2}{2\lambda_1} + \sigma^2\textstyle\sum_{k=1}^N \tfrac{L_\Psi\beta_k^2}{\Gamma_k}\right),$$

which, in view of the definition of $\bar{x}_R$, then implies that

$$\mathbb{E}[\Psi(\bar{x}_R) - \Psi(x^*)] = \textstyle\sum_{k=1}^N \tfrac{\Gamma_k^{-1}\beta_k(1-L_\Psi\beta_k)}{\sum_{\tau=1}^N \Gamma_\tau^{-1}\beta_\tau(1-L_\Psi\beta_\tau)}\mathbb{E}_{\xi_{[N]}}[\Psi(\bar{x}_k) - \Psi(x^*)]$$

$$\leq \frac{\sum_{k=1}^{N} \beta_k (1-L_\Psi \beta_k)\left[(2\lambda_1)^{-1}\|x_0-x\|^2 + L_\Psi \sigma^2 \sum_{j=1}^{k} \Gamma_j^{-1} \beta_j^2\right]}{\sum_{k=1}^{N} \Gamma_k^{-1} \beta_k (1-L_\Psi \beta_k)}.$$

■

We now add a few remarks about the results obtained in Theorem 6.12. First, note that similar to the deterministic case, we can use the assumption in (6.4.31) instead of the one in (6.4.68). Second, the expectations in (6.4.67), (6.4.70), and (6.4.71) are taken with respect to one more random variable R in addition to ξ coming from the SFO. Specifically, the output of Algorithm 6.4 is chosen randomly from the generated trajectory $\{(\underline{x}_1, \bar{x}_1), \ldots, (\underline{x}_N, \bar{x}_N)\}$ according to (6.4.63), as mentioned earlier in this subsection. Third, the probabilities $\{p_k\}$ depend on the choice of $\{\alpha_k\}$, $\{\beta_k\}$, and $\{\lambda_k\}$.

Below, we specialize the results obtained in Theorem 6.12 for some particular selections of $\{\alpha_k\}$, $\{\beta_k\}$, and $\{\lambda_k\}$.

Corollary 6.16. *The following statements hold for Algorithm 6.4 when applied to problem (6.4.1) under Assumption 16.*

(a) If $\{\alpha_k\}$ and $\{\lambda_k\}$ in the RSAGD method are set to (6.4.32) and (6.4.33), respectively, $\{p_k\}$ is set to (6.4.66), $\{\beta_k\}$ is set to

$$\beta_k = \min\left\{\frac{8}{21L_\Psi}, \frac{\tilde{D}}{\sigma\sqrt{N}}\right\}, \quad k \geq 1 \tag{6.4.74}$$

for some $\tilde{D} > 0$, and an iteration limit $N \geq 1$ is given, then we have

$$\mathbb{E}[\|\nabla \Psi(\underline{x}_R)\|^2] \leq \frac{21L_\Psi[\Psi(x_0)-\Psi^*]}{4N} + \frac{2\sigma}{\sqrt{N}}\left(\frac{\Psi(x_0)-\Psi^*}{\tilde{D}} + L_\Psi \tilde{D}\right) =: \mathscr{U}_N. \tag{6.4.75}$$

(b) Assume that $\Psi(\cdot)$ is convex and that an optimal solution x^* exists for problem (6.4.1). If $\{\alpha_k\}$ is set to (6.4.32), $\{p_k\}$ is set to (6.4.69), $\{\beta_k\}$ and $\{\lambda_k\}$ are set to

$$\beta_k = \min\left\{\frac{1}{2L_\Psi}, \left(\frac{\tilde{D}^2}{L_\Psi^2 \sigma^2 N^3}\right)^{\frac{1}{4}}\right\} \tag{6.4.76}$$

$$\text{and} \quad \lambda_k = \frac{kL_\Psi \beta_k^2}{2}, \quad k \geq 1, \tag{6.4.77}$$

for some $\tilde{D} > 0$, and an iteration limit $N \geq 1$ is given, then we have

$$\mathbb{E}[\|\nabla \Psi(\underline{x}_R)\|^2] \leq \frac{96L_\Psi^2\|x_0-x^*\|^2}{N(N+1)(N+2)} + \frac{2L_\Psi^{\frac{1}{2}}\sigma^{\frac{3}{2}}}{N^{\frac{3}{4}}}\left(\frac{6\|x_0-x^*\|^2}{\tilde{D}^{\frac{3}{2}}} + \tilde{D}^{\frac{1}{2}}\right), \tag{6.4.78}$$

$$\mathbb{E}[\Psi(\bar{x}_R) - \Psi(x^*)] \leq \frac{48L_\Psi\|x_0-x^*\|^2}{N(N+1)} + \frac{2\sigma}{\sqrt{N}}\left(\frac{6\|x_0-x^*\|^2}{\tilde{D}} + \tilde{D}\right). \tag{6.4.79}$$

Proof. We first show part (a). It follows from (6.4.33), (6.4.40), and (6.4.74) that

$$C_k \geq 1 - \tfrac{21}{16} L_\Psi \beta_k \geq \tfrac{1}{2} > 0 \ \text{ and } \ \lambda_k C_k \geq \tfrac{\beta_k}{2}.$$

Also by (6.4.33), (6.4.38), (6.4.39), and (6.4.74), we have

$$\lambda_k^2\left[1+\tfrac{(\lambda_k-\beta_k)^2}{\alpha_k\Gamma_k\lambda_k^2}\left(\Sigma_{\tau=k}^N\Gamma_\tau\right)\right]\le\lambda_k^2\left[1+\tfrac{1}{\alpha_k\Gamma_k\lambda_k^2}\left(\tfrac{\alpha_k\beta_k}{4}\right)^2\tfrac{2}{k}\right]=\lambda_k^2+\tfrac{\beta_k^2}{8}$$
$$\le\left[\left(1+\tfrac{\alpha_k}{4}\right)^2+\tfrac{1}{8}\right]\beta_k^2\le 2\beta_k^2$$

for any $k\ge1$. These observations together with (6.4.67) then imply that

$$\begin{aligned}\mathbb{E}[\|\nabla\Psi(\underline{x}_R)\|^2]&\le\tfrac{2}{\sum_{k=1}^N\beta_k}\left(\Psi(x_0)-\Psi^*+L_\Psi\sigma^2\Sigma_{k=1}^N\beta_k^2\right)\\&\le\tfrac{2[\Psi(x_0)-\Psi^*]}{N\beta_1}+2L_\Psi\sigma^2\beta_1\\&\le\tfrac{2[\Psi(x_0)-\Psi^*]}{N}\left\{\tfrac{21L_\Psi}{8}+\tfrac{\sigma\sqrt{N}}{\tilde D}\right\}+\tfrac{2L_\Psi\tilde D\sigma}{\sqrt{N}},\end{aligned}$$

which implies (6.4.74).

We now show part (b). It can be easily checked that (6.4.15) and (6.4.68) hold in view of (6.4.76) and (6.4.77). By (6.4.38) and (6.4.76), we have

$$\Sigma_{k=1}^N\Gamma_k^{-1}\beta_k(1-L_\Psi\beta_k)\ge\tfrac{1}{2}\Sigma_{k=1}^N\Gamma_k^{-1}\beta_k=\tfrac{\beta_1}{2}\Sigma_{k=1}^N\Gamma_k^{-1},\tag{6.4.80}$$
$$\Sigma_{k=1}^N\Gamma_k^{-1}=\Sigma_{k=1}^N\tfrac{k+k^2}{2}=\tfrac{N(N+1)(N+2)}{6}.\tag{6.4.81}$$

Using these observations, (6.4.38), (6.4.70), (6.4.76), and (6.4.77), we have

$$\begin{aligned}\mathbb{E}[\|\nabla\Psi(\underline{x}_R)\|^2]&\le\tfrac{2}{\beta_1\sum_{k=1}^N\Gamma_k^{-1}}\left(\tfrac{\|x_0-x^*\|^2}{L_\Psi\beta_1^2}+L_\Psi\sigma^2\beta_1^2\Sigma_{k=1}^N\Gamma_k^{-1}\right)\\&=\tfrac{2\|x_0-x^*\|^2}{L_\Psi\beta_1^3\sum_{k=1}^N\Gamma_k^{-1}}+2L_\Psi\sigma^2\beta_1\le\tfrac{12\|x_0-x^*\|^2}{L_\Psi N(N+1)(N+2)}\beta_1^3+2L_\Psi\sigma^2\beta_1\\&\le\tfrac{96L_\Psi^2\|x_0-x^*\|^2}{N(N+1)(N+2)}+\tfrac{2L_\Psi^{\frac{1}{2}}\sigma^{\frac{3}{2}}}{N^{\frac{3}{4}}}\left(\tfrac{6\|x_0-x^*\|^2}{\tilde D^{\frac{3}{2}}}+\tilde D^{\frac{1}{2}}\right).\end{aligned}$$

Also observe that by (6.4.76), we have $1-L_\Psi\beta_k\le1$ for any $k\ge1$. Using this observation, (6.4.71), (6.4.76), (6.4.80), and (6.4.81), we obtain

$$\begin{aligned}\mathbb{E}[\Psi(\bar x_R)-\Psi(x^*)]&\le\tfrac{2}{\sum_{k=1}^N\Gamma_k^{-1}}\left[N(2\lambda_1)^{-1}\|x_0-x^*\|^2+L_\Psi\sigma^2\beta_1^2\Sigma_{k=1}^N\Sigma_{j=1}^k\Gamma_j^{-1}\right]\\&\le\tfrac{12\|x_0-x^*\|^2}{N(N+1)}L_\Psi\beta_1^2+\tfrac{12L_\Psi\sigma^2\beta_1^2}{N(N+1)(N+2)}\Sigma_{k=1}^N\tfrac{k(k+1)(k+2)}{6}\\&=\tfrac{12\|x_0-x^*\|^2}{N(N+1)L_\Psi\beta_1^2}+\tfrac{L_\Psi\sigma^2\beta_1^2(N+3)}{2}\\&\le\tfrac{48L_\Psi\|x_0-x^*\|^2}{N(N+1)}+\tfrac{2\sigma}{N^{\frac{1}{2}}}\left(\tfrac{6\|x_0-x^*\|^2}{\tilde D}+\tilde D\right),\end{aligned}$$

where the equality follows from the fact that $\sum_{k=1}^{N} k(k+1)(k+2) = N(N+1)(N+2)(N+3)/4$. ■

We now add a few remarks about the results obtained in Corollary 6.16. First, note that the stepsizes $\{\beta_k\}$ in the above corollary depend on the parameter $\tilde{D}$. While the RSAGD method converges for any $\tilde{D} > 0$, by minimizing the RHS of (6.4.75) and (6.4.79), the optimal choices of $\tilde{D}$ would be $\sqrt{[\Psi(x_0) - \Psi(x^*)]/L_\Psi}$ and $\sqrt{6}\|x_0 - x^*\|$, respectively, for solving nonconvex and convex smooth SP problems. With such selections for $\tilde{D}$, the bounds in (6.4.75), (6.4.78), and (6.4.79), respectively, reduce to

$$\mathbb{E}[\|\nabla\Psi(\underline{x}_R)\|^2] \le \frac{21L_\Psi[\Psi(x_0)-\Psi^*]}{4N} + \frac{4\sigma[L_\Psi(\Psi(x_0)-\Psi^*)]^{\frac{1}{2}}}{\sqrt{N}}, \tag{6.4.82}$$

$$\mathbb{E}[\|\nabla\Psi(\underline{x}_R)\|^2] \le \frac{96L_\Psi^2\|x_0-x^*\|^2}{N^3} + \frac{4(\sqrt{6}L_\Psi\|x_0-x^*\|)^{\frac{1}{2}}\sigma^{\frac{3}{2}}}{N^{\frac{3}{4}}}, \tag{6.4.83}$$

and

$$\mathbb{E}[\Psi(\bar{x}_R) - \Psi(x^*)] \le \frac{48L_\Psi\|x_0-x^*\|^2}{N^2} + \frac{4\sqrt{6}\|x_0-x^*\|\sigma}{\sqrt{N}}. \tag{6.4.84}$$

It should be noted, however, that such optimal choices of $\tilde{D}$ are usually not available, and that one needs to replace $\Psi(x_0) - \Psi(x^*)$ or $\|x_0 - x^*\|$ in the aforementioned optimal selections of $\tilde{D}$ with their respective upper bounds in practice. Second, the rate of convergence of the RSAGD algorithm in (6.4.75) for general nonconvex problems is the same as that of the RSGD method for smooth nonconvex SP problems (see Sect. 6.1). However, if the problem is convex, then the complexity of the RSAGD algorithm will be significantly better than the latter algorithm. More specifically, in view of (6.4.84), the RSAGD is an optimal method for smooth stochastic optimization, while the rate of convergence of the RSGD method is only nearly optimal. Moreover, in view of (6.4.78), if $\Psi(\cdot)$ is convex, then the number of iterations performed by the RSAGD algorithm to find an ε-solution of (6.4.1), i.e., a point $\bar{x}$ such that $\mathbb{E}[\|\nabla\Psi(\bar{x})\|^2] \le \varepsilon$, can be bounded by

$$\mathcal{O}\left\{\left(\frac{1}{\varepsilon^{\frac{1}{3}}} + \frac{\sigma^2}{\varepsilon^{\frac{4}{3}}}\right)(L_\Psi\|x_0 - x^*\|)^{\frac{2}{3}}\right\}.$$

In addition to the aforementioned expected complexity results of the RSAGD method, we can establish their associated large deviation properties. For example, by Markov's inequality and (6.4.75), we have

$$\text{Prob}\left\{\|\nabla\Psi(\underline{x}_R)\|^2 \ge \lambda\mathscr{U}_N\right\} \le \tfrac{1}{\lambda} \ \forall\lambda > 0, \tag{6.4.85}$$

which implies that the total number of calls to the $\mathscr{SO}$ performed by the RSAGD method for finding an (ε,Λ)-*solution* of problem (6.4.1), i.e., a point $\bar{x}$ satisfying $\text{Prob}\{\|\nabla\Psi(\bar{x})\|^2 \le \varepsilon\} \ge 1 - \Lambda$ for some $\varepsilon > 0$ and $\Lambda \in (0,1)$, after disregarding a few constant factors, can be bounded by

$$\mathcal{O}\left\{\frac{1}{\Lambda\varepsilon}+\frac{\sigma^2}{\Lambda^2\varepsilon^2}\right\}. \tag{6.4.86}$$

To improve the dependence of the above bound on the confidence level Λ, we can design a variant of the RSAGD method which has two phases: optimization and post-optimization phase. The optimization phase consists of independent runs of the RSAGD method to generate a list of candidate solutions and the post-optimization phase then selects a solution from the generated candidate solutions in the optimization phase (see Sect. 6.1 for more details).

6.4.2.2 Minimization of Nonconvex Stochastic Composite Functions

In this subsection, we consider the stochastic composite problem (6.4.3), which satisfies both Assumptions 16 and 17. Our goal is to show that under the above assumptions, we can choose the same aggressive stepsize policy in the RSAGD method no matter if the objective function $\Psi(\cdot)$ in (6.4.3) is convex or not.

We will modify the RSAGD method in Algorithm 6.4 by replacing the stochastic gradient $\nabla\Psi(\underline{x}_k,\xi_k)$ with

$$\bar{G}_k = \tfrac{1}{m_k}\Sigma_{i=1}^{m_k} G(\underline{x}_k,\xi_{k,i}) \tag{6.4.87}$$

for some $m_k \geq 1$, where $G(\underline{x}_k,\xi_{k,i}), i=1,\ldots,m_k$ are the stochastic gradients returned by the i-th call to the SFO at iteration k. The modified RSAGD algorithm is formally described as follows.

Algorithm 6.5 The RSAGD algorithm for stochastic composite optimization

Replace (6.4.64) and (6.4.65), respectively, in Step 2 of Algorithm 6.4 by

$$x_k = \text{argmin}_{u\in\mathbb{R}^n}\left\{\langle\bar{G}_k,u\rangle+\tfrac{1}{2\lambda_k}\|u-x_{k-1}\|^2+\chi(u)\right\}, \tag{6.4.88}$$

$$\bar{x}_k = \text{argmin}_{u\in\mathbb{R}^n}\left\{\langle\bar{G}_k,u\rangle+\tfrac{1}{2\beta_k}\|u-\underline{x}_k\|^2+\chi(u)\right\}, \tag{6.4.89}$$

where $\bar{G}_k$ is defined in (6.4.87) for some $m_k \geq 1$.

A few remarks about the above RSAGD algorithm are in place. First, note that by calling the SFO multiple times at each iteration, we can obtain a better estimator for $\nabla\Psi(\underline{x}_k)$ than the one obtained by using one call to the SFO as in Algorithm 6.4. More specifically, under Assumption 16, we have

$$\begin{aligned}\mathbb{E}[\bar{G}_k] &= \tfrac{1}{m_k}\Sigma_{i=1}^{m_k}\mathbb{E}[G(\underline{x}_k,\xi_{k,i})] = \nabla\Psi(\underline{x}_k),\\ \mathbb{E}[\|\bar{G}_k-\nabla\Psi(\underline{x}_k)\|^2] &= \tfrac{1}{m_k^2}\mathbb{E}\left[\|\Sigma_{i=1}^{m_k}[G(\underline{x}_k,\xi_{k,i})-\nabla\Psi(\underline{x}_k)]\|^2\right] \leq \tfrac{\sigma^2}{m_k},\end{aligned} \tag{6.4.90}$$

where the last inequality follows from (6.2.40). Thus, by increasing m_k, we can decrease the error existing in the estimation of $\nabla\Psi(\underline{x}_k)$. We will discuss the appropriate choice of m_k later in this subsection. Second, since we do not have access to $\nabla\Psi(\underline{x}_k)$, we cannot compute the exact gradient mapping, i.e., $P_X(\underline{x}_k, \nabla\Psi(\underline{x}_k), \beta_k)$ as the one used in Sect. 6.4.1.2 for composite optimization. However, by (6.4.43) and (6.4.88), we can compute an approximate stochastic gradient mapping given by $P_X(\underline{x}_k, \bar{G}_k, \beta_k)$. Indeed, by (6.4.45) and (6.4.90), we have

$$\mathbb{E}[\|P_X(\underline{x}_k, \nabla\Psi(\underline{x}_k), \beta_k) - P_X(\underline{x}_k, \bar{G}_k, \beta_k)\|^2] \leq \mathbb{E}[\|\bar{G}_k - \nabla\Psi(\underline{x}_k)\|^2] \leq \tfrac{\sigma^2}{m_k}. \quad (6.4.91)$$

Finally, it is worth mentioning that although several SGD-type algorithms have been developed for convex programming with $m_k = 1$, the mini-batch SGD method in Algorithm 6.5 (i.e., $m_k > 1$) is more attractive when computing the projection subproblems (6.4.88) and (6.4.89) is more expensive than calling the stochastic first-order oracle.

We are ready to describe the main convergence properties of Algorithm 6.5 for solving nonconvex stochastic composite problems.

Theorem 6.13. *Suppose that $\{\alpha_k\}$, $\{\beta_k\}$, $\{\lambda_k\}$, and $\{p_k\}$ in Algorithm 6.5 satisfy (6.4.14), (6.4.15), and (6.4.69). Then under Assumptions 16 and 17, we have*

$$\begin{aligned}\mathbb{E}[\|P_X(\underline{x}_R, \nabla\Psi(\underline{x}_R), \beta_R)\|^2] \leq 8\left[\textstyle\sum_{k=1}^N \Gamma_k^{-1}\beta_k(1 - L_\Psi\beta_k)\right]^{-1} \\ \Big[\tfrac{\|x_0 - x^*\|^2}{2\lambda_1} + \tfrac{L_f}{\Gamma_N}(\|x^*\|^2 + M^2) \\ + \sigma^2 \textstyle\sum_{k=1}^N \tfrac{\beta_k\left(4 + (1 - L_\Psi\beta_k)^2\right)}{4\Gamma_k(1 - L_\Psi\beta_k)m_k}\Big],\end{aligned} \quad (6.4.92)$$

where the expectation is taken with respect to R and $\xi_{k,i}$, $k = 1,..,N$, $i = 1,...,m_k$. If, in addition, $L_f = 0$, then we have

$$\begin{aligned}\mathbb{E}[\Phi(\bar{x}_R) - \Phi(x^*)] \leq \left[\textstyle\sum_{k=1}^N \Gamma_k^{-1}\beta_k(1 - L_\Psi\beta_k)\right]^{-1} \Big[\textstyle\sum_{k=1}^N \beta_k(1 - L_\Psi\beta_k) \\ \Big(\tfrac{\|x_0 - x^*\|^2}{2\lambda_1} + \sigma^2 \textstyle\sum_{j=1}^k \tfrac{\beta_j(4 + (1 - L_\Psi\beta_j)^2)}{4\Gamma_j(1 - L_\Psi\beta_j)m_j}\Big)\Big],\end{aligned} \quad (6.4.93)$$

where $\Phi(x) \equiv \Psi(x) + \chi(x)$.

Proof. Denoting $\bar{\delta}_k \equiv \bar{G}_k - \nabla\Psi(\underline{x}_k)$ and $\bar{\delta}_{[k]} \equiv \{\bar{\delta}_1, \ldots, \bar{\delta}_k\}$ for any $k \geq 1$, similar to (6.4.54) and (6.4.55), we have

$$\begin{aligned}\langle \nabla\Psi(\underline{x}_k) + \bar{\delta}_k, x_k - x\rangle + \chi(x_k) \leq \chi(x) + \tfrac{1}{2\lambda_k}\big[\|x_{k-1} - x\|^2 - \|x_k - x\|^2 \\ - \|x_k - x_{k-1}\|^2\big],\end{aligned} \quad (6.4.94)$$

$$\begin{aligned}\langle \nabla\Psi(\underline{x}_k) + \bar{\delta}_k, \bar{x}_k - x\rangle + \chi(\bar{x}_k) \leq \chi(x) + \tfrac{1}{2\beta_k}\big[\|\underline{x}_k - x\|^2 \\ - \|\bar{x}_k - x\|^2 - \|\bar{x}_k - \underline{x}_k\|^2\big]\end{aligned} \quad (6.4.95)$$

for any $x \in \mathbb{R}^n$. By using the above relations and similar arguments used to obtain (6.4.56) in the proof of Theorem 6.11, we obtain

$$\langle \nabla\Psi(\underline{x}_k) + \bar{\delta}_k, \bar{x}_k - \alpha_k x - (1 - \alpha_k)\bar{x}_{k-1}\rangle + \chi(\bar{x}_k) \leq (1 - \alpha_k)\chi(\bar{x}_{k-1}) + \alpha_k\chi(x)$$

$$+\tfrac{\alpha_k}{2\lambda_k}\left[\|x_{k-1}-x\|^2-\|x_k-x\|^2\right]+\tfrac{\alpha_k(\lambda_k\alpha_k-\beta_k)}{2\beta_k\lambda_k}\|x_k-x_{k-1}\|^2-\tfrac{1}{2\beta_k}\|\bar{x}_k-\underline{x}_k\|^2$$
$$\leq (1-\alpha_k)\chi(\bar{x}_{k-1})+\alpha_k\chi(x)+\tfrac{\alpha_k}{2\lambda_k}\left[\|x_{k-1}-x\|^2-\|x_k-x\|^2\right]-\tfrac{1}{2\beta_k}\|\bar{x}_k-\underline{x}_k\|^2, \quad (6.4.96)$$

where the last inequality follows from the assumption that $\alpha_k\lambda_k \leq \beta_k$. Combining the above relation with (6.4.52) and (6.4.53), and using the definition $\Phi(x) \equiv \Psi(x)+\chi(x)$, we have

$$\begin{aligned}
\Phi(\bar{x}_k) &\leq (1-\alpha_k)\Phi(\bar{x}_{k-1})+\alpha_k\Phi(x)-\tfrac{1}{2}\left(\tfrac{1}{\beta_k}-L_\Psi\right)\|\bar{x}_k-\underline{x}_k\|^2\\
&\quad+\langle\bar{\delta}_k,\alpha_k(x-x_{k-1})+\underline{x}_k-\bar{x}_k\rangle\\
&\quad+\tfrac{\alpha_k}{2\lambda_k}\left[\|x_{k-1}-x\|^2-\|x_k-x\|^2\right]+\tfrac{L_f\alpha_k}{2}\|\underline{x}_k-x\|^2\\
&\quad+\tfrac{L_f\alpha_k^2(1-\alpha_k)}{2}\|\bar{x}_{k-1}-x_{k-1}\|^2\\
&\leq (1-\alpha_k)\Phi(\bar{x}_{k-1})+\alpha_k\Phi(x)+\langle\bar{\delta}_k,\alpha_k(x-x_{k-1})\rangle\\
&\quad-\tfrac{1}{4}\left(\tfrac{1}{\beta_k}-L_\Psi\right)\|\bar{x}_k-\underline{x}_k\|^2+\tfrac{\beta_k\|\bar{\delta}_k\|^2}{1-L_\Psi\beta_k}\\
&\quad+\tfrac{\alpha_k}{2\lambda_k}\left[\|x_{k-1}-x\|^2-\|x_k-x\|^2\right]+\tfrac{L_f\alpha_k}{2}\|\underline{x}_k-x\|^2\\
&\quad+\tfrac{L_f\alpha_k^2(1-\alpha_k)}{2}\|\bar{x}_{k-1}-x_{k-1}\|^2,
\end{aligned}$$

where the last inequality follows from the Young's inequality. Subtracting $\Phi(x)$ from both sides of the above inequality, rearranging the terms, and using Lemma 3.17 and (6.4.30), we obtain

$$\begin{aligned}
&\tfrac{\Phi(\bar{x}_N)-\Phi(x)}{\Gamma_N}+\textstyle\sum_{k=1}^N\tfrac{1-L_\Psi\beta_k}{4\beta_k\Gamma_k}\|\bar{x}_k-\underline{x}_k\|^2 \leq \tfrac{\|x_0-x\|^2}{2\lambda_1}+\textstyle\sum_{k=1}^N\tfrac{\alpha_k}{\Gamma_k}\langle\bar{\delta}_k,x-x_{k-1}\rangle\\
&+\tfrac{L_f}{2}\textstyle\sum_{k=1}^N\tfrac{\alpha_k}{\Gamma_k}\left[\|\underline{x}_k-x\|^2+\alpha_k(1-\alpha_k)\|\bar{x}_{k-1}-x_{k-1}\|^2\right]\\
&\qquad+\textstyle\sum_{k=1}^N\tfrac{\beta_k\|\bar{\delta}_k\|^2}{\Gamma_k(1-L_\Psi\beta_k)} \quad \forall x\in\mathbb{R}^n.
\end{aligned}$$

Letting $x=x^*$ in the above inequality, and using (6.4.21) and (6.4.58), we have

$$\begin{aligned}
\tfrac{\Phi(\bar{x}_N)-\Phi(x^*)}{\Gamma_N}+\textstyle\sum_{k=1}^N\tfrac{1-L_\Psi\beta_k}{4\beta_k\Gamma_k}\|\bar{x}_k-\underline{x}_k\|^2 &\leq \tfrac{\|x_0-x^*\|^2}{2\lambda_1}+\textstyle\sum_{k=1}^N\tfrac{\alpha_k}{\Gamma_k}\langle\bar{\delta}_k,x^*-x_{k-1}\rangle\\
&\quad+\tfrac{L_f}{\Gamma_N}(\|x^*\|^2+M^2)+\textstyle\sum_{k=1}^N\tfrac{\beta_k\|\bar{\delta}_k\|^2}{\Gamma_k(1-L_\Psi\beta_k)}.
\end{aligned}$$

Taking expectation from both sides of the above inequality, noting that under Assumption 16, $\mathbb{E}[\langle\bar{\delta}_k,x^*-x_{k-1}\rangle|\bar{\delta}_{[k-1]}]=0$, and using (6.4.90) and the definition of the gradient mapping in (6.4.43), we conclude

$$\begin{aligned}
&\tfrac{\mathbb{E}_{\bar{\delta}_{[N]}}[\Phi(\bar{x}_N)-\Phi(x^*)]}{\Gamma_N}+\textstyle\sum_{k=1}^N\tfrac{\beta_k[1-L_\Psi\beta_k]}{4\Gamma_k}\mathbb{E}_{\bar{\delta}_{[N]}}[\|P_X(\underline{x}_k,\bar{G}_k,\beta_k)\|^2]\\
&\leq \tfrac{\|x_0-x^*\|^2}{2\lambda_1}+\tfrac{L_f}{\Gamma_N}(\|x^*\|^2+M^2)+\sigma^2\textstyle\sum_{k=1}^N\tfrac{\beta_k}{\Gamma_k(1-L_\Psi\beta_k)m_k},
\end{aligned}$$

which, together with the fact that $\mathbb{E}_{\bar{\delta}_{[N]}}[\|P_X(\underline{x}_k,\bar{G}_k,\beta_k)\|^2] \geq \mathbb{E}_{\bar{\delta}_{[N]}}[\|P_X(\underline{x}_k,\nabla\Psi(\underline{x}_k),\beta_k)\|^2]/2-\sigma^2/m_k$ due to (6.4.91), then imply that

$$\begin{aligned}
&\tfrac{\mathbb{E}_{\bar{\delta}_{[N]}}[\Phi(\bar{x}_N)-\Phi(x)]}{\Gamma_N}+\textstyle\sum_{k=1}^N \tfrac{\beta_k(1-L_\Psi\beta_k)}{8\Gamma_k}\mathbb{E}_{\bar{\delta}_{[N]}}[\|P_X(\underline{x}_k,\nabla\Psi(\underline{x}_k),\beta_k)\|^2]\\
&\leq \tfrac{\|x_0-x^*\|^2}{2\lambda_1}+\tfrac{L_f}{\Gamma_N}(\|x^*\|^2+M^2)+\sigma^2\left(\textstyle\sum_{k=1}^N \tfrac{\beta_k}{\Gamma_k(1-L_\Psi\beta_k)m_k}+\textstyle\sum_{k=1}^N \tfrac{\beta_k(1-L_\Psi\beta_k)}{4\Gamma_k m_k}\right)\\
&= \tfrac{\|x_0-x^*\|^2}{2\lambda_1}+\tfrac{L_f}{\Gamma_N}(\|x^*\|^2+M^2)+\sigma^2\textstyle\sum_{k=1}^N \tfrac{\beta_k\left[4+(1-L_\Psi\beta_k)^2\right]}{4\Gamma_k(1-L_\Psi\beta_k)m_k}.
\end{aligned} \tag{6.4.97}$$

Since the above relation is similar to the relation (6.4.73), the rest of proof is also similar to the last part of the proof for Theorem 6.12 and hence the details are skipped. ■

Theorem 6.13 shows that by using the RSAGD method in Algorithm 6.5, we can have a unified treatment and analysis for stochastic composite problem (6.4.3), no matter it is convex or not. In the next result, we specialize the results obtained in Theorem 6.13 for some particular selections of $\{\alpha_k\}$, $\{\beta_k\}$, and $\{\lambda_k\}$.

Corollary 6.17. *Suppose that the stepsizes $\{\alpha_k\}$, $\{\beta_k\}$, and $\{\lambda_k\}$ in Algorithm 6.5 are set to (6.4.32) and (6.4.35), respectively, and $\{p_k\}$ is set to (6.4.69). Also assume that an optimal solution x^* exists for problem (6.4.3). Then under Assumptions 16 and 17, for any $N \geq 1$, we have*

$$\begin{aligned}
\mathbb{E}[\|P_X(\underline{x}_R,\nabla\Psi(\underline{x}_R),\beta_R)\|^2] \leq 96L_\Psi \Big[&\tfrac{4L_\Psi\|x_0-x^*\|^2}{N(N+1)(N+2)}+\tfrac{L_f}{N}(\|x^*\|^2+M^2)\\
&+\tfrac{2\sigma^2}{L_\Psi N(N+1)(N+2)}\textstyle\sum_{k=1}^N\tfrac{k(k+1)}{m_k}\Big].
\end{aligned} \tag{6.4.98}$$

If, in addition, $L_f=0$, then for any $N \geq 1$, we have

$$\mathbb{E}[\Phi(\bar{x}_R)-\Phi(x^*)] \leq \tfrac{12L_\Psi\|x_0-x^*\|^2}{N(N+1)}+\tfrac{4\sigma^2}{L_\Psi N(N+1)(N+2)}\textstyle\sum_{k=1}^N\sum_{j=1}^k\tfrac{j(j+1)}{m_j}. \tag{6.4.99}$$

Proof. Similar to Corollary 6.14(b), we can easily show that (6.4.14) and (6.4.15) hold. By (6.4.92), (6.4.32), (6.4.35), (6.4.38), and (6.4.41), we have

$$\begin{aligned}
\mathbb{E}[\|P_X(\underline{x}_R,\nabla\Psi(\underline{x}_R),\beta_R)\|^2] \leq \tfrac{192L_\Psi}{N(N+1)(N+2)}\Big[&2L_\Psi\|x_0-x^*\|^2\\
&+\tfrac{N(N+1)L_f}{2}(\|x^*\|^2+M^2)+\sigma^2\textstyle\sum_{k=1}^N\tfrac{17k(k+1)}{32L_\Psi m_k}\Big],
\end{aligned}$$

which clearly implies (6.4.98). By (6.4.93), (6.4.32), (6.4.35), (6.4.38), and (6.4.41), we have

$$\mathbb{E}[\Phi(\bar{x}_R)-\Phi(x^*)] \leq \tfrac{24L_\Psi}{N(N+1)(N+2)}\left[\tfrac{N}{2}\|x_0-x^*\|^2+\tfrac{\sigma^2}{4L_\Psi}\textstyle\sum_{k=1}^N\sum_{j=1}^k\tfrac{17j(j+1)}{32L_\Psi m_j}\right],$$

which implies (6.4.99). ■

Note that all the bounds in the above corollary depend on $\{m_k\}$ and they may not converge to zero for all values of $\{m_k\}$. In particular, if $\{m_k\}$ is set to a positive integer constant, then the last terms in (6.4.98) and (6.4.99), unlike the other terms, will not vanish as the algorithm advances. On the other hand, if $\{m_k\}$ is very big, then each iteration of Algorithm 6.5 will be expensive due to the computation of stochastic gradients. Next result provides an appropriate selection of $\{m_k\}$.

Corollary 6.18. *Suppose that the stepsizes $\{\alpha_k\}$, $\{\beta_k\}$, and $\{\lambda_k\}$ in Algorithm 6.5 are set to (6.4.32) and (6.4.35), respectively, and $\{p_k\}$ is set to (6.4.69). Also assume that an optimal solution x^* exists for problem (6.4.3), an iteration limit $N \geq 1$ is given, and*

$$m_k = \left\lceil \tfrac{\sigma^2}{L_\Psi \tilde{D}^2} \min\left\{ \tfrac{k}{L_f}, \tfrac{k(k+1)N}{L_\Psi} \right\} \right\rceil, \quad k = 1,2,\ldots,N \tag{6.4.100}$$

for some parameter $\tilde{D}$. Then under Assumptions 16 and 17, we have

$$\mathbb{E}[\|P_X(\underline{x}_R, \nabla\Psi(\underline{x}_R), \beta_R)\|^2] \leq 96 L_\Psi \left[\tfrac{4L_\Psi(\|x_0-x^*\|^2+\tilde{D}^2)}{N(N+1)(N+2)} + \tfrac{L_f(\|x^*\|^2+M^2+2\tilde{D}^2)}{N} \right]. \tag{6.4.101}$$

If, in addition, $L_f = 0$, then

$$\mathbb{E}[\Phi(\bar{x}_R) - \Phi(x^*)] \leq \tfrac{2L_\Psi}{N(N+1)} \left(6\|x_0 - x^*\|^2 + \tilde{D}^2\right). \tag{6.4.102}$$

Proof. By (6.4.100), we have

$$\begin{aligned} \tfrac{\sigma^2}{L_\Psi} \textstyle\sum_{k=1}^N \tfrac{k(k+1)}{m_k} &\leq \tilde{D}^2 \textstyle\sum_{k=1}^N k(k+1) \max\left\{ \tfrac{L_f}{k}, \tfrac{L_\Psi}{k(k+1)N} \right\} \\ &\leq \tilde{D}^2 \textstyle\sum_{k=1}^N k(k+1) \left\{ \tfrac{L_f}{k} + \tfrac{L_\Psi}{k(k+1)N} \right\} \leq \tilde{D}^2 \left[\tfrac{L_f N(N+3)}{2} + L_\Psi \right], \end{aligned}$$

which together with (6.4.98) imply (6.4.101). If $L_f = 0$, then due to (6.4.100), we have

$$m_k = \left\lceil \tfrac{\sigma^2 k(k+1)N}{L_\Psi^2 \tilde{D}^2} \right\rceil, \quad k = 1,2,\ldots,N. \tag{6.4.103}$$

Using this observation, we have

$$\tfrac{\sigma^2}{L_\Psi} \textstyle\sum_{k=1}^N \sum_{j=1}^k \tfrac{j(j+1)}{m_j} \leq \tfrac{L_\Psi \tilde{D}^2 (N+1)}{2},$$

which, in view of (6.4.99), then implies (6.4.102). ■

We now add a few remarks about the results obtained in Corollary 6.18. First, we conclude from (6.4.101) and (6.4.44) that by running Algorithm 6.5 for at most

$$\mathcal{O}\left\{ \left[\tfrac{L_\Psi^2(\|x_0-x^*\|^2+\tilde{D}^2)}{\varepsilon} \right]^{\frac{1}{3}} + \tfrac{L_f L_\Psi(M^2+\|x^*\|^2+\tilde{D}^2)}{\varepsilon} \right\}$$

iterations, we have $-\nabla\Psi(\bar{x}_R) \in \partial\chi(\bar{x}_R) + B(\sqrt{\varepsilon})$. Also at the k-th iteration of this algorithm, the SFO is called m_k times and hence the total number of calls to the SFO equals to $\sum_{k=1}^N m_k$. Now, observe that by (6.4.100), we have

$$\sum_{k=1}^{N} m_k \leq \sum_{k=1}^{N} \left(1 + \frac{k\sigma^2}{L_f L_\Psi \tilde{D}^2}\right) \leq N + \frac{\sigma^2 N^2}{L_f L_\Psi \tilde{D}^2}. \tag{6.4.104}$$

Using these two observations, we conclude that the total number of calls to the SFO performed by Algorithm 6.5 to find an ε-stationary point of problem (6.4.3) i.e., a point $\bar{x}$ satisfying $-\nabla\Psi(\bar{x}) \in \partial\chi(\bar{x}) + B(\sqrt{\varepsilon})$ for some $\varepsilon > 0$, can be bounded by

$$\mathcal{O}\left\{ \left[\frac{L_\Psi^2(\|x_0-x^*\|^2+\tilde{D}^2)}{\varepsilon}\right]^{\frac{1}{3}} + \frac{L_f L_\Psi(M^2+\|x^*\|^2+\tilde{D}^2)}{\varepsilon} + \left[\frac{L_\Psi^{\frac{1}{2}}(\|x_0-x^*\|^2+\tilde{D}^2)\sigma^3}{L_f^{\frac{3}{2}}\tilde{D}^3\varepsilon}\right]^{\frac{2}{3}} \right.$$
$$\left. + \frac{L_f L_\Psi(M^2+\|x^*\|^2+\tilde{D}^2)^2\sigma^2}{\tilde{D}^2\varepsilon^2} \right\}. \tag{6.4.105}$$

Second, note that there are various choices for the parameter $\tilde{D}$ in the definition of m_k. While Algorithm 6.5 converges for any $\tilde{D}$, an optimal choice would be $\sqrt{\|x^*\|^2 + M^2}$ for solving composite nonconvex SP problems, if the last term in (6.4.105) is the dominating one. Third, due to (6.4.102) and (6.4.103), it can be easily shown that when $L_f = 0$, Algorithm 6.5 possesses an optimal complexity for solving convex SP problems which is similar to the one obtained in the Sect. 6.4.2.1 for smooth problems. Fourth, note that the definition of $\{m_k\}$ in Corollary 6.18 depends on the iteration limit N. In particular, due to (6.4.100), we may call the SFO many times (depending on N) even at the beginning of Algorithm 6.5. In the next result, we specify a different choice for $\{m_k\}$ which is independent of N. However, the following result is slightly weaker than the one in (6.4.101) when $L_f = 0$.

Corollary 6.19. *Suppose that the stepsizes $\{\alpha_k\}$, $\{\beta_k\}$, and $\{\lambda_k\}$ in Algorithm 6.5 are set to (6.4.32) and (6.4.35), respectively, and $\{p_k\}$ is set to (6.4.69). Also assume that an optimal solution x^* exists for problem (6.4.3), and*

$$m_k = \left\lceil \frac{\sigma^2 k}{L_\Psi \tilde{D}^2} \right\rceil, \quad k = 1, 2, \ldots \tag{6.4.106}$$

for some parameter $\tilde{D}$. Then under Assumptions 16 and 17, for any $N \geq 1$, we have

$$\mathbb{E}[\|P_X(\underline{x}_R, \nabla\Psi(\underline{x}_R), \beta_R)\|^2] \leq 96 L_\Psi \left[\frac{4L_\Psi\|x_0-x^*\|^2}{N(N+1)(N+2)} + \frac{L_f(\|x^*\|^2+M^2)+2\tilde{D}^2}{N}\right]. \tag{6.4.107}$$

Proof. Observe that by (6.4.106), we have

$$\frac{\sigma^2}{L_\Psi}\sum_{k=1}^{N}\frac{k(k+1)}{m_k} \leq \tilde{D}^2\sum_{k=1}^{N}(k+1) \leq \frac{\tilde{D}^2 N(N+3)}{2}.$$

Using this observation and (6.4.98), we obtain (6.4.107). ■

Using Markov's inequality, (6.4.104), (6.4.106), and (6.4.107), we conclude that the total number of calls to the SFO performed by Algorithm 6.5 for finding an *(ε, Λ)-solution* of problem (6.4.3), i.e., a point $\bar{x}$ satisfying

$\text{Prob}\{\|P_X(\bar{x}, \nabla\Psi(\bar{x}), c)\|^2 \le \varepsilon\} \ge 1 - \Lambda$ for any $c > 0$, some $\varepsilon > 0$ and $\Lambda \in (0,1)$, can be bounded by (6.4.86) after disregarding a few constant factors. We can also design a two-phase method for improving the dependence of this bound on the confidence level Λ.

Note that in this section, we focus on the Euclidean setting by assuming that $\|\cdot\|$ is the Euclidean norm. It should be noted that if the problem is deterministic, we can easily extend our results to the non-Euclidean setting by modifying (6.4.27) and (6.4.54) and using some existing results on prox-mapping discussed in Sect. 6.2.1. Similar extensions to the non-Euclidean setting for stochastic problems, however, are more complicated, mainly because the variance reduction inequality in (6.4.90) requires an inner product norm. One can possibly obtain a similar relation as in (6.4.90) if the associated norm is given by $\|\cdot\|_p$ for $p \ge 2$. However, such a relation does not necessarily hold when $p < 2$. Another possibility is to derive complexity results by noting that all the norms in $\mathbb{R}^n$ are equivalent. However, such complexity results will have more complicated dependence on the dimension of the problem in general.

6.5 Nonconvex Variance-Reduced Mirror Descent

In this section, we consider the following nonconvex finite-sum problem

$$\Psi^* := \min_{x \in X} \{\Psi(x) := f(x) + h(x)\}, \tag{6.5.1}$$

where X is a closed convex set in Euclidean space $\mathbb{R}^n$, f is the average of m smooth but possibly nonconvex component functions f_i, i.e., $f(x) = \sum_{i=1}^m f_i(x)/m$, and h is a simple convex function with known structure, but possibly nonsmooth (e.g., $h(x) = \|x\|_1$ or $h(x) \equiv 0$). We assume that for $\forall$ i=1,2,...,m, $\exists L_i > 0$, s.t.

$$\|\nabla f_i(x) - \nabla f_i(y)\|_* \le L_i \|x - y\|, \ \forall x, y \in X.$$

Clearly, f has Lipschitz continuous gradients with constant

$$L_f \le L \equiv \tfrac{1}{m} \textstyle\sum_{i=1}^m L_i. \tag{6.5.2}$$

Throughout this section, we assume that Ψ is bounded below over X, i.e., Ψ^* is finite. Observe that the problem is similar to (5.3.1), the difference is that f_i are possibly nonconvex.

Our goal in this section is to adapt the variance reduction techniques in Sect. 5.3 into the nonconvex mirror descent method in Sect. 6.2.2 and demonstrate that the resulting algorithm can significantly save the number of gradient evaluations of f_i for solving nonconvex finite-sum optimization problems. We will modify the basic scheme of this algorithm to solve an important class stochastic optimization problems where f is given by an expectation function. Similar to the previous section, we assume that $\|\cdot\|$ is the Euclidean norm.

6.5.1 Basic Scheme for Deterministic Problems

We first focus on the basic case when the number of terms m is fixed. Similar to the variance-reduced mirror descent method in Sect. 5.3, the nonconvex variance-reduced mirror descent method (see Algorithm 6.6) will compute a full gradient for every T iterations. However, different from the variance-reduced mirror descent for solving convex finite-sum problems, the full gradient $\nabla f(\tilde{x})$ will not directly participate in the computation of the gradient estimator G_k. Instead the gradient estimator G_k will be computed in a recursive manner based on G_{k-1}. Another difference from the convex variance-reduced mirror descent method exist in that a mini-batch of sample of size b is used to define G_k. It should be noted, however, that the original algorithmic scheme of the variance-reduced mirror descent method could still be applied to the nonconvex finite-sum problems, even though it would not exhibit the best possible rate of convergence in terms of the total number of required gradient evaluations.

Algorithm 6.6 A nonconvex variance-reduced mirror descent method

Input: $x_1, \gamma, T, \{\theta_t\}$ and probability distribution $Q = \{q_1, \ldots, q_m\}$ on $\{1, \ldots, m\}$.
for $k = 1, 2, \ldots, N$ **do**
  if k % T == 1 **then**
    Set $G_k = \nabla f(x_k)$.
  else
    Generate i.i.d. samples I_b of size b according to Q.
    Set $G_k = \frac{1}{b}\sum_{i\in I_b}(\nabla f_i(x_k) - \nabla f_i(x_{k-1}))/(q_i m) + G_{k-1}$.
  end if
  Set $x_{k+1} = \operatorname{argmin}_{x\in X}\{\gamma[\langle G_k, x\rangle + h(x)] + V(x_k, x)\}$.
end for
Output x_R, where R is uniformly distributed over $\{1, \ldots, N\}$.

In order to facilitate the analysis of the algorithm, we will group the iteration indices $k = 1, 2, \ldots$ into different epochs given by

$$\{\{1,2,\ldots,T\},\{T+1,T+2,\ldots,2T\},\ldots,\{sT+1,sT+2\ldots,(s+1)T\},\ldots\}.$$

In other words, except for the last epoch, each epoch s, $s \geq 0$, consists of T iterations starting from $sT+1$ to $(s+1)T$, and the last epoch consists of the remaining iterations. For a given iteration index $k = sT + t$, we will always use the index k and the pair (s,t) interchangeably. For notational convenience, we also denote $(s, T+1) == (s+1, 1)$. Sometimes we will simply denote (s,t) by t if the epoch s is clear from the context.

We will use the generalized projected gradient $P_X(x_k, \nabla f(x_k), \gamma)$ defined in (6.2.7) to evaluate the solution quality at a given search point $x_k \in X$. Moreover, replacing $\nabla f(x_k)$ with its estimator G_k, we then obtain the stochastic generalized projected gradient $P_X(x_k, G_k, \gamma)$. For notation convenience, we simply denote

$$g_{X,k} \equiv P_X(x_k, \nabla f(x_k), \gamma) \text{ and } \tilde{g}_{X,k} \equiv P_X(x_k, G_k, \gamma).$$

Let us denote $\delta_k \equiv G_k - \nabla f(x_k)$. Then by Corollary 6.1, we have

$$\|g_{X,k} - \tilde{g}_{X,k}\| \leq \|\delta_k\|. \tag{6.5.3}$$

We first provide a bound on the size of $\|\delta_k\|$.

Lemma 6.10. *Let L be defined in (6.5.2) and suppose that the probabilities q_i are set to*

$$q_i = \tfrac{L_i}{mL} \tag{6.5.4}$$

for $i = 1,\ldots,m$. If the iteration index k (or equivalently $(s,\ t)$) represents the t-th iteration at the s-epoch, then

$$\mathbb{E}[\|\delta_k\|^2] \equiv \mathbb{E}[\|\delta_{(s,t)}\|^2] \leq \tfrac{L^2}{b}\textstyle\sum_{i=2}^{t}\mathbb{E}[\|x_{(s,i)} - x_{(s,i-1)}\|^2]. \tag{6.5.5}$$

Proof. Consider the s-th epoch. For simplicity let us denote $\delta_t \equiv \delta_{s,t}$ and $x_t \equiv x_{(s,t)}$. It is easy to see that for the first iteration in epoch s, we have $\delta_1 = 0$. Note that by the definition of δ_t, we have

$$\begin{aligned}
\mathbb{E}[\|\delta_t\|^2] &= \mathbb{E}[\|\tfrac{1}{b}\textstyle\sum_{i\in I_b}[\nabla f_i(x_t) - \nabla f_i(x_{t-1})]/(q_i m) + G_{t-1} - \nabla f(x_t)\|^2] \\
&= \mathbb{E}[\|\tfrac{1}{b}\textstyle\sum_{i\in I_b}[\nabla f_i(x_t) - \nabla f_i(x_{t-1})]/(q_i m) - [\nabla f(x_t) - \nabla f(x_{t-1})] + \delta_{t-1}]\|^2.
\end{aligned}$$

Let us denote $\zeta_i = [\nabla f_i(x_t) - \nabla f_i(x_{t-1})]/(q_i m)$. By taking the conditional expectation w.r.t. $i \in I_b$, we have $\mathbb{E}[\zeta_i] = \nabla f(x_t) - \nabla f(x_{t-1})$, which together with the above identity then imply that

$$\begin{aligned}
\mathbb{E}[\|\delta_t\|^2] &= \mathbb{E}[\|\tfrac{1}{b}\textstyle\sum_{i\in I_b}[\zeta_i - (\nabla f(x_t) - \nabla f(x_{t-1})\|^2] + \mathbb{E}[\|\delta_{t-1}\|^2] \\
&\leq \tfrac{1}{b^2}\textstyle\sum_{i\in I_b}\mathbb{E}[\|\zeta_i\|^2] + \mathbb{E}[\|\delta_{t-1}\|^2] \\
&= \tfrac{1}{b^2}\textstyle\sum_{i\in I_b}\mathbb{E}[\tfrac{1}{m^2 q_i}\|\nabla f_i(x_t) - \nabla f_i(x_{t-1})\|^2] + \mathbb{E}[\|\delta_{t-1}\|^2 \\
&\leq \tfrac{1}{b}\textstyle\sum_{j=1}^{m}\tfrac{L_j^2}{m^2 q_j}\|x_t - x_{t-1}\|^2 + \mathbb{E}[\|\delta_{t-1}\|^2 \\
&= \tfrac{L^2}{b}\|x_t - x_{t-1}\|^2 + \mathbb{E}[\|\delta_{t-1}\|^2.
\end{aligned} \tag{6.5.6}$$

The result then follows by applying the above inequality inductively. ■

Now we are ready to prove the main convergence properties of the nonconvex variance-reduced mirror descent method.

Theorem 6.14. *Assume that the probabilities q_i are set to (6.5.4) and that*

$$\gamma = 1/L \quad and \quad b = 17T. \tag{6.5.7}$$

Then for any $k \geq 1$,

$$\mathbb{E}[\Psi(x_{k+1})] + \tfrac{1}{8L}\textstyle\sum_{j=1}^{k}\mathbb{E}[\|g_{X,j}\|^2] \leq \mathbb{E}[\Psi(x_1)],\ \forall k \geq 1. \tag{6.5.8}$$

As a consequence,

$$\mathbb{E}[\|g_{X,R}\|^2] \le \tfrac{8L}{N}[\Psi(x_1) - \Psi^*]. \tag{6.5.9}$$

Proof. Using the smoothness of f and Lemma 6.4, we have for any $k \ge 1$,

$$\begin{aligned}
f(x_{k+1}) &\le f(x_k) + \langle \nabla f(x_k), x_{k+1} - x_k\rangle + \tfrac{L}{2}\|x_{k+1} - x_k\|^2 \\
&= f(x_k) + \langle G_k, x_{k+1} - x_k\rangle + \tfrac{L}{2}\|x_{k+1} - x_k\|^2 - \langle \delta_k, x_{k+1} - x_k\rangle \\
&= f(x_k) - \gamma\langle G_k, \tilde{g}_{X,k}\rangle + \tfrac{L\gamma^2}{2}\|\tilde{g}_{X,k}\|^2 + \gamma\langle \delta_k, \tilde{g}_{X,k}\rangle \\
&\le f(x_k) - \gamma\Big[\|\tilde{g}_{X,k}\|^2 + \tfrac{1}{\gamma}(h(x_{k+1} - h(x_k))\Big] + \tfrac{L\gamma^2}{2}\|\tilde{g}_{X,k}\|^2 + \gamma\langle \delta_k, \tilde{g}_{X,k}\rangle.
\end{aligned}$$

Rearranging the terms in the above inequality and applying the Cauchy–Schwarz inequality, we obtain

$$\begin{aligned}
\Psi(x_{k+1}) &\le \Psi(x_k) - \gamma\|\tilde{g}_{X,k}\|^2 + \tfrac{L\gamma^2}{2}\|\tilde{g}_{X,k}\|^2 + \gamma\langle \delta_k, \tilde{g}_{X,k}\rangle \\
&\le \Psi(x_k) - \gamma\Big(1 - \tfrac{L\gamma}{2} - \tfrac{q}{2}\Big)\|\tilde{g}_{X,k}\|^2 + \tfrac{\gamma}{2q}\|\delta_k\|^2
\end{aligned} \tag{6.5.10}$$

for any $q > 0$. Observe that by (6.5.3),

$$\|g_{X,k}\|^2 = \|g_{X,k} - \tilde{g}_{X,k} + \tilde{g}_{X,k}\|^2 \le 2\|\delta_k\|^2 + 2\|\tilde{g}_{X,k}\|^2.$$

Multiplying the above inequality for any $p > 0$ and adding it to (6.5.10), we have

$$\Psi(x_{k+1}) + p\|g_{X,k}\|^2 \le \Psi(x_k) - \Big[\gamma(1 - \tfrac{L\gamma}{2}) - 2p\Big]\|\tilde{g}_{X,k}\|^2 + (\tfrac{\gamma}{2q} + 2p)\|\delta_k\|^2.$$

We now show possible progress made by each epoch of the nonconvex variance-reduced mirror-descent method. Using (7.4.4) and the fact that $x_{s,t} - x_{s,t-1} = -\gamma\tilde{g}_{X,t}$, $t = 1, \ldots, T+1$ (here we use the notation that $(s, T+1) = (s+1, 1)$), we have

$$\mathbb{E}[\|\delta_{(s,t)}\|^2] \le \tfrac{L^2}{b}\Sigma_{i=2}^t\|x_{s,i} - x_{s,i-1}\|^2 = \tfrac{\gamma^2L^2}{b}\Sigma_{i=2}^t\|\tilde{g}_{X,(s,i)}\|^2.$$

Combining the above two inequalities, we obtain

$$\begin{aligned}
\mathbb{E}[\Psi(x_{s,t+1})] + p\mathbb{E}[\|g_{X,(s,t)}\|^2] &\le \mathbb{E}[\Psi(x_{s,t})] - \Big[\gamma(1 - \tfrac{L\gamma}{2}) - 2p\Big]\mathbb{E}[\|\tilde{g}_{X,(s,t)}\|^2] \\
&\quad + (\tfrac{\gamma}{2q} + 2p)\tfrac{\gamma^2L^2}{b}\Sigma_{i=2}^t\mathbb{E}[\|\tilde{g}_{X,(s,i)}\|^2].
\end{aligned} \tag{6.5.11}$$

Taking the telescope sum of the above inequalities, we have for any $t = 1, \ldots, T$

$$\begin{aligned}
&\mathbb{E}[\Psi(x_{s,t+1})] + p\Sigma_{j=1}^t\mathbb{E}[\|g_{X,(s,j)}\|^2] \le \mathbb{E}[\Psi(x_{(s,1)})] \\
&\quad - \Big[\gamma(1 - \tfrac{L\gamma}{2}) - 2p\Big]\Sigma_{j=1}^t\mathbb{E}[\|\tilde{g}_{X,(s,j)}\|^2] + (\tfrac{\gamma}{2q} + 2p)\tfrac{\gamma^2L^2}{b}\Sigma_{j=1}^t\Sigma_{i=2}^j\mathbb{E}[\|\tilde{g}_{X,(s,i)}\|^2] \\
&\le \mathbb{E}[\Psi(x_1)] - \Big[\gamma(1 - \tfrac{L\gamma}{2}) - 2p\Big]\Sigma_{j=1}^t\mathbb{E}[\|\tilde{g}_{X,(s,j)}\|^2] \\
&\quad + (\tfrac{\gamma}{2q} + 2p)\tfrac{\gamma^2L^2(t-1)}{b}\Sigma_{j=2}^t\mathbb{E}[\|\tilde{g}_{X,(s,j)}\|^2]
\end{aligned}$$

$$\leq \mathbb{E}[\Psi(x_{(s,1)}] - \left[\gamma(1-\tfrac{L\gamma}{2}) - 2p - (\tfrac{\gamma}{2q}+2p)\tfrac{\gamma^2 L^2 (t-1)}{b}\right] \textstyle\sum_{j=1}^{t} \mathbb{E}[\|\tilde{g}_{X,j}\|^2], \quad (6.5.12)$$

for any $p > 0$ and $q > 0$. Fixing $p = 1/(8L)$ and $q = 1/8$ in the above inequality, and using the facts that $\gamma = 1/L$, $b = 21T$, and $1 \leq t \leq T$, we observe

$$\gamma(1-\tfrac{L\gamma}{2}) - 2p - (\tfrac{\gamma}{2q}+2p)\tfrac{\gamma^2 L^2 (t-1)}{b} = \tfrac{1}{4L} - \tfrac{17(T-1)}{4Lb} > 0.$$

Then for any epochs $s \geq 0$ and $1 \leq t \leq T$, we have

$$\mathbb{E}[\Psi(x_{(s,t+1)})] + \tfrac{1}{8L}\textstyle\sum_{j=1}^{t} \mathbb{E}[\|g_{X,(s,i)}\|^2] \leq \mathbb{E}[\Psi(x_{(s,1)})]. \quad (6.5.13)$$

Therefore, (6.5.8) easily follows by summing up the first k inequalities in the above form. In addition, (6.5.9) follows from (6.5.8) (with $k = N$) due to the definition of the random variable R and the fact that $\Psi(x_{N+1}) \geq \Psi^*$. ■

We are now ready to provide a bound on the total number of gradient evaluations required by the nonconvex variance-reduced mirror descent method.

Corollary 6.20. *Assume that the probabilities q_i are set to (6.5.4) and that γ and b are set to (6.5.7). In addition if $T = \sqrt{m}$, then the total number of gradient evaluations required by the nonconvex variance-reduced mirror descent method to find a solution $\bar{x} \in X$ s.t. $\mathbb{E}[\|P_X(x_k, \nabla f(x_k), \gamma)\|^2] \leq \varepsilon$ can be bounded by*

$$\mathscr{O}(m + \tfrac{\sqrt{m}L[\Psi(x_1)-\Psi^*]}{\varepsilon}).$$

Proof. Clearly, the total number of gradient evaluations will be bounded by

$$(m + bT)\left\lceil \tfrac{N}{T} \right\rceil = (m + 17T^2)\left\lceil \tfrac{N}{T} \right\rceil.$$

By (6.5.9), the total number of iterations performed by this method will be bounded by $N = \frac{8L}{\varepsilon}[\Psi(x_1) - \Psi^*]$. Our result then follows from these observations and the assumption that $T = \sqrt{m}$. ■

6.5.2 Generalization for Stochastic Optimization Problems

In this section, we still consider problem (6.5.1), but with f given by

$$f(x) = \mathbb{E}[F(x,\xi)], \quad (6.5.14)$$

where ξ is a random vector supported on $\Xi \subseteq \mathbb{R}^d$ for some $d \geq 1$. We make the following assumptions throughout this subsection.

- $F(x,\xi)$ is a smooth function with Lipschitz constant L for any $\xi \in \Xi$ almost surely.
- It is possible to generate a realization $\xi \in \Xi$, and to compute $\nabla F(x,\xi)$ and $\nabla F(y,\xi)$ for any given two point $x, y \in X$.

- For any x, we have $\mathbb{E}[\nabla F(x,\xi)] = \nabla f(x)$ and

$$\mathbb{E}[\|\nabla F(x,\xi) - \nabla f(x)\|^2] \le \sigma^2. \tag{6.5.15}$$

We observe that the above assumptions are much stronger than those required for the RSGD and RSMD methods.

Algorithm 6.7 A nonconvex variance-reduced mirror descent method

---

Input: $x_1, \gamma, T, \{\theta_t\}$, sample sizes b and m, and epoch index $s = 0$.
for $k = 1, 2, \ldots, N$ **do**
  if k % T == 1 **then**
    Generate an i.i.d. sample $H^s = \{\xi_1^s, \ldots, \xi_m^s\}$ for the random variable ξ.
    Set $G_k = \frac{1}{m}\sum_1^m \nabla F(x_k, \xi_i^s)$.
    Set $s \leftarrow s+1$.
  else
    Generate an i.i.d. sample $I^k = \{\xi_1^k, \ldots, \xi_b^k\}$ for the random variable ξ.
    Set $G_k = \frac{1}{b}\sum_{i=1}^b (\nabla F(x_k, \xi_i^k) - \nabla F(x_{k-1}, \xi_i^k)) + G_{k-1}$.
  end if
  Set $x_{k+1} = \mathrm{argmin}_{x \in X} \{\gamma[\langle G_k, x\rangle + h(x)] + V(x_k, x)\}$.
end for
Output x_R, where R is uniformly distributed over $\{1, \ldots, N\}$.

Similar to the previous section, we will first need to provide a bound on the size of $\delta_k = G_k - \nabla f(x_k)$.

Lemma 6.11. *If the iteration index k (or equivalently (s,t)) represents the t-th iteration at the s-epoch, then*

$$\mathbb{E}[\|\delta_k\|^2] \equiv \mathbb{E}[\|\delta_{(s,t)}\|^2] \le \tfrac{L^2}{b}\textstyle\sum_{i=2}^t \mathbb{E}[\|x_{(s,i)} - x_{(s,i-1)}\|^2] + \tfrac{\sigma^2}{m}. \tag{6.5.16}$$

Proof. Similar to (6.5.6), we can show that

$$\mathbb{E}[\|\delta_{(s,t)}\|^2] = \tfrac{L^2}{b}\|x_{(s,t)} - x_{(s,t-1)}\|^2 + \mathbb{E}[\|\delta_{(s,t-1)}\|^2.$$

Moreover, it can be easily shown that $\mathbb{E}[\|\delta_{(s,0)}\|^2] \le \sigma^2/m$. Combining these two inequalities, we obtain the result. ∎

Theorem 6.15. *Assume that γ and b are set to (6.5.7). We have*

$$\mathbb{E}[\Psi(x_{k+1})] + \tfrac{1}{8L}\textstyle\sum_{j=1}^k \mathbb{E}[\|g_{X,j}\|^2] \le \mathbb{E}[\Psi(x_1)] + \tfrac{21k\sigma^2}{4Lm}, \forall k \ge 1,$$
$$\mathbb{E}[\|g_{X,R}\|^2] \le \tfrac{8L}{N}[\Psi(x_1) - \Psi^*] + \tfrac{21\sigma^2}{4m}.$$

Proof. Using (6.5.16) and an argument similar to the one used in the proof of (6.5.11), we can show that

$$\mathbb{E}[\Psi(x_{s,t+1})] + p\mathbb{E}[\|g_{X,(s,t)}\|^2] \le \mathbb{E}[\Psi(x_{s,t})] - \left[\gamma(1 - \tfrac{L\gamma}{2}) - 2p\right]\mathbb{E}[\|\tilde{g}_{X,(s,t)}\|^2]$$

$$+\left(\tfrac{\gamma}{2q}+2p\right)\left[\tfrac{\gamma^2L^2}{b}\textstyle\sum_{i=2}^t\mathbb{E}[\|\tilde{g}_{X,(s,i)}\|^2]+\tfrac{\sigma^2}{m}\right]. \tag{6.5.17}$$

Therefore, similar to (6.5.12),

$$\begin{aligned}&\mathbb{E}[\Psi(x_{s,t+1})]+p\textstyle\sum_{j=1}^t\mathbb{E}[\|g_{X,(s,j)}\|^2]\leq\mathbb{E}[\Psi(x_1)]\\&-\left[\gamma(1-\tfrac{L\gamma}{2})-2p-(\tfrac{\gamma}{2q}+2p)\tfrac{\gamma^2L^2(t-1)}{b}\right]\textstyle\sum_{j=1}^t\mathbb{E}[\|\tilde{g}_{X,j}\|^2]+t(\tfrac{\gamma}{2q}+2p)\tfrac{\sigma^2}{m},\end{aligned}$$

for any $p>0$ and $q>0$. Fixing $p=1/(8L)$ and $q=1/8$ in the above inequality, and using the facts that $\gamma=1/L$, $b=17T$, and $1\leq t\leq T$, we observe

$$\gamma(1-\tfrac{L\gamma}{2})-2p-(\tfrac{\gamma}{2q}+2p)\tfrac{\gamma^2L^2(t-1)}{b}=\tfrac{1}{4L}-\tfrac{17(T-1)}{4Lb}>0.$$

Then for any epochs $s\geq 0$ and $1\leq t\leq T$, we have

$$\mathbb{E}[\Psi(x_{(s,t+1)})]+\tfrac{1}{8L}\textstyle\sum_{j=1}^t\mathbb{E}[\|g_{X,(s,i)}\|^2]\leq\mathbb{E}[\Psi(x_{(s,1)})]+\tfrac{17t\sigma^2}{4Lm}. \tag{6.5.18}$$

Applying these inequalities inductively, we have

$$\begin{aligned}\mathbb{E}[\Psi(x_{k+1})]+\tfrac{1}{8L}\textstyle\sum_{j=1}^k\mathbb{E}[\|g_{X,j}\|^2]&\leq\mathbb{E}[\Psi(x_1)]+\tfrac{17L\sigma^2}{4}\left(\textstyle\sum_{i=0}^{s-1}\tfrac{T}{m}+\tfrac{t}{m}\right)\\&=\mathbb{E}[\Psi(x_1)]+\tfrac{17k\sigma^2}{4Lm}.\end{aligned}$$

■

We are now ready to provide a bound on the total number of gradient evaluations required by the nonconvex variance-reduced mirror descent method for solving stochastic optimization problems.

Corollary 6.21. *Assume that γ and b are set to (6.5.7). For a given accuracy $\varepsilon>0$, if $m=\sigma^2/\varepsilon^2$ and $T=\sqrt{m}$, then the total number of gradient evaluations required by the nonconvex variance-reduced mirror descent method to find a solution $\bar{x}\in X$ s.t. $\mathbb{E}[\|P_X(x_k,\nabla f(x_k),\gamma)\|^2]\leq\varepsilon$ can be bounded by*

$$\mathcal{O}\left(\tfrac{L\sigma[\Psi(x_1)-\Psi^*]}{\varepsilon^{3/2}}\right). \tag{6.5.19}$$

Proof. It follows from Theorem 6.15 that the total number of iterations N and the sample size m should be bounded by $\mathcal{O}(\frac{L}{\varepsilon}[\Psi(x_1)-\Psi^*])$ and $\mathcal{O}(\sigma^2/\varepsilon)$, respectively. Clearly, the total number of gradient evaluations will be bounded by

$$(m+bT)\left\lceil\tfrac{N}{T}\right\rceil=(m+17T^2)\left\lceil\tfrac{N}{T}\right\rceil=\mathcal{O}(\sqrt{m}N),$$

which implies the bound in (6.5.19). ■

We observe that the above complexity bound is much better than the one obtained RSMD method (see Sect. 6.2) due to the special structure information we assumed for the problem. In particular, we need to assume that the function $F(x,\xi)$ is smooth for every ξ almost surely. On the other hand, the RSMD only requires f to be smooth. There are many cases when F is nonsmooth but f becomes smooth. The second assumption that we rely on is the possibility to fix a random variable ξ when computing gradients at different search points. This assumption is satisfied in many applications in machine learning, but not necessarily in some other applications, e.g., simulation or stochastic dynamic optimization, where the random variable may depend on the decision variables.

6.6 Randomized Accelerated Proximal-Point Methods

In this section, we discuss a different class of acceleration methods based on proximal point methods for solving nonconvex optimization problems. In these methods, we first transfer the original nonconvex optimization problem into a series of strongly convex optimization problems, and then apply accelerated methods for solving them.

We consider two classes of nonconvex optimization problems that are widely used in machine learning. The first class of problems intends to minimize the summation of many terms:

$$\min_{x\in X}\{f(x) := \tfrac{1}{m}\textstyle\sum_{i=1}^m f_i(x)\}, \tag{6.6.1}$$

where $X \subseteq \mathbb{R}^n$ is a closed convex set, and $f_i : X \to \mathbb{R}$, $i = 1,\ldots,m$, are nonconvex smooth functions with L-Lipschitz continuous gradients over X, i.e., for some $L \geq 0$,

$$\|\nabla f_i(x_1) - \nabla f_i(x_2)\| \leq L\|x_1 - x_2\|, \quad \forall x_1, x_2 \in X. \tag{6.6.2}$$

Moreover, we assume that there exists $0 < \mu \leq L$ such that (s.t.)

$$f_i(x_1) - f_i(x_2) - \langle \nabla f_i(x_2), x_1 - x_2\rangle \geq -\tfrac{\mu}{2}\|x_1 - x_2\|^2, \quad \forall x_1, x_2 \in X. \tag{6.6.3}$$

Clearly, (6.6.2) implies (6.6.3) (with $\mu = L$). While in the classical nonlinear programming setting one only assumes (6.6.2), by using both conditions (6.6.2) and (6.6.3) we can explore more structural information for the design of solution methods of problem (6.6.1). This class of problems cover, for example, the nonconvex composite problem we discussed in Sect. 6.4 as a special case. In this section, we intend to develop more efficient algorithms to solve problems where the condition number L/μ associated with problem (6.6.1) is large. Some applications of these problems can be found in variable selection in statistics with $f(x) = \frac{1}{m}\sum_{i=1}^m h_i(x) + \rho p(x)$, where h_i's are smooth convex functions, p is a nonconvex function, and $\rho > 0$ is a relatively small penalty parameter. Note that some examples of the nonconvex penalties are given by minimax concave penalty (MCP) or smoothly clipped absolute deviation (SCAD). It can be shown that the condition number for these problems is usually larger than m.

In addition to (6.6.1), we consider an important class of nonconvex multi-block optimization problems with linearly coupled constraints, i.e.,

$$\min_{x_i\in X_i} \textstyle\sum_{i=1}^m f_i(x_i)$$

$$\text{s.t. } \sum_{i=1}^{m} A_i x_i = b. \tag{6.6.4}$$

Here $X_i \subseteq \mathbb{R}^{d_i}$ are closed convex sets, $A_i \subseteq \mathbb{R}^{n\times d_i}$, $b \subseteq \mathbb{R}^n$, $f_i : X_i \to \mathbb{R}$ satisfy, for some $\mu \geq 0$,

$$f_i(x) - f_i(y) - \langle \nabla f_i(y), x-y\rangle \geq -\tfrac{\mu}{2}\|x-y\|^2, \quad \forall x,y \in X_i, \tag{6.6.5}$$

and $f_m : \mathbb{R}^n \to \mathbb{R}$ has L-Lipschitz continuous gradients, i.e., $\exists L \geq 0$ s.t.

$$\|\nabla f_m(x) - \nabla f_m(y)\| \leq L\|x-y\|, \quad \forall x,y \in \mathbb{R}^n. \tag{6.6.6}$$

Moreover, we assume that $X_m = \mathbb{R}^n$ and A_m is invertible. In other words, we make the structural assumption that one of the blocks equals the dimension of the variable in order to guarantee the strong concavity of the Lagrangian dual of the subproblem. Moreover, we assume that it is relatively easy to compute A_m^{-1} (e.g., A_m is the identity matrix, sparse or symmetric diagonally dominant) to simplify the statement and analysis of the algorithm (see Remark 6.1 for more discussions). Problem of this type arises naturally in compressed sensing and distributed optimization. For instance, consider the compressed sensing problem via nonconvex shrinkage penalties: $\min_{x_i \in X_i} \{p(x) : Ax = b\}$, where $A \in \mathbb{R}^{n\times d}$ is a big sensing matrix with $d >> n$, and $p(x) = \sum_{i=1}^{m} p_i(x_i)$ is a nonconvex and separable penalty function. Since it is easy to find an invertible submatrix in A, w.l.o.g, we assume that the last n columns of A forms an invertible matrix. We can then view this problem as a special case of (6.6.4) by grouping the last n components of x into block x_m, and dividing the remaining $d-n$ components into another $m-1$ blocks.

6.6.1 *Nonconvex Finite-Sum Problems*

In this section, we develop a randomized accelerated proximal gradient (RapGrad) method for solving the nonconvex finite-sum optimization problem in (6.6.1) and demonstrate that it can significantly improve the rates of convergence for solving these problems, especially when their objective functions have large condition numbers. We will describe this algorithm and establish its convergence in Sects. 6.6.1.1 and 6.6.1.2, respectively.

6.6.1.1 The RapGrad Algorithm

The basic idea of RapGrad is to solve problem (6.6.1) iteratively by using the proximal-point type method. More specifically, given a current search point $\bar{x}^{\ell-1}$ at the l-th iteration, we will employ a randomized accelerated gradient (RaGrad) obtained by properly modifying the randomized primal–dual gradient method in Sect. 5.1, to approximately solve

$$\min_{x \in X} \tfrac{1}{m} \textstyle\sum_{i=1}^m f_i(x) + \tfrac{3\mu}{2} \|x - \bar{x}^{\ell-1}\|^2 \tag{6.6.7}$$

to compute a new search point $\bar{x}^\ell$.

The algorithmic schemes for RapGrad and RaGrad are described in Algorithm 6.8 and Algorithm 6.9, respectively. While it seems that we can directly apply the randomized primal–dual gradient method (or other fast randomized incremental gradient method) to solve (6.6.7) since it is strongly convex due to (6.6.3), a direct application of these methods would require us to compute the full gradient from time to time whenever a new subproblem needs to be solved. Moreover, a direct application of these existing first-order methods to solve (6.6.7) would result in some extra logarithmic factor ($\log(1/\varepsilon)$) in the final complexity bound. Therefore, we employed the RaGrad method to solve (6.6.7), which differs from the original randomized primal–dual gradient method in the following several aspects. First, different from the randomized primal–dual gradient method, the design and analysis of RaGrad do not involve the conjugate functions of f_i's, but only first-order information (function values and gradients). Such an analysis enables us to build a relation between successive search points $\bar{x}^\ell$, as well as the convergence of the sequences $\underline{\bar{x}}_i^\ell$ where the gradients $\bar{y}_i^\ell$ are computed. With these relations at hand, we can determine the number of iterations s required by Algorithm 6.9 to ensure the overall RapGrad Algorithm to achieve an accelerated rate of convergence.

Second, the original randomized primal–dual gradient method requires the computation of only one randomly selected gradient at each iteration, and does not require the computation of full gradients from time to time. However, it is unclear whether a full pass of all component functions is required whenever we solve a new proximal subproblem (i.e., $\bar{x}^{\ell-1}$ changes at each iteration). It turns out that by properly initializing a few intertwined primal and gradient sequences in RaGrad using information obtained from previous subproblems, we will compute full gradient only once for the very first time when this method is called, and do not need to compute full gradients any more when solving all other subproblems throughout the RapGrad method. Indeed, the output y_s^i of RaGrad (Algorithm 6.9) represents the gradients of ψ_i at the search points $\underline{x}_i^s$. By using the strong convexity of the objective functions, we will be able to show that all the search points $\underline{x}_s^i$, $i = 1, \ldots, m$, will converge, similar to the search point x^s, to the optimal solution of the subproblem in (6.6.8) (see Lemma 6.13 below). Therefore, we can use y_s^i to approximate $\nabla \psi_i(x^s)$ and thus remove the necessity of computing the full gradient of x^s when solving the next subproblem.

Algorithm 6.8 RapGrad for nonconvex finite-sum optimization

Let $\bar{x}^0 \in X$, and set $\underline{x}_i^0 = \bar{x}^0$, $\bar{y}_i^0 = \nabla f_i(\bar{x}^0)$, $i = 1,\ldots,m$.
for $\ell = 1,\ldots,k$ **do**
Set $x^{-1} = x^0 = \bar{x}^{\ell-1}$, $\underline{x}_i^0 = \underline{\bar{x}}_i^{\ell-1}$, and $y_i^0 = \bar{y}_i^{\ell-1}$, $i = 1,\ldots,m$.
Run RaGrad (cf., Algorithm 6.9) with input x^{-1}, x^0, $\underline{x}_i^0$, y_i^0, $i = 1,\ldots,m$, and s to solve the following subproblem

$$\min_{x\in X} \tfrac{1}{m}\textstyle\sum_{i=1}^m \psi_i(x) + \varphi(x) \tag{6.6.8}$$

to obtain output x^s, $\underline{x}_i^s$, y_i^s, $i = 1,\ldots,m$, where $\psi_i(x) \equiv \psi_i^\ell(x) := f_i(x) + \mu\|x - \bar{x}^{\ell-1}\|^2$, $i = 1,\ldots,m$, and $\varphi(x) \equiv \varphi^\ell(x) := \frac{\mu}{2}\|x - \bar{x}^{\ell-1}\|^2$.
Set $\bar{x}^\ell = x^s$, $\underline{\bar{x}}_i^\ell = \underline{x}_i^s$ and $\bar{y}_i^\ell = y_i^s + 2\mu(\bar{x}^{\ell-1} - \bar{x}^\ell)$, $i = 1,\ldots,m$ (note $\bar{y}_i^\ell = \nabla\psi_i^{\ell+1}(\underline{\bar{x}}_i^\ell)$ always holds).
end for
return $\bar{x}^{\hat{\ell}}$ for some random $\hat{\ell} \in [k]$.

Algorithm 6.9 RaGrad for iteratively solving subproblem (6.6.8)

Input $x^{-1} = x^0 \in X$, $\underline{x}_i^0 \in X$, y_i^0, $i = 1,\ldots,m$, number of iterations s. Assume nonnegative parameters $\{\alpha_t\}$, $\{\tau_t\}$, $\{\eta_t\}$ are given.
for $t = 1,\ldots,s$ **do**
1. Generate a random variable i_t uniformly distributed over $[m]$.
2. Update x^t and y^t according to

$$\tilde{x}^t = \alpha_t(x^{t-1} - x^{t-2}) + x^{t-1}. \tag{6.6.9}$$

$$\underline{x}_i^t = \begin{cases} (1+\tau_t)^{-1}(\tilde{x}^t + \tau_t \underline{x}_i^{t-1}), & i = i_t, \\ \underline{x}_i^{t-1}, & i \neq i_t, \end{cases} \tag{6.6.10}$$

$$y_i^t = \begin{cases} \nabla\psi_i(\underline{x}_i^t), & i = i_t, \\ y_i^{t-1}, & i \neq i_t, \end{cases} \tag{6.6.11}$$

$$\tilde{y}_i^t = m(y_i^t - y_i^{t-1}) + y_i^{t-1}, \quad \forall i = 1,\ldots,m \tag{6.6.12}$$

$$x^t = \operatorname{argmin}_{x\in X}\ \varphi(x) + \langle \tfrac{1}{m}\textstyle\sum_{i=1}^m \tilde{y}_i^t, x\rangle + \eta_t V_\varphi(x, x^{t-1}). \tag{6.6.13}$$

end for
return x^s, $\underline{x}_i^s$, and y_i^s, $i = 1,\ldots,m$.

Before establishing the convergence of the RapGrad method, we first need to define an approximate stationary point for problem (6.6.1) suitable for the analysis of proximal-point type methods. A point $x \in X$ is called an approximate stationary point if it sits within a small neighborhood of a point $\hat{x} \in X$ which approximately satisfies the first-order optimality condition.

Definition 6.1. A point $x \in X$ is called an (ε, δ)-solution of (6.6.1) if there exists some $\hat{x} \in X$ such that

$$[d(\nabla f(\hat{x}), -N_X(\hat{x}))]^2 \leq \varepsilon \quad \text{and} \quad \|x - \hat{x}\|^2 \leq \delta.$$

A stochastic (ε, δ)-solution of (6.6.1) is one such that

$$\mathbb{E}[d(\nabla f(\hat{x}), -N_X(\hat{x}))]^2 \leq \varepsilon \quad \text{and} \quad \mathbb{E}\|x - \hat{x}\|^2 \leq \delta.$$

Here, $d(x,Z) := \inf_{z\in Z}\|x-z\|$ denotes the distance from x to set Z, and $N_X(\hat{x}) := \{x \in \mathbb{R}^n | \langle x, y-\hat{x}\rangle \le 0 \text{ for all } y \in X\}$ denotes the normal cone of X at $\hat{x}$.

To have a better understanding of the above definition, let us consider the unconstrained problem (6.6.1), i.e., $X = \mathbb{R}^n$. Suppose that $x \in X$ is an (ε,δ)-solution with $\delta = \varepsilon/L^2$. Then there exists $\hat{x} \in X$ s.t. $\|\nabla f(\hat{x})\|^2 \le \varepsilon$ and $\|x-\hat{x}\|^2 \le \varepsilon/L^2$, which implies that

$$\begin{aligned}\|\nabla f(x)\|^2 &= \|\nabla f(x) - \nabla f(\hat{x}) + \nabla f(\hat{x})\|^2 \le 2\|\nabla f(x) - \nabla f(\hat{x})\|^2 + 2\|\nabla f(\hat{x})\|^2 \\ &\le 2L^2\|x-\hat{x}\|^2 + 2\|\nabla f(\hat{x})\|^2 \le 4\varepsilon. \end{aligned} \tag{6.6.14}$$

Moreover, if X is a compact set and $x \in X$ is an (ε,δ)-solution, we can bound the so-called Wolfe gap (see also Sect. 7.1.1) as follows:

$$\begin{aligned}\text{gap}(x) &:= \max_{z\in X}\langle \nabla f(x), x-z\rangle \\ &= \max_{z\in X}\langle \nabla f(x) - \nabla f(\hat{x}), x-z\rangle + \max_{z\in X}\langle \nabla f(\hat{x}), x-\hat{x}\rangle + \max_{z\in X}\langle \nabla f(\hat{x}), \hat{x}-z\rangle \\ &\le L\sqrt{\delta}D_X + \sqrt{\delta}\|\nabla f(\hat{x})\| + \sqrt{\varepsilon}D_X, \end{aligned} \tag{6.6.15}$$

where $D_X := \max_{x_1,x_2\in X}\|x_1-x_2\|$. In comparison with the two well-known criterions in (6.6.14) and (6.6.15), the criterion given in Definition 6.1 seems to be applicable to a wider class of problems and is particularly suitable for proximal-point type methods.

We are now ready to state the main convergence properties for RapGrad. The proof of this result is more involved and hence is put into Sect. 6.6.1.2 separately.

Theorem 6.16. *Let the iterates $\bar{x}^\ell$, $\ell = 1,\ldots,k$, be generated by Algorithm 6.8 and $\hat{\ell}$ be randomly selected from $[k]$. Suppose that in Algorithm 6.9, the number of iterations $s = \lceil -\log \widetilde{M}/\log\alpha\rceil$ with*

$$\widetilde{M} := 6\left(5 + \tfrac{2L}{\mu}\right)\max\left\{\tfrac{6}{5}, \tfrac{L^2}{\mu^2}\right\}, \quad \alpha = 1 - \frac{2}{m\left(\sqrt{1+16c/m}+1\right)}, \quad c = 2 + \tfrac{L}{\mu}, \tag{6.6.16}$$

and other parameters are set to

$$\alpha_t = \alpha, \quad \gamma_t = \alpha^{-t}, \quad \tau_t = \tfrac{1}{m(1-\alpha)} - 1, \quad \text{and} \quad \eta_t = \tfrac{\alpha}{1-\alpha}, \quad \forall t = 1,\ldots,s. \tag{6.6.17}$$

Then we have

$$\begin{aligned}\mathbb{E}\left[d\left(\nabla f(x_*^{\hat{\ell}}), -N_X(x_*^{\hat{\ell}})\right)\right]^2 &\le \tfrac{36\mu}{k}[f(\bar{x}^0) - f(x^*)], \\ \mathbb{E}\|\bar{x}^{\hat{\ell}} - x_*^{\hat{\ell}}\|^2 &\le \tfrac{4\mu}{kL^2}[f(\bar{x}^0) - f(x^*)],\end{aligned}$$

where x^* and x_*^ℓ denote the optimal solutions to problem (6.6.1) *and the ℓ-th subproblem* (6.6.7), *respectively.*

Theorem 6.16 guarantees, in expectation, the existence of an approximate stationary point $x_*^{\hat{\ell}}$, which is the optimal solution to the $\hat{\ell}$-th subproblem. Though $x_*^{\hat{\ell}}$ is unknown to us, we can output the computable solution $\bar{x}^{\hat{\ell}}$ since it is close enough to $x_*^{\hat{\ell}}$. Moreover, its quality can be directly measured by (6.6.14) and (6.6.15) under certain important circumstances.

In view of Theorem 6.16, we can bound the total number of gradient evaluations required by RapGrad to yield a stochastic (ε, δ)-solution of (6.6.1). Indeed, observe that the full gradient is computed only once in the first outer loop, and that for each subproblem (6.6.1), we only need to compute s gradients with

$$s = \left\lceil -\tfrac{\log \widetilde{M}}{\log \alpha} \right\rceil \sim \mathcal{O}\left(\left(m + \sqrt{m\tfrac{L}{\mu}}\right) \log\left(\tfrac{L}{\mu}\right)\right).$$

Hence, the total number of gradient evaluations performed by RapGrad can be bounded by

$$N(\varepsilon, \delta) := \mathcal{O}\left(m + \mu\left(m + \sqrt{m\tfrac{L}{\mu}}\right) \log\left(\tfrac{L}{\mu}\right) \cdot \max\left\{\tfrac{1}{\delta L^2}, \tfrac{1}{\varepsilon}\right\} D^0\right),$$

where $D^0 := f(\bar{x}^0) - f(x^*)$. As a comparison, the batch version of this algorithm, obtained by viewing $\frac{1}{m}\sum_{i=1}^m f_i(x)$ as a single component, would update all the $\underline{x}_i^t$ and y_i^t for $i = 1, \ldots, m$, in (6.6.10) and (6.6.11) at each iteration, and hence would require

$$\hat{N}(\varepsilon, \delta) := \mathcal{O}\left(m\sqrt{L\mu} \log\left(\tfrac{L}{\mu}\right) \cdot \max\left\{\tfrac{1}{\delta L^2}, \tfrac{1}{\varepsilon}\right\} D^0\right)$$

gradient evaluations to compute an (ε, δ)-solution of (6.6.1). For problems with $L/\mu \geq m$, RapGrad can potentially save the total number of gradient computations up to a factor of $\mathcal{O}(\sqrt{m})$ gradient evaluations than its batch counterpart as well as other deterministic batch methods. It is also interesting to compare RapGrad with those variance-reduced stochastic algorithms (see Sect. 6.5). For simplicity, consider for the case when $\delta = \varepsilon/L^2$ and $X \equiv \mathbb{R}^n$. In this case, the complexity bound of RapGrad, given by $\mathcal{O}(\sqrt{mL\mu}/\varepsilon)$, is smaller than the complexity bound $\mathcal{O}(\sqrt{m}L/\varepsilon)$ by a factor of $\mathcal{O}(L^{\frac{1}{2}}/\mu^{\frac{1}{2}})$, which must be greater than $\mathcal{O}(m^{\frac{1}{2}})$ due to $L/\mu \geq m$.

Theorem 6.16 only shows the convergence of RapGrad in expectation. Similarly to the nonconvex stochastic gradient descent methods, we can establish and then further improve the convergence of RapGrad with overwhelming probability by using a two-phase procedure. More specifically, we can compute a short list of candidate solutions in the optimization phase by either taking a few independent runs of RapGrad or randomly selecting a few solutions from the trajectory of RapGrad, and then choose the best solution, e.g., in terms of either (6.6.14) or (6.6.15), in the post-optimization phase.

6.6.1.2 Convergence Analysis for RapGrad

In this section, we will first develop the convergence results for Algorithm 6.9 applied to the convex finite-sum subproblem (6.6.8), and then using them to establish the convergence of RapGrad. Observe that the component functions ψ_i and φ in (6.6.8) satisfy:

1. $\frac{\mu}{2}\|x-y\|^2 \le \psi_i(x) - \psi_i(y) - \langle \nabla \psi_i(y), x-y\rangle \le \frac{\hat{L}}{2}\|x-y\|^2, \quad \forall x,y \in X, \quad i = 1,\ldots,m,$
2. $\varphi(x) - \varphi(y) - \langle \nabla \varphi(y), x-y\rangle \ge \frac{\mu}{2}\|x-y\|^2, \ \forall x,y \in X,$

where $\hat{L} = L + 2\mu$.

We first state some simple relations about the iterations generated by Algorithm 6.9.

Lemma 6.12. *Let* $\hat{\underline{x}}_i^t = (1+\tau_t)^{-1}(\tilde{x}^t + \tau_t \underline{x}_i^{t-1})$, *for* $i = 1,\ldots,m$, $t = 1,\ldots,s$.

$$\mathbb{E}_{i_t}[\psi(\hat{\underline{x}}_i^t)] = m\psi(\underline{x}_i^t) - (m-1)\psi(\underline{x}_i^{t-1}), \tag{6.6.18}$$

$$\mathbb{E}_{i_t}[\nabla\psi(\hat{\underline{x}}_i^t)] = m\nabla\psi(\underline{x}_i^t) - (m-1)\nabla\psi(\underline{x}_i^{t-1}) = \mathbb{E}_{i_t}[\tilde{y}_i^t]. \tag{6.6.19}$$

Proof. By the definition of $\hat{\underline{x}}_i^t$, it is easy to see that $\mathbb{E}_{i_t}[\underline{x}_i^t] = \frac{1}{m}\hat{\underline{x}}_i^t + \frac{m-1}{m}\underline{x}_i^{t-1}$, thus $\mathbb{E}_{i_t}[\psi_i(\underline{x}_i^t)] = \frac{1}{m}\psi_i(\hat{\underline{x}}_i^t) + \frac{m-1}{m}\psi_i(\underline{x}_i^{t-1})$, and $\mathbb{E}_{i_t}[\nabla\psi_i(\underline{x}_i^t)] = \frac{1}{m}\nabla\psi_i(\hat{\underline{x}}_i^t) + \frac{m-1}{m}\nabla\psi_i(\underline{x}_i^{t-1})$, which combined with the fact $\tilde{y}_i^t = m(y_i^t - y_i^{t-1}) + y_i^{t-1}$ gives us the desired relations. ∎

Lemma 6.13 below describes an important result about Algorithm 6.9, which improves Proposition 5.1 by showing the convergence of $\underline{x}_i^s$.

Lemma 6.13. *Let the iterates* x^t *and* y^t, *for* $t = 1,\ldots,s$, *be generated by Algorithm 6.9 and* x^* *be an optimal solution of* (6.6.8). *If the parameters in Algorithm 6.9 satisfy for all* $t = 1,\ldots,s-1$,

$$\alpha_{t+1}\gamma_{t+1} = \gamma_t, \tag{6.6.20}$$

$$\gamma_{t+1}[m(1+\tau_{t+1}) - 1] \le m\gamma_t(1+\tau_t), \tag{6.6.21}$$

$$\gamma_{t+1}\eta_{t+1} \le \gamma_t(1+\eta_t), \tag{6.6.22}$$

$$\tfrac{\eta_s\mu}{4} \ge \tfrac{(m-1)^2\hat{L}}{m^2\tau_s}, \tag{6.6.23}$$

$$\tfrac{\eta_t\mu}{2} \ge \tfrac{\alpha_{t+1}\hat{L}}{\tau_{t+1}} + \tfrac{(m-1)^2\hat{L}}{m^2\tau_t}, \tag{6.6.24}$$

$$\tfrac{\eta_s\mu}{4} \ge \tfrac{\hat{L}}{m(1+\tau_s)}, \tag{6.6.25}$$

then we have

$$\begin{aligned}\mathbb{E}_s\Big[\gamma_s(1+\eta_s)V_\varphi(x^*,x^s) + \textstyle\sum_{i=1}^m \frac{\mu\gamma_s(1+\tau_s)}{4}\|\underline{x}_i^s - x^*\|^2\Big] &\le \gamma_1\eta_1\mathbb{E}_s V_\varphi(x^*,x^0)\\ &+ \textstyle\sum_{i=1}^m \frac{\gamma_1[(1+\tau_1)-1/m]\hat{L}}{2}\mathbb{E}_s\|\underline{x}_i^0 - x^*\|^2,\end{aligned}$$

where $\mathbb{E}_s[X]$ *denotes the expectation of a random variable* X *on* $i_1,\ldots,i_s$.

Proof. By convexity of ψ and optimality of x^*, we have

$$Q_t := \varphi(x^t) + \psi(x^*) + \langle \nabla\psi(x^*), x^t - x^*\rangle$$

$$
\begin{aligned}
&- \left[\varphi(x^*) + \tfrac{1}{m}\textstyle\sum_{i=1}^m (\psi_i(\hat{\underline{x}}_i^t) + \langle \nabla \psi_i(\hat{\underline{x}}_i^t), x^* - \hat{\underline{x}}_i^t\rangle)\right] \\
&\geq \varphi(x^t) + \psi(x^*) + \langle \nabla \psi(x^*), x^t - x^*\rangle - [\varphi(x^*) + \psi(x^*)] \\
&= \varphi(x^t) - \varphi(x^*) + \langle \nabla \psi(x^*), x^t - x^*\rangle \qquad (6.6.26)\\
&\geq \langle \nabla \varphi(x^*) + \nabla \psi(x^*), x^t - x^*\rangle \geq 0. \qquad (6.6.27)
\end{aligned}
$$

For notation convenience, let $\Psi(x,z) := \psi(x) - \psi(z) - \langle \nabla \psi(z), x - z\rangle$,

$$
\begin{aligned}
Q_t &= \varphi(x^t) - \varphi(x^*) + \langle \tfrac{1}{m}\textstyle\sum_{i=1}^m \tilde{y}_i^t, x^t - x^*\rangle + \delta_1^t + \delta_2^t, \qquad (6.6.28)\\
\delta_1^t &:= \psi(x^*) - \langle \nabla \psi(x^*), x^*\rangle - \tfrac{1}{m}\textstyle\sum_{i=1}^m [\psi_i(\hat{\underline{x}}_i^t) - \langle \nabla \psi_i(\hat{\underline{x}}_i^t), \hat{\underline{x}}_i^t\rangle \\
&\quad + \langle \nabla \psi_i(\hat{\underline{x}}_i^t) - \nabla \psi_i(x^*), \tilde{x}^t\rangle] \\
&= \tfrac{1}{m}\textstyle\sum_{i=1}^m [\tau_t \Psi(\underline{x}_i^{t-1}, x^*) - (1+\tau_t)\Psi(\hat{\underline{x}}_i^t, x^*) - \tau_t \Psi(\underline{x}_i^{t-1}, \hat{\underline{x}}_i^t)], \qquad (6.6.29)\\
\delta_2^t &:= \tfrac{1}{m}\textstyle\sum_{i=1}^m [\langle \nabla \psi_i(\hat{\underline{x}}_i^t) - \nabla \psi_i(x^*), \tilde{x}^t\rangle - \langle \tilde{y}_i^t - \nabla \psi_i(x^*), x^t\rangle + \langle \tilde{y}_i^t - \nabla \psi_i(\hat{\underline{x}}_i^t), x^*\rangle].
\end{aligned}
$$

In view of (6.6.19), we have

$$
\begin{aligned}
\mathbb{E}_{i_t}[\delta_2^t] &= \tfrac{1}{m}\textstyle\sum_{i=1}^m \mathbb{E}_{i_t}\left[\langle \nabla \psi_i(\hat{\underline{x}}_i^t) - \nabla \psi_i(x^*), \tilde{x}^t\rangle - \langle \tilde{y}_i^t - \nabla \psi_i(x^*), x^t\rangle \right.\\
&\quad \left. + \langle \tilde{y}_i^t - \nabla \psi_i(\hat{\underline{x}}_i^t), x^*\rangle\right] \\
&= \tfrac{1}{m}\textstyle\sum_{i=1}^m \mathbb{E}_{i_t}\langle \tilde{y}_i^t - \nabla \psi_i(x^*), \tilde{x}^t - x^t\rangle. \qquad (6.6.30)
\end{aligned}
$$

Multiplying each Q_t by a nonnegative γ_t and summing them up, we obtain

$$
\begin{aligned}
\mathbb{E}_s[\textstyle\sum_{t=1}^s \gamma_t Q_t] &\leq \mathbb{E}_s\{\textstyle\sum_{t=1}^s \gamma_t[\eta_t V_\varphi(x^*, x^{t-1}) - (1+\eta_t)V_\varphi(x^*, x^t) - \eta_t V_\varphi(x^t, x^{t-1})]\} \\
&\quad + \mathbb{E}_s\left\{\textstyle\sum_{t=1}^s \sum_{i=1}^m \left[\gamma_t(1+\tau_t - \tfrac{1}{m})\Psi(\underline{x}_i^{t-1}, x^*) - \gamma_t(1+\tau_t)\Psi(\underline{x}_i^t, x^*)\right]\right\} \\
&\quad + \mathbb{E}_s\left\{\textstyle\sum_{t=1}^s \sum_{i=1}^m \gamma_t \left[\tfrac{1}{m}\langle \tilde{y}_i^t - \nabla \psi_i(x^*), x^t - x^t\rangle - \tau_t \Psi(\underline{x}_i^{t-1}, \underline{x}_i^t))\right]\right\} \\
&\leq \mathbb{E}_s\left[\gamma_1 \eta_1 V_\varphi(x^*, x^0) - (1+\eta_s)V_\varphi(x^*, x^s)\right] \\
&\quad 1 + \mathbb{E}_s\{\textstyle\sum_{i=1}^m [\gamma_1(1+\tau_1 - \tfrac{1}{m})\Psi(\underline{x}_i^0, x^*) \\
&\quad - \gamma_s(1+\tau_s)\Psi(\underline{x}_i^s, x^*)]\} - \mathbb{E}_s[\textstyle\sum_{t=1}^s \gamma_t \delta_t], \qquad (6.6.31)
\end{aligned}
$$

where

$$
\begin{aligned}
\delta_t &:= \eta_t V_\varphi(x^t, x^{t-1}) - \textstyle\sum_{i=1}^m \left[\tfrac{1}{m}\langle \tilde{y}_i^t - \nabla \psi_i(x^*), \tilde{x}^t - x^t\rangle - \tau_t \Psi(\underline{x}_i^{t-1}, \underline{x}_i^t)\right] \\
&= \eta_t V_\varphi(x^t, x^{t-1}) - \tfrac{1}{m}\textstyle\sum_{i=1}^m \langle \tilde{y}_i^t - \nabla \psi_i(x^*), \tilde{x}^t - x^t\rangle + \tau_t \Psi(\underline{x}_{i_t}^{t-1}, \underline{x}_{i_t}^t), \qquad (6.6.32)
\end{aligned}
$$

the first inequality follows from (6.6.28), (6.6.29), (6.6.45), (6.6.30), and Lemma 6.12, and the second inequality is implied by (6.6.21) and (6.6.22).

By the definition of $\tilde{x}$ in (6.6.9), we have

$$
\begin{aligned}
&\tfrac{1}{m}\textstyle\sum_{i=1}^m \langle \tilde{y}_i^t - \nabla \psi_i(x^*), \tilde{x}^t - x^t\rangle \\
&= \tfrac{1}{m}\textstyle\sum_{i=1}^m [\langle \tilde{y}_i^t - \nabla \psi_i(x^*), x^{t-1} - x^t\rangle - \alpha_t \langle \tilde{y}_i^t - \nabla \psi_i(x^*), x^{t-2} - x^{t-1}\rangle] \\
&= \tfrac{1}{m}\textstyle\sum_{i=1}^m [\langle \tilde{y}_i^t - \nabla \psi_i(x^*), x^{t-1} - x^t\rangle - \alpha_t \langle \tilde{y}_i^{t-1} - \nabla \psi_i(x^*), x^{t-2}
\end{aligned}
$$

$$
\begin{aligned}
&\quad -x^{t-1}\rangle - \alpha_t \langle \tilde{y}_i^t - \tilde{y}_i^{t-1}, x^{t-2} - x^{t-1}\rangle] \\
&= \tfrac{1}{m}\textstyle\sum_{i=1}^m [\langle \tilde{y}_i^t - \nabla\psi_i(x^*), x^{t-1} - x^t\rangle - \alpha_t \langle \tilde{y}_i^{t-1} - \nabla\psi_i(x^*), x^{t-2} - x^{t-1}\rangle] \\
&\quad - \alpha_t \langle \nabla\psi_{i_t}(\underline{x}_{i_t}^t) - \nabla\psi_{i_t}(\underline{x}_{i_t}^{t-1}), x^{t-2} - x^{t-1}\rangle \\
&\quad - (1-\tfrac{1}{m})\alpha_t \langle \nabla\psi_{i_{t-1}}(\underline{x}_{i_{t-1}}^{t-2}) - \nabla\psi_{i_{t-1}}(\underline{x}_{i_{t-1}}^{t-1}), x^{t-2} - x^{t-1}\rangle. \qquad (6.6.33)
\end{aligned}
$$

From the relation (6.6.24) and the fact $x^{-1} = x^0$, we have

$$
\begin{aligned}
&\textstyle\sum_{t=1}^s \gamma_t \tfrac{1}{m}\sum_{i=1}^m [\langle \tilde{y}_i^t - \nabla\psi_i(x^*), x^{t-1} - x^t\rangle - \alpha_t \langle \tilde{y}_i^{t-1} - \nabla\psi_i(x^*), x^{t-2} - x^{t-1}\rangle] \\
&= \gamma_s \tfrac{1}{m}\textstyle\sum_{i=1}^m \langle \tilde{y}_i^s - \nabla\psi_i(x^*), x^{s-1} - x^s\rangle \\
&= \gamma_s \tfrac{1}{m}\textstyle\sum_{i=1}^m \langle \nabla\psi_i(\underline{x}_i^s) - \nabla\psi_i(x^*), x^{s-1} - x^s\rangle \\
&\quad + \gamma_s \textstyle\sum_{i=1}^m (1-\tfrac{1}{m})\langle \nabla\psi_i(\underline{x}_i^s) - \nabla\psi_i(\underline{x}_i^{s-1}), x^{s-1} - x^s\rangle \\
&= \gamma_s \tfrac{1}{m}\textstyle\sum_{i=1}^m \langle \nabla\psi_i(\underline{x}_i^s) - \nabla\psi_i(x^*), x^{s-1} - x^s\rangle \\
&\quad + \gamma_s (1-\tfrac{1}{m}) \langle \nabla\psi_{i_s}(\underline{x}_{i_s}^s) - \nabla\psi_{i_s}(\underline{x}_{i_s}^{s-1}), x^{s-1} - x^s\rangle. \qquad (6.6.34)
\end{aligned}
$$

Now we are ready to bound the last term in (6.6.31) as follows:

$$
\begin{aligned}
&\sum_{t=1}^s \gamma_t \delta_t \\
&\overset{(a)}{=} \textstyle\sum_{t=1}^s \gamma_t [\eta_t V_\varphi(x^t, x^{t-1}) - \tfrac{1}{m}\sum_{i=1}^m \langle \tilde{y}_i^t - \nabla\psi_i(x^*), \tilde{x}^t - x^t\rangle + \tau_t \Psi(\underline{x}_{i_t}^{t-1}, \underline{x}_{i_t}^t)] \\
&\overset{(b)}{=} \textstyle\sum_{t=1}^s \gamma_t \Big[\eta_t V_\varphi(x^t, x^{t-1}) + \alpha_t \langle \nabla\psi_{i_t}(\underline{x}_{i_t}^t) - \nabla\psi_{i_t}(\underline{x}_{i_t}^{t-1}), x^{t-2} - x^{t-1}\rangle \\
&\quad + (1-\tfrac{1}{m})\alpha_t \langle \nabla\psi_{i_{t-1}}(\underline{x}_{i_{t-1}}^{t-2}) - \nabla\psi_{i_{t-1}}(\underline{x}_{i_{t-1}}^{t-1})\rangle, x^{t-2} - x^{t-1} + \tau_t \Psi(\underline{x}_{i_t}^{t-1}, \underline{x}_{i_t}^t)\Big] \\
&\quad + \gamma_s \tfrac{1}{m}\textstyle\sum_{i=1}^m \langle \nabla\psi_i(\underline{x}_i^s) - \nabla\psi_i(x^*), x^{s-1} - x^s\rangle \\
&\quad - \gamma_s (1-\tfrac{1}{m})\langle \nabla\psi_{i_s}(\underline{x}_{i_s}^s) - \nabla\psi_{i_s}(\underline{x}_{i_s}^{s-1}), x^{s-1} - x^s\rangle \\
&\overset{(c)}{\geq} \textstyle\sum_{t=1}^s \gamma_t \Big[\tfrac{\eta_t r}{2}\|x^t - x^{t-1}\|^2 + \alpha_t \langle \nabla\psi_{i_t}(\underline{x}_{i_t}^t) - \nabla\psi_{i_t}(\underline{x}_{i_t}^{t-1}), x^{t-2} - x^{t-1}\rangle \\
&\quad + (1-\tfrac{1}{m})\alpha_t \langle \nabla\psi_{i_{t-1}}(\underline{x}_{i_{t-1}}^{t-2}) - \nabla\psi_{i_{t-1}}(\underline{x}_{i_{t-1}}^{t-1}), x^{t-2} - x^{t-1}\rangle \\
&\quad + \tfrac{\tau_t}{2\hat{L}}\|\nabla\psi_{i_t}(\underline{x}_{i_t}^t) - \nabla\psi_{i_t}(\underline{x}_{i_t}^{t-1})\|^2\Big] \\
&\quad - \tfrac{\gamma_s}{m}\textstyle\sum_{i=1}^m \langle x^{s-1} - x^s, \nabla\psi_i(\underline{x}_i^s) - \nabla\psi_i(x^*)\rangle \\
&\quad - \gamma_s (1-\tfrac{1}{m}) \langle x^{s-1} - x^s, \nabla\psi_{i_s}(\underline{x}_{i_s}^s) - \nabla\psi_{i_s}(\underline{x}_{i_s}^{s-1})\rangle, \qquad (6.6.35)
\end{aligned}
$$

where (a) follows from the definition δ_t in (8.3.8), (b) follows relations (6.6.33) and (6.6.34), and (c) follows from the fact that $V_\varphi(x^t, x^{t-1}) \geq \frac{r}{2}\|x^t - x^{t-1}\|^2$, $\Psi(\underline{x}_{i_t}^{t-1}, \underline{x}_{i_t}^t) \geq \frac{1}{2\hat{L}}\|\nabla\psi_{i_t}(\underline{x}_{i_t}^{t-1}) - \nabla\psi_{i_t}(\underline{x}_{i_t}^t)\|^2$.

By properly regrouping the terms on the right-hand side of (6.6.35), we have

$$
\begin{aligned}
&\textstyle\sum_{t=1}^{s}\gamma_t\delta_t \\
&\geq \gamma_s\left[\tfrac{\eta_s r}{4}\|x^s-x^{s-1}\|^2-\tfrac{1}{m}\textstyle\sum_{i=1}^{m}\langle\nabla\psi_i(\underline{x}_i^s)-\nabla\psi_i(x^*),x^{s-1}-x^s\rangle\right] \\
&\quad+\gamma_s\left[\tfrac{\eta_s r}{4}\|x^s-x^{s-1}\|^2-(1-\tfrac{1}{m})\langle\nabla\psi_{i_s}(\underline{x}_{i_s}^s)-\nabla\psi_{i_s}(\underline{x}_{i_s}^{s-1}),x^{s-1}-x^s\rangle\right. \\
&\quad\left.+\tfrac{\tau_s}{4\hat{L}}\|\nabla\psi_{i_s}(\underline{x}_{i_s}^s)-\nabla\psi_{i_s}(\underline{x}_{i_s}^{s-1})\|^2\right] \\
&\quad+\textstyle\sum_{t=2}^{s}\gamma_t\left[\alpha_t\langle\nabla\psi_{i_t}(\underline{x}_{i_t}^t)-\nabla\psi_{i_t}(\underline{x}_{i_t}^{t-1}),x^{t-2}-x^{t-1}\rangle\right. \\
&\qquad\qquad\qquad\left.+\tfrac{\tau_t}{4\hat{L}}\|\nabla\psi_{i_t}(\underline{x}_{i_t}^t)-\nabla\psi_{i_t}(\underline{x}_{i_t}^{t-1})\|^2\right] \\
&\quad+\textstyle\sum_{t=2}^{s}\left[\gamma_t(1-\tfrac{1}{m})\alpha_t\langle\nabla\psi_{i_{t-1}}(\underline{x}_{i_{t-1}}^{t-1})-\nabla\psi_{i_{t-1}}(\underline{x}_{i_{t-1}}^{t-2}),x^{t-2}-x^{t-1}\rangle\right. \\
&\quad\left.+\tfrac{\tau_{t-1}\gamma_{t-1}}{4\hat{L}}\|\nabla\psi_{i_{t-1}}(\underline{x}_{i_{t-1}}^{t-1})-\nabla\psi_{i_{t-1}}(\underline{x}_{i_{t-1}}^{t-2})\|^2\right]+\textstyle\sum_{t=2}^{s}\tfrac{\gamma_{t-1}\eta_{t-1}r}{2}\|x^{t-1}-x^{t-2}\|^2 \\
&\overset{(a)}{\geq}\gamma_s\left[\tfrac{\eta_s r}{4}\|x^s-x^{s-1}\|^2-\tfrac{1}{m}\textstyle\sum_{i=1}^{m}\langle\nabla\psi_i(\underline{x}_i^s)-\nabla\psi_i(x^*),x^{s-1}-x^s\rangle\right] \\
&\quad+\gamma_s\left(\tfrac{\eta_s r}{4}-\tfrac{(m-1)^2\hat{L}}{m^2\tau_s}\right)\|x^s-x^{s-1}\|^2 \\
&\quad+\textstyle\sum_{t=2}^{s}\left(\tfrac{\gamma_{t-1}\eta_{t-1}r}{2}-\tfrac{\gamma_t\alpha_t^2\hat{L}}{\tau_t}-\tfrac{(m-1)^2\gamma_t^2\alpha_t^2\hat{L}}{m^2\gamma_{t-1}\tau_{t-1}}\right)\|x^s-x^{s-1}\|^2 \\
&\overset{(b)}{\geq}\gamma_s\left[\tfrac{\eta_s r}{4}\|x^s-x^{s-1}\|^2-\tfrac{1}{m}\textstyle\sum_{i=1}^{m}\langle\nabla\psi_i(\underline{x}_i^s)-\nabla\psi_i(x^*),x^{s-1}-x^s\rangle\right],
\end{aligned}
$$

where (a) follows from the simple relation that $b\langle u,v\rangle-a\|v\|^2/2\leq b^2\|u\|^2/(2a)$, $\forall a>0$ and (b) follows from (6.6.20), (6.6.23), and (6.6.24). By using the above inequality, (6.6.26) and (6.6.31), we obtain

$$
\begin{aligned}
0&\leq\mathbb{E}_s\left[\gamma_1\eta_1V_\varphi(x^*,x^0)-\gamma_s(1+\eta_s)V_\varphi(x^*,x^s)\right]-\tfrac{\gamma_s\eta_s r}{4}\mathbb{E}_s\|x^s-x^{s-1}\|^2 \\
&\quad+\gamma_s\mathbb{E}_s\left[\tfrac{1}{m}\textstyle\sum_{i=1}^{m}\langle\nabla\psi_i(\underline{x}_i^s)-\nabla\psi_i(x^*),x^{s-1}-x^s\rangle\right] \\
&\quad+\mathbb{E}_s\left\{\textstyle\sum_{i=1}^{m}[\gamma_1(1+\tau_1-\tfrac{1}{m})\Psi(\underline{x}_i^0,x^*)-\gamma_s(1+\tau_s)\Psi(\underline{x}_i^s,x^*)]\right\} \\
&\overset{(a)}{\leq}\mathbb{E}_s[\gamma_1\eta_1V_\varphi(x^*,x^0)-\gamma_s(1+\eta_s)V_\varphi(x^*,x^s)]-\tfrac{\gamma_s\eta_s r}{4}\mathbb{E}_s\|x^s-x^{s-1}\|^2 \\
&\quad+\mathbb{E}_s\left\{\textstyle\sum_{i=1}^{m}[\gamma_1(1+\tau_1-\tfrac{1}{m})\Psi(\underline{x}_i^0,x^*)-\tfrac{\gamma_s(1+\tau_s)}{2}\Psi(\underline{x}_i^s,x^*)]\right\} \\
&\quad-\gamma_s\tfrac{1}{m}\textstyle\sum_{i=1}^{m}\mathbb{E}_s\left[\tfrac{m(1+\tau_s)}{4\hat{L}}\|\nabla\psi_i(\underline{x}_i^s)-\nabla\psi_i(x^*)\|^2-\langle\nabla\psi_i(\underline{x}_i^s)\right. \\
&\qquad\qquad\qquad\qquad\left.-\nabla\psi_i(x^*),x^{s-1}-x^s\rangle\right] \\
&\overset{(b)}{\leq}\mathbb{E}_s[\gamma_1\eta_1V_\varphi(x^*,x^0)-\gamma_s(1+\eta_s)V_\varphi(x^*,x^s)]-\gamma_s\left[\tfrac{\eta_s r}{4}-\tfrac{\hat{L}}{m(1+\tau_s)}\right]\mathbb{E}_s\|x^s-x^{s-1}\|^2 \\
&\quad+\textstyle\sum_{i=1}^{m}\mathbb{E}_s[\gamma_1(1+\tau_1-\tfrac{1}{m})\Psi(\underline{x}_i^0,x^*)-\tfrac{\gamma_s(1+\tau_s)}{2}\Psi(\underline{x}_i^s,x^*)] \\
&\overset{(c)}{\leq}\mathbb{E}_s\left[\gamma_1\eta_1V_\varphi(x^*,x^0)-\gamma_s(1+\eta_s)V_\varphi(x^*,x^s)\right] \\
&\quad+\tfrac{\gamma_1[(1+\tau_1)-\frac{1}{m}]\hat{L}}{2}\textstyle\sum_{i=1}^{m}\mathbb{E}_s\|\underline{x}_i^0-x^*\|^2-\tfrac{\mu\gamma_s(1+\tau_s)}{4}\textstyle\sum_{i=1}^{m}\mathbb{E}_s\|\underline{x}_i^s-x^*\|^2,
\end{aligned}
\tag{6.6.36}
$$

where (a) follows from $\Psi(\underline{x}_i^0, x^*) \geq \frac{1}{2\hat{L}}\|\nabla\psi_{i_t}(\underline{x}_i^0) - \nabla\psi_{i_t}(x^*)\|^2$; (b) follows from the simple relation that $b\langle u, v\rangle - a\|v\|^2/2 \leq b^2\|u\|^2/(2a)$, $\forall a > 0$; and (c) follows from (6.6.25), strong convexity of ψ_i and Lipschitz continuity of $\nabla\psi_i$. This completes the proof. ■

With the help of Lemma 6.13, we now establish the main convergence properties of Algorithm 6.9.

Theorem 6.17. *Let x^* be an optimal solution of* (6.6.8), *and suppose that the parameters $\{\alpha_t\}$, $\{\tau_t\}$, $\{\eta_t\}$, and $\{\gamma_t\}$ are set as in* (6.6.16) *and* (6.6.17). *If $\varphi(x) = \frac{\mu}{2}\|x - z\|^2$, for some $z \in X$, then, for any $s \geq 1$, we have*

$$\begin{aligned} \mathbb{E}_s\left[\|x^* - x^s\|^2\right] &\leq \alpha^s(1 + 2\tfrac{\hat{L}}{\mu})\,\mathbb{E}_s\left[\|x^* - x^0\|^2 + \tfrac{1}{m}\textstyle\sum_{i=1}^m \|\underline{x}_i^0 - x^0\|^2\right], \\ \mathbb{E}_s\left[\tfrac{1}{m}\textstyle\sum_{i=1}^m \|\underline{x}_i^s - x^s\|^2\right] &\leq 6\alpha^s(1 + 2\tfrac{\hat{L}}{\mu})\,\mathbb{E}_s\left[\|x^* - x^0\|^2 + \tfrac{1}{m}\textstyle\sum_{i=1}^m \|\underline{x}_i^0 - x^0\|^2\right]. \end{aligned}$$

Proof. It is easy to check that (6.6.16) and (6.6.17) satisfy conditions (6.6.20), (6.6.21), (6.6.22) (6.6.23), (6.6.24), and (6.6.25). Then by Lemma 6.13, we have

$$\mathbb{E}_s\left[V_\varphi(x^*, x^s) + \textstyle\sum_{i=1}^m \tfrac{\mu}{4m}\|\underline{x}_i^s - x^*\|^2\right] \leq \alpha^s \mathbb{E}_s\left[V_\varphi(x^*, x^0) + \textstyle\sum_{i=1}^m \tfrac{\hat{L}}{2m}\|\underline{x}_i^0 - x^*\|^2\right]. \tag{6.6.37}$$

Since $\varphi(x) = \frac{\mu}{2}\|x - z\|^2$, we have $V_\varphi(x^*, x^s) = \frac{\mu}{2}\|x^* - x^s\|^2$, and $V_\varphi(x^0, x^s) = \frac{\mu}{2}\|x^* - x^0\|^2$. Plugging into (6.6.37), we obtain the following two relations:

$$\begin{aligned} \mathbb{E}_s\left[\|x^* - x^s\|^2\right] &\leq \alpha^s \mathbb{E}_s\left[\|x^* - x^0\|^2 + \textstyle\sum_{i=1}^m \tfrac{\hat{L}}{mr}\|\underline{x}_i^0 - x^*\|^2\right] \\ &\leq \alpha^s \mathbb{E}_s\left[\|x^* - x^0\|^2 + \textstyle\sum_{i=1}^m \tfrac{\hat{L}}{mr}(2\|\underline{x}_i^0 - x^0\|^2 + 2\|x^0 - x^*\|^2)\right] \\ &= \alpha^s\,\mathbb{E}_s\left[(1 + 2\tfrac{\hat{L}}{\mu})\|x^* - x^0\|^2 + \textstyle\sum_{i=1}^m \tfrac{2\hat{L}}{m\mu}\|\underline{x}_i^0 - x^0\|^2\right] \\ &\leq \alpha^s(1 + 2\tfrac{\hat{L}}{\mu})\,\mathbb{E}_s\left[\|x^* - x^0\|^2 + \textstyle\sum_{i=1}^m \tfrac{1}{m}\|\underline{x}_i^0 - x^0\|^2\right], \\ \mathbb{E}_s\left[\tfrac{1}{m}\textstyle\sum_{i=1}^m \|\underline{x}_i^s - x^*\|^2\right] &\leq 2\alpha^s \mathbb{E}_s\left[\|x^* - x^0\|^2 + \textstyle\sum_{i=1}^m \tfrac{\hat{L}}{m\mu}\|\underline{x}_i^0 - x^*\|^2\right] \\ &\leq 2\alpha^s(1 + \tfrac{2\hat{L}}{\mu})\mathbb{E}_s\left[\|x^* - x^0\|^2 + \tfrac{1}{m}\textstyle\sum_{i=1}^m \|\underline{x}_i^0 - x^0\|^2\right]. \end{aligned}$$

In view of the above two relations, we have

$$\begin{aligned} \mathbb{E}_s\left[\tfrac{1}{m}\textstyle\sum_{i=1}^m \|\underline{x}_i^s - x^s\|^2\right] &\leq \mathbb{E}_s\left[\tfrac{1}{m}\textstyle\sum_{i=1}^m 2(\|\underline{x}_i^s - x^*\|^2 + \|x^* - x^s\|^2)\right] \\ &= 2\mathbb{E}_s\left[\tfrac{1}{m}\textstyle\sum_{i=1}^m \|\underline{x}_i^s - x^*\|^2\right] + 2\mathbb{E}_s\|x^* - x^s\|^2 \\ &\leq 6\alpha^s(1 + 2\tfrac{\hat{L}}{\mu})\,\mathbb{E}_s\left[\|x^* - x^0\|^2 + \tfrac{1}{m}\textstyle\sum_{i=1}^m \|\underline{x}_i^0 - x^0\|^2\right]. \end{aligned}$$

■

In view of Theorem 6.17, Algorithm 6.9 applied to subproblem (6.6.8) exhibits a fast linear rate of convergence. Actually, as shown below we do not need to solve the subproblem too accurately, and a constant number of iterations of Algorithm 6.9 for each subproblem is enough to guarantee the convergence of Algorithm 6.8.

Lemma 6.14. *Let the number of inner iterations $s \ge \lceil -\log(7M/6)/\log\alpha \rceil$ with $M := 6(5+2L/\mu)$ be given. Also let the iterates $\bar{x}^\ell$, $\ell = 1,\ldots,k$, be generated by Algorithm 6.8, and $\hat{\ell}$ be randomly selected from $[k]$. Then*

$$\mathbb{E}\|x_*^{\hat{\ell}} - \bar{x}^{\hat{\ell}-1}\|^2 \le \tfrac{4(1-M\alpha^s)}{k\mu(6-7M\alpha^s)}[f(\bar{x}^0) - f(x^*)],$$
$$\mathbb{E}\|x_*^{\hat{\ell}} - \bar{x}^{\hat{\ell}}\|^2 \le \tfrac{2M\alpha^s}{3k\mu(6-7M\alpha^s)}[f(\bar{x}^0) - f(x^*)],$$

where x^* and x_*^ℓ are the optimal solutions to problem (6.6.1) *and the ℓ-th subproblem* (6.6.7)*, respectively.*

Proof. According to Theorem 6.17 (with $\hat{L} = 2\mu + L$), we have, for $\ell \ge 1$,

$$\begin{aligned}
\mathbb{E}\|x_*^\ell - \bar{x}^\ell\|^2 &\le \alpha^s(5+\tfrac{2L}{\mu})\mathbb{E}\left[\|x_*^\ell - \bar{x}^{\ell-1}\|^2 + \textstyle\sum_{i=1}^m \tfrac{1}{m}\|\underline{x}_i^{\ell-1} - \bar{x}^{\ell-1}\|^2\right]\\
&\le \tfrac{M\alpha^s}{6}\,\mathbb{E}\left[\|x_*^\ell - \bar{x}^{\ell-1}\|^2 + \textstyle\sum_{i=1}^m \tfrac{1}{m}\|\underline{x}_i^{\ell-1} - \bar{x}^{\ell-1}\|^2\right], \quad (6.6.38)\\
\mathbb{E}\left[\tfrac{1}{m}\textstyle\sum_{i=1}^m \|\underline{x}_i^\ell - \bar{x}^\ell\|^2\right] &\le 4\alpha^s(5+\tfrac{2L}{\mu})\,\mathbb{E}\left[\|x_*^\ell - \bar{x}^{\ell-1}\|^2 + \textstyle\sum_{i=1}^m \tfrac{1}{m}\|\underline{x}_i^{\ell-1} - \bar{x}^{\ell-1}\|^2\right]\\
&\le M\alpha^s\,\mathbb{E}\left[\|x_*^\ell - \bar{x}^{\ell-1}\|^2 + \textstyle\sum_{i=1}^m \tfrac{1}{m}\|\underline{x}_i^{\ell-1} - \bar{x}^{\ell-1}\|^2\right]. \quad (6.6.39)
\end{aligned}$$

By induction on (6.6.39) and noting $\underline{x}_i^0 = \bar{x}^0$, $i = 1,\ldots,m$, we have

$$\mathbb{E}\left[\tfrac{1}{m}\textstyle\sum_{i=1}^m \|\underline{x}_i^\ell - \bar{x}^\ell\|^2\right] \le \textstyle\sum_{j=1}^\ell (M\alpha^s)^{\ell-j+1}\mathbb{E}\|x_*^j - \bar{x}^{j-1}\|^2.$$

In view of the above relation and (6.6.38), for $\ell \ge 2$, we have

$$\mathbb{E}\|x_*^\ell - \bar{x}^\ell\|^2 \le \tfrac{M\alpha^s}{6}\,\mathbb{E}\left[\|x_*^\ell - \bar{x}^{\ell-1}\|^2 + \textstyle\sum_{j=1}^{\ell-1}(M\alpha^s)^{\ell-j}\|x_*^j - \bar{x}^{j-1}\|^2\right].$$

Summing up both sides of the above inequality from $\ell = 1$ to k, we then obtain

$$\begin{aligned}
&\textstyle\sum_{\ell=1}^k \mathbb{E}\|x_*^\ell - \bar{x}^\ell\|^2\\
&\le \tfrac{M\alpha^s}{6}\mathbb{E}\left[\|x_*^1 - \bar{x}^0\|^2 + \textstyle\sum_{\ell=2}^k\left(\|x_*^\ell - \bar{x}^{\ell-1}\|^2 + \textstyle\sum_{j=1}^{\ell-1}(M\alpha^s)^{\ell-j}\|x_*^j - \bar{x}^{j-1}\|^2\right)\right]\\
&= \tfrac{M\alpha^s}{6}\,\mathbb{E}\left[\|x_*^k - \bar{x}^{k-1}\|^2 + \textstyle\sum_{\ell=1}^{k-1}\left(\tfrac{1}{1-M\alpha^s} - \tfrac{(M\alpha^s)^{k+1-\ell}}{1-M\alpha^s}\right)\|x_*^\ell - \bar{x}^{\ell-1}\|^2\right]\\
&\le \tfrac{M\alpha^s}{6(1-M\alpha^s)}\textstyle\sum_{\ell=1}^k \mathbb{E}\|x_*^\ell - \bar{x}^{\ell-1}\|^2. \qquad (6.6.40)
\end{aligned}$$

Using the fact that x_*^ℓ is optimal to the ℓ-th subproblem, and letting $x_*^0 = \bar{x}^0$ (x_*^0 is a free variable), we have

$$\textstyle\sum_{\ell=1}^k[\psi^\ell(x_*^\ell) + \varphi^\ell(x_*^\ell)] \le \textstyle\sum_{\ell=1}^k[\psi^\ell(x_*^{\ell-1}) + \varphi^\ell(x_*^{\ell-1})],$$

which, in view of the definition of ψ^ℓ and φ^ℓ, then implies that

$$\textstyle\sum_{\ell=1}^k \mathbb{E}[f(x_*^\ell) + \tfrac{3\mu}{2}\|x_*^\ell - \bar{x}^{\ell-1}\|^2] \le \textstyle\sum_{\ell=1}^k \mathbb{E}[f(x_*^{\ell-1}) + \tfrac{3\mu}{2}\|x_*^{\ell-1} - \bar{x}^{\ell-1}\|^2]. \quad (6.6.41)$$

Combining (6.6.40) and (6.6.41), we obtain

$$\begin{aligned}
&\tfrac{3\mu}{2}\textstyle\sum_{\ell=1}^k \mathbb{E}\|x_*^\ell - \bar{x}^{\ell-1}\|^2\\
&\le \textstyle\sum_{\ell=1}^k \mathbb{E}\{f(x_*^{\ell-1}) - f(x_*^\ell)\} + \tfrac{3\mu}{2}\sum_{\ell=1}^k \mathbb{E}\|x_*^{\ell-1} - \bar{x}^{\ell-1}\|^2\\
&\le \textstyle\sum_{\ell=1}^k \mathbb{E}\{f(x_*^{\ell-1}) - f(x_*^\ell)\} + \tfrac{3\mu}{2}\sum_{\ell=1}^k \mathbb{E}\|x_*^{\ell} - \bar{x}^{\ell}\|^2\\
&\le \textstyle\sum_{\ell=1}^k \mathbb{E}\{f(x_*^{\ell-1}) - f(x_*^\ell)\} + \tfrac{3\mu}{2}\tfrac{M\alpha^s}{6(1-M\alpha^s)}\sum_{\ell=1}^k \mathbb{E}\|x_*^{\ell} - \bar{x}^{\ell-1}\|^2. \qquad (6.6.42)
\end{aligned}$$

Using (6.6.42), (6.6.40), and the condition on s, we have

$$\textstyle\sum_{\ell=1}^k \mathbb{E}\|x_*^\ell - \bar{x}^{\ell-1}\|^2 \le \tfrac{4(1-M\alpha^s)}{\mu(6-7M\alpha^s)}[f(\bar{x}^0) - f(x^*)],$$
$$\textstyle\sum_{\ell=1}^k \mathbb{E}\|x_*^\ell - \bar{x}^{\ell}\|^2 \le \tfrac{2M\alpha^s}{3\mu(6-7M\alpha^s)}[f(\bar{x}^0) - f(x^*)].$$

Our results then immediately follow since $\hat{\ell}$ is chosen randomly in $[k]$. ■

Now we are ready to prove Theorem 6.16 using all the previous results we have developed.

Proof of Theorem 6.16. By the optimality condition of the $\hat{\ell}$-th subproblem (6.6.7),

$$\nabla\psi^{\hat{\ell}}(x_*^{\hat{\ell}}) + \nabla\varphi^{\hat{\ell}}(x_*^{\hat{\ell}}) \in -N_X(x_*^{\hat{\ell}}). \qquad (6.6.43)$$

From the definition of $\psi^{\hat{\ell}}$ and $\varphi^{\hat{\ell}}$, we have

$$\nabla f(x_*^{\hat{\ell}}) + 3\mu(x_*^{\hat{\ell}} - \bar{x}^{\hat{\ell}-1}) \in -N_X(x_*^{\hat{\ell}}). \qquad (6.6.44)$$

From the optimality condition of (6.6.13), we obtain

$$\begin{aligned}
&\varphi(x^t) - \varphi(x^*) + \langle \tfrac{1}{m}\textstyle\sum_{i=1}^m \tilde{y}_i^t, x^t - x^*\rangle\\
&\quad\le \eta_t V_\varphi(x^*, x^{t-1}) - (1+\eta_t)V_\varphi(x^*, x^t) - \eta_t V_\varphi(x^t, x^{t-1}). \qquad (6.6.45)
\end{aligned}$$

Using the above relation and Lemma 6.14, we have

$$\begin{aligned}
\mathbb{E}\|\bar{x}^{\hat{\ell}-1} - x_*^{\hat{\ell}}\|^2 &\le \tfrac{4(1-M\alpha^s)}{k\mu(6-7M\alpha^s)}[f(\bar{x}^0) - f(x^*)] \le \tfrac{4}{k\mu}[f(\bar{x}^0) - f(x^*)],\\
\mathbb{E}\left[d\left(\nabla f(x_*^{\hat{\ell}}), -N_X(x_*^{\hat{\ell}})\right)\right]^2 &\le \mathbb{E}\|3\mu(\bar{x}^{\hat{\ell}-1} - x_*^{\hat{\ell}})\|^2 \le \tfrac{36\mu}{k}[f(\bar{x}^0) - f(x^*)],\\
\mathbb{E}\|\bar{x}^{\hat{\ell}} - x_*^{\hat{\ell}}\|^2 &\le \tfrac{2M\alpha^s}{3k\mu(6-7M\alpha^s)}[f(\bar{x}^0) - f(x^*)] \le \tfrac{4M\alpha^s}{k\mu}[f(\bar{x}^0) - f(x^*)]\\
&\le \tfrac{4\mu}{kL^2}[f(\bar{x}^0) - f(x^*)].
\end{aligned}$$

■

6.6.2 Nonconvex Multi-Block Problems

In this section, we present a randomized accelerated proximal dual (RapDual) algorithm for solving the nonconvex multi-block optimization problem in (6.6.4) and show the potential advantages in terms of the total number of block updates.

As mentioned earlier, we assume the inverse of the last block of the constraint matrix is easily computable. Hence, denoting $\mathbf{A}_i = A_m^{-1}A_i$, $i = 1, \ldots, m-1$ and $\mathbf{b} = A_m^{-1}$, we can reformulate problem (6.6.4) as

$$\min_{\mathbf{x}\in X,\ x_m\in\mathbb{R}^n} f(\mathbf{x}) + f_m(x_m),$$
$$\text{s.t.}\ \ \mathbf{A}\mathbf{x} + x_m = \mathbf{b}, \tag{6.6.46}$$

where $f(\mathbf{x}) := \sum_{i=1}^{m-1} f_i(x_i)$, $X = X_1 \times \ldots \times X_{m-1}$, $\mathbf{A} = [\mathbf{A}_1, \ldots, \mathbf{A}_{m-1}]$, and $\mathbf{x} = (x_1, \ldots, x_{m-1})$. It should be noted that except for some special cases, the computation of A_m^{-1} requires up to $\mathcal{O}(n^3)$ arithmetic operations, which will be a one-time computational cost added on top of the overall computational cost of our algorithm (see Remark 6.1 below for more discussions).

One may also reformulate problem (6.6.46) in the form of (6.6.1) and directly apply Algorithm 6.8 to solve it. More specifically, substituting x_m with $\mathbf{b} - \mathbf{A}\mathbf{x}$ in the objective function of (6.6.46), we obtain

$$\min_{\mathbf{x}\in X} \sum_{i=1}^{m-1} f_i(B_i\mathbf{x}) + f_m(\mathbf{b} - \mathbf{A}\mathbf{x}), \tag{6.6.47}$$

where $B_i = (\mathbf{0}, \ldots, I, \ldots, \mathbf{0})$ with the i-th block given a $d_i \times d_i$ identity matrix and hence $x_i = B_i\mathbf{x}$. However, this method will be inefficient since we enlarge the dimension of each f_i from d_i to $\sum_{i=1}^{m-1} d_i$ and as a result, every block has to be updated in each iteration. One may also try to apply a nonconvex randomized block mirror descent method in Sect. 6.3 to solve the above reformulation. However, such methods do not apply to the case when f_i are both nonconvex and nonsmooth. This motivates us to design the new RapDual method which requires to update only a single block at a time, applies to the case when f_i is nonsmooth, and achieves an accelerated rate of convergence when f_i is smooth.

6.6.2.1 The RapDual Algorithm

The main idea of RapDual is similar to the one used to design the RapGrad method introduced in Sect. 6.6.1.1. Given the proximal points $\bar{\mathbf{x}}^{\ell-1}$ and $\bar{x}_m^{\ell-1}$ from the previous iteration, we define a new proximal subproblem as

$$\min_{\mathbf{x}\in X, x_m\in\mathbb{R}^n} \psi(\mathbf{x}) + \psi_m(x_m)$$
$$\text{s.t.}\ \ \mathbf{A}\mathbf{x} + x_m = \mathbf{b}, \tag{6.6.48}$$

where $\psi(\mathbf{x}) := f(\mathbf{x}) + \mu\|\mathbf{x} - \bar{\mathbf{x}}^{\ell-1}\|^2$ and $\psi_m(x_m) := f_m(x_m) + \mu\|x_m - x_m^{\ell-1}\|^2$. Obviously, RaGrad does not apply directly to this type of subproblem. In this subsection, we present a new randomized algorithm, named the randomized accelerated dual (RaDual) method to solve the subproblem in (6.6.48), which will be iteratively called by the RapDual method to solve problem (6.6.46).

RaDual (cf. Algorithm 6.11) can be viewed as a randomized primal–dual type method. Indeed, by the method of multipliers and Fenchel conjugate duality, we have

$$\begin{aligned}
&\min_{\mathbf{x}\in X, x_m\in\mathbb{R}^n} \{\psi(\mathbf{x}) + \psi_m(x_m) + \max_{y\in\mathbb{R}^n} \langle \textstyle\sum_{i=1}^m \mathbf{A}_i x_i - \mathbf{b}, y\rangle\} \\
&= \min_{\mathbf{x}\in X} \{\psi(\mathbf{x}) + \max_{y\in\mathbb{R}^n} [\langle \mathbf{A}\mathbf{x} - \mathbf{b}, y\rangle + \min_{x_m\in\mathbb{R}^n} \{\psi_m(x_m) + \langle x_m, y\rangle\}]\} \\
&= \min_{\mathbf{x}\in X} \{\psi(\mathbf{x}) + \max_{y\in\mathbb{R}^n} [\langle \mathbf{A}\mathbf{x} - \mathbf{b}, y\rangle - h(y)]\},
\end{aligned} \tag{6.6.49}$$

where $h(y) := -\min_{x_m\in\mathbb{R}^n}\{\psi_m(x_m) + \langle x_m, y\rangle\} = \psi_m^*(-y)$. Observe that the above saddle point problem is both strongly convex in $\mathbf{x}$ and strongly concave in y. Indeed, $\psi(\mathbf{x})$ is strongly convex due to the added proximal term. Moreover, since ψ_m has $\hat{L}$-Lipschitz continuous gradients, $h(y) = \psi_m^*(-y)$ is $1/\hat{L}$-strongly convex. Using the fact that h is strongly convex, we can see that (6.6.52)–(6.6.53) in Algorithm 6.11 is equivalent to a dual mirror-descent step with a properly chosen distance generating function $V_h(y, y^{t-1})$. Specifically,

$$\begin{aligned}
y^t &= \operatorname{argmin}_{y\in\mathbb{R}^n} h(y) + \langle -\mathbf{A}\tilde{\mathbf{x}}^t + \mathbf{b}, y\rangle + \tau_t V_h(y, y^{t-1}) \\
&= \operatorname{argmax}_{y\in\mathbb{R}^n} \langle (\mathbf{A}\tilde{\mathbf{x}}^t - \mathbf{b} + \tau_t \nabla h(y^{t-1}))/(1+\tau_t), y\rangle - h(y) \\
&= \nabla h^*[(\mathbf{A}\tilde{\mathbf{x}}^t - \mathbf{b} + \tau_t \nabla h(y^{t-1}))/(1+\tau_t)].
\end{aligned}$$

If we set $g^0 = \nabla h(y^0) = -x_m^0$, then it is easy to see by induction that $g^t = (\tau_t g^{t-1} + \mathbf{A}\tilde{\mathbf{x}}^t - \mathbf{b})/(1+\tau_t)$, and $y^t = \nabla h^*(g^t)$ for all $t \ge 1$. Moreover, $h^*(g) = \max_{y\in\mathbb{R}^n}\langle g, y\rangle - h(y) = \max_{y\in\mathbb{R}^n}\langle g, y\rangle - \psi_m^*(-y) = \psi_m(-g)$, thus $y^t = -\nabla\psi_m(-g^t)$ is the negative gradient of ψ_m at point $-g^t$. Therefore, Algorithm 6.11 does not explicitly depend on the function h, even though the above analysis does.

Each iteration of Algorithm 6.11 updates only a randomly selected block i_t in (6.6.54), making it especially favorable when the number of blocks m is large. However, similar difficulty as mentioned in Sect. 6.6.1.1 also appears when we integrate this algorithm with proximal-point type method to yield the final RapDual method in Algorithm 6.10. First, Algorithm 6.11 also keeps a few intertwined primal and dual sequences, thus we need to carefully decide the input and output of Algorithm 6.11 so that information from previous iterations of RapDual is fully used. Second, the number of iterations performed by Algorithm 6.11 to solve each subproblem plays a vital role in the convergence rate of RapDual, which should be carefully predetermined.

Algorithm 6.10 describes the basic scheme of RapDual. At the beginning, all the blocks are initialized using the output from solving the previous subproblem. Note that x_m^0 is used to initialize g, which further helps compute the dual variable y

without using the conjugate function h of ψ_m. We will derive the convergence result for Algorithm 6.11 in terms of primal variables and construct relations between successive search points $(\mathbf{x}^\ell, x_m^l)$, which will be used to prove the final convergence of RapDual.

Algorithm 6.10 RapDual for nonconvex multi-block optimization

Compute A_m^{-1} and reformulate problem (6.6.4) as (6.6.46).
Let $\bar{\mathbf{x}}^0 \in X, \bar{x}_m^0 \in \mathbb{R}^n$, such that $\mathbf{A}\bar{\mathbf{x}}^0 + \bar{x}_m^0 = \mathbf{b}$, and $\bar{y}^0 = -\nabla f_m(\bar{x}_m^0)$.
for $\ell = 1, \ldots, k$ **do**
Set $\mathbf{x}^{-1} = \mathbf{x}^0 = \bar{\mathbf{x}}^{\ell-1}$, $x_m^0 = \bar{x}_m^{\ell-1}$.
Run Algorithm 6.11 with input $\mathbf{x}^{-1}$, $\mathbf{x}^0$, x_m^0, and s to solve the following subproblem

$$\begin{aligned} \min_{\mathbf{x} \in X, x_m \in \mathbb{R}^n} \quad & \psi(\mathbf{x}) + \psi_m(x_m) \\ \text{s.t.} \quad & \mathbf{A}\mathbf{x} + x_m = \mathbf{b}, \end{aligned} \tag{6.6.50}$$

to compute output $(\mathbf{x}^s, x_m^s)$, where $\psi(\mathbf{x}) \equiv \psi^\ell(\mathbf{x}) := f(\mathbf{x}) + \mu\|\mathbf{x} - \bar{\mathbf{x}}^{\ell-1}\|^2$ and $\psi_m(x) \equiv \psi_m^\ell(x) := f_m(x_m) + \mu\|x_m - \bar{x}_m^{\ell-1}\|^2$.
Set $\bar{\mathbf{x}}^\ell = \mathbf{x}^s$, $\bar{x}_m^\ell = x_m^s$.

end forreturn $(\bar{\mathbf{x}}^{\hat{\ell}}, \bar{x}_m^{\hat{\ell}})$ for some random $\hat{\ell} \in [k]$.

Algorithm 6.11 RaDual for solving subproblem (6.6.48)

Let $\mathbf{x}^{-1} = \mathbf{x}^0 \in X, x_m \in \mathbb{R}^n$, number of iterations s and nonnegative parameters $\{\alpha_t\}$, $\{\tau_t\}$, $\{\eta_t\}$ be given. Set $g^0 = -x_m^0$.
for $t = 1, \ldots, s$ **do**
1. Generate a random variable i_t uniformly distributed over $[m-1]$.
2. Update x^t and y^t according to

$$\tilde{\mathbf{x}}^t = \alpha_t(\mathbf{x}^{t-1} - \mathbf{x}^{t-2}) + \mathbf{x}^{t-1}, \tag{6.6.51}$$

$$g^t = (\tau_t g^{t-1} + \mathbf{A}\tilde{\mathbf{x}}^t - \mathbf{b})/(1 + \tau_t), \tag{6.6.52}$$

$$y^t = \text{argmin}_{y \in \mathbb{R}^n} h(y) + \langle -\mathbf{A}\tilde{\mathbf{x}}^t + \mathbf{b}, y \rangle + \tau_t V_h(y, y^{t-1}) = -\nabla \psi_m(-g^t), \tag{6.6.53}$$

$$x_i^t = \begin{cases} \text{argmin}_{x_i \in X_i} \psi_i(x_i) + \langle \mathbf{A}_i^\top y^t, x_i \rangle + \frac{\eta_t}{2}\|x_i - x_i^{t-1}\|^2, & i = i_t, \\ x_i^{t-1}, & i \neq i_t. \end{cases} \tag{6.6.54}$$

end for
Compute $x_m^s = \text{argmin}_{x_m \in \mathbb{R}^n}\{\psi_m(x_m) + \langle x_m, y^s \rangle\}$.
return $(\mathbf{x}^s, x_m^s)$.

We first define an approximate stationary point for problem (6.6.4) before establishing the convergence of RapDual.

Definition 6.2. A point $(x, x_m) \in X \times \mathbb{R}^n$ is called an $(\varepsilon, \delta, \sigma)$-solution of (6.6.4) if there exists some $\hat{x} \in X$, and $\lambda \in \mathbb{R}^n$ such that

$$\left[d(\nabla f(\hat{x})+\mathbf{A}^\top\lambda,-N_X(\hat{x}))\right]^2 \le \varepsilon,\ \|\nabla f_m(x_m)+\lambda\|^2 \le \varepsilon,$$
$$\|x-\hat{x}\|^2 \le \delta,\ \|\mathbf{A}x+x_m-\mathbf{b}\|^2 \le \sigma.$$

A stochastic counterpart is one that satisfies

$$\mathbb{E}\left[d(\nabla f(\hat{x})+\mathbf{A}^\top\lambda,-N_X(\hat{x}))\right]^2 \le \varepsilon,\ \mathbb{E}\|\nabla f_m(x_m)+\lambda\|^2 \le \varepsilon,$$
$$\mathbb{E}\|x-\hat{x}\|^2 \le \delta,\ \mathbb{E}\|\mathbf{A}x+x_m-\mathbf{b}\|^2 \le \sigma.$$

Consider the unconstrained problem with $X=\mathbb{R}^{\sum_{i=1}^{m-1}d_i}$. If $(\mathbf{x},x_m)\in X\times\mathbb{R}^n$ is an $(\varepsilon,\delta,\sigma)$-solution with $\delta=\varepsilon/L^2$, then exists some $\hat{\mathbf{x}}\in X$ such that $\|\nabla f(\hat{\mathbf{x}})\|^2\le\varepsilon$ and $\|\mathbf{x}-\hat{\mathbf{x}}\|^2\le\delta$. By similar argument in (6.6.14), we obtain $\|\nabla f(\mathbf{x})\|^2\le 4\varepsilon$. Besides, the definition of a $(\varepsilon,\delta,\sigma)$-solution guarantees $\|\nabla f_m(x_m)+\lambda\|^2\le\varepsilon$ and $\|\mathbf{A}\mathbf{x}+x_m-\mathbf{b}\|^2\le\sigma$, which altogether justify that $(\mathbf{x},x_m)$ is a reasonably good solution. The following result shows the convergence of RapDual to find such an approximate solution. Its proof is involved and will be postponed to Sect. 6.6.2.2.

Theorem 6.18. *Let the iterates $(\mathbf{x}^\ell,x_m^\ell)$ for $\ell=1,\ldots,k$ be generated by Algorithm 6.10 and $\hat{\ell}$ be randomly selected from $[k]$. Suppose that in Algorithm 6.11, number of iterations $s=\left\lceil -\log\widehat{M}/\log\alpha\right\rceil$ with*

$$\widehat{M}=(2+\tfrac{L}{\mu})\cdot\max\left\{2,\tfrac{L^2}{\mu^2}\right\}\ \textit{and}\ \alpha=1-\tfrac{2}{(m-1)(\sqrt{1+8c}+1)},\tag{6.6.55}$$

where

$$c=\tfrac{\bar{A}^2}{\mu\bar{\mu}}=\tfrac{(2\mu+L)\bar{A}^2}{\mu}\quad\textit{and}\quad\bar{A}=\max_{i\in[m-1]}\|\mathbf{A}_i\|,$$

and that other parameters are set to

$$\alpha_t=(m-1)\alpha,\gamma_t=\alpha^{-t},\tau_t=\tfrac{\alpha}{1-\alpha},\ \textit{and}\ \eta_t=\frac{\left(\alpha-\frac{m-2}{m-1}\right)\mu}{1-\alpha},\forall t=1,\ldots,s.\tag{6.6.56}$$

Then there exists some $\lambda^*\in\mathbb{R}^n$ such that

$$\mathbb{E}\left[d(\nabla f(\mathbf{x}_*^{\hat{\ell}})+\mathbf{A}^\top\lambda^*,-N_X(\mathbf{x}_*^{\hat{\ell}}))\right]^2\le\tfrac{8\mu}{k}\left\{f(\bar{\mathbf{x}}^0)+f_m(\bar{x}_m^0)-[f(\mathbf{x}^*)+f_m(x_m^*)]\right\},$$
$$\mathbb{E}\|\nabla f_m(x_m^{\hat{\ell}})+\lambda^*\|^2\le\tfrac{34\mu}{k}\left\{f(\bar{\mathbf{x}}^0)+f_m(\bar{x}_m^0)-[f(\mathbf{x}^*)+f_m(x_m^*)]\right\},$$
$$\mathbb{E}\|\mathbf{x}^{\hat{\ell}}-\mathbf{x}_*^{\hat{\ell}}\|^2\le\tfrac{2\mu}{kL^2}\left\{f(\bar{\mathbf{x}}^0)+f_m(\bar{x}_m^0)-[f(\mathbf{x}^*)+f_m(x_m^*)]\right\},$$
$$\mathbb{E}\|\mathbf{A}\mathbf{x}^{\hat{\ell}}+x_m^{\hat{\ell}}-\mathbf{b}\|^2\le\tfrac{2(\|\mathbf{A}\|^2+1)\mu}{kL^2}\left\{f(\bar{\mathbf{x}}^0)+f_m(\bar{x}_m^0)-[f(\mathbf{x}^*)+f_m(x_m^*)]\right\},$$

where $(\mathbf{x}^*,x_m^*)$ and $(\mathbf{x}_*^\ell,x_{m*}^\ell)$ denote the optimal solutions to (6.6.4) *and the ℓ-th subproblem* (6.6.48), *respectively.*

Theorem 6.18 ensures that our output solution $(\mathbf{x}^{\hat{\ell}},x_m^{\hat{\ell}})$ is close enough to an unknown approximate stationary point $(\mathbf{x}_*^{\hat{\ell}},x_{m*}^{\hat{\ell}})$. According to Theorem 6.18, we can bound the complexity of RapDual to compute a stochastic $(\varepsilon,\delta,\sigma)$-solution of (6.6.4) in terms of block updates in (6.6.54). Note that for each subproblem (6.6.48), we only need to update s primal blocks with

$$s = \left\lceil -\tfrac{\log \hat{M}}{\log \alpha} \right\rceil \sim \mathscr{O}\left(m\bar{A}\sqrt{\tfrac{L}{\mu}}\log\left(\tfrac{L}{\mu}\right)\right).$$

Let $\mathscr{D}^0 := f(\bar{\mathbf{x}}^0) + f_m(\bar{x}_m^0) - [f(\mathbf{x}^*) + f_m(x_m^*)]$. It can be seen that the total number of primal block updates required to obtain a stochastic $(\varepsilon,\delta,\sigma)$-solution can be bounded by

$$N(\varepsilon,\delta,\sigma) := \mathscr{O}\left(m\bar{A}\sqrt{L\mu}\log\left(\tfrac{L}{\mu}\right)\cdot \max\left\{\tfrac{1}{\varepsilon}, \tfrac{1}{\delta L^2}, \tfrac{\|\mathbf{A}\|^2}{\sigma L^2}\right\}\mathscr{D}^0\right). \tag{6.6.57}$$

As a comparison, the batch version of this algorithm would update all the x_i^t for $i = 1,\ldots,m$, in (6.6.54), and thus would require

$$\hat{N}(\varepsilon,\delta,\sigma) := \mathscr{O}\left(m\|\mathbf{A}\|\sqrt{L\mu}\log\left(\tfrac{L}{\mu}\right)\cdot \max\left\{\tfrac{1}{\varepsilon}, \tfrac{1}{\delta L^2}, \tfrac{\|\mathbf{A}\|^2}{\sigma L^2}\right\}\mathscr{D}^0\right).$$

primal block updates to obtain an $(\varepsilon,\delta,\sigma)$-solution of (6.6.4). Therefore, the benefit of randomization comes from the difference between $\|\mathbf{A}\|$ and $\bar{A}$. Obviously we always have $\|\mathbf{A}\| > \bar{A}$, and the relative gap between $\|\mathbf{A}\|$ and $\bar{A}$ can be large when all the matrix blocks have close norms. In the case when all the blocks are identical, i.e., $\mathbf{A}_1 = \mathbf{A}_2 = \ldots = \mathbf{A}_{m-1}$, we immediately have $\|\mathbf{A}\| = \sqrt{m-1}\bar{A}$, which means that RapDual can potentially save the number of primal block updates by a factor of $\mathscr{O}(\sqrt{m})$ than its batch counterpart.

It is also interesting to compare RapDual with the nonconvex randomized block mirror descent method in Sect. 6.3. To compare these methods, let us assume that f_i is smooth with $\bar{L}$-Lipschitz continuous gradient for some $\bar{L} \ge \mu$ for any $i = 1,\ldots,m$. Also let us assume that $\sigma > \|\mathbf{A}\|^2\varepsilon/L^2$, $\delta = \varepsilon/L^2$, and $X = \mathbb{R}^{\sum_{i=1}^{m-1} d_i}$. Then, after disregarding some constant factors, the bound in (6.6.57) reduces to $\mathscr{O}(m\bar{A}\sqrt{L\mu}\mathscr{D}^0/\varepsilon)$, which is always smaller than the $\mathscr{O}(m(\bar{L}+L\bar{A}^2)\mathscr{D}^0/\varepsilon)$ complexity bound implied by Corollary 6.12.

Remark 6.1. In this section, we assume that A_m^{-1} is easily computable. One natural question is whether we can avoid the computation of A_m^{-1} by directly solving (6.6.4) instead of its reformulation (6.6.46). To do so, we can iteratively solve the following saddle-point subproblems in place of the ones in (6.6.49):

$$\begin{aligned}
&\min_{\mathbf{x}\in X, x_m\in\mathbb{R}^n} \{\psi(\mathbf{x}) + \psi_m(x_m) + \max_{y\in\mathbb{R}^n} \langle \textstyle\sum_{i=1}^m A_i x_i - b, y\rangle\} \\
&= \min_{\mathbf{x}\in X}\{\psi(\mathbf{x}) + \max_{y\in\mathbb{R}^n}[\langle \mathbf{A}\mathbf{x} - b, y\rangle + \min_{x_m\in\mathbb{R}^n}\{\psi_m(x_m) + \langle A_m x_m, y\rangle\}]\} \\
&= \min_{\mathbf{x}\in X}\{\psi(\mathbf{x}) + \max_{y\in\mathbb{R}^n}[\langle \mathbf{A}\mathbf{x} - b, y\rangle - \tilde{h}(y)]\},
\end{aligned} \tag{6.6.58}$$

where $A := [A_1, \ldots, A_{m-1}]$ and $\tilde{h}(y) := \max_{x_m \in \mathbb{R}^n} \{-\psi_m(x_m) - \langle A_m x_m, y \rangle\}$. Instead of keeping $\tilde{h}(y)$ in the projection subproblem (6.6.53) as we did for $h(y)$, we need to linearize it at each iteration by computing its gradients $\nabla \tilde{h}(y^{t-1}) = -A_m^T \bar{x}_m(y^{t-1})$, where $\bar{x}_m(y^{t-1}) = \operatorname{argmax}_{x_m \in \mathbb{R}^n} \{-\psi_m(x_m) - \langle A_m x_m, y^{t-1} \rangle\}$. Note that the latter optimization problem can be solved by using an efficient first-order method due to the smoothness and strong concavity of its objective function. As a result, we will be able to obtain a similar rate of convergence to RapDual without computing A_m^{-1}. However, the statement and analysis of the algorithm will be much more complicated than RapDual in its current form. ■

6.6.2.2 Convergence Analysis for RapDual

In this section, we first show the convergence of Algorithm 6.11 for solving the convex multi-block subproblem (6.6.50) with

1. $\psi_i(x) - \psi_i(y) - \langle \nabla \psi_i(y), x - y \rangle \geq \frac{\mu}{2}\|x - y\|^2, \ \forall x, y \in X_i, \quad i = 1, \ldots, m-1,$
2. $\frac{\mu}{2}\|x - y\|^2 \leq \psi_m(x) - \psi_m(y) - \langle \nabla \psi_m(y), x - y \rangle \leq \frac{\hat{L}}{2}\|x - y\|^2, \ \forall x, y \in \mathbb{R}^n.$

Some simple relations about the iterations generated by the Algorithm 6.11 are characterized in the following lemma, and the proof follows directly from the definition of $\hat{\mathbf{x}}$ in (6.6.59), thus has been omitted.

Lemma 6.15. *Let* $\hat{\mathbf{x}}^0 = \mathbf{x}^0$ *and* $\hat{\mathbf{x}}^t$ *for* $t = 1, \ldots, s$ *be defined as follows:*

$$\hat{\mathbf{x}}^t = \operatorname{argmin}_{\mathbf{x} \in X} \ \psi(\mathbf{x}) + \langle \mathbf{A}^\top y^t, \mathbf{x} \rangle + \frac{\eta_t}{2}\|\mathbf{x} - \mathbf{x}^{t-1}\|^2, \tag{6.6.59}$$

where $\mathbf{x}^t$ *and* y^t *are obtained from* (6.6.53)–(6.6.54), *then we have*

$$\mathbb{E}_{i_t}\left\{\|\mathbf{x} - \hat{\mathbf{x}}^t\|^2\right\} = \mathbb{E}_{i_t}\left\{(m-1)\|\mathbf{x} - \mathbf{x}^t\|^2 - (m-2)\|\mathbf{x} - \mathbf{x}^{t-1}\|^2\right\}, \tag{6.6.60}$$

$$\mathbb{E}_{i_t}\left\{\|\hat{\mathbf{x}}^t - \mathbf{x}^{t-1}\|^2\right\} = \mathbb{E}_{i_t}\left\{(m-1)\|x_{i_t}^t - x_{i_t}^{t-1}\|^2\right\}. \tag{6.6.61}$$

The following lemma 6.16 builds some connections between the input and output of Algorithm 6.11 in terms of both primal and dual variables.

Lemma 6.16. *Let the iterates* $\mathbf{x}^t$ *and* y^t *for* $t = 1, \ldots, s$ *be generated by Algorithm 6.11 and* $(\mathbf{x}^*, y^*)$ *be a saddle point of* (6.6.49). *Assume that the parameters in Algorithm 6.11 satisfy for all* $t = 1, \ldots, s-1$

$$\alpha_{t+1} = (m-1)\tilde{\alpha}_{t+1}, \tag{6.6.62}$$

$$\gamma_t = \gamma_{t+1}\tilde{\alpha}_{t+1}, \tag{6.6.63}$$

$$\gamma_{t+1}\left((m-1)\eta_{t+1} + (m-2)\mu\right) \leq (m-1)\gamma_t(\eta_t + \mu), \tag{6.6.64}$$

$$\gamma_{t+1}\tau_{t+1} \leq \gamma_t(\tau_t + 1), \tag{6.6.65}$$

$$2(m-1)\tilde{\alpha}_{t+1}\bar{A}^2 \leq \bar{\mu}\eta_t\tau_{t+1}, \tag{6.6.66}$$

where $\bar{A} = \max_{i \in [m-1]} \|\mathbf{A}_i\|$. *Then we have*

$$\mathbb{E}_s\left\{\tfrac{\gamma_1((m-1)\eta_1+(m-2)\mu)}{2}\|\mathbf{x}^*-\mathbf{x}^0\|^2 - \tfrac{(m-1)\gamma_s(\eta_s+\mu)}{2}\|\mathbf{x}^*-\mathbf{x}^s\|^2\right\}$$
$$+\mathbb{E}_s\left\{\gamma_1\tau_1 V_h(y^*,y^0) - \tfrac{\gamma_s(\tau_s+1)\bar{\mu}}{2}V_h(y^*,y^s)\right\} \geq 0. \qquad (6.6.67)$$

Proof. For any $t \geq 1$, since $(\mathbf{x}^*, y^*)$ is a saddle point of (6.6.49), we have

$$\psi(\hat{\mathbf{x}}^t) - \psi(\mathbf{x}^*) + \langle \mathbf{A}\hat{\mathbf{x}}^t - \mathbf{b}, y^*\rangle - \langle \mathbf{A}\mathbf{x}^* - \mathbf{b}, y^t\rangle + h(y^*) - h(y^t) \geq 0.$$

For nonnegative γ_t, we further obtain

$$\mathbb{E}_s\left\{\textstyle\sum_{t=1}^s \gamma_t\left[\psi(\hat{\mathbf{x}}^t) - \psi(\mathbf{x}^*) + \langle \mathbf{A}\hat{\mathbf{x}}^t - \mathbf{b}, y^*\rangle - \langle \mathbf{A}\mathbf{x}^* - \mathbf{b}, y^t\rangle + h(y^*) - h(y^t)\right]\right\} \geq 0. \qquad (6.6.68)$$

According to optimality conditions of (6.6.59) and (6.6.53) respectively, and strongly convexity of ψ and h we obtain

$$\begin{aligned}
&\psi(\hat{\mathbf{x}}^t) - \psi(\mathbf{x}^*) + \tfrac{\mu}{2}\|\mathbf{x}^* - \hat{\mathbf{x}}^t\|^2 + \langle \mathbf{A}^\top y^t, \hat{\mathbf{x}}^t - \mathbf{x}^*\rangle \\
&\quad \leq \tfrac{\eta_t}{2}\left[\|\mathbf{x}^* - \mathbf{x}^{t-1}\|^2 - \|\mathbf{x}^* - \hat{\mathbf{x}}^t\|^2 - \|\hat{\mathbf{x}}^t - \mathbf{x}^{t-1}\|^2\right], \\
&h(y^t) - h(y^*) + \langle -\mathbf{A}\tilde{\mathbf{x}}^t + \mathbf{b}, y^t - y^*\rangle \\
&\quad \leq \tau_t V_h(y^*, y^{t-1}) - (\tau_t + 1)V_h(y^*, y^t) - \tau_t V_h(y^t, y^{t-1}).
\end{aligned}$$

Combining the above two inequalities with relation (6.6.68), we have

$$\begin{aligned}
&\mathbb{E}_s\left\{\textstyle\sum_{t=1}^k \left[\tfrac{\gamma_t\eta_t}{2}\|\mathbf{x}^* - \mathbf{x}^{t-1}\|^2 - \tfrac{\gamma_t(\eta_t+\mu)}{2}\|\mathbf{x}^* - \hat{\mathbf{x}}^t\|^2 - \tfrac{\gamma_t\eta_t}{2}\|\hat{\mathbf{x}}^t - \mathbf{x}^{t-1}\|^2\right]\right\} \\
&+\mathbb{E}_s\left\{\textstyle\sum_{t=1}^s \gamma_t\left[\tau_t V_h(y^*, y^{t-1}) - (\tau_t+1)V_h(y^*, y^t) - \tau_t V_h(y^t, y^{t-1})\right]\right\} \\
&+\mathbb{E}_s\left[\textstyle\sum_{t=1}^s \gamma_t\langle \mathbf{A}(\hat{\mathbf{x}}^t - \tilde{\mathbf{x}}^t), y^* - y^t\rangle\right] \geq 0.
\end{aligned}$$

Observe that for $t \geq 1$,

$$\mathbb{E}_{i_t}\left\{\langle \mathbf{A}(\hat{\mathbf{x}}^t - \tilde{\mathbf{x}}^t), y^*\rangle\right\} = \mathbb{E}_{i_t}\left\{\langle \mathbf{A}((m-1)\mathbf{x}^t - (m-2)\mathbf{x}^{t-1} - \tilde{\mathbf{x}}^t), y^*\rangle\right\}.$$

Applying this and the results (6.6.60), (6.6.61) in Lemma 6.15, we further have

$$\begin{aligned}
0 &\leq \mathbb{E}_s\left\{\textstyle\sum_{t=1}^s \left[\tfrac{\gamma_t((m-1)\eta_t+(m-2)\mu)}{2}\|\mathbf{x}^* - \mathbf{x}^{t-1}\|^2 - \tfrac{(m-1)\gamma_t(\eta_t+\mu)}{2}\|\mathbf{x}^* - \mathbf{x}^t\|^2\right]\right\} \\
&\quad +\mathbb{E}_s\left\{\textstyle\sum_{t=1}^s \left[\gamma_t\tau_t V_h(y^*, y^{t-1}) - \gamma_t(\tau_t+1)V_h(y^*, y^t)\right]\right\} + \mathbb{E}_s\left\{\textstyle\sum_{t=1}^s \gamma_t\delta_t\right\} \\
&\leq \mathbb{E}_s\left[\tfrac{\gamma_1((m-1)\eta_1+(m-2)\mu)}{2}\|\mathbf{x}^* - \mathbf{x}^0\|^2 - \tfrac{(m-1)\gamma_s(\eta_s+\mu)}{2}\|\mathbf{x}^* - \mathbf{x}^s\|^2\right] \\
&\quad +\mathbb{E}_s[\gamma_1\tau_1 V_h(y^*, y^0) - \gamma_s(\tau_s+1)V_h(y^*, y^s)] + \mathbb{E}_s[\textstyle\sum_{t=1}^s \gamma_t\delta_t],
\end{aligned} \qquad (6.6.69)$$

where

$$\begin{aligned}
\delta_t = &-\tfrac{(m-1)\eta_t}{2}\|x_{i_t}^t - x_{i_t}^{t-1}\|^2 - \tau_t V_h(y^t, y^{t-1}) \\
&+\langle \mathbf{A}((m-1)\mathbf{x}^t - (m-2)\mathbf{x}^{t-1} - \tilde{\mathbf{x}}^t), y^* - y^t\rangle.
\end{aligned}$$

and the second inequality follows from (6.6.64) and (6.6.65).

By (6.6.62) and the definition of $\tilde{x}^t$ in (6.6.51) we have:

$$\begin{aligned}\Sigma_{t=1}^s \gamma_t \delta_t =&\Sigma_{t=1}^s \left[-\tfrac{(m-1)\gamma_t\eta_t}{2}\|x_{i_t}^t - x_{i_t}^{t-1}\|^2 - \gamma_t\tau_t V_h(y^t, y^{t-1})\right]\\ &+ \Sigma_{t=1}^s \gamma_t(m-1)\langle \mathbf{A}(\mathbf{x}^t - \mathbf{x}^{t-1}), y^* - y^t\rangle \\ &- \Sigma_{t=1}^s \gamma_t(m-1)\tilde{\alpha}_t\langle \mathbf{A}(\mathbf{x}^{t-1} - \mathbf{x}^{t-2}), y^* - y^{t-1}\rangle\\ &- \Sigma_{t=1}^s \gamma_t(m-1)\tilde{\alpha}_t\langle \mathbf{A}(\mathbf{x}^{t-1} - \mathbf{x}^{t-2}), y^{t-1} - y^t\rangle\end{aligned} \tag{6.6.70}$$

$$\begin{aligned}=&\Sigma_{t=1}^s \left[-\tfrac{(m-1)\gamma_t\eta_t}{2}\|x_{i_t}^t - x_{i_t}^{t-1}\|^2 - \gamma_t\tau_t V_h(y^t, y^{t-1})\right]\\ &+ \gamma_s(m-1)\langle \mathbf{A}(\mathbf{x}^s - \mathbf{x}^{s-1}), y^* - y^s\rangle\\ &- \Sigma_{t=1}^s \gamma_t(m-1)\tilde{\alpha}_t\langle \mathbf{A}(\mathbf{x}^{t-1} - \mathbf{x}^{t-2}), y^{t-1} - y^t\rangle,\end{aligned} \tag{6.6.71}$$

where the second equality follows from (6.6.63) and the fact that $x^0 = x^{-1}$.

Since $\langle \mathbf{A}(\mathbf{x}^{t-1} - \mathbf{x}^{t-2}), y^{t-1} - y^t\rangle = \langle \mathbf{A}_{t-1}(x_{i_{t-1}}^{t-1} - x_{i_{t-1}}^{t-2}), y^{t-1} - y^t\rangle \le \|\mathbf{A}_{i_{t-1}}\|\|x_{i_{t-1}}^{t-1} - x_{i_{t-1}}^{t-2}\|\|y^t - y^{t-1}\|$ and $V_h(y^t, y^{t-1}) \ge \frac{\bar{\mu}}{2}\|y^t - y^{t-1}\|^2$, from (6.6.71) we have

$$\begin{aligned}\Sigma_{t=1}^s \gamma_t\delta_t \le&\Sigma_{t=1}^s \left[-\tfrac{(m-1)\gamma_t\eta_t}{2}\|x_{i_t}^t - x_{i_t}^{t-1}\|^2 - \tfrac{g_t\tau_t\bar{\mu}}{2}\|y^t, y^{t-1}\|^2\right]\\ &+ \gamma_s(m-1)\langle \mathbf{A}(\mathbf{x}^s - \mathbf{x}^{s-1}), y^* - y^s\rangle\\ &- \Sigma_{t=1}^s \gamma_t(m-1)\tilde{\alpha}_t\|\mathbf{A}_{i_{t-1}}\|\|x_{i_{t-1}}^{t-1} - x_{i_{t-1}}^{t-2}\|\|y^t - y^{t-1}\|\\ \overset{(a)}{=}& \gamma_s(m-1)\langle \mathbf{A}(\mathbf{x}^s - \mathbf{x}^{s-1}), y^* - y^s\rangle - \tfrac{(m-1)\gamma_s\eta_s}{2}\|x_{i_s}^s - x_{i_s}^{s-1}\|^2\\ &+ \Sigma_{t=2}^s\Big[\gamma_t(m-1)\tilde{\alpha}_t\|\mathbf{A}_{i_{t-1}}\|\|x_{i_{t-1}}^{t-1} - x_{i_{t-1}}^{t-2}\|\|y^t - y^{t-1}\|\\ &\quad - \tfrac{(m-1)\gamma_{t-1}\eta_{t-1}}{2}\|x_{i_{t-1}}^{t-1} - x_{i_{t-1}}^{t-2}\|^2 - \tfrac{\bar{\mu}\gamma_t\tau_t}{2}\|y^t - y^{t-1}\|^2\Big]\\ \overset{(b)}{\le}& \gamma_s(m-1)\langle \mathbf{A}(\mathbf{x}^s - \mathbf{x}^{s-1}), y^* - y^s\rangle - \tfrac{(m-1)\gamma_s\eta_s}{2}\|x_{i_s}^s - x_{i_s}^{s-1}\|^2\\ &+ \Sigma_{t=2}^s\left(\tfrac{\gamma_t^2(m-1)^2\tilde{\alpha}_t^2\bar{A}^2}{2(m-1)\gamma_{t-1}\eta_{t-1}} - \tfrac{\bar{\mu}\gamma_t\tau_t}{2}\right)\|y^t - y^{t-1}\|^2\\ \overset{(c)}{=}& \gamma_s(m-1)\langle \mathbf{A}(\mathbf{x}^s - \mathbf{x}^{s-1}), y^* - y^s\rangle - \tfrac{(m-1)\gamma_s\eta_s}{2}\|x_{i_s}^s - x_{i_s}^{s-1}\|^2,\end{aligned}$$

where (a) follows from regrouping the terms; (b) follows from the definition $\bar{A} = \max_{i\in[m-1]}\|\mathbf{A}_i\|$ and the simple relation that $b\langle u, v\rangle - a\|v\|^2/2 \le b^2\|u\|^2/(2a), \forall a > 0$; and (c) follows from (6.6.63) and (6.6.66).

By combining the relation above with (6.6.69), we obtain

$$\begin{aligned}0 \le& \mathbb{E}_s\left[\tfrac{\gamma_1((m-1)\eta_1 + (m-2)\mu)}{2}\|\mathbf{x}^* - \mathbf{x}^0\|^2 - \tfrac{(m-1)\gamma_s(\eta_s+\mu)}{2}\|\mathbf{x}^* - \mathbf{x}^s\|^2\right]\\ &+ \mathbb{E}_s\left[\gamma_1\tau_1 V_h(y^*, y^0) - \gamma_s(\tau_s + 1)V_h(y^*, y^s)\right]\\ &+ \mathbb{E}_s\left[\gamma_s(m-1)\langle \mathbf{A}(\mathbf{x}^s - \mathbf{x}^{s-1}), y^* - y^s\rangle - \tfrac{(m-1)\gamma_s\eta_s}{2}\|x_{i_s}^s - x_{i_s}^{s-1}\|^2\right].\end{aligned} \tag{6.6.72}$$

Notice the fact that

$$
\begin{aligned}
&\mathbb{E}_s\left[\tfrac{\gamma_s(\tau_s+1)}{2}V_h(y^*,y^s)-\gamma_s(m-1)\langle \mathbf{A}(\mathbf{x}^s-\mathbf{x}^{s-1}),y^*-y^s\rangle\right.\\
&\qquad\qquad\qquad\qquad\left.+\tfrac{(m-1)\gamma_s\eta_s}{2}\|x_{i_s}^s-x_{i_s}^{s-1}\|^2\right]\\
&=\mathbb{E}_s\left[\tfrac{\gamma_s(\tau_s+1)}{2}V_h(y^*,y^s)-\gamma_s(m-1)\langle \mathbf{A}(x_{i_s}^s-x_{i_s}^{s-1}),y^*-y^s\rangle\right.\\
&\qquad\qquad\qquad\qquad\left.+\tfrac{(m-1)\gamma_s\eta_s}{2}\|x_{i_s}^s-x_{i_s}^{s-1}\|^2\right]\\
&\ge\gamma_s\mathbb{E}_s\left[\tfrac{(\tau_s+1)\bar{\mu}}{4}\|y^*-y^s\|^2-(m-1)\bar{A}\|x_{i_s}^s-x_{i_s}^{s-1}\|\|y^*-y^s\|\right.\\
&\qquad\qquad\qquad\qquad\left.+\tfrac{(m-1)\eta_s}{2}\|x_{i_s}^s-x_{i_s}^{s-1}\|^2\right]\\
&\ge\gamma_s\mathbb{E}_s\left(\sqrt{\tfrac{(m-1)(\tau_s+1)\bar{\mu}\eta_s}{2}}-(m-1)\bar{A}\right)\|x_{i_s}^s-x_{i_s}^{s-1}\|\|y^*-y^s\|\ge 0. \qquad (6.6.73)
\end{aligned}
$$

In view of (6.6.72) and (6.6.73), we complete the proof. ■

Now we present the main convergence result of Algorithm 6.11 in Theorem 6.19, which eliminates the dependence on dual variables and relates directly the successive searching points of RapDual.

Theorem 6.19. *Let (x^*,y^*) be a saddle point of* (6.6.49), *and suppose that the parameters $\{\alpha_t\}$, $\{\tau_t\}$, $\{\eta_t\}$, and $\{\gamma_t\}$ are set as in* (6.6.55) *and* (6.6.56), *and $\tilde{\alpha}_t=\alpha$. Then, for any $s\ge 1$, we have*

$$
\mathbb{E}_s\left\{\|\mathbf{x}^s-\mathbf{x}^*\|^2+\|x_m^s-x_m^*\|^2\right\}\le\alpha^s M(\|\mathbf{x}^0-\mathbf{x}^*\|^2+\|x_m^0-x_m^*\|^2),
$$

where $x_m^*=\operatorname{argmin}_{x_m\in\mathbb{R}^n}\{\psi_m(x_m)+\langle x_m,y^*\rangle\}$ and $M=2\hat{L}/\mu$.

Proof. It is easy to check that (6.6.55) and (6.6.56) satisfy conditions (6.6.62), (6.6.63), (6.6.64), (6.6.65), and (6.6.66) when $\mu,\bar{\mu}>0$. Then we have

$$
\begin{aligned}
&\mathbb{E}_s\left\{\tfrac{(m-1)\gamma_s(\eta_s+\mu)}{2}\|\mathbf{x}^s-\mathbf{x}^*\|^2+\tfrac{\gamma_s(\tau_s+1)\bar{\mu}}{2}V_h(y^s,y^*)\right\}\\
&\le\tfrac{\gamma_1((m-1)\eta_1+(m-2)\mu)}{2}\|\mathbf{x}^0-\mathbf{x}^*\|^2+\gamma_1\tau_1V_h(y^0,y^*).
\end{aligned}
$$

Therefore, by plugging in those values in (6.6.55) and (6.6.56), we have

$$
\mathbb{E}_s\left[\mu\|\mathbf{x}^s-\mathbf{x}^*\|^2+V_h(y^s,y^*)\right]\le\alpha^s\left[\mu\|\mathbf{x}^0-\mathbf{x}^*\|^2+2V_h(y^0,y^*)\right], \qquad (6.6.74)
$$

Since $h(y)$ has $1/\mu$-Lipschitz continuous gradients and is $1/L$-strongly convex, we obtain

$$
\begin{aligned}
V_h(y^s,y^*)&\ge\tfrac{\mu}{2}\|\nabla h(y^s)-\nabla h(y^*)\|^2=\tfrac{\mu}{2}\|-x_m^s+x_m^*\|^2, &(6.6.75)\\
V_h(y^0,y^*)&\le\tfrac{\hat{L}}{2}\|\nabla h(y^0)-\nabla h(y^*)\|^2=\tfrac{\hat{L}}{2}\|-x_m^0+x_m^*\|^2. &(6.6.76)
\end{aligned}
$$

Combining (6.6.74), (6.6.75), and (6.6.76), we have

$$
\mathbb{E}_s\left\{\|\mathbf{x}^s-\mathbf{x}^*\|^2+\|x_m^s-x_m^*\|^2\right\}\le\alpha^s M(\|\mathbf{x}^0-\mathbf{x}^*\|^2+\|x_m^0-x_m^*\|^2).
$$

■

The above theorem shows that subproblem (6.6.50) can be solved efficiently by Algorithm 6.11 with a linear rate of convergence. In fact, we need not solve it too accurately. With a fixed and relatively small number of iterations s Algorithm 6.11 can still converge, as shown by the following lemma.

Lemma 6.17. *Let the inner iteration number $s \geq \lceil -\log M / \log \alpha \rceil$ with $M = 4 + 2L/\mu$ be given. Also the iterates $(\mathbf{x}^\ell, x_m^\ell)$ for $\ell = 1, \ldots, k$ be generated by Algorithm 6.10 and $\hat{\ell}$ be randomly selected from $[k]$. Then*

$$\begin{aligned}
&\mathbb{E}\left(\|\mathbf{x}_*^\ell - \bar{\mathbf{x}}^{\ell-1}\|^2 + \|x_{m^*}^\ell - \bar{x}_m^{\ell-1}\|^2\right) \\
&\quad \leq \tfrac{1}{k\mu(1-M\alpha^s)} \left\{f(\bar{\mathbf{x}}^0) + f_m(\bar{x}_m^0) - [f(\mathbf{x}^*) + f_m(x_m^*)]\right\}, \\
&\mathbb{E}\left(\|\mathbf{x}_*^\ell - \bar{\mathbf{x}}^\ell\|^2 + \|x_{m^*}^\ell - \bar{x}_m^\ell\|^2\right) \\
&\quad \leq \tfrac{M\alpha^s}{k\mu(1-M\alpha^s)} \left\{f(\bar{\mathbf{x}}^0) + f_m(\bar{x}_m^0) - [f(\mathbf{x}^*) + f_m(x_m^*)]\right\},
\end{aligned}$$

where $(\mathbf{x}^*, x_m^*)$ and $(\mathbf{x}_*^\ell, x_{m^*}^\ell)$ are the optimal solutions to (6.6.4) *and the ℓ-th subproblem* (6.6.48), *respectively.*

Proof. According to Theorem 6.19, we have

$$\mathbb{E}\left(\|\bar{\mathbf{x}}^\ell - \mathbf{x}_*^\ell\|^2 + \|\bar{x}_m^\ell - x_{m^*}^\ell\|^2\right) \leq \alpha^s M(\|\bar{\mathbf{x}}^{\ell-1} - \mathbf{x}_*^\ell\|^2 + \|\bar{x}_m^{\ell-1} - x_{m^*}^\ell\|^2). \quad (6.6.77)$$

Let us denote $(\mathbf{x}_*^0, x_{m^*}^0) = (\bar{\mathbf{x}}^0, \bar{x}_m^0)$ and by selection, it is feasible to subproblem (6.6.48) when $\ell = 1$. Since $(\mathbf{x}_*^\ell, x_{m^*}^\ell)$ is optimal and $(\mathbf{x}_*^{\ell-1}, x_{m^*}^{\ell-1})$ is feasible to the ℓ-th subproblem, we have

$$\psi^\ell(\mathbf{x}_*^\ell) + \psi_m^\ell(x_{m^*}^\ell) \leq \psi^\ell(\mathbf{x}_*^{\ell-1}) + \psi_m^\ell(x_{m^*}^{\ell-1}).$$

Plugging in the definition of ψ^ℓ and ψ_m^ℓ in the above inequality, and summing up from $\ell = 1$ to k, we have

$$\begin{aligned}
&\Sigma_{\ell=1}^k [f(\mathbf{x}_*^\ell) + f_m(x_{m^*}^\ell) + \mu(\|\mathbf{x}_*^\ell - \bar{\mathbf{x}}^{\ell-1}\|^2 + \|x_{m^*}^\ell - \bar{x}_m^{\ell-1}\|^2)] \\
&\quad \leq \Sigma_{\ell=1}^k [f(\mathbf{x}_*^{\ell-1}) + f_m(x_{m^*}^{\ell-1}) + \mu(\|\mathbf{x}_*^{\ell-1} - \bar{\mathbf{x}}^{\ell-1}\|^2 + \|x_{m^*}^{\ell-1} - \bar{x}_m^{\ell-1}\|^2)]. \quad (6.6.78)
\end{aligned}$$

Combining (6.6.77) and (6.6.78) and noticing that $(\mathbf{x}_*^0, x_{m^*}^0) = (\bar{\mathbf{x}}^0, \bar{x}_m^0)$, we have

$$\begin{aligned}
&\mu \Sigma_{\ell=1}^k \mathbb{E}(\|\mathbf{x}_*^\ell - \bar{\mathbf{x}}^{\ell-1}\|^2 + \|x_{m^*}^\ell - \bar{x}_m^{\ell-1}\|^2) \quad (6.6.79) \\
&\leq \Sigma_{\ell=1}^k \{f(\mathbf{x}_*^{\ell-1}) + f_m(x_{m^*}^{\ell-1}) - [f(\mathbf{x}_*^\ell) + f_m(x_{m^*}^\ell)]\} \\
&\quad + \mu \Sigma_{\ell=1}^k \mathbb{E}(\|\mathbf{x}_*^\ell - \bar{\mathbf{x}}^\ell\|^2 + \|x_{m^*}^\ell - \bar{x}_m^\ell\|^2) \\
&\leq f(\bar{\mathbf{x}}^0) + f_m(\bar{x}_m^0) - [f(\mathbf{x}^*) + f_m(x_m^*)] \\
&\quad + \mu\alpha^s M \Sigma_{\ell=1}^k \mathbb{E}(\|\mathbf{x}_*^\ell - \bar{\mathbf{x}}^{\ell-1}\|^2 + \|x_{m^*}^\ell - \bar{x}_m^{\ell-1}\|^2). \quad (6.6.80)
\end{aligned}$$

In view of (6.6.79) and (6.6.77), we have

$$\begin{aligned}
&\textstyle\sum_{\ell=1}^k \mathbb{E}\left(\|\mathbf{x}_*^\ell - \bar{\mathbf{x}}^{\ell-1}\|^2 + \|x_{m^*}^\ell - \bar{x}_m^{\ell-1}\|^2\right)\\
&\quad\le \tfrac{1}{\mu(1-M\alpha^s)}\left\{f(\bar{\mathbf{x}}^0) + f_m(\bar{x}_m^0) - [f(\mathbf{x}^*) + f_m(x_m^*)]\right\}\\
&\textstyle\sum_{\ell=1}^k \mathbb{E}\left(\|\mathbf{x}_*^\ell - \bar{\mathbf{x}}^{\ell}\|^2 + \|x_{m^*}^\ell - \bar{x}_m^{\ell}\|^2\right)\\
&\quad\le \tfrac{M\alpha^s}{\mu(1-M\alpha^s)}\left\{f(\bar{\mathbf{x}}^0) + f_m(\bar{x}_m^0) - [f(\mathbf{x}^*) + f_m(x_m^*)]\right\},
\end{aligned}$$

which, in view of the fact that $\hat{\ell}$ is chosen randomly in $[k]$, implies our results. ■

Now we are ready to prove the results in Theorem 6.18 with all the results proved above.

Proof of Theorem 6.18. By the optimality condition of the $\hat{\ell}$-th subproblem (6.6.48), there exists some λ^* such that

$$\begin{aligned}
&\nabla \psi^{\hat{\ell}}(\mathbf{x}_*^{\hat{\ell}}) + \mathbf{A}^\top \lambda^* \in -N_X(\mathbf{x}_*^{\hat{\ell}}),\\
&\nabla \psi_m^{\hat{\ell}}(x_{m^*}^{\hat{\ell}}) + \lambda^* = 0,\\
&\mathbf{A}\mathbf{x}_*^{\hat{\ell}} + x_{m^*}^{\hat{\ell}} = \mathbf{b}.
\end{aligned} \tag{6.6.81}$$

Plugging in the definition of $\psi^{\hat{\ell}}$ and $\psi_m^{\hat{\ell}}$, we have

$$\nabla f^{\hat{\ell}}(\mathbf{x}_*^{\hat{\ell}}) + 2\mu(\mathbf{x}_*^{\hat{\ell}} - \bar{\mathbf{x}}^{\hat{\ell}-1}) + \mathbf{A}^\top \lambda^* \in -N_X(\mathbf{x}_*^{\hat{\ell}}), \tag{6.6.82}$$

$$\nabla f_m^{\hat{\ell}}(x_{m^*}^{\hat{\ell}}) + 2\mu(x_{m^*}^{\hat{\ell}} - \bar{x}_m^{\hat{\ell}-1}) + \lambda^* = 0. \tag{6.6.83}$$

Now we are ready to evaluate the quality of the solution $(\bar{\mathbf{x}}^{\hat{\ell}}, \bar{x}_m^{\hat{\ell}})$. In view of (6.6.82) and Lemma 6.17, we have

$$\begin{aligned}
\mathbb{E}\left[d(\nabla f^{\hat{\ell}}(\mathbf{x}_*^{\hat{\ell}}) + \mathbf{A}^\top\lambda^*, -N_X(\mathbf{x}_*^{\hat{\ell}}))\right]^2 &\le \mathbb{E}\|2\mu(\mathbf{x}_*^{\hat{\ell}} - \bar{\mathbf{x}}^{\hat{\ell}-1})\|^2\\
&\le \tfrac{4\mu}{k(1-M\alpha^s)}\left\{f(\bar{\mathbf{x}}^0) + f_m(\bar{x}_m^0)\right.\\
&\qquad\left. - [f(\mathbf{x}^*) + f_m(x_m^*)]\right\}\\
&\le \tfrac{8\mu}{k}\left\{f(\bar{\mathbf{x}}^0) + f_m(\bar{x}_m^0) - [f(\mathbf{x}^*) + f_m(x_m^*)]\right\}.
\end{aligned}$$

Similarly, due to (6.6.83) and Lemma 6.17, we have

$$\begin{aligned}
\mathbb{E}\|\nabla f_m^{\hat{\ell}}(x_m^{\hat{\ell}}) + \lambda^*\|^2 &= \mathbb{E}\|\nabla f_m^{\hat{\ell}}(x_m^{\hat{\ell}}) - \nabla f_m^{\hat{\ell}}(x_{m^*}^{\hat{\ell}}) - 2\mu(x_{m^*}^{\hat{\ell}} - x_m^{\hat{\ell}-1})\|^2\\
&\le 2\mathbb{E}\{\|\nabla f_m^{\hat{\ell}}(x_m^{\hat{\ell}}) - \nabla f_m^{\hat{\ell}}(x_{m^*}^{\hat{\ell}})\|^2 + 4\mu^2\|x_{m^*}^{\hat{\ell}} - x_m^{\hat{\ell}-1}\|^2\}\\
&\le \mathbb{E}\left\{18\mu^2\|x_{m^*}^{\hat{\ell}} - \bar{x}_m^{\hat{\ell}}\|^2 + 8\mu^2\|x_{m^*}^{\hat{\ell}} - \bar{x}_m^{\hat{\ell}-1}\|^2\right\}\\
&\le \mu^2(8 + 18M\alpha^s)\mathbb{E}\{\|\mathbf{x}^{\hat{\ell}} - \bar{\mathbf{x}}^{\hat{\ell}-1}\|^2 + \|x_{m^*}^{\hat{\ell}} - \bar{x}_m^{\hat{\ell}-1}\|^2\}\\
&\le \tfrac{2\mu(4+9M\alpha^s)}{k(1-M\alpha^s)}\mathbb{E}\left\{f(\bar{\mathbf{x}}^0) + f_m(\bar{x}_m^0) - [f(\mathbf{x}^*) + f_m(x_m^*)]\right\}\\
&\le \tfrac{34\mu}{k}\left\{f(\bar{\mathbf{x}}^0) + f_m(\bar{x}_m^0) - [f(\mathbf{x}^*) + f_m(x_m^*)]\right\}.
\end{aligned}$$

By Lemma 6.17 we have

$$\begin{aligned}\mathbb{E}\|\mathbf{x}^{\hat{\ell}} - \mathbf{x}_*^{\hat{\ell}}\|^2 &\le \tfrac{M\alpha^s}{k\mu(1-M\alpha^s)} \left\{f(\bar{\mathbf{x}}^0) + f_m(x_m^0) - [f(\mathbf{x}^*) + f_m(x_m^*)]\right\} \\ &\le \tfrac{2M\alpha^s}{k\mu} \left\{f(\bar{\mathbf{x}}^0) + f_m(\bar{x}_m^0) - [f(\mathbf{x}^*) + f_m(x_m^*)]\right\} \\ &\le \tfrac{2\mu}{kL^2} \left\{f(\bar{\mathbf{x}}^0) + f_m(\bar{x}_m^0) - [f(\mathbf{x}^*) + f_m(x_m^*)]\right\}.\end{aligned}$$

Combining (6.6.81) and Lemma 6.17, we have

$$\begin{aligned}\mathbb{E}\|\mathbf{A}\bar{\mathbf{x}}^{\hat{\ell}} + \bar{x}_m^{\hat{\ell}} - \mathbf{b}\|^2 &= \mathbb{E}\|\mathbf{A}(\bar{\mathbf{x}}^{\hat{\ell}} - \mathbf{x}_*^{\hat{\ell}}) + \bar{x}_m^{\hat{\ell}} - x_{m*}^{\hat{\ell}}\|^2 \\ &\le 2\mathbb{E}\{\|\mathbf{A}\|^2\|\bar{\mathbf{x}}^{\hat{\ell}} - \mathbf{x}_*^{\hat{\ell}}\|^2 + \|\bar{x}_m^{\hat{\ell}} - x_{m*}^{\hat{\ell}}\|^2\} \\ &\le 2(\|\mathbf{A}\|^2 + 1)\mathbb{E}\{\|\bar{\mathbf{x}}^{\hat{\ell}} - \mathbf{x}_*^{\hat{\ell}}\|^2 + \|\bar{x}_m^{\hat{\ell}} - x_{m*}^{\hat{\ell}}\|^2\} \\ &\le \tfrac{2(\|\mathbf{A}\|^2+1)M\alpha^s}{k\mu(1-M\alpha^s)} \left\{f(\bar{\mathbf{x}}^0) + f_m(\bar{x}_m^0) - [f(\mathbf{x}^*) + f_m(x_m^*)]\right\} \\ &\le \tfrac{2(\|\mathbf{A}\|^2+1)\mu}{kL^2} \left\{f(\bar{\mathbf{x}}^0) + f_m(\bar{x}_m^0) - [f(\mathbf{x}^*) + f_m(x_m^*)]\right\}.\end{aligned}$$

■

We observe that the main ideas of this RapDual, i.e., using proximal point method to transform the nonconvex multi-block problem into a series of convex subproblems, and using randomized dual method to them, can be applied for solving much more general multi-block optimization problems for which there does not exist an invertible block. In this more general case, the saddle-point subproblems will only be strongly convex in the primal space, but not in the dual space. Therefore, the complexity of solving the subproblem will only be sublinear, and as a consequence, the overall complexity will be much worse than $\mathcal{O}(1/\varepsilon)$. It is also worth noting that the proposed RapGrad and RapDual implicitly assume that the parameter μ is known, or a (tight) upper bound of it can be obtained.

6.7 Exercises and Notes

1. Consider the problem of $\min_{x\in X} f(x)$, where $X \subseteq \mathbb{R}^n$ is a closed convex set and f is a convex function. Assume that we apply either the gradient descent or accelerated gradient descent method for solving

$$\min_{x\in X}\{f(x) + \tfrac{\mu}{2}\|x - x_0\|^2,$$

for a given initial point $x_0 \in X$ and a small enough $\mu > 0$. Please establish the best possible complexity bound for these methods to find a stationary point so that the size of the projected gradient is small.

2. Establish the convergence of the RSGF method if the random variable u in (6.1.49) is not a Gaussian random variable, but a uniform random variable over the standard Euclidean ball.

3. Develop a stochastic gradient-free mirror descent method similar to 2-RSMD-V for which only stochastic zeroth-order information is available and establish its complexity bounds.

Notes. The random stochastic gradient descent method for nonconvex optimization problems was first studied by Ghadimi and Lan in [40]. Ghadimi, Lan, and Zhang generalized this method for solving nonconvex stochastic composite problems in [42]. Dang and Lan developed the randomized block coordinate descent method for solving nonconvex problems in [28]. The convergence of accelerated gradient descent method for solving nonconvex and stochastic optimization problems was first established by Ghaidimi and Lan in [41]. The gradient estimator used in the nonconvex variance-reduced mirror descent method was first introduced in [103] in convex optimization and analyzed for nonconvex optimization in [34, 106, 137]. A different estimator based on nested variance reduction technique was developed in [142]. Section 7.4 further generalizes the analysis to nonconvex mirror descent setting. The RapGrad and RapDual methods for nonconvex finite-sum and multi-block optimization were first introduced by Lan and Yang in [68]. It is worth noting that there exists active research on the geometry of a few special classes of nonconvex optimization problems, e.g., the low-rank matrix problems and dictionary learning, as well as on the study of a few nonconvex optimization algorithms for solving these problems.

Chapter 7
Projection-Free Methods

In this chapter, we present conditional gradient type methods that have attracted much attention in both machine learning and optimization community recently. These methods call a linear optimization (LO) oracle to minimize a series of linear functions over the feasible set. We will introduce the classic conditional gradient (a.k.a. Frank–Wolfe method) and a few of its variants. We will also discuss the conditional gradient sliding (CGS) algorithm which can skip the computation of gradients from time to time, and as a result, can achieve the optimal complexity bounds in terms of not only the number of calls to the LO oracle, but also the number of gradient evaluations. Extension of these methods for solving nonconvex optimization problems will also be discussed.

7.1 Conditional Gradient Method

In this section, we study a new class of optimization algorithms, referred to as linear-optimization-based convex programming (LCP) methods, for solving large-scale convex programming (CP) problems. Specifically, consider the CP problem of

$$f^* := \min_{x \in X} f(x), \tag{7.1.1}$$

where $X \subseteq \mathbb{R}^n$ is a convex compact set and $f : X \to \mathbb{R}$ is a closed convex function. The LCP methods solve problem (7.1.1) by iteratively calling a linear optimization (LO) oracle, which, for a given input vector $p \in \mathbb{R}^n$, computes the solution of subproblems given in the form of

$$\mathrm{Argmin}_{x \in X} \langle p, x \rangle. \tag{7.1.2}$$

In particular, if p is computed based on first-order information, then we call these algorithms first-order LCP methods. Clearly, the difference between first-order LCP methods and the more general first-order methods exists in the restrictions on the

G. Lan, *First-Order and Stochastic Optimization Methods for Machine Learning*,
Springer Series in the Data Sciences, https://doi.org/10.1007/978-3-030-39568-1_7

format of subproblems. For example, in the subgradient (mirror) descent method, we solve the projection (or prox-mapping) subproblems given in the form of

$$\text{argmin}_{x\in X} \{\langle p,x\rangle + d(x)\}. \tag{7.1.3}$$

Here $d : X \to \mathbb{R}$ is a certain strongly convex function (e.g., $d(x) = \|x\|_2^2/2$).

The development of LCP methods has recently regained some interests from both machine learning and optimization community mainly for the following reasons.

- *Low iteration cost.* In many cases, the solution of the linear subproblem (7.1.2) is much easier to solve than the nonlinear subproblem (7.1.3). For example, if X is a spectrahedron given by $X = \{x \in \mathbb{R}^{n\times n} : \text{Tr}(x) = 1, x \succeq 0\}$, the solution of (7.1.2) can be much faster than that of (7.1.3).
- *Simplicity.* The CndG method is simple to implement since it does not require the selection of the distance function $d(x)$ in (7.1.3) and the fine-tuning of stepsizes, which are required in most other first-order methods (with exceptions to some extent for a few level-type first-order methods as discussed earlier in Sect. 3.9).
- *Structural properties for the generated solutions.* The output solutions of the CndG method may have certain desirable structural properties, e.g., sparsity and low rank, as they can often be written as the convex combination of a small number of extreme points of X.

In this section, we first establish the rate of convergence of the classic conditional gradient method and its variants, in terms of the number of calls to the LO oracle, for solving different classes of CP problems under an LO oracle as follows.

(a) f is a smooth convex function satisfying

$$\|f'(x) - f'(y)\|_* \le L\|x-y\|, \forall x,y \in X. \tag{7.1.4}$$

(b) f is a special nonsmooth function with f given by

$$f(x) = \max_{y\in Y} \{\langle Ax,y\rangle - \hat{f}(y)\}. \tag{7.1.5}$$

Here $Y \subseteq \mathbb{R}^m$ is a convex compact set, $A : \mathbb{R}^n \to \mathbb{R}^m$ a linear operator, and $\hat{f} : Y \to \mathbb{R}$ is a simple convex function.

(c) f is a general nonsmooth Lipschitz continuous convex function such that

$$|f(x) - f(y)| \le M\|x-y\|,\ \forall x,y \in X, \tag{7.1.6}$$

(d) We also discuss the possibility to improve the complexity of the CndG method under strong convexity assumption about $f(\cdot)$ and with an enhanced LO oracle.

We then present a few new LCP methods, namely the primal averaging CndG (PA-CndG) and primal–dual averaging CndG (PDA-CndG) algorithms, for solving large-scale CP problems under an LO oracle. These methods are obtained by replacing the projection subproblems with linear optimization subproblems in the accelerated gradient descent method.

Finally, we show that to solve CP problems under an LO oracle is fundamentally more difficult than to solve CP problems without such restrictions, by establishing a series of lower complexity bounds for solving different classes of CP problems under an LO oracle. We then show that the number of calls to the LO oracle performed by the aforementioned LCP methods do not seem to be improvable in general.

To fix a notation, let $X \in \mathbb{R}^n$ and $Y \in \mathbb{R}^m$ be given convex compact sets. Also let $\|\cdot\|_X$ and $\|\cdot\|_Y$ be the norms (not necessarily associated with inner product) in $\mathbb{R}^n$ and $\mathbb{R}^m$, respectively. For the sake of simplicity, we often skip the subscripts in the norms $\|\cdot\|_X$ and $\|\cdot\|_Y$. For a given norm $\|\cdot\|$, we denote its conjugate by $\|s\|_* = \max_{\|x\|\le 1}\langle s,x\rangle$. We use $\|\cdot\|_1$ and $\|\cdot\|_2$, respectively, to denote the regular l_1 and l_2 norms. Let $A : \mathbb{R}^n \to \mathbb{R}^m$ be a given linear operator, we use $\|A\|$ to denote its operator norm given by $\|A\| := \max_{\|x\|\le 1}\|Ax\|$. Let $f : X \to \mathbb{R}$ be a convex function, we denote its linear approximation at x by

$$l_f(x;y) := f(x) + \langle f'(x), y-x\rangle. \tag{7.1.7}$$

Clearly, if f satisfies (7.1.4), then

$$f(y) \le l_f(x;y) + \tfrac{L}{2}\|y-x\|^2, \quad \forall x,y \in X. \tag{7.1.8}$$

Notice that the constant L in (7.1.4) and (7.1.8) depends on $\|\cdot\|$.

7.1.1 Classic Conditional Gradient

Our goal in this section is to establish the rate of convergence for the classic CndG method and its variants for solving different classes of CP problems in terms of the number of calls to the LO oracle.

7.1.1.1 Smooth Convex Problems Under an LO Oracle

The classic CndG method is one of the earliest iterative algorithms to solve problem (7.1.1). The basic scheme of this algorithm is stated as follows.

Algorithm 7.1 The classic conditional gradient (CndG) method

Let $x_0 \in X$ be given. Set $y_0 = x_0$.
for $k = 1, \ldots$ **do**
  Call the LO oracle to compute $x_k \in \mathrm{Argmin}_{x\in X}\langle f'(y_{k-1}), x\rangle$.
  Set $y_k = (1-\alpha_k)y_{k-1} + \alpha_k x_k$ for some $\alpha_k \in [0,1]$.
end for

In order to guarantee the convergence of the classic CndG method, we need to properly specify the stepsizes α_k used in the definition of y_k. There are two popular options for selecting α_k: one is to set

$$\alpha_k = \tfrac{2}{k+1}, \quad k = 1,2,\ldots, \tag{7.1.9}$$

and the other is to compute α_k by solving a one-dimensional minimization problem:

$$\alpha_k = \mathrm{argmin}_{\alpha\in[0,1]} f((1-\alpha)y_{k-1} + \alpha x_k), \quad k = 1,2,\ldots. \tag{7.1.10}$$

We now formally describe the convergence properties of the above classic CndG method. Observe that we state explicitly in Theorem 7.1 how the rate of convergence associated with this algorithm depends on distance between the previous iterate y_{k-1} and the output of the LO oracle, i.e., $\|x_k - y_{k-1}\|$. Also observe that, given a candidate solution $\bar{x} \in X$, we use the function optimality gap $f(\bar{x}) - f^*$ as a termination criterion for the algorithm. In Sect. 7.2, we will show that the CndG method also exhibits the same rate of convergence in terms of a stronger termination criterion, i.e., the Wolfe gap given by $\max_{x\in X}\langle f'(\bar{x}), \bar{x} - x\rangle$. The following quantity will be used for our convergence analysis.

$$\Gamma_k := \begin{cases} 1, & k = 1, \\ (1-\gamma_k)\,\Gamma_{k-1}, & k \geq 2. \end{cases} \tag{7.1.11}$$

Theorem 7.1. *Let $\{x_k\}$ be the sequence generated by the classic CndG method applied to problem (7.1.1) with the stepsize policy in (7.1.9) or (7.1.10). If $f(\cdot)$ satisfies (7.1.4), then for any $k = 1,2,\ldots$,*

$$f(y_k) - f^* \leq \tfrac{2L}{k(k+1)} \textstyle\sum_{i=1}^k \|x_i - y_{i-1}\|^2. \tag{7.1.12}$$

Proof. Let Γ_k be defined in (7.1.11) with

$$\gamma_k := \tfrac{2}{k+1}. \tag{7.1.13}$$

It is easy to check that

$$\Gamma_k = \tfrac{2}{k(k+1)} \quad \text{and} \quad \tfrac{\gamma_k^2}{\Gamma_k} \leq 2, \quad k = 1,2,\ldots. \tag{7.1.14}$$

Denoting $\tilde{y}_k = (1-\gamma_k)y_{k-1} + \gamma_k x_k$, we conclude from from (7.1.9) (or (7.1.10)) and the definition of y_k in Algorithm 7.1 that $f(y_k) \leq f(\tilde{y}_k)$. It also follows from the definition of $\tilde{y}_k$ that $\tilde{y}_k - y_{k-1} = \gamma_k(x_k - y_{k-1})$. Letting $l_f(x;y)$ be defined in (7.1.7) and using these two observations, (7.1.8), the definition of x_k, and the convexity of $f(\cdot)$, we have

$$\begin{aligned} f(y_k) &\leq f(\tilde{y}_k) \leq l_f(y_{k-1};\tilde{y}_k) + \tfrac{L}{2}\|y_k - y_{k-1}\|^2 \\ &= (1-\gamma_k)f(y_{k-1}) + \gamma_k l_f(y_{k-1};x_k) + \tfrac{L}{2}\gamma_k^2\|x_k - y_{k-1}\|^2 \\ &\leq (1-\gamma_k)f(y_{k-1}) + \gamma_k l_f(y_{k-1};x) + \tfrac{L}{2}\gamma_k^2\|x_k - y_{k-1}\|^2, \\ &\leq (1-\gamma_k)f(y_{k-1}) + \gamma_k f(x) + \tfrac{L}{2}\gamma_k^2\|x_k - y_{k-1}\|^2, \ \forall x \in X. \end{aligned} \tag{7.1.15}$$

Subtracting $f(x)$ from both sides of the above inequality, we obtain

$$f(y_k) - f(x) \le (1-\gamma_k)[f(y_{k-1}) - f(x)] + \tfrac{L}{2}\gamma_k^2 \|x_k - y_{k-1}\|^2, \tag{7.1.16}$$

which, in view of Lemma 3.17, then implies that

$$\begin{aligned} f(y_k) - f(x) &\le \Gamma_k(1-\gamma_1)[f(y_0) - f(x)] + \tfrac{\Gamma_k L}{2}\textstyle\sum_{i=1}^k \tfrac{\gamma_i^2}{\Gamma_i}\|x_i - y_{i-1}\|^2 \\ &\le \tfrac{2L}{k(k+1)}\textstyle\sum_{i=1}^k \|x_i - y_{i-1}\|^2, \ k = 1,2,\ldots, \end{aligned} \tag{7.1.17}$$

where the last inequality follows from the fact that $\gamma_1 = 1$ and (7.1.14). ■

We now add a few remarks about the results obtained in Theorem 7.1. Let us denote

$$\bar{D}_X \equiv \bar{D}_{X,\|\cdot\|} := \max_{x,y\in X} \|x - y\|. \tag{7.1.18}$$

First, note that by (7.1.12) and (7.1.18), we have, for any $k = 1,\ldots,$

$$f(y_k) - f^* \le \tfrac{2L}{k+1}\bar{D}_X^2.$$

Hence, the number of iterations required by the classic CndG method to find an ε-solution of problem (7.1.1) is bounded by

$$\mathcal{O}(1)\tfrac{L\bar{D}_X^2}{\varepsilon}. \tag{7.1.19}$$

Second, although the CndG method does not require the selection of the norm $\|\cdot\|$, the iteration complexity of this algorithm, as stated in (7.1.19), does depend on $\|\cdot\|$ as the two constants, i.e., $L \equiv L_{\|\cdot\|}$ and $\bar{D}_X \equiv \bar{D}_{X,\|\cdot\|}$, depend on $\|\cdot\|$. However, since the result in (7.1.19) holds for an arbitrary $\|\cdot\|$, the iteration complexity of the classic CndG method to solve problem (7.1.1) can actually be bounded by

$$\mathcal{O}(1)\inf_{\|\cdot\|}\left\{\tfrac{L_{\|\cdot\|}\bar{D}_{X,\|\cdot\|}^2}{\varepsilon}\right\}. \tag{7.1.20}$$

For example, if X is a simplex, a widely accepted strategy to accelerate gradient type methods is to set $\|\cdot\| = \|\cdot\|_1$ and $d(x) = \sum_{i=1}^n x_i \log x_i$ in (7.1.3), in order to obtain (nearly) dimension-independent complexity results, which only grow mildly with the increase of the dimension of the problem. On the other hand, the classic CndG method does automatically adjust to the geometry of the feasible set X in order to obtain such scalability to high-dimensional problems.

Third, observe that the rate of convergence in (7.1.12) depends on $\|x_k - y_{k-1}\|$ which usually does not vanish as k increases. For example, suppose $\{y_k\} \to x^*$ (this is true if x^* is a unique optimal solution of (7.1.1)), the distance $\{\|x_k - y_{k-1}\|\}$ does not necessarily converge to zero unless x^* is an extreme point of X. In these cases, the summation $\sum_{i=1}^k \|x_i - y_{i-1}\|^2$ increases linearly with respect to k. We will discuss some techniques in Sect. 7.1.2 that might help to improve this situation.

7.1.1.2 Bilinear Saddle Point Problems Under an LO Oracle

In this subsection, we show that the CndG method, after incorporating some proper modification, can be used to solve the bilinear saddle point problem with the objective function f given by (7.1.5).

Since f given by (7.1.5) is nonsmooth in general, we cannot directly apply the CndG method. However, recall that the function $f(\cdot)$ in (7.1.5) can be closely approximated by a class of smooth convex functions. More specifically, for a given strongly convex function $\omega : Y \to \mathbb{R}$ such that

$$\omega(y) \geq \omega(x) + \langle \omega'(x), y - x \rangle + \tfrac{\sigma_\omega}{2}\|y-x\|^2, \forall x, y \in Y, \tag{7.1.21}$$

let us denote $c_\omega := \mathrm{argmin}_{y \in Y} \omega(y)$, $W(y) := \omega(y) - \omega(c_\omega) - \langle \nabla\omega(c_\omega), y - c_\omega \rangle$ and

$$D_Y \equiv D_{Y,\omega} := [\max_{y \in Y} W(y)]^{1/2}. \tag{7.1.22}$$

Then the function $f(\cdot)$ in (7.1.5) can be closely approximated by

$$f_\eta(x) := \max_y \left\{ \langle Ax, y \rangle - \hat{f}(y) - \eta\,[V(y) - D_Y^2] : \; y \in Y \right\}. \tag{7.1.23}$$

Indeed, by definition we have $0 \leq V(y) \leq D_Y^2$ and hence, for any $\eta \geq 0$,

$$f(x) \leq f_\eta(x) \leq f(x) + \eta\, D_Y^2, \quad \forall x \in X. \tag{7.1.24}$$

Moreover, it can be shown that $f_\eta(\cdot)$ is differentiable and its gradients are Lipschitz continuous with the Lipschitz constant given by (see Lemma 3.7)

$$\mathscr{L}_\eta := \tfrac{\|A\|^2}{\eta \sigma_v}. \tag{7.1.25}$$

In view of this result, we modify the CndG method by replacing the gradient $f'(y_k)$ in Algorithm 7.1 with the gradient $f'_{\eta_k}(y_k)$ for some $\eta_k > 0$. Observe that in the original smoothing scheme in Sect. 3.5, we first need to define the smooth approximation function f_η in (7.1.23) by specifying in advance the smoothing parameter η and then apply a smooth optimization method to solve the approximation problem. The specification of η usually requires explicit knowledge of $\bar{D}_X$, D_Y^2, and the target accuracy ε given a priori. However, by using a different analysis, we show that one can use variable smoothing parameters η_k and thus does not need to know the target accuracy ε in advance. In addition, wrong estimation on $\bar{D}_X$ and D_Y^2 only affects the rate of convergence of the modified CndG method by a constant factor. Our analysis relies on a slightly different construction of $f_\eta(\cdot)$ in (7.1.23) and the following simple observation.

Lemma 7.1. *Let $f_\eta(\cdot)$ be defined in (7.1.23) and $\eta_1 \geq \eta_2 \geq 0$ be given. Then, we have $f_{\eta_1}(x) \geq f_{\eta_2}(x)$ for any $x \in X$.*

Proof. The result directly follows from the definition of $f_\eta(\cdot)$ in (7.1.23) and the fact that $V(y) - D_Y^2 \le 0$. ■

We are now ready to describe the main convergence properties of this modified CndG method to solve the bilinear saddle point problems.

Theorem 7.2. *Let $\{x_k\}$ and $\{y_k\}$ be the two sequences generated by the CndG method with $f'(y_k)$ replaced by $f_{\eta_k}(y_k)$, where $f_\eta(\cdot)$ is defined in (7.1.5). If the stepsizes α_k, $k = 1,2,\ldots$, are set to (7.1.9) or (7.1.10), and $\{\eta_k\}$ satisfies*

$$\eta_1 \ge \eta_2 \ge \ldots, \tag{7.1.26}$$

then we have, for any $k = 1,2,\ldots$,

$$f(y_k) - f^* \le \tfrac{2}{k(k+1)} \left[\textstyle\sum_{i=1}^k \left(i\eta_i D_Y^2 + \tfrac{\|A\|^2}{\sigma_\nu \eta_i}\|x_i - y_{i-1}\|^2\right)\right]. \tag{7.1.27}$$

In particular, if

$$\eta_k = \tfrac{\|A\|\bar{D}_X}{D_Y\sqrt{\sigma_\nu k}}, \tag{7.1.28}$$

then we have, for any $k = 1,2,\ldots$,

$$f(y_k) - f^* \le \tfrac{2\sqrt{2}\|A\|\bar{D}_X D_Y}{\sqrt{\sigma_\nu k}}, \tag{7.1.29}$$

where $\bar{D}_X$ and D_Y are defined in (7.1.18) and (7.1.22), respectively.

Proof. Let Γ_k and γ_k be defined in (7.1.11) and (7.1.13), respectively. Similar to (7.1.16), we have, for any $x \in X$,

$$\begin{aligned}
f_{\eta_k}(y_k) &\le (1-\gamma_k)[f_{\eta_k}(y_{k-1})] + \gamma_k f_{\eta_k}(x) + \tfrac{L_{\eta_k}}{2}\gamma_k^2\|x_k - y_{k-1}\|^2\\
&\le (1-\gamma_k)[f_{\eta_{k-1}}(y_{k-1})] + \gamma_k f_{\eta_k}(x) + \tfrac{L_{\eta_k}}{2}\gamma_k^2\|x_k - y_{k-1}\|^2\\
&\le (1-\gamma_k)[f_{\eta_{k-1}}(y_{k-1})] + \gamma_k[f(x) + \eta_k D_Y^2] + \tfrac{L_{\eta_k}}{2}\gamma_k^2\|x_k - y_{k-1}\|^2,
\end{aligned}$$

where the second inequality follows from (7.1.26) and Lemma 7.1, and the third inequality follows from (7.1.24). Now subtracting $f(x)$ from both sides of the above inequality, we obtain, $\forall x \in X$,

$$\begin{aligned}
f_{\eta_k}(y_k) - f(x) &\le (1-\gamma_k)[f_{\eta_{k-1}}(y_{k-1}) - f(x)] + \gamma_k\eta_k D_Y^2 + \tfrac{L_{\eta_k}}{2}\gamma_k^2\|x_k - y_{k-1}\|^2\\
&\le (1-\gamma_k)[f_{\eta_{k-1}}(y_{k-1}) - f(x)] + \gamma_k\eta_k D_Y^2 + \tfrac{\|A\|^2\gamma_k^2}{2\sigma_\nu\eta_k}\|x_k - y_{k-1}\|^2,
\end{aligned}$$

which, in view of Lemma 3.17, (7.1.13), and (7.1.14), then implies that, $\forall x \in X$,

$$f_{\eta_k}(y_k) - f(x) \le \tfrac{2}{k(k+1)} \left[\textstyle\sum_{i=1}^k \left(i\eta_i D_Y^2 + \tfrac{\|A\|^2}{\sigma_\nu \eta_i}\|x_i - y_{i-1}\|^2\right)\right], \quad \forall k \ge 1.$$

Our result in (7.1.27) then immediately follows from (7.1.24) and the above inequality. Now it is easy to see that the selection of η_k in (7.1.28) satisfies (7.1.26).

By (7.1.27) and (7.1.28), we have

$$\begin{aligned} f(y_k) - f^* &\leq \tfrac{2}{k(k+1)} \left[\Sigma_{i=1}^k \left(i\eta_i D_Y^2 + \tfrac{\|A\|^2}{\sigma_v \eta_i} \bar{D}_X^2 \right) \right] \\ &= \tfrac{4\|A\| \bar{D}_X D_{v,Y}}{k(k+1)\sqrt{\sigma_v}} \Sigma_{i=1}^k \sqrt{i} \leq \tfrac{8\sqrt{2}\|A\| \bar{D}_X D_Y}{3\sqrt{\sigma_v k}}, \end{aligned}$$

where the last inequality follows from the fact that

$$\Sigma_{i=1}^k \sqrt{i} \leq \int_0^{k+1} t dt \leq \tfrac{2}{3}(k+1)^{3/2} \leq \tfrac{2\sqrt{2}}{3}(k+1)\sqrt{k}. \tag{7.1.30}$$

■

A few remarks about the results obtained in Theorem 7.2 are in order. First, observe that the specification of η_k in (7.1.28) requires the estimation of a few problem parameters, including $\|A\|$, $\bar{D}_X$, D_Y, and σ_v. However, wrong estimation on these parameters will only result in the increase on the rate of convergence of the modified CndG method by a constant factor. For example, if $\eta_k = 1/\sqrt{k}$ for any $k \geq 1$, then (7.1.27) reduces to

$$f(y_k) - f^* \leq \tfrac{8\sqrt{2}}{3\sqrt{k}} \left(D_Y^2 + \tfrac{\|A\|^2 \bar{D}_X^2}{\sigma_v} \right).$$

It is worth noting that similar adaptive smoothing schemes can also be used when one applies the accelerated gradient descent method to solve the bilinear saddle point problems. Second, suppose that the norm $\|\cdot\|$ in the dual space associated with Y is an inner product norm and $v(y) = \|y\|^2/2$. Also let us denote

$$\bar{D}_Y \equiv \bar{D}_{Y,\|\cdot\|} := \max_{x,y \in Y} \|x - y\|. \tag{7.1.31}$$

In this case, by the definitions of $\bar{D}_Y$ and D_Y in (7.1.31) and (7.1.22), we have $D_Y \leq \bar{D}_Y$. Using this observation and (7.1.29), we conclude that the number of iterations required by the modified CndG method to solve $\mathscr{F}^0_{\|A\|}(X,Y)$ can be bounded by

$$\mathscr{O}(1) \left(\tfrac{\|A\| \bar{D}_X \bar{D}_Y}{\varepsilon} \right)^2.$$

7.1.1.3 General Nonsmooth Problems Under an LO Oracle

In this subsection, we present a randomized CndG method and establish its rate of convergence for solving general nonsmooth CP problems under an LO oracle.

The basic idea is to approximate the general nonsmooth CP problems by using the convolution-based smoothing. The intuition underlying such an approach is that convolving two functions yields a new function that is at least as smooth as the smoother one of the original two functions. In particular, let μ denote the density of a random variable with respect to Lebesgue measure and consider the function f_μ given by

$$f_\mu(x) := (f * \mu)(x) = \int_{\mathbb{R}^n} f(y)\mu(x-y)d(y) = \mathbb{E}_\mu[x+Z],$$

where Z is a random variable with density μ. Since μ is a density with respect to Lebesgue measure, f_μ is differentiable. The above convolution-based smoothing technique has been extensively studied in stochastic optimization, e.g., [11, 32, 59, 102, 119]. For the sake of simplicity, we assume throughout this subsection that $\|\cdot\| = \|\cdot\|_2$ and Z is uniformly distributed over a certain Euclidean ball. The following result is known in the literature (see, e.g., [32]).

Lemma 7.2. *Let ξ be uniformly distributed over the l_2-ball $\mathscr{B}_2(0,1) := \{x \in \mathbb{R}^n : \|x\|_2 \le 1\}$ and $u > 0$ is given. Suppose that (7.1.6) holds for any $x, y \in X + u\mathscr{B}_2(0,1)$. Then, the following statements hold for the function $f_u(\cdot)$ given by*

$$f_u(x) := \mathbb{E}[f(x+u\xi)]. \tag{7.1.32}$$

(a) $f(x) \le f_u(x) \le f(x) + Mu$;
(b) $f_u(x)$ has $M\sqrt{n}/u$-Lipschitz continuous gradient with respect to $\|\cdot\|_2$;
(c) $\mathbb{E}[f'(x+u\xi)] = f'_u(x)$ and $\mathbb{E}[\|f'(x+u\xi) - f'_u(x)\|^2] \le M^2$;
(d) If $u_1 \ge u_2 \ge 0$, then $f_{u_1}(x) \ge f_{u_2}(x)$ for any $x \in X$.

In view of the above result, we can apply the CndG method directly to $\min_{x\in X} f_u(x)$ for a properly chosen μ in order to solve the original problem (7.1.1). The only problem is that we cannot compute the gradient of $f_u(\cdot)$ exactly. To address this issue, we will generate an i.i.d. random sample $(\xi_1, \ldots, \xi_T)$ for some $T > 0$ and approximate the gradient $f'_\mu(x)$ by $\tilde{f}'_u(x) := \frac{1}{T}\sum_{t=1}^T f'(x, u\xi_t)$. After incorporating the aforementioned randomized smoothing scheme, the CndG method exhibits the following convergence properties for solving general nonsmooth convex optimization problems.

Theorem 7.3. *Let $\{x_k\}$ and $\{y_k\}$ be the two sequences generated by the classic CndG method with $f'(y_{k-1})$ replaced by the average of the sampled gradients, i.e.,*

$$\tilde{f}'_{u_k}(y_{k-1}) := \tfrac{1}{T_k}\textstyle\sum_{t=1}^{T_k} f'(y_{k-1} + u_k\xi_t), \tag{7.1.33}$$

where f_u is defined in (7.1.32) and $\{\xi_1, \ldots, \xi_{T_k}\}$ is an i.i.d. sample of ξ. If the stepsizes α_k, $k = 1, 2, \ldots$, are set to (7.1.9) or (7.1.10), and $\{u_k\}$ satisfies

$$u_1 \ge u_2 \ge \ldots, \tag{7.1.34}$$

then we have

$$\mathbb{E}[f(y_k)] - f(x) \le \tfrac{2M}{k(k+1)}\left[\textstyle\sum_{i=1}^k \left(\tfrac{i}{\sqrt{T_i}}\bar{D}_X + iu_i + \tfrac{\sqrt{n}}{u_i}\bar{D}_X^2\right)\right], \tag{7.1.35}$$

where M is given by (7.1.6). In particular, if

$$T_k = k \quad \text{and} \quad u_k = \tfrac{n^{1/4}\bar{D}_X}{\sqrt{k}}, \tag{7.1.36}$$

then

$$\mathbb{E}[f(y_k)] - f(x) \leq \frac{4(1+2n^{1/4})M\bar{D}_X}{3\sqrt{k}}, k = 1, 2, \ldots. \quad (7.1.37)$$

Proof. Let γ_k be defined in (7.1.13), similar to (7.1.15), we have

$$\begin{aligned} f_{u_k}(y_k) &\leq (1-\gamma_k) f_{u_k}(y_{k-1}) + \gamma_k l_{f_{u_k}}(x_k; y_{k-1}) + \tfrac{M\sqrt{n}}{2u_k}\gamma_k^2 \|x_k - y_{k-1}\|^2 \\ &\leq (1-\gamma_k) f_{u_{k-1}}(y_{k-1}) + \gamma_k l_{f_{u_k}}(x_k; y_{k-1}) \\ &\quad + \tfrac{M\sqrt{n}}{2u_k}\gamma_k^2 \|x_k - y_{k-1}\|^2, \end{aligned} \quad (7.1.38)$$

where the last inequality follows from the fact that $f_{u_{k-1}}(y_{k-1}) \geq f_{u_k}(y_{k-1})$ due to Lemma 7.2(d). Let us denote $\delta_k := f'_{u_k}(y_{k-1}) - \tilde{f}'_{u_k}(y_{k-1})$. Noting that by definition of x_k and the convexity of $f_{u_k}(\cdot)$,

$$\begin{aligned} l_{f_{u_k}}(x_k; y_{k-1}) &= f_{u_k}(y_{k-1}) + \langle f'_{u_k}(y_{k-1}, x_k - y_{k-1}\rangle \\ &= f_{u_k}(y_{k-1}) + \langle \tilde{f}'_{u_k}(y_{k-1}), x_k - y_{k-1}\rangle + \langle \delta_k, x_k - y_{k-1}\rangle \\ &\leq f_{u_k}(y_{k-1}) + \langle \tilde{f}'_{u_k}(y_{k-1}), x - y_{k-1}\rangle + \langle \delta_k, x_k - y_{k-1}\rangle \\ &= f_{u_k}(y_{k-1}) + \langle f'_{u_k}(y_{k-1}), x - y_{k-1}\rangle + \langle \delta_k, x_k - x\rangle \\ &\leq f_{u_k}(x) + \|\delta_k\|\bar{D}_X \leq f(x) + \|\delta_k\|\bar{D}_X + Mu_k, \ \ \forall x \in X, \end{aligned}$$

where the last inequality follows from Lemma 7.2(a), we conclude from (7.1.38) that, $\forall x \in X$,

$$f_{u_k}(y_k) \leq (1-\gamma_k) f_{u_{k-1}}(y_{k-1}) + \gamma_k [f(x) + \|\delta_k\|\bar{D}_X + Mu_k] + \tfrac{M\sqrt{n}}{2u_k}\gamma_k^2 \|x_k - y_{k-1}\|^2,$$

which implies that

$$f_{u_k}(y_k) - f(x) \leq (1-\gamma_k)[f_{u_{k-1}}(y_{k-1}) - f(x)] + \gamma_k [\|\delta_k\|\bar{D}_X + Mu_k] + \tfrac{M\sqrt{n}}{2u_k}\gamma_k^2 \bar{D}_X^2.$$

Noting that by Jensen's inequality and Lemma 7.2(c),

$$\{\mathbb{E}[\|\delta_k\|]\}^2 \leq \mathbb{E}[\|\delta_k\|^2] = \tfrac{1}{T_k^2}\textstyle\sum_{t=1}^{T_k} \mathbb{E}[\|f'(y_{k-1} + u_k\xi_k) - f'_{u_k}(y_{k-1})\|^2] \leq \tfrac{M^2}{T_k}, \quad (7.1.39)$$

we conclude from the previous inequality that

$$\begin{aligned} \mathbb{E}[f_{u_k}(y_k) - f(x)] \leq (1-\gamma_k)\mathbb{E}[f_{u_{k-1}}(y_{k-1}) - f(x)] + \tfrac{\gamma_k}{\sqrt{T_k}} M\bar{D}_X + M\gamma_k u_k \\ + \tfrac{M\sqrt{n}}{2u_k}\gamma_k^2 \bar{D}_X^2, \end{aligned}$$

which, in view of Lemma 3.17, (7.1.13), and (7.1.14), then implies that, $\forall x \in X$,

$$\mathbb{E}[f_{u_k}(y_k) - f(x)] \leq \tfrac{2}{k(k+1)} \left[\textstyle\sum_{i=1}^k \left(\tfrac{i}{\sqrt{T_i}} M\bar{D}_X + Miu_i + \tfrac{M\sqrt{n}}{u_i}\bar{D}_X^2\right)\right].$$

The result in (7.1.35) follows directly from Lemma 7.2(a) and the above inequality. Using (7.1.30), (7.1.35), and (7.1.36), we can easily verify that the bound in (7.1.37) holds. ■

We now add a few remarks about the results obtained in Theorem 7.3. First, note that in order to obtain the result in (7.1.37), we need to set $T_k = k$. This implies that at the k-th iteration of the randomized CndG method in Theorem 7.3, we need to take an i.i.d. sample $\{\xi_1,\ldots,\xi_k\}$ of ξ and compute the corresponding gradients $\{f'(y_{k-1},\xi_1),\ldots,f'(y_{k-1},\xi_k)\}$. Also note that from the proof of the above result, we can recycle the generated samples $\{\xi_1,\ldots,\xi_k\}$ for usage in subsequent iterations.

Second, observe that we can apply the randomized CndG method to solve the bilinear saddle point problems with f given by (7.1.5). In comparison with the smoothing CndG method in Sect. 7.1.1.2, we do not need to solve the subproblems given in the form of (7.1.23), but to solve the subproblems

$$\max_y \left\{\langle A(x+\xi_i), y\rangle - \hat{f}(y) : y \in Y\right\},$$

in order to compute $f'(y_{k-1},\xi_i)$, $i = 1,\ldots,k$, at the k-th iteration. In particular, if $\hat{f}(y) = 0$, then we only need to solve linear optimization subproblems over the set Y. To the best of our knowledge, this is the only optimization algorithm that requires linear optimization in both primal and dual spaces.

Third, in view of (7.1.37), the number of iterations (calls to the LO oracle) required by the randomized CndG method to find a solution $\bar{x}$ such that $\mathbb{E}[f(\bar{x}) - f^*] \leq \varepsilon$ can be bounded by

$$N_\varepsilon := \mathcal{O}(1)\frac{\sqrt{n}M^2\bar{D}_X^2}{\varepsilon^2}, \tag{7.1.40}$$

and that the total number of subgradient evaluations can be bounded by

$$\Sigma_{k=1}^{N_\varepsilon} T_k = \Sigma_{k=1}^{N_\varepsilon} k = \mathcal{O}(1)N_\varepsilon^2.$$

7.1.1.4 Strongly Convex Problems Under an Enhanced LO Oracle

In this subsection, we assume that the objective function $f(\cdot)$ in (7.1.1) is smooth and strongly convex, i.e., in addition to (7.1.4), it also satisfies

$$f(y) - f(x) - \langle f'(x), y - x\rangle \geq \tfrac{\mu}{2}\|y-x\|^2, \ \forall x,y \in X. \tag{7.1.41}$$

It is known that the optimal complexity for the general first-order methods to solve this class of problems is given by

$$\mathcal{O}(1)\sqrt{\tfrac{L}{\mu}}\max\left(\log\tfrac{\mu\bar{D}_X}{\varepsilon}, 1\right).$$

On the other hand, the number of calls to the LO oracle for the CndG method is given by $\mathcal{O}(L\bar{D}_X^2/\varepsilon)$.

Our goal in this subsection is to show that, under certain stronger assumptions on the LO oracle, we can somehow "improve" the complexity of the CndG method for solving these strongly convex problems. More specifically, we assume throughout this subsection that we have access to an enhanced LO oracle, which can solve optimization problems given in the form of

$$\min\{\langle p,x\rangle : x\in X, \|x-x_0\|\le R\} \tag{7.1.42}$$

for some given $x_0 \in X$. For example, we can assume that the norm $\|\cdot\|$ is chosen such that problem (7.1.42) is relatively easy to solve. In particular, if X is a polytope, we can set $\|\cdot\| = \|\cdot\|_\infty$ or $\|\cdot\| = \|\cdot\|_1$ and then the complexity to solve (7.1.42) will be comparable to the one to solve (7.1.2). Note, however, that such a selection of $\|\cdot\|$ will possibly increase the value of the condition number given by L/μ. Using similar technique in Sect. 4.2.3, we present a shrinking CndG method under the above assumption on the enhanced LO oracle.

Algorithm 7.2 The shrinking conditional gradient (CndG) method

Let $p_0 \in X$ be given. Set $R_0 = \bar{D}_X$.
for $t = 1,\ldots$ **do**
  Set $y_0 = p_{t-1}$.
  for $k = 1,\ldots,8L/\mu$ **do**
    Call the enhanced LO oracle to compute $x_k \in \mathrm{Argmin}_{x\in X_{t-1}} \langle f'(y_{k-1}),x\rangle$, where $X_{t-1} := \{x\in X : \|x-p_{t-1}\| \le R_{t-1}\}$.
    Set $y_k = (1-\alpha_k)y_{k-1} + \alpha_k x_k$ for some $\alpha_k \in [0,1]$.
  end for
  Set $p_t = y_k$ and $R_t = R_{t-1}/\sqrt{2}$;
end for

Note that an outer (resp., inner) iteration of the above shrinking CndG method occurs whenever t (resp., k) increases by 1. Observe also that the feasible set X_t will be reduced at every outer iteration t. The following result summarizes the convergence properties for this algorithm.

Theorem 7.4. *Suppose that conditions (7.1.4) and (7.1.41) hold. If the stepsizes $\{\alpha_k\}$ in the shrinking CndG method are set to (7.1.9) or (7.1.10), then the number of calls to the enhanced* LO *oracle performed by this algorithm to find an ε-solution of problem (7.1.1) can be bounded by*

$$\tfrac{8L}{\mu}\left\lceil \max\left(\log\tfrac{\mu R_0}{\varepsilon},1\right)\right\rceil. \tag{7.1.43}$$

Proof. Denote $K \equiv 8L/\mu$. We first claim that $x^* \in X_t$ for any $t \ge 0$. This relation is obviously true for $t=0$ since $\|y_0 - x^*\| \le R_0 = \bar{D}_X$. Now suppose that $x^* \in X_{t-1}$ for some $t\ge 1$. Under this assumption, relation (7.1.17) holds with $x = x^*$ for inner iterations $k = 1,\ldots,K$ performed at the t-th outer iteration. Hence, we have

$$f(y_k) - f(x^*) \le \tfrac{2L}{k(k+1)}\textstyle\sum_{i=1}^k \|x_i - y_{i-1}\|^2 \le \tfrac{2L}{k+1}R_{t-1}^2,\ \ k=1,\ldots,K. \tag{7.1.44}$$

Letting $k = K$ in the above relation, and using the facts that $p_t = y_K$ and $f(y_K) - f^* \geq \mu \|y_K - x^*\|^2/2$, we conclude that

$$\|p_t - x^*\|^2 \leq \tfrac{2}{\mu}[f(p_t) - f^*] = \tfrac{2}{\mu}[f(y_K) - f^*] \leq \tfrac{4L}{\mu(K+1)} R_{t-1}^2 \leq \tfrac{1}{2} R_{t-1}^2 = R_t^2, \quad (7.1.45)$$

which implies that $x^* \in X_t$. We now provide a bound on the total number of calls to the LO oracle (i.e., the total number of inner iterations) performed by the shrinking CndG method. It follows from (7.1.45) and the definition of R_t that

$$f(p_t) - f^* \leq \tfrac{\mu}{2} R_t^2 = \tfrac{\mu}{2} \tfrac{R_0}{2^{t-1}}, \quad t = 1, 2, \ldots.$$

Hence the total number of outer iterations performed by the shrinking CndG method for finding an ε-solution of (7.1.1) is bounded by $\lceil \max(\log \mu R_0/\varepsilon, 1) \rceil$. This observation, in view of the fact that K inner iterations are performed at each outer iteration t, then implies that the total number of inner iterations is bounded by (7.1.43). ■

7.1.2 New Variants of Conditional Gradient

Our goal in this section is to present a few new LCP methods, obtained by replacing the projection (prox-mapping) subproblems with linear optimization subproblems in the accelerated gradient descent method. We will demonstrate that these methods can exhibit faster rate of convergence under certain circumstances than the original CndG method. Throughout this section, we focus on smooth CP problems (i.e., (7.1.4) holds). However, the developed algorithms can be easily modified to solve saddle point problems, general nonsmooth CP problems, and strongly convex problems, by using similar ideas to those described in Sect. 7.1.1.

7.1.2.1 Primal Averaging CndG Method

In this subsection, we present a new LCP method, obtained by incorporating a primal averaging step into the CndG method. This algorithm is formally described as follows.

Algorithm 7.3 The primal averaging conditional gradient (PA-CndG) method

---

Let $x_0 \in X$ be given. Set $y_0 = x_0$.
for $k = 1, \ldots$ **do**
  Set $z_{k-1} = \frac{k-1}{k+1} y_{k-1} + \frac{2}{k+1} x_{k-1}$ and $p_k = f'(z_{k-1})$.
  Call the LO oracle to compute $x_k \in \text{Argmin}_{x \in X} \langle p_k, x \rangle$.
  Set $y_k = (1 - \alpha_k) y_{k-1} + \alpha_k x_k$ for some $\alpha_k \in [0, 1]$.
end for

It can be easily seen that the PA-CndG method stated above differs from the classic CndG method in the way that the search direction p_k is defined. In particular, while p_k is set to $f'(x_{k-1})$ in the classic CndG algorithm, the search direction p_k in PA-CndG is given by $f'(z_{k-1})$ for some $z_{k-1} \in \text{Conv}\{x_0, x_1, \dots, x_{k-1}\}$. In other words, we will need to "average" the primal sequence $\{x_k\}$ before calling the LO oracle to update the iterates. It is worth noting that the PA-CndG method can be viewed as a variant of the accelerated gradient descent method, obtained by replacing the projection (or prox-mapping) subproblem with a simpler linear optimization subproblem.

By properly choosing the stepsize parameter α_k, we have the following convergence results for the PA-CndG method described above.

Theorem 7.5. *Let $\{x_k\}$ and $\{y_k\}$ be the sequences generated by the PA-CndG method applied to problem (7.1.1) with the stepsize policy in (7.1.9) or (7.1.10). Then we have*

$$f(y_k) - f^* \le \tfrac{2L}{k(k+1)} \textstyle\sum_{i=1}^k \|x_i - x_{i-1}\|^2 \le \tfrac{2L\bar{D}_X^2}{k+1}, \ k = 1, 2, \dots, \tag{7.1.46}$$

where L is given by (7.1.8).

Proof. Let γ_k and Γ_k be defined in (7.1.11) and (7.1.13), respectively. Denote $\tilde{y}_k = (1-\gamma_k)y_{k-1} + \gamma_k x_k$. It can be easily seen from (7.1.9) (or (7.1.10)) and the definition of y_k in Algorithm 7.3 that $f(y_k) \le f(\tilde{y}_k)$. Also by definition, we have $z_{k-1} = (1-\gamma_k)y_{k-1} + \gamma_k x_{k-1}$ and hence

$$\tilde{y}_k - z_{k-1} = \gamma_k (x_k - x_{k-1}).$$

Letting $l_f(\cdot,\cdot)$ be defined in (7.1.7), and using the previous two observations, (7.1.8), the definition of x_k in Algorithm 7.3, and the convexity of $f(\cdot)$, we obtain

$$\begin{aligned} f(y_k) &\le f(\tilde{y}_k) \le l_f(z_{k-1}; \tilde{y}_k) + \tfrac{L}{2}\|\tilde{y}_k - z_{k-1}\|^2 \\ &= (1-\gamma_k) l_f(z_{k-1}; y_{k-1}) + \gamma_k l_f(z_{k-1}; x_k) + \tfrac{L}{2}\gamma_k^2 \|x_k - x_{k-1}\|^2 \\ &\le (1-\gamma_k) f(y_{k-1}) + \gamma_k l_f(z_{k-1}; x) + \tfrac{L}{2}\gamma_k^2 \|x_k - x_{k-1}\|^2 \\ &\le (1-\gamma_k) f(y_{k-1}) + \gamma_k f(x) + \tfrac{L}{2}\gamma_k^2 \|x_k - x_{k-1}\|^2. \end{aligned} \tag{7.1.47}$$

Subtracting $f(x)$ from both sides of the above inequality, we have

$$f(y_k) - f(x) \le (1-\gamma_k)[f(y_{k-1}) - f(x)] + \tfrac{L}{2}\gamma_k^2 \|x_k - x_{k-1}\|^2,$$

which, in view of Lemma 3.17, (7.1.14), and the fact that $\gamma_1 = 1$, then implies that, $\forall x \in X$,

$$\begin{aligned} f(y_k) - f(x) &\le \Gamma_k (1-\gamma_1)[f(y_0) - f(x)] + \tfrac{\Gamma_k L}{2} \textstyle\sum_{i=1}^k \tfrac{\gamma_i^2}{\Gamma_i} \|x_i - x_{i-1}\|^2 \\ &\le \tfrac{2L}{k(k+1)} \textstyle\sum_{i=1}^k \|x_i - x_{i-1}\|^2, \ k = 1, 2, \dots. \end{aligned}$$

■

We now add a few remarks about the results obtained in Theorem 7.5. First, similar to (7.1.19), we can easily see that the number of iterations required by the PA-CndG method to find an ε-solution of problem (7.1.1) is bounded by $\mathscr{O}(1)L\bar{D}_X^2/\varepsilon$. In addition, since the selection of $\|\cdot\|$ is arbitrary, the iteration complexity of this method can also be bounded by (7.1.20).

Second, while the rate of convergence for the CndG method (cf. (7.1.12)) depends on $\|x_k - y_{k-1}\|$, the one for the PA-CndG method depends on $\|x_k - x_{k-1}\|$, i.e., the distance between the output of the LO oracle in two consecutive iterations. Clearly, the distance $\|x_k - x_{k-1}\|$ will depend on the geometry of X and the difference between p_k and p_{k-1}. Let γ_k be defined in (7.1.13) and suppose that α_k is set to (7.1.9) (i.e., $\alpha_k = \gamma_k$). Observe that by definitions of z_k and y_k in Algorithm 7.3, we have

$$\begin{aligned} z_k - z_{k-1} &= (y_k - y_{k-1}) + \gamma_{k+1}(x_k - y_k) - \gamma_k(x_{k-1} - y_{k-1}) \\ &= \alpha_k(x_k - y_{k-1}) + \gamma_{k+1}(x_k - y_k) - \gamma_k(x_{k-1} - y_{k-1}) \\ &= \gamma_k(x_k - y_{k-1}) + \gamma_{k+1}(x_k - y_k) - \gamma_k(x_{k-1} - y_{k-1}), \end{aligned}$$

which implies that $\|z_k - z_{k-1}\| \le 3\gamma_k \bar{D}_X$. Using this observation, (7.1.4), and the definition of p_k, we have

$$\|p_k - p_{k-1}\|_* = \|f'(z_{k-1}) - f'(z_{k-2})\|_* \le 3\gamma_{k-1}L\bar{D}_X. \qquad (7.1.48)$$

Hence, the difference between p_k and p_{k-1} vanishes as k increases. By exploiting this fact, we establish in Corollary 7.1 certain necessary conditions about the LO oracle, under which the rate of convergence of the PA-CndG algorithm can be improved.

Corollary 7.1. *Let $\{y_k\}$ be the sequence generated by the PA-CndG method applied to problem (7.1.1) with the stepsize policy in (7.1.9). Suppose that the* LO *oracle satisfies*

$$\|x_k - x_{k-1}\| \le Q\|p_k - p_{k-1}\|_*^\rho, \ \ k \ge 2, \qquad (7.1.49)$$

for some $\rho \in (0,1]$ and $Q > 0$. Then we have, for any $k \ge 1$,

$$f(y_k) - f^* \le \mathscr{O}(1) \begin{cases} Q^2 L^{2\rho+1}\bar{D}_X^{2\rho} / \left[(1-2\rho)k^{2\rho+1}\right], & \rho \in (0, 0.5), \\ Q^2L^2\bar{D}_X \log(k+1)/k^2, & \rho = 0.5, \\ Q^2L^{2\rho+1}\bar{D}_X^{2\rho} / \left[(2\rho-1)k^2\right], & \rho \in (0.5, 1]. \end{cases} \qquad (7.1.50)$$

Proof. Let γ_k be defined in (7.1.13). By (7.1.48) and (7.1.49), we have

$$\|x_k - x_{k-1}\| \le Q\|p_k - p_{k-1}\|_*^\rho \le Q(3\gamma_k L\bar{D}_X)^\rho$$

for any $k \ge 2$. The result follows by plugging the above bound into (7.1.46) and noting that

$$\Sigma_{i=1}^{k}(i+1)^{-2\rho} \leq \begin{cases} \frac{(k+1)^{-2\rho+1}}{1-2\rho}, & \rho \in (0,0.5), \\ \log(k+1), & \rho = 0.5, \\ \frac{1}{2\rho-1}, & \rho \in (0.5,1]. \end{cases}$$

■

The bound obtained in (7.1.50) provides some interesting insights on the relation between first-order LCP methods and the general optimal first-order methods for CP. More specifically, if the LO oracle satisfies the Hölder's continuity condition (7.1.49) for some $\rho \in (0.5,1]$, then we can obtain an $\mathcal{O}(1/k^2)$ rate of convergence for the PA-CndG method for solving smooth convex optimization problems.

While these assumptions on the LO oracle seem to quite strong, we provide some examples below for which the LO oracle satisfies (7.1.49).

Example 7.1. Suppose X is given by $\{x \in \mathbb{R}^n : \|Bx\| \leq 1\}$ and $f = \|Ax - b\|_2^2$. Moreover, the system $Ax - b$ is overdetermined. Then it can be seen that condition (7.1.49) will be satisfied.

It is possible to generalize the above example for more general convex sets (e.g., strongly convex sets) and for more general convex functions satisfying certain growth conditions.

7.1.2.2 Primal–Dual Averaging CndG Methods

Our goal in this subsection is to present another new LCP method, namely the primal–dual averaging CndG method, obtained by introducing a different acceleration scheme into the CndG method. This algorithm is formally described as follows.

Algorithm 7.4 The primal–dual averaging conditional gradient (PDA-CndG) method

---

Let $x_0 \in X$ be given and set $y_0 = x_0$.
for $k = 1, \ldots$ **do**
  Set $z_{k-1} = \frac{k-1}{k+1} y_{k-1} + \frac{2}{k+1} x_{k-1}$.
  Set $p_k = \Theta_k^{-1} \Sigma_{i=1}^{k} [\theta_i f'(z_{i-1})]$, where $\theta_i \geq 0$ are given and $\Theta_k = \Sigma_{i=1}^{k} \theta_i$.
  Call the LO oracle to compute $x_k \in \mathrm{Argmin}_{x \in X} \langle p_k, x \rangle$.
  Set $y_k = (1 - \alpha_k) y_{k-1} + \alpha_k x_k$ for some $\alpha_k \in [0,1]$.
end for

While the input vector p_k to the LO oracle is set to $f'(z_{k-1})$ in the PA-CndG method in the previous subsection, the vector p_k in the PDA-CndG method is defined as a weighted average of $f'(z_{i-1})$, $i = 1, \ldots, k$, for some properly chosen weights θ_i, $i = 1, \ldots, k$. This algorithm can also be viewed as the projection-free version of an ∞-memory variant of the accelerated gradient descent method.

Note that by convexity of f, the function $\Psi_k(x)$ given by

$$\Psi_k(x) := \begin{cases} 0, & k = 0, \\ \Theta_k^{-1} \sum_{i=1}^k \theta_i l_f(z_{i-1}; x), & k \geq 1, \end{cases} \tag{7.1.51}$$

underestimates $f(x)$ for any $x \in X$. In particular, by the definition of x_k in Algorithm 7.4, we have

$$\Psi_k(x_k) \leq \Psi_k(x) \leq f(x), \ \ \forall x \in X, \tag{7.1.52}$$

and hence $\Psi_k(x_k)$ provides a lower bound on the optimal value f^* of problem (7.1.1). In order to establish the convergence of the PDA-CndG method, we first need to show a simple technical result about $\Psi_k(x_k)$.

Lemma 7.3. *Let $\{x_k\}$ and $\{z_k\}$ be the two sequences computed by the PDA-CndG method. We have*

$$\theta_k l_f(z_{k-1}; x_k) \leq \Theta_k \Psi_k(x_k) - \Theta_{k-1} \Psi_{k-1}(x_{k-1}), \ \ k = 1, 2, \ldots, \tag{7.1.53}$$

where $l_f(\cdot;\cdot)$ and $\Psi_k(\cdot)$ are defined in (7.1.7) and (7.1.51), respectively.

Proof. It can be easily seen from (7.1.51) and the definition of x_k in Algorithm 7.4 that $x_k \in \text{Argmin}_{x \in X} \Psi_k(x)$ and hence that $\Psi_{k-1}(x_{k-1}) \leq \Psi_{k-1}(x_k)$. Using the previous observation and (7.1.51), we obtain

$$\begin{aligned} \Theta_k \Psi_k(x_k) &= \sum_{i=1}^k \theta_i l_f(z_{i-1}; x_i) = \theta_k l_f(z_{k-1}; x_k) + \sum_{i=1}^{k-1} \theta_i l_f(z_{i-1}; x_i) \\ &= \theta_k l_f(z_{k-1}; x_k) + \Theta_{k-1} \Psi_{k-1}(x_k) \\ &\geq \theta_k l_f(z_{k-1}; x_k) + \Theta_{k-1} \Psi_{k-1}(x_{k-1}). \end{aligned}$$

■

We are now ready to establish the main convergence properties of the PDA-CndG method.

Theorem 7.6. *Let $\{x_k\}$ and $\{y_k\}$ be the two sequences generated by the PDA-CndG method applied to problem (7.1.1) with the stepsize policy in (7.1.9) or (7.1.10). Also let $\{\gamma_k\}$ be defined in (7.1.13). If the parameters θ_k are chosen such that*

$$\theta_k \Theta_k^{-1} = \gamma_k, \ \ k = 1, 2, \ldots, \tag{7.1.54}$$

then, we have

$$f(y_k) - f^* \leq f(y_k) - \Psi_k(x_k) \leq \tfrac{2L}{k(k+1)} \sum_{i=1}^k \|x_i - x_{i-1}\|^2 \leq \tfrac{2L\bar{D}_X^2}{k+1} \tag{7.1.55}$$

for any $k = 1, 2, \ldots$, where L is given by (7.1.8).

Proof. Denote $\tilde{y}_k = (1 - \gamma_k) y_{k-1} + \gamma_k x_k$. It follows from (7.1.9) (or (7.1.10)) and the definition of y_k that $f(y_k) \leq f(\tilde{y}_k)$. Also noting that, by definition, we have $z_{k-1} = (1 - \gamma_k) y_{k-1} + \gamma_k x_{k-1}$ and hence

$$\tilde{y}_k - z_{k-1} = \gamma_k (x_k - x_{k-1}).$$

Using these two observations, (7.1.8), the definitions of x_k in Algorithm 7.4, the convexity of f, and (7.1.53), we obtain

$$\begin{aligned} f(y_k) &\le f(\tilde{y}_k) \le l_f(z_{k-1};\tilde{y}_k) + \tfrac{L}{2}\|\tilde{y}_k - z_{k-1}\|^2 \\ &= (1-\gamma_k) l_f(z_{k-1};y_{k-1}) + \gamma_k l_f(z_{k-1};x_k) + \tfrac{L}{2}\gamma_k^2 \|x_k - x_{k-1}\|^2 \\ &= (1-\gamma_k) f(y_{k-1}) + \gamma_k l_f(x_k;z_{k-1}) + \tfrac{L}{2}\gamma_k^2 \|x_k - x_{k-1}\|^2 \\ &\le (1-\gamma_k) f(y_{k-1}) + \gamma_k \theta_k^{-1}[\Theta_k \Psi_k(x_k) - \Theta_{k-1}\Psi_{k-1}(x_{k-1})] \\ &\quad + \tfrac{L}{2}\gamma_k^2 \|x_k - x_{k-1}\|^2. \end{aligned} \tag{7.1.56}$$

Also, using (7.1.54) and the fact that $\Theta_{k-1} = \Theta_k - \theta_k$, we have

$$\begin{aligned} \gamma_k \theta_k^{-1}[\Theta_k \Psi_k(x_k) - \Theta_{k-1}\Psi_{k-1}(x_{k-1})] &= \Psi_k(x_k) - \Theta_{k-1}\Theta_k^{-1}\Psi_{k-1}(x_{k-1}) \\ &= \Psi_k(x_k) - \left(1 - \theta_k \Theta_k^{-1}\right)\Psi_{k-1}(x_{k-1}) \\ &= \Psi_k(x_k) - (1-\gamma_k)\Psi_{k-1}(x_{k-1}). \end{aligned}$$

Combining the above two relations and rearranging the terms, we obtain

$$f(y_k) - \Psi_k(x_k) \le (1-\gamma_k)\left[f(y_{k-1}) - \Psi_{k-1}(x_{k-1})\right] + \tfrac{L}{2}\gamma_k^2 \|x_k - x_{k-1}\|^2,$$

which, in view of Lemma 3.17, (7.1.13), and (7.1.14), then implies that

$$f(y_k) - \Psi_k(x_k) \le \tfrac{2L}{k(k+1)} \textstyle\sum_{i=1}^k \|x_i - x_{i-1}\|^2.$$

Our result then immediately follows from (7.1.52) and the above inequality. ■

We now add a few remarks about the results obtained in Theorem 7.6. First, observe that we can simply set $\theta_k = k$, $k = 1, 2, \ldots$ in order to satisfy (7.1.54). Second, in view of the discussion after Theorem 7.5, the rate of convergence for the PDA-CndG method is exactly the same as the one for the PA-CndG method. In addition, its rate of convergence is invariant of the selection of the norm $\|\cdot\|$ (see (7.1.20)). Third, according to (7.1.55), we can compute an online lower bound $\Psi_k(x_k)$ on the optimal value f^*, and terminate the PDA-CndG method based on the optimality gap $f(y_k) - \Psi_k(x_k)$.

Similar to the PA-CndG method, the rate of convergence of the PDA-CndG method depends on $x_k - x_{k-1}$, which in turn depends on the geometry of X and the input vectors p_k and p_{k-1} to the LO oracle. One can easily check the closeness between p_k and p_{k-1}. Indeed, by the definition of p_k, we have $p_k = \Theta_k^{-1}[(1 - \theta_k)p_{k-1} + \theta_k f_k'(z_{k-1})$ and hence

$$p_k - p_{k-1} = \Theta_k^{-1}\theta_k[p_{k-1} + f_k'(z_{k-1})] = \gamma_k[p_{k-1} + f_k'(z_{k-1})], \tag{7.1.57}$$

where the last inequality follows from (7.1.54). Noting that by (7.1.8), we have $\|f'(x)\|_* \le \|f'(x^*)\|_* + L\bar{D}_X$ for any $x \in X$ and hence that $\|p_k\|_* \le \|f'(x^*)\|_* + L\bar{D}_X$ due to the definition of p_k. Using these observations, we obtain

$$\|p_k - p_{k-1}\|_* \leq 2\gamma_k[\|f'(x_*)\|_* + L\bar{D}_X], \;\; k \geq 1. \tag{7.1.58}$$

Hence, under certain continuity assumptions on the LO oracle, we can obtain a result similar to Corollary 7.1. Note that both stepsize policies in (7.1.9) and (7.1.10) can be used in this result.

Corollary 7.2. *Let $\{y_k\}$ be the sequences generated by the PDA-CndG method applied to problem (7.1.1) with the stepsize policy in (7.1.9) or (7.1.10). Assume that (7.1.54) holds. Also suppose that the* LO *oracle satisfies (7.1.49) for some $\rho \in (0,1]$ and $Q > 0$. Then we have, for any $k \geq 1$,*

$$f(y_k) - f^* \leq \mathcal{O}(1) \begin{cases} LQ^2\left[\|f'(x_*)\|_* + L\,\bar{D}_X\right]^{2\rho} / \left[(1-2\rho)\,k^{2\rho+1}\right], & \rho \in (0,0.5), \\ LQ^2\left[\|f'(x_*)\|_* + L\,\bar{D}_X\right]\log(k+1)/k^2, & \rho = 0.5, \\ LQ^2\left[\|f'(x_*)\|_* + L\,\bar{D}_X\right]^{2\rho} / \left[(2\rho-1)\,k^2\right], & \rho \in (0.5,1]. \end{cases} \tag{7.1.59}$$

Similar to Corollary 7.1, Corollary 7.2 also helps to build some connections between LCP methods and the more general optimal first-order method.

7.1.3 Lower Complexity Bound

Our goal in this section is to establish a few lower complexity bounds for solving different classes of CP problems under an LO oracle. More specifically, we first introduce a generic LCP algorithm in Sect. 7.1.3.1 and then present a few lower complexity bounds for these types of algorithms to solve different smooth and non-smooth CP problems in Sects. 7.1.3.2 and 7.1.3.3, respectively.

7.1.3.1 A Generic LCP Algorithm

The LCP algorithms solve problem (7.1.1) iteratively. In particular, at the k-th iteration, these algorithms perform a call to the LO oracle in order to update the iterates by minimizing a given linear function $\langle p_k, x\rangle$ over the feasible region X. A generic framework for these types of algorithms is described as follows.

Algorithm 7.5 A generic LCP algorithm

Let $x_0 \in X$ be given.
for $k = 1, 2, \ldots,$ **do**
  Define the linear function $\langle p_k, \cdot\rangle$.
  Call the LO oracle to compute $x_k \in \text{Argmin}_{x\in X}\langle p_k, x\rangle$.
  Output $y_k \in \text{Conv}\{x_0, \ldots, x_k\}$.
end for

Observe that the above LCP algorithm can be quite general. First, there are no restrictions regarding the definition of the linear function $\langle p_k, \cdot \rangle$. For example, if f is a smooth function, then p_k can be defined as the gradient computed at some feasible solution or a linear combination of some previously computed gradients. If f is nonsmooth, we can define p_k as the gradient computed for a certain approximation function of f. We can also consider the situation when some random noise or second-order information is incorporated into the definition of p_k. Second, the output solution y_k is written as a convex combination of $x_0, \ldots, x_k$, and thus can be different from any points in $\{x_k\}$. Algorithm 7.5 covers, as certain special cases, the classic CndG method and several new LCP methods in Sects. 7.1.1 and 7.1.2.

It is interesting to observe the difference between the above LCP algorithm and the general first-order methods for CP. On one hand, the LCP algorithm can only solve linear, rather than nonlinear subproblems (e.g., projection or prox-mapping) to update iterates. On the other hand, the LCP algorithm allows more flexibility in the definitions of the search direction p_k and the output solution y_k.

7.1.3.2 Lower Complexity Bounds for Smooth Minimization

In this subsection, we consider a class of smooth CP problems, which consist of any CP problems given in the form of (7.1.1) with f satisfying assumption (7.1.4). Our goal is to derive a lower bound on the number of calls to the LO oracle required by any LCP methods for solving this class of problems.

In the same vein as the classic complexity analysis for CP, we assume that the LO oracle used in the LCP algorithm is *resisting*, implying that: (i) the LCP algorithm does not know how the solution of (7.1.2) is computed; and (ii) in the worst case, the LO oracle provides the least amount of information for the LCP algorithm to solve problem (7.1.1). Using this assumption, we will construct a class of worst-case instances for smooth convex optimization and establish a lower bound on the number of iterations required by any LCP algorithms to solve these instances.

Theorem 7.7. *Let $\varepsilon > 0$ be a given target accuracy. The number of iterations required by any LCP methods to solve smooth convex optimization problems, in the worst case, cannot be smaller than*

$$\left\lceil \min\left\{\tfrac{n}{2}, \tfrac{L\bar{D}_X^2}{4\varepsilon}\right\}\right\rceil - 1, \tag{7.1.60}$$

where $\bar{D}_X$ is given by (7.1.18).

Proof. Consider the CP problem of

$$f_0^* := \min_{x \in X_0} \left\{ f_0(x) := \tfrac{L}{2} \Sigma_{i=1}^n (x^{(i)})^2 \right\}, \tag{7.1.61}$$

where $X_0 := \left\{ x \in \mathbb{R}^n : \Sigma_{i=1}^n x^{(i)} = D, x^{(i)} \geq 0 \right\}$ for some $D > 0$. It can be easily seen that the optimal solution x^* and the optimal value f_0^* for problem (7.1.61) are given by

$$x^* = \left(\tfrac{D}{n}, \ldots, \tfrac{D}{n}\right) \text{ and } f_0^* = \tfrac{LD^2}{n}. \tag{7.1.62}$$

Clearly, this class of problems belong to $\mathscr{F}^{1,1}_{L,\|\cdot\|}(X)$ with $\|\cdot\| = \|\cdot\|_2$.

Without loss of generality, we assume that the initial point is given by $x_0 = De_1$ where $e_1 = (1,0,\ldots,0)$ is the unit vector. Otherwise, for an arbitrary $x_0 \in X_0$, we can consider a similar problem given by

$$\begin{aligned}
\min\nolimits_x \ & \left(x^{(1)}\right)^2 + \textstyle\sum_{i=2}^n \left(x^{(i)} - x_0^{(i)}\right)^2 \\
\text{s.t.} \ & x^{(1)} + \textstyle\sum_{i=2}^n \left(x^{(i)} - x_0^{(i)}\right) = D \\
& x^{(1)} \geq 0 \\
& x^{(i)} - x_0^{(i)} \geq 0, i = 2, \ldots, n,
\end{aligned}$$

and adapt our following argument to this problem without much modification.

Now suppose that problem (7.1.61) is to be solved by an LCP algorithm. At the k-th iteration, this algorithm will call the LO oracle to compute a new search point x_k based on the input vector p_k, $k = 1, \ldots$. We assume that the LO oracle is resisting in the sense that it always outputs an extreme point $x_k \in \{De_1, De_2, \ldots, De_n\}$ such that

$$x_k \in \operatorname{Argmin}_{x \in X_0} \langle p_k, x \rangle.$$

Here e_i, $i = 1, \ldots, n$, denotes the i-th unit vector in $\mathbb{R}^n$. In addition, whenever x_k is not uniquely defined, it breaks the tie arbitrarily. Let us denote $x_k = De_{p_k}$ for some $1 \leq p_k \leq n$. By definition, we have $y_k \in D\text{Conv}\{x_0, x_1, \ldots, x_k\}$ and hence

$$y_k \in D\text{Conv}\{e_1, e_{p_1}, e_{p_2}, \ldots, e_{p_k}\}. \tag{7.1.63}$$

Suppose that totally q unit vectors from the set $\{e_1, e_{p_1}, e_{p_2}, \ldots, e_{p_k}\}$ are linearly independent for some $1 \leq q \leq k+1 \leq n$. Without loss of generality, assume that the vectors e_1, e_{p_1}, e_{p_2}, $\ldots$, $e_{p_{q-1}}$ are linearly independent. Therefore, we have

$$\begin{aligned}
f_0(y_k) &\geq \min_x \left\{ f_0(x) : x \in D\text{Conv}\{e_1, e_{p_1}, e_{p_2}, \ldots, e_{p_k}\} \right\} \\
&= \min_x \left\{ f_0(x) : x \in D\text{Conv}\{e_1, e_{p_1}, e_{p_2}, \ldots, e_{p_{q-1}}\} \right\} \\
&= \tfrac{LD^2}{q} \geq \tfrac{LD^2}{k+1},
\end{aligned}$$

where the second identity follows from the definition of f_0 in (7.1.61). The above inequality together with (7.1.62) then imply that

$$f_0(y_k) - f_0^* \geq \tfrac{LD^2}{k+1} - \tfrac{LD^2}{n} \tag{7.1.64}$$

for any $k = 1, \ldots, n-1$. Let us denote

$$\bar{K} := \left\lceil \min\left\{ \tfrac{n}{2}, \tfrac{L\bar{D}^2_{X_0}}{4\varepsilon} \right\} \right\rceil - 1.$$

By the definition of $\bar{D}_X$ and X_0, and the fact that $\|\cdot\| = \|\cdot\|_2$, we can easily see that $\bar{\bar{D}}_{X_0} = \sqrt{2}D$ and hence that

$$\bar{K} = \left\lceil \tfrac{1}{2}\min\left\{n, \tfrac{LD^2}{\varepsilon}\right\}\right\rceil - 1.$$

Using (7.1.64) and the above identity, we conclude that, for any $1 \le k \le \bar{K}$,

$$\begin{aligned} f_0(y_k) - f_0^* &\ge \tfrac{LD^2}{\bar{K}+1} - \tfrac{LD^2}{n} \ge \frac{2LD^2}{\min\{n, LD^2/\varepsilon\}} - \frac{LD^2}{n} \\ &= \frac{LD^2}{\min\{n, LD^2/\varepsilon\}} + \left(\frac{LD^2}{\min\{n, LD^2/\varepsilon\}} - \frac{LD^2}{n}\right) \ge \frac{LD^2}{LD^2/\varepsilon} + \left(\tfrac{LD^2}{n} - \tfrac{LD^2}{n}\right) = \varepsilon. \end{aligned}$$

Our result then immediately follows since (7.1.61) is a special class of problems in $\mathscr{F}^{1,1}_{L,\|\cdot\|}(X)$. ∎

We now add a few remarks about the results obtained in Theorem 7.7. First, it can be easily seen from (7.1.60) that if $n \ge L\bar{D}_X^2/(2\varepsilon)$, then the number of calls to the LO oracle required by any LCP methods for solving smooth convex optimization problems, in the worst case, cannot be smaller than $\mathscr{O}(1)L\bar{D}_X^2/\varepsilon$. Second, it is worth noting that the objective function f_0 in (7.1.61) is actually strongly convex. Hence, the performance of the LCP methods cannot be improved by assuming strong convexity when n is sufficiently large (see Sect. 7.1.1.4 for more discussions). This is in sharp contrast to the general first-order methods whose complexity for solving strongly convex problems depends on $\log(1/\varepsilon)$.

Comparing (7.1.60) with a few complexity bounds we obtained for the CndG, PA-CndG, and PDA-CndG methods, we conclude that these algorithms achieve an optimal bound on the number of calls to the LO oracle for solving smooth convex optimization if n is sufficiently large. It should be noted, however, that the lower complexity bound in (7.1.60) was established for the number of calls to the LO oracles. It is possible that we can achieve better complexity than $\mathscr{O}(1)L\bar{D}_X^2/\varepsilon$ in terms of the number of gradient computations for smooth convex optimization. We will discuss this issue in more detail in Sect. 7.2.

7.1.3.3 Lower Complexity Bounds for Nonsmooth Minimization

In this subsection, we consider two classes of nonsmooth CP problems. The first one is a general class of nonsmooth CP problems which consist of any CP problems given in the form of (7.1.1) with f satisfying (7.1.6). The second one is a special class of bilinear saddle-point problems, composed of all CP problems (7.1.1) with f given by (7.1.5). Our goal in this subsection is to derive the lower complexity bounds for any LCP algorithms to solve these two classes of nonsmooth CP problems.

It can be seen that if $f(\cdot)$ is given by (7.1.5), then

$$\|f'(x)\|_* \le \|A\|\bar{D}_Y, \;\; \forall x \in X,$$

where $\bar{D}_Y$ is given by (7.1.18). Hence, the saddle point problems with f given by (7.1.5) are a special class of nonsmooth CP problems.

Theorem 7.8 below provides lower complexity bounds for solving these two classes of nonsmooth CP problems by using LCP algorithms.

Theorem 7.8. *Let $\varepsilon > 0$ be a given target accuracy. Then, the number of iterations required by any LCP methods to solve the problem classes with f satisfying (7.1.6) and with f given by (7.1.5), respectively, cannot be smaller than*

$$\tfrac{1}{4}\min\left\{n, \tfrac{M^2\bar{D}_X^2}{2\varepsilon^2}\right\} - 1 \tag{7.1.65}$$

and

$$\tfrac{1}{4}\min\left\{n, \tfrac{\|A\|^2\bar{D}_X^2\bar{D}_Y^2}{2\varepsilon^2}\right\} - 1, \tag{7.1.66}$$

where $\bar{D}_X$ and $\bar{D}_Y$ are defined in (7.1.18) and (7.1.31), respectively.

Proof. We first show the bound in (7.1.65). Consider the CP problem of

$$\hat{f}_0^* := \min_{x\in X_0}\left\{\hat{f}(x) := M\left(\textstyle\sum_{i=1}^n x_i^2\right)^{1/2}\right\}, \tag{7.1.67}$$

where $X_0 := \left\{x \in \mathbb{R}^n : \sum_{i=1}^n x^{(i)} = D, x^{(i)} \geq 0\right\}$ for some $D > 0$. It can be easily seen that the optimal solution x^* and the optimal value f_0^* for problem (7.1.67) are given by

$$x^* = \left(\tfrac{D}{n},\ldots,\tfrac{D}{n}\right) \text{ and } \hat{f}_0^* = \tfrac{MD}{\sqrt{n}}. \tag{7.1.68}$$

Clearly, this class of problems satisfies (7.1.6) with $\|\cdot\| = \|\cdot\|_2$. Now suppose that problem (7.1.61) is to be solved by an arbitrary LCP method. Without loss of generality, we assume that the initial point is given by $x_0 = De_1$ where $e_1 = (1,0,\ldots,0)$ is the unit vector. Assume that the LO oracle is resisting in the sense that it always outputs an extreme point solution. By using an argument similar to the one used in the proof of (7.1.63), we can show that

$$y_k \in D\mathrm{Conv}\{e_1, e_{p_1}, e_{p_2}, \ldots, e_{p_k}\},$$

where e_{p_i}, $i = 1,\ldots,k$, are the unit vectors in $\mathbb{R}^n$. Suppose that totally q unit vectors in the set $\{e_1, e_{p_1}, e_{p_2}, \ldots, e_{p_k}\}$ are linearly independent for some $1 \leq q \leq k+1 \leq n$. We have

$$\hat{f}_0(y_k) \geq \min_x\left\{\hat{f}_0(x) : x \in D\mathrm{Conv}\{e_1, e_{p_1}, e_{p_2}, \ldots, e_{p_k}\}\right\} = \tfrac{MD}{\sqrt{q}} \geq \tfrac{MD}{\sqrt{k+1}},$$

where the identity follows from the definition of $\hat{f}_0$ in (7.1.67). The above inequality together with (7.1.68) then imply that

$$\hat{f}_0(y_k) - \hat{f}_0^* \geq \tfrac{MD}{\sqrt{k+1}} - \tfrac{LD^2}{\sqrt{n}} \tag{7.1.69}$$

for any $k = 1,\ldots,n-1$. Let us denote

$$\bar{K} := \tfrac{1}{4}\left\lceil \min\left\{n, \tfrac{M^2\bar{D}^2_{X_0}}{2\varepsilon^2}\right\}\right\rceil - 1.$$

Using the above definition, (7.1.69) and the fact that $\bar{D}_{X_0} = \sqrt{2}D$, we conclude that

$$\hat{f}_0(y_k) - \hat{f}_0^* \geq \tfrac{MD}{\sqrt{\bar{K}+1}} - \tfrac{MD}{n} \geq \tfrac{2MD}{\min\left\{\sqrt{n}, \tfrac{MD}{\varepsilon}\right\}} - \tfrac{MD}{\sqrt{n}} \geq \varepsilon$$

for any $1 \leq k \leq \bar{K}$. Our result in (7.1.65) then immediately follows since (7.1.67) is a special class of problems with f satisfying (7.1.6).

In order to prove the lower complexity bound in (7.1.66), we consider a class of saddle point problems given in the form of

$$\min_{x\in X_0} \max_{\|y\|_2\leq\tilde{D}} M\langle x,y\rangle. \tag{7.1.70}$$

Clearly, these problems belong to $\mathscr{S}_{\|A\|}(X,Y)$ with $A = MI$. Noting that problem (7.1.70) is equivalent to

$$\min_{x\in X_0} M\tilde{D}\left(\Sigma_{i=1}^n x_i^2\right)^{1/2},$$

we can show the lower complexity bound in (7.1.66) by using an argument similar to the one used in the proof of bound (7.1.65). ∎

Observe that the lower complexity bound in (7.1.65) is in the same order of magnitude as the one for general first-order methods to solve these problems. However, the bound in (7.1.65) holds not only for first-order LCP methods, but also for any other LCP methods, including those based on higher-order information.

In view of Theorem 7.8 and the discussions after Theorem 7.2, the CndG method when coupled with smoothing technique is optimal for solving bilinear saddle point problems with f given by (7.1.5), when n is sufficiently large.

In addition, from (7.1.65) and the discussion after Theorem 7.3, we conclude that the complexity bound in (7.1.40) on the number of calls to the LO oracle is nearly optimal for general nonsmooth convex optimization due to the following facts: (i) the above result is in the same order of magnitude as (7.1.65) with an additional factor of $\sqrt{n}$; and (ii) the termination criterion is in terms of expectation.

7.2 Conditional Gradient Sliding Method

In the previous section, we show that the number of calls to the LO oracle performed by LCP methods cannot be smaller than $\mathscr{O}(1/\varepsilon)$ for solving smooth convex optimization problems. Moreover, it has been shown that the CndG method and a few of its variants can find an ε-solution of (7.1.1) (i.e., a point $\bar{x} \in X$ s.t. $f(\bar{x}) - f^* \leq \varepsilon$)

in at most $\mathcal{O}(1/\varepsilon)$ iterations. Note that each iteration of the CndG method requires one call to the LO oracle and one gradient evaluation. Therefore, the CndG method requires totally $\mathcal{O}(1/\varepsilon)$ gradient evaluations returned by the first-order (FO) oracle. Since the aforementioned $\mathcal{O}(1/\varepsilon)$ bound on gradient evaluations is significantly worse than the optimal $\mathcal{O}(1/\sqrt{\varepsilon})$ bound for smooth convex optimization, a natural question is whether one can further improve the $\mathcal{O}(1/\varepsilon)$ complexity bound associated with the CndG method.

Our main goal in this section is to show that although the number of calls to the LO oracle cannot be improved for the LCP methods in general, we can substantially improve their complexity bounds in terms of the number of gradient evaluations. To this end, we present a new LCP algorithm, referred to as the conditional gradient sliding (CGS) method, which can skip the computation for the gradient of f from time to time while still maintaining the optimal bound on the number of calls to the LO oracle. Our development has been leveraged on the basic idea of applying the CndG method to the subproblems of the accelerated gradient descent method, rather than to the original CP problem in (7.1.1) itself. As a result, the same first-order information of f will be used throughout a large number of CndG iterations. Moreover, the accuracy of the approximate solutions to these subproblems is measured by the first-order optimality condition (or Wolfe gap), which allows us to establish the convergence of an inexact version of the accelerated gradient descent method.

This section proceeds as follows. First, we show that if f is a smooth convex function satisfying (7.1.4), then the number of calls to the FO and LO oracles, respectively, can be bounded by $\mathcal{O}(1/\sqrt{\varepsilon})$ and $\mathcal{O}(1/\varepsilon)$. Moreover, if f is smooth and strongly convex, then the number of calls to the FO oracle can be significantly reduced to $\mathcal{O}(\log 1/\varepsilon)$ while the number of calls to the LO oracle remains the same. It should be noted that these improved complexity bounds were obtained without enforcing any stronger assumptions on the LO oracle or the feasible set X.

Second, we consider the stochastic case where one can only have access to a stochastic first-order oracle (SFO) of f, which upon requests returns unbiased estimators for the gradient of f. By developing a stochastic counterpart of the CGS method, i.e., the SCGS algorithm, we show that the number of calls to the SFO and LO oracles, respectively, can be optimally bounded by $\mathcal{O}(1/\varepsilon^2)$ and $\mathcal{O}(1/\varepsilon)$ when f is smooth. In addition, if f is smooth and strongly convex, then the former bound can be significantly reduced to $\mathcal{O}(1/\varepsilon)$.

Third, we generalize the CGS and SCGS algorithms to solve an important class of nonsmooth CP problems that can be closely approximated by a class of smooth functions. By incorporating an adaptive smoothing technique into the conditional gradient sliding algorithms, we show that the number of gradient evaluations and calls to the LO oracle can bound optimally by $\mathcal{O}(1/\varepsilon)$ and $\mathcal{O}(1/\varepsilon^2)$, respectively.

7.2.1 Deterministic Conditional Gradient Sliding

Our goal in this subsection is to present a new LCP method, namely the conditional gradient sliding (CGS) method, which can skip the computation for the gradient of f from time to time when performing linear optimization over the feasible region X. More specifically, we introduce the CGS method for smooth convex problems in Sect. 7.2.1.1 and generalize it for smooth and strongly convex problems in Sect. 7.2.1.2.

7.2.1.1 Smooth Convex Optimization

The basic scheme of the CGS method is obtained by applying the classic conditional gradient (CndG) method to solve the projection subproblems existing in the accelerated gradient descent (AGD) method approximately. By properly specifying the accuracy for solving these subproblems, we will show that the resulting CGS method can achieve the optimal bounds on the number of calls to the FO and LO oracles for solving problem (7.1.1).

The CGS method is formally described as follows.

Algorithm 7.6 The conditional gradient sliding (CGS) method

Input: Initial point $x_0 \in X$ and iteration limit N.
Let $\beta_k \in \mathbb{R}_{++}, \gamma_k \in [0,1]$, and $\eta_k \in \mathbb{R}_+$, $k = 1,2,\ldots$, be given and set $y_0 = x_0$.
for $k = 1,2,\ldots,N$ **do**

$$z_k = (1-\gamma_k)y_{k-1} + \gamma_k x_{k-1}, \tag{7.2.1}$$

$$x_k = \mathrm{CndG}(f'(z_k), x_{k-1}, \beta_k, \eta_k), \tag{7.2.2}$$

$$y_k = (1-\gamma_k)y_{k-1} + \gamma_k x_k. \tag{7.2.3}$$

end for
Output: y_N.

procedure $u^+ = \mathrm{CndG}(g,\, u,\, \beta,\, \eta)$
1. Set $u_1 = u$ and $t = 1$.
2. Let v_t be an optimal solution for the subproblem of

$$V_{g,u,\beta}(u_t) := \max_{x\in X} \langle g + \beta(u_t - u), u_t - x\rangle. \tag{7.2.4}$$

3. If $V_{g,u,\beta}(u_t) \le \eta$, set $u^+ = u_t$ and **terminate** the procedure.
4. Set $u_{t+1} = (1-\alpha_t)u_t + \alpha_t v_t$ with

$$\alpha_t = \min\left\{1, \tfrac{\langle \beta(u-u_t) - g, v_t - u_t\rangle}{\beta\|v_t - u_t\|^2}\right\}. \tag{7.2.5}$$

5 Set $t \leftarrow t+1$ and go to step 2.
end procedure

Clearly, the most crucial step of the CGS method is to update the search point x_k by calling the CndG procedure in (7.2.2). Denoting $\phi(x) := \langle g, x\rangle + \beta\|x-u\|^2/2$, the CndG procedure can be viewed as a specialized version of the classical conditional gradient method applied to $\min_{x\in X}\phi(x)$. In particular, it can be easily seen that $V_{g,u,\beta}(u_t)$ in (7.2.4) is equivalent to $\max_{x\in X}\langle\phi'(u_t), u_t - x\rangle$, which is often called the Wolfe gap, and the CndG procedure terminates whenever $V_{g,u,\beta}(u_t)$ is smaller than the pre-specified tolerance η. In fact, this procedure is slightly simpler than the generic conditional gradient method in that the selection of α_t in (7.2.5) explicitly solves

$$\alpha_t = \mathrm{argmin}_{\alpha\in[0,1]}\phi((1-\alpha)u_t + \alpha v_t). \tag{7.2.6}$$

In view of the above discussion, we can easily see that x_k obtained in (7.2.2) is an approximate solution for the projection subproblem

$$\min_{x\in X}\left\{\phi_k(x) := \langle f'(z_k), x\rangle + \tfrac{\beta_k}{2}\|x - x_{k-1}\|^2\right\} \tag{7.2.7}$$

such that

$$\langle\phi_k'(x_k), x_k - x\rangle = \langle f'(z_k) + \beta_k(x_k - x_{k-1}), x_k - x\rangle \le \eta_k, \ \ \forall x\in X, \tag{7.2.8}$$

for some $\eta_k \ge 0$.

Clearly, problem (7.2.7) is equivalent to $\min_{x\in X}\beta_k/2\|x - x_{k-1} + f'(z_k)/\beta_k\|^2$ after completing the square, and it admits explicit solutions in some special cases, e.g., when X is a standard Euclidean ball. However, we focus on the case where (7.2.7) is solved iteratively by calling the LO oracle.

We now add a few comments about the main CGS method. First, similar to the accelerated gradient method, the above CGS method maintains the updating of three intertwined sequences, namely $\{x_k\}$, $\{y_k\}$, and $\{z_k\}$, in each iteration. The main difference between CGS and the original AGD exists in the computation of x_k. More specifically, x_k in the original AGD method is set to the exact solution of (7.2.7) (i.e., $\eta_k = 0$ in (7.2.8)), while the subproblem in (7.2.7) is only solved approximately for the CGS method (i.e., $\eta_k > 0$ in (7.2.8)).

Second, we say that an inner iteration of the CGS method occurs whenever the index t in the CndG procedure increments by 1. Accordingly, an outer iteration of CGS occurs whenever k increases by 1. While we need to call the FO oracle to compute the gradient $f'(z_k)$ in each outer iteration, the gradient $\phi_k'(p_t)$ used in the CndG subroutine is given explicitly by $f'(z_k) + \beta_k(p - x_{k-1})$. Hence, the main cost per each inner iteration of the CGS method is to call the LO oracle to solve linear optimization problem in (7.2.4). As a result, the total number of outer and inner iterations performed by the CGS algorithm are equivalent to the total number of calls to the FO and LO oracles, respectively.

Third, observe that the above CGS method is conceptual only since we have not specified a few parameters, including $\{\beta_k\}$, $\{\gamma_k\}$, and $\{\eta_k\}$, used in this algorithm yet. We will come back to this issue after establishing some important convergence properties for the above generic CGS algorithm.

Theorem 7.9 describes the main convergence properties of the above CGS method. More specifically, both Theorem 7.9(a) and (b) show the convergence of the AGD method when the projection subproblem is approximately solved according to (7.2.8), while Theorem 7.9(c) states the convergence of the CndG procedure by using the Wolfe gap as the termination criterion.

Observe that the following quantity will be used in the convergence analysis of the CGS algorithm:

$$\Gamma_k := \begin{cases} 1 & k = 1 \\ \Gamma_{k-1}(1-\gamma_k) & k \geq 2. \end{cases} \tag{7.2.9}$$

Theorem 7.9. *Let Γ_k be defined in (7.2.9). Suppose that $\{\beta_k\}$ and $\{\gamma_k\}$ in the CGS algorithm satisfy*

$$\gamma_1 = 1 \text{ and } L\gamma_k \leq \beta_k, \; k \geq 1. \tag{7.2.10}$$

(a) If

$$\tfrac{\beta_k \gamma_k}{\Gamma_k} \geq \tfrac{\beta_{k-1}\gamma_{k-1}}{\Gamma_{k-1}}, \; k \geq 2, \tag{7.2.11}$$

then for any $x \in X$ and $k \geq 1$,

$$f(y_k) - f(x^*) \leq \tfrac{\beta_k \gamma_k}{2} \bar{D}_X^2 + \Gamma_k \Sigma_{i=1}^k \tfrac{\eta_i \gamma_i}{\Gamma_i}. \tag{7.2.12}$$

where x^* is an arbitrary optimal solution of (7.1.1) and $\bar{D}_X$ is defined in (7.1.18).

(b) If

$$\tfrac{\beta_k \gamma_k}{\Gamma_k} \leq \tfrac{\beta_{k-1}\gamma_{k-1}}{\Gamma_{k-1}}, \; k \geq 2, \tag{7.2.13}$$

then for any $x \in X$ and $k \geq 1$,

$$f(y_k) - f(x^*) \leq \tfrac{\beta_1 \Gamma_k}{2} \|x_0 - x^*\|^2 + \Gamma_k \Sigma_{i=1}^k \tfrac{\eta_i \gamma_i}{\Gamma_i}. \tag{7.2.14}$$

(c) Under the assumptions in either part (a) or (b), the number of inner iterations performed at the k-th outer iteration can be bounded by

$$T_k := \left\lceil \tfrac{6\beta_k \bar{D}_X^2}{\eta_k} \right\rceil, \; \forall k \geq 1. \tag{7.2.15}$$

Proof. We first show part (a). Note that by (7.2.1) and (7.2.3), we have $y_k - z_k = \gamma_k(x_k - x_{k-1})$. By using this observation, (7.1.8), and (7.2.3) we have

$$\begin{aligned} f(y_k) &\leq l_f(z_k; y_k) + \tfrac{L}{2}\|y_k - z_k\|^2 \\ &= (1-\gamma_k) l_f(z_k; y_{k-1}) + \gamma_k l_f(z_k; x_k) + \tfrac{L\gamma_k^2}{2}\|x_k - x_{k-1}\|^2 \\ &= (1-\gamma_k) l_f(z_k; y_{k-1}) + \gamma_k l_f(z_k; x_k) + \tfrac{\beta_k \gamma_k}{2}\|x_k - x_{k-1}\|^2 \\ &\quad - \tfrac{\gamma_k}{2}(\beta_k - L\gamma_k)\|x_k - x_{k-1}\|^2 \\ &\leq (1-\gamma_k) f(y_{k-1}) + \gamma_k l_f(z_k; x_k) + \tfrac{\beta_k \gamma_k}{2}\|x_k - x_{k-1}\|^2, \end{aligned} \tag{7.2.16}$$

where the last inequality follows from the convexity of $f(\cdot)$ and (7.2.10). Also observe that by (7.2.8), we have

$$\langle f'(z_k) + \beta_k(x_k - x_{k-1}), x_k - x \rangle \le \eta_k, \ \ \forall x \in X,$$

which implies that

$$\begin{aligned}\tfrac{1}{2}\|x_k - x_{k-1}\|^2 &= \tfrac{1}{2}\|x_{k-1} - x\|^2 - \langle x_{k-1} - x_k, x_k - x \rangle - \tfrac{1}{2}\|x_k - x\|^2 \\ &\le \tfrac{1}{2}\|x_{k-1} - x\|^2 + \tfrac{1}{\beta_k}\langle f'(z_k), x - x_k \rangle - \tfrac{1}{2}\|x_k - x\|^2 + \tfrac{\eta_k}{\beta_k}. \quad (7.2.17)\end{aligned}$$

Combining (7.2.16) and (7.2.17), we obtain

$$\begin{aligned}f(y_k) &\le (1-\gamma_k)f(y_{k-1}) + \gamma_k l_f(z_k; x) + \tfrac{\beta_k\gamma_k}{2}\left(\|x_{k-1} - x\|^2 - \|x_k - x\|^2\right) + \eta_k\gamma_k \\ &\le (1-\gamma_k)f(y_{k-1}) + \gamma_k f(x) + \tfrac{\beta_k\gamma_k}{2}\left(\|x_{k-1} - x\|^2 - \|x_k - x\|^2\right) + \eta_k\gamma_k, \quad (7.2.18)\end{aligned}$$

where the last inequality follows from the convexity of $f(\cdot)$. Subtracting $f(x)$ from both sides of the above inequality, we have

$$\begin{aligned}f(y_k) - f(x) \le (1-\gamma_k)[f(y_{k-1}) - f(x)] + \tfrac{\beta_k\gamma_k}{2}\left(\|x_{k-1} - x\|^2 - \|x_k - x\|^2\right) \\ + \eta_k\gamma_k, \ \ \forall x \in X.\end{aligned}$$

which, in view of Lemma 3.17, then implies that

$$\begin{aligned}f(y_k) - f(x) \le \tfrac{\Gamma_k(1-\gamma_1)}{\Gamma_1}[f(y_0) - f(x)] \\ + \Gamma_k \textstyle\sum_{i=1}^k \tfrac{\beta_i\gamma_i}{2\Gamma_i}\left(\|x_{i-1} - x\|^2 - \|x_i - x\|^2\right) + \Gamma_k \textstyle\sum_{i=1}^k \tfrac{\eta_i\gamma_i}{\Gamma_i}. \quad (7.2.19)\end{aligned}$$

Our result in part (a) then immediately follows from the above inequality, the assumption that $\gamma_1 = 1$, and the fact that

$$\begin{aligned}&\textstyle\sum_{i=1}^k \tfrac{\beta_i\gamma_i}{\Gamma_i}\left(\|x_{i-1} - x\|^2 - \|x_i - x\|^2\right) \\ &= \tfrac{\beta_1\gamma_1}{\Gamma_1}\|x_0 - x\|^2 + \textstyle\sum_{i=2}^k \left(\tfrac{\beta_i\gamma_i}{\Gamma_i} - \tfrac{\beta_{i-1}\gamma_{i-1}}{\Gamma_{i-1}}\right)\|x_{i-1} - x\|^2 - \tfrac{\beta_k\gamma_k}{\Gamma_k}\|x_k - x\|^2 \\ &\le \tfrac{\beta_1\gamma_1}{\Gamma_1}\bar{D}_X^2 + \textstyle\sum_{i=2}^k \left(\tfrac{\beta_i\gamma_i}{\Gamma_i} - \tfrac{\beta_{i-1}\gamma_{i-1}}{\Gamma_{i-1}}\right)\bar{D}_X^2 = \tfrac{\beta_k\gamma_k}{\Gamma_k}\bar{D}_X^2, \quad (7.2.20)\end{aligned}$$

where the inequality follows from the third assumption in (7.2.11) and the definition of $\bar{D}_X$ in (7.1.18).

Similarly, part (b) follows from (7.2.19), the assumption that $\gamma_1 = 1$, and the fact that

$$\textstyle\sum_{i=1}^k \tfrac{\beta_i\gamma_i}{\Gamma_i}\left(\|x_{i-1} - x\|^2 - \|x_i - x\|^2\right) \le \tfrac{\beta_1\gamma_1}{\Gamma_1}\|x_0 - x\|^2 - \tfrac{\beta_k\gamma_k}{\Gamma_k}\|x_k - x\|^2 \le \beta_1\|x_0 - x\|^2, \quad (7.2.21)$$

due to the assumptions in (7.2.10) and (7.2.13).

Now we show that part (c) holds. Let us denote $\phi \equiv \phi_k$ and $\phi^* \equiv \min_{x\in X} \phi(x)$. Also let us denote

$$\lambda_t := \tfrac{2}{t} \text{ and } \Lambda_t = \tfrac{2}{t(t-1)}. \tag{7.2.22}$$

It then follows from the above definitions that

$$\Lambda_{t+1} = \Lambda_t(1-\lambda_{t+1}), \ \forall t \geq 2. \tag{7.2.23}$$

Let us define $\bar{u}_{t+1} := (1-\lambda_{t+1})u_t + \lambda_{t+1}v_t$. Clearly we have $\bar{u}_{t+1} - u_t = \lambda_{t+1}(v_t - u_t)$. Observe that $u_{t+1} = (1-\alpha_t)u_t + \alpha_t v_t$ and α_t is an optimal solution of (7.2.6), and hence that $\phi(u_{t+1}) \leq \phi(\bar{u}_{t+1})$. Using this observation, (7.1.8), and the fact that ϕ has Lipschitz continuous gradients, we have

$$\begin{aligned}\phi(u_{t+1}) &\leq \phi(\bar{u}_{t+1}) \leq l_\phi(u_t, \bar{u}_{t+1}) + \tfrac{\beta}{2}\|\bar{u}_{t+1} - u_t\|^2 \\ &\leq (1-\lambda_{t+1})\phi(u_t) + \lambda_{t+1} l_\phi(u_t, v_t) + \tfrac{\beta\lambda_{t+1}^2}{2}\|v_t - u_t\|^2.\end{aligned} \tag{7.2.24}$$

Also observe that by (7.1.7) and the fact that v_t solves (7.2.4), we have

$$l_\phi(u_t, v_t) = \phi(u_t) + \langle \phi'(u_t), v_t - u_t\rangle \leq \phi(u_t) + \langle \phi'(u_t), x - u_t\rangle \leq \phi(x)$$

for any $x \in X$, where the last inequality follows from the convexity of $\phi(\cdot)$. Combining the above two inequalities and rearranging the terms, we obtain

$$\phi(u_{t+1}) - \phi(x) \leq (1-\lambda_{t+1})[\phi(u_t) - \phi(x)] + \tfrac{\beta\lambda_{t+1}^2}{2}\|v_t - u_t\|^2, \ \forall x \in X,$$

which, in view of Lemma 3.17, then implies that, for any $x \in X$ and $t \geq 1$,

$$\begin{aligned}\phi(u_{t+1}) - \phi(x) &\leq \Lambda_{t+1}(1-\lambda_2)[\phi(u_1) - \phi(x)] + \Lambda_{t+1}\beta\textstyle\sum_{j=1}^t \tfrac{\lambda_{j+1}^2}{2\Lambda_{j+1}}\|v_j - u_j\|^2 \\ &\leq \tfrac{2\beta\bar{D}_X^2}{t+1},\end{aligned} \tag{7.2.25}$$

where the last inequality easily follows from (7.2.22) and the definition of $\bar{D}_X$ in (7.1.18). Now, let the gap function $V_{g,u,\beta}$ be defined in (7.2.4). Also let us denote $\Delta_j = \phi(u_j) - \phi^*$. It then follows from (7.1.7), (7.2.4), and (7.2.24) that that for any $j = 1, \ldots, t$,

$$\begin{aligned}\lambda_{j+1}V_{g,u,\beta}(u_j) &\leq \phi(u_j) - \phi(u_{j+1}) + \tfrac{\beta\lambda_{j+1}^2}{2}\|v_j - u_j\|^2 \\ &= \Delta_j - \Delta_{j+1} + \tfrac{\beta\lambda_{j+1}^2}{2}\|v_j - u_j\|^2.\end{aligned}$$

Dividing both sides of the above inequality by Λ_{j+1} and summing up the resulting inequalities, we obtain

$$\begin{aligned}\textstyle\sum_{j=1}^t \tfrac{\lambda_{j+1}}{\Lambda_{j+1}} V_{g,u,\beta}(u_j) &\leq -\tfrac{1}{\Lambda_{t+1}}\Delta_{t+1} + \textstyle\sum_{j=2}^t \left(\tfrac{1}{\Lambda_{j+1}} - \tfrac{1}{\Lambda_j}\right)\Delta_j + \Delta_1 \\ &\quad + \textstyle\sum_{j=1}^t \tfrac{\beta\lambda_{j+1}^2}{2\Lambda_{j+1}}\|v_j - u_j\|^2\end{aligned}$$

$$\le \sum_{j=2}^{t}\left(\frac{1}{\Lambda_{j+1}}-\frac{1}{\Lambda_j}\right)\Delta_j+\Delta_1$$
$$+\sum_{j=1}^{t}\frac{\beta\lambda_{j+1}^2}{2\Lambda_{j+1}}\bar{D}_X^2 \le \sum_{j=1}^{t} j\Delta_j + t\beta\bar{D}_X^2,$$

where the last inequality follows from the definitions of λ_t and Λ_t in (7.2.22). Using the above inequality and the bound on Δ_j given in (7.2.25), we conclude that

$$\min_{j=1,\ldots,t} V_{g,u,\beta}(u_j)\sum_{j=1}^{t}\frac{\lambda_{j+1}}{\Lambda_{j+1}} \le \sum_{j=1}^{t}\frac{\lambda_{j+1}}{\Lambda_{j+1}}V_{g,u,\beta}(u_j) \le 3t\beta\bar{D}_X^2,$$

which, in view of the fact that $\sum_{j=1}^{t}\lambda_{j+1}/\Lambda_{j+1} = t(t+1)/2$, then clearly implies that

$$\min_{j=1,\ldots,t} V_{g,u,\beta}(u_j) \le \frac{6\beta\bar{D}_X^2}{t+1}, \quad \forall t\ge 1, \tag{7.2.26}$$

from which part (c) immediately follows. ■

Clearly, there exist various options to specify the parameters $\{\beta_k\}$, $\{\gamma_k\}$, and $\{\eta_k\}$ so as to guarantee the convergence of the CGS method. In the following corollaries, we provide two different parameter settings for $\{\beta_k\}$, $\{\gamma_k\}$, and $\{\eta_k\}$, which lead to optimal complexity bounds on the total number of calls to the FO and LO oracles for smooth convex optimization.

Corollary 7.3. *If $\{\beta_k\}$, $\{\gamma_k\}$, and $\{\eta_k\}$ in the CGS method are set to*

$$\beta_k = \frac{3L}{k+1}, \quad \gamma_k = \frac{3}{k+2}, \quad \text{and} \quad \eta_k = \frac{L\bar{D}_X^2}{k(k+1)}, \quad \forall k\ge 1, \tag{7.2.27}$$

then for any $k\ge 1$,

$$f(y_k)-f(x^*) \le \frac{15L\bar{D}_X^2}{2(k+1)(k+2)}. \tag{7.2.28}$$

As a consequence, the total number of calls to the FO *and* LO *oracles performed by the CGS method for finding an ε-solution of (7.1.1) can be bounded by $\mathcal{O}\left(\sqrt{L\bar{D}_X^2/\varepsilon}\right)$ and $\mathcal{O}\left(L\bar{D}_X^2/\varepsilon\right)$, respectively.*

Proof. We first show part (a). It can be easily seen from (7.2.27) that (7.2.10) holds. Also note that by (7.2.27), we have

$$\Gamma_k = \frac{6}{k(k+1)(k+2)}, \tag{7.2.29}$$

and

$$\frac{\beta_k\gamma_k}{\Gamma_k} = \frac{9L}{(k+1)(k+2)}\frac{k(k+1)(k+2)}{6} = \frac{3Lk}{2},$$

which implies that (7.2.11) is satisfied. It then follows from Theorem 7.9(a), (7.2.27), and (7.2.29) that

$$f(y_k)-f(x^*) \le \frac{9L\bar{D}_X^2}{2(k+1)(k+2)} + \frac{6}{k(k+1)(k+2)}\sum_{i=1}^{k}\frac{\eta_i\gamma_i}{\Gamma_i} = \frac{15L\bar{D}_X^2}{2(k+1)(k+2)},$$

which implies that the total number of outer iterations performed by the CGS method for finding an ε-solution can be bounded by $N = \sqrt{15L\bar{D}_X^2/(2\varepsilon)}$. Moreover, it follows from the bound in (7.2.15) and (7.2.27) that the total number of inner iterations can be bounded by

$$\Sigma_{k=1}^N T_k \le \Sigma_{k=1}^N \left(\tfrac{6\beta_k \bar{D}_X^2}{\eta_k} + 1\right) = 18\Sigma_{k=1}^N k + N = 9N^2 + 10N,$$

which implies that the total number of inner iterations is bounded by $\mathscr{O}(L\bar{D}_X^2/\varepsilon)$. ■

Observe that in the above result, the number of calls to the LO oracle is not improvable in terms of their dependence on ε, L, and $\bar{D}_X$ for LCP methods. Similarly, the number of calls to the FO oracle is also optimal in terms of its dependence on ε and L. It should be noted, however, that we can potentially improve the latter bound in terms of its dependence on $\bar{D}_X$. Indeed, by using a different parameter setting, we show in Corollary 7.4 a slightly improved bound on the number of calls to the FO oracle which only depends on the distance from the initial point to the set of optimal solutions, rather than the diameter $\bar{D}_X$. This result will play an important role for the analysis of the CGS method for solving strongly convex problems. The disadvantage of using this parameter setting is that we need to fix the number of iterations N in advance.

Corollary 7.4. *Suppose that there exists an estimate $D_0 \ge \|x_0 - x^*\|$ and that the outer iteration limit $N \ge 1$ is given. If*

$$\beta_k = \tfrac{2L}{k}, \quad \gamma_k = \tfrac{2}{k+1}, \quad \eta_k = \tfrac{2LD_0^2}{Nk}, \tag{7.2.30}$$

for any $k \ge 1$, then

$$f(y_N) - f(x^*) \le \tfrac{6LD_0^2}{N(N+1)}. \tag{7.2.31}$$

As a consequence, the total number of calls to the FO *and* LO *oracles performed by the CGS method for finding an ε-solution of (7.1.1), respectively, can be bound by*

$$\mathscr{O}\left(D_0\sqrt{\tfrac{L}{\varepsilon}}\right) \tag{7.2.32}$$

and

$$\mathscr{O}\left(\tfrac{L\bar{D}_X^2}{\varepsilon} + D_0\sqrt{\tfrac{L}{\varepsilon}}\right). \tag{7.2.33}$$

Proof. It can be easily seen from the definition of γ_k in (7.2.30) and Γ_k in (7.2.9) that

$$\Gamma_k = \tfrac{2}{k(k+1)}. \tag{7.2.34}$$

Using the previous identity and (7.2.30), we have $\beta_k\gamma_k/\Gamma_k = 2L$, which implies that (7.2.13) holds. It then follows from (7.2.14), (7.2.30), and (7.2.34) that

$$f(y_N) - f(x^*) \le \Gamma_N\left(LD_0^2 + \Sigma_{i=1}^N \tfrac{\eta_i\gamma_i}{\Gamma_i}\right) = \Gamma_N\left(LD_0^2 + \Sigma_{i=1}^N i\eta_i\right) = \tfrac{6LD_0^2}{N(N+1)}.$$

Moreover, it follows from the bound in (7.2.15) and (7.2.30) that the total number of inner iterations can be bounded by

$$\sum_{k=1}^{N} T_k \le \sum_{k=1}^{N} \left(\frac{6\beta_k \bar{D}_X^2}{\eta_k} + 1 \right) = \frac{6N^2 \bar{D}_X^2}{D_0^2} + N.$$

The complexity bounds in (7.2.32) and (7.2.33) then immediately follow from the previous two inequalities. ■

In view of the classic complexity theory for convex optimization, the bound on the total number of calls to the FO oracle in (7.2.32) is optimal for smooth convex optimization. Moreover, in view of the complexity results established in the previous section, the total number of calls to the LO oracle in (7.2.33) is not improvable for a wide class of LCP methods. To the best of our knowledge, the CGS method is the first algorithm in the literature that can achieve these two optimal bounds at the same time.

Remark 7.1. Observe that in this section, we have assumed that the Euclidean distance function $\|x - x_{k-1}\|^2$ has been used in the subproblem (7.2.7). However, one can also replace it with the more general Bregman distance

$$V(x, x_{k-1}) := \omega(x) - [\omega(x_{k-1}) + \langle \omega'(x_{k-1}), x - x_{k-1} \rangle]$$

and relax the assumption that the norms are associated with the inner product, where ω is a strongly convex function. We can show similar complexity results as those in Corollaries 7.3 and 7.4 under the following assumptions: (i) ω is a smooth convex function with Lipschitz continuous gradients; and (ii) in the CndG subroutine, the objective function in (7.2.4) and the stepsizes α_t in (7.2.5) are replaced by $g + \beta[\omega'(u_t) - \omega'(u)]$ and $2/(t+1)$, respectively. However, if ω is nonsmooth (e.g., the entropy function), then we cannot obtain these results since the CndG subroutine cannot be directly applied to the modified subproblem. ■

7.2.1.2 Strongly Convex Optimization

In this subsection, we assume that the objective function f is not only smooth (i.e., (7.1.8) holds), but also strongly convex, that is, $\exists \mu > 0$ s.t.

$$f(y) - f(x) - \langle f'(x), y - x \rangle \ge \tfrac{\mu}{2} \|y - x\|^2, \quad \forall x, y \in X. \tag{7.2.35}$$

Our goal is to show that a linear rate of convergence, in terms of the number of calls to the FO oracle, can be obtained by only performing linear optimization over the feasible region X. In contrast with the shrinking conditional gradient method in the previous section, here we do not need to enforce any additional assumptions on the LO oracle. We also show that the total number of calls to the LO oracle is bounded by $\mathcal{O}(L\bar{D}_X^2/\varepsilon)$, which has been shown to be optimal for strongly convex optimization.

We are now ready to formally describe the CGS method for solving strongly convex problems, which is obtained by properly restarting the CGS method in Algorithm 7.6.

Algorithm 7.7 The CGS method for strongly convex problems

Input: Initial point $p_0 \in X$ and an estimate $\delta_0 > 0$ satisfying $f(p_0) - f(x^*) \leq \delta_0$.
for $s = 1, 2, \ldots$
Call the CGS method in Algorithm 7.6 with input

$$x_0 = p_{s-1} \quad \text{and} \quad N = \left\lceil 2\sqrt{\tfrac{6L}{\mu}} \right\rceil, \tag{7.2.36}$$

and parameters

$$\beta_k = \tfrac{2L}{k}, \; \gamma_k = \tfrac{2}{k+1}, \; \text{and} \; \eta_k = \eta_{s,k} := \tfrac{8L\delta_0 2^{-s}}{\mu N k}, \tag{7.2.37}$$

and let p_s be its output solution.
end for

In Algorithm 7.7, we restart the CGS method for smooth optimization (i.e., Algorithm 7.6) every $\lceil 2\sqrt{6L/\mu} \rceil$ iterations. We say that a phase of the above CGS algorithm occurs whenever s increases by 1. Observe that $\{\eta_k\}$ decrease by a factor of 2 as s increments by 1, while $\{\beta_k\}$ and $\{\gamma_k\}$ remain the same. The following theorem shows the convergence of the above variant of the CGS method.

Theorem 7.10. *Assume (7.2.35) holds and let $\{p_s\}$ be generated by Algorithm 7.7. Then,*

$$f(p_s) - f(x^*) \leq \delta_0 2^{-s}, \; s \geq 0. \tag{7.2.38}$$

As a consequence, the total number of calls to the FO *and* LO *oracles performed by this algorithm for finding an ε-solution of problem (7.1.1) can be bounded by*

$$\mathcal{O}\left\{ \sqrt{\tfrac{L}{\mu}} \left\lceil \log_2 \max\left(1, \tfrac{\delta_0}{\varepsilon}\right) \right\rceil \right\} \tag{7.2.39}$$

and

$$\mathcal{O}\left\{ \tfrac{L\bar{D}_X^2}{\varepsilon} + \sqrt{\tfrac{L}{\mu}} \left\lceil \log_2 \max\left(1, \tfrac{\delta_0}{\varepsilon}\right) \right\rceil \right\}, \tag{7.2.40}$$

respectively.

Proof. We prove (7.2.38) by using induction. This inequality holds obviously when $s = 0$ due to our assumption on δ_0. Now suppose that (7.2.38) holds before the s-th phase starts, i.e.,

$$f(p_{s-1}) - f(x^*) \leq \delta_0 2^{-s+1}.$$

Using the above relation and the strong convexity of f, we have

$$\|p_{s-1} - x^*\|^2 \leq \tfrac{2}{\mu}[f(p_{s-1}) - f(x^*)] \leq \tfrac{4\delta_0 2^{-s}}{\mu}.$$

Hence, by comparing the parameter settings in (7.2.37) with those in (7.2.30), we can easily see that Corollary 7.4 holds with $x_0 = p_{s-1}$, $y_N = p_s$, and $D_0^2 = 4\delta_0 2^{-s}/\mu$, which implies that

$$f(y_s) - f(x^*) \leq \frac{6LD_0^2}{N(N+1)} = \frac{24L\delta_0 2^{-s}}{\mu N(N+1)} \leq \delta_0 2^{-s},$$

where the last inequality follows from the definition of N in (7.2.36). In order to show the bounds in (7.2.39) and (7.2.40), it suffices to consider the case when $\delta_0 > \varepsilon$ (otherwise, the results are obvious). Let us denote

$$S := \left\lceil \log_2 \max\left(\frac{\delta_0}{\varepsilon}, 1\right)\right\rceil. \tag{7.2.41}$$

By (7.2.38), an ε-solution of (7.1.1) can be found at the s-th phase for some $1 \leq s \leq S$. Since the number of calls to the FO oracle in each phase is bounded by N, the total number of calls to the FO oracle performed by Algorithm 7.7 is clearly bounded by NS, which is bounded by (7.2.39). Now, let $T_{s,k}$ denote the number of calls to LO oracle required at the k-th outer iteration in s-th phase. It follows from Theorem 7.9(c) that

$$T_{s,k} \leq \frac{6\beta_k \bar{D}_X^2}{\eta_{k,s}} + 1 \leq \frac{3\mu \bar{D}_X^2 2^s N}{2\delta_0} + 1.$$

Therefore, the total number of calls to the LO can be bounded by

$$\begin{aligned}
\Sigma_{s=1}^S \Sigma_{k=1}^N T_{s,k} &\leq \Sigma_{s=1}^S \Sigma_{k=1}^N \frac{3\mu \bar{D}_X^2 2^s N}{2\delta_0} + NS = \frac{3\mu \bar{D}_X^2 N^2}{2\delta_0} \Sigma_{s=1}^S 2^s + NS \\
&\leq \frac{3\mu \bar{D}_X^2 N^2}{2\delta_0} 2^{S+1} + NS \\
&\leq \frac{6}{\varepsilon} \mu \bar{D}_X^2 N^2 + NS,
\end{aligned} \tag{7.2.42}$$

which is bounded by (7.2.40) due to the definitions of N and S in (7.2.36) and (7.2.41), respectively. ■

In view of the classic complexity theory for convex optimization, the bound on the total number of calls to FO oracle in (7.2.39) is optimal for strongly convex optimization. Moreover, in view of the complexity results established in the previous section, the bound on the total number of calls to the LO oracle in (7.2.40) is also not improvable for a wide class of linear-optimization based convex programming methods. CGS is the first optimization method that can achieve these two bounds simultaneously.

7.2.2 Stochastic Conditional Gradient Sliding Method

7.2.2.1 The Algorithm and the Main Convergence Results

In this section, we still consider smooth convex optimization problems satisfying (7.1.4). However, here we only have access to the stochastic first-order information about f. More specifically, we assume that f is represented by a stochastic first-order (SFO) oracle, which, for a given search point $z_k \in X$, outputs a vector $G(z_k, \xi_k)$ s.t.

$$\mathbb{E}\left[G(z_k, \xi_k)\right] = f'(z_k), \tag{7.2.43}$$

$$\mathbb{E}\left[\|G(z_k, \xi_k) - f'(z_k)\|_*^2\right] \le \sigma^2. \tag{7.2.44}$$

Our goal in this section is to present a stochastic conditional gradient type algorithm that can achieve the optimal bounds on the number of calls to SFO and LO oracles.

The stochastic CGS (SCGS) method is obtained by simply replacing the exact gradients in Algorithm 7.6 with an unbiased estimator computed by the SFO oracle. The algorithm is formally described as follows.

Algorithm 7.8 The stochastic conditional gradient sliding method

---

This algorithm is the same as Algorithm 7.6 except that (7.2.2) is replaced by

$$x_k = \text{CndG}(g_k, x_{k-1}, \beta_k, \eta_k). \tag{7.2.45}$$

Here,

$$g_k := \tfrac{1}{B_k}\textstyle\sum_{j=1}^{B_k} G(z_k, \xi_{k,j}) \tag{7.2.46}$$

and $G(z_k, \xi_{k,j})$, $j = 1, \ldots, B_k$, are stochastic gradients computed by the SFO at z_k.

In the above stochastic CGS method, the parameters $\{B_k\}$ denote the batch sizes used to compute g_k. It can be easily seen from (7.2.43), (7.2.44), and (7.2.46) that

$$\mathbb{E}[g_k - f'(z_k)] = 0 \ \text{ and } \ \mathbb{E}[\|g_k - f'(z_k)\|_*^2] \le \tfrac{\sigma^2}{B_k} \tag{7.2.47}$$

and hence g_k is an unbiased estimator of $f'(z_k)$. Indeed, letting $S_{B_k} = \sum_{j=1}^{B_k}(G(z_k, \xi_{k,j}) - f'(z_k))$, from (7.2.43) and (7.2.44), we have

$$\begin{aligned}
\mathbb{E}\left[\|S_{B_k}\|_*^2\right] &= \mathbb{E}\left[\|S_{B_k-1} + G(z_k, \xi_{k,B_k}) - f'(z_k)\|_*^2\right] \\
&= \mathbb{E}\left[\|S_{B_k-1}\|_*^2 + 2\langle S_{B_k-1}, G(z_k, \xi_{k,B_k}) - f'(z_k)\rangle \right. \\
&\qquad \left. + \|G(z_k, \xi_{k,B_k}) - f'(z_k)\|_*^2\right] \\
&= \mathbb{E}\left[\|S_{B_k-1}\|_*^2\right] + \mathbb{E}\left[\|G(z_k, \xi_{k,B_k}) - f'(z_k)\|_*^2\right] = \ldots \\
&= \textstyle\sum_{j=1}^{B_k} \mathbb{E}\left[\|G(z_k, \xi_{k,j}) - f'(z_k)\|_*^2\right] \le B_k \sigma^2.
\end{aligned}$$

Note that by (7.2.46), we have

$$g_k - f'(z_k) = \tfrac{1}{B_k}\sum_{j=1}^{B_k} G(z_k,\xi_{k,j}) - f'(z_k) = \tfrac{1}{B_k}\sum_{j=1}^{B_k}\left[G(z_k,\xi_{k,j}) - f'(z_k)\right] = \tfrac{1}{B_k} S_{B_k}.$$

Therefore, the second relationship in (7.2.47) immediately follows. Since the algorithm is stochastic, we will establish the complexity for finding a stochastic ε-solution, i.e., a point $\bar{x} \in X$ s.t. $\mathbb{E}[f(\bar{x}) - f(x^*)] \le \varepsilon$, as well as a stochastic (ε,Λ)-solution, i.e., a point $\bar{x} \in X$ s.t. $\text{Prob}\{f(\bar{x}) - f(x^*) \le \varepsilon\} \ge 1-\Lambda$ for some $\varepsilon > 0$ and $\Lambda \in (0,1)$.

Observe that the above SCGS method is conceptual only as we have not yet specified the parameters $\{B_k\}$, $\{\beta_k\}$, $\{\gamma_k\}$, and $\{\eta_k\}$. We will come back to this issue after establishing the main convergence properties for this algorithm.

Theorem 7.11. *Let Γ_k and $\bar{D}_X$ be defined in (7.2.9) and (7.1.18), respectively. Also assume that $\{\beta_k\}$ and $\{\gamma_k\}$ satisfy (7.2.10) and (7.2.11).*

(a) Under assumptions (7.2.43) and (7.2.44), we have

$$\mathbb{E}[f(y_k) - f(x^*)] \le \mathscr{C}_e := \tfrac{\beta_k\gamma_k}{2}\bar{D}_X^2 + \Gamma_k\sum_{i=1}^{k}\left[\tfrac{\eta_i\gamma_i}{\Gamma_i} + \tfrac{\gamma_i\sigma^2}{2\Gamma_i B_i(\beta_i - L\gamma_i)}\right], \quad \forall k \ge 1, \quad (7.2.48)$$

where x^* is an arbitrary optimal solution of (7.1.1).

(b) If (7.2.13) (rather than (7.2.11)) is satisfied, then the results in part (a) still hold by replacing $\beta_k\gamma_k\bar{D}_X^2$ with $\beta_1\Gamma_k\|x_0 - x^*\|^2$ in the first term of $\mathscr{C}_e$ in (7.2.48).

(c) Under the assumptions in part (a) or (b), the number of inner iterations performed at the k-th outer iterations is bounded by (7.2.15).

Proof. Let us denote $\delta_{k,j} = G(z_k,\xi_{k,j}) - f'(z_k)$ and $\delta_k \equiv g_k - f'(z_k) = \sum_{j=1}^{B_k}\delta_{k,j}/B_k$. Note that by (7.2.16) and (7.2.17) (with $f'(z_k)$ replaced by g_k), we have

$$\begin{aligned}
f(y_k) &\le (1-\gamma_k)f(y_{k-1}) + \gamma_k l_f(z_k,x_k) + \gamma_k\langle g_k, x - x_k\rangle \\
&\quad + \tfrac{\beta_k\gamma_k}{2}\left[\|x_{k-1}-x\|^2 - \|x_k - x\|^2\right] + \eta_k\gamma_k - \tfrac{\gamma_k}{2}(\beta_k - L\gamma_k)\|x_k - x_{k-1}\|^2 \\
&= (1-\gamma_k)f(y_{k-1}) + \gamma_k l_f(z_k,x) + \gamma_k\langle\delta_k, x - x_k\rangle \\
&\quad + \tfrac{\beta_k\gamma_k}{2}\left[\|x_{k-1}-x\|^2 - \|x_k - x\|^2\right] + \eta_k\gamma_k - \tfrac{\gamma_k}{2}(\beta_k - L\gamma_k)\|x_k - x_{k-1}\|^2.
\end{aligned}$$

Using the above inequality and the fact that

$$\begin{aligned}
&\langle\delta_k, x - x_k\rangle - \tfrac{1}{2}(\beta_k - L\gamma_k)\|x_k - x_{k-1}\|^2 \\
&\quad = \langle\delta_k, x - x_{k-1}\rangle + \langle\delta_k, x_{k-1} - x_k\rangle - \tfrac{1}{2}(\beta_k - L\gamma_k)\|x_k - x_{k-1}\|^2 \\
&\quad \le \langle\delta_k, x - x_{k-1}\rangle + \tfrac{\|\delta_k\|_*^2}{2(\beta_k - L\gamma_k)},
\end{aligned}$$

we obtain

$$\begin{aligned}
f(y_k) &\le (1-\gamma_k)f(y_{k-1}) + \gamma_k f(x) + \tfrac{\beta_k\gamma_k}{2}\left[\|x_{k-1} - x\|^2 - \|x_k - x\|^2\right] + \eta_k\gamma_k \\
&\quad + \gamma_k\langle\delta_k, x - x_{k-1}\rangle + \tfrac{\gamma_k\|\delta_k\|_*^2}{2(\beta_k - L\gamma_k)}, \quad \forall x \in X. \qquad (7.2.49)
\end{aligned}$$

Subtracting $f(x)$ from both sides of (7.2.49) and using Lemma 3.17, we have

$$\begin{aligned} f(y_k) - f(x) &\leq \Gamma_k(1-\gamma_1)\left[f(y_0) - f(x)\right] \\ &\quad + \Gamma_k \sum_{i=1}^k \left\{ \tfrac{\beta_i \gamma_i}{2\Gamma_i} \left[\|x_{k-1} - x\|^2 - \|x_k - x\|^2\right] + \tfrac{\eta_i \gamma_i}{\Gamma_i} \right\} \\ &\quad + \Gamma_k \sum_{i=1}^k \tfrac{\gamma_i}{\Gamma_i} \left[\langle \delta_i, x - x_{i-1} \rangle + \tfrac{\|\delta_i\|_*^2}{2(\beta_i - L\gamma_i)} \right] \\ &\leq \tfrac{\beta_k \gamma_k}{2} \bar{D}_X^2 + \Gamma_k \sum_{i=1}^k \tfrac{\eta_i \gamma_i}{\Gamma_i} \\ &\quad + \Gamma_k \sum_{i=1}^k \tfrac{\gamma_i}{\Gamma_i} \left[\sum_{j=1}^{B_i} B_i^{-1} \langle \delta_{i,j}, x - x_{i-1} \rangle + \tfrac{\|\delta_i\|_*^2}{2(\beta_i - L\gamma_i)} \right], \end{aligned} \tag{7.2.50}$$

where the last inequality follows from (7.2.20) and the fact that $\gamma_1 = 1$. Note that by our assumptions on the SFO, the random variables $\delta_{i,j}$ are independent of the search point x_{i-1} and hence $\mathbb{E}[\langle \delta_{i,j}, x^* - x_{i-1} \rangle] = 0$. In addition, relation (7.2.47) implies that $\mathbb{E}[\|\delta_i\|_*^2] \leq \sigma^2 / B_i$. Using the previous two observations and taking expectation on both sides of (7.2.50) (with $x = x^*$), we obtain (7.2.48).

Part (b) follows similarly from the bound in (7.2.21) and (7.2.50), and the proof of part (c) is exactly the same as that of Theorem 7.9(c). ∎

Now we provide a set of parameters $\{\beta_k\}, \{\gamma_k\}, \{\eta_k\}$, and $\{B_k\}$ which lead to optimal bounds on the number of calls to the SFO and LO oracles.

Corollary 7.5. *Suppose that $\{\beta_k\}, \{\gamma_k\}$, $\{\eta_k\}$, and $\{B_k\}$ in the SCGS method are set to*

$$\beta_k = \tfrac{4L}{k+2}, \ \ \gamma_k = \tfrac{3}{k+2}, \ \ \eta_k = \tfrac{L\bar{D}_X^2}{k(k+1)}, \ \ \textit{and} \ \ B_k = \left\lceil \tfrac{\sigma^2 (k+2)^3}{L^2 \bar{D}_X^2} \right\rceil, \ \ k \geq 1. \tag{7.2.51}$$

Under assumptions (7.2.43) and (7.2.44), we have

$$\mathbb{E}\left[f(y_k) - f(x^*)\right] \leq \tfrac{6L\bar{D}_X^2}{(k+2)^2} + \tfrac{9L\bar{D}_X^2}{2(k+1)(k+2)}, \ \ \forall k \geq 1. \tag{7.2.52}$$

As a consequence, the total number of calls to the SFO *and* LO *oracles performed by the SCGS method for finding a stochastic ε-solution of (7.1.1), respectively, can be bounded by*

$$\mathcal{O}\left\{ \sqrt{\tfrac{L\bar{D}_X^2}{\varepsilon}} + \tfrac{\sigma^2 \bar{D}_X^2}{\varepsilon^2} \right\} \ \ \textit{and} \ \ \mathcal{O}\left\{ \tfrac{L\bar{D}_X^2}{\varepsilon} \right\}. \tag{7.2.53}$$

Proof. It can be easily seen from (7.2.51) that (7.2.10) holds. Also by (7.2.51), Γ_k is given by (7.2.29) and hence

$$\tfrac{\beta_k \gamma_k}{\Gamma_k} = \tfrac{2Lk(k+1)}{k+2},$$

which implies that (7.2.11) holds. It can also be easily checked from (7.2.29) and (7.2.51) that

$$\sum_{i=1}^k \tfrac{\eta_i \gamma_i}{\Gamma_i} \leq \tfrac{kL\bar{D}_X^2}{2}, \ \ \sum_{i=1}^k \tfrac{\gamma_i}{\Gamma_i B_i (\beta_i - L\gamma_i)} \leq \tfrac{kL\bar{D}_X^2}{2\sigma^2}.$$

Using the bound in (7.2.48), we obtain (7.2.52), which implies that the total number of outer iterations can be bounded by

$$\mathscr{O}\left(\sqrt{\tfrac{L\bar{D}_X^2}{\varepsilon}}\right)$$

under the assumptions (7.2.43) and (7.2.44). The bounds in (7.2.53) then immediately follow from this observation and the fact that the number of calls to the SFO and LO oracles are bounded by

$$\textstyle\sum_{k=1}^N B_k \le \sum_{k=1}^N \frac{\sigma^2(k+2)^3}{L^2\bar{D}_X^2} + N \le \frac{\sigma^2(N+3)^4}{4L^2\bar{D}_X^2} + N,$$
$$\textstyle\sum_{k=1}^N T_k \le \sum_{k=1}^N \left(\frac{6\beta_k\bar{D}_X^2}{\eta_k} + 1\right) \le 12N^2 + 13N.$$

■

Now we give a different set of parameters $\{\beta_k\}, \{\gamma_k\}, \{\eta_k\}$, and $\{B_k\}$, which can slightly improve the bounds on the number of calls to the SFO in terms of its dependence on $\bar{D}_X$.

Corollary 7.6. *Suppose that there exists an estimate D_0 s.t. $\|x_0 - x^*\| \le D_0 \le \bar{D}_X$. Also assume that the outer iteration limit $N \ge 1$ is given. If*

$$\beta_k = \tfrac{3L}{k}, \ \gamma_k = \tfrac{2}{k+1}, \ \eta_k = \tfrac{2LD_0^2}{Nk}, \ \textit{and} \ B_k = \left\lceil \tfrac{\sigma^2 N(k+1)^2}{L^2 D_0^2} \right\rceil, \ k \ge 1. \tag{7.2.54}$$

Under assumptions (7.2.43) and (7.2.44),

$$\mathbb{E}[f(y_N) - f(x^*)] \le \tfrac{8LD_0^2}{N(N+1)}, \ \forall N \ge 1. \tag{7.2.55}$$

As a consequence, the total number of calls to the SFO *and* LO *oracles performed by the SCGS method for finding a stochastic ε-solution of (7.1.1), respectively, can be bounded by*

$$\mathscr{O}\left\{\sqrt{\tfrac{LD_0^2}{\varepsilon}} + \tfrac{\sigma^2 D_0^2}{\varepsilon^2}\right\} \ \textit{and} \ \mathscr{O}\left\{\tfrac{L\bar{D}_X^2}{\varepsilon}\right\}. \tag{7.2.56}$$

Proof. It can be easily seen from (7.2.54) that (7.2.10) holds. Also by (7.2.54), Γ_k is given by (7.2.34) and hence

$$\tfrac{\beta_k\gamma_k}{\Gamma_k} = 3L,$$

which implies that (7.2.13) holds. It can also be easily checked from (7.2.34) and (7.2.54) that

$$\textstyle\sum_{i=1}^N \frac{\eta_i\gamma_i}{\Gamma_i} \le 2LD_0^2, \ \sum_{i=1}^N \frac{\gamma_i}{\Gamma_i B_i(\beta_i - L\gamma_i)} \le \sum_{i=1}^N \frac{i(i+1)}{LB_i} \le \frac{LD_0^2}{\sigma^2}.$$

Using the bound in (7.2.48) (with $\beta_k\gamma_k\bar{D}_X^2$ replaced by $\beta_1\Gamma_k D_0^2$ in the definition of $\mathscr{C}_e$), we obtain (7.2.55), which implies that the total number of outer iterations can be bounded by

$$\mathscr{O}\left(\sqrt{\tfrac{LD_0^2}{\varepsilon}}\right)$$

under the assumptions (7.2.43) and (7.2.44). The bounds in (7.2.56) then immediately follow from this observation and the fact that the total number calls to the SFO and LO are bounded by

$$\sum_{k=1}^{N} B_k \leq N \sum_{k=1}^{N} \frac{\sigma^2 (k+1)^2}{L^2 D_0^2} + N \leq \frac{\sigma^2 N(N+1)^3}{3L^2 D_0^2} + N,$$
$$\sum_{k=1}^{N} T_k \leq \sum_{k=1}^{N} \frac{6\beta_k \bar{D}_X^2}{\eta_k} + N \leq \frac{9N^2 \bar{D}_X^2}{D_0^2} + N.$$

■

According to the complexity bounds in Corollaries 7.5 and 7.6, the total number of calls to the SFO oracle can be bounded by $\mathcal{O}(1/\varepsilon^2)$, which is optimal in view of the classic complexity theory for stochastic convex optimization. Moreover, the total number of calls to the LO oracle can be bounded by $\mathcal{O}(1/\varepsilon)$, which is the same as the CGS method for deterministic smooth convex optimization and hence not improvable for a wide class of LCP methods.

In view of the results in Corollary 7.6, we can present an optimal algorithm for solving stochastic strongly convex problems, similar to the deterministic case.

Algorithm 7.9 The stochastic CGS method for solving strongly convex problems

---

Input: Initial point $p_0 \in X$ and an estimate $\delta_0 > 0$ satisfying $f(p_0) - f(x^*) \leq \delta_0$.
for $s = 1, 2, \ldots$

Call the stochastic CGS method in Algorithm 7.8 with input

$$x_0 = p_{s-1} \text{ and } N = \left\lceil 4\sqrt{\tfrac{2L}{\mu}} \right\rceil, \tag{7.2.57}$$

and parameters

$$\beta_k = \tfrac{3L}{k},\ \gamma_k = \tfrac{2}{k+1},\ \eta_k = \eta_{s,k} := \tfrac{8L\delta_0 2^{-s}}{\mu N k},\ \text{and } B_k = B_{s,k} := \left\lceil \tfrac{\mu \sigma^2 N(k+1)^2}{4L^2 \delta_0 2^{-s}} \right\rceil, \tag{7.2.58}$$

and let p_s be its output solution.
end for

The main convergence properties of Algorithm 7.9 are described as follows.

Theorem 7.12. *Assume that (7.2.35) holds and let $\{p_s\}$ be generated by Algorithm 7.9. Then,*

$$\mathbb{E}[f(p_s) - f(x^*)] \leq \delta_0 2^{-s},\ s \geq 0. \tag{7.2.59}$$

As a consequence, the total number of calls to the SFO *and* LO *oracles performed by this algorithm for finding a stochastic ε-solution of problem (7.1.1) can be bounded by*

$$\mathcal{O}\left\{ \tfrac{\sigma^2}{\mu\varepsilon} + \sqrt{\tfrac{L}{\mu}} \left\lceil \log_2 \max\left(1, \tfrac{\delta_0}{\varepsilon}\right) \right\rceil \right\} \tag{7.2.60}$$

and

$$\mathcal{O}\left\{ \tfrac{L\bar{D}_X^2}{\varepsilon} + \sqrt{\tfrac{L}{\mu}} \left\lceil \log_2 \max\left(1, \tfrac{\delta_0}{\varepsilon}\right) \right\rceil \right\}, \tag{7.2.61}$$

respectively.

Proof. In view of Corollary 7.6, (7.2.59) can be proved in a way similar to (7.2.38). It now remains to show the bounds in (7.2.60) and (7.2.61), respectively, for the total number of calls to the SFO and LO oracles. It suffices to consider the case when $\delta_0 > \varepsilon$, since otherwise the results are obvious. Let us denote

$$S := \left\lceil \log_2 \max\left(\tfrac{\delta_0}{\varepsilon}, 1\right)\right\rceil. \tag{7.2.62}$$

By (7.2.59), a stochastic ε-solution of (7.1.1) can be found at the s-th phase for some $1 \le s \le S$. Since the number of calls to the SFO oracle in each phase is bounded by N, the total number of calls to the SFO oracle can be bounded by

$$\begin{aligned}\textstyle\sum_{s=1}^{S}\sum_{k=1}^{N} B_k &\le \textstyle\sum_{s=1}^{S}\sum_{k=1}^{N}\left(\tfrac{\mu\sigma^2 N(k+1)^2}{4L^2\delta_0 2^{-s}}+1\right)\\ &\le \tfrac{\mu\sigma^2 N(N+1)^3}{12L^2\delta_0}\textstyle\sum_{s=1}^{S}2^s + SN \le \tfrac{\mu\sigma^2 N(N+1)^3}{3L^2\varepsilon} + SN.\end{aligned}$$

Moreover, let $T_{s,k}$ denote the number of calls to LO oracle required at the k-th outer iteration in s-th phase of the stochastic CGS method. It follows from Theorem 7.9(c) that

$$T_{s,k} \le \tfrac{6\beta_k \bar{D}_X^2}{\eta_{k,s}} + 1 \le \tfrac{9\mu\bar{D}_X^2 2^s N}{4\delta_0} + 1.$$

Therefore, the total number of calls to the LO oracle can be bounded by

$$\begin{aligned}\textstyle\sum_{s=1}^{S}\sum_{k=1}^{N} T_{s,k} &\le \textstyle\sum_{s=1}^{S}\sum_{k=1}^{N}\tfrac{9\mu\bar{D}_X^2 2^s N}{4\delta_0} + NS = \tfrac{9}{4}\mu\bar{D}_X^2 N^2\delta_0^{-1}\textstyle\sum_{s=1}^{S}2^s + NS\\ &\le \tfrac{9}{\varepsilon}\mu\bar{D}_X^2 N^2 + NS,\end{aligned}$$

which is bounded by (7.2.40) due to the definitions of N and S in (7.2.57) and (7.2.62), respectively. ■

According to Theorem 7.12, the total number of calls to the SFO oracle can be bounded by $\mathcal{O}(1/\varepsilon)$, which is optimal in view of the classic complexity theory for strongly convex stochastic optimization. Moreover, the total number of calls to the LO oracle can be bounded by $\mathcal{O}(1/\varepsilon)$, which is the same as the deterministic CGS method for strongly convex optimization and not improvable for a wide class of LCP methods discussed in the previous section.

7.2.2.2 The Large Deviation Results

For the sake of simplicity, in this subsection we only consider smooth convex optimization problems rather than strongly convex problems. In order to develop some large deviation results associated with the aforementioned optimal complexity bounds, we need to make some assumptions about the objective function values and its estimator, $F(x,\xi)$, given by the SFO. More specifically, we assume that

$$\mathbb{E}[F(x,\xi)] = f(x), \text{ and } \mathbb{E}\left[\exp\left\{(F(x,\xi)-f(x))^2/M^2\right\}\right] \le \exp\{1\} \quad (7.2.63)$$

for some $M \geq 0$.

We now propose a variant of the SCGS method which has some desirable large deviation properties. Similar to the 2-RSPG algorithm in Sect. 6.2, this method consists of two phases: an optimization phase and a post-optimization phase. In the optimization phase, we restart the SCGS algorithm for a certain number of times to generate a list of candidate solutions, and in the post-optimization phase, we choose a solution $\hat{x}$ from this list according to a certain rule.

Algorithm 7.10 A two phase SCGS (2-SCGS) algorithm

Input: Initial point $x_0 \in X$, number of restart times S, iteration limit N, and the sample size K in the post-optimization phase.

Optimization phase:

for $s = 1,\ldots,S$ **do**

Call the SCGS algorithm with iteration limit N, and the initial point x_{s-1}, where $x_s = x_{N_s}$, $s = 1,\ldots,S$, are the outputs of the s-th run of the SCGS algorithm.

end for

Let $\{\bar{x}_s = x_{N_s}, s = 1,\ldots,S\}$, be the output list of candidate solutions.

Post-optimization phase:

Choose a solution $\hat{x}$ from the candidate list $\{\bar{x}_1,\ldots,\bar{x}_S\}$ such that

$$\hat{x} = \operatorname{argmin}_{s=1,\ldots,S}\{\hat{f}(\bar{x}_s)\}, \quad (7.2.64)$$

where $\hat{f}(x) = \frac{1}{K}\Sigma_{j=1}^{K} F(x,\xi_j)$.

Now we are ready to state the large deviation results obtained for the above 2-SCGS algorithm.

Theorem 7.13. *Assuming that $\{\beta_k\}$ and $\{\gamma_k\}$ satisfy (7.2.10) and (7.2.11), under assumption (7.2.63), we have*

$$\operatorname{Prob}\left\{f(\hat{x}) - f(x^*) \geq \tfrac{2\sqrt{2}(1+\lambda)M}{\sqrt{K}} + 2\mathscr{C}_e\right\} \leq S\exp\left\{-\lambda^2/3\right\} + 2^{-S}, \quad (7.2.65)$$

where $\hat{x}$ is the output of the 2-SCGS algorithm, x^* is an arbitrary optimal solution of (7.1.1), and $\mathscr{C}_e$ is defined in (7.2.48).

Proof. It follows from the definition of $\hat{x}$ in (7.2.64) that

$$\begin{aligned}
\hat{f}(\hat{x}) - f(x^*) &= \min_{s=1,\ldots,S} \hat{f}(\bar{x}_s) - f(x^*) \\
&= \min_{s=1,\ldots,S} \{\hat{f}(\bar{x}_s) - f(\bar{x}_s) + f(\bar{x}_s) - f(x^*)\} \\
&\leq \min_{s=1,\ldots,S} \{|\hat{f}(\bar{x}_s) - f(\bar{x}_s)| + f(\bar{x}_s) - f(x^*)\} \\
&\leq \max_{s=1,\ldots,S} |\hat{f}(\bar{x}_s) - f(\bar{x}_s)| + \min_{s=1,\ldots,S} \{f(\bar{x}_s) - f(x^*)\},
\end{aligned}$$

which implies that

$$\begin{aligned} f(\hat{x})-f(x^*) &= f(\hat{x})-\hat{f}(\hat{x})+\hat{f}(\hat{x})-f(x^*) \\ &\le f(\hat{x})-\hat{f}(\hat{x})+\max_{s=1,\ldots,S}|\hat{f}(\bar{x}_s)-f(\bar{x}_s)|+\min_{s=1,\ldots,S}\{f(\bar{x}_s)-f(x^*)\} \\ &\le 2\max_{s=1,\ldots,S}|\hat{f}(\bar{x}_s)-f(\bar{x}_s)|+\min_{s=1,\ldots,S}\{f(\bar{x}_s)-f(x^*)\}. \end{aligned} \tag{7.2.66}$$

Note that by the Markov's inequality and (7.2.48), we obtain

$$\text{Prob}\{f(\bar{x}_s)-f(x^*)\ge 2\mathscr{C}_e\}\le \tfrac{\mathbb{E}[f(\bar{x}_s)-f(x^*)]}{2\mathscr{C}_e}\le \tfrac{1}{2},\ \forall s=1,\ldots,S. \tag{7.2.67}$$

Let E_s be the event that $f(\bar{x}_s)-f(x^*)\ge 2\mathscr{C}_e$, note that due to the boundedness of X, and the above observation, we have

$$\text{Prob}\left\{E_s|\bigcap_{j=1}^{s-1}E_j\right\}\le \tfrac{1}{2}, s=1,\ldots,S,$$

which then implies that

$$\begin{aligned} &\text{Prob}\left\{\min_{s=1,\ldots,S}[f(\bar{x}_s)-f(x^*)]\ge 2\mathscr{C}_e\right\} \\ &=\text{Prob}\left\{\bigcap_{s=1}^{S}E_s\right\}=\prod_{s=1}^{S}\text{Prob}\left\{E_s|\bigcap_{j=1}^{s-1}E_j\right\}\le 2^{-S}. \end{aligned} \tag{7.2.68}$$

By assumption (7.2.63) and Lemma 4.1, it is clear that

$$\begin{aligned} &\text{Prob}\left\{|\Sigma_{j=1}^{K}[F(\bar{x}_s,\xi_j)-f(\bar{x}_s)]|\ge\sqrt{2}(1+\lambda)\sqrt{KM^2}\right\} \\ &\qquad\le\exp\left\{-\lambda^2/3\right\},\ s=1,\ldots,S, \end{aligned}$$

which implies

$$\text{Prob}\left\{|\hat{f}(\bar{x}_s)-f(\bar{x}_s)|\ge\tfrac{\sqrt{2}(1+\lambda)M}{\sqrt{K}}\right\}\le\exp\left\{-\lambda^2/3\right\},\ s=1,\ldots,S.$$

Therefore, we obtain

$$\text{Prob}\left\{\max_{s=1,\ldots,S}|\hat{f}(\bar{x}_s)-f(\bar{x}_s)|\ge\tfrac{\sqrt{2}(1+\lambda)M}{\sqrt{K}}\right\}\le S\exp\left\{-\lambda^2/3\right\}. \tag{7.2.69}$$

Our result in (7.2.65) directly follows from (7.2.66), (7.2.68), and (7.2.69). ∎

Now we state a set of parameters S, N, and K, and the associated bounds on the number of calls to the SFO and LO oracles.

Corollary 7.7. *Suppose that parameters $\{\beta_k\}$, $\{\gamma_k\}$, $\{\eta_k\}$, and $\{B_k\}$ in the 2-SCGS method are set as in (7.2.51) for each run of SCGS algorithm. Let $\varepsilon>0$ and $\Lambda\in$*

$(0,1)$ *be given, parameters S, N, and K are set to*

$$S(\Lambda) := \lceil \log_2(2/\Lambda) \rceil, \; N(\varepsilon) := \left\lceil \sqrt{\tfrac{42L\bar{D}_X^2}{\varepsilon}} \right\rceil, \; \text{and} \; K(\varepsilon,\Lambda) := \left\lceil \tfrac{32(1+\lambda)^2 M^2}{\varepsilon^2} \right\rceil, \tag{7.2.70}$$

where $\lambda = \sqrt{3\ln(2S/\Lambda)}$, *then the total number of calls to the* SFO *and* LO *oracles performed by the 2-SCGS method in the optimization phase to compute a stochastic* (ε,Λ)*-solution of the problem (7.1.1), respectively, can be bounded by*

$$\mathcal{O}\left\{\sqrt{\tfrac{L\bar{D}_X^2}{\varepsilon}} \log_2 \tfrac{2}{\Lambda} + \tfrac{\sigma^2 \bar{D}_X^2}{\varepsilon^2} \log_2 \tfrac{2}{\Lambda}\right\} \; \text{and} \; \mathcal{O}\left\{\tfrac{L\bar{D}_X^2}{\varepsilon} \log_2 \tfrac{2}{\Lambda}\right\}. \tag{7.2.71}$$

Proof. By Corollary 7.5, we have

$$\mathcal{C}_e \leq \tfrac{21L\bar{D}_X^2}{2(N+1)^2},$$

together with the definition of S, N, and K in (7.2.70), (7.2.65), and $\lambda = \sqrt{3\ln(2S/\Lambda)}$, we have

$$\text{Prob}\,\{f(\hat{x}) - f(x^*) \geq \varepsilon\} \leq \Lambda,$$

i.e., $\hat{x}$ is a stochastic (ε,Λ)-solution of problem (7.1.1). Moreover, we obtain from Corollary 7.5 that the bounds for the number of calls to the SFO and LO oracles for each run of SCGS algorithm as (7.2.53), which immediately implies the bounds in (7.2.71), as we restart the SCGS algorithm in 2-SCGS method S times. ■

7.2.3 Generalization to Saddle Point Problems

In this section, we consider an important class of saddle point problems with f given in the form of:

$$f(x) = \max_{y \in Y} \left\{ \langle Ax, y \rangle - \hat{f}(y) \right\}, \tag{7.2.72}$$

where $A : \mathbb{R}^n \to \mathbb{R}^m$ denotes a linear operator, $Y \in \mathbb{R}^m$ is a convex compact set, and $\hat{f} : Y \to \mathbb{R}$ is a simple convex function. Since the objective function f given in (7.2.72) is nonsmooth, we cannot directly apply the CGS method presented in the previous section. However, as discussed in the previous section, the function $f(\cdot)$ in (7.2.72) can be closely approximated by a class of smooth convex functions

$$f_\tau(x) := \max_y \left\{ \langle Ax, y \rangle - \hat{f}(y) - \tau\,[V(y) - D_Y^2] : \; y \in Y \right\} \tag{7.2.73}$$

for some $\tau > 0$.

In this subsection, we assume that the feasible region Y and the function $\hat{f}$ are simple enough, so that the subproblem in (7.2.73) is easy to solve, and as a result, the major computational cost for computing the gradient of f_τ exists in the evaluation of the linear operator A and its adjoint operator A^T. Our goal is to present a variant of the CGS method, which can achieve the optimal bounds on the number of calls to the LO oracle and the number of evaluations for the linear operator A and A^T.

Algorithm 7.11 The CGS method for solving saddle point problems

This algorithm is the same as Algorithm 7.6 except that (7.2.2) is replaces by

$$x_k = \text{CndG}(f'_{\tau_k}(z_k), x_{k-1}, \beta_k, \eta_k), \tag{7.2.74}$$

for some $\tau_k \geq 0$.

We are now ready to describe the main convergence properties of this modified CGS method to solve the saddle point problem in (7.1.1)–(7.2.72).

Theorem 7.14. *Suppose that $\tau_1 \geq \tau_2 \geq \ldots \geq 0$. Also assume that $\{\beta_k\}$ and $\{\gamma_k\}$ satisfy (7.2.10) (with L replaced by L_{τ_k}) and (7.2.11). Then,*

$$f(y_k) - f(x^*) \leq \tfrac{\beta_k \gamma_k}{2} \bar{D}_X^2 + \Gamma_k \textstyle\sum_{i=1}^k \tfrac{\gamma_i}{\Gamma_i} \left(\eta_i + \tau_i D_Y^2\right), \quad \forall k \geq 1, \tag{7.2.75}$$

where x^* is an arbitrary optimal solution of (7.1.1)–(7.2.72). Moreover, the number of inner iterations performed at the k-th outer iteration can be bounded by (7.2.15).

Proof. First, observe that by the definition of $f_\tau(\cdot)$ in (7.2.73), and the facts that $V(y) - D_Y^2 \leq 0$ and $\tau_{k-1} \geq \tau_k$, we have

$$f_{\tau_{k-1}}(x) \geq f_{\tau_k}(x) \ \ \forall x \in X, \ \forall k \geq 1. \tag{7.2.76}$$

Applying relation (7.2.18) to f_{τ_k} and using (7.2.76), we obtain

$$\begin{aligned}
f_{\tau_k}(y_k) &\leq (1-\gamma_k) f_{\tau_k}(y_{k-1}) + \gamma_k f_{\tau_k}(x) + \tfrac{\beta_k \gamma_k}{2}(\|x_{k-1}-x\|^2 - \|x_k - x\|^2) + \eta_k \gamma_k \\
&\leq (1-\gamma_k) f_{\tau_{k-1}}(y_{k-1}) + \gamma_k \left[f(x) + \tau_k D_Y^2\right] \\
&\quad + \tfrac{\beta_k \gamma_k}{2}(\|x_{k-1}-x\|^2 - \|x_k - x\|^2) + \eta_k \gamma_k
\end{aligned}$$

for any $x \in X$, where the second inequality follows from (7.1.24) and (7.2.76). Subtracting $f(x)$ from both sides of the above inequality, we have

$$\begin{aligned}
f_{\tau_k}(y_k) - f(x) &\leq (1-\gamma_k)\left[f_{\tau_{k-1}}(y_{k-1}) - f(x)\right] + \tfrac{\beta_k \gamma_k}{2}(\|x_{k-1}-x\|^2 - \|x_k - x\|^2) \\
&\quad + \eta_k \gamma_k + \gamma_k \tau_k D_Y^2
\end{aligned}$$

for any $x \in X$, which, in view of Lemma 3.17 and (7.2.20), then implies that

$$\begin{aligned}
f_{\tau_k}(y_k) - f(x) &\leq \Gamma_k \textstyle\sum_{i=1}^k \tfrac{\beta_i \gamma_i}{2\Gamma_i}(\|x_{i-1}-x\|^2 - \|x_i - x\|^2) + \Gamma_k \textstyle\sum_{i=1}^k \tfrac{\gamma_i}{\Gamma_i}\left(\eta_i + \tau_i D_Y^2\right) \\
&\leq \tfrac{\beta_k \gamma_k}{2} \bar{D}_X^2 + \Gamma_k \textstyle\sum_{i=1}^k \tfrac{\gamma_i}{\Gamma_i}\left(\eta_i + \tau_i D_Y^2\right).
\end{aligned} \tag{7.2.77}$$

Our result in (7.2.75) then immediately follows from the above relation and the fact that $f_{\tau_k}(y_k) \geq f(y_k)$ due to (7.1.24). The last part of our claim easily follows from Theorem 7.9(c). ■

We now provide two sets of parameters for $\{\beta_k\}, \{\gamma_k\}, \{\eta_k\}$, and $\{\tau_k\}$ which can guarantee the optimal convergence of the above variant of CGS method for saddle point optimization.

Corollary 7.8. *Assume the outer iteration limit $N \geq 1$ is given. If*

$$\tau_k \equiv \tau = \frac{2\|A\|\bar{D}_X}{D_Y\sqrt{\sigma_v N}}, \quad k \geq 1, \tag{7.2.78}$$

and $\{\beta_k\}$, $\{\gamma_k\}$, and $\{\eta_k\}$ used in Algorithm 7.11 are set to

$$\beta_k = \frac{3\mathcal{L}_{\tau_k}}{k+1}, \quad \gamma_k = \frac{3}{k+2}, \text{and } \eta_k = \frac{\mathcal{L}_{\tau_k}\bar{D}_X^2}{k^2}, \quad k \geq 1, \tag{7.2.79}$$

then the number of linear operator evaluations (for A and A^T) and the number of calls to the LO *oracle performed by Algorithm 7.11 for finding an ε-solution of problem (7.1.1)–(7.2.72), respectively, can be bounded by*

$$\mathcal{O}\left\{\frac{\|A\|\bar{D}_X D_Y}{\sqrt{\sigma_v}\varepsilon}\right\} \quad \text{and} \quad \mathcal{O}\left\{\frac{\|A\|^2\bar{D}_X^2 D_Y^2}{\sigma_v \varepsilon^2}\right\}. \tag{7.2.80}$$

Proof. Observe that Γ_k is given by (7.2.29) due to the definition of γ_k in (7.2.79). By (7.2.29) and (7.2.79), we have

$$\frac{\beta_k}{\gamma_k} = \frac{\mathcal{L}_\tau(k+2)}{k+1} \geq \mathcal{L}_\tau,$$

and

$$\frac{\beta_k\gamma_k}{\Gamma_k} = \frac{3\mathcal{L}_\tau k}{2} \geq \frac{\beta_{k-1}\gamma_{k-1}}{\Gamma_{k-1}}.$$

The above results indicate that the assumptions in Theorem 7.14 are satisfied. It then follows from Theorem 7.14, (7.2.78), and (7.2.79) that

$$\begin{aligned} f(y_N) - f(x^*) &\leq \frac{9\mathcal{L}_\tau\bar{D}_X^2}{2(N+1)(N+2)} + \frac{6}{N(N+1)(N+2)}\sum_{i=1}^N \left[\frac{\mathcal{L}_\tau\bar{D}_X^2}{i^2} + \frac{2\|A\|\bar{D}_X D_Y}{\sqrt{\sigma_v N}}\right]\frac{i(i+1)}{2} \\ &\leq \frac{9\|A\|\bar{D}_X D_Y}{4\sqrt{\sigma_v}(N+2)} + \frac{15\|A\|\bar{D}_X D_Y}{\sqrt{\sigma_v}N(N+1)(N+2)}\sum_{i=1}^N N \leq \frac{69\|A\|\bar{D}_X D_Y}{4\sqrt{\sigma_v}(N+2)}, \end{aligned}$$

where the second inequality follows from the definition of $\mathcal{L}_\tau$ in (8.1.93). Moreover, it follows from (7.2.15) and (7.2.79) that the total number of calls to the LO oracle can be bounded by

$$\sum_{k=1}^N T_k \leq \sum_{k=1}^N \left(\frac{18\mathcal{L}_{\tau_k}\bar{D}_X^2}{k+1}\frac{k^2}{\mathcal{L}_{\tau_k}\bar{D}_X^2} + 1\right) \leq \frac{18(N+1)N}{2} + N \leq 9N^2 + 10N.$$

The bounds in (7.2.80) then immediately follow from the previous two conclusions. ■

In the above result, we used a static smoothing technique, in which we need to fix the number of outer iterations N in advance for obtaining a constant τ_k in (7.2.78). We now state a dynamic parameter setting for τ_k so that the number of outer iterations N need not be given a priori.

Corollary 7.9. *Suppose that parameter $\{\tau_k\}$ is now set to*

$$\tau_k = \frac{2\|A\|\bar{D}_X}{D_Y\sqrt{\sigma_v}k}, \quad k \geq 1, \tag{7.2.81}$$

and the parameters $\{\beta_k\}$, $\{\gamma_k\}$, and $\{\eta_k\}$ used in Algorithm 7.11 are set as in (7.2.79). Then, the number of linear operator evaluations (for A and A^T) and the number of calls to the LO *oracle performed by Algorithm 7.11 for finding an ε-solution of problem (7.1.1)–(7.2.72), respectively, can also be bounded by (7.2.80).*

Proof. Note that γ_k is defined in (7.2.79), and hence that Γ_k is given by (7.2.29). We have

$$\frac{\beta_k}{\gamma_k} \geq \mathcal{L}_{\tau_k},$$

and

$$\frac{\beta_k \gamma_k}{\Gamma_k} = \frac{3\mathcal{L}_{\tau_k} k}{2} = \frac{3\|A\|D_Y k^2}{4\sqrt{\sigma_v}\bar{D}_X} \geq \frac{\beta_{k-1}\gamma_{k-1}}{\Gamma_{k-1}}.$$

Therefore, the assumptions in Theorem 7.14 are satisfied. It then follows from Theorem 7.14, (7.2.79), and (7.2.81) that

$$\begin{aligned}
f(y_k) - f(x^*) &\leq \frac{9\mathcal{L}_{\tau_k}\bar{D}_X^2}{2(k+1)(k+2)} + \frac{6}{k(k+1)(k+2)}\sum_{i=1}^{k}\left[\frac{\mathcal{L}_{\tau_i}\bar{D}_X^2}{i^2} + \frac{2\|A\|\bar{D}_X D_Y}{\sqrt{\sigma_v}i}\right]\frac{i(i+1)}{2} \\
&\leq \frac{9\|A\|\bar{D}_X D_Y k}{4\sqrt{\sigma_v}(k+1)(k+2)} + \frac{15\|A\|\bar{D}_X D_Y}{\sqrt{\sigma_v}k(k+1)(k+2)}\sum_{i=1}^{k} i \leq \frac{39\|A\|\bar{D}_X D_Y}{4(k+2)\sqrt{\sigma_v}},
\end{aligned}$$

where the second inequality follows from the definition of $\mathcal{L}_{\tau_k}$ in (8.1.93). Similar to the proof in Corollary 7.8, we can show that the total number of calls to the LO oracle in N outer iterations can be bounded by $\mathcal{O}(N^2)$. The bounds in (7.2.80) then immediately follow. ■

Observe that the $\mathcal{O}(1/\varepsilon)$ bound on the total number of operator evaluations is not improvable for solving the saddle point problems in (7.1.1)–(7.2.72). Moreover, the $\mathcal{O}(1/\varepsilon^2)$ bound on the total number of calls to the LO is also optimal for the LCP methods for solving the saddle point problems in (7.1.1)–(7.2.72).

We now turn our attention to stochastic saddle point problems for which only stochastic gradients of f_τ are available. In particular, we consider the situation when the original objective function f in (7.1.1) is given by

$$f(x) = \mathbb{E}\left[\max_{y \in Y}\langle A_\xi x, y\rangle - \hat{f}(y,\xi)\right], \tag{7.2.82}$$

where $\hat{f}(\cdot,\xi)$ is simple concave function for all $\xi \in \Xi$ and A_ξ is a random linear operator such that

$$\mathbb{E}\left[\|A_\xi\|^2\right] \leq L_A^2 \tag{7.2.83}$$

We can solve this stochastic saddle point problem by replacing (7.2.74) with

$$x_k = \mathrm{CndG}(g_k, x_{k-1}, \beta_k, \eta_k) \;\text{ where }\; g_k = \tfrac{1}{B_k}\textstyle\sum_{j=1}^{B_k} F'(z_k, \xi_j) \tag{7.2.84}$$

for some $\tau_k \geq 0$ and $B_k \geq 1$. By properly specifying $\{\beta_k\}$, $\{\eta_k\}$, $\{\tau_k\}$, and $\{B_k\}$, we can show that the number of linear operator evaluations (for A_ξ and A_ξ^T) and the number of calls to the LO performed by this variant of CGS method for finding a stochastic ε-solution of problem (7.1.1)–(7.2.82) can be bounded by

$$\mathcal{O}\left\{\tfrac{L_A^2 \bar{D}_X^2 D_Y^2}{\sigma_v \varepsilon^2}\right\}. \tag{7.2.85}$$

This result can be proved by combining the techniques in Sect. 7.2.2 and those in Theorem 7.14. However, we skip the details of these developments for the sake of simplicity.

7.3 Nonconvex Conditional Gradient Method

In this section, we consider the conditional gradient method applied to solve the following nonconvex optimization problem

$$f^* \equiv \min_{x \in X} f(x). \tag{7.3.1}$$

Here $X \subseteq \mathbb{R}^n$ is a compact convex set and f is differentiable but not necessarily convex. Moreover, we assume that the gradients of f satisfy

$$\|\nabla f(x) - \nabla f(y)\|_* \leq L\|x - y\|, \; \forall x, y \in X \tag{7.3.2}$$

for a given norm $\|\cdot\|$ in $\mathbb{R}^n$, where $\|\cdot\|_*$ denotes the conjugate norm of $\|\cdot\|$.

For a given $\bar{x} \in X$, we evaluate its accuracy using the Wolfe gap given by

$$\mathrm{gap}(\bar{x}) := \max_{x \in X} \langle \nabla f(\bar{x}), \bar{x} - x \rangle. \tag{7.3.3}$$

Clearly, $\bar{x} \in X$ satisfies the first-order optimality condition for (7.3.1) if and only if $\mathrm{gap}(\bar{x}) = 0$.

We study the convergence behavior of the conditional gradient method presented in Algorithm 7.1 for solving (7.3.1).

Theorem 7.15. *Let $\{y_t\}_{t=0}^k$ be generated by the conditional gradient method in Algorithm 7.1 applied to (7.3.1). Then we have*

$$\min_{t=1,\ldots,k} \text{gap}(y_{t-1}) \leq \tfrac{1}{\sum_{t=1}^k \alpha_t} \left[f(y_0) - f^* + \tfrac{L\bar{D}_X^2}{2} \sum_{t=1}^k \alpha_t^2 \right]. \tag{7.3.4}$$

In particular, if k is given in advance and $\alpha_t = \theta/\sqrt{k}$, $t = 1,\ldots,k$, for some $\theta > 0$, then

$$\min_{t=1,\ldots,k} \text{gap}(y_{t-1}) \leq \tfrac{1}{\sqrt{k}} \left[\tfrac{f(y_0)-f^*}{\theta} + \tfrac{\theta L\bar{D}_X^2}{2} \right], \tag{7.3.5}$$

where $\bar{D}_X := \max_{x,y\in X} \|x-y\|$.

Proof. Using the smoothness property of f, the fact that $y_k = (1-\alpha_k)y_{k-1} + \alpha_k x_k$, we have

$$\begin{aligned} f(y_k) &\leq f(y_{k-1}) + \langle \nabla f(y_{k-1}), y_k - y_{k-1} \rangle + \tfrac{L}{2}\|y_k - y_{k-1}\|^2 \\ &= f(y_{k-1}) + \alpha_k \langle f(y_{k-1}), x_k - y_{k-1} \rangle + \tfrac{L\alpha_k^2}{2}\|x_k - y_{k-1}\|^2 \\ &\leq f(y_{k-1}) + \alpha_k \langle f(y_{k-1}), x_k - y_{k-1} \rangle + \tfrac{L\alpha_k^2}{2}\bar{D}_X^2 \end{aligned}$$

for any $k \geq 1$ Summing up the above inequalities and rearranging the terms, we conclude that

$$\begin{aligned} \textstyle\sum_{t=1}^k \alpha_t \text{gap}(y_{t-1}) &= \textstyle\sum_{t=1}^k \left(\alpha_t \langle f(y_{k-1}), x_k - y_{k-1} \rangle \right) \\ &\leq f(y_0) - f(y_k) + \tfrac{L\bar{D}_X^2}{2} \textstyle\sum_{t=1}^k \alpha_t^2 \\ &\leq f(y_0) - f^* + \tfrac{L\bar{D}_X^2}{2} \textstyle\sum_{t=1}^k \alpha_t^2, \end{aligned}$$

which clearly implies (7.3.4). ■

In view of (7.3.5), the best stepsize policy for α_t would be

$$\alpha_t = \tfrac{\theta}{\sqrt{k}}, t = 1,\ldots,k, \text{with } \theta = \sqrt{\tfrac{2[f(y_0)-f^*]}{L\bar{D}_X^2}}.$$

In this case, we have

$$\min_{t=1,\ldots,k} \text{gap}(y_{t-1}) \leq \tfrac{1}{\sqrt{k}} \sqrt{2[f(y_0) - f^*]L\bar{D}_X^2}.$$

7.4 Stochastic Nonconvex Conditional Gradient

In this section, we consider the following nonconvex finite-sum problem

$$f^* := \min_{x\in X} \{f(x)\}, \tag{7.4.1}$$

where X is a closed compact set in Euclidean space $\mathbb{R}^n$, and f can be given as the average of m smooth but possibly nonconvex component functions f_i, i.e., $f(x) =$

$\sum_{i=1}^{m} f_i(x)/m$, or given as an expectation function, i.e., $f(x) = \mathbb{E}[F(x,\xi)]$ for some random variable $\xi \subseteq \Xi$. Our goal is to develop projection-free stochastic methods for solving these problems.

7.4.1 Basic Scheme for Finite-Sum Problems

We first focus on the basic case when the number of terms m is fixed. Our goal is to develop a variance-reduced conditional gradient method and establish its convergence properties.

This method computes a full gradient for every T iterations and use it to recursively define a gradient estimator G_k, which will be used in the linear optimization problem.

Algorithm 7.12 Nonconvex variance-reduced conditional gradient for finite-sum problems

Input: $x_1, T, \{\alpha_k\}$ and probability distribution $Q = \{q_1, \ldots, q_m\}$ on $\{1, \ldots, m\}$.
for $k = 1, 2, \ldots, N$ **do**
  if k % T == 1 **then**
    Set $G_k = \nabla f(x_k)$.
  else
    Generate i.i.d. samples I_b of size b according to Q.
    Set $G_k = \frac{1}{b}\sum_{i \in I_b}(\nabla f_i(x_k) - \nabla f_i(x_{k-1}))/(q_i m) + G_{k-1}$.
  end if
  Set $y_k = \text{argmin}_{x \in X}\langle G_k, x\rangle$.
  Set $x_{k+1} = (1-\alpha_k)x_k + \alpha_k y_k$.
end for
Output x_R, where R is a random variable s.t. to

$$\text{Prob}\{R = k\} = \frac{\alpha_k}{\sum_{k=1}^{N}\alpha_k}, k = 1, \ldots, N.$$

In order to facilitate the analysis of the algorithm, we will group the iteration indices $k = 1, 2, \ldots$ into different epochs given by

$$\{\{1,2,\ldots,T\},\{T+1,T+2,\ldots,2T\},\ldots,\{sT+1,sT+2\ldots,(s+1)T\},\ldots\}.$$

In other words, except for the last epoch, each epoch s, $s \geq 0$, consists of T iterations starting from $sT+1$ to $(s+1)T$, and the last epoch consists of the remaining iterations. For a given iteration index $k = sT+t$, we will always use the index k and the pair (s,t) interchangeably. For notational convenience, we also denote $(s, T+1) == (s+1, 1)$. Sometimes we will simply denote (s,t) by t if the epoch s is clear from the context.

For a given $\bar{x} \in X$, we evaluate its accuracy using the Wolfe gap given by (7.3.3). Denoting $\delta_k \equiv G_k - \nabla f(x_k)$, we can easily see that

$$\text{gap}(x_k) \le \max_{x\in X}\langle G_k, x_k - x\rangle + \|\delta_k\|\bar{D}_X, \tag{7.4.2}$$

where $\bar{D}_X := \max_{x,y\in X}\|x-y\|$.

We first provide a bound on the size of $\|\delta_k\|$. Its proof is skipped since it is similar to that of Lemma 6.10.

Lemma 7.4. *Let L be defined in (6.5.2) and suppose that the probabilities q_i are set to*

$$q_i = \tfrac{L_i}{mL} \tag{7.4.3}$$

for $i = 1,\ldots,m$. If the iteration index k (or equivalently (s, t)) represents the t-th iteration at the s-th epoch, then

$$\mathbb{E}[\|\delta_k\|^2] \equiv \mathbb{E}[\|\delta_{s,t}\|^2] \le \tfrac{L^2}{b}\textstyle\sum_{i=2}^t \mathbb{E}[\|x_{s,i} - x_{s,i-1}\|^2]. \tag{7.4.4}$$

Now we are ready to prove the main convergence properties of the nonconvex variance-reduced conditional gradient method.

Theorem 7.16. *If the probabilities q_i are set to (7.4.3) and the batch size $b \ge T$, then*

$$\mathbb{E}[\text{gap}(x_R)] \le \tfrac{f(x_1)-f^*}{\sum_{k=1}^N \alpha_k} + \tfrac{L\bar{D}_X^2}{\sum_{k=1}^N \alpha_k}\left[\tfrac{3}{2}\textstyle\sum_{k=1}^N \alpha_k^2 + \sum_{s=0}^S\left(\sum_{j=1}^T \alpha_{s,j}\max_{j=2,\ldots,T}\alpha_{s,j}\right)\right],$$

where $S = \lfloor N/T \rfloor$.

Proof. Using the smoothness property of f and the fact that $x_{k+1} = (1-\alpha_k)x_k + \alpha_k y_k$, we have

$$\begin{aligned} f(x_{k+1}) &\le f(x_k) + \langle \nabla f(x_k), x_{k+1} - x_k\rangle + \tfrac{L}{2}\|x_{k+1} - x_k\|^2 \\ &= f(x_k) + \alpha_k\langle G_k, y_k - x_k\rangle + \alpha_k\langle \delta_k, y_k - x_k\rangle + \tfrac{L\alpha_k^2}{2}\|y_k - x_k\|^2 \\ &\le f(x_k) + \alpha_k\langle G_k, y_k - x_k\rangle + \tfrac{1}{2L}\|\delta_k\|^2 + L\alpha_k^2\|y_k - x_k\|^2 \\ &= f(x_k) - \alpha_k\max_{x\in X}\langle G_k, x_k - x\rangle + \tfrac{1}{2L}\|\delta_k\|^2 + L\alpha_k^2\|y_k - x_k\|^2 \end{aligned} \tag{7.4.5}$$

for any $k \ge 1$, where the second inequality follows from the Cauchy–Schwarz inequality and the last inequality follows from the definition of y_k. Note that by (7.4.4) and the definition of x_{k+1},

$$\begin{aligned} \mathbb{E}[\|\delta_k\|^2] = \mathbb{E}[\|\delta_{s,t}\|^2] &\le \tfrac{L^2}{b}\textstyle\sum_{i=2}^t \mathbb{E}[\|x_{s,i} - x_{s,i-1}\|^2] \\ &= \tfrac{L^2}{b}\textstyle\sum_{i=2}^t \alpha_{s,i}^2\|y_{s,i} - x_{s,i}\|^2 \\ &\le \tfrac{L^2\bar{D}_X^2}{b}\textstyle\sum_{i=2}^t \alpha_{s,i}^2. \end{aligned}$$

Combining the above two inequalities with (7.4.2), we have for any iteration t at epoch s,

$$\mathbb{E}[f(x_{s,t+1})] \le \mathbb{E}[f(x_{s,t})] - \alpha_{s,t}\mathbb{E}[\max_{x\in X}\langle G_{s,t}, x_{s,t}-x\rangle] + \tfrac{L\bar{D}_X^2}{2b}\textstyle\sum_{i=2}^t \alpha_{s,i}^2 + L\alpha_{s,t}^2\bar{D}_X^2$$
$$\le \mathbb{E}[f(x_{s,t})] - \alpha_{s,t}\mathbb{E}[\text{gap}(x_{s,t})] + L\bar{D}_X^2\left[\alpha_{s,t}(\tfrac{1}{b}\textstyle\sum_{i=2}^t \alpha_{s,i}^2)^{1/2} + \tfrac{1}{2b}\textstyle\sum_{i=2}^t \alpha_{s,i}^2 + \alpha_{s,t}^2\right]. \tag{7.4.6}$$

Summing up these inequalities, we conclude that for any $t = 1,\dots,T$,

$$\mathbb{E}[f(x_{s,t+1})] \le \mathbb{E}[f(x_{s,1})] - \textstyle\sum_{j=1}^t \alpha_{s,j}\mathbb{E}[\text{gap}(x_{s,j})]$$
$$+ L\bar{D}_X^2 \textstyle\sum_{j=1}^t \left[\alpha_{s,j}(\tfrac{1}{b}\textstyle\sum_{i=2}^j \alpha_{s,i}^2)^{1/2} + \tfrac{1}{2b}\textstyle\sum_{i=2}^j \alpha_{s,i}^2 + \alpha_{s,j}^2\right].$$

Observing

$$\tfrac{1}{2b}\textstyle\sum_{j=1}^t\sum_{i=2}^j \alpha_{s,i}^2 = \tfrac{1}{2b}\textstyle\sum_{j=2}^t (t-j+1)\alpha_{s,j}^2 \le \tfrac{t-1}{2b}\textstyle\sum_{j=2}^t \alpha_{s,j}^2 \le \tfrac{1}{2}\textstyle\sum_{j=2}^t \alpha_{s,j}^2$$
$$\textstyle\sum_{j=1}^t \alpha_{s,j}(\tfrac{1}{b}\textstyle\sum_{i=2}^j \alpha_{s,i}^2)^{1/2} \le \tfrac{1}{\sqrt{b}}\left(\max_{j=2,\dots,T}\alpha_{s,j}\right)\textstyle\sum_{j=1}^t \sqrt{j-1}\,\alpha_{s,j}$$
$$\le \textstyle\sum_{j=1}^t \alpha_{s,j} \max_{j=2,\dots,T}\alpha_{s,j},$$

we conclude that for any $t = 1,\dots,T$,

$$\textstyle\sum_{j=1}^t \alpha_{s,j}\mathbb{E}[\text{gap}(x_{s,j})] \le \mathbb{E}[f(x_{s,1})] - \mathbb{E}[f(x_{s,t+1})]$$
$$+ L\bar{D}_X^2\left[\tfrac{3}{2}\textstyle\sum_{j=1}^t \alpha_{s,j}^2 + \textstyle\sum_{j=1}^t \alpha_{s,j}\max_{j=2,\dots,T}\alpha_{s,j}\right].$$

Noting $S = \lfloor N/T \rfloor$ and $J = N\%T$, and taking telescope sum over epochs $s = 0,\dots,S$, we obtain

$$\textstyle\sum_{k=1}^N \alpha_k \mathbb{E}[\text{gap}(x_k)]$$
$$\le f(x_1) - \mathbb{E}[f(x_{N+1})] + L\bar{D}_X^2\left[\tfrac{3}{2}\textstyle\sum_{k=1}^N \alpha_k^2 + \textstyle\sum_{s=0}^S\left(\textstyle\sum_{j=1}^T \alpha_{s,j}\max_{j=2,\dots,T}\alpha_{s,j}\right)\right]$$
$$\le f(x_1) - f^* + L\bar{D}_X^2\left[\tfrac{3}{2}\textstyle\sum_{k=1}^N \alpha_k^2 + \textstyle\sum_{s=0}^S\left(\textstyle\sum_{j=1}^T \alpha_{s,j}\max_{j=2,\dots,T}\alpha_{s,j}\right)\right].$$

Our result immediately follows from the above inequality and the selection of the random variable R. ■

We are now ready to specify stepsizes α_k and establish the bounds on the total number of gradient evaluations and linear optimization required by the nonconvex variance-reduced conditional gradient method.

Corollary 7.10. *Assume that the probabilities q_i are set to (7.4.3) and that*

$$b = T = \sqrt{m}. \tag{7.4.7}$$

If N is given and

$$\alpha_k = \alpha := \tfrac{1}{\sqrt{N}}, \tag{7.4.8}$$

then the total number of linear oracles and gradient evaluations required by the nonconvex variance-reduced conditional gradient method to find a solution $\bar{x} \in X$ *s.t.* $\mathbb{E}[\text{gap}(\bar{x})] \le \varepsilon$ *can be bounded by*

$$\mathscr{O}\left\{\tfrac{1}{\varepsilon^2}\left[f(x_1) - f^* + L\bar{D}_X^2\right]^2\right\} \tag{7.4.9}$$

and

$$\mathscr{O}\left\{m + \tfrac{\sqrt{m}}{\varepsilon^2}\left[f(x_1) - f^* + L\bar{D}_X^2\right]^2\right\} \tag{7.4.10}$$

respectively.

Proof. Denote $S = \lfloor N/T \rfloor$. Observe that by (7.4.8), $\sum_{k=1}^N \alpha_k = N\alpha$, $\sum_{k=1}^N \alpha_k^2 = N\alpha^2$, and

$$\Sigma_{s=0}^S \left(\Sigma_{j=1}^T \alpha_{s,j} \max_{j=2,\ldots,T} \alpha_{s,j}\right) = \Sigma_{s=0}^S \Sigma_{j=1}^T \alpha^2 \le 2N\alpha^2.$$

Using these observations in Theorem 7.16, we conclude that

$$\mathbb{E}[\text{gap}(x_R)] \le \tfrac{f(x_1) - f^*}{N\alpha} + \tfrac{7L\bar{D}_X^2\alpha}{2} = \tfrac{1}{\sqrt{N}}\left[f(x_1) - f^* + \tfrac{7L\bar{D}_X^2}{2}\right].$$

Hence, the total number of linear oracles will be bounded by (7.4.9). Moreover, the total number of gradient evaluations will be bounded by

$$(m + bT)\left\lceil \tfrac{N}{T} \right\rceil = (2m)\left\lceil \tfrac{N}{T} \right\rceil = 2m + \sqrt{m}N,$$

and thus by (7.4.10). ■

In practice, it makes sense to choose a nonuniform distribution to select the output solution x_R. We provide such an option below with increasing α_k so as to put more weight on the iterations generated later by the algorithm.

Corollary 7.11. *Assume that the probabilities* q_i *and batch size* b *are set to (7.4.3) and (7.4.7), respectively. If* N *is given and*

$$\alpha_k = \alpha := \tfrac{k^{1/4}}{N^{3/4}}, \tag{7.4.11}$$

then the total number of linear oracles and gradient evaluations required by the nonconvex variance-reduced conditional gradient method to find a solution $\bar{x} \in X$ *s.t.* $\mathbb{E}[\text{gap}(\bar{x})] \le \varepsilon$ *can be bounded by (7.4.9) and (7.4.10), respectively.*

Proof. Denote $S = \lfloor N/T \rfloor$. Observe that by (7.4.11),

$$\begin{aligned}
&\Sigma_{k=1}^N \alpha_k = \tfrac{1}{N^{3/4}} \Sigma_{k=1}^N k^{1/4} \ge \tfrac{4}{5}\sqrt{N},\\
&\Sigma_{k=1}^N \alpha_k^2 = \tfrac{1}{N^{3/2}} \Sigma_{k=1}^N k^{1/2} \le \tfrac{2(N+1)^{3/2}}{3N^{3/2}} \le \tfrac{4\sqrt{2}}{3},\\
&\Sigma_{s=0}^S \left(\Sigma_{j=1}^T \alpha_{s,j} \max_{j=2,\ldots,T} \alpha_{s,j}\right) = \Sigma_{s=0}^S \Sigma_{j=1}^T \alpha_{s,j}\alpha_{s,T} \le \Sigma_{s=0}^S \Sigma_{j=1}^T \alpha_{s,T}
\end{aligned}$$

$$= \frac{T^{3/2}}{N^{3/2}} \sum_{s=0}^{S} (s+1)^{1/2} \le \frac{2T^{3/2}}{3N^{3/2}} (S+2)^{3/2} \le 2\sqrt{3}.$$

Using these observations in Theorem 7.16, we conclude that

$$\mathbb{E}[\text{gap}(x_R)] \le \frac{5}{2\sqrt{N}} \left[\tfrac{1}{2} f(x_1) - f^* + (\sqrt{2}+\sqrt{3}) L\bar{D}_X^2\right].$$

Hence, the total number of linear oracles will be bounded by (7.4.9). Moreover, the total number of gradient evaluations will be bounded by

$$(m + bT) \left\lceil \tfrac{N}{T} \right\rceil = (2m) \left\lceil \tfrac{N}{T} \right\rceil = 2m + \sqrt{m}N,$$

and thus by (7.4.10). ■

In view of the results in Corollaries 7.12 and 7.11, the nonconvex variance-reduced conditional gradient method can save up to a factor of $\sqrt{m}$ gradient evaluations without increasing the number of calls to the linear oracle than the deterministic nonconvex variance-reduced conditional gradient method. Observe that both stepsize policies in (7.4.8) and (7.4.11) require us to fix the number of iterations N a priori. It is possible to relax this assumption, e.g., by setting $\alpha_k = 1/\sqrt{k}$. However, this stepsize policy will result in a slightly worse rate of convergence result up to some logarithmic factors than those in Corollaries 7.12 and 7.11.

7.4.2 Generalization for Stochastic Optimization Problems

In this section, we still consider problem (7.4.1), but with f given by

$$f(x) = \mathbb{E}[F(x,\xi)], \tag{7.4.12}$$

where ξ is a random vector supported on $\Xi \subseteq \mathbb{R}^d$ for some $d \ge 1$. We make the following assumptions throughout this subsection.

- $F(x,\xi)$ is a smooth function with Lipschitz constant L for any $\xi \in \Xi$ almost surely.
- It is possible to generate a realization $\xi \in \Xi$, and to compute $\nabla F(x,\xi)$ and $\nabla F(y,\xi)$ for any given two point $x, y \in X$ for a fixed realization ξ.
- For any x, we have $\mathbb{E}[\nabla F(x,\xi)] = \nabla f(x)$ and

$$\mathbb{E}[\|\nabla F(x,\xi) - \nabla f(x)\|^2] \le \sigma^2. \tag{7.4.13}$$

These assumptions are the same as those for the nonconvex variance-reduced mirror-descent method in Sect. 7.4, but much stronger than those required for the RSMD method in Sect. 6.2.

Algorithm 7.13 Nonconvex variance-reduced conditional gradient for stochastic problems

Input: $x_1, T, \{\alpha_k\}$ and sample size m.
for $k = 1, 2, \ldots, N$ **do**
  if k % T == 1 **then**
    Generate an i.i.d. sample $H^s = \{\xi_1^s, \ldots, \xi_m^s\}$ for the random variable ξ.
    Set $G_k = \frac{1}{m}\sum_{i=1}^m \nabla F(x_k, \xi_i^s)$.
    Set $s \leftarrow s+1$.
  else
    Generate an i.i.d. sample $I^k = \{\xi_1^k, \ldots, \xi_b^k\}$ for the random variable ξ.
    Set $G_k = \frac{1}{b}\sum_{i=1}^b (\nabla F(x_k, \xi_i^k) - \nabla F(x_{k-1}, \xi_i^k)) + G_{k-1}$.
  end if
  Set $y_k = \mathrm{argmin}_{x \in X} \langle G_k, x \rangle$.
  Set $x_{k+1} = (1-\alpha_k)x_k + \alpha_k y_k$.
end for
Output x_R, where R is a random variable s.t. to

$$\mathrm{Prob}\{R = k\} = \frac{\alpha_k}{\sum_{k=1}^N \alpha_k}, k = 1, \ldots, N.$$

Similar to the previous section, we will first need to provide a bound on the size of $\delta_k = G_k - \nabla f(x_k)$. The proof of this result is almost identical to that of Lemma 6.11 and hence its details are skipped.

Lemma 7.5. *If the iteration index k (or equivalently (s,t)) represents the t-th iteration at the s-epoch, then*

$$\mathbb{E}[\|\delta_k\|^2] \equiv \mathbb{E}[\|\delta_{(s,t)}\|^2] \le \tfrac{L^2}{b}\textstyle\sum_{i=2}^t \mathbb{E}[\|x_{(s,i)} - x_{(s,i-1)}\|^2] + \tfrac{\sigma^2}{m}. \tag{7.4.14}$$

Theorem 7.17. *If the probabilities q_i are set to (7.4.3) and the batch size $b \ge T$, then*

$$\begin{aligned}\mathbb{E}[\mathrm{gap}(x_R)] \le \tfrac{f(x_1)-f^*}{\sum_{k=1}^N \alpha_k} + \tfrac{L\bar{D}_X^2}{\sum_{k=1}^N \alpha_k}\left[\tfrac{3}{2}\textstyle\sum_{k=1}^N \alpha_k^2 + \sum_{s=0}^S \left(\sum_{j=1}^T \alpha_{s,j} \max_{j=2,\ldots,T} \alpha_{s,j}\right)\right] \\ + \tfrac{N\sigma^2}{2Lm\sum_{k=1}^N \alpha_k},\end{aligned}$$

where $S = \lfloor N/T \rfloor$.

Proof. The result can be proved similar to Theorem 7.16 after we replace (6.5.16) with (7.4.14). ∎

We are now ready to specify stepsizes α_k and establish the bounds on the total number of gradient evaluations and calls to the linear optimization oracle required by the nonconvex variance-reduced conditional gradient method.

Corollary 7.12. *Assume that b and T are set to (7.4.7). If*

$$\alpha_k = \alpha := \left[\left(\tfrac{1}{N} + \tfrac{\sigma^2}{Lm}\right)\tfrac{1}{L\bar{D}_X^2}\right]^{1/2} \tag{7.4.15}$$

for some fixed in advance iterations count N, then we have

$$\mathbb{E}[\mathrm{gap}(x_R)] \le \tfrac{f(x_1)-f^*}{\sqrt{N}} + \tfrac{7L\bar{D}_X^2}{2\sqrt{N}} + \tfrac{4\sigma\bar{D}_X}{\sqrt{m}}.$$

As a consequence, the total number of linear oracles and gradient evaluations required by the nonconvex variance-reduced conditional gradient method to find a solution $\bar{x} \in X$ s.t. $\mathbb{E}[\mathrm{gap}(\bar{x})] \le \varepsilon$ can be bounded by

$$\mathcal{O}\left\{\left(\tfrac{f(x_1)-f^*+L\bar{D}_X^2}{\varepsilon}\right)^2\right\} \tag{7.4.16}$$

and

$$\mathcal{O}\left\{\left(\tfrac{\sigma\bar{D}_X}{\varepsilon}\right)^2 + \tfrac{\sigma\bar{D}_X}{\varepsilon}\left(\tfrac{f(x_1)-f^*+L\bar{D}_X^2}{\varepsilon}\right)^2\right\}, \tag{7.4.17}$$

respectively.

Proof. Denote $S = \lfloor N/T \rfloor$. Observe that by (7.4.15), $\Sigma_{k=1}^N \alpha_k = N\alpha$, $\Sigma_{k=1}^N \alpha_k^2 = N\alpha^2$, and

$$\Sigma_{s=0}^S \left(\Sigma_{j=1}^T \alpha_{s,j} \max_{j=2,\ldots,T} \alpha_{s,j}\right) = \Sigma_{s=0}^S \Sigma_{j=1}^T \alpha^2 \le 2N\alpha^2.$$

Using these observations in Theorem 7.17, we conclude that

$$\begin{aligned}\mathbb{E}[\mathrm{gap}(x_R)] &\le \tfrac{f(x_1)-f^*}{N\alpha} + \tfrac{7L\bar{D}_X^2\alpha}{2} + \tfrac{\sigma^2}{2Lm\alpha}\\ &= \tfrac{f(x_1)-f^*}{\sqrt{N}} + \tfrac{7L\bar{D}_X^2}{2\sqrt{N}} + \tfrac{4\sigma\bar{D}_X}{\sqrt{m}}.\end{aligned}$$

Now if we choose

$$m = \mathcal{O}\left\{\left(\tfrac{\sigma\bar{D}_X}{\varepsilon}\right)^2\right\},$$

then an ε-solution will be found in

$$N = \mathcal{O}\left\{\left(\tfrac{f(x_1)-f^*+L\bar{D}_X^2}{\varepsilon}\right)^2\right\}$$

iterations. Hence, the total number of linear oracles will be bounded by (7.4.16). Moreover, the total number of gradient evaluations will be bounded by

$$(m+bT)\left\lceil \tfrac{N}{T} \right\rceil = (2m)\left\lceil \tfrac{N}{T} \right\rceil = 2m + \sqrt{m}N,$$

and thus by (7.4.17). ∎

One can also choose a nonuniform distribution to select the output solution x_R, similar to the deterministic case in Corollary 7.11. We leave this as an exercise.

7.5 Stochastic Nonconvex Conditional Gradient Sliding

We have so far discussed different types of termination criterions for solving nonconvex optimization problems given in the form of (7.4.1), including one based on Wolfe gap and the other based on projected gradient. In this section, we first compare these two criterions and then present a stochastic nonconvex conditional gradient sliding method for solving problem (7.4.1), which can potentially outperform the stochastic nonconvex conditional gradient method in the previous section in terms of the latter criterion based on projected gradient.

7.5.1 Wolfe Gap vs Projected Gradient

Recall that for a given search point $\bar{x} \in X$, the projected gradient $g_X(\bar{x})$ associated with problem (7.4.1) is given by (see (6.2.7))

$$g_X(\bar{x}) \equiv P_X(x, \nabla f(\bar{x}), \gamma) := \tfrac{1}{\gamma}(\bar{x} - \bar{x}^+), \tag{7.5.1}$$

where

$$\bar{x}^+ = \operatorname{argmin}_{u \in X} \left\{ \langle \nabla f(\bar{x}), u \rangle + \tfrac{1}{\gamma} V(\bar{x}, u) \right\}, \tag{7.5.2}$$

where V denotes the prox-function (or Bregman distance) associated with the distance generating function ν. We assume that ν has L_ν-Lipschitz gradients and modulus 1. It then follows from Lemma 6.3 that if

$$\|g_X(\bar{x})\| \le \varepsilon,$$

then

$$-\nabla f(\bar{x}^+) \in N_X(\bar{x}^+) + B(\varepsilon(\gamma L + L_\nu)), \tag{7.5.3}$$

which, in view of the definitions of the normal cone in (6.2.8) and the Wolfe gap in (7.3.3), then clearly imply that

$$\text{gap}(\bar{x}^+) \le \varepsilon(\gamma L + L_\nu).$$

Now suppose that we have a solution $\bar{x}$ satisfying $\text{gap}(\bar{x}) \le \varepsilon$. It can be easily seen that

$$-\nabla f(\bar{x}) \in N_X(\bar{x}) + B(\varepsilon). \tag{7.5.4}$$

Observe that one nice feature about the definition of $\text{gap}(\bar{x})$ is that it does not rely on the selection of the norm. Now let us provide a bound on the size of projected gradient for $\bar{x}$. By the optimality condition of (7.5.2), we have

$$\langle \gamma \nabla f(\bar{x}) + \nabla v(\bar{x}^+) - \nabla v(\bar{x}), x - \bar{x}^+ \rangle \geq 0, \forall x \in X.$$

Letting $x = \bar{x}$ in the above inequality, we have

$$\begin{aligned}\langle \gamma \nabla f(\bar{x}), \bar{x}^+ - \bar{x} \rangle &\geq \langle \nabla v(\bar{x}) - \nabla v(\bar{x}^+), \bar{x} - \bar{x}^+ \rangle \\ &\geq \|\bar{x} - \bar{x}^+\|^2 = \gamma^2 g_X(\bar{x}),\end{aligned}$$

which implies that

$$\|g_X(\bar{x})\|^2 \leq \langle \nabla f(\bar{x}), \bar{x}^+ - \bar{x} \rangle \leq \text{gap}(\bar{x}).$$

In other words, if $\text{gap}(\bar{x}) \leq \varepsilon$, in general we can only guarantee that

$$\|g_X(\bar{x})\| \leq \sqrt{\varepsilon}.$$

Therefore, it appears that the projected gradient is a stronger termination criterion than the Wolfe gap, even though they both imply that the gradient $\nabla f(\bar{x})$ (or $\nabla f(\bar{x}^+)$) falls within a small neighborhood of the normal cone $N_X(\bar{x})$ (or $N_X(\bar{x}^+)$) with similar magnitude of perturbation (see (7.5.3) and (7.5.4)).

7.5.2 Projection-Free Method to Drive Projected Gradient Small

Suppose that our goal indeed is to find a solution of problem (7.4.1) with small projected gradient, i.e., a solution $\bar{x} \in X$ s.t. $\mathbb{E}[\|g_X(\bar{x})\|^2] \leq \varepsilon$. First consider the finite-sum case when f is given by the average of m components. If we apply the nonconvex variance-reduced conditional gradient method, then the total number of stochastic gradients and calls to the linear optimization oracle will be bounded by

$$\mathcal{O}\left\{m + \tfrac{\sqrt{m}}{\varepsilon^2}\right\} \text{ and } \mathcal{O}\left\{\tfrac{1}{\varepsilon^2}\right\},$$

respectively. On the other hand, if we apply the nonconvex variance-reduced mirror descent method in Sect. 7.4, then the total number of stochastic gradient will be bounded by $\mathcal{O}(m + \sqrt{m}/\varepsilon)$. Therefore, the number of stochastic gradients required by the nonconvex variance-reduced conditional gradient method can be worse than that by the nonconvex variance-reduced mirror descent method up to a factor of $\mathcal{O}(1/\varepsilon)$. The same situation happens for the stochastic case when f is given in the form of expectation. If we apply the nonconvex variance-reduced conditional gradient method, then the total number of stochastic gradients and calls to the linear optimization oracle will be bounded by

$$\mathcal{O}\left\{\tfrac{1}{\varepsilon^3}\right\} \text{ and } \mathcal{O}\left\{\tfrac{1}{\varepsilon^2}\right\},$$

respectively. However, the total number of stochastic gradient required by the nonconvex variance-reduced mirror descent method will be bounded by $\mathcal{O}(1/\varepsilon^{3/2})$. Therefore, the total number of stochastic gradients required by the nonconvex variance-reduced conditional gradient method can be worse than that by the nonconvex variance-reduced mirror descent method up to a factor of $\mathcal{O}(1/\varepsilon^{3/2})$.

Our goal in this section is to present the nonconvex stochastic conditional gradient sliding method for solving problem (7.4.1) and show that it can substantially reduce the total number of required stochastic gradients than the nonconvex variance-reduced conditional gradient method, but without increasing the total number of calls to the linear oracle.

Similar to the conditional gradient sliding method, the basic idea of the nonconvex stochastic conditional gradient sliding method is to apply the conditional gradient method for solving the projection subproblem existing in the nonconvex variance-reduced mirror descent method in Sect. 7.4. We formally state this algorithm as follows.

Nonconvex Stochastic Conditional Gradient Sliding for Finite-Sum Problems
Replace the definition of x_{k+1} in Algorithm 7.13 by

$$x_{k+1} = \text{CndG}(G_k, x_k, 1/\gamma, \eta), \tag{7.5.5}$$

where the procedure CndG is defined in the conditional gradient sliding method (see Algorithm 7.6).

In (7.5.5), we call the classical conditional gradient method to approximately solve the projection subproblem

$$\min_{x \in X} \left\{ \phi_k(x) := \langle G_k, x \rangle + \tfrac{1}{2\gamma} \|x - x_k\|^2 \right\} \tag{7.5.6}$$

such that

$$\langle \phi_k'(x_{k+1}), x_{k+1} - x \rangle = \langle G_k + (x_{k+1} - x_k)/\gamma, x_{k+1} - x \rangle \le \eta, \ \ \forall x \in X, \tag{7.5.7}$$

for some $\eta \ge 0$.

Theorem 7.18. *Suppose that the probabilities q_i are set to (7.4.3). If*

$$b = 10T \quad \textit{and} \quad \gamma = \tfrac{1}{L}, \tag{7.5.8}$$

then we have

$$\mathbb{E}[\|g_{X,k}\|^2] \le \tfrac{16L}{N}[f(x_1) - f^*] + 24L\eta. \tag{7.5.9}$$

Proof. Let us denote $\bar{x}_{k+1} := \text{argmin}_{x \in X} \phi_k(x)$, $\hat{x}_{k+1} := \text{argmin}\langle \nabla f(x_k), x \rangle + \frac{1}{2\gamma}\|x - x_k\|^2$, and $\tilde{g}_k \equiv \frac{1}{\gamma}(x_k - x_{k+1})$. It follows from (7.5.7) with $x = x_k$ that

$$\tfrac{1}{\gamma}\|x_k - x_{k+1}\|^2 \le \langle G_k, x_k - x_{k+1} \rangle + \eta.$$

Using the above relation and the smoothness of f, we have for any $k \ge 1$,

$$
\begin{aligned}
f(x_{k+1}) &\leq f(x_k) + \langle \nabla f(x_k), x_{k+1} - x_k \rangle + \tfrac{L}{2}\|x_{k+1} - x_k\|^2 \\
&= f(x_k) + \langle G_k, x_{k+1} - x_k \rangle + \tfrac{L}{2}\|x_{k+1} - x_k\|^2 - \langle \delta_k, x_{k+1} - x_k \rangle \\
&\leq f(x_k) - \tfrac{1}{\gamma}\|x_k - x_{k+1}\|^2 + \eta_k + \tfrac{L}{2}\|x_{k+1} - x_k\|^2 - \langle \delta_k, x_{k+1} - x_k \rangle \\
&\leq f(x_k) - \left(\tfrac{1}{\gamma} - \tfrac{L}{2} - \tfrac{q}{2}\right)\|x_k - x_{k+1}\|^2 + \tfrac{1}{2q}\|\delta_k\|^2 + \eta_k,
\end{aligned}
$$

for any $q > 0$. Using the definition of $\tilde{g}_k$ in the above relation, we have

$$
f(x_{k+1}) \leq f(x_k) - \gamma\left(1 - \tfrac{L\gamma}{2} - \tfrac{q\gamma}{2}\right)\|\tilde{g}_{X,k}\|^2 + \tfrac{1}{2q}\|\delta_k\|^2 + \eta_k. \tag{7.5.10}
$$

Moreover, by (7.5.7) with $x = \bar{x}_{k+1}$ and the strong convexity of ϕ_k, we obtain

$$
\tfrac{1}{2\gamma}\|x_{k+1} - \bar{x}_{k+1}\|^2 \leq \langle \phi_k'(x_{k+1}), x_{k+1} - \bar{x}_{k+1} \rangle \leq \eta_k. \tag{7.5.11}
$$

Using the simple observation that

$$
g_{X,k} = (x_k - \hat{x}_{k+1})/\gamma = [(x_k - x_{k+1}) + (x_{k+1} - \bar{x}_{k+1}) + (\bar{x}_{k+1} - \hat{x}_{k+1})]/\gamma,
$$

we conclude from (6.2.11) and (7.5.11) that

$$
\begin{aligned}
\|g_{X,k}\|^2 &\leq 2\|\tilde{g}_{X,k}\|^2 + 4\|(x_{k+1} - \bar{x}_{k+1})/\gamma\|^2 + 4\|(\bar{x}_{k+1} - \hat{x}_{k+1})/\gamma\|^2 \\
&\leq 2\|\tilde{g}_{X,k}\|^2 + \tfrac{8\eta_k}{\gamma} + 4\|\delta_k\|^2.
\end{aligned}
$$

Multiplying the above inequality for any $p > 0$ and adding it to (7.5.10), we have

$$
\begin{aligned}
f(x_{k+1}) + p\|g_{X,k}\|^2 \leq f(x_k) &- \left[\gamma\left(1 - \tfrac{L\gamma}{2} - \tfrac{q\gamma}{2}\right) - 2p\right]\|\tilde{g}_{X,k}\|^2 \\
&+ (4p + \tfrac{1}{2q})\|\delta_k\|^2 + (1 + \tfrac{8p}{\gamma})\eta_k.
\end{aligned}
$$

Now using an argument similar to (6.5.12), we can show that for any epoch s of the nonconvex stochastic conditional gradient sliding method,

$$
\begin{aligned}
&\mathbb{E}[f(x_{s,t+1})] + p\Sigma_{j=1}^t \mathbb{E}[\|g_{X,(s,j)}\|^2] \\
&\leq \mathbb{E}[f(x_{s,1})] - \left[\gamma\left(1 - \tfrac{L\gamma}{2} - \tfrac{q\gamma}{2}\right) - 2p - (4p + \tfrac{1}{2q})\tfrac{\gamma^2 L^2 (t-1)}{b}\right]\Sigma_{j=1}^t \mathbb{E}[\|\tilde{g}_{X,(s,j)}\|^2] \\
&\quad + (1 + \tfrac{8p}{\gamma})\Sigma_{j=1}^t \eta_{s,j}.
\end{aligned} \tag{7.5.12}
$$

Fixing $\gamma = 1/L$, $p = 1/(16L)$, $q = L/2$, and $b = 10T$ in the above inequality, and observing

$$
\gamma\left(1 - \tfrac{L\gamma}{2} - \tfrac{q\gamma}{2}\right) - 2p - (4p + \tfrac{1}{2q})\tfrac{\gamma^2 L^2 (t-1)}{b} = \tfrac{1}{8L} - \tfrac{5(t-1)}{4Lb} > 0, \forall t = 1, \ldots, T,
$$

we have

$$
\mathbb{E}[f(x_{(s,t+1)})] + \tfrac{1}{16L}\Sigma_{j=1}^t \mathbb{E}[\|g_{X,(s,j)}\|^2] \leq \mathbb{E}[f(x_{(s,1)})] + \tfrac{3}{2}\Sigma_{j=1}^t \eta_{s,j}. \tag{7.5.13}
$$

Therefore, by summing up the first N inequalities in the above form we obtain

$$\mathbb{E}[f(x_{N+1})] + \tfrac{1}{16L}\textstyle\sum_{k=1}^{N}\mathbb{E}[\|g_{X,k}\|^2] \le f(x_1) + \tfrac{3}{2}\textstyle\sum_{k=1}^{N}\eta_k.$$

The result then follows from the above inequality, the definition of the random variable R, and the fact that $f(x_{N+1}) \ge f^*$. ∎

Using the above result, we can bound the total number of stochastic gradients and calls to the linear optimization oracle.

Corollary 7.13. *Suppose that the probabilities* q_i *are set to (7.4.3) and that* b *and* γ *are set to (7.5.8) with* $T = \sqrt{m}$. *Then the total number of stochastic gradients and calls to the linear optimization oracle performed by the nonconvex stochastic conditional gradient sliding method to find a solution* $\bar{x} \in X$ *s.t.* $\mathbb{E}[g_X(\bar{x})] \le \varepsilon$ *will be bounded by*

$$\mathcal{O}\left\{m + \tfrac{\sqrt{m}L}{\varepsilon}[f(x_1) - f^*]\right\} \tag{7.5.14}$$

and

$$\mathcal{O}\left\{\tfrac{L^3\bar{D}_X^2[f(x_1)-f^*]}{\varepsilon^2}\right\}, \tag{7.5.15}$$

respectively.

Proof. Assume that $\eta = \varepsilon/(48L)$. Clearly, by Theorem 7.18, the total number of iterations N will be bounded by

$$\tfrac{32L}{\varepsilon}[f(x_1) - f^*].$$

Therefore, the total number of gradient evaluations will be bounded by

$$(m + bT)\lceil \tfrac{N}{T} \rceil \le 11m(\tfrac{N}{\sqrt{m}} + 1)\rceil,$$

which is bounded by (7.5.14). Moreover, in view of Theorem 7.9(c), the number of calls to the linear optimization oracle performed at each iteration can be bounded by

$$\left\lceil \frac{6\bar{D}_X^2}{\gamma\eta} \right\rceil$$

and hence the total number of calls to the linear optimization oracle will be bounded by

$$N\left\lceil \frac{6\bar{D}_X^2}{\gamma\eta} \right\rceil,$$

which is bounded by (7.5.15). ∎

We can develop a similar stochastic nonconvex conditional gradient sliding method for solving stochastic optimization problem with the objective function given in the form of expectation. We leave this as an exercise for the readers.

7.6 Exercises and Notes

1. Try to provide a game interpretation for the conditional gradient method, similar to the game interpretation for the accelerated gradient method discussed in Sect. 3.4.
2. Similar to the conditional gradient sliding method, try to solve the subproblems in the primal–dual method in Sect. 3.6 by using the conditional gradient method and establish the rate of convergence of the resulting algorithm.
3. Similar to the conditional gradient sliding method, try to solve the subproblems in the mirror-prox method in Sect. 3.8 by using the conditional gradient method and establish the rate of convergence of the resulting algorithm.

Notes. The conditional gradient method was first introduced by Frank and Wolfe in [35]. The variants of the conditional gradient method obtained by replacing the projection with linear optimization in the accelerated gradient descent method was first introduced by Lan in [64]. Lan [64] also discussed nonsmooth conditional gradient method and the low complexity bounds on the number of calls to the linear optimization oracles for solving different classes of convex optimization problems. Lan and Zhou [69] introduced the conditional gradient methods, which was the first class of optimization algorithms that can achieve the lower complexity bound for linear optimization oracles while maintaining the optimal rate of convergence in terms of the number of calls to the first-order oracle. The complexity of nonconvex conditional gradient and conditional gradient sliding methods was analyzed in [55] and [111], respectively. Lan developed the materials on stochastic nonconvex conditional gradient in Sects. 7.4 and 7.5 when writing the book at the end of 2018 and early 2019, before realizing that some results in Sect. 7.4 were developed in [125]. It is worth mentioning that the best complexity result so far, in terms of gradient computation, for variance-reduced conditional gradient methods was reported by Reddi et al. [115] even though their algorithm requires more memory than the one presented in Sect. 7.4. Conditional gradient type methods have attracted a lot of interest in both optimization and machine learning community recently (see, e.g., [1, 4, 6, 24, 25, 36, 43, 45, 47, 51–53, 81, 124]). In particular, Jaggi et al. [51–53] revisited the classical Frank–Wolfe method and improved its analysis, which helped to popularize this method in machine learning and computer science community.

Chapter 8
Operator Sliding and Decentralized Optimization

In this chapter, we will further explore the structure properties for solving optimization problems. We will identify potential bottlenecks for solving these problems and develop new techniques that can skip expensive operations from time to time. More specifically, we first consider a class of composite optimization problems whose objective function is given by the summation of a general smooth and nonsmooth component, and present the gradient sliding (GS) algorithm, which can skip the computation of the gradient for the smooth component from time to time. We then discuss an accelerated gradient sliding (AGS) method for minimizing the summation of two smooth convex functions with different Lipschitz constants and show that the AGS method can skip the gradient computation for one of these smooth components without slowing down the overall optimal rate of convergence. The AGS method can further improve the complexity for solving an important class of bilinear saddle point problems. In addition, we present a new class of decentralized first-order methods for nonsmooth and stochastic optimization problems defined over multiagent networks. These methods can skip the inter-node communications while agents solve the primal subproblems iteratively through linearizations of their local objective functions.

8.1 Gradient Sliding for Composite Optimization

In this section, we consider a class of composite convex programming (CP) problems given in the form of

$$\Psi^* \equiv \min_{x \in X} \{\Psi(x) := f(x) + h(x) + \mathscr{X}(x)\}. \tag{8.1.1}$$

Here, $X \subseteq \mathbb{R}^n$ is a closed convex set, $\mathscr{X}$ is a relatively simple convex function, and $f : X \to \mathbb{R}$ and $h : X \to \mathbb{R}$, respectively, are general smooth and nonsmooth convex functions satisfying

$$f(x) \le f(y) + \langle \nabla f(y), x - y\rangle + \tfrac{L}{2}\|x - y\|^2, \forall x, y \in X, \tag{8.1.2}$$

G. Lan, *First-Order and Stochastic Optimization Methods for Machine Learning*, Springer Series in the Data Sciences, https://doi.org/10.1007/978-3-030-39568-1_8

$$h(x) \le h(y) + \langle h'(y), x - y \rangle + M\|x - y\|, \forall x, y \in X, \tag{8.1.3}$$

for some $L > 0$ and $M > 0$, where $h'(x) \in \partial h(x)$. Composite problem of this type appears in many data analysis applications, where either f or h corresponds to a certain data fidelity term, while the other components in Ψ denote regularization terms used to enforce certain structural properties for the obtained solutions.

Throughout this section, we assume that one can access the first-order information of f and h separately. More specifically, in the deterministic setting, we can compute the exact gradient $\nabla f(x)$ and a subgradient $h'(x) \in \partial h(x)$ for any $x \in X$. We also consider the stochastic situation where only a stochastic subgradient of the nonsmooth component h is available. The main goal of this section to provide a better theoretical understanding on how many number of gradient evaluations of ∇f and subgradient evaluations of h' are needed in order to find a certain approximate solution of (8.1.1).

Most existing first-order methods for solving (8.1.1) require the computation of both ∇f and h' in each iteration. In particular, since the objective function Ψ in (8.1.1) is nonsmooth, these algorithms would require $\mathcal{O}(1/\varepsilon^2)$ first-order iterations, and hence $\mathcal{O}(1/\varepsilon^2)$ evaluations for both ∇f and h' to find an ε-solution of (8.1.1), i.e., a point $\bar{x} \in X$ s.t. $\Psi(\bar{x}) - \Psi^* \le \varepsilon$. Much research effort has been directed to reducing the impact of the Lipschitz constant L on the aforementioned complexity bounds for composite optimization. For example, we show in Sect. 4.2 that the number of evaluations for ∇f and h' required to find an ε-solution of (8.1.1) can be bounded by

$$\mathcal{O}\left(\sqrt{\tfrac{L_f}{\varepsilon}} + \tfrac{M^2}{\varepsilon^2}\right). \tag{8.1.4}$$

It is also shown in Sect. 4.2 that similar bounds hold for the stochastic case where only unbiased estimators for ∇f and h' are available. It is observed in Sect. 4.2 that such a complexity bound is not improvable if one can only access the first-order information for the summation of f and h all together.

Note, however, that it is unclear whether the complexity bound in (8.1.4) is optimal if one does have access to the first-order information of f and h separately. In particular, one would expect that the number of evaluations for ∇f can be bounded by $\mathcal{O}(1/\sqrt{\varepsilon})$, if the nonsmooth term h in (8.1.1) does not appear. However, it is unclear whether such a bound still holds for the more general composite problem in (8.1.1) without significantly increasing the bound in (8.1.4) on the number of subgradient evaluations for h'. It should be pointed out that in many applications the bottleneck of first-order methods exists in the computation of ∇f rather than that of h'. To motivate our discussion, let us mention a few such examples.

(a) In many inverse problems, we need to enforce certain block sparsity (e.g., total variation and overlapped group Lasso) by solving the problem of $\min_{x \in \mathbb{R}^n} \|Ax - b\|_2^2 + r(Bx)$. Here $A : \mathbb{R}^n \to \mathbb{R}^m$ is a given linear operator, $b \in \mathbb{R}^m$ denotes the collected observations, $r : \mathbb{R}^p \to \mathbb{R}$ is a relatively simple nonsmooth convex function (e.g., $r = \|\cdot\|_1$), and $B : \mathbb{R}^n \to \mathbb{R}^p$ is a very sparse matrix. In this case, evaluating the gradient of $\|Ax-b\|^2$ requires $\mathcal{O}(mn)$ arithmetic operations, while the computation of $r'(Bx)$ only needs $\mathcal{O}(n+p)$ arithmetic operations.

(b) In many machine learning problems, we need to minimize a regularized loss function given by $\min_{x\in\mathbb{R}^n} \mathbb{E}_\xi[l(x,\xi)] + q(Bx)$. Here $l : \mathbb{R}^n \times \mathbb{R}^d \to \mathbb{R}$ denotes a certain simple loss function, ξ is a random variable with unknown distribution, q is a certain smooth convex function, and $B : \mathbb{R}^n \to \mathbb{R}^p$ is a given linear operator. In this case, the computation of the stochastic subgradient for the loss function $\mathbb{E}_\xi[l(x,\xi)]$ requires only $\mathcal{O}(n+d)$ arithmetic operations, while evaluating the gradient of $q(Bx)$ needs $\mathcal{O}(np)$ arithmetic operations.

(c) In some cases, the computation of ∇f involves a black-box simulation procedure, the solution of an optimization problem, or a partial differential equation, while the computation of h' is given explicitly.

In all these cases mentioned above, it is desirable to reduce the number of gradient evaluations of ∇f to improve the overall efficiency for solving the composite problem (8.1.1).

In this section, we first present a new class of first-order methods, namely the gradient sliding algorithms, and show that the number of gradient evaluations for ∇f required by these algorithms to find an ε-solution of (8.1.1) can be significantly reduced from (8.1.4) to

$$\mathcal{O}\left(\sqrt{\tfrac{L}{\varepsilon}}\right), \tag{8.1.5}$$

while the total number of subgradient evaluations for h' is still bounded by (8.1.4). The basic scheme of these algorithms is to skip the computation of ∇f from time to time so that only $\mathcal{O}(1/\sqrt{\varepsilon})$ gradient evaluations are needed in the $\mathcal{O}(1/\varepsilon^2)$ iterations required to solve (8.1.1). Similar to the conditional gradient sliding method in Sect. 7.2, such an algorithmic framework originated from the simple idea of incorporating an iterative procedure to solve the subproblems in the aforementioned accelerated proximal gradient methods, although the analysis of these gradient sliding algorithms appears to be more technical and involved.

We then consider the stochastic case where the nonsmooth term h is represented by a stochastic oracle (SFO), which, for a given search point $u_t \in X$, outputs a vector $H(u_t,\xi_t)$ such that (s.t.)

$$\mathbb{E}[H(u_t,\xi_t)] = h'(u_t) \in \partial h(u_t), \tag{8.1.6}$$

$$\mathbb{E}[\|H(u_t,\xi_t) - h'(u_t)\|_*^2] \le \sigma^2, \tag{8.1.7}$$

where ξ_t is a random vector independent of the search points u_t. Note that $H(u_t,\xi_t)$ is referred to as a stochastic subgradient of h at u_t and its computation is often much cheaper than the exact subgradient h'. Based on the gradient sliding techniques, we develop a new class of stochastic gradient descent type algorithms and show that the total number gradient evaluations of ∇f required by these algorithms to find a stochastic ε-solution of (8.1.1), i.e., a point $\bar{x} \in X$ s.t. $\mathbb{E}[\Psi(\bar{x}) - \Psi^*] \le \varepsilon$, can still be bounded by (8.1.5), while the total number of stochastic subgradient evaluations can be bounded by

$$\mathcal{O}\left(\sqrt{\tfrac{L}{\varepsilon}} + \tfrac{M^2+\sigma^2}{\varepsilon^2}\right).$$

We also establish large-deviation results associated with these complexity bounds under certain "light-tail" assumptions on the stochastic subgradients returned by the SFO.

Finally, we generalize the gradient sliding algorithms for solving two important classes of composite problems given in the form of (8.1.1), but with f satisfying additional or alternative assumptions. We first assume that f is not only smooth, but also strongly convex, and show that the number of evaluations for ∇f and h' can be significantly reduced from $\mathcal{O}(1/\sqrt{\varepsilon})$ and $\mathcal{O}(1/\varepsilon^2)$, respectively, to $\mathcal{O}(\log(1/\varepsilon))$ and $\mathcal{O}(1/\varepsilon)$. We then consider the case when f is nonsmooth, but can be closely approximated by a class of smooth functions. By incorporating the smoothing scheme in Sect. 3.5 into the gradient sliding algorithms, we show that the number of gradient evaluations can be bounded by $\mathcal{O}(1/\varepsilon)$, while the optimal $\mathcal{O}(1/\varepsilon^2)$ bound on the number of subgradient evaluations of h' is still retained.

8.1.1 Deterministic Gradient Sliding

In this section, we consider the gradient sliding method for solving the deterministic problem in (8.1.1) where exact subgradients of h are available.

Let us provide a brief review on the proximal gradient methods from which the proposed gradient sliding algorithms originate, and point out a few problems associated with these existing algorithms when applied to solve problem (8.1.1).

We start with the simplest proximal gradient method which works for the case when the nonsmooth component h does not appear or is relatively simple (e.g., h is affine). Let $V(x,u)$ be the prox-function associated with the distance generating function ν with modulus 1 (see Sect. 3.2). For a given $x \in X$, let

$$m_\Psi(x,u) := l_f(x,u) + h(u) + \chi(u), \quad \forall u \in X, \tag{8.1.8}$$

where

$$l_f(x;y) := f(x) + \langle \nabla f(x), y - x \rangle. \tag{8.1.9}$$

Clearly, by the convexity of f and (8.1.2), we have

$$m_\Psi(x,u) \le \Psi(u) \le m_\Psi(x,u) + \tfrac{L}{2}\|u-x\|^2 \le m_\Psi(x,u) + LV(x,u)$$

for any $u \in X$, where the last inequality follows from the strong convexity of ν. Hence, $m_\Psi(x,u)$ is a good approximation of $\Psi(u)$ when u is "close" enough to x. In view of this observation, we update the search point $x_k \in X$ at the k-th iteration of the proximal gradient method by

$$x_k = \text{argmin}_{u\in X}\left\{l_f(x_{k-1},u) + h(u) + \chi(u) + \beta_k V(x_{k-1},u)\right\}, \tag{8.1.10}$$

Here, $\beta_k > 0$ is a parameter which determines how well we "trust" the proximity between $m_\Psi(x_{k-1},u)$ and $\Psi(u)$. In particular, a larger value of β_k implies less confidence on $m_\Psi(x_{k-1},u)$ and results in a smaller step moving from x_{k-1} to x_k. It can be shown that the number of iterations required by the proximal gradient method for finding an ε-solution of (8.1.1) can be bounded by $\mathcal{O}(1/\varepsilon)$ (see Sect. 3.1).

The efficiency of the above proximal gradient method can be significantly improved by incorporating a multi-step acceleration scheme. The basic idea of this scheme is to introduce three closely related search sequences, namely $\{\underline{x}_k\}$, $\{x_k\}$, and $\{\bar{x}_k\}$, which will be used to build the model m_Ψ, control the proximity between m_Ψ and Ψ, and compute the output solution, respectively. More specifically, these three sequences are updated according to

$$\underline{x}_k = (1-\gamma_k)\bar{x}_{k-1} + \gamma_k x_{k-1}, \tag{8.1.11}$$

$$x_k = \operatorname{argmin}_{u\in X}\left\{\Phi_k(u) := l_f(\underline{x}_k,u) + h(u) + \chi(u) + \beta_k V(x_{k-1},u)\right\}, \tag{8.1.12}$$

$$\bar{x}_k = (1-\gamma_k)\bar{x}_{k-1} + \gamma_k x_k, \tag{8.1.13}$$

where $\beta_k \geq 0$ and $\gamma_k \in [0,1]$ are given parameters for the algorithm. Clearly, (8.1.11)–(8.1.13) reduce to (8.1.10), if $\bar{x}_0 = x_0$ and γ_k is set to 1. However, by properly specifying β_k and γ_k, e.g., $\beta_k = 2L/k$ and $\gamma_k = 2/(k+2)$, one can show that the above accelerated gradient descent method can find an ε-solution of (8.1.1) in at most $\mathcal{O}(1/\sqrt{\varepsilon})$ iterations (see Sect. 3.3 for the analysis of the scheme in (8.1.11)–(8.1.13)). Since each iteration of this algorithm requires only one evaluation of ∇f, the total number of gradient evaluations of ∇f can also be bounded by $\mathcal{O}(1/\sqrt{\varepsilon})$.

One crucial problem associated with the aforementioned proximal gradient type methods is that the subproblems (8.1.10) and (8.1.12) are difficult to solve when h is a general nonsmooth convex function. To address this issue, one can possibly apply an enhanced accelerated gradient method introduced in Sect. 4.2. This algorithm is obtained by replacing $h(u)$ in (8.1.12) with

$$l_h(\underline{x}_k;u) := h(\underline{x}_k) + \langle h'(\underline{x}_k), u - \underline{x}_k\rangle \tag{8.1.14}$$

for some $h'(\underline{x}_k) \in \partial h(\underline{x}_k)$. As a result, the subproblems in this algorithm become easier to solve. Moreover, with a proper selection of $\{\beta_k\}$ and $\{\gamma_k\}$, this approach can find an ε-solution of (8.1.1) in at most

$$\mathcal{O}\left\{\sqrt{\frac{LV(x_0,x^*)}{\varepsilon}} + \frac{M^2 V(x_0,x^*)}{\varepsilon^2}\right\} \tag{8.1.15}$$

iterations. Since each iteration requires one computation of ∇f and h', the total number of evaluations for f and h' is bounded by $\mathcal{O}(1/\varepsilon^2)$. This bound in (8.1.15) is not improvable if one can only compute the subgradient of the composite function $f(x)+h(x)$ as a whole. However, as mentioned earlier, we do have access to separate first-order information about f and h in many applications. One interesting problem is whether we can further improve the performance of proximal gradient type methods in the latter case.

By presenting the gradient sliding method, we show that one can significantly reduce the number of gradient evaluations for ∇f required to solve (8.1.1), while maintaining the optimal bound on the total number of subgradient evaluations for h'. The basic idea of the GS method is to incorporate an iterative procedure to approximately solve the subproblem (8.1.12) in the accelerated proximal gradient methods. A critical observation in our development of the GS method is that one needs to compute a pair of closely related approximate solutions of problem (8.1.12). One of them will be used in place of x_k in (8.1.11) to construct the model m_Ψ, while the other one will be used in place of x_k in (8.1.13) to compute the output solution $\bar{x}_k$. Moreover, we show that such a pair of approximation solutions can be obtained by applying a simple subgradient projection type subroutine. We now formally describe this algorithm as follows.

Algorithm 8.1 The gradient sliding (GS) algorithm

Input: Initial point $x_0 \in X$ and iteration limit N.
Let $\beta_k \in \mathbb{R}_{++}, \gamma_k \in \mathbb{R}_+$, and $T_k \in \mathcal{N}$, $k = 1,2,\ldots$, be given and set $\bar{x}_0 = x_0$.
for $k = 1,2,\ldots,N$ **do**
1. Set $\underline{x}_k = (1-\gamma_k)\bar{x}_{k-1} + \gamma_k x_{k-1}$, and let $g_k(\cdot) \equiv l_f(\underline{x}_k, \cdot)$ be defined in (8.1.9).
2. Set

$$(x_k, \tilde{x}_k) = \mathrm{PS}(g_k, x_{k-1}, \beta_k, T_k); \tag{8.1.16}$$

3. Set $\bar{x}_k = (1-\gamma_k)\bar{x}_{k-1} + \gamma_k \tilde{x}_k$.
end for
Output: $\bar{x}_N$.

The PS (prox-sliding) procedure called at step 2 is stated as follows.
procedure $(x^+, \tilde{x}^+) = \mathrm{PS}(g, x, \beta, T)$
Let the parameters $p_t \in \mathbb{R}_{++}$ and $\theta_t \in [0,1]$, $t = 1,\ldots$, be given. Set $u_0 = \tilde{u}_0 = x$.
for $t = 1,2,\ldots,T$ **do**

$$u_t = \mathrm{argmin}_{u\in X} \{g(u) + l_h(u_{t-1}, u) + \beta V(x,u) + \beta p_t V(u_{t-1}, u) + \chi(u)\}, \tag{8.1.17}$$

$$\tilde{u}_t = (1-\theta_t)\tilde{u}_{t-1} + \theta_t u_t. \tag{8.1.18}$$

end for
Set $x^+ = u_T$ and $\tilde{x}^+ = \tilde{u}_T$.
end procedure

Observe that when supplied with an affine function $g(\cdot)$, prox-center $x \in X$, parameter β, and sliding period T, the PS procedure computes a pair of approximate solutions $(x^+, \tilde{x}^+) \in X \times X$ for the problem of:

$$\mathrm{argmin}_{u\in X} \{\Phi(u) := g(u) + h(u) + \beta V(x,u) + \chi(u)\}. \tag{8.1.19}$$

Clearly, problem (8.1.19) is equivalent to (8.1.12) when the input parameters are set to (8.1.16). Since the same affine function $g(\cdot) = l_f(\underline{x}_{k-1}, \cdot)$ has been used throughout the T iterations of the PS procedure, we skip the computation of the gradients of f when performing the T projection steps in (8.1.17). This differs from the accelerated gradient method in Sect. 4.2, where one needs to compute $\nabla f + h'$ in each projection step.

It should also be noted that there has been some related work on the accelerated gradient methods with inexact solution of the proximal mapping step (8.1.19). The results basically state that the approximation error at each step has to decrease very fast to maintain the accelerated convergence rate. Since (8.1.19) is strongly convex, one can apply the subgradient method to solve it efficiently. However, one needs to carefully deal with some difficulties in this intuitive approach. First, one has to define an appropriate termination criterion for solving (8.1.19). It turns out that using the natural function optimality gap as the termination criterion for this subproblem could not lead to the desirable convergence rates, and we need to use in the GS algorithm a special termination criterion defined by the summation of the function optimality gap and the distance to the optimal solution (see (8.1.21) below). Second, even though (8.1.19) is strongly convex, it is nonsmooth and the strong convexity modulus decreases as the number of iterations increases. Hence, one has to carefully determine the specification of these nested (accelerated) subgradient algorithms. Third, one important modification that we incorporated in the GS method is to use two different approximate solutions in the two interpolation updates in the accelerated gradient methods. Otherwise, one could not obtain the optimal complexity bounds on the computation of both ∇f and h'.

A few more remarks about the above GS algorithm are in order. First, we say that an outer iteration of the GS algorithm occurs whenever k in Algorithm 8.1 increments by 1. Each outer iteration of the GS algorithm involves the computation of the gradient $\nabla f(\underline{x}_{k-1})$ and a call to the PS procedure to update x_k and $\tilde{x}_k$. Second, the PS procedure solves problem (8.1.19) iteratively. Each iteration of this procedure consists of the computation of subgradient $h'(u_{t-1})$ and the solution of the projection subproblem (8.1.17), which is assumed to be relatively easy to solve (see Sect. 5.1.1). For notational convenience, we refer to an iteration of the PS procedure as an inner iteration of the GS algorithm. Third, the GS algorithm described above is conceptual only since we have not yet specified the selection of $\{\beta_k\}$, $\{\gamma_k\}$, $\{T_k\}$, $\{p_t\}$, and $\{\theta_t\}$. We will return to this issue after establishing some convergence properties of the generic GS algorithm described above.

We first present a result which summarizes some important convergence properties of the PS procedure.

Proposition 8.1. *If* $\{p_t\}$ *and* $\{\theta_t\}$ *in the* PS *procedure satisfy*

$$\theta_t = \tfrac{P_{t-1}-P_t}{(1-P_t)P_{t-1}} \quad \text{with} \quad P_t := \begin{cases} 1, & t=0, \\ p_t(1+p_t)^{-1}P_{t-1}, & t \geq 1, \end{cases} \tag{8.1.20}$$

then, for any $t \geq 1$ and $u \in X$,

$$\beta(1-P_t)^{-1}V(u_t,u) + [\Phi(\tilde{u}_t) - \Phi(u)] \leq P_t(1-P_t)^{-1}\left[\beta V(u_0,u) + \tfrac{M^2}{2\beta}\textstyle\sum_{i=1}^t (p_i^2 P_{i-1})^{-1}\right], \quad (8.1.21)$$

where Φ is defined in (8.1.19).

Proof. By (8.1.3) and the definition of l_h in (8.1.14), we have $h(u_t) \leq l_h(u_{t-1},u_t) + M\|u_t - u_{t-1}\|$. Adding $g(u_t) + \beta V(x,u_t) + \chi(u_t)$ to both sides of this inequality and using the definition of Φ in (8.1.19), we obtain

$$\Phi(u_t) \leq g(u_t) + l_h(u_{t-1},u_t) + \beta V(x,u_t) + \chi(u_t) + M\|u_t - u_{t-1}\|. \quad (8.1.22)$$

Now applying Lemma 3.5 to (8.1.17), we obtain

$$\begin{aligned}
&g(u_t) + l_h(u_{t-1},u_t) + \beta V(x,u_t) + \chi(u_t) + \beta p_t V(u_{t-1},u_t)\\
&\leq g(u) + l_h(u_{t-1},u) + \beta V(x,u) + \chi(u) + \beta p_t V(u_{t-1},u) - \beta(1+p_t)V(u_t,u)\\
&\leq g(u) + h(u) + \beta V(x,u) + \chi(u) + \beta p_t V(u_{t-1},u) - \beta(1+p_t)V(u_t,u)\\
&= \Phi(u) + \beta p_t V(u_{t-1},u) - \beta(1+p_t)V(u_t,u),
\end{aligned}$$

where the second inequality follows from the convexity of h. Moreover, by the strong convexity of ν,

$$-\beta p_t V(u_{t-1},u_t) + M\|u_t - u_{t-1}\| \leq -\tfrac{\beta p_t}{2}\|u_t - u_{t-1}\|^2 + M\|u_t - u_{t-1}\| \leq \tfrac{M^2}{2\beta p_t},$$

where the last inequality follows from the simple fact that $-at^2/2 + bt \leq b^2/(2a)$ for any $a > 0$. Combining the previous three inequalities, we conclude that

$$\Phi(u_t) - \Phi(u) \leq \beta p_t V(u_{t-1},u) - \beta(1+p_t)V(u_t,u) + \tfrac{M^2}{2\beta p_t}.$$

Dividing both sides by $1 + p_t$ and rearranging the terms, we obtain

$$\beta V(u_t,u) + \tfrac{\Phi(u_t) - \Phi(u)}{1+p_t} \leq \tfrac{\beta p_t}{1+p_t}V(u_{t-1},u) + \tfrac{M^2}{2\beta(1+p_t)p_t},$$

which, in view of the definition of P_t in (8.1.20) and Lemma 3.17 (with $k = t$, $w_k = 1/(1+p_t)$ and $W_k = P_t$), then implies that

$$\begin{aligned}
\tfrac{\beta}{P_t}V(u_t,u) + \textstyle\sum_{i=1}^t \tfrac{\Phi(u_i) - \Phi(u)}{P_i(1+p_i)} &\leq \beta V(u_0,u) + \tfrac{M^2}{2\beta}\textstyle\sum_{i=1}^t \tfrac{1}{P_i(1+p_i)p_i}\\
&= \beta V(u_0,u) + \tfrac{M^2}{2\beta}\textstyle\sum_{i=1}^t (p_i^2 P_{i-1})^{-1}, \quad (8.1.23)
\end{aligned}$$

where the last identity also follows from the definition of P_t in (8.1.20). Also note that by the definition of $\tilde{u}_t$ in the PS procedure and (8.1.20), we have

$$\tilde{u}_t = \tfrac{P_t}{1-P_t}\left(\tfrac{1-P_{t-1}}{P_{t-1}}\tilde{u}_{t-1} + \tfrac{1}{P_t(1+p_t)}u_t\right).$$

Applying this relation inductively and using the fact that $P_0 = 1$, we can easily see that

$$\begin{aligned}\tilde{u}_t &= \tfrac{P_t}{1-P_t}\left[\tfrac{1-P_{t-2}}{P_{t-2}}\tilde{u}_{t-2} + \tfrac{1}{P_{t-1}(1+p_{t-1})}u_{t-1} + \tfrac{1}{P_t(1+p_t)}u_t\right]\\ &= \ldots = \tfrac{P_t}{1-P_t}\textstyle\sum_{i=1}^{t}\tfrac{1}{P_i(1+p_i)}u_i,\end{aligned}$$

which, in view of the convexity of Φ, then implies that

$$\Phi(\tilde{u}_t) - \Phi(u) \le \tfrac{P_t}{1-P_t}\textstyle\sum_{i=1}^{t}\tfrac{\Phi(u_i)-\Phi(u)}{P_i(1+p_i)}. \tag{8.1.24}$$

Combining the above inequality with (8.1.23) and rearranging the terms, we obtain (8.1.21). ■

Setting u to be the optimal solution of (8.1.19), we can see that both x_k and $\tilde{x}_k$ are approximate solutions of (8.1.19) if the right-hand side (RHS) of (8.1.21) is small enough. With the help of this result, we can establish an important recursion from which the convergence of the GS algorithm easily follows.

Proposition 8.2. *Suppose that $\{p_t\}$ and $\{\theta_t\}$ in the* PS *procedure satisfy (8.1.20). Also assume that $\{\beta_k\}$ and $\{\gamma_k\}$ in the GS algorithm satisfy*

$$\gamma_1 = 1 \quad and \quad \beta_k - L\gamma_k \ge 0, \;\; k \ge 1. \tag{8.1.25}$$

Then for any $u \in X$ and $k \ge 1$,

$$\begin{aligned}\Psi(\bar{x}_k) - \Psi(u) \le &(1-\gamma_k)[\Psi(\bar{x}_{k-1}) - \Psi(u)] + \gamma_k(1-P_{T_k})^{-1}\\ &\left[\beta_k V(x_{k-1},u) - \beta_k V(x_k,u) + \tfrac{M^2 P_{T_k}}{2\beta_k}\textstyle\sum_{i=1}^{T_k}(p_i^2 P_{i-1})^{-1}\right].\end{aligned} \tag{8.1.26}$$

Proof. First, notice that by the definition of $\bar{x}_k$ and $\underline{x}_k$, we have $\bar{x}_k - \underline{x}_k = \gamma_k(\tilde{x}_k - x_{k-1})$. Using this observation, (8.1.2), the definition of l_f in (8.1.9), and the convexity of f, we obtain

$$\begin{aligned}f(\bar{x}_k) &\le l_f(\underline{x}_k,\bar{x}_k) + \tfrac{L}{2}\|\bar{x}_k - \underline{x}_k\|^2\\ &= (1-\gamma_k)l_f(\underline{x}_k,\bar{x}_{k-1}) + \gamma_k l_f(\underline{x}_k,\tilde{x}_k) + \tfrac{L\gamma_k^2}{2}\|\tilde{x}_k - x_{k-1}\|^2\\ &\le (1-\gamma_k)f(\bar{x}_{k-1}) + \gamma_k\left[l_f(\underline{x}_k,\tilde{x}_k) + \beta_k V(x_{k-1},\tilde{x}_k)\right]\\ &\quad - \gamma_k\beta_k V(x_{k-1},\tilde{x}_k) + \tfrac{L\gamma_k^2}{2}\|\tilde{x}_k - x_{k-1}\|^2\\ &\le (1-\gamma_k)f(\bar{x}_{k-1}) + \gamma_k\left[l_f(\underline{x}_k,\tilde{x}_k) + \beta_k V(x_{k-1},\tilde{x}_k)\right]\\ &\quad - \left(\gamma_k\beta_k - L\gamma_k^2\right)V(x_{k-1},\tilde{x}_k)\\ &\le (1-\gamma_k)f(\bar{x}_{k-1}) + \gamma_k\left[l_f(\underline{x}_k,\tilde{x}_k) + \beta_k V(x_{k-1},\tilde{x}_k)\right],\end{aligned} \tag{8.1.27}$$

where the third inequality follows from the strong convexity of v and the last inequality follows from (8.1.25). By the convexity of h and χ, we have

$$h(\bar{x}_k)+\chi(\bar{x}_k) \le (1-\gamma_k)[h(\bar{x}_{k-1})+\chi(\bar{x}_{k-1})]+\gamma_k[h(\tilde{x}_k)+\chi(\tilde{x}_k)]. \tag{8.1.28}$$

Adding up the previous two inequalities, and using the definitions of Ψ in (8.1.1) and Φ_k in (8.1.12), we have

$$\Psi(\bar{x}_k) \le (1-\gamma_k)\Psi(\bar{x}_{k-1})+\gamma_k\Phi_k(\tilde{x}_k).$$

Subtracting $\Psi(u)$ from both sides of the above inequality, we obtain

$$\Psi(\bar{x}_k)-\Psi(u) \le (1-\gamma_k)[\Psi(\bar{x}_{k-1})-\Psi(u)]+\gamma_k[\Phi_k(\tilde{x}_k)-\Psi(u)]. \tag{8.1.29}$$

Also note that by the definition of Φ_k in (8.1.12) and the convexity of f,

$$\Phi_k(u) \le f(u)+h(u)+\chi(u)+\beta_k V(x_{k-1},u) = \Psi(u)+\beta_k V(x_{k-1},u), \ \forall u \in X. \tag{8.1.30}$$

Combining these two inequalities (i.e., replacing the third $\Psi(u)$ in (8.1.29) by $\phi_k(u)-\beta_k V(x_{k-1},u)$), we obtain

$$\begin{aligned}\Psi(\bar{x}_k)-\Psi(u) \le (1-\gamma_k)&[\Psi(\bar{x}_{k-1})-\Psi(u)]\\ &+\gamma_k[\Phi_k(\tilde{x}_k)-\Phi_k(u)+\beta_k V(x_{k-1},u)].\end{aligned} \tag{8.1.31}$$

Now, in view of the definition of Φ_k in (8.1.12) and the origin of $(x_k,\tilde{x}_k)$ in (8.1.16), we can apply Proposition 1 with $\phi=\phi_k$, $u_0=x_{k-1}$, $u_t=x_k$, $\tilde{u}_t=\tilde{x}_k$, and $\beta=\beta_k$, and conclude that for any $u \in X$ and $k \ge 1$,

$$\begin{aligned}&\tfrac{\beta_k}{1-P_{T_k}}V(x_k,u)+[\Phi_k(\tilde{x}_k)-\Phi_k(u)] \le\\ &\qquad \tfrac{P_{T_k}}{1-P_{T_k}}\left[\beta_k V(x_{k-1},u)+\tfrac{M^2}{2\beta_k}\textstyle\sum_{i=1}^{T_k}(p_i^2 P_{i-1})^{-1}\right].\end{aligned}$$

Plugging the above bound on $\Phi_k(\tilde{x}_k)-\Phi_k(u)$ into (8.1.31), we obtain (8.1.26). ■

We are now ready to establish the main convergence properties of the GS algorithm. Note that the following quantity will be used in our analysis of this algorithm.

$$\Gamma_k = \begin{cases} 1, & k=1,\\ (1-\gamma_k)\Gamma_{k-1}, & k \ge 2. \end{cases} \tag{8.1.32}$$

Theorem 8.1. *Assume that $\{p_t\}$ and $\{\theta_t\}$ in the* PS *procedure satisfy (8.1.20), and also that $\{\beta_k\}$ and $\{\gamma_k\}$ in the GS algorithm satisfy (8.1.25).*

(a) If for any $k \ge 2$,

$$\tfrac{\gamma_k\beta_k}{\Gamma_k(1-P_{T_k})} \le \tfrac{\gamma_{k-1}\beta_{k-1}}{\Gamma_{k-1}(1-P_{T_{k-1}})}, \tag{8.1.33}$$

then we have, for any $N \ge 1$,

$$\begin{aligned}\Psi(\bar{x}_N)-\Psi(x^*) \le \mathcal{B}_d(N) := &\tfrac{\Gamma_N\beta_1}{1-P_{T_1}}V(x_0,x^*)\\ &+\tfrac{M^2\Gamma_N}{2}\textstyle\sum_{k=1}^{N}\sum_{i=1}^{T_k}\tfrac{\gamma_k P_{T_k}}{\Gamma_k\beta_k(1-P_{T_k})p_i^2 P_{i-1}},\end{aligned} \tag{8.1.34}$$

where $x^* \in X$ is an arbitrary optimal solution of problem (8.1.1), and P_t and Γ_k are defined in (8.1.20) and (8.1.32), respectively.

(b) *If X is compact, and for any $k \ge 2$,*

$$\frac{\gamma_k \beta_k}{\Gamma_k(1-P_{T_k})} \ge \frac{\gamma_{k-1}\beta_{k-1}}{\Gamma_{k-1}(1-P_{T_{k-1}})}, \tag{8.1.35}$$

then (8.1.34) still holds by simply replacing the first term in the definition of $\mathscr{B}_d(N)$ with $\gamma_N \beta_N \bar{V}(x^*)/(1-P_{T_N})$, where $\bar{V}(u) = \max_{x\in X} V(x,u)$.

Proof. We conclude from (8.1.26) and Lemma 3.17 that

$$\begin{aligned}
\Psi(\bar{x}_N) - \Psi(u) &\le \Gamma_N \tfrac{1-\gamma_1}{\Gamma_1}[\Psi(\bar{x}_0) - \Psi(u)] \\
&\quad + \Gamma_N \textstyle\sum_{k=1}^N \tfrac{\beta_k \gamma_k}{\Gamma_k(1-P_{T_k})}[V(x_{k-1},u) - V(x_k,u)] \\
&\quad + \tfrac{M^2 \Gamma_N}{2} \textstyle\sum_{k=1}^N \sum_{i=1}^{T_k} \tfrac{\gamma_k P_{T_k}}{\Gamma_k \beta_k (1-P_{T_k}) p_i^2 P_{i-1}} \\
&= \Gamma_N \textstyle\sum_{k=1}^N \tfrac{\beta_k \gamma_k}{\Gamma_k(1-P_{T_k})}[V(x_{k-1},u) - V(x_k,u)] \\
&\quad + \tfrac{M^2 \Gamma_N}{2} \textstyle\sum_{k=1}^N \sum_{i=1}^{T_k} \tfrac{\gamma_k P_{T_k}}{\Gamma_k \beta_k (1-P_{T_k}) p_i^2 P_{i-1}},
\end{aligned} \tag{8.1.36}$$

where the last identity follows from the fact that $\gamma_1 = 1$. Now it follows from (8.1.33) that

$$\begin{aligned}
&\textstyle\sum_{k=1}^N \tfrac{\beta_k \gamma_k}{\Gamma_k(1-P_{T_k})}[V(x_{k-1},u) - V(x_k,u)] \\
&\quad \le \tfrac{\beta_1 \gamma_1}{\Gamma_1(1-P_{T_1})} V(x_0,u) - \tfrac{\beta_N \gamma_N}{\Gamma_N(1-P_{T_N})} V(x_N,u) \le \tfrac{\beta_1}{1-P_{T_1}} V(x_0,u),
\end{aligned} \tag{8.1.37}$$

where the last inequality follows from the facts that $\gamma_1 = \Gamma_1 = 1$, $P_{T_N} \le 1$, and $V(x_N,u) \ge 0$. The result in part (a) then clearly follows from the previous two inequalities with $u = x^*$. Moreover, using (8.1.35) and the fact $V(x_k,u) \le \bar{V}(u)$, we conclude that

$$\begin{aligned}
&\textstyle\sum_{k=1}^N \tfrac{\beta_k \gamma_k}{\Gamma_k(1-P_{T_k})}[V(x_{k-1},u) - V(x_k,u)] \\
&\quad \le \tfrac{\beta_1}{1-P_{T_1}} \bar{V}(u) - \textstyle\sum_{k=2}^N \left[\tfrac{\beta_{k-1}\gamma_{k-1}}{\Gamma_{k-1}(1-P_{T_{k-1}})} - \tfrac{\beta_k \gamma_k}{\Gamma_k(1-P_{T_k})}\right] \bar{V}(u) \\
&\quad = \tfrac{\gamma_N \beta_N}{\Gamma_N(1-P_{T_N})} \bar{V}(u).
\end{aligned} \tag{8.1.38}$$

Part (b) then follows from the above observation and (8.1.36) with $u = x^*$. ∎

Clearly, there are various options for specifying the parameters $\{p_t\}$, $\{\theta_t\}$, $\{\beta_k\}$, $\{\gamma_k\}$, and $\{T_k\}$ to guarantee the convergence of the GS algorithm. Below we provide a few such selections which lead to the best possible rate of convergence for solving problem (8.1.1). In particular, Corollary 8.1(a) provides a set of such parameters for the case when the feasible region X is unbounded and the iteration limit N is given a priori, while the one in Corollary 8.1(b) works only for the case when X is compact, but does not require N to be given in advance.

Corollary 8.1. *Assume that $\{p_t\}$ and $\{\theta_t\}$ in the* PS *procedure are set to*

$$p_t = \tfrac{t}{2} \quad and \quad \theta_t = \tfrac{2(t+1)}{t(t+3)}, \ \forall t \geq 1. \tag{8.1.39}$$

(a) If N is fixed a priori, and $\{\beta_k\}$, $\{\gamma_k\}$, and $\{T_k\}$ are set to

$$\beta_k = \tfrac{2L}{\nu k}, \ \gamma_k = \tfrac{2}{k+1}, \ and \ T_k = \left\lceil \tfrac{M^2 N k^2}{\tilde{D} L^2} \right\rceil \tag{8.1.40}$$

for some $\tilde{D} > 0$, then

$$\Psi(\bar{x}_N) - \Psi(x^*) \leq \tfrac{2L}{N(N+1)} \left[3V(x_0, x^*) + 2\tilde{D}\right], \ \forall N \geq 1. \tag{8.1.41}$$

(b) If X is compact, and $\{\beta_k\}$, $\{\gamma_k\}$, and $\{T_k\}$ are set to

$$\beta_k = \frac{9L(1-P_{T_k})}{2(k+1)}, \ \gamma_k = \tfrac{3}{k+2}, \ and \ T_k = \left\lceil \frac{M^2(k+1)^3}{\tilde{D} L^2} \right\rceil, \tag{8.1.42}$$

for some $\tilde{D} > 0$, then

$$\Psi(\bar{x}_N) - \Psi(x^*) \leq \tfrac{L}{(N+1)(N+2)} \left(\tfrac{27\bar{V}(x^*)}{2} + \tfrac{8\tilde{D}}{3} \right), \ \forall N \geq 1. \tag{8.1.43}$$

Proof. We first show part (a). By the definitions of P_t and p_t in (8.1.20) and (8.1.39), we have

$$P_t = \tfrac{tP_{t-1}}{t+2} = \ldots = \tfrac{2}{(t+1)(t+2)}. \tag{8.1.44}$$

Using the above identity and (8.1.39), we can easily see that the condition in (8.1.20) holds. It also follows from (8.1.44) and the definition of T_k in (8.1.40) that

$$P_{T_k} \leq P_{T_{k-1}} \leq \ldots \leq P_{T_1} \leq \tfrac{1}{3}. \tag{8.1.45}$$

Now, it can be easily seen from the definition of β_k and γ_k in (8.1.40) that (8.1.25) holds. It also follows from (8.1.32) and (8.1.40) that

$$\Gamma_k = \tfrac{2}{k(k+1)}. \tag{8.1.46}$$

By (8.1.40), (8.1.45), and (8.1.46), we have

$$\tfrac{\gamma_k \beta_k}{\Gamma_k(1-P_{T_k})} = \tfrac{2L}{1-P_{T_k}} \leq \tfrac{2L}{1-P_{T_{k-1}}} = \tfrac{\gamma_{k-1}\beta_{k-1}}{\Gamma_{k-1}(1-P_{T_{k-1}})},$$

from which (8.1.33) follows. Now, by (8.1.44) and the fact that $p_t = t/2$, we have

$$\textstyle\sum_{i=1}^{T_k} \tfrac{1}{p_i^2 P_{i-1}} = 2\sum_{i=1}^{T_k} \tfrac{i+1}{i} \leq 4T_k, \tag{8.1.47}$$

which, together with (8.1.40) and (8.1.46), then imply that

$$\sum_{i=1}^{T_k} \frac{\gamma_k P_{T_k}}{\Gamma_k \beta_k (1-P_{T_k}) p_i^2 P_{i-1}} \leq \frac{4\gamma_k P_{T_k} T_k}{\Gamma_k \beta_k (1-P_{T_k})} = \frac{4k^2}{L(T_k+3)}. \tag{8.1.48}$$

Using this observation, (8.1.34), (8.1.45), and (8.1.46), we have

$$\begin{aligned}\mathscr{B}_d(N) &\leq \frac{4LV(x_0,x^*)}{N(N+1)(1-P_{T_1})} + \frac{4M^2}{LN(N+1)} \sum_{k=1}^{N} \frac{k^2}{T_k+3} \\ &\leq \frac{6LV(x_0,x^*)}{N(N+1)} + \frac{4M^2}{LN(N+1)} \sum_{k=1}^{N} \frac{k^2}{T_k+3},\end{aligned}$$

which, in view of Theorem 8.1(a) and the definition of T_k in (8.1.40), then clearly implies (8.1.41).

Now let us show that part (b) holds. It follows from (8.1.45), and the definition of β_k and γ_k in (8.1.42) that

$$\beta_k \geq \frac{3L}{k+1} \geq L\gamma_k \tag{8.1.49}$$

and hence that (8.1.25) holds. It also follows from (8.1.32) and (8.1.42) that

$$\Gamma_k = \frac{6}{k(k+1)(k+2)}, \quad k \geq 1, \tag{8.1.50}$$

and hence that

$$\frac{\gamma_k \beta_k}{\Gamma_k (1-P_{T_k})} = \frac{k(k+1)}{2} \frac{9L}{2(k+1)} = \frac{9Lk}{4}, \tag{8.1.51}$$

which implies that (8.1.35) holds. Using (8.1.42), (8.1.45), (8.1.47), and (8.1.49), we have

$$\begin{aligned}\sum_{i=1}^{T_k} \frac{\gamma_k P_{T_k}}{\Gamma_k \beta_k (1-P_{T_k}) p_i^2 P_{i-1}} &\leq \frac{4\gamma_k P_{T_k} T_k}{\Gamma_k \beta_k (1-P_{T_k})} = \frac{4k(k+1)^2 P_{T_k} T_k}{9L(1-P_{T_k})^2} \\ &= \frac{8k(k+1)^2 (T_k+1)(T_k+2)}{9LT_k (T_k+3)^2} \leq \frac{8k(k+1)^2}{9LT_k}.\end{aligned} \tag{8.1.52}$$

Using this observation, (8.1.42), (8.1.50), and Theorem 8.1(b), we conclude that

$$\begin{aligned}\Psi(\bar{x}_N) - \Psi(x^*) &\leq \frac{\gamma_N \beta_N \bar{V}(x^*)}{(1-P_{T_N})} + \frac{M^2 \Gamma_N}{2} \sum_{k=1}^{N} \frac{8k(k+1)^2}{9LT_k} \\ &\leq \frac{\gamma_N \beta_N \bar{V}(x^*)}{(1-P_{T_N})} + \frac{8L\tilde{D}}{3(N+1)(N+2)} \\ &\leq \frac{L}{(N+1)(N+2)} \left(\frac{27\bar{V}(x^*)}{2} + \frac{8\tilde{D}}{3} \right).\end{aligned}$$

■

Observe that by (8.1.18) and (8.1.44), when the selection of $p_t = t/2$, the definition of $\tilde{u}_t$ in the PS procedure can be simplified as

$$\tilde{u}_t = \frac{(t+2)(t-1)}{t(t+3)} \tilde{u}_{t-1} + \frac{2(t+1)}{t(t+3)} u_t.$$

In view of Corollary 8.1, we can establish the complexity of the GS algorithm for finding an ε-solution of problem (8.1.1).

Corollary 8.2. *Suppose that $\{p_t\}$ and $\{\theta_t\}$ are set to (8.1.39). Also assume that there exists an estimate $D_X > 0$ s.t.*

$$V(x,y) \le D_X^2, \ \ \forall x,y \in X. \tag{8.1.53}$$

If $\{\beta_k\}$, $\{\gamma_k\}$, and $\{T_k\}$ are set to (8.1.40) with $\tilde{D} = 3D_X^2/2$ for some $N > 0$, then the total number of evaluations for ∇f and h' can be bounded by

$$\mathcal{O}\left(\sqrt{\tfrac{LD_X^2}{\varepsilon}}\right) \tag{8.1.54}$$

and

$$\mathcal{O}\left\{\tfrac{M^2D_X^2}{\varepsilon^2} + \sqrt{\tfrac{LD_X^2}{\varepsilon}}\right\}, \tag{8.1.55}$$

respectively. Moreover, the above two complexity bounds also hold if X is bounded, and $\{\beta_k\}$, $\{\gamma_k\}$, and $\{T_k\}$ are set to (8.1.42) with $\tilde{D} = 81D_X^2/16$.

Proof. In view of Corollary 8.1(a), if $\{\beta_k\}$, $\{\gamma_k\}$, and $\{T_k\}$ are set to (8.1.40), the total number of outer iterations (or gradient evaluations) performed by the GS algorithm to find an ε-solution of (8.1.1) can be bounded by

$$N \le \sqrt{\tfrac{L}{\varepsilon}\left[3V(x_0,x^*) + 2\tilde{D}\right]} \le \sqrt{\tfrac{6LD_X^2}{\varepsilon}}. \tag{8.1.56}$$

Moreover, using the definition of T_k in (8.1.40), we conclude that the total number of inner iterations (or subgradient evaluations) can be bounded by

$$\textstyle\sum_{k=1}^N T_k \le \sum_{k=1}^N \left(\tfrac{M^2Nk^2}{\tilde{D}L^2} + 1\right) \le \tfrac{M^2N(N+1)^3}{3\tilde{D}L^2} + N = \tfrac{2M^2N(N+1)^3}{9D_X^2L^2} + N,$$

which, in view of (8.1.56), then clearly implies the bound in (8.1.55). Using Corollary 8.1(b) and similar arguments, we can show that the complexity bounds (8.1.54) and (8.1.55) also hold when X is bounded, and $\{\beta_k\}$, $\{\gamma_k\}$, and $\{T_k\}$ are set to (8.1.42) . ■

In view of Corollary 8.2, the GS algorithm can achieve the optimal complexity bound for solving problem (8.1.1) in terms of the number of evaluations for both ∇f and h'.

It is also worth noting that we can relax the requirement on D_X in (8.1.53) to $V(x_0,x^*) \le D_X^2$ or $\max_{x\in X} V(x,x^*) \le D_X^2$, respectively, when the stepsize policies in (8.1.40) or in (8.1.42) are used. Accordingly, we can tighten the complexity bounds in (8.1.54) and (8.1.55) by a constant factor.

8.1.2 Stochastic Gradient Sliding

We now consider the situation when the computation of stochastic subgradients of h is much easier than that of exact subgradients. This situation happens, for example, when h is given in the form of an expectation or as the summation of many nonsmooth components. By presenting a stochastic gradient sliding (SGS) method, we show that similar complexity bounds as in Sect. 8.1.1 for solving problem (8.1.1) can still be obtained in expectation or with high probability, but the iteration cost of the SGS method can be substantially smaller than that of the GS method.

More specifically, we assume that the nonsmooth component h is represented by a stochastic oracle (SFO) satisfying (8.1.6) and (8.1.7). Sometimes, we augment (8.1.7) by a "light-tail" assumption:

$$\mathbb{E}[\exp(\|H(u,\xi)-h'(u)\|_*^2/\sigma^2)] \le \exp(1). \tag{8.1.57}$$

It can be easily seen that (8.1.57) implies (8.1.7) by Jensen's inequality.

The stochastic gradient sliding (SGS) algorithm is obtained by simply replacing the exact subgradients in the PS procedure with the stochastic subgradients returned by the SFO. This algorithm is formally described as follows.

Algorithm 8.2 The stochastic gradient sliding (SGS) algorithm

The algorithm is the same as GS except that the identity (8.1.17) in the PS procedure is replaced by

$$u_t = \text{argmin}_{u\in X}\left\{g(u)+\langle H(u_{t-1},\xi_{t-1}),u\rangle+\beta V(x,u)+\beta p_t V(u_{t-1},u)+\chi(u)\right\}. \tag{8.1.58}$$

The above modified PS procedure is called the SPS (stochastic PS) procedure.

We add a few remarks about the above SGS algorithm. First, in this algorithm, we assume that the exact gradient of f will be used throughout the T_k inner iterations. This is different from the stochastic accelerated gradient method in Sect. 4.2, where one needs to compute ∇f at each subgradient projection step. Second, let us denote

$$\tilde{l}_h(u_{t-1},u) := h(u_{t-1})+\langle H(u_{t-1},\xi_{t-1}),u-u_{t-1}\rangle. \tag{8.1.59}$$

It can be easily seen that (8.1.58) is equivalent to

$$u_t = \text{argmin}_{u\in X}\left\{g(u)+\tilde{l}_h(u_{t-1},u)+\beta V(x,u)+\beta p_t V(u_{t-1},u)+\chi(u)\right\}. \tag{8.1.60}$$

This problem reduces to (8.1.17) if there is no stochastic noise associated with the SFO, i.e., $\sigma=0$ in (8.1.7). Third, note that we have not provided the specification of $\{\beta_k\}$, $\{\gamma_k\}$, $\{T_k\}$, $\{p_t\}$, and $\{\theta_t\}$ in the SGS algorithm. Similar to Sect. 8.1.1, we will return to this issue after establishing some convergence properties about the generic SPS procedure and SGS algorithm.

The following result describes some important convergence properties of the SPS procedure.

Proposition 8.3. *Assume that $\{p_t\}$ and $\{\theta_t\}$ in the* SPS *procedure satisfy (8.1.20). Then for any $t \geq 1$ and $u \in X$,*

$$\beta(1-P_t)^{-1}V(u_t,u) + [\Phi(\tilde{u}_t) - \Phi(u)] \leq \beta P_t(1-P_t)^{-1}V(u_{t-1},u) + P_t(1-P_t)^{-1}\Sigma_{i=1}^{t}(p_iP_{i-1})^{-1}\left[\tfrac{(M+\|\delta_i\|_*)^2}{2\beta p_i} + \langle\delta_i, u - u_{i-1}\rangle\right], \tag{8.1.61}$$

where Φ is defined in (8.1.19),

$$\delta_t := H(u_{t-1},\xi_{t-1}) - h'(u_{t-1}), \;\; \textit{and} \;\; h'(u_{t-1}) = \mathbb{E}[H(u_{t-1},\xi_{t-1})]. \tag{8.1.62}$$

Proof. Let $\tilde{l}_h(u_{t-1},u)$ be defined in (8.1.59). Clearly, we have $\tilde{l}_h(u_{t-1},u) - l_h(u_{t-1},u) = \langle\delta_t, u - u_{t-1}\rangle$. Using this observation and (8.1.22), we obtain

$$\begin{aligned}
\Phi(u_t) &\leq g(u) + l_h(u_{t-1},u_t) + \beta V(x,u_t) + \chi(u_t) + M\|u_t - u_{t-1}\| \\
&= g(u) + \tilde{l}_h(u_{t-1},u_t) - \langle\delta_t, u_t - u_{t-1}\rangle + \beta V(x,u_t) + \chi(u_t) + M\|u_t - u_{t-1}\| \\
&\leq g(u) + \tilde{l}_h(u_{t-1},u_t) + \beta V(x,u_t) + \chi(u_t) + (M + \|\delta_t\|_*)\|u_t - u_{t-1}\|,
\end{aligned}$$

where the last inequality follows from the Cauchy–Schwarz inequality. Now applying Lemma 3.5 to (8.1.58), we obtain

$$\begin{aligned}
& g(u_t) + \tilde{l}_h(u_{t-1},u_t) + \beta V(x,u_t) + \beta p_t V(u_{t-1},u_t) + \chi(u_t) \\
&\leq g(u) + \tilde{l}_h(u_{t-1},u) + \beta V(x,u) + \beta p_t V(u_{t-1},u) + \chi(u) - \beta(1+p_t)V(u_t,u) \\
&= g(u) + l_h(u_{t-1},u) + \langle\delta_t, u - u_{t-1}\rangle \\
&\qquad + \beta V(x,u) + \beta p_t V(u_{t-1},u) + \chi(u) - \beta(1+p_t)V(u_t,u) \\
&\leq \Phi(u) + \beta p_t V(u_{t-1},u) - \beta(1+p_t)V(u_t,u) + \langle\delta_t, u - u_{t-1}\rangle,
\end{aligned}$$

where the last inequality follows from the convexity of h and (8.1.19). Moreover, by the strong convexity of ν,

$$\begin{aligned}
& -\beta p_t V(u_{t-1},u_t) + (M + \|\delta_t\|_*)\|u_t - u_{t-1}\| \\
&\qquad \leq -\tfrac{\beta p_t}{2}\|u_t - u_{t-1}\|^2 + (M + \|\delta_t\|_*)\|u_t - u_{t-1}\| \leq \tfrac{(M+\|\delta_t\|_*)^2}{2\beta p_t},
\end{aligned}$$

where the last inequality follows from the simple fact that $-at^2/2 + bt \leq b^2/(2a)$ for any $a > 0$. Combining the previous three inequalities, we conclude that

$$\Phi(u_t) - \Phi(u) \leq \beta p_t V(u_{t-1},u) - \beta(1+p_t)V(u_t,u) + \tfrac{(M+\|\delta_t\|_*)^2}{2\beta p_t} + \langle\delta_t, u - u_{t-1}\rangle.$$

Now dividing both sides of the above inequality by $1+p_t$ and rearranging the terms, we obtain

$$\beta V(u_t,u)+\tfrac{\Phi(u_t)-\Phi(u)}{1+p_t}\le\tfrac{\beta p_t}{1+p_t}V(u_{t-1},u)+\tfrac{(M+\|\delta_t\|_*)^2}{2\beta(1+p_t)p_t}+\tfrac{\langle\delta_t,u-u_{t-1}\rangle}{1+p_t},$$

which, in view of Lemma 3.17, then implies that

$$\begin{aligned}\tfrac{\beta}{P_t}V(u_t,u)+\textstyle\sum_{i=1}^t\tfrac{\Phi(u_i)-\Phi(u)}{P_i(1+p_i)}&\le\beta V(u_0,u)\\&+\textstyle\sum_{i=1}^t\left[\tfrac{(M+\|\delta_i\|_*)^2}{2\beta P_i(1+p_i)p_i}+\tfrac{\langle\delta_i,u-u_{i-1}\rangle}{P_i(1+p_i)}\right].\end{aligned}\tag{8.1.63}$$

The result then immediately follows from the above inequality and (8.1.24). ■

It should be noted that the search points $\{u_t\}$ generated by different calls to the SPS procedure in different outer iterations of the SGS algorithm are distinct from each other. To avoid ambiguity, we use $u_{k,t}$, $k\ge1$, and $t\ge0$ to denote the search points generated by the SPS procedure in the k-th outer iteration. Accordingly, we use

$$\delta_{k,t-1}:=H(u_{k,t-1},\xi_{t-1})-h'(u_{k,t-1}),\ \ k\ge1,t\ge1,\tag{8.1.64}$$

to denote the stochastic noises associated with the SFO. Then, by (8.1.61), the definition of Φ_k in (8.1.12), and the origin of $(x_k,\tilde{x}_k)$ in the SGS algorithm, we have

$$\begin{aligned}&\beta_k(1-P_{T_k})^{-1}V(x_k,u)+[\Phi_k(\tilde{x}_k)-\Phi_k(u)]\le\beta_kP_{T_k}(1-P_{T_k})^{-1}V(x_{k-1},u)+\\&\quad P_{T_k}(1-P_{T_k})^{-1}\textstyle\sum_{i=1}^{T_k}\tfrac{1}{p_iP_{i-1}}\left[\tfrac{(M+\|\delta_{k,i-1}\|_*)^2}{2\beta_kp_i}+\langle\delta_{k,i-1},u-u_{k,i-1}\rangle\right]\end{aligned}\tag{8.1.65}$$

for any $u\in X$ and $k\ge1$.

With the help of (8.1.65), we are now ready to establish the main convergence properties of the SGS algorithm.

Theorem 8.2. *Suppose that $\{p_t\}$, $\{\theta_t\}$, $\{\beta_k\}$, and $\{\gamma_k\}$ in the SGS algorithm satisfy (8.1.20) and (8.1.25).*

(a) If relation (8.1.33) holds, then under Assumptions (8.1.6) and (8.1.7), we have, for any $N\ge1$,

$$\begin{aligned}\mathbb{E}[\Psi(\bar{x}_N)-\Psi(x^*)]\le\tilde{\mathscr{B}}_d(N):=&\tfrac{\Gamma_N\beta_1}{1-P_{T_1}}V(x_0,u)\\&+\Gamma_N\textstyle\sum_{k=1}^N\sum_{i=1}^{T_k}\tfrac{(M^2+\sigma^2)\gamma_kP_{T_k}}{\beta_k\Gamma_k(1-P_{T_k})p_i^2P_{i-1}},\end{aligned}\tag{8.1.66}$$

where x^* is an arbitrary optimal solution of (8.1.1), and P_t and Γ_k are defined in (8.1.18) and (8.1.32), respectively.

(b) If in addition, X is compact and Assumption (8.1.57) holds, then

$$\mathrm{Prob}\left\{\Psi(\bar{x}_N)-\Psi(x^*)\ge\tilde{\mathscr{B}}_d(N)+\lambda\mathscr{B}_p(N)\right\}\le\exp\{-2\lambda^2/3\}+\exp\{-\lambda\},\tag{8.1.67}$$

for any $\lambda > 0$ and $N \geq 1$, where

$$\tilde{\mathscr{B}}_p(N) := \sigma\Gamma_N \left\{2\bar{V}(x^*)\sum_{k=1}^N \sum_{i=1}^{T_k} \left[\frac{\gamma_k P_{T_k}}{\Gamma_k(1-P_{T_k})p_i P_{i-1}}\right]^2\right\}^{1/2} + \Gamma_N \sum_{k=1}^N \sum_{i=1}^{T_k} \frac{\sigma^2 \gamma_k P_{T_k}}{\beta_k \Gamma_k (1-P_{T_k}) p_i^2 P_{i-1}}. \tag{8.1.68}$$

(c) *If X is compact and relation (8.1.35) (instead of (8.1.33)) holds, then both parts (a) and (b) still hold by replacing the first term in the definition of $\tilde{\mathscr{B}}_d(N)$ with $\gamma_N \beta_N \bar{V}(x^*)/(1-P_{T_N})$.*

Proof. Using (8.1.31) and (8.1.65), we have

$$\Psi(\bar{x}_k) - \Psi(u) \leq (1-\gamma_k)[\Psi(\bar{x}_{k-1}) - \Psi(u)] + \gamma_k \left\{\frac{\beta_k}{1-P_{T_k}}[V(x_{k-1},u) - V(x_k,u)] + \frac{P_{T_k}}{1-P_{T_k}} \sum_{i=1}^{T_k} \frac{1}{p_i P_{i-1}} \left[\frac{(M + \|\delta_{k,i-1}\|_*)^2}{2\beta_k p_i} + \langle \delta_{k,i-1}, u - u_{k,i-1}\rangle\right]\right\}.$$

Using the above inequality and Lemma 3.17, we conclude that

$$\begin{aligned}\Psi(\bar{x}_N) - \Psi(u) &\leq \Gamma_N(1-\gamma_1)[\Psi(\bar{x}_0) - \Psi(u)] \\ &+ \Gamma_N \sum_{k=1}^N \frac{\beta_k \gamma_k}{\Gamma_k(1-P_{T_k})}[V(x_{k-1},u) - V(x_k,u)] + \Gamma_N \sum_{k=1}^N \frac{\gamma_k P_{T_k}}{\Gamma_k(1-P_{T_k})} \\ &\sum_{i=1}^{T_k} \frac{1}{p_i P_{i-1}} \left[\frac{(M + \|\delta_{k,i-1}\|_*)^2}{2\beta_k p_i} + \langle \delta_{k,i-1}, u - u_{k,i-1}\rangle\right].\end{aligned}$$

The above relation, in view of (8.1.37) and the fact that $\gamma_1 = 1$, then implies that

$$\Psi(\bar{x}_N) - \Psi(u) \leq \frac{\beta_k}{1-P_{T_1}} V(x_0,u) + \Gamma_N \sum_{k=1}^N \frac{\gamma_k P_{T_k}}{\Gamma_k(1-P_{T_k})} \sum_{i=1}^{T_k} \frac{1}{p_i P_{i-1}} \left[\frac{M^2 + \|\delta_{k,i-1}\|_*^2}{\beta_k p_i} + \langle \delta_{k,i-1}, u - u_{k,i-1}\rangle\right]. \tag{8.1.69}$$

We now provide bounds on the RHS of (8.1.69) in expectation or with high probability.

We first show part (a). Note that by our assumptions on the SFO, the random variable $\delta_{k,i-1}$ is independent of the search point $u_{k,i-1}$ and hence $\mathbb{E}[\langle \Delta_{k,i-1}, x^* - u_{k,i}\rangle] = 0$. In addition, Assumption 8.1.7 implies that $\mathbb{E}[\|\delta_{k,i-1}\|_*^2] \leq \sigma^2$. Using the previous two observations and taking expectation on both sides of (8.1.69) (with $u = x^*$), we obtain (8.1.66).

We now show that part (b) holds. Note that by our assumptions on the SFO and the definition of $u_{k,i}$, the sequence $\{\langle \delta_{k,i-1}, x^* - u_{k,i-1}\rangle\}_{k\geq 1, 1\leq i\leq T_k}$ is a martingale-difference sequence. Denoting

$$\alpha_{k,i} := \frac{\gamma_k P_{T_k}}{\Gamma_k(1-P_{T_k}) p_i P_{i-1}},$$

and using the large-deviation theorem for martingale-difference sequence and the fact that

$$\begin{aligned}
&\mathbb{E}\left[\exp\left\{\alpha_{k,i}^2\langle\delta_{k,i-1},x^*-u_{k,i}\rangle^2/\left(2\alpha_{k,i}^2\bar{V}(x^*)\sigma^2\right)\right\}\right]\\
&\le\mathbb{E}\left[\exp\left\{\alpha_{k,i}^2\|\delta_{k,i-1}\|_*^2\|x^*-u_{k,i}\|^2/\left(2\bar{V}(x^*)\sigma^2\right)\right\}\right]\\
&\le\mathbb{E}\left[\exp\left\{\|\delta_{k,i-1}\|_*^2V(u_{k,i},x^*)/\left(\bar{V}(x^*)\sigma^2\right)\right\}\right]\\
&\le\mathbb{E}\left[\exp\left\{\|\delta_{k,i-1}\|_*^2/\sigma^2\right\}\right]\le\exp\{1\},
\end{aligned}$$

we conclude that

$$\begin{aligned}
&\mathrm{Prob}\left\{\textstyle\sum_{k=1}^N\sum_{i=1}^{T_k}\alpha_{k,i}\langle\delta_{k,i-1},x^*-u_{k,i-1}\rangle>\lambda\sigma\sqrt{2\bar{V}(x^*)\sum_{k=1}^N\sum_{i=1}^{T_k}\alpha_{k,i}^2}\right\}\\
&\le\exp\{-\lambda^2/3\},\forall\lambda>0.
\end{aligned}\tag{8.1.70}$$

Now let

$$S_{k,i}:=\frac{\gamma_kP_{T_k}}{\beta_k\Gamma_k(1-P_{T_k})p_i^2P_{i-1}}$$

and $S:=\sum_{k=1}^N\sum_{i=1}^{T_k}S_{k,i}$. By the convexity of exponential function, we have

$$\begin{aligned}
&\mathbb{E}\left[\exp\left\{\tfrac{1}{S}\textstyle\sum_{k=1}^N\sum_{i=1}^{T_k}S_{k,i}\|\delta_{k,i}\|_*^2/\sigma^2\right\}\right]\\
&\le\mathbb{E}\left[\tfrac{1}{S}\textstyle\sum_{k=1}^N\sum_{i=1}^{T_k}S_i\exp\left\{\|\delta_{k,i}\|_*^2/\sigma^2\right\}\right]\le\exp\{1\},
\end{aligned}$$

where the last inequality follows from Assumption 8.1.57. Therefore, by Markov's inequality, for all $\lambda>0$,

$$\begin{aligned}
&\mathrm{Prob}\left\{\textstyle\sum_{k=1}^N\sum_{i=1}^{T_k}S_{k,i}\|\delta_{k,i-1}\|_*^2>(1+\lambda)\sigma^2\sum_{k=1}^N\sum_{i=1}^{T_k}S_{k,i}\right\}\\
&=\mathrm{Prob}\left\{\exp\left\{\tfrac{1}{S}\textstyle\sum_{k=1}^N\sum_{i=1}^{T_k}S_{k,i}\|\delta_{k,i-1}\|_*^2/\sigma^2\right\}\ge\exp\{1+\lambda\}\right\}\le\exp\{-\lambda\}.
\end{aligned}\tag{8.1.71}$$

Our result now directly follows from (8.1.69), (8.1.70), and (8.1.71). The proof of part (c) is very similar to parts (a) and (b) in view of the bound in (8.1.38), and hence the details are skipped. ■

We now provide some specific choices for the parameters $\{\beta_k\}$, $\{\gamma_k\}$, $\{T_k\}$, $\{p_t\}$, and $\{\theta_t\}$ used in the SGS algorithm. In particular, while the stepsize policy in Corollary 8.3(a) requires the number of iterations N given a priori, such an assumption is not needed in Corollary 8.3(b) given that X is bounded. However, in order to provide some large-deviation results associated with the rate of convergence for the SGS algorithm (see (8.1.74) and (8.1.77) below), we need to assume the boundedness of X in both Corollary 8.3(a) and Corollary 8.3(b).

Corollary 8.3. *Assume that $\{p_t\}$ and $\{\theta_t\}$ in the* SPS *procedure are set to (8.1.39).*

(a) If N is given a priori, $\{\beta_k\}$ and $\{\gamma_k\}$ are set to (8.1.40), and $\{T_k\}$ is given by

$$T_k = \left\lceil \frac{N(M^2+\sigma^2)k^2}{\tilde{D}L^2} \right\rceil \tag{8.1.72}$$

for some $\tilde{D} > 0$. Then under Assumptions (8.1.6) and (8.1.7), we have

$$\mathbb{E}\left[\Psi(\bar{x}_N) - \Psi(x^*)\right] \le \frac{2L}{N(N+1)}\left[3V(x_0,x^*) + 4\tilde{D}\right], \ \ \forall N \ge 1. \tag{8.1.73}$$

If in addition, X is compact and Assumption (8.1.57) holds, then

$$\begin{aligned} &\text{Prob}\left\{\Psi(\bar{x}_N) - \Psi(x^*) \ge \frac{2L}{N(N+1)}\left[3V(x_0,x^*) + 4(1+\lambda)\tilde{D} + \frac{4\lambda\sqrt{\tilde{D}\bar{V}(x^*)}}{\sqrt{3}}\right]\right\} \\ &\quad \le \exp\left\{-2\lambda^2/3\right\} + \exp\left\{-\lambda\right\}, \ \forall \lambda > 0, \forall N \ge 1. \end{aligned} \tag{8.1.74}$$

(b) If X is compact, $\{\beta_k\}$ and $\{\gamma_k\}$ are set to (8.1.42), and $\{T_k\}$ is given by

$$T_k = \left\lceil \frac{(M^2+\sigma^2)(k+1)^3}{\tilde{D}L^2} \right\rceil \tag{8.1.75}$$

for some $\tilde{D} > 0$. Then under Assumptions (8.1.6) and (8.1.7), we have

$$\mathbb{E}\left[\Psi(\bar{x}_N) - \Psi(x^*)\right] \le \frac{L}{(N+1)(N+2)}\left[\frac{27\bar{V}(x^*)}{2} + \frac{16\tilde{D}}{3}\right], \ \ \forall N \ge 1. \tag{8.1.76}$$

If in addition, Assumption (8.1.57) holds, then

$$\begin{aligned} &\text{Prob}\left\{\Psi(\bar{x}_N) - \Psi(x^*) \ge \frac{L}{N(N+2)}\left[\frac{27\bar{V}(x^*)}{2} + \frac{8}{3}(2+\lambda)\tilde{D} + \frac{12\lambda\sqrt{2\tilde{D}\bar{V}(x^*)}}{\sqrt{3}}\right]\right\} \\ &\quad \le \exp\left\{-2\lambda^2/3\right\} + \exp\left\{-\lambda\right\}, \ \forall \lambda > 0, \forall N \ge 1. \end{aligned} \tag{8.1.77}$$

Proof. We first show part (a). It can be easily seen from (8.1.46) that (8.1.25) holds. Moreover, using (8.1.40), (8.1.45), and (8.1.46), we can easily see that (8.1.33) holds. By (8.1.45), (8.1.46), (8.1.48), (8.1.66), and (8.1.72), we have

$$\begin{aligned} \tilde{\mathscr{B}}_d(N) &\le \frac{4LV(x_0,x^*)}{N(N+1)(1-P_{T_1})} + \frac{8(M^2+\sigma^2)}{LN(N+1)}\sum_{k=1}^N \frac{k^2}{T_k+3} \\ &\le \frac{6L}{N(N+1)} + \frac{8(M^2+\sigma^2)}{LN(N+1)}\sum_{k=1}^N \frac{k^2}{T_k+3} \\ &\le \frac{2L}{N(N+1)}\left[3V(x_0,x^*) + 4\tilde{D}\right], \end{aligned} \tag{8.1.78}$$

which, in view of Theorem 8.2(a), then clearly implies (8.1.73). Now observe that by the definition of γ_k in (8.1.40) and relation (8.1.46),

$$\begin{aligned} \sum_{i=1}^{T_k}\left[\frac{\gamma_k P_{T_k}}{\Gamma_k(1-P_{T_k})p_i P_{i-1}}\right]^2 &= \left(\frac{2k}{T_k(T_k+3)}\right)^2 \sum_{i=1}^{T_k}(i+1)^2 \\ &= \left(\frac{2k}{T_k(T_k+3)}\right)^2 \frac{(T_k+1)(T_k+2)(2T_k+3)}{6} \le \frac{8k^2}{3T_k}, \end{aligned}$$

which together with (8.1.46), (8.1.48), and (8.1.68) then imply that

$$\begin{aligned}\tilde{\mathscr{B}}_p(N) &\le \tfrac{2\sigma}{N(N+1)}\left[2\bar{V}(x^*)\textstyle\sum_{k=1}^N \tfrac{8k^2}{3T_k}\right]^{1/2} + \tfrac{8\sigma^2}{LN(N+1)}\textstyle\sum_{k=1}^N \tfrac{k^2}{T_k+3}\\ &\le \tfrac{2\sigma}{N(N+1)}\left[\tfrac{16\tilde{D}L^2\bar{V}(x^*)}{3(M^2+\sigma^2)}\right]^{1/2} + \tfrac{8\tilde{D}L\sigma^2}{N(N+1)(M^2+\sigma^2)}\\ &\le \tfrac{8L}{N(N+1)}\left(\tfrac{\sqrt{\tilde{D}\bar{V}(x^*)}}{\sqrt{3}} + \tilde{D}\right).\end{aligned}$$

Using the above inequality, (8.1.78), and Theorem 8.2(b), we obtain (8.1.74).

We now show that part (b) holds. Note that P_t and Γ_k are given by (8.1.44) and (8.1.50), respectively. It then follows from (8.1.49) and (8.1.51) that both (8.1.25) and (8.1.35) hold. Using (8.1.52), the definitions of γ_k and β_k in (8.1.42), (8.1.75), and Theorem 8.2(c), we conclude that

$$\begin{aligned}\mathbb{E}\left[\Psi(\bar{x}_N)-\Psi(x^*)\right] &\le \tfrac{\gamma_N\beta_N\bar{V}(x^*)}{(1-P_{T_N})} + \Gamma_N(M^2+\sigma^2)\textstyle\sum_{k=1}^N\sum_{i=1}^{T_k}\tfrac{\gamma_k P_{T_k}}{\beta_k\Gamma_k(1-P_{T_k})p_i^2P_{i-1}}\\ &\le \tfrac{\gamma_N\beta_N\bar{V}(x^*)}{(1-P_{T_N})} + \tfrac{16L\tilde{D}}{3(N+1)(N+2)}\\ &\le \tfrac{L}{(N+1)(N+2)}\left(\tfrac{27\bar{V}(x^*)}{2}+\tfrac{16\tilde{D}}{3}\right).\end{aligned} \tag{8.1.79}$$

Now observe that by the definition of γ_k in (8.1.42), the fact that $p_t = t/2$, (8.1.44), and (8.1.50), we have

$$\begin{aligned}\textstyle\sum_{i=1}^{T_k}\left[\tfrac{\gamma_k P_{T_k}}{\Gamma_k(1-P_{T_k})p_iP_{i-1}}\right]^2 &= \left(\tfrac{k(k+1)}{T_k(T_k+3)}\right)^2\textstyle\sum_{i=1}^{T_k}(i+1)^2\\ &= \left(\tfrac{k(k+1)}{T_k(T_k+3)}\right)^2\tfrac{(T_k+1)(T_k+2)(2T_k+3)}{6} \le \tfrac{8k^4}{3T_k},\end{aligned}$$

which together with (8.1.50), (8.1.52), and (8.1.68) then imply that

$$\begin{aligned}\tilde{\mathscr{B}}_p(N) &\le \tfrac{6}{N(N+1)(N+2)}\left[\sigma\left(2\bar{V}(x^*)\textstyle\sum_{k=1}^N\tfrac{8k^4}{3T_k}\right)^{1/2} + \tfrac{4\sigma^2}{9L}\textstyle\sum_{k=1}^N\tfrac{k(k+1)^2}{T_k}\right]\\ &= \tfrac{6}{N(N+1)(N+2)}\left[\sigma\left(\tfrac{8\bar{V}(x^*)\tilde{D}L^2N(N+1)}{3(M^2+\sigma^2)}\right)^{1/2} + \tfrac{4\sigma^2L\tilde{D}N}{9(M^2+\sigma^2)}\right]\\ &\le \tfrac{6L}{N(N+2)}\left(\tfrac{2\sqrt{2\bar{V}(x^*)\tilde{D}}}{\sqrt{3}}+\tfrac{4\tilde{D}}{9}\right).\end{aligned}$$

The relation in (8.1.77) then immediately follows from the above inequality, (8.1.79), and Theorem 8.2(c). ■

Corollary 8.4 below states the complexity of the SGS algorithm for finding a stochastic ε-solution of (8.1.1), i.e., a point $\bar{x}\in X$ s.t. $\mathbb{E}[\Psi(\bar{x})-\Psi^*]\le\varepsilon$ for some $\varepsilon>0$, as well as a stochastic (ε,Λ)-solution of (8.1.1), i.e., a point $\bar{x}\in X$ s.t. $\text{Prob}\{\Psi(\bar{x})-\Psi^*\le\varepsilon\}>1-\Lambda$ for some $\varepsilon>0$ and $\Lambda\in(0,1)$. Since this result follows as an immediate consequence of Corollary 8.3, we skipped the details of its proof.

Corollary 8.4. *Suppose that* $\{p_t\}$ *and* $\{\theta_t\}$ *are set to (8.1.39). Also assume that there exists an estimate* $D_X > 0$ *s.t. (8.1.53) holds.*

(a) If $\{\beta_k\}$ *and* $\{\gamma_k\}$ *are set to (8.1.40), and* $\{T_k\}$ *is given by (8.1.72) with* $\tilde{D} = 3D_X^2/(4)$ *for some* $N > 0$*, then the number of evaluations for* ∇f *and* h'*, respectively, required by the SGS algorithm to find a stochastic* ε*-solution of (8.1.1) can be bounded by*

$$\mathscr{O}\left(\sqrt{\tfrac{LD_X^2}{\varepsilon}}\right) \tag{8.1.80}$$

and

$$\mathscr{O}\left\{\tfrac{(M^2+\sigma^2)D_X^2}{\varepsilon^2} + \sqrt{\tfrac{LD_X^2}{\varepsilon}}\right\}. \tag{8.1.81}$$

(b) If in addition, Assumption (8.1.57) holds, then the number of evaluations for ∇f *and* h'*, respectively, required by the SGS algorithm to find a stochastic* (ε, Λ)*-solution of (8.1.1) can be bounded by*

$$\mathscr{O}\left\{\sqrt{\tfrac{LD_X^2}{\varepsilon}\max\left(1, \log\tfrac{1}{\Lambda}\right)}\right\} \tag{8.1.82}$$

and

$$\mathscr{O}\left\{\tfrac{M^2D_X^2}{\varepsilon^2}\max\left(1, \log^2\tfrac{1}{\Lambda}\right) + \sqrt{\tfrac{LD_X^2}{\varepsilon}\max\left(1, \log\tfrac{1}{\Lambda}\right)}\right\}. \tag{8.1.83}$$

(c) The above bounds in parts (a) and (b) still hold if X *is bounded,* $\{\beta_k\}$ *and* $\{\gamma_k\}$ *are set to (8.1.42), and* $\{T_k\}$ *is given by (8.1.75) with* $\tilde{D} = 81D_X^2/(32)$.

Observe that both bounds in (8.1.80) and (8.1.81) on the number of evaluations for ∇f and h' are essentially not improvable. In fact, it is interesting to note that only $\mathscr{O}(1/\sqrt{\varepsilon})$ gradient evaluations is required by this stochastic approximation type algorithm applied to solve the composite problem in (8.1.1).

8.1.3 Strongly Convex and Structured Nonsmooth Problems

We intend to show that the gradient sliding techniques developed in Sects. 8.1.1 and 8.1.2 can be further generalized to some other important classes of CP problems. More specifically, we first study in Sect. 8.1.3.1 the composite CP problems in (8.1.1) with f being strongly convex, and then consider in Sect. 8.1.3.2 the case where f is a special nonsmooth function given in a bilinear saddle point form. Throughout this subsection, we assume that the nonsmooth component h is represented by a SFO. It is clear that our discussion covers also the deterministic composite problems as certain special cases by setting $\sigma = 0$ in (8.1.7) and (8.1.57).

8.1.3.1 Strongly Convex Optimization

In this section, we assume that the smooth component f in (8.1.1) is strongly convex, i.e., $\exists \mu > 0$ such that

$$f(x) \geq f(y) + \langle \nabla f(y), x-y \rangle + \mu V(y,x), \quad \forall x, y \in X. \tag{8.1.84}$$

One way to solve these strongly convex composite problems is to apply the aforementioned stochastic accelerated gradient descent method (or accelerated stochastic approximation algorithm) which would require $\mathcal{O}(1/\varepsilon)$ evaluations for ∇f and h' to find an ε-solution of (8.1.1) (Sect. 4.2). However, we will show in this subsection that this bound on the number of evaluations for ∇f can be significantly reduced to $\mathcal{O}(\log(1/\varepsilon))$, by properly restarting the SGS algorithm in Sect. 8.1.2. This multi-phase stochastic gradient sliding (M-SGS) algorithm is formally described as follows.

Algorithm 8.3 The multi-phase stochastic gradient sliding (M-SGS) algorithm

Input: Initial point $y_0 \in X$, iteration limit N_0, and an initial estimate Δ_0 s.t. $\Psi(y_0) - \Psi^* \leq \Delta_0$.
for $s = 1, 2, \ldots, S$ **do**
Run the SGS algorithm with $x_0 = y_{s-1}$, $N = N_0$, $\{p_t\}$, and $\{\theta_t\}$ in (8.1.39), $\{\beta_k\}$ and $\{\gamma_k\}$ in (8.1.40), and $\{T_k\}$ in (8.1.72) with $\tilde{D} = \Delta_0/(\mu 2^s)$, and let y_s be its output solution.
end for
Output: y_S.

We now establish the main convergence properties of the M-SGS algorithm described above.

Theorem 8.3. *If* $N_0 = \left\lceil 2\sqrt{5L/(\mu)} \right\rceil$ *in the M-SGS algorithm, then*

$$\mathbb{E}[\Psi(y_s) - \Psi^*] \leq \tfrac{\Delta_0}{2^s}, \quad s \geq 0. \tag{8.1.85}$$

As a consequence, the total number of evaluations for ∇f *and* H*, respectively, required by the M-SGS algorithm to find a stochastic* ε*-solution of (8.1.1) can be bounded by*

$$\mathcal{O}\left(\sqrt{\tfrac{L}{\mu}} \log_2 \max\left\{\tfrac{\Delta_0}{\varepsilon}, 1\right\}\right) \tag{8.1.86}$$

and

$$\mathcal{O}\left(\tfrac{M^2+\sigma^2}{\mu\varepsilon} + \sqrt{\tfrac{L}{\mu}} \log_2 \max\left\{\tfrac{\Delta_0}{\varepsilon}, 1\right\}\right). \tag{8.1.87}$$

Proof. We show (8.1.85) by induction. Note that (8.1.85) clearly holds for $s = 0$ by our assumption on Δ_0. Now assume that (8.1.85) holds at phase $s-1$, i.e., $\Psi(y_{s-1}) - \Psi^* \leq \Delta_0/2^{(s-1)}$ for some $s \geq 1$. In view of Corollary 8.3 and the definition of y_s, we have

$$\mathbb{E}[\Psi(y_s) - \Psi^* | y_{s-1}] \leq \tfrac{2L}{N_0(N_0+1)} \left[3V(y_{s-1}, x^*) + 4\tilde{D}\right]$$

$$\leq \tfrac{2L}{N_0^2}\left[\tfrac{3}{\mu}(\Psi(y_{s-1}) - \Psi^*) + 4\tilde{D}\right].$$

where the second inequality follows from the strong convexity of Ψ. Now taking expectation on both sides of the above inequality w.r.t. y_{s-1}, and using the induction hypothesis and the definition of $\tilde{D}$ in the M-SGS algorithm, we conclude that

$$\mathbb{E}[\Psi(y_s) - \Psi^*] \leq \tfrac{2L}{N_0^2}\tfrac{5\Delta_0}{\mu 2^{s-1}} \leq \tfrac{\Delta_0}{2^s},$$

where the last inequality follows from the definition of N_0. Now, by (8.1.85), the total number of phases performed by the M-SGS algorithm can be bounded by $S = \lceil \log_2 \max\left\{\frac{\Delta_0}{\varepsilon}, 1\right\}\rceil$. Using this observation, we can easily see that the total number of gradient evaluations of ∇f is given by $N_0 S$, which is bounded by (8.1.86). Now let us provide a bound on the total number of stochastic subgradient evaluations of h'. Without loss of generality, let us assume that $\Delta_0 > \varepsilon$. Using the previous bound on S and the definition of T_k, the total number of stochastic subgradient evaluations of h' can be bounded by

$$\begin{aligned}
\textstyle\sum_{s=1}^{S}\sum_{k=1}^{N_0} T_k &\leq \textstyle\sum_{s=1}^{S}\sum_{k=1}^{N_0}\left(\tfrac{\mu N_0(M^2+\sigma^2)k^2}{\Delta_0 L^2}2^s + 1\right)\\
&\leq \textstyle\sum_{s=1}^{S}\left[\tfrac{\mu N_0(M^2+\sigma^2)}{3\Delta_0 L^2}(N_0+1)^3 2^s + N_0\right]\\
&\leq \tfrac{\mu N_0(N_0+1)^3(M^2+\sigma^2)}{3\Delta_0 L^2}2^{S+1} + N_0 S\\
&\leq \tfrac{4\mu N_0(N_0+1)^3(M^2+\sigma^2)}{3\varepsilon L^2} + N_0 S.
\end{aligned}$$

This observation, in view of the definition of N_0, then clearly implies the bound in (8.1.87). ■

We now add a few remarks about the results obtained in Theorem 8.3. First, the M-SGS algorithm possesses optimal complexity bounds in terms of the number of gradient evaluations for ∇f and subgradient evaluations for h', while existing algorithms only exhibit optimal complexity bounds on the number of stochastic subgradient evaluations (see Sect. 4.2). Second, in Theorem 8.3, we only establish the optimal convergence of the M-SGS algorithm in expectation. It is also possible to establish the optimal convergence of this algorithm with high probability by making use of the light-tail assumption in (8.1.57) and a domain shrinking procedure.

8.1.3.2 Structured Nonsmooth Problems

Our goal in this subsection is to further generalize the gradient sliding algorithms to the situation when f is nonsmooth, but can be closely approximated by a certain smooth convex function.

More specifically, we assume that f is given in the form of

$$f(x) = \max_{y \in Y} \langle Ax, y \rangle - J(y), \tag{8.1.88}$$

where $A : \mathbb{R}^n \to \mathbb{R}^m$ denotes a linear operator, Y is a closed convex set, and $J : Y \to \mathbb{R}$ is a relatively simple, proper, convex, and lower semi-continuous (l.s.c.) function (i.e., problem (8.1.91) below is easy to solve). Observe that if J is the convex conjugate of some convex function F and $Y \equiv \mathscr{Y}$, then problem (8.1.1) with f given in (8.1.88) can be written equivalently as

$$\min_{x \in X} h(x) + F(Ax).$$

Similar to the previous subsection, we focus on the situation when h is represented by a SFO. Stochastic composite problems in this form have wide applications in machine learning, for example, to minimize the regularized loss function of

$$\min_{x \in X} \mathbb{E}_{\xi}[l(x, \xi)] + F(Ax),$$

where $l(\cdot, \xi)$ is a convex loss function for any $\xi \in \Xi$ and $F(Kx)$ is a certain regularization.

Since f in (8.1.88) is nonsmooth, we cannot directly apply the gradient sliding methods developed in the previous sections. However, the function $f(\cdot)$ in (8.1.88) can be closely approximated by a class of smooth convex functions. More specifically, for a given strongly convex function $\omega : Y \to \mathbb{R}$ such that

$$\omega(y) \geq \omega(x) + \langle \nabla \omega(x), y - x \rangle + \tfrac{1}{2} \|y - x\|^2, \forall x, y \in Y, \tag{8.1.89}$$

let us denote $c_\omega := \mathrm{argmin}_{y \in Y} \omega(y)$, $W(y) \equiv W(c_\omega, y) := \omega(y) - \omega(c_\omega) - \langle \nabla \omega(c_\omega), y - c_\omega \rangle$, and

$$D_Y := [\max_{y \in Y} W(y)]^{1/2}. \tag{8.1.90}$$

Then the function $f(\cdot)$ in (8.1.88) can be closely approximated by

$$f_\eta(x) := \max_{y} \{\langle Ax, y \rangle - J(y) - \eta\, W(y) : \ y \in Y\}. \tag{8.1.91}$$

Indeed, by definition we have $0 \leq W(y) \leq D_Y^2$ and hence, for any $\eta \geq 0$,

$$f(x) - \eta D_Y^2 \leq f_\eta(x) \leq f(x), \quad \forall x \in X. \tag{8.1.92}$$

Moreover, $f_\eta(\cdot)$ is differentiable and its gradients are Lipschitz continuous with the Lipschitz constant given by

$$\mathscr{L}_\eta := \tfrac{\|A\|^2}{\eta}. \tag{8.1.93}$$

We are now ready to present a smoothing stochastic gradient sliding (S-SGS) method and study its convergence properties.

Theorem 8.4. *Let $(\bar{x}_k, x_k)$ be the search points generated by a smoothing stochastic gradient sliding (S-SGS) method, which is obtained by replacing f with $f_\eta(\cdot)$ in the definition of g_k in the SGS method. Suppose that $\{p_t\}$ and $\{\theta_t\}$ in the* SPS *procedure are set to (8.1.39). Also assume that $\{\beta_k\}$ and $\{\gamma_k\}$ are set to (8.1.40) and that T_k is given by (8.1.72) with $\tilde{D} = 3D_X^2/4$ for some $N \geq 1$, where D_X is given by (8.1.53). If*

$$\eta = \frac{2\sqrt{3}\|A\|D_X}{ND_Y},$$

then the total number of outer iterations and inner iterations performed by the S-SGS algorithm to find an ε-solution of (8.1.1) can be bounded by

$$\mathcal{O}\left(\frac{\|A\|D_X D_Y}{\varepsilon}\right) \tag{8.1.94}$$

and

$$\mathcal{O}\left\{\frac{(M^2+\sigma^2)\|A\|^2 V(x_0,x^*)}{\varepsilon^2} + \frac{\|A\|D_Y\sqrt{V(x_0,x^*)}}{\varepsilon}\right\}, \tag{8.1.95}$$

respectively.

Proof. Let us denote $\Psi_\eta(x) = f_\eta(x) + h(x) + \chi(x)$. In view of (8.1.73) and (8.1.93), we have

$$\begin{aligned}
\mathbb{E}[\Psi_\eta(\bar{x}_N) - \Psi_\eta(x)] &\leq \frac{2L_\eta}{N(N+1)}\left[3V(x_0,x) + 4\tilde{D}\right] \\
&= \frac{2\|A\|^2}{\eta N(N+1)}\left[3V(x_0,x) + 4\tilde{D}\right], \quad \forall x \in X, N \geq 1.
\end{aligned}$$

Moreover, it follows from (8.1.92) that

$$\Psi_\eta(\bar{x}_N) - \Psi_\eta(x) \geq \Psi(\bar{x}_N) - \Psi(x) - \eta D_Y^2.$$

Combining the above two inequalities, we obtain

$$\mathbb{E}[\Psi(\bar{x}_N) - \Psi(x)] \leq \frac{2\|A\|^2}{\eta N(N+1)}\left[3V(x_0,x) + 4\tilde{D}\right] + \eta D_Y^2, \quad \forall x \in X,$$

which implies that

$$\mathbb{E}[\Psi(\bar{x}_N) - \Psi(x^*)] \leq \frac{2\|A\|^2}{\eta N(N+1)}\left[3D_X^2 + 4\tilde{D}\right] + \eta D_Y^2. \tag{8.1.96}$$

Plugging the value of $\tilde{D}$ and η into the above bound, we can easily see that

$$\mathbb{E}[\Psi(\bar{x}_N) - \Psi(x^*)] \leq \frac{4\sqrt{3}\|A\|D_X D_Y}{N}, \quad \forall x \in X, N \geq 1.$$

It then follows from the above relation that the total number of outer iterations to find an ε-solution of problem (8.1.88) can be bounded by

$$\bar{N}(\varepsilon) = \frac{4\sqrt{3}\|A\|D_X D_Y}{\varepsilon}.$$

Now observe that the total number of inner iterations is bounded by

$$\sum_{k=1}^{\bar{N}(\varepsilon)} T_k = \sum_{k=1}^{\bar{N}(\varepsilon)} \left\lceil \frac{(M^2+\sigma^2)\bar{N}(\varepsilon)k^2}{\tilde{D}L_\eta^2} + 1 \right\rceil = \sum_{k=1}^{\bar{N}(\varepsilon)} \left\lceil \frac{(M^2+\sigma^2)\bar{N}(\varepsilon)k^2}{\tilde{D}L_\eta^2} + 1 \right\rceil.$$

Combining these two observations, we conclude that the total number of inner iterations is bounded by (8.4). ■

In view of Theorem 8.4, by using the smoothing SGS algorithm, we can significantly reduce the number of outer iterations, and hence the number of times to access the linear operator A and A^T, from $\mathcal{O}(1/\varepsilon^2)$ to $\mathcal{O}(1/\varepsilon)$ in order to find an ε-solution of (8.1.1), while still maintaining the optimal bound on the total number of stochastic subgradient evaluations for h'. It should be noted that, by using the result in Theorem 8.2(b), we can show that the aforementioned savings on the access to the linear operator A and A^T also hold with overwhelming probability under the light-tail assumption in (8.1.57) associated with the SFO oracle.

8.2 Accelerated Gradient Sliding

In this section, we show that one can skip gradient computations without slowing down the convergence of gradient descent type methods for solving certain structured convex programming (CP) problems. To motivate our study, let us first consider the following classic bilinear saddle point problem (SPP):

$$\psi^* := \min_{x\in X}\left\{\psi(x) := f(x) + \max_{y\in Y}\langle Ax, y\rangle - J(y)\right\}. \tag{8.2.1}$$

Here, $X \subseteq \mathbb{R}^n$ and $Y \subseteq \mathbb{R}^m$ are closed convex sets, $A : \mathbb{R}^n \to \mathbb{R}^m$ is a linear operator, J is a relatively simple convex function, and $f : X \to \mathbb{R}$ is a continuously differentiable convex function satisfying

$$0 \le f(x) - l_f(u,x) \le \tfrac{L}{2}\|x-u\|^2,\ \forall x,u \in X, \tag{8.2.2}$$

for some $L > 0$, where $l_f(u,x) := f(u) + \langle \nabla f(u), x-u\rangle$ denotes the first-order Taylor expansion of f at u. Observe that problem (8.2.1) is different from the problem discussed in Sect. 8.1.3.2, whose objective function consists of a general nonsmooth convex function h (rather than a smooth convex function f), although both problems contain a structured nonsmooth component given in the form of $\max_{y\in Y}\langle Ax, y\rangle - J(y)$.

Since ψ is a nonsmooth convex function, traditional nonsmooth optimization methods, e.g., the subgradient method, would require $\mathcal{O}(1/\varepsilon^2)$ iterations to find an ε-solution of (8.2.1), i.e., a point $\bar{x} \in X$ s.t. $\psi(\bar{x}) - \psi^* \le \varepsilon$. As discussed in Sect. 3.6, we can approximate ψ by a smooth convex function

$$\psi_\rho^* := \min_{x\in X}\left\{\psi_\rho(x) := f(x) + h_\rho(x)\right\}, \tag{8.2.3}$$

with

$$h_\rho(x) := \max_{y \in Y} \langle Ax, y \rangle - J(y) - \rho W(y_0, y) \tag{8.2.4}$$

for some $\rho > 0$, where $y_0 \in Y$ and $W(y_0, \cdot)$ is a strongly convex function. By properly choosing ρ and applying the optimal gradient method to (8.2.3), one can compute an ε-solution of (8.2.1) in at most

$$\mathcal{O}\left(\sqrt{\tfrac{L}{\varepsilon}} + \tfrac{\|A\|}{\varepsilon}\right) \tag{8.2.5}$$

iterations. Such complexity bounds can also be achieved by primal–dual type methods and their equivalent form as the alternating direction method of multipliers (Sects. 3.6-3.7).

One problem associated with the smoothing scheme and the related methods mentioned above is that each iteration of these methods require both the computation of ∇f and the evaluation of the linear operators (A and A^T). As a result, the total number of gradient and linear operator evaluations will both be bounded by $\mathcal{O}(1/\varepsilon)$. However, in many applications the computation of ∇f is often much more expensive than the evaluation of the linear operators A and A^T. This happens, for example, when the linear operator A is sparse (e.g., total variation, overlapped group lasso and graph regularization), while f involves a more expensive data-fitting term. In Sect. 8.1, we considered some similar situation and proposed a gradient sliding (GS) algorithm to minimize a class of composite problems whose objective function is given by the summation of a general smooth and nonsmooth component. We show that one can skip the computation of the gradient for the smooth component from time to time, while still maintaining the $\mathcal{O}(1/\varepsilon^2)$ iteration complexity bound. More specifically, by applying the GS method to problem (8.2.1), we can show that the number of gradient evaluations of ∇f will be bounded by

$$\mathcal{O}\left(\sqrt{\tfrac{L}{\varepsilon}}\right), \tag{8.2.6}$$

which is significantly better than (8.2.5). Unfortunately, the total number of evaluations for the linear operators A and A^T will be bounded by

$$\mathcal{O}\left(\sqrt{\tfrac{L}{\varepsilon}} + \tfrac{\|A\|^2}{\varepsilon^2}\right), \tag{8.2.7}$$

which is much worse than (8.2.5). An important question is whether one can still preserve the optimal $\mathcal{O}(1/\varepsilon)$ complexity bound in (8.2.5) for solving (8.2.1) by utilizing only $\mathcal{O}(1/\sqrt{\varepsilon})$ gradient computations of ∇f to find an ε-solution of (8.2.1). If so, we could be able to keep the total number of iterations relatively small, but significantly reduce the total number of required gradient computations.

In order to address the aforementioned issues associated with existing solution methods for (8.2.1), we introduce in this section a different approach to exploit the structural information of (8.2.1). First, instead of concentrating solely on nonsmooth optimization as in Sect. 8.1, we study the following smooth composite optimization problem:

$$\phi^* := \min_{x \in X} \{\phi(x) := f(x) + h(x)\}. \tag{8.2.8}$$

Here f and h are smooth convex functions satisfying (8.2.2) and

$$0 \leq h(x) - l_h(u,x) \leq \tfrac{M}{2}\|x-u\|^2, \ \forall x, u \in X, \tag{8.2.9}$$

respectively. It is worth noting that problem (8.2.8) can be viewed as special cases of (8.2.1) or (8.2.3) (with $J = h^*$ being a strongly convex function, $Y = \mathbb{R}^n$, $A = I$, and $\rho = 0$). Under the assumption that $M \geq L$, we present a novel accelerated gradient sliding (AGS) method which can skip the computation of ∇f from time to time. We show that the total number of required gradient evaluations of ∇f and ∇h, respectively, can be bounded by

$$\mathscr{O}\left(\sqrt{\tfrac{L}{\varepsilon}}\right) \quad \text{and} \quad \mathscr{O}\left(\sqrt{\tfrac{M}{\varepsilon}}\right) \tag{8.2.10}$$

to find an ε-solution of (8.2.8). Observe that the above complexity bounds are sharper than the complexity bound obtained by the accelerated gradient method (see Sect. 3.3) for smooth convex optimization, which is given by

$$\mathscr{O}\left(\sqrt{\tfrac{L+M}{\varepsilon}}\right).$$

In particular, for the AGS method, the Lipschitz constant M associated with ∇h does not affect at all the number of gradient evaluations of ∇f. Clearly, the higher ratio of M/L will potentially result in more savings on the gradient computation of ∇f. Moreover, if f is strongly convex with modulus μ, then the above two complexity bounds in (8.2.10) can be significantly reduced to

$$\mathscr{O}\left(\sqrt{\tfrac{L}{\mu}}\log\tfrac{1}{\varepsilon}\right) \quad \text{and} \quad \mathscr{O}\left(\sqrt{\tfrac{M}{\mu}}\log\tfrac{1}{\varepsilon}\right), \tag{8.2.11}$$

respectively, which also improves the accelerated gradient descent method applied to (8.2.8) in terms of the number gradient evaluations of ∇f. Observe that in the classic black-box setting, the complexity bounds in terms of gradient evaluations of ∇f and ∇h are intertwined, and a larger Lipschitz constant M will result in more gradient evaluations of ∇f, even though there is no explicit relationship between ∇f and M. In our development, we break down the black-box assumption by assuming that we have separate access to ∇f and ∇h rather than $\nabla \phi$ as a whole. To the best of our knowledge, these types of separate complexity bounds as in (8.2.10) and (8.2.11) have never been obtained before for smooth convex optimization.

Second, we apply the above AGS method to the smooth approximation problem (8.2.3) in order to solve the aforementioned bilinear SPP in (8.2.1). By choosing the smoothing parameter properly, we show that the total number of gradient evaluations of ∇f and operator evaluations of A (and A^T) for finding an ε-solution of (8.2.1) can be bounded by

$$\mathscr{O}\left(\sqrt{\tfrac{L}{\varepsilon}}\right) \quad \text{and} \quad \mathscr{O}\left(\tfrac{\|A\|}{\varepsilon}\right), \tag{8.2.12}$$

respectively. In comparison with the original smoothing scheme and other existing methods for solving (8.2.1), this method can provide significant savings on the number of gradient computations of ∇f without increasing the complexity bound on the number of operator evaluations of A and A^T. In comparison with the GS method in Sect. 3.3, this method can reduce the number of operator evaluations of A and A^T from $\mathscr{O}(1/\varepsilon^2)$ to $\mathscr{O}(1/\varepsilon)$. Moreover, if f is strongly convex with modulus μ, the above two bounds will be significantly reduced to

$$\mathscr{O}\left(\sqrt{\tfrac{L}{\mu}}\log\tfrac{1}{\varepsilon}\right) \quad \text{and} \quad \mathscr{O}\left(\tfrac{\|A\|}{\sqrt{\varepsilon}}\right), \tag{8.2.13}$$

respectively. To the best of our knowledge, this is the first time that these tight complexity bounds were obtained for solving the classic bilinear saddle point problem (8.2.1).

It should be noted that, even though the idea of skipping the computation of ∇f is similar to Sect. 3.3, the AGS method presented in this section significantly differs from the GS method in Sect. 3.3. In particular, each iteration of GS method consists of one accelerated gradient iteration together with a bounded number of subgradient iterations. On the other hand, each iteration of the AGS method is composed of an accelerated gradient iteration nested with a few other accelerated gradient iterations to solve a different subproblem. The development of the AGS method seems to be more technical than GS and its convergence analysis is also nontrivial.

8.2.1 Composite Smooth Optimization

In this subsection, we present an accelerated gradient sliding (AGS) algorithm for solving the smooth composite optimization problem in (8.2.8) and discuss its convergence properties. Our main objective is to show that the AGS algorithm can skip the evaluation of ∇f from time to time and achieve better complexity bounds in terms of gradient computations than the classical optimal first-order methods applied to (8.2.8) (e.g., the accelerated gradient descent method in Sect. 3.3). Without loss of generality, throughout this section we assume that $M \geq L$ in (8.2.2) and (8.2.9).

The AGS method evolves from the gradient sliding (GS) algorithm in Sect. 8.1, which was designed to solve a class of composite convex optimization problems with the objective function given by the summation of a smooth and nonsmooth component. The basic idea of the GS method is to keep the nonsmooth term inside the projection (or proximal mapping) in the accelerated gradient method and then to apply a few subgradient descent iterations to solve the projection subproblem. Inspired by GS, we suggest to keep the smooth term h that has a larger Lipschitz constant in the proximal mapping in the accelerated gradient method, and then to apply a few accelerated gradient iterations to solve this smooth subproblem. As a consequence, the proposed AGS method involves two nested loops (i.e., outer and inner iterations), each of which consists of a set of modified accelerated gradient descent iterations (see Algorithm 8.4). At the k-th outer iteration, we first build a linear approximation $g_k(u) = l_f(\underline{x}_k, u)$ of f at the search point $\underline{x}_k \in X$ and then call the ProxAG procedure in (8.2.18) to compute a new pair of search points $(x_k, \tilde{x}_k) \in X \times X$. Let $V(x,u)$ be the prox-function associated with the distance generating function ν with modulus 1 so that

$$V(x,u) \geq \tfrac{1}{2}\|x-u\|^2 \;\; \forall x, y \in X. \tag{8.2.14}$$

The ProxAG procedure can be viewed as a subroutine to compute a pair of approximate solutions to

$$\min_{u \in X} g_k(u) + h(u) + \beta V(x_{k-1}, u), \tag{8.2.15}$$

where $g_k(\cdot)$ is defined in (8.2.17), and x_{k-1} is called the prox-center at the k-th outer iteration. It is worth mentioning that there are two essential differences associated with the steps (8.2.16)–(8.2.20) from the standard accelerated gradient iterations. First, we use two different search points, i.e., x_k and $\bar{x}_k$, respectively, to update $\underline{x}_k$ to compute the linear approximation and $\bar{x}_k$ to compute the output solution in (8.2.19). Second, we employ two parameters, i.e., γ_k and λ_k, to update $\underline{x}_k$ and $\bar{x}_k$, respectively, rather than just one single parameter.

The ProxAG procedure in Algorithm 8.4 performs T_k inner accelerated gradient iterations to solve (8.2.15) with certain properly chosen starting points $\tilde{u}_0$ and u_0. It should be noted, however, that the accelerated gradient iterations in (8.2.20)–(8.2.22) also differ from the standard accelerated gradient iterations in the sense that the definition of the search point $\underline{u}_t$ involves a fixed search point $\bar{x}$. Since each inner iteration of the ProxAG procedure requires one evaluation of ∇h and no evaluation of ∇f, the number of gradient evaluations of ∇h will be greater than that of ∇f as long as $T_k > 1$. On the other hand, if $\lambda_k \equiv \gamma_k$ and $T_k \equiv 1$ in the AGS method, and $\alpha_t \equiv 1$, and $p_t \equiv q_t \equiv 0$ in the ProxAG procedure, then (8.2.18) becomes

$$x_k = \tilde{x}_k = \text{argmin}_{u \in X} g_k(u) + l_h(\underline{x}_k, u) + \beta_k V(x_{k-1}, u).$$

In this case, the AGS method reduces to a variant of the accelerated gradient descent method.

Our goal in the remaining part of this section is to establish the convergence of the AGS method and to provide theoretical guidance to specify quite a few parameters, including $\{\gamma_k\}$, $\{\beta_k\}$, $\{T_k\}$, $\{\lambda_k\}$, $\{\alpha_t\}$, $\{p_t\}$, and $\{q_t\}$, used in the generic statement of this algorithm. In particular, we will provide upper bounds on the number of outer and inner iterations, corresponding to the number of gradient evaluations of ∇f and ∇h, respectively, performed by the AGS method to find an ε-solution to (8.2.8).

Algorithm 8.4 Accelerated gradient sliding (AGS) algorithm for solving (8.2.8)

Choose $x_0 \in X$. Set $\bar{x}_0 = x_0$.
for $k = 1,\ldots,N$ **do**

$$\underline{x}_k = (1-\gamma_k)\bar{x}_{k-1} + \gamma_k x_{k-1}, \tag{8.2.16}$$

$$g_k(\cdot) = l_f(\underline{x}_k,\cdot), \tag{8.2.17}$$

$$(x_k,\tilde{x}_k) = ProxAG(g_k,\bar{x}_{k-1},x_{k-1},\lambda_k,\beta_k,T_k), \tag{8.2.18}$$

$$\bar{x}_k = (1-\lambda_k)\bar{x}_{k-1} + \lambda_k \tilde{x}_k. \tag{8.2.19}$$

end for
Output $\bar{x}_N$.

procedure $(x^+,\tilde{x}^+) = ProxAG(g,\bar{x},x,\lambda,\beta,\gamma,T)$
Set $\tilde{u}_0 = \bar{x}$ and $u_0 = x$.
for $t = 1,\ldots,T$ **do**

$$\underline{u}_t = (1-\lambda)\bar{x} + \lambda(1-\alpha_t)\tilde{u}_{t-1} + \lambda\alpha_t u_{t-1}, \tag{8.2.20}$$

$$u_t = \mathrm{argmin}_{u\in X} g(u) + l_h(\underline{u}_t,u) + \beta V(x,u) + (\beta p_t + q_t)V(u_{t-1},u), \tag{8.2.21}$$

$$\tilde{u}_t = (1-\alpha_t)\tilde{u}_{t-1} + \alpha_t u_t, \tag{8.2.22}$$

end for
Output $x^+ = u_T$ and $\tilde{x}^+ = \tilde{u}_T$.
end procedure

We will first study the convergence properties of the ProxAG procedure from which the convergence of the AGS method immediately follows. In our analysis, we measure the quality of the output solution computed at the k-th call to the ProxAG procedure by

$$Q_k(x,u) := g_k(x) - g_k(u) + h(x) - h(u). \tag{8.2.23}$$

Indeed, if x^* is an optimal solution to (8.2.8), then $Q_k(x,x^*)$ provides a linear approximation for the function optimality gap $\phi(x) - \phi(x^*) = f(x) - f(x^*) + h(x) - h(x^*)$ obtained by replacing f with g_k. The following result describes some relationship between $\phi(x)$ and $Q_k(\cdot,\cdot)$.

Lemma 8.1. *For any $u \in X$, we have*

$$\begin{aligned} &\phi(\bar{x}_k) - \phi(u) \\ \le &(1-\gamma_k)[\phi(\bar{x}_{k-1}) - \phi(u)] + Q_k(\bar{x}_k,u) - (1-\gamma_k)Q_k(\bar{x}_{k-1},u) \\ &+ \tfrac{L}{2}\|\bar{x}_k - \underline{x}_k\|^2. \end{aligned} \tag{8.2.24}$$

Proof. By (8.2.2), (8.2.8), (8.2.17), and the convexity of $f(\cdot)$, we have

$$
\begin{aligned}
&\phi(\bar{x}_k) - (1-\gamma_k)\phi(\bar{x}_{k-1}) - \gamma_k\phi(u)\\
\le& g_k(\bar{x}_k) + \tfrac{L}{2}\|\bar{x}_k - \underline{x}_k\|^2 + h(\bar{x}_k)\\
&- (1-\gamma_k)f(\bar{x}_{k-1}) - (1-\gamma_k)h(\bar{x}_{k-1}) - \gamma_k f(u) - \gamma_k h(u)\\
\le& g_k(\bar{x}_k) + \tfrac{L}{2}\|\bar{x}_k - \underline{x}_k\|^2 + h(\bar{x}_k)\\
&- (1-\gamma_k)g_k(\bar{x}_{k-1}) - (1-\gamma_k)h(\bar{x}_{k-1}) - \gamma_k g_k(u) - \gamma_k h(u)\\
=& Q_k(\bar{x}_k, u) - (1-\gamma_k)Q_k(\bar{x}_{k-1}, u) + \tfrac{L}{2}\|\bar{x}_k - \underline{x}_k\|^2.
\end{aligned}
$$

□

■

We need to derive some useful equalities for our convergence analysis. Let $\{\alpha_t\}$ be the parameters used in the ProxAG procedure (see (8.2.20) and (8.2.22)) and consider the sequence $\{\Lambda_t\}_{t\ge 1}$ defined by

$$
\Lambda_t = \begin{cases} 1 & t = 1, \\ (1-\alpha_t)\Lambda_{t-1} & t > 1. \end{cases} \tag{8.2.25}
$$

By Lemma 3.17, we have

$$
1 = \Lambda_t \left[\tfrac{1-\alpha_1}{\Lambda_1} + \textstyle\sum_{i=1}^t \tfrac{\alpha_i}{\Lambda_i}\right] = \Lambda_t(1-\alpha_1) + \Lambda_t \textstyle\sum_{i=1}^t \tfrac{\alpha_i}{\Lambda_i}, \tag{8.2.26}
$$

where the last identity follows from the fact that $\Lambda_1 = 1$ in (8.2.25). Similarly, applying Lemma 3.17 to the recursion $\tilde{u}_t = (1-\alpha_t)\tilde{u}_{t-1} + \alpha_t u_t$ in (8.2.22), we have

$$
\tilde{u}_t = \Lambda_t \left[(1-\alpha_1)\tilde{u}_0 + \textstyle\sum_{i=1}^t \tfrac{\alpha_i}{\Lambda_i} u_i\right]. \tag{8.2.27}
$$

In view of (8.2.26) and the fact that $\tilde{u}_0 = \bar{x}$ in the description of the ProxAG procedure, the above relation indicates that $\tilde{u}_t$ is a convex combination of $\bar{x}$ and $\{u_i\}_{i=1}^t$.

With the help of the above technical results, we are now ready to derive some important convergence properties for the ProxAG procedure in terms of the error measure $Q_k(\cdot,\cdot)$. For the sake of notational convenience, when we work on the k-th call to the ProxAG procedure, we drop the subscript k in (8.2.23) and just denote

$$
Q(x,u) := g(x) - g(u) + h(x) - h(x). \tag{8.2.28}
$$

In a similar vein, we also define

$$
\underline{x} := (1-\gamma)\bar{x} + \gamma x \text{ and } \bar{x}^+ := (1-\lambda)\bar{x} + \lambda\tilde{x}^+. \tag{8.2.29}
$$

Comparing the above notations with (8.2.16) and (8.2.19), we can observe that $\underline{x}$ and $\bar{x}^+$, respectively, represent $\underline{x}_k$ and $\bar{x}_k$ in the k-th call to the ProxAG procedure.

Lemma 8.2. *Consider the k-th call to the ProxAG procedure in Algorithm 8.4 and let Λ_t and $\bar{x}^+$ be defined in (8.2.25) and (8.2.29) respectively. If the parameters* *satisfy*

$$\lambda \leq 1, \Lambda_T(1-\alpha_1) = 1 - \tfrac{\gamma}{\lambda}, \text{ and } \beta p_t + q_t \geq \lambda M \alpha_t, \tag{8.2.30}$$

then

$$Q(\bar{x}^+, u) - (1-\gamma)Q(\bar{x}, u) \leq \Lambda_T \textstyle\sum_{t=1}^{T} \frac{\Upsilon_t(u)}{\Lambda_t}, \ \forall u \in X, \tag{8.2.31}$$

where

$$\Upsilon_t(u) := \lambda \beta \alpha_t [V(x,u) - V(x,u_t) + p_t V(u_{t-1}, u) - (1+p_t) V(u_t, u)] \tag{8.2.32}$$

$$+ \lambda \alpha_t q_t [V(u_{t-1}, u) - V(u_t, u)]. \tag{8.2.33}$$

Proof. Let us fix any arbitrary $u \in X$ and denote

$$v := (1-\lambda)\bar{x} + \lambda u, \text{ and } \bar{u}_t := (1-\lambda)\bar{x} + \lambda \tilde{u}_t. \tag{8.2.34}$$

Our proof consists of two major parts. We first prove that

$$Q(\bar{x}^+, u) - (1-\gamma)Q(\bar{x}, u) \leq Q(\bar{u}_T, v) - \left(1 - \tfrac{\lambda}{\gamma}\right) Q(\bar{u}_0, v), \tag{8.2.35}$$

and then estimate the right-hand side of (8.2.35) through the following recurrence property:

$$Q(\bar{u}_t, v) - (1-\alpha_t) Q(\bar{u}_{t-1}, v) \leq \Upsilon_t(u). \tag{8.2.36}$$

The result in (8.2.31) then follows as an immediate consequence of (8.2.35) and (8.2.36). Indeed, by Lemma 3.17 applied to (8.2.36) (with $k = t$, $C_k = \Lambda_t$, $c_k = \alpha_t$, $\delta_k = Q(\bar{u}_t, v)$, and $B_k = \Upsilon_t(u)$), we have

$$\begin{aligned} Q(\bar{u}_T, v) \leq & \Lambda_T \left[\tfrac{1-\alpha_1}{\Lambda_1} Q(\bar{u}_0, v) - \textstyle\sum_{t=1}^{T} \frac{\Upsilon_t(u)}{\Lambda_t} \right] \\ = & \left(1 - \tfrac{\lambda}{\gamma}\right) Q(\bar{u}_0, v) - \Lambda_T \textstyle\sum_{t=1}^{T} \frac{\Upsilon_t(u)}{\Lambda_t}, \end{aligned}$$

where the last inequality follows from (8.2.30) and the fact that $\Lambda_1 = 1$ in (8.2.25). The above relation together with (8.2.35) then clearly imply (8.2.31).

We start with the first part of the proof regarding (8.2.35). By (8.2.28) and the linearity of $g(\cdot)$, we have

$$\begin{aligned} & Q(\bar{x}^+, u) - (1-\gamma)Q(\bar{x}, u) \\ = & g(\bar{x}^+ - (1-\gamma)\bar{x} - \gamma u) + h(\bar{x}^+) - (1-\gamma)h(\bar{x}) - \gamma h(u) \\ = & g(\bar{x}^+ - \bar{x} + \gamma(\bar{x} - u)) + h(\bar{x}^+) - h(\bar{x}) + \gamma(h(\bar{x}) - h(u)). \end{aligned} \tag{8.2.37}$$

Now, noting that by the relation between u and v in (8.2.34), we have

$$\gamma(\bar{x} - u) = \tfrac{\gamma}{\lambda}(\lambda \bar{x} - \lambda u) = \tfrac{\gamma}{\lambda}(\bar{x} - v). \tag{8.2.38}$$

In addition, by (8.2.34) and the convexity of $h(\cdot)$, we obtain

$$\tfrac{\gamma}{\lambda}[h(v)-(1-\lambda)h(\overline{x})-\lambda h(u)]\le 0,$$

or equivalently,

$$\gamma(h(\overline{x})-h(u))\le\tfrac{\gamma}{\lambda}(h(\overline{x})-h(v)). \tag{8.2.39}$$

Applying (8.2.38) and (8.2.39) to (8.2.37), and using the definition of $Q(\cdot,\cdot)$ in (8.2.28), we obtain

$$Q(\overline{x}^+,u)-(1-\gamma)Q(\overline{x},u)\le Q(\overline{x}^+,v)-\left(1-\tfrac{\lambda}{\gamma}\right)Q(\overline{x},v).$$

Noting that $\tilde{u}_0=\overline{x}$ and $\tilde{x}=\tilde{u}_T$ in the description of the ProxAG procedure, by (8.2.29) and (8.2.34) we have $\overline{x}^+=\overline{u}_T$ and $\overline{u}_0=\overline{x}$. Therefore, the above relation is equivalent to (8.2.35), and we conclude the first part of the proof.

For the second part of the proof regarding (8.2.36), first observe that by the definition of $Q(\cdot,\cdot)$ in (8.2.28), the convexity of $h(\cdot)$, and (8.2.9),

$$\begin{aligned}
&Q(\overline{u}_t,v)-(1-\alpha_t)Q(\overline{u}_{t-1},v)\\
=&\lambda\alpha_t(g(u_t)-g(u))+h(\overline{u}_t)-(1-\alpha_t)h(\overline{u}_{t-1})-\alpha_t h(v)\\
\le&\lambda\alpha_t(g(u_t)-g(u))+l_h(\underline{u}_t,\overline{u}_t)+\tfrac{M}{2}\|\overline{u}_t-\underline{u}_t\|^2\\
&-(1-\alpha_t)l_h(\underline{u}_t,\overline{u}_{t-1})-\alpha_t l_h(\underline{u}_t,v)\\
=&\lambda\alpha_t(g(u_t)-g(u))+l_h(\underline{u}_t,\overline{u}_t-(1-\alpha_t)\overline{u}_{t-1}-\alpha_t v)+\tfrac{M}{2}\|\overline{u}_t-\underline{u}_t\|^2.
\end{aligned} \tag{8.2.40}$$

Also note that by (8.2.20), (8.2.22), and (8.2.34),

$$\begin{aligned}
&\overline{u}_t-(1-\alpha_t)\overline{u}_{t-1}-\alpha_t v=(\overline{u}_t-\overline{u}_{t-1})+\alpha_t(\overline{u}_{t-1}-v)\\
=&\lambda(\tilde{u}_t-\tilde{u}_{t-1})+\lambda\alpha_t(\tilde{u}_{t-1}-u)=\lambda(\tilde{u}_t-(1-\alpha_t)\tilde{u}_{t-1})-\lambda\alpha_t u\\
=&\lambda\alpha_t(u_t-u).
\end{aligned}$$

By a similar argument as the above, we have

$$\overline{u}_t-\underline{u}_t=\lambda(\tilde{u}_t-(1-\alpha_t)\tilde{u}_{t-1})-\lambda\alpha_t u_{t-1}=\lambda\alpha_t(u_t-u_{t-1}). \tag{8.2.41}$$

Using the above two identities in (8.2.40), we have

$$\begin{aligned}
&Q(\overline{u}_t,v)-(1-\alpha_t)Q(\overline{u}_{t-1},v)\\
\le&\lambda\alpha_t\left[g(u_t)-g(u)+l_h(\underline{u}_t,u_t)-l_h(\underline{u}_t,u)+\tfrac{M\lambda\alpha_t}{2}\|u_t-u_{t-1}\|^2\right].
\end{aligned}$$

Moreover, it follows from Lemma 3.5 applied to (8.2.21) that

$$\begin{aligned}
&g(u_t)-g(u)+l_h(\underline{u}_t,u_t)-l_h(\underline{u}_t,u)\\
\le&\beta(V(x,u)-V(u_t,u)-V(x,u_t))\\
&+(\beta p_t+q_t)(V(u_{t-1},u)-V(u_t,u)-V(u_{t-1},u_t)).
\end{aligned} \tag{8.2.42}$$

Also by (8.2.14) and (8.2.30), we have

$$\tfrac{M\lambda\alpha_t}{2}\|u_t-u_{t-1}\|^2 \le \tfrac{M\lambda\alpha_t}{2}V(u_{t-1},u_t) \le (\beta p_t+q_t)V(u_{t-1},u_t). \tag{8.2.43}$$

Combining the above three relations, we conclude (8.2.36). □

■

In the following proposition, we provide certain sufficient conditions under which the right-hand side of (8.2.31) can be properly bounded. As a consequence, we obtain a recurrence relation for the ProxAG procedure in terms of $Q(\bar{x}_k,u)$.

Proposition 8.4. *Consider the k-th call to the ProxAG procedure. If* (8.2.30) *holds,*

$$\tfrac{\alpha_t q_t}{\Lambda_t} = \tfrac{\alpha_{t+1}q_{t+1}}{\Lambda_{t+1}} \text{ and } \tfrac{\alpha_t(1+p_t)}{\Lambda_t} = \tfrac{\alpha_{t+1}p_{t+1}}{\Lambda_{t+1}} \tag{8.2.44}$$

for any $1\le t\le T-1$*, then we have*

$$\begin{aligned}&Q(\overline{x}^+,u)-(1-\gamma)Q(\bar{x},u)\\ \le&\lambda\alpha_T[\beta(1+p_T)+q_T]\left[V(x,u)-V(x^+,u)\right]-\tfrac{\beta}{2\gamma}\|\overline{x}^+-\underline{x}\|^2,\end{aligned} \tag{8.2.45}$$

where $\overline{x}^+$ *and* $\underline{x}$ *are defined in* (8.2.29).

Proof. To prove the proposition it suffices to estimate the right-hand side of (8.2.31). We make three observations regarding the terms in (8.2.31) and (8.2.32). First, by (8.2.26),

$$\lambda\beta\Lambda_T\textstyle\sum_{t=1}^T\tfrac{\alpha_t}{\Lambda_t}V(x,u)=\lambda\beta(1-\Lambda_T(1-\alpha_1))V(x,u).$$

Second, by (8.2.14), (8.2.26), (8.2.27), (8.2.30), and the fact that $\tilde{u}_0=\bar{x}$ and $\tilde{x}^+=\tilde{u}_T$ in the ProxAG procedure, we have

$$\begin{aligned}\lambda\beta\Lambda_T\textstyle\sum_{t=1}^T\tfrac{\alpha_t}{\Lambda_t}V(x,u_t)&\ge\tfrac{\gamma\beta}{2}\cdot\tfrac{\Lambda_T}{(1-\Lambda_T(1-\alpha_1))}\textstyle\sum_{t=1}^T\tfrac{\alpha_t}{\Lambda_t}\|x-u_t\|^2\\&\ge\tfrac{\gamma\beta}{2}\left\|x-\tfrac{\Lambda_T}{1-\Lambda_T(1-\alpha_1)}\textstyle\sum_{i=1}^T\tfrac{\alpha_t}{\Lambda_t}u_t\right\|^2\\&=\tfrac{\gamma\beta}{2}\left\|x-\tfrac{\tilde{u}_T-\Lambda_T(1-\alpha_1)\tilde{u}_0}{1-\Lambda_T(1-\alpha_1)}\right\|^2\\&=\tfrac{\gamma\beta}{2}\left\|x-\tfrac{\lambda}{\gamma}\tilde{u}_T-\left(1-\tfrac{\lambda}{\gamma}\right)\tilde{u}_0\right\|^2\\&=\tfrac{\beta}{2\gamma}\left\|\gamma x-\lambda\tilde{x}^+-(\gamma-\lambda)\bar{x}\right\|^2\\&=\tfrac{\beta}{2\gamma}\|\underline{x}-\overline{x}^+\|^2,\end{aligned}$$

where the last equality follows from (8.2.29). Third, by (8.2.44), the fact that $\Lambda_1=1$ in (8.2.25), and the relations that $u_0=x$ and $u_T=x^+$ in the ProxAG procedure, we have

$$\begin{aligned}
&\lambda\beta\Lambda_T\textstyle\sum_{t=1}^{T}\frac{\alpha_t}{\Lambda_t}[p_tV(u_{t-1},u)-(1+p_t)V(u_t,u)]\\
&+\lambda\Lambda_T\textstyle\sum_{t=1}^{T}\frac{\alpha_t q_t}{\Lambda_t}[V(u_{t-1},u)-V(u_t,u)]\\
=&\lambda\beta\Lambda_T\left[\alpha_1p_1V(u_0,u)-\textstyle\sum_{i=1}^{T-1}\left(\frac{\alpha_t(1+p_t)}{\Lambda_t}-\frac{\alpha_{t+1}p_{t+1}}{\Lambda_{t+1}}\right)V(u_t,u)\right.\\
&\left.-\frac{\alpha_T(1+p_T)}{\Lambda_T}V(u_T,u)\right]+\lambda\alpha_Tq_T[V(u_0,u)-V(u_T,u)]\\
=&\lambda\beta\left[\Lambda_T\alpha_1p_1V(u_0,u)-\alpha_T(1+p_T)V(u_T,u)\right]+\lambda\alpha_Tq_T[V(u_0,u)-V(u_T,u)]\\
=&\lambda\beta\left[\Lambda_T\alpha_1p_1V(x,u)-\alpha_T(1+p_T)V(x^+,u)\right]+\lambda\alpha_Tq_T[V(x,u)-V(x^+,u)].
\end{aligned}$$

Using the above three observations in (8.2.31), we have

$$\begin{aligned}
&Q(\overline{x}^+,u)-(1-\gamma)Q(\overline{x},u)\\
\le&\lambda\beta\left[(1-\Lambda_T(1-\alpha_1)+\Lambda_T\alpha_1p_1)V(x,u)-\alpha_T(1+p_T)V(x^+,u)\right]\\
&+\lambda\alpha_Tq_T[V(x,u)-V(x^+,u)]-\tfrac{\beta}{2\gamma}\|\underline{x}-\overline{x}^+\|^2.
\end{aligned}$$

Comparing the above equation with (8.2.45), it now remains to show that

$$\alpha_T(1+p_T)=\Lambda_T\alpha_1p_1+1-\Lambda_T(1-\alpha_1). \tag{8.2.46}$$

By (8.2.26), the last relation in (8.2.44), and the fact that $\Lambda_1=1$, we have

$$\frac{\alpha_{t+1}p_{t+1}}{\Lambda_{t+1}}=\frac{\alpha_tp_t}{\Lambda_t}+\frac{\alpha_t}{\Lambda_t}=\ldots=\frac{\alpha_1p_1}{\Lambda_1}+\textstyle\sum_{i=1}^{t}\frac{\alpha_i}{\Lambda_i}=\alpha_1p_1+\frac{1-\Lambda_t(1-\alpha_1)}{\Lambda_t}.$$

Using the second relation in (8.2.44) to the above equation, we have

$$\frac{\alpha_t(1+p_t)}{\Lambda_t}=\alpha_1p_1+\frac{1-\Lambda_t(1-\alpha_1)}{\Lambda_t},$$

which implies $\alpha_t(1+p_t)=\Lambda_t\alpha_1p_1+1-\Lambda_t(1-\alpha_1)$ for any $1\le t\le T$. □

■

With the help of the above proposition and Lemma 8.1, we are now ready to establish the convergence of the AGS method. Note that the following sequence will be used in the analysis of the AGS method:

$$\Gamma_k=\begin{cases}1 & k=1\\ (1-\gamma_k)\Gamma_{k-1} & k>1.\end{cases} \tag{8.2.47}$$

Theorem 8.5. *Suppose that* (8.2.30) *and* (8.2.44) *hold. If*

$$\gamma_1=1 \text{ and } \beta_k\ge L\gamma_k, \tag{8.2.48}$$

then

$$\begin{aligned}
&\phi(\overline{x}_k)-\phi(u)\\
\le&\Gamma_k\textstyle\sum_{i=1}^{k}\frac{\lambda_i\alpha_{T_i}(\beta_i(1+p_{T_i})+q_{T_i})}{\Gamma_i}(V(x_{i-1},u)-V(x_i,u)),
\end{aligned} \tag{8.2.49}$$

where Γ_k is defined in (8.2.47).

Proof. It follows from Proposition 8.4 that for all $u \in X$,

$$\begin{aligned}&Q_k(\overline{x}_k,u)-(1-\gamma_k)Q_k(\overline{x}_{k-1},u)\\ \le&\lambda_k\alpha_{T_k}(\beta_k(1+p_{T_k})+q_{T_k})(V(x_{k-1},u)-V(x_k,u))-\tfrac{\beta_k}{2\gamma_k}\|\overline{x}_k-\underline{x}_k\|^2.\end{aligned}$$

Substituting the above bound to (8.2.24) in Lemma 8.1, and using (8.2.48), we have

$$\begin{aligned}&\phi(\overline{x}_k)-\phi(u)\\ \le&(1-\gamma_k)[\phi(\overline{x}_{k-1})-\phi(u)]\\ &+\lambda_k\alpha_{T_k}(\beta_k(1+p_{T_k})+q_{T_k})(V(x_{k-1},u)-V(x_k,u)),\end{aligned}$$

which, in view of Lemma 3.17 (with $c_k=\gamma_k$, $C_k=\Gamma_k$, and $\delta_k=\phi(\overline{x}_k)-\phi(u)$), then implies that

$$\begin{aligned}&\phi(\overline{x}_k)-\phi(u)\\ \le&\Gamma_k\Big[\tfrac{1-\gamma_1}{\Gamma_1}(\phi(\overline{x}_0)-\phi(u))\\ &+\textstyle\sum_{i=1}^k\tfrac{\lambda_i\alpha_{T_i}(\beta_i(1+p_{T_i})+q_{T_i})}{\Gamma_i}(V(x_{i-1},u)-V(x_i,u))\Big]\\ =&\Gamma_k\textstyle\sum_{i=1}^k\tfrac{\lambda_i\alpha_{T_i}(\beta_i(1+p_{T_i})+q_{T_i})}{\Gamma_i}(V(x_{i-1},u)-V(x_i,u)),\end{aligned}$$

where the last equality follows from the fact that $\gamma_1=1$ in (8.2.48). □

■

There are many possible selections of parameters that satisfy the assumptions of the above theorem. In the following corollaries we describe two different ways to specify the parameters of Algorithm 8.4 that lead to the optimal complexity bounds in terms of the number of gradient evaluations of ∇f and ∇h.

Corollary 8.5. *Consider problem* (8.2.8) *with the Lipschitz constants in* (8.2.2) *and* (8.2.9) *satisfying $M \ge L$. Suppose that the parameters of Algorithm 8.4 are set to*

$$\begin{aligned}&\gamma_k=\tfrac{2}{k+1},\ T_k\equiv T:=\left\lceil\sqrt{\tfrac{M}{L}}\right\rceil,\\ &\lambda_k=\begin{cases}1 & k=1,\\ \frac{\gamma_k(T+1)(T+2)}{T(T+3)} & k>1,\end{cases}\quad \text{and } \beta_k=\tfrac{3L\gamma_k}{k\lambda_k}.\end{aligned}\tag{8.2.50}$$

Also assume that the parameters in the first call to the ProxAG procedure ($k=1$) are set to

$$\alpha_t=\tfrac{2}{t+1},\ \ p_t=\tfrac{t-1}{2},\ \text{and } q_t=\tfrac{6M}{t},\tag{8.2.51}$$

and the parameters in the remaining calls to the ProxAG procedure ($k>1$) are set to

$$\alpha_t=\tfrac{2}{t+2},\ \ p_t=\tfrac{t}{2},\ \text{and } q_t=\tfrac{6M}{k(t+1)}.\tag{8.2.52}$$

Then the numbers of gradient evaluations of ∇f and ∇h performed by the AGS method to compute an ε-solution of (8.2.8) *can be bounded by*

$$N_f := \sqrt{\frac{30LV(x_0,x^*)}{\varepsilon}} \tag{8.2.53}$$

and

$$N_h := \sqrt{\frac{30MV(x_0,x^*)}{\varepsilon}} + \sqrt{\frac{30LV(x_0,x^*)}{\varepsilon}} \tag{8.2.54}$$

respectively, where x^* is a solution to (8.2.8).

Proof. Let us start with verification of (8.2.30), (8.2.44), and (8.2.48) for the purpose of applying Theorem 8.5. We will consider the first call to the ProxAG procedure ($k = 1$) and the remaining calls ($k > 1$) separately.

When $k = 1$, by (8.2.50) we have $\lambda_1 = \gamma_1 = 1$, and $\beta_1 = 3L$, hence (8.2.48) holds immediately. By (8.2.51) we can observe that $\Lambda_t = 2/(t(t+1))$ satisfies (8.2.25), and that

$$\tfrac{\alpha_t q_t}{\Lambda_t} \equiv 6M, \text{ and } \tfrac{\alpha_t(1+p_t)}{\Lambda_t} = \tfrac{t(t+1)}{2} = \tfrac{\alpha_{t+1}p_{t+1}}{\Lambda_{t+1}},$$

hence (8.2.44) holds. In addition, by (8.2.50) and (8.2.51) we have $\lambda = \gamma = 1$ and $\alpha_1 = 1$ in (8.2.30), and that

$$\beta p_t + q_t \geq q_t = \tfrac{6M}{t} > \tfrac{2M}{t+1} = \lambda M \alpha_t.$$

Therefore (8.2.30) also holds.

For the case when $k > 1$, we can observe from (8.2.52) that $\Lambda_t = 6/(t+1)(t+2)$ satisfies (8.2.25), $\alpha_t q_t/\Lambda_t \equiv 2M/(k)$, and that

$$\tfrac{\alpha_t(1+p_t)}{\Lambda_t} = \tfrac{(t+1)(t+2)}{6} = \tfrac{\alpha_{t+1}p_{t+1}}{\Lambda_{t+1}}.$$

Therefore (8.2.44) holds. Also, from (8.2.50) and noting that $k, T \geq 1$, we have

$$\tfrac{3}{k} > \tfrac{3\gamma_k}{2} = \tfrac{3\lambda_k}{2}\left(1 - \tfrac{2}{(T+1)(T+2)}\right) \geq \tfrac{3\lambda_k}{2}\left(1 - \tfrac{2}{2\cdot 3}\right) = \lambda_k. \tag{8.2.55}$$

Applying the above relation to the definition of β_k in (8.2.50) we have (8.2.48). It now suffices to verify (8.2.30) in order to apply Theorem 8.5. Applying (8.2.50), (8.2.52), (8.2.55), and noting that $k \geq 2$ and that $\Lambda_T = 6/(T+1)(T+2)$ with $T \geq 1$, we can verify in (8.2.30) that

$$\begin{aligned}
\lambda &= \tfrac{\gamma(T+1)(T+2)}{T(T+3)} = \tfrac{2}{k+1}\left(1 + \tfrac{2}{T(T+3)}\right) \leq \tfrac{2}{3}\left(1 + \tfrac{2}{1\cdot 4}\right) = 1,\\
\Lambda_T(1-\alpha_1) &= \tfrac{2}{(T+1)(T+2)} = 1 - \tfrac{T(T+3)}{(T+1)(T+2)} = 1 - \tfrac{\gamma}{\lambda},\\
\beta p_t + q_t &> q_t = \tfrac{2M}{t+1} \cdot \tfrac{3}{k} > \tfrac{2\lambda M}{t+1} \geq \lambda M \alpha_t.
\end{aligned}$$

Therefore, the conditions in (8.2.30) are satisfied.

We are now ready to apply Theorem 8.5. In particular, noting that $\alpha_t(1+p_t) \equiv 1$ from (8.2.51) and (8.2.52), we obtain from (8.2.49) (with $u = x^*$) that

$$\phi(\bar{x}_k) - \phi^* \leq \Gamma_k \textstyle\sum_{i=1}^k \xi_i(V(x_{i-1},x^*) - V(x_i,x^*)), \tag{8.2.56}$$

where

$$\xi_i := \tfrac{\lambda_i(\beta_i + \alpha_{T_i} q_{T_i})}{\Gamma_i}. \tag{8.2.57}$$

Substituting (8.2.50) and (8.2.51) to (8.2.57), and noting that $\Gamma_i = 2/(i(i+1))$ by (8.2.47), we have

$$\begin{aligned}
\xi_1 =& \beta_1 + \alpha_T q_T = 3L + \tfrac{12M}{T(T+1)}, \text{ and} \\
\xi_i =& \tfrac{\lambda_i \beta_i}{\Gamma_i} + \tfrac{\lambda_i \alpha_{T_i} q_{T_i}}{\Gamma_i} = \tfrac{3L\gamma_i}{i\Gamma_i} + \tfrac{\gamma_i}{\Gamma_i} \tfrac{(T_i+1)(T_i+2)}{T_i(T_i+3)} \tfrac{2}{T_i+2} \tfrac{6M}{i(T_i+1)} \\
\equiv& 3L + \tfrac{12M}{T(T+3)}, \forall i > 1.
\end{aligned}$$

Applying the above two results regarding ξ_i to (8.2.56), and noting that $\xi_1 > \xi_2$, we have

$$\begin{aligned}
&\phi(\bar{x}_k) - \phi^* \\
\leq& \Gamma_k \left[\xi_1(V(x_0,x^*) - V(x_1,x^*)) + \textstyle\sum_{i=2}^k \xi_i(V(x_{i-1},x^*) - V(x_i,x^*)) \right] \\
=& \Gamma_k \left[\xi_1(V(x_0,x^*) - V(x_1,x^*)) + \xi_2(V(x_1,x^*) - V(x_k,x^*)) \right] \\
\leq& \Gamma_k \xi_1 V(x_0,x^*) \\
=& \tfrac{2}{k(k+1)} \left(3L + \tfrac{12M}{T(T+1)} \right) V(x_0,x^*) \\
\leq& \tfrac{30L}{k(k+1)} V(x_0,x^*),
\end{aligned}$$

where the last inequality is due to the fact that $T \geq \sqrt{M/L}$.

From the above inequality, the number of calls to the ProxAG procedure for computing an ε-solution of (8.2.8) is bounded by N_f in (8.2.53). This is also the bound for the number of gradient evaluations of ∇f. Moreover, the number of gradient evaluations of ∇h is bounded by

$$TN_f \leq \left(\sqrt{\tfrac{M}{L}} + 1 \right) N_f = \sqrt{\tfrac{30MV(x_0,x^*)}{\varepsilon}} + \sqrt{\tfrac{30LV(x_0,x^*)}{\varepsilon}} = N_h. \tag{8.2.58}$$

□

■

In the above corollary, the constant factors in (8.2.53) and (8.2.54) are both given by $\sqrt{30}$. In the following corollary, we provide a slightly different set of parameters for Algorithm 8.4 that results in a tighter constant factor for (8.2.53).

Corollary 8.6. *Consider problem* (8.2.8) *with the Lipschitz constants in* (8.2.2) *and* (8.2.9) *satisfying $M \geq L$. Suppose that the parameters in the first call to the ProxAG procedure ($k = 1$) are set to*

$$\alpha_t = \tfrac{2}{t+1},\ p_t = \tfrac{t-1}{2},\ \text{and}\ q_t = \tfrac{7LT(T+1)}{4t}, \tag{8.2.59}$$

and that the parameters in the k-th call ($k > 1$) are set to

$$p_t \equiv p := \sqrt{\tfrac{M}{L}},\ \alpha_t \equiv \alpha := \tfrac{1}{p+1},\ \text{and}\ q_t \equiv 0. \tag{8.2.60}$$

If the other parameters in Algorithm 8.4 satisfy

$$\begin{aligned} &\gamma_k = \tfrac{2}{k+1}, T_k := \begin{cases} \left\lceil \sqrt{\tfrac{8M}{7L}} \right\rceil, & k = 1 \\ \left\lceil \tfrac{\ln(3)}{-\ln(1-\alpha)} \right\rceil, & k > 1, \end{cases} \\ &\lambda_k := \begin{cases} 1, & k = 1 \\ \tfrac{\gamma_k}{1-(1-\alpha)^{T_k}}, & k > 1, \end{cases} \ \text{and}\ \beta_k := \begin{cases} L, & k = 1 \\ \tfrac{9L\gamma_k}{2k\lambda_k}, & k > 1, \end{cases} \end{aligned} \tag{8.2.61}$$

where α is defined in (8.2.60)*, then the numbers of gradient evaluations of ∇f and ∇h performed by the AGS method to find an ε-solution to problem* (8.2.8) *can be bounded by*

$$N_f := 3\sqrt{\tfrac{LV(x_0,x^*)}{\varepsilon}} \tag{8.2.62}$$

and

$$\begin{aligned} N_h :=& (1+\ln 3)N_f \left(\sqrt{\tfrac{M}{L}}+1\right) \\ \leq& 7\left(\sqrt{\tfrac{MV(x_0,x^*)}{\varepsilon}} + \sqrt{\tfrac{LV(x_0,x^*)}{\varepsilon}}\right), \end{aligned} \tag{8.2.63}$$

respectively.

Proof. Let us verify (8.2.30), (8.2.48), and (8.2.44) first, so that we could apply Theorem 8.5. We consider the case when $k = 1$ first. By the definition of γ_k and β_k in (8.2.61), it is clear that (8.2.48) is satisfied when $k = 1$. Also, by (8.2.59) we have that $\Lambda_t = 2/(t(t+1))$ in (8.2.25),

$$\tfrac{\alpha_t q_t}{\Lambda_t} \equiv \tfrac{7LT_1(T_1+1)}{4},\ \text{and}\ \tfrac{\alpha_t(1+p_t)}{\Lambda_t} = \tfrac{t(t+1)}{2} = \tfrac{\alpha_{t+1}p_{t+1}}{\Lambda_{t+1}},$$

hence (8.2.44) also holds. Moreover, by (8.2.59) and (8.2.61), we can verify in (8.2.30) that

$$\lambda = \gamma = 1, \Lambda_{T_1}(1-\alpha_1) = 0 = 1 - \tfrac{\gamma}{\lambda},$$

and

$$\beta p_t + q_t \geq q_t > \tfrac{7LT^2}{4t} = \tfrac{8M}{4t} > M\alpha_t.$$

Therefore the relations in (8.2.30) are all satisfied.

Now we consider the case when $k > 1$. By (8.2.25) and (8.2.60), we observe that $\Lambda_t = (1-\alpha)^{t-1}$ for all $t \geq 1$. Moreover, from the definition of T_k in (8.2.61), we can also observe that

$$(1-\alpha)^{T_k} \leq \tfrac{1}{3}.$$

Four relations can be derived based on the aforementioned two observations, (8.2.60), and (8.2.61). First,

$$\tfrac{\alpha_t q_t}{\Lambda_t} \equiv 0, \ \tfrac{\alpha_t(1+p_t)}{\Lambda_t} = \tfrac{1}{(1-\alpha)^{t-1}} = \tfrac{\alpha_{t+1}p_{t+1}}{\Lambda_{t+1}},$$

which verifies (8.2.44). Second,

$$\beta_k = \tfrac{9L(1-(1-\alpha)^{T_k})}{2k} \geq \tfrac{3L}{k} > L\gamma_k,$$

which leads to (8.2.48). Third, noting that $k \geq 2$, we have

$$\tfrac{\gamma_k}{1-\Lambda_{T_k}(1-\alpha)} = \lambda_k = \tfrac{\gamma_k}{1-(1-\alpha)^{T_k}} \leq \tfrac{3\gamma_k}{2} = \tfrac{3}{k+1} \leq 1.$$

Fourth,

$$\begin{aligned}
\tfrac{\beta_k p}{\lambda_k M\alpha} &= \tfrac{9L\gamma_k p(p+1)}{2k\lambda_k^2 M} = \tfrac{9Lp(p+1)\left(1-(1-\alpha)^{T_k}\right)^2}{2k\gamma_k M} \\
&= \tfrac{9(k+1)}{4k} \cdot \left(\tfrac{Lp(p+1)}{M}\right) \cdot \left(1-(1-\alpha)^{T_k}\right)^2 \\
&> \tfrac{9}{4} \cdot 1 \cdot \tfrac{4}{9} = 1.
\end{aligned}$$

The last two relations imply that (8.2.30) holds.

Summarizing the above discussions regarding both the cases $k = 1$ and $k > 1$, applying Theorem 8.5, and noting that $\alpha_t(1+p_t) \equiv 1$, we have

$$\phi(\bar{x}_k) - \phi(u) \leq \Gamma_k \textstyle\sum_{i=1}^k \xi_i(V(x_{i-1},u) - V(x_i,u)), \ \forall u \in X, \tag{8.2.64}$$

where

$$\xi_i := \tfrac{\lambda_i(\beta_i + \alpha_{T_i} q_{T_i})}{\Gamma_i}.$$

It should be observed from the definition of γ_k in (8.2.61) that $\Gamma_i := 2/(i(i+1))$ satisfies (8.2.47). Using this observation, applying (8.2.59), (8.2.60), and (8.2.61) to the above equation we have

$$\xi_1 = \beta_1 + \alpha_{T_1} q_{T_1} = L + \tfrac{7L}{2} = \tfrac{9L}{2}$$

and

$$\xi_i = \frac{\lambda_i \beta_i}{\Gamma_i} \equiv \frac{9L}{2}, \ \forall i > 1.$$

Therefore, (8.2.64) becomes

$$\begin{aligned} \phi(\bar{x}_k) - \phi(u) \leq & \frac{9L}{k(k+1)}(V(x_0,u) - V(x_k,u)) \\ \leq & \frac{9L}{k(k+1)} V(x_0,u). \end{aligned} \tag{8.2.65}$$

Setting $u = x^*$ in the above inequality, we observe that the number of calls to the ProxAG procedure for computing an ε-solution of (8.2.8) is bounded by N_f in (8.2.62). This is also the bound for the number of gradient evaluations of ∇f. Moreover, by (8.2.60), (8.2.61), and (8.2.62) we conclude that the number of gradient evaluations of ∇h is bounded by

$$\begin{aligned} \Sigma_{k=1}^{N_f} T_k = & T_1 + \Sigma_{k=2}^{N_f} T_k \leq \left(\sqrt{\tfrac{8M}{7L}} + 1\right) + (N_f - 1)\left(\tfrac{\ln 3}{-\ln(1-\alpha)} + 1\right) \\ \leq & \left(\sqrt{\tfrac{8M}{7L}} + 1\right) + (N_f - 1)\left(\tfrac{\ln 3}{\alpha} + 1\right) \\ = & \left(\sqrt{\tfrac{8M}{7L}} + 1\right) + (N_f - 1)\left(\left(\sqrt{\tfrac{M}{L}} + 1\right)\ln 3 + 1\right) \\ < & (1 + \ln 3) N_f \left(\sqrt{\tfrac{M}{L}} + 1\right) \\ < & 7\left(\sqrt{\tfrac{MV(x_0,x^*)}{\varepsilon}} + \sqrt{\tfrac{LV(x_0,x^*)}{\varepsilon}}\right). \end{aligned}$$

Here the second inequity is from the property of logarithm functions that $-\ln(1-\alpha) \geq \alpha$ for $\alpha \in [0,1)$. □

■

Since $M \geq L$ in (8.2.2) and (8.2.9), the results obtained in Corollaries 8.5 and 8.6 indicate that the number of gradient evaluations of ∇f and ∇h that Algorithm 8.4 requires for computing an ε-solution of (8.2.8) can be bounded by $\mathcal{O}(\sqrt{L/\varepsilon})$ and $\mathcal{O}(\sqrt{M/\varepsilon})$, respectively. Such a result is particularly useful when M is significantly larger, e.g., $M = \mathcal{O}(L/\varepsilon)$, since the number of gradient evaluations of ∇f would not be affected at all by the large Lipschitz constant of the whole problem. It is interesting to compare the above result with the best known so-far complexity bound under the traditional black-box oracle assumption. If we treat problem (8.2.8) as a general smooth convex optimization and study its oracle complexity, i.e., under the assumption that there exists an *oracle* that outputs $\nabla\phi(x)$ for any test point x (and $\nabla\phi(x)$ only), it has been shown that the number of calls to the oracle cannot be smaller than $\mathcal{O}(\sqrt{(L+M)/\varepsilon})$ for computing an ε-solution. Under such "single oracle" assumption, the complexity bounds in terms of gradient evaluations of ∇f and ∇h are intertwined, and a larger Lipschitz constant M will result in more gradient evaluations of ∇f, even though there is no explicit relationship between ∇f and M. However, the results in Corollaries 8.5 and 8.6 suggest that we can study the oracle complexity of problem (8.2.8) based on the assumption of *two separate oracles*: one oracle $\mathscr{O}_f$ to compute ∇f for any test point x, and the other one $\mathscr{O}_h$ to compute $\nabla h(y)$ for any test point y. In particular, these two oracles do not have to be called at the same time, and hence it is possible to obtain separate complexity bounds $\mathscr{O}(\sqrt{L/\varepsilon})$ and $\mathscr{O}(\sqrt{M/\varepsilon})$ on the number of calls to $\mathscr{O}_f$ and $\mathscr{O}_h$, respectively.

We now consider a special case of (8.2.8) where f is strongly convex. More specifically, we assume that there exists $\mu > 0$ such that

$$\mu V(u,x) \le f(x) - l_f(u,x) \le \tfrac{L}{2}\|x-u\|^2, \ \forall x,u \in X. \tag{8.2.66}$$

Under the above assumption, we develop a multi-stage AGS algorithm that can skip computation of ∇f from time to time, and compute an ε-solution of (8.2.8) with

$$\mathscr{O}\left(\sqrt{\tfrac{L}{\mu}}\log\tfrac{1}{\varepsilon}\right) \tag{8.2.67}$$

gradient evaluations of ∇f (see Algorithm 8.5). It should be noted that, under the traditional black-box setting, where one could only access $\nabla\psi(x)$ for each inquiry x, the number of evaluations of $\nabla\psi(x)$ required to compute an ε-solution is bounded by

$$\mathscr{O}\left(\sqrt{\tfrac{L+M}{\mu}}\log\tfrac{1}{\varepsilon}\right). \tag{8.2.68}$$

Algorithm 8.5 The multi-stage accelerated gradient sliding (M-AGS) algorithm

Choose $v_0 \in X$, accuracy ε, iteration limit N_0, and initial estimate Δ_0 such that $\phi(v_0) - \phi^* \le \Delta_0$.
for $s = 1,\ldots,S$ **do**
  Run the AGS algorithm with $x_0 = v_{s-1}$, $N = N_0$, and parameters in Corollary 8.6, and let $v_s = \overline{x}_N$.
end for
Output v_S.

Theorem 8.6 below describes the main convergence properties of the M-AGS algorithm.

Theorem 8.6. *Suppose that $M \ge L$ in* (8.2.9) *and* (8.2.66). *If the parameters in Algorithm 8.5 are set to*

$$N_0 = 3\sqrt{\tfrac{2L}{\mu}} \text{ and } S = \log_2 \max\left\{\tfrac{\Delta_0}{\varepsilon}, 1\right\}, \tag{8.2.69}$$

then its output v_S must be an ε-solution of (8.2.1). *Moreover, the total number of gradient evaluations of ∇f and ∇h performed by Algorithm 8.5 can be bounded by*

$$N_f := 3\sqrt{\tfrac{2L}{\mu}}\log_2 \max\left\{\tfrac{\Delta_0}{\varepsilon}, 1\right\} \tag{8.2.70}$$

and

$$\begin{aligned} N_h :=& (1+\ln 3)N_f\left(\sqrt{\tfrac{M}{L}}+1\right) \\ <& 9\left(\sqrt{\tfrac{L}{\mu}}+\sqrt{\tfrac{M}{\mu}}\right)\log_2 \max\left\{\tfrac{\Delta_0}{\varepsilon}, 1\right\}, \end{aligned} \tag{8.2.71}$$

respectively.

Proof. With input $x_0 = v_{s-1}$ and $N = N_0$, we conclude from (8.2.65) in the proof of Corollary 8.6 (with $u = x^*$ a solution to (8.2.8)) that

$$\phi(\bar{x}_N) - \phi^* \le \tfrac{9L}{N_0(N_0+1)} V(x_0, x^*) \le \tfrac{\mu}{2} V(x_0, x^*),$$

where the last inequality follows from (8.2.69). Using the facts that the input of the AGS algorithm is $x_0 = v_{s-1}$ and that the output is set to $v_s = \bar{x}_N$, we conclude

$$\phi(v_s) - \phi^* \le \tfrac{\mu}{2} V(v_{s-1}, x^*) \le \tfrac{1}{2}(\phi(v_{s-1}) - \phi^*),$$

where the last inequality is due to the strong convexity of $\phi(\cdot)$. It then follows from the above relation, the definition of Δ_0 in Algorithm 8.5, and (8.2.69) that

$$\phi(v_S) - \phi^* \le \tfrac{1}{2^S}(\phi(v_0) - \phi^*) \le \tfrac{\Delta_0}{2^S} \le \varepsilon.$$

Comparing Algorithms 8.4 and 8.5, we can observe that the total number of gradient evaluations of ∇f in Algorithm 8.5 is bounded by $N_0 S$, and hence we have (8.2.70). Moreover, comparing (8.2.62) and (8.2.63) in Corollary 8.6, we conclude (8.2.71).

□

■

In view of Theorem 8.6, the total number of gradient evaluations of ∇h required by the M-AGS algorithm to compute an ε-solution of (8.2.8) is the same as the traditional result (8.2.68). However, by skipping the gradient evaluations of ∇f from time to time in the M-AGS algorithm, the total number of gradient evaluations of ∇f is improved from (8.2.68) to (8.2.67). Such an improvement becomes more significant as the ratio M/L increases.

8.2.2 Composite Bilinear Saddle Point Problems

Our goal in this section is to show the advantages of the AGS method when applied to our motivating problem, i.e., the composite bilinear saddle point problem in (8.2.1). In particular, we show in Sect. 8.2.2.1 that the AGS algorithm can be used to solve (8.2.1) by incorporating the smoothing technique discussed in Sect. 3.6 and derive new complexity bounds in terms of the number of gradient computations of

∇f and operator evaluations of A and A^T. Moreover, we demonstrate in Sect. 8.2.2.2 that even more significant saving on gradient computation of ∇f can be obtained when f is strongly convex in (8.2.1) by incorporating the multi-stage AGS method.

8.2.2.1 Saddle Point Problems

Our goal in this section is to extend the AGS algorithm from composite smooth optimization to nonsmooth optimization. By incorporating the smoothing technique in Sect. 3.6, we can apply AGS to solve the composite saddle point problem (8.2.1). Throughout this section, we assume that the dual feasible set Y in (8.2.1) is bounded, i.e., there exists $y_0 \in Y$ such that

$$D_Y := [\max_{v \in Y} W(y_0, v)]^{1/2} \tag{8.2.72}$$

is finite, where $W(\cdot,\cdot)$ is the prox-function associated with Y with modulus 1.

Let ψ_ρ be the smooth approximation of ψ defined in (8.2.3). It can be easily shown that

$$\psi_\rho(x) \le \psi(x) \le \psi_\rho(x) + \rho D_Y^2, \ \forall x \in X. \tag{8.2.73}$$

Therefore, if $\rho = \varepsilon/(2D_Y^2)$, then an $(\varepsilon/2)$-solution to (8.2.3) is also an ε-solution to (8.2.1). Moreover, it follows that problem (8.2.3) is given in the form of (8.2.8) (with $h(x) = h_\rho(x)$) and satisfies (8.2.9) with $M = \|A\|^2/\rho$. Using these observations, we are ready to summarize the convergence properties of the AGS algorithm for solving problem (8.2.1).

Proposition 8.5. *Let $\varepsilon > 0$ be given and assume that $2\|A\|^2 D_Y^2 > \varepsilon L$. If we apply the AGS method in Algorithm 8.4 to problem* (8.2.3) *(with $h = h_\rho$ and $\rho = \varepsilon/(2D_Y^2)$), in which the parameters are set to* (8.2.59)–(8.2.61) *with $M = \|A\|^2/(\rho)$, then the total number of gradient evaluations of ∇f and linear operator evaluations of A (and A^T) in order to find an ε-solution of* (8.2.1) *can be bounded by*

$$N_f := 3\left(\sqrt{\frac{2LV(x_0,x^*)}{\varepsilon}}\right) \tag{8.2.74}$$

and

$$N_A := 14\left(\sqrt{\frac{2LV(x_0,x^*)}{\varepsilon}} + \frac{2\|A\|D_Y\sqrt{V(x_0,x^*)}}{\varepsilon}\right), \tag{8.2.75}$$

respectively.

Proof. By (8.2.73) we have $\psi_\rho^* \le \psi^*$ and $\psi(x) \le \psi_\rho(x) + \rho D_Y^2$ for all $x \in X$, and hence

$$\psi(x) - \psi^* \le \psi_\rho(x) - \psi_\rho^* + \rho D_Y^2, \ \forall x \in X.$$

Using the above relation and the fact that $\rho = \varepsilon/(2D_Y^2)$ we conclude that if $\psi_\rho(x) - \psi_\rho^* \le \varepsilon/2$, then x is an ε-solution to (8.2.1). To finish the proof, it suffices to consider the complexity of AGS for computing an $\varepsilon/2$-solution of (8.2.3). By Corollary 8.6, the total number of gradient evaluations of ∇f is bounded by (8.2.74). Note that the evaluation of ∇h_ρ is equivalent to 2 evaluations of linear operators: one computation of form Ax for computing the maximizer $y^*(x)$ for problem (8.2.4), and one computation of form $A^T y^*(x)$ for computing $\nabla h_\rho(x)$. Using this observation, and substituting $M = \|A\|^2/\rho$ in (8.2.63), we conclude (8.2.75). □

■

According to Proposition 8.5, the total number of gradient evaluations of ∇f and linear operator evaluations of both A and A^T are bounded by

$$\mathcal{O}\left(\sqrt{\tfrac{L}{\varepsilon}}\right) \tag{8.2.76}$$

and

$$\mathcal{O}\left(\sqrt{\tfrac{L}{\varepsilon}} + \tfrac{\|A\|}{\varepsilon}\right) \tag{8.2.77}$$

respectively, for computing an ε-solution of the saddle point problem (8.2.1). Therefore, if $L \le \mathcal{O}(\|A\|^2/\varepsilon)$, then the number of gradient evaluations of ∇f will not be affected by the dominating term $\mathcal{O}(\|A\|/\varepsilon)$. This result significantly improves the best known so-far complexity results for solving the bilinear saddle point problem (8.2.1). Specifically, it improves the complexity regarding number of gradient computations of ∇f from $\mathcal{O}(1/\varepsilon)$ associated with the smoothing technique or primal–dual type methods to $\mathcal{O}(1/\sqrt{\varepsilon})$, and also improves the complexity regarding operator evaluations involving A from $\mathcal{O}(1/\varepsilon^2)$ associated with the gradient sliding methods to $\mathcal{O}(1/\varepsilon)$.

8.2.2.2 Strongly Convex Composite Saddle Point Problems

In this subsection, we still consider problem (8.2.1), but assume that f is strongly convex (i.e., (8.2.66) holds). In this case, it has been shown previously that $\mathcal{O}(\|A\|/\sqrt{\varepsilon})$ first-order iterations, each one of them involving the computation of ∇f, and the evaluation of A and A^T, are needed in order to compute an ε-solution of (8.2.1) (e.g., Sect. 3.6). However, we demonstrate in this subsection that the complexity with respect to the gradient evaluation of ∇f can be significantly improved from $\mathcal{O}(1/\sqrt{\varepsilon})$ to $\mathcal{O}(\log(1/\varepsilon))$.

Such an improvement can be achieved by properly restarting the AGS method applied to solve a series of smooth optimization problem of form (8.2.3), in which the smoothing parameter ρ changes over time. The proposed multi-stage AGS algorithm with dynamic smoothing is stated in Algorithm 8.6.

Algorithm 8.6 The multi-stage AGS algorithm with dynamic smoothing

Choose $v_0 \in X$, accuracy ε, smoothing parameter ρ_0, iteration limit N_0, and initial estimate Δ_0 of (8.2.1) such that $\psi(v_0) - \psi^* \le \Delta_0$.
for $s = 1, \ldots, S$ **do**
Run the AGS algorithm to problem (8.2.3) with $\rho = 2^{-s/2}\rho_0$ (where $h = h_\rho$ in AGS). In the AGS algorithm, set $x_0 = v_{s-1}$, $N = N_0$, and parameters in Corollary 8.6, and let $v_s = \overline{x}_N$.
end for
Output v_S.

Theorem 8.7 describes the main convergence properties of Algorithm 8.6.

Theorem 8.7. *Let $\varepsilon > 0$ be given and suppose that the Lipschitz constant L in* (8.2.66) *satisfies*

$$D_Y^2\|A\|^2 \max\left\{\sqrt{\tfrac{15\Delta_0}{\varepsilon}}, 1\right\} \ge 2\Delta_0 L.$$

If the parameters in Algorithm 8.6 are set to

$$N_0 = 3\sqrt{\tfrac{2L}{\mu}},\ S = \log_2 \max\left\{\tfrac{15\Delta_0}{\varepsilon}, 1\right\},\ \textit{and}\ \rho_0 = \tfrac{4\Delta_0}{D_Y^2 2^{S/2}}, \tag{8.2.78}$$

then the output v_S of this algorithm must be an ε-solution (8.2.1). *Moreover, the total number of gradient evaluations of ∇f and operator evaluations involving A and A^T performed by Algorithm 8.6 can be bounded by*

$$N_f := 3\sqrt{\tfrac{2L}{\mu}} \log_2 \max\left\{\tfrac{15\Delta_0}{\varepsilon}, 1\right\} \tag{8.2.79}$$

and

$$N_A := 18\sqrt{\tfrac{L}{\mu}} \log_2 \max\left\{\tfrac{15\Delta_0}{\varepsilon}, 1\right\} + \tfrac{56 D_Y\|A\|}{\sqrt{\mu\Delta_0}} \cdot \max\left\{\sqrt{\tfrac{15\Delta_0}{\varepsilon}}, 1\right\}, \tag{8.2.80}$$

respectively.

Proof. Suppose that x^* is an optimal solution to (8.2.1). By (8.2.65) in the proof of Corollary 8.6, in the s-th stage of Algorithm 8.6 (calling AGS with input $x_0 = v_{s-1}$, output $v_s = \overline{x}_N$, and iteration number $N = N_0$), we have

$$\begin{aligned}
&\psi_\rho(v_s) - \psi_\rho(x^*) = \psi_\rho(\overline{x}_N) - \psi_\rho(x^*)\\
&\le \tfrac{9L}{N_0(N_0+1)} V(x_0, x^*) \le \tfrac{\mu}{2} V(x_0, x^*) = \tfrac{\mu}{2} V(v_{s-1}, x^*),
\end{aligned}$$

where the last two inequalities follow from (8.2.78), respectively. Moreover, by (8.2.73) we have $\psi(v_s) \le \psi_\rho(v_s) + \rho D_Y^2$ and $\psi^* = \psi(x^*) \ge \psi_\rho(x^*)$, hence

$$\psi(v_s) - \psi^* \le \psi_\rho(v_s) - \psi_\rho(x^*) + \rho D_Y^2.$$

Combing the above two equations and using the strong convexity of $\psi(\cdot)$, we have

$$\begin{aligned}&\psi(v_s)-\psi^*\le \tfrac{\mu}{2}V(v_{s-1},x^*)+\rho D_Y^2\\ &\le\tfrac{1}{2}[\psi(v_{s-1})-\psi^*]+\rho D_Y^2=\tfrac{1}{2}[\psi(v_{s-1})-\psi^*]+2^{-s/2}\rho_0 D_Y^2,\end{aligned}$$

where the last equality is due to the selection of ρ in Algorithm 8.6. Reformulating the above relation as

$$2^s[\psi(v_s)-\psi^*]\le 2^{s-1}[\psi(v_{s-1})-\psi^*]+2^{s/2}\rho_0 D_Y^2,$$

and summing the above inequalities from $s=1,\ldots,S$, we have

$$\begin{aligned}&2^S(\psi(v_S)-\psi^*)\\ &\le\Delta_0+\rho_0 D_Y^2\Sigma_{s=1}^S 2^{s/2}=\Delta_0+\rho_0 D_Y^2\tfrac{\sqrt{2}(2^{S/2}-1)}{\sqrt{2}-1}<\Delta_0+\tfrac{7}{2}\rho_0 D_Y^2 2^{S/2}=15\Delta_0,\end{aligned}$$

where the first inequality follows from the fact that $\psi(v_0)-\psi^*\le\Delta_0$ and the last equality is due to (8.2.78). By (8.2.78) and the above result, we have $\psi(v_S)-\psi^*\le\varepsilon$. Comparing the descriptions of Algorithms 8.4 and 8.6, we can clearly see that the total number of gradient evaluations of ∇f in Algorithm 8.6 is given $N_0 S$, hence we have (8.2.79).

To complete the proof it suffices to estimate the total number of operator evaluations involving A and A^T. Note that in the s-th stage of Algorithm 8.6, the number of operator evaluations involving A is equivalent to twice the number of evaluations of ∇h_ρ in the AGS algorithm, which, in view of (8.2.63) in Corollary 8.6, is given by

$$\begin{aligned}&2(1+\ln 3)N\left(\sqrt{\tfrac{M}{L}}+1\right)\\ &=2(1+\ln 3)N\left(\sqrt{\tfrac{\|A\|^2}{\rho L}}+1\right)=2(1+\ln 3)N_0\left(\sqrt{\tfrac{2^{s/2}\|A\|^2}{\rho_0 L}}+1\right),\end{aligned}$$

where we used the relation $M=\|A\|^2/\rho$ (see Sect. 8.2.2.1) in the first equality and relations $\rho=2^{-s/2}\rho_0$ and $N=N_0$ from Algorithm 8.6 in the last equality. It then follows from the above result and (8.2.78) that the total number of operator evaluations involving A in Algorithm 8.6 can be bounded by

$$\begin{aligned}&\Sigma_{s=1}^S 2(1+\ln 3)N_0\left(\sqrt{\tfrac{2^{s/2}\|A\|^2}{\rho_0 L}}+1\right)\\ &=2(1+\ln 3)N_0 S+\tfrac{2(1+\ln 3)N_0\|A\|}{\sqrt{\rho_0 L}}\Sigma_{s=1}^S 2^{s/4}\\ &=2(1+\ln 3)N_0 S+\tfrac{3\sqrt{2}(1+\ln 3)D_Y\|A\|2^{S/4}}{\sqrt{\mu\Delta_0}}\cdot\tfrac{2^{1/4}(2^{S/4}-1)}{2^{1/4}-1}\\ &<2(1+\ln 3)N_0 S+\tfrac{56 D_Y\|A\|}{\sqrt{\mu\Delta_0}}\cdot 2^{S/2}\end{aligned}$$

$$<18\sqrt{\tfrac{L}{\mu}}\log_2 \max\left\{\tfrac{15\Delta_0}{\varepsilon},1\right\}+\tfrac{56D_Y\|A\|}{\sqrt{\mu\Delta_0}}\cdot\max\left\{\sqrt{\tfrac{15\Delta_0}{\varepsilon}},1\right\}.$$

□

■

By Theorem 8.7, the total number of operator evaluations involving A performed by Algorithm 8.6 to compute an ε-solution of (8.2.8) can be bounded by

$$\mathcal{O}\left(\sqrt{\tfrac{L}{\mu}}\log\tfrac{1}{\varepsilon}+\tfrac{\|A\|}{\sqrt{\varepsilon}}\right),$$

which matches with the best-known complexity result (e.g., Sect. 3.6). However, the total number of gradient evaluations of ∇f is now bounded by

$$\mathcal{O}\left(\sqrt{\tfrac{L}{\mu}}\log\tfrac{1}{\varepsilon}\right),$$

which drastically improves existing results from $\mathcal{O}(1/\sqrt{\varepsilon})$ to $\mathcal{O}(\log(1/\varepsilon))$.

8.3 Communication Sliding and Decentralized Optimization

In this section, we consider the following decentralized optimization problem which is cooperatively solved by the network of m agents:

$$\begin{aligned} f^* := \min_x\ & f(x) := \textstyle\sum_{i=1}^m f_i(x) \\ \text{s.t. }\ & x\in X, \quad X := \cap_{i=1}^m X_i, \end{aligned} \tag{8.3.1}$$

where $f_i : X_i \to \mathbb{R}$ is a convex and possibly nonsmooth objective function of agent i. Note that f_i and X_i are private and only known to agent i. Throughout this section, we assume the feasible set X is nonempty.

In this section, we also consider the situation where one can only have access to noisy first-order information (function values and subgradients) of the functions f_i, $i=1,\ldots,m$. This happens, for example, when the function f_i's are given in the form of expectation, i.e.,

$$f_i(x) := \mathbb{E}_{\xi_i}[F_i(x;\xi_i)], \tag{8.3.2}$$

where the random variable ξ_i models a source of uncertainty and the distribution $\mathbb{P}(\xi_i)$ is not known in advance. As a special case of (8.3.2), f_i may be given as the summation of many components, i.e.,

$$f_i(x) := \textstyle\sum_{j=1}^l f_i^j(x), \tag{8.3.3}$$

where $l \geq 1$ is a large number. Stochastic optimization problem of this type has great potential of applications in data analysis, especially in machine learning. In particular, problem (8.3.2) corresponds to the minimization of generalized risk and is particularly useful for dealing with online (streaming) data distributed over a network, while problem (8.3.3) aims at the collaborative minimization of empirical risk.

Currently the dominant approach to solve (8.3.1) is to collect all agents' private data on a server (or cluster) and to apply centralized machine learning techniques. However, this centralization scheme would require agents to submit their private data to the service provider without much control on how the data will be used, in addition to incurring high setup cost related to the transmission of data to the service provider. Decentralized optimization provides a viable approach to deal with these data privacy related issues. Each network agent i is associated with the local objective function $f_i(x)$ and all agents intend to cooperatively minimize the system objective $f(x)$ as the sum of all local objective f_i's in the absence of full knowledge about the global problem and network structure. A necessary feature in decentralized optimization is, therefore, that the agents must communicate with their neighboring agents to propagate the distributed information to every location in the network.

Many of the current studies on optimization over networks have been focused on incremental gradient methods (see Sect. 5.2). All of these incremental methods are not fully decentralized in a sense that they require a special star network topology in which the existence of a central authority is necessary for operation. To consider a more general distributed network topology without a central authority, one can possibly generalize the subgradient descent methods by requiring each node to compute a local subgradient and followed by the communication with neighboring agents iteratively. However, subgradient methods converge slowly, achieving a rate of convergence as $\mathcal{O}(1/\varepsilon^2)$ to obtain an ε-optimal solution, i.e., a point $\hat{x} \in X$, s.t., $\mathbb{E}[f(\hat{x}) - f^*] \le \varepsilon$. While the subgradient computation at each step can be inexpensive, due to the fact that one iteration in decentralized optimization is equivalent to at least one communication round among agents, these methods can incur a significant latency for solving (8.3.1). In fact, CPUs in these days can read and write the memory at over 10–100 GB per second whereas communication over TCP/IP is about 100 MB per second. Therefore, the gap between intra-node computation and inter-node communication is about three orders of magnitude. The communication start-up cost itself is also not negligible as it usually takes a few milliseconds. Improvements on communication complexity can be obtained when the objective function (8.3.1) is smooth and/or strongly convex.

Besides subgradient based methods, another well-known type of decentralized algorithms relies on dual methods, where at each step for a fixed dual variable, the primal variables are solved to minimize some local Lagrangian related function, then the dual variables associated with the consistency constraints are updated accordingly. In particular, decentralized alternating direction method of multipliers (ADMM) algorithms have received much attention recently (see Sect. 3.7). For relatively simple convex functions f_i, the decentralized ADMM has been shown to require $\mathcal{O}(1/\varepsilon)$ communications. An improved $\mathcal{O}(\log 1/\varepsilon)$ complexity bound on communication rounds can be achieved for decentralized ADMM if stronger assumptions, i.e., smoothness and strong convexity, are imposed on f_i. Although dual type methods usually require fewer numbers of iterations (hence, fewer communication rounds) than the subgradient based methods, the local Lagrangian minimization problem associated with each agent cannot be solved efficiently in many cases, especially when the problem is constrained.

While decentralized algorithms for solving deterministic optimization problems have been extensively studied during the past few years, there exists only limited research on decentralized stochastic optimization, for which only noisy gradient information of functions f_i, $i = 1,\ldots,m$, in (8.3.1) can be easily computed. Existing decentralized stochastic first-order methods for problem (8.3.1) require $\mathcal{O}(1/\varepsilon^2)$ inter-node communications and intra-node gradient computations to obtain an ε-optimal solution for solving general convex problems. When the objective functions are strongly convex, multiagent mirror descent method for decentralized stochastic optimization can achieve an $\mathcal{O}(1/\varepsilon)$ complexity bound. All these previous works in decentralized stochastic optimization suffered from high communication costs due to the coupled scheme for stochastic subgradient evaluation and communication, i.e., each evaluation of stochastic subgradient will incur one round of communication.

Inspired by the gradient sliding methods in Sect. 8.1, the main goal of this section is to develop dual based decentralized algorithms for solving (8.3.1) which are communication efficient and have local subproblems approximately solved by each agent through the utilization of (noisy) first-order information of f_i. More specifically, we will provide a theoretical understanding on how many rounds of inter-node communications and intra-node (stochastic) subgradient computations of f_i are required in order to find a certain approximate solution of (8.3.1) in which f_i's are convex or strongly convex, but not necessarily smooth, and their exact first-order information is not necessarily computable.

More specifically, we first introduce a new decentralized primal–dual type method, called decentralized communication sliding (DCS), where the agents can skip communications while solving their local subproblems iteratively through successive linearizations of their local objective functions. We show that agents can still find an ε-optimal solution in $\mathcal{O}(1/\varepsilon)$ (resp., $\mathcal{O}(1/\sqrt{\varepsilon})$) communication rounds while maintaining the $\mathcal{O}(1/\varepsilon^2)$ (resp., $\mathcal{O}(1/\varepsilon)$) bound on the total number of intra-node subgradient evaluations when the objective functions are general convex (resp., strongly convex). The bounds on the subgradient evaluations are actually comparable to those optimal complexity bounds required for centralized nonsmooth optimization under certain conditions on the target accuracy, and hence are not improvable in general.

We then present a stochastic decentralized communication sliding method, denoted by SDCS, for solving stochastic optimization problems and show complexity bounds similar to those of DCS on the total number of required communication rounds and stochastic subgradient evaluations. In particular, only $\mathcal{O}(1/\varepsilon)$ (resp., $\mathcal{O}(1/\sqrt{\varepsilon})$) communication rounds are required while agents perform up to $\mathcal{O}(1/\varepsilon^2)$ (resp., $\mathcal{O}(1/\varepsilon)$) stochastic subgradient evaluations for general convex (resp., strongly convex) functions. Only requiring the access to stochastic subgradient at each iteration, SDCS is particularly efficient for solving problems with f_i given in the form of (8.3.2) and (8.3.3). In the former case, SDCS requires only one realization of the random variable at each iteration and provides a communication-efficient way to deal with streaming data and decentralized machine learning. In the latter case, each iteration of SDCS requires only one randomly selected component, leading up to a factor of $\mathscr{O}(l)$ savings on the total number of subgradient computations over DCS.

To fix notation, all vectors are viewed as column vectors, and for a vector $x \in \mathbb{R}^d$, we use $x^\top$ to denote its transpose. For a stacked vector of x_i's, we often use $(x_1,\ldots,x_m)$ to represent the column vector $[x_1^\top,\ldots,x_m^\top]^\top$. We denote by $\mathbf{0}$ and $\mathbf{1}$ the vector of all zeros and ones whose dimensions vary from the context. The cardinality of a set S is denoted by $|S|$. We use I_d to denote the identity matrix in $\mathbb{R}^{d\times d}$. We use $A\otimes B$ for matrices $A \in \mathbb{R}^{n_1\times n_2}$ and $B \in \mathbb{R}^{m_1\times m_2}$ to denote their Kronecker product of size $\mathbb{R}^{n_1m_1\times n_2m_2}$. For a matrix $A \in \mathbb{R}^{n\times m}$, we use A_{ij} to denote the entry of i-th row and j-th column. For any $m \geq 1$, the set of integers $\{1,\ldots,m\}$ is denoted by $[m]$.

8.3.1 Problem Formulation

In Sects. 8.3.1.1 and 8.3.1.2 we introduce the saddle point reformulation of (8.3.1) and define appropriate gap functions which will be used for the convergence analysis of our algorithms. Moreover, in Sect. 8.3.1.3 we provide a brief review on the distance generating function and prox-function under the decentralized setting.

8.3.1.1 Problem Formulation

Consider a multiagent network system whose communication is governed by an undirected graph $\mathscr{G} = (\mathscr{N},\mathscr{E})$, where $\mathscr{N} = [m]$ indexes the set of agents, and $\mathscr{E} \subseteq \mathscr{N}\times\mathscr{N}$ represents the pairs of communicating agents. If there exists an edge from agent i to j which we denote by (i,j), agent i may send its information to agent j and vice versa. Thus, each agent $i \in \mathscr{N}$ can directly receive (resp., send) information only from (resp., to) the agents in its neighborhood

$$N_i = \{j \in \mathscr{N} \mid (i,j) \in \mathscr{E}\} \cup \{i\}, \tag{8.3.4}$$

where we assume that there always exists a self-loop (i,i) for all agents $i \in \mathscr{N}$. Then, the associated Laplacian $L \in \mathbb{R}^{m\times m}$ of $\mathscr{G}$ is $L := D - A$ where D is the diagonal degree matrix, and $A \in \mathbb{R}^{m\times m}$ is the adjacency matrix with the property that $A_{ij} = 1$ if and only if $(i,j) \in \mathscr{E}$ and $i \neq j$, i.e.,

$$L_{ij} = \begin{cases} |N_i| - 1 & \text{if } i = j \\ -1 & \text{if } i \neq j \text{ and } (i,j) \in \mathscr{E} \\ 0 & \text{otherwise.} \end{cases} \tag{8.3.5}$$

We consider a reformulation of problem (8.3.1) which will be used in the development of our decentralized algorithms. We introduce an individual copy x_i of the decision variable x for each agent $i \in \mathscr{N}$ and impose the constraint $x_i = x_j$ for all pairs $(i, j) \in \mathscr{E}$. The transformed problem can be written compactly by using the Laplacian matrix L:

$$\begin{aligned} &\min_{\mathbf{x}} F(\mathbf{x}) := \textstyle\sum_{i=1}^m f_i(x_i) \\ &\text{s.t. } \mathbf{L}\mathbf{x} = \mathbf{0}, \quad x_i \in X_i, \text{ for all } i = 1, \ldots, m, \end{aligned} \tag{8.3.6}$$

where $\mathbf{x} = (x_1, \ldots, x_m) \in X_1 \times \ldots \times X_m$, $F : X_1 \times \ldots \times X_m \to \mathbb{R}$, and $\mathbf{L} = L \otimes I_d \in \mathbb{R}^{md \times md}$. The constraint $\mathbf{L}\mathbf{x} = \mathbf{0}$ is a compact way of writing $x_i = x_j$ for all agents i and j which are connected by an edge. By construction,$\mathbf{L}$ is symmetric positive semidefinite and its null space coincides with the "agreement" subspace, i.e., $\mathbf{L}\mathbf{1} = \mathbf{0}$ and $\mathbf{1}^\top \mathbf{L} = \mathbf{0}$. To ensure each node gets information from every other node, we need the following assumption.

Assumption 18 *The graph $\mathscr{G}$ is connected.*

Under Assumption 18, problems (8.3.1) and (8.3.6) are equivalent. We let Assumption 18 be a blanket assumption for the rest of this section.

We next consider a reformulation of the problem (8.3.6) as a saddle point problem. By the method of Lagrange multipliers, problem (8.3.6) is equivalent to the following saddle point problem:

$$\min_{\mathbf{x} \in \mathbf{X}} \left[F(\mathbf{x}) + \max_{\mathbf{y} \in \mathbb{R}^{md}} \langle \mathbf{L}\mathbf{x}, \mathbf{y} \rangle \right], \tag{8.3.7}$$

where $\mathbf{X} := X_1 \times \ldots \times X_m$ and $\mathbf{y} = (y_1, \ldots, y_m) \in \mathbb{R}^{md}$ are the Lagrange multipliers associated with the constraints $\mathbf{L}\mathbf{x} = \mathbf{0}$. We assume that there exists an optimal solution $\mathbf{x}^* \in \mathbf{X}$ of (8.3.6) and that there exists $\mathbf{y}^* \in \mathbb{R}^{md}$ such that $(\mathbf{x}^*, \mathbf{y}^*)$ is a saddle point of (8.3.7). In fact, since our objective function $F(\mathbf{x})$ is convex, strong duality holds if constraint qualification (CQ) condition holds. In particular, CQ condition states that there exists $\bar{\mathbf{x}} \in \mathbf{X}$ such that $\mathbf{L}\bar{\mathbf{x}} = \mathbf{0}$, which is implied by the assumption that there exists an optimal solution to (8.3.6).

8.3.1.2 Gap Functions: Termination Criteria

Given a pair of feasible solutions $\mathbf{z} = (\mathbf{x}, \mathbf{y})$ and $\bar{\mathbf{z}} = (\bar{\mathbf{x}}, \bar{\mathbf{y}})$ of (8.3.7), we define the *primal-dual gap function* $Q(\mathbf{z}; \bar{\mathbf{z}})$ by

$$Q(\mathbf{z}; \bar{\mathbf{z}}) := F(\mathbf{x}) + \langle \mathbf{L}\mathbf{x}, \bar{\mathbf{y}} \rangle - [F(\bar{\mathbf{x}}) + \langle \mathbf{L}\bar{\mathbf{x}}, \mathbf{y} \rangle]. \tag{8.3.8}$$

Sometimes we also use the notations $Q(\mathbf{z}; \bar{\mathbf{z}}) := Q(\mathbf{x}, \mathbf{y}; \bar{\mathbf{x}}, \bar{\mathbf{y}})$ or $Q(\mathbf{z}; \bar{\mathbf{z}}) := Q(\mathbf{x}, \mathbf{y}; \bar{\mathbf{z}}) = Q(\mathbf{z}; \bar{\mathbf{x}}, \bar{\mathbf{y}})$. One can easily see that $Q(\mathbf{z}^*; \mathbf{z}) \le 0$ and $Q(\mathbf{z}; \mathbf{z}^*) \ge 0$ for all $\mathbf{z} \in \mathbf{X} \times \mathbb{R}^{md}$, where $\mathbf{z}^* = (\mathbf{x}^*, \mathbf{y}^*)$ is a saddle point of (8.3.7). For compact sets $\mathbf{X} \subset \mathbb{R}^{md}$, $Y \subset \mathbb{R}^{md}$, the gap function

$$\sup_{\bar{\mathbf{z}} \in \mathbf{X} \times Y} Q(\mathbf{z}; \bar{\mathbf{z}}) \tag{8.3.9}$$

measures the accuracy of the approximate solution $\mathbf{z}$ to the saddle point problem (8.3.7).

However, the saddle point formulation (8.3.7) of our problem of interest (8.3.1) may have an unbounded feasible set. We adopt the perturbation-based termination criterion and propose a modified version of the gap function in (8.3.9). More specifically, we define

$$g_Y(\mathbf{s}, \mathbf{z}) := \sup_{\bar{\mathbf{y}} \in Y} Q(\mathbf{z}; \mathbf{x}^*, \bar{\mathbf{y}}) - \langle \mathbf{s}, \bar{\mathbf{y}} \rangle, \tag{8.3.10}$$

for any closed set $Y \subseteq \mathbb{R}^{md}$, $\mathbf{z} \in \mathbf{X} \times \mathbb{R}^{md}$, and $\mathbf{s} \in \mathbb{R}^{md}$. If $Y = \mathbb{R}^{md}$, we omit the subscript Y and simply use the notation $g(\mathbf{s}, \mathbf{z})$.

This perturbed gap function allows us to bound the objective function value and the feasibility separately. We first define the following terminology.

Definition 8.1. A point $\mathbf{x} \in \mathbf{X}$ is called an (ε, δ)-solution of (8.3.6) if

$$F(\mathbf{x}) - F(\mathbf{x}^*) \leq \varepsilon \text{ and } \|\mathbf{L}\mathbf{x}\| \leq \delta. \tag{8.3.11}$$

We say that $\mathbf{x}$ has primal residual ε and feasibility residual δ.

Similarly, a stochastic (ε, δ)-solution of (8.3.6) can be defined as a random point $\hat{\mathbf{x}} \in \mathbf{X}$ s.t. $\mathbb{E}[F(\hat{\mathbf{x}}) - F(\mathbf{x}^*)] \leq \varepsilon$ and $\mathbb{E}[\|\mathbf{L}\hat{\mathbf{x}}\|] \leq \delta$ for some $\varepsilon, \delta > 0$. Note that for problem (8.3.6), the feasibility residual measures the disagreement among the local copies x_i, for $i \in \mathcal{N}$.

In the following proposition, we establish the relationship between the perturbed gap function (8.3.10) and the approximate solutions to problem (8.3.6). Although the proposition was originally developed for deterministic cases, the extension of this to stochastic cases is straightforward.

Proposition 8.6. *For any $Y \subset \mathbb{R}^{md}$ such that $\mathbf{0} \in Y$, if $g_Y(\mathbf{L}\mathbf{x}, \mathbf{z}) \leq \varepsilon < \infty$ and $\|\mathbf{L}\mathbf{x}\| \leq \delta$, where $\mathbf{z} = (\mathbf{x}, \mathbf{y}) \in \mathbf{X} \times \mathbb{R}^{md}$, then $\mathbf{x}$ is an (ε, δ)-solution of* (8.3.6). *In particular, when $Y = \mathbb{R}^{md}$, for any $\mathbf{s}$ such that $g(\mathbf{s}, \mathbf{z}) \leq \varepsilon < \infty$ and $\|\mathbf{s}\| \leq \delta$, we always have* $\mathbf{s} = \mathbf{L}\mathbf{x}$.

Proof. By (8.3.8) and (8.3.10), for all $\mathbf{s} \in \mathbb{R}^{md}$ and $Y \subseteq \mathbb{R}^{md}$, we have

$$\begin{aligned} g_Y(\mathbf{s}, \mathbf{z}) &= \sup_{\tilde{y} \in Y} [F(\mathbf{x}) - \langle \tilde{y}, \mathbf{L}\mathbf{x} \rangle] - F(\mathbf{x}^*) + \langle \mathbf{s}, \tilde{y} \rangle \\ &= F(\mathbf{x}) - F(\mathbf{x}^*) + \sup_{\tilde{y} \in Y} \langle -\tilde{y}, \mathbf{L}\mathbf{x} - \mathbf{s} \rangle. \end{aligned}$$

From the above we see that if $g_Y(\mathbf{L}\mathbf{x}, \mathbf{z}) = F(\mathbf{x}) - F(\mathbf{x}^*) \leq \varepsilon$ and $\|\mathbf{L}\mathbf{x}\| \leq \delta$, then (w, z) is an (ε, δ)-solution. In addition, if $Y = \mathbb{R}^{md}$, we can also see that $g(\mathbf{s}, \mathbf{z}) = \infty$ if $\mathbf{s} \neq \mathbf{L}\mathbf{x}$, hence $g(\mathbf{s}, \mathbf{z}) < \infty$ implies that $\mathbf{s} = \mathbf{L}\mathbf{x}$. ∎

8.3.1.3 Prox-Function

We assume that the individual constraint set X_i for each agent in problem (8.3.1) are equipped with norm $\|\cdot\|_{X_i}$, and their associated prox-functions associated with the distance generating function ω_i are given by $V_i(\cdot,\cdot)$. Moreover, we assume that each $V_i(\cdot,\cdot)$ shares the same strongly convex modulus $\nu = 1$, i.e.,

$$V_i(x_i,u_i) \geq \tfrac{1}{2}\|x_i - u_i\|_{X_i}^2, \quad \forall x_i, u_i \in X_i,\ i = 1,\ldots,m. \tag{8.3.12}$$

We define the norm associated with the primal feasible set $\mathbf{X} = X_1 \times \ldots \times X_m$ of (8.3.7) as follows:[1]

$$\|\mathbf{x}\|^2 \equiv \|\mathbf{x}\|_{\mathbf{X}}^2 := \textstyle\sum_{i=1}^m \|x_i\|_{X_i}^2, \tag{8.3.13}$$

where $\mathbf{x} = (x_1,\ldots,x_m) \in \mathbf{X}$ for any $x_i \in X_i$. Therefore, the corresponding prox-function $\mathbf{V}(\cdot,\cdot)$ can be defined as

$$\mathbf{V}(\mathbf{x},\mathbf{u}) := \textstyle\sum_{i=1}^m V_i(x_i,u_i),\ \forall \mathbf{x},\mathbf{u} \in \mathbf{X}. \tag{8.3.14}$$

Note that by (8.3.12) and (8.3.13), it can be easily seen that

$$\mathbf{V}(\mathbf{x},\mathbf{u}) \geq \tfrac{1}{2}\|\mathbf{x}-\mathbf{u}\|^2,\ \forall \mathbf{x},\mathbf{u} \in \mathbf{X}. \tag{8.3.15}$$

Throughout this section, we endow the dual space where the multipliers $\mathbf{y}$ of (8.3.7) reside with the standard Euclidean norm $\|\cdot\|_2$, since the feasible region of $\mathbf{y}$ is unbounded. For simplicity, we often write $\|\mathbf{y}\|$ instead of $\|\mathbf{y}\|_2$ for a dual multiplier $\mathbf{y} \in \mathbb{R}^{md}$.

Given the prox-function V_i, we assume that the objective functions associated with agent i satisfy

$$\mu V_i(y,x) \leq f_i(x) - f_i(y) - \langle f_i'(y), x - y\rangle \leq M\|x-y\|,\ \ \forall x,y \in X_i, \tag{8.3.16}$$

for some $M,\mu \geq 0$ and $f_i'(y) \in \partial f_i(y)$, where $\partial f_i(y)$ denotes the subdifferential of f_i at y, and $X_i \subseteq \mathbb{R}^d$ is a closed convex constraint set of agent i. Clearly, f_i's are strongly convex if $\mu > 0$.

8.3.2 Decentralized Communication Sliding

In this section, we introduce a primal–dual algorithmic framework, namely, the decentralized communication sliding (DCS) method, for solving the saddle point problem (8.3.7) in a decentralized fashion. Moreover, we will establish complexity bounds on the required number of inter-node communication rounds as well as

---

[1] We can define the norm associated with $\mathbf{X}$ in a more general way, e.g., $\|\mathbf{x}\|^2 := \sum_{i=1}^m p_i\|x_i\|_{X_i}^2$, $\forall \mathbf{x} = (x_1,\ldots,x_m) \in \mathbf{X}$, for some $p_i > 0$, $i = 1,\ldots,m$. Accordingly, the prox-function $\mathbf{V}(\cdot,\cdot)$ can be defined as $\mathbf{V}(\mathbf{x},\mathbf{u}) := \sum_{i=1}^m p_i V_i(x_i,u_i)$, $\forall \mathbf{x},\mathbf{u} \in \mathbf{X}$. This setting gives us flexibility to choose p_i's based on the information of individual X_i's, and the possibility to further refine the convergence results.

the total number of required subgradient evaluations. Throughout this section, we consider the deterministic case where exact subgradients of f_i's are available.

8.3.2.1 The DCS Algorithm

The basic scheme of the DCS algorithm is inspired by the primal–dual method in Sect. 3.6. When applied to our saddle point reformulation defined in (8.3.7), for any given initial points $\mathbf{x}^0 = \mathbf{x}^{-1} \in \mathbf{X}$ and $\mathbf{y}^0 \in \mathbb{R}^{md}$, and certain nonnegative parameters $\{\alpha_k\}$, $\{\tau_k\}$ and $\{\eta_k\}$, the primal–dual method updates $(\mathbf{x}^k, \mathbf{y}^k)$ according to

$$\tilde{\mathbf{x}}^k = \alpha_k(\mathbf{x}^{k-1} - \mathbf{x}^{k-2}) + \mathbf{x}^{k-1}, \tag{8.3.17}$$

$$\mathbf{y}^k = \mathrm{argmin}_{\mathbf{y}\in\mathbb{R}^{md}} \, \langle -\mathbf{L}\tilde{\mathbf{x}}^k, \mathbf{y}\rangle + \tfrac{\tau_k}{2}\|\mathbf{y} - \mathbf{y}^{k-1}\|^2, \tag{8.3.18}$$

$$\mathbf{x}^k = \mathrm{argmin}_{\mathbf{x}\in\mathbf{X}} \left\{ \Phi^k(\mathbf{x}) := \langle \mathbf{L}\mathbf{y}^k, \mathbf{x}\rangle + F(\mathbf{x}) + \eta_k \mathbf{V}(\mathbf{x}^{k-1}, \mathbf{x}) \right\}. \tag{8.3.19}$$

In each iteration of the primal–dual method, only the computation of the matrix-vector products $\mathbf{L}\tilde{\mathbf{x}}^k$ and $\mathbf{L}\mathbf{y}^k$ will involve the communication among different agents, while the other computations such as updating of $\tilde{\mathbf{x}}^k, \mathbf{y}^k$ and $\mathbf{x}^k$ can be performed separately by each agent. Under the assumption that the subproblem (8.3.19) can be easily solved, we can show that by properly choosing the algorithmic parameters α_k, τ_k, and η_k one can find an ε-solution, i.e., a point $\bar{\mathbf{x}} \in \mathbf{X}$ such that $F(\bar{\mathbf{x}}) - F(\mathbf{x}^*) \le \varepsilon$ and $\|\mathbf{L}\bar{\mathbf{x}}\| \le \varepsilon$, within $\mathcal{O}(1/\varepsilon)$ iterations. This implies that one can find such an ε-solution in $\mathcal{O}(1/\varepsilon)$ rounds of communication, which already improves the existing $\mathcal{O}(1/\varepsilon^2)$ communication complexity for decentralized nonsmooth optimization. However, such a communication complexity bound is not quite meaningful because F is a general nonsmooth convex function and it is often difficult to solve the primal subproblem (8.3.19) explicitly.

One natural way to address this issue is to approximately solve (8.3.19) through an iterative subgradient descent method. Inside this iterative subgradient descent method, we do not need to recompute the matrix-vector products $\mathbf{L}\tilde{\mathbf{x}}^k$ and $\mathbf{L}\mathbf{y}^k$, and hence no communication cost is involved. However, a straightforward pursuit of this approach, i.e., to solve the subproblem accurately enough at each iteration, does not necessarily yield the best complexity bound in terms of the total number of subgradient computations. To achieve the best possible complexity bounds in terms of both subgradient computation and communication, the proposed DCS method (along with its analysis) is in fact more complicated than the aforementioned inexact primal–dual method in the following two aspects. First, while in most inexact first-order methods one usually computes only one approximate solution of the subproblems, in the proposed DCS method we need to generate a pair of closely related approximate solutions $\mathbf{x}^k = (x_1^k, \ldots, x_m^k)$ and $\hat{\mathbf{x}}^k = (\hat{x}_1^k, \ldots, \hat{x}_m^k)$ to the subproblem in (8.3.19). Second, we need to modify the primal–dual method in a way such that one of these sequence (i.e.,$\{\hat{\mathbf{x}}^k\}$) will be used in the extrapolation step in (8.3.17), while the other sequence $\{\mathbf{x}^k\}$ will act as the prox-center in $\mathbf{V}(\mathbf{x}^{k-1}, \mathbf{x})$ (see (8.3.19)).

Algorithm 8.7 DCS from agent i's perspective

Let $x_i^0 = x_i^{-1} = \hat{x}_i^0 \in X_i$, $y_i^0 \in \mathbb{R}^d$ for $i \in [m]$ and the nonnegative parameters $\{\alpha_k\}$, $\{\tau_k\}$, $\{\eta_k\}$, and $\{T_k\}$ be given.
for $k = 1, \ldots, N$ **do**
Update $z_i^k = (\hat{x}_i^k, y_i^k)$ according to

$$\tilde{x}_i^k = \alpha_k(\hat{x}_i^{k-1} - x_i^{k-2}) + x_i^{k-1}, \tag{8.3.20}$$

$$v_i^k = \sum_{j \in N_i} L_{ij} \tilde{x}_j^k, \tag{8.3.21}$$

$$y_i^k = \mathrm{argmin}_{y_i \in \mathbb{R}^d} \langle -v_i^k, y_i \rangle + \tfrac{\tau_k}{2} \|y_i - y_i^{k-1}\|^2 = y_i^{k-1} + \tfrac{1}{\tau_k} v_i^k, \tag{8.3.22}$$

$$w_i^k = \sum_{j \in N_i} L_{ij} y_j^k, \tag{8.3.23}$$

$$(x_i^k, \hat{x}_i^k) = \mathrm{CS}(f_i, X_i, V_i, T_k, \eta_k, w_i^k, x_i^{k-1}). \tag{8.3.24}$$

end forreturn $z_i^N = \left(\sum_{k=1}^N \theta_k\right)^{-1} \sum_{k=1}^N \theta_k z_i^k$

The CS (Communication-Sliding) procedure called at (8.3.24) is stated as follows.
procedure: $(x, \hat{x}) = \mathrm{CS}(\phi, U, V, T, \eta, w, x)$
Let $u^0 = \hat{u}^0 = x$ and the parameters $\{\beta_t\}$ and $\{\lambda_t\}$ be given.
for $t = 1, \ldots, T$ **do**

$$h^{t-1} = \phi'(u^{t-1}) \in \partial\phi(u^{t-1}), \tag{8.3.25}$$

$$u^t = \mathrm{argmin}_{u \in U} \left[\langle w + h^{t-1}, u\rangle + \eta V(x, u) + \eta \beta_t V(u^{t-1}, u)\right]. \tag{8.3.26}$$

end for
Set

$$\hat{u}^T := \left(\sum_{t=1}^T \lambda_t\right)^{-1} \sum_{t=1}^T \lambda_t u^t. \tag{8.3.27}$$

Set $x = u^T$ and $\hat{x} = \hat{u}^T$.
end procedure

We formally describe our DCS method in Algorithm 8.7. An outer iteration of the DCS algorithm occurs whenever the index k in Algorithm 8.7 is incremented by 1. More specifically, each primal estimate x_i^0 is locally initialized from some arbitrary point in X_i, and x_i^{-1} and $\hat{x}_i^0$ are also set to be the same value. At each time step $k \geq 1$, each agent $i \in \mathcal{N}$ computes a local prediction $\tilde{x}_i^k$ using these three previous primal iterates (ref. (8.3.20)), and sends it to all of the nodes in its neighborhood, i.e., to all agents $j \in N_i$. In (8.3.21)–(8.3.22), each agent i then calculates the neighborhood disagreement v_i^k using the messages received from agents in N_i, and updates the dual subvector y_i^k. Then, another round of communication occurs in (8.3.23) when calculating w_i^k based on these updated dual variables. Therefore, each outer iteration k involves two communication rounds, one for the primal estimates and the other for the dual variables. Lastly, each agent i approximately solves the proximal projection subproblem (8.3.19), i.e.,

$$\mathrm{argmin}_{u \in U} \langle w, u \rangle + \phi(u) + \eta V(x, u) \tag{8.3.28}$$

with $u = x_i$, $U = X_i$, $w = w_i^k$, $\phi = f_i$, $\eta = \eta_k$, and $V = V_i$, by calling the CS procedure for $T = T_k$ iterations in (8.3.24).

Each iteration performed by the CS procedure, referred to as an inner iteration of the DCS method, is equivalent to a subgradient descent step applied to (8.3.28). More specifically, each inner iteration consists of the computation of the subgradient $\phi'(u^{t-1})$ in (8.3.25) and the solution of the projection subproblem in (8.3.26). Note that the objective function of (8.3.26) consists of two parts: (1) the inner product of u and the summation of w and the current subgradient $\phi'(u^{t-1})$; and (2) two Bregman distances requiring that the new iterate lies near x and u^{t-1}. By using the definition of Bregman distance, we can see that (8.3.26) is equivalent to

$$u^t = \text{argmin}_{u \in U} \left[\langle w + h^{t-1} - \eta \nabla \omega(x) - \eta \beta_t \nabla \omega(u^{t-1}), u \rangle + \eta (1 + \beta_t) \omega(u) \right].$$

Similar to mirror-descent type methods, we assume that this problem is easy to solve. Also observe that the same dual information $w = w_i^k$ (see (8.3.23)) has been used throughout the $T = T_k$ iterations of the CS procedure, and hence no additional communication is required within the procedure, which explains the name of the DCS method.

Observe that the DCS method, in spirit, has been inspired by the gradient sliding method (Sect. 8.1). However, the gradient sliding method focuses on how to save gradient evaluations for solving certain structured convex optimization problems, rather than how to save communication rounds (or matrix-vector products) for decentralized optimization, and its algorithmic scheme is also quite different from the DCS method. It should also be noted that the description of the algorithm is only conceptual at this moment since we have not specified the parameters $\{\alpha_k\}$, $\{\eta_k\}$, $\{\tau_k\}$, $\{T_k\}$, $\{\beta_t\}$, and $\{\lambda_t\}$ yet. We will later instantiate this generic algorithm when we state its convergence properties.

8.3.2.2 Convergence of DCS on General Convex Functions

We now establish the main convergence properties of the DCS algorithm. More specifically, we provide in Lemma 8.3 an estimate on the gap function defined in (8.3.8) together with stepsize policies which work for the general nonsmooth convex case with $\mu = 0$ (cf. (8.3.16)). The proof of this lemma can be found in Sect. 8.3.5.

Lemma 8.3. *Let the iterates* $(\hat{\mathbf{x}}^k, \mathbf{y}^k)$, $k = 1, \ldots, N$ *be generated by Algorithm 8.7 and* $\hat{\mathbf{z}}^N$ *be defined as* $\hat{\mathbf{z}}^N := \left(\sum_{k=1}^N \theta_k\right)^{-1} \sum_{k=1}^N \theta_k (\hat{\mathbf{x}}^k, \mathbf{y}^k)$. *If the objective* f_i, $i = 1, \ldots, m$, *are general nonsmooth convex functions, i.e.,* $\mu = 0$ *and* $M > 0$, *let the parameters* $\{\alpha_k\}$, $\{\theta_k\}$, $\{\eta_k\}$, $\{\tau_k\}$, *and* $\{T_k\}$ *in Algorithm 8.7 satisfy*

$$\theta_k \tfrac{(T_k+1)(T_k+2)\eta_k}{T_k(T_k+3)} \le \theta_{k-1} \tfrac{(T_{k-1}+1)(T_{k-1}+2)\eta_{k-1}}{T_{k-1}(T_{k-1}+3)}, \quad k = 2, \ldots, N, \tag{8.3.29}$$

$$\alpha_k \theta_k = \theta_{k-1}, \quad k = 2, \ldots, N, \tag{8.3.30}$$

$$\theta_k \tau_k = \theta_1 \tau_1, \quad k = 2, \ldots, N, \tag{8.3.31}$$

$$\alpha_k \|\mathbf{L}\|^2 \leq \eta_{k-1}\tau_k, \quad k = 2,\ldots,N, \tag{8.3.32}$$

$$\theta_N \|\mathbf{L}\|^2 \leq \theta_1 \tau_1 \eta_N, \tag{8.3.33}$$

and the parameters $\{\lambda_t\}$ *and* $\{\beta_t\}$ *in the CS procedure of Algorithm 8.7 be set to*

$$\lambda_t = t+1, \quad \beta_t = \tfrac{t}{2}, \quad \forall t \geq 1. \tag{8.3.34}$$

Then, we have for all $\mathbf{z} := (\mathbf{x},\mathbf{y}) \in \mathbf{X} \times \mathbb{R}^{md}$,

$$Q(\hat{\mathbf{z}}^N;\mathbf{z}) \leq \left(\textstyle\sum_{k=1}^N \theta_k\right)^{-1} \Big[\tfrac{(T_1+1)(T_1+2)\theta_1\eta_1}{T_1(T_1+3)} \mathbf{V}(\mathbf{x}^0,\mathbf{x}) + \tfrac{\theta_1\tau_1}{2}\|\mathbf{y}^0\|^2 + \langle \hat{\mathbf{s}}, \mathbf{y}\rangle + \textstyle\sum_{k=1}^N \tfrac{4mM^2\theta_k}{(T_k+3)\eta_k}\Big], \tag{8.3.35}$$

where $\hat{\mathbf{s}} := \theta_N \mathbf{L}(\hat{\mathbf{x}}^N - \mathbf{x}^{N-1}) + \theta_1\tau_1(\mathbf{y}^N - \mathbf{y}^0)$ *and* Q *is defined in* (8.3.8). *Furthermore, for any saddle point* $(\mathbf{x}^*,\mathbf{y}^*)$ *of* (8.3.7), *we have*

$$\begin{aligned} &\tfrac{\theta_N}{2}\left(1 - \tfrac{\|\mathbf{L}\|^2}{\eta_N\tau_N}\right) \max\{\eta_N\|\hat{\mathbf{x}}^N - \mathbf{x}^{N-1}\|^2, \tau_N\|\mathbf{y}^* - \mathbf{y}^N\|^2\} \\ &\leq \tfrac{(T_1+1)(T_1+2)\theta_1\eta_1}{T_1(T_1+3)} \mathbf{V}(\mathbf{x}^0,\mathbf{x}^*) + \tfrac{\theta_1\tau_1}{2}\|\mathbf{y}^* - \mathbf{y}^0\|^2 + \textstyle\sum_{k=1}^N \tfrac{4mM^2\theta_k}{\eta_k(T_k+3)}. \end{aligned} \tag{8.3.36}$$

In the following theorem, we provide a specific selection of $\{\alpha_k\}$, $\{\theta_k\}$, $\{\eta_k\}$, $\{\tau_k\}$, and $\{T_k\}$ satisfying (8.3.29)–(8.3.33). Using Lemma 8.3 and Proposition 8.6, we also establish the complexity of the DCS method for computing an (ε,δ)-solution of problem (8.3.6) when the objective functions are general convex.

Theorem 8.8. *Let* $\mathbf{x}^*$ *be an optimal solution of (8.3.6), the parameters* $\{\lambda_t\}$ *and* $\{\beta_t\}$ *in the CS procedure of Algorithm 8.7 be set to* (8.3.34), *and suppose that* $\{\alpha_k\}$, $\{\theta_k\}$, $\{\eta_k\}$, $\{\tau_k\}$, *and* $\{T_k\}$ *are set to*

$$\alpha_k = \theta_k = 1, \ \eta_k = 2\|\mathbf{L}\|, \ \tau_k = \|\mathbf{L}\|, \text{ and } T_k = \left\lceil \tfrac{mM^2N}{\|\mathbf{L}\|^2\tilde{D}} \right\rceil, \quad \forall k = 1,\ldots,N, \tag{8.3.37}$$

for some $\tilde{D} > 0$. *Then, for any* $N \geq 1$, *we have*

$$F(\hat{\mathbf{x}}^N) - F(\mathbf{x}^*) \leq \tfrac{\|\mathbf{L}\|}{N}\left[3\mathbf{V}(\mathbf{x}^0,\mathbf{x}^*) + \tfrac{1}{2}\|\mathbf{y}^0\|^2 + 2\tilde{D}\right] \tag{8.3.38}$$

and

$$\|\mathbf{L}\hat{\mathbf{x}}^N\| \leq \tfrac{\|\mathbf{L}\|}{N}\left[3\sqrt{6\mathbf{V}(\mathbf{x}^0,\mathbf{x}^*) + 4\tilde{D}} + 4\|\mathbf{y}^* - \mathbf{y}^0\|\right], \tag{8.3.39}$$

where $\hat{\mathbf{x}}^N = \frac{1}{N}\sum_{k=1}^N \hat{\mathbf{x}}^k$, *and* y^* *is an arbitrary dual optimal solution.*

Proof. It is easy to check that (8.3.37) satisfies conditions (8.3.29)–(8.3.33). Particularly,

$$\tfrac{(T_1+1)(T_1+2)}{T_1(T_1+3)} = 1 + \tfrac{2}{T_1^2+3T_1} \leq \tfrac{3}{2}.$$

Therefore, by plugging in these values to (8.3.35), we have

$$Q(\hat{\mathbf{z}}^N;\mathbf{x}^*,\mathbf{y}) \le \tfrac{\|\mathbf{L}\|}{N}\left[3\mathbf{V}(\mathbf{x}^0,\mathbf{x}^*)+\tfrac{1}{2}\|\mathbf{y}^0\|^2+2\tilde{D}\right]+\tfrac{1}{N}\langle\hat{\mathbf{s}},\mathbf{y}\rangle. \tag{8.3.40}$$

Letting $\hat{\mathbf{s}}^N=\frac{1}{N}\hat{\mathbf{s}}$, then from (8.3.36), we have

$$\begin{aligned}\|\hat{\mathbf{s}}^N\| &\le \tfrac{\|\mathbf{L}\|}{N}\left[\|\hat{\mathbf{x}}^N-\mathbf{x}^{N-1}\|+\|\mathbf{y}^N-\mathbf{y}^*\|+\|\mathbf{y}^*-\mathbf{y}^0\|\right]\\ &\le \tfrac{\|\mathbf{L}\|}{N}\left[3\sqrt{6\mathbf{V}(\mathbf{x}^0,\mathbf{x}^*)+\|\mathbf{y}^*-\mathbf{y}^0\|^2+4\tilde{D}}+\|\mathbf{y}^*-\mathbf{y}^0\|\right].\end{aligned}$$

Furthermore, by (8.3.40), we have

$$g(\hat{\mathbf{s}}^N,\hat{\mathbf{z}}^N) \le \tfrac{\|\mathbf{L}\|}{N}\left[3\mathbf{V}(\mathbf{x}^0,\mathbf{x}^*)+\tfrac{1}{2}\|\mathbf{y}^0\|^2+2\tilde{D}\right].$$

Applying Proposition 8.6 to the above two inequalities, the results in (8.3.38) and (8.3.39) follow immediately. ■

We now make some remarks about the results obtained in Theorem 8.8. First, even though one can choose any $\tilde{D}>0$ (e.g., $\tilde{D}=1$) in (8.3.37), the best selection of $\tilde{D}$ would be $\mathbf{V}(\mathbf{x}^0,\mathbf{x}^*)$ so that the first and third terms in (8.3.40) are about the same order. In practice, if there exists an estimate $D_{\mathbf{X}}>0$ s.t.

$$\mathbf{V}(\mathbf{x}_1,\mathbf{x}_2)\le D_{\mathbf{X}}^2,\ \forall \mathbf{x}_1,\mathbf{x}_2\in\mathbf{X}, \tag{8.3.41}$$

then we can set $\tilde{D}=D_{\mathbf{X}}^2$.

Second, the complexity of the DCS method directly follows from (8.3.38) and (8.3.39). For simplicity, let us assume that X is bounded, $\tilde{D}=D_{\mathbf{X}}^2$ and $\mathbf{y}^0=\mathbf{0}$. We can see that the total number of inter-node communication rounds and intra-node subgradient evaluations required by each agent for finding an (ε,δ)-solution of (8.3.6) can be bounded by

$$\mathcal{O}\left\{\|\mathbf{L}\|\max\left(\tfrac{D_{\mathbf{X}}^2}{\varepsilon},\tfrac{D_{\mathbf{X}}+\|\mathbf{y}^*\|}{\delta}\right)\right\} \text{ and } \mathcal{O}\left\{mM^2\max\left(\tfrac{D_{\mathbf{X}}^2}{\varepsilon^2},\tfrac{D_{\mathbf{X}}^2+\|\mathbf{y}^*\|^2}{D_{\mathbf{X}}^2\delta^2}\right)\right\}, \tag{8.3.42}$$

respectively. In particular, if ε and δ satisfy

$$\tfrac{\varepsilon}{\delta}\le\tfrac{D_{\mathbf{X}}^2}{D_{\mathbf{X}}+\|\mathbf{y}^*\|}, \tag{8.3.43}$$

then the previous two complexity bounds in (8.3.42), respectively, reduce to

$$\mathcal{O}\left\{\tfrac{\|\mathbf{L}\|D_{\mathbf{X}}^2}{\varepsilon}\right\} \text{ and } \mathcal{O}\left\{\tfrac{mM^2D_{\mathbf{X}}^2}{\varepsilon^2}\right\}. \tag{8.3.44}$$

Third, it is interesting to compare DCS with the centralized mirror descent method (Sect. 3.2) applied to (8.3.1). In the worst case, the Lipschitz constant of f in (8.3.1) can be bounded by $M_f\le mM$, and each iteration of the method will incur m subgradient evaluations. Hence, the total number of subgradient evaluations performed by the mirror descent method for finding an ε-solution of (8.3.1), i.e., a point $\bar{x} \in X$ such that $f(\bar{x}) - f^* \leq \varepsilon$, can be bounded by

$$\mathcal{O}\left\{\frac{m^3 M^2 D_{\tilde{X}}^2}{\varepsilon^2}\right\}, \tag{8.3.45}$$

where $D_{\tilde{X}}^2$ characterizes the diameter of X, i.e., $D_{\tilde{X}}^2 := \max_{x_1,x_2 \in X} V(x_1,x_2)$. Noting that $D_{\tilde{X}}^2/D_{\mathbf{X}}^2 = \mathcal{O}(1/m)$, and that the second bound in (8.3.44) states only the number of subgradient evaluations for each agent in the DCS method, we conclude that the total number of subgradient evaluations performed by DCS is comparable to the classic mirror descent method as long as (8.3.43) holds and hence not improvable in general.

Finally, observe that the parameter setting (8.3.37) requires the knowledge of the norm of Laplacian matrix $\mathbf{L}$, i.e., $\|\mathbf{L}\| = \max_{\|x\| \leq 1}\{\|\mathbf{L}\mathbf{x}\|_2\}$. If we use l_2-norm for the primal space, $\|\mathbf{L}\|$ will be the maximum eigenvalue of L. We can estimate it using power iteration method or simply bound it by the maximum degree of the graph. If we use l_1-norm in the primal space, then $\|\mathbf{L}\|$ will be the $L_{1,2}$-norm for $\|\mathbf{L}\|$, i.e., $\|\mathbf{L}\| = \|\mathbf{L}\|_{1,2} = \left(\sum_{i=1}^{md} \|\mathbf{L}_i\|_1^2\right)^{1/2} = 2\sqrt{d\sum_{j=1}^{m} \deg_j^2}$, where $\mathbf{L}_i$'s denote the row vectors of $\mathbf{L}$ and $\deg_j$ denotes the degree of node j. The estimation of $\|\mathbf{L}\|$ will involve a few rounds of communication; however, these initial setup costs are independent of the target accuracy ε of the solution. It should also be noted that the number of inner iterations T_k given in (8.3.37) is fixed as a constant in order to achieve the best complexity bounds. In practice, it is reasonable to choose T_k dynamically so that a smaller number of inner iterations will be performed in the first few outer iterations. One simple strategy would be to set

$$T_k = \min\left(ck, \left\lceil \frac{mM^2 N}{\|\mathbf{L}\|^2 \tilde{D}} \right\rceil\right)$$

for some constant $c > 0$. While theoretically such a selection of T_k will result in slightly worse complexity bounds (up to an $\mathcal{O}(\log(1/\varepsilon))$ factor) in terms of subgradient computations and communication rounds, it may improve the practical performance of the DCS method especially in the beginning of the execution of this method.

8.3.2.3 Boundedness of $\|\mathbf{y}^*\|$

In this subsection, we will provide a bound on the optimal dual multiplier $\mathbf{y}^*$. By doing so, we show that the complexity of DCS algorithm (as well as the stochastic DCS algorithm in Sect. 8.3.3) only depends on the parameters for the primal problem along with the smallest nonzero eigenvalue of $\mathbf{L}$ and the initial point $\mathbf{y}^0$, even though these algorithms are intrinsically primal–dual type methods.

Theorem 8.9. *Suppose that f_i's are Lipschitz continuous, i.e., the subgradients of f_i are bounded by a constant M_f w.r.t. $\|\cdot\|_2$. Let $\mathbf{x}^*$ be an optimal solution of* (8.3.6). *Then there exists an optimal dual multiplier $\mathbf{y}^*$ for* (8.3.7) *s.t.*

$$\|\mathbf{y}^*\|_2 \leq \frac{\sqrt{m}M_f}{\tilde{\sigma}_{min}(\mathbf{L})}, \tag{8.3.46}$$

where $\tilde{\sigma}_{min}(\mathbf{L})$ *denotes the smallest nonzero eigenvalue of* $\mathbf{L}$.

Proof. Since we only relax the linear constraints in problem (8.3.6) to obtain the Lagrange dual problem (8.3.7), it follows from the strong Lagrange duality and the existence of $\mathbf{x}^*$ to (8.3.6) that an optimal dual multiplier $\mathbf{y}^*$ for problem (8.3.7) must exist. It is clear that

$$\mathbf{y}^* = \mathbf{y}_N^* + \mathbf{y}_C^*,$$

where $\mathbf{y}_N^*$ and $\mathbf{y}_C^*$ denote the projections of $\mathbf{y}^*$ over the null space and the column space of $\mathbf{L}^T$, respectively.

We consider two cases. Case (1) $\mathbf{y}_C^* = \mathbf{0}$. Since $\mathbf{y}_N^*$ belongs to the null space of $\mathbf{L}^T$, $\mathbf{L}^T\mathbf{y}^* = \mathbf{L}^T\mathbf{y}_N^* = \mathbf{0}$, which implies that for any $c \in \mathbb{R}$, $c\mathbf{y}^*$ is also an optimal dual multiplier of (8.3.7). Therefore, (8.3.46) clearly holds, because we can scale $\mathbf{y}^*$ to an arbitrary small vector.

Case (2) $\mathbf{y}_C^* \neq \mathbf{0}$. Using the fact that $\mathbf{L}^T\mathbf{y}^* = \mathbf{L}^T\mathbf{y}_C^*$ and the definition of a saddle point of (8.3.7), we conclude that $\mathbf{y}_C^*$ is also an optimal dual multiplier of (8.3.7). Since $\mathbf{y}_C^*$ in the column space of $\mathbf{L}$, we have

$$\begin{aligned}\|\mathbf{L}^T\mathbf{y}_C^*\|_2^2 = (\mathbf{y}_C^*)^T\mathbf{L}\mathbf{L}^T\mathbf{y}_C^* = (\mathbf{y}_C^*)^T\mathbf{U}^T\Lambda\mathbf{U}\mathbf{y}_C^* &\geq \tilde{\lambda}_{\min}(\mathbf{L}\mathbf{L}^T)\|\mathbf{U}\mathbf{y}_C^*\|_2^2\\ &= \tilde{\sigma}_{\min}^2(\mathbf{L})\|\mathbf{y}_C^*\|_2^2,\end{aligned}$$

where $\mathbf{U}$ is an orthonormal matrix whose rows consist of the eigenvectors of $\mathbf{L}\mathbf{L}^T$, Λ is the diagonal matrix whose diagonal elements are the corresponding eigenvalues, $\tilde{\lambda}_{min}(\mathbf{L}\mathbf{L}^T)$ denotes the smallest nonzero eigenvalue of $\mathbf{L}\mathbf{L}^T$, and $\tilde{\sigma}_{min}(\mathbf{L})$ denotes the smallest nonzero eigenvalue of $\mathbf{L}$. In particular,

$$\|\mathbf{y}_C^*\|_2 \leq \frac{\|\mathbf{L}^T\mathbf{y}_C^*\|_2}{\tilde{\sigma}_{\min}(\mathbf{L})}. \tag{8.3.47}$$

Moreover, if we denote the saddle point problem defined in (8.3.7) as follows:

$$\mathscr{L}(\mathbf{x},\mathbf{y}) := F(\mathbf{x}) + \langle \mathbf{L}\mathbf{x}, \mathbf{y}\rangle.$$

By the definition of a saddle point of (8.3.7), we have $\mathscr{L}(\mathbf{x}^*,\mathbf{y}_C^*) \leq \mathscr{L}(\mathbf{x},\mathbf{y}_C^*)$, i.e.,

$$F(\mathbf{x}^*) - F(\mathbf{x}) \leq \langle -\mathbf{L}^T\mathbf{y}_C^*, \mathbf{x} - \mathbf{x}^*\rangle.$$

Hence, from the definition of subgradients, we conclude that $-\mathbf{L}^T\mathbf{y}_C^* \in \partial F(\mathbf{x}^*)$, which together with the fact that f_i's are Lipschitz continuous implies that

$$\|\mathbf{L}^T\mathbf{y}_C^*\|_2 = \|(f_1'(x_1^*), f_2'(x_2^*), \ldots, f_m'(x_m^*))\|_2 \leq \sqrt{m}M_f.$$

Our result in (8.3.46) follows immediately from the above relation, (8.3.47) and the fact that $\mathbf{y}_C^*$ is also an optimal dual multiplier of (8.3.7). ■

Observe that our bound for the dual multiplier $\mathbf{y}^*$ in (8.3.46) contains only the primal information. Given an initial dual multiplier $\mathbf{y}^0$, this result can be used to provide an upper bound on $\|\mathbf{y}^0 - \mathbf{y}^*\|$ in Theorems 8.8–8.12 throughout this section. Note also that we can assume $\mathbf{y}^0 = 0$ to simplify these complexity bounds.

8.3.2.4 Convergence of DCS on Strongly Convex Functions

In this subsection, we assume that the objective functions f_i's are strongly convex (i.e., $\mu > 0$ (8.3.16)).

We next provide in Lemma 8.4 an estimate on the gap function defined in (8.3.8) together with stepsize policies which work for the strongly convex case. The proof of this lemma can be found in Sect. 8.3.5.

Lemma 8.4. *Let the iterates $(\hat{\mathbf{x}}^k, \mathbf{y}^k)$, $k = 1, \ldots, N$ be generated by Algorithm 8.7 and $\hat{\mathbf{z}}^N$ be defined as $\hat{\mathbf{z}}^N := \left(\sum_{k=1}^N \theta_k\right)^{-1} \sum_{k=1}^N \theta_k (\hat{\mathbf{x}}^k, \mathbf{y}^k)$. If the objective f_i, $i = 1, \ldots, m$ are strongly convex functions, i.e., $\mu, M > 0$, let the parameters $\{\alpha_k\}$, $\{\theta_k\}$, $\{\eta_k\}$, and $\{\tau_k\}$ in Algorithm 8.7 satisfy* (8.3.30)–(8.3.33) *and*

$$\theta_k \eta_k \leq \theta_{k-1}(\mu + \eta_{k-1}), \quad k = 2, \ldots, N, \tag{8.3.48}$$

and the parameters $\{\lambda_t\}$ and $\{\beta_t\}$ in the CS procedure of Algorithm 8.7 be set to

$$\lambda_t = t, \quad \beta_t^{(k)} = \tfrac{(t+1)\mu}{2\eta_k} + \tfrac{t-1}{2}, \quad \forall t \geq 1. \tag{8.3.49}$$

Then, we have for all $\mathbf{z} \in \mathbf{X} \times \mathbb{R}^{md}$

$$\begin{aligned} Q(\hat{\mathbf{z}}^N; \mathbf{z}) \leq \left(\textstyle\sum_{k=1}^N \theta_k\right)^{-1} \Big[&\theta_1 \eta_1 \mathbf{V}(\mathbf{x}^0, \mathbf{x}) + \tfrac{\theta_1 \tau_1}{2} \|\mathbf{y}^0\|^2 + \langle \hat{\mathbf{s}}, \mathbf{y} \rangle \\ &+ \textstyle\sum_{k=1}^N \sum_{t=1}^{T_k} \tfrac{2mM^2 \theta_k}{T_k(T_k+1)} \tfrac{t}{(t+1)\mu + (t-1)\eta_k} \Big], \end{aligned} \tag{8.3.50}$$

where $\hat{\mathbf{s}} := \theta_N \mathbf{L}(\hat{\mathbf{x}}^N - \mathbf{x}^{N-1}) + \theta_1 \tau_1 (\mathbf{y}^N - \mathbf{y}^0)$ and Q is defined in (8.3.8)*. Furthermore, for any saddle point $(\mathbf{x}^*, \mathbf{y}^*)$ of* (8.3.7)*, we have*

$$\begin{aligned} &\tfrac{\theta_N}{2} \left(1 - \tfrac{\|\mathbf{L}\|^2}{\eta_N \tau_N}\right) \max\{\eta_N \|\hat{\mathbf{x}}^N - \mathbf{x}^{N-1}\|^2, \tau_N \|\mathbf{y}^* - \mathbf{y}^N\|^2\} \\ &\leq \theta_1 \eta_1 \mathbf{V}(\mathbf{x}^0, \mathbf{x}^*) + \tfrac{\theta_1 \tau_1}{2} \|\mathbf{y}^* - \mathbf{y}^0\|^2 + \textstyle\sum_{k=1}^N \sum_{t=1}^{T_k} \tfrac{2mM^2 \theta_k}{T_k(T_k+1)} \tfrac{t}{(t+1)\mu + (t-1)\eta_k}. \end{aligned} \tag{8.3.51}$$

In the following theorem, we provide a specific selection of $\{\alpha_k\}$, $\{\theta_k\}$, $\{\eta_k\}$, $\{\tau_k\}$, and $\{T_k\}$ satisfying (8.3.30)–(8.3.33) and (8.3.48). Also, by using Lemma 8.4 and Proposition 8.6, we establish the complexity of the DCS method for computing an (ε, δ)-solution of problem (8.3.6) when the objective functions are strongly convex. The choice of variable stepsizes rather than using constant stepsizes will accelerate its convergence rate.

Theorem 8.10. *Let $\mathbf{x}^*$ be an optimal solution of (8.3.6), the parameters $\{\lambda_t\}$ and $\{\beta_t\}$ in the CS procedure of Algorithm 8.7 be set to* (8.3.49) *and suppose that $\{\alpha_k\}$, $\{\theta_k\}$, $\{\eta_k\}$, $\{\tau_k\}$, and $\{T_k\}$ are set to*

$$\alpha_k = \tfrac{k}{k+1},\ \theta_k = k+1,\ \eta_k = \tfrac{k\mu}{2},\ \tau_k = \tfrac{4\|\mathbf{L}\|^2}{(k+1)\mu},\ and$$
$$T_k = \left\lceil \sqrt{\tfrac{2m}{\tilde{D}}}\tfrac{MN}{\mu}\max\left\{\sqrt{\tfrac{2m}{\tilde{D}}}\tfrac{4M}{\mu}, 1\right\}\right\rceil, \tag{8.3.52}$$

$\forall k = 1,\ldots,N$, for some $\tilde{D} > 0$. Then, for any $N \geq 2$, we have

$$F(\hat{\mathbf{x}}^N) - F(\mathbf{x}^*) \leq \tfrac{2}{N(N+3)}\left[\mu\mathbf{V}(\mathbf{x}^0,\mathbf{x}^*) + \tfrac{2\|\mathbf{L}\|^2}{\mu}\|\mathbf{y}^0\|^2 + 2\mu\tilde{D}\right], \tag{8.3.53}$$

and

$$\|\mathbf{L}\hat{\mathbf{x}}^N\| \leq \tfrac{8\|\mathbf{L}\|}{N(N+3)}\left[3\sqrt{2\tilde{D} + \mathbf{V}(\mathbf{x}^0,\mathbf{x}^*)} + \tfrac{7\|\mathbf{L}\|}{\mu}\|\mathbf{y}^* - \mathbf{y}^0\|\right], \tag{8.3.54}$$

where $\hat{\mathbf{x}}^N = \frac{2}{N(N+3)}\sum_{k=1}^N (k+1)\hat{\mathbf{x}}^k$, and y^* is an arbitrary dual optimal solution.

Proof. It is easy to check that (8.3.52) satisfies conditions (8.3.30)–(8.3.33) and (8.3.48). Moreover, we have

$$\begin{aligned}
\textstyle\sum_{k=1}^N \sum_{t=1}^{T_k} \tfrac{2mM^2\theta_k}{T_k(T_k+1)} \tfrac{t}{(t+1)\mu + (t-1)\eta_k} &= \textstyle\sum_{k=1}^N \tfrac{2mM^2\theta_k}{T_k(T_k+1)\mu}\sum_{t=1}^{T_k} \tfrac{2t}{2(t+1)+(t-1)k} \\
&\leq \textstyle\sum_{k=1}^N \tfrac{2mM^2\theta_k}{T_k(T_k+1)\mu}\left(\tfrac{1}{2} + \sum_{t=2}^{T_k}\tfrac{2t}{(t-1)(k+1)}\right) \\
&\leq \textstyle\sum_{k=1}^N \tfrac{mM^2(k+1)}{T_k(T_k+1)\mu} + \sum_{k=1}^N \tfrac{8mM^2(T_k-1)}{T_k(T_k+1)\mu} \leq 2\mu\tilde{D}.
\end{aligned}$$

Therefore, by plugging in these values to (8.3.50), we have

$$Q(\hat{\mathbf{z}}^N;\mathbf{x}^*,\mathbf{y}) \leq \tfrac{2}{N(N+3)}\left[\mu\mathbf{V}(\mathbf{x}^0,\mathbf{x}^*) + \tfrac{2\|\mathbf{L}\|^2}{\mu}\|\mathbf{y}^0\|^2 + 2\mu\tilde{D} + \langle\hat{\mathbf{s}},\mathbf{y}\rangle\right]. \tag{8.3.55}$$

Furthermore, from (8.3.51), we have for $N \geq 2$

$$\begin{aligned}
\|\hat{\mathbf{x}}^N - \mathbf{x}^{N-1}\|^2 &\leq \tfrac{8}{\mu(N+1)(N-1)}\left[\mu\mathbf{V}(\mathbf{x}^0,\mathbf{x}^*) + \tfrac{2\|\mathbf{L}\|^2}{\mu}\|\mathbf{y}^0 - \mathbf{y}^*\|^2 + 2\mu\tilde{D}\right], \\
\|\mathbf{y}^* - \mathbf{y}^N\|^2 &\leq \tfrac{N\mu}{(N-1)\|\mathbf{L}\|^2}\left[\mu\mathbf{V}(\mathbf{x}^0,\mathbf{x}^*) + \tfrac{2\|\mathbf{L}\|^2}{\mu}\|\mathbf{y}^0 - \mathbf{y}^*\|^2 + 2\mu\tilde{D}\right].
\end{aligned} \tag{8.3.56}$$

Let $\mathbf{s}^N := \frac{2}{N(N+3)}\hat{\mathbf{s}}$, then by using (8.3.56), we have for $N \geq 2$

$$\begin{aligned}
\|\mathbf{s}^N\| &\leq \tfrac{2}{N(N+3)}\left[(N+1)\|\mathbf{L}\|\|\hat{\mathbf{x}}^N - \mathbf{x}^{N-1}\| + \tfrac{4\|\mathbf{L}\|^2}{\mu}\|\mathbf{y}^N - \mathbf{y}^*\| + \tfrac{4\|\mathbf{L}\|^2}{\mu}\|\mathbf{y}^* - \mathbf{y}^0\|\right] \\
&\leq \tfrac{8\|\mathbf{L}\|}{N(N+3)}\left[3\sqrt{2\tilde{D} + \mathbf{V}(\mathbf{x}^0,\mathbf{x}^*) + \tfrac{2\|\mathbf{L}\|^2}{\mu^2}\|\mathbf{y}^0 - \mathbf{y}^*\|^2} + \tfrac{\|\mathbf{L}\|}{\mu}\|\mathbf{y}^* - \mathbf{y}^0\|\right] \\
&\leq \tfrac{8\|\mathbf{L}\|}{N(N+3)}\left[3\sqrt{2\tilde{D} + \mathbf{V}(\mathbf{x}^0,\mathbf{x}^*)} + \tfrac{7\|\mathbf{L}\|}{\mu}\|\mathbf{y}^* - \mathbf{y}^0\|\right].
\end{aligned}$$

From (8.3.55), we further have

$$g(\hat{\mathbf{s}}^N,\hat{\mathbf{z}}^N) \leq \tfrac{2}{N(N+3)}\left[\mu\mathbf{V}(\mathbf{x}^0,\mathbf{x}^*) + \tfrac{2\|\mathbf{L}\|^2}{\mu}\|\mathbf{y}^0\|^2 + 2\mu\tilde{D}\right].$$

Applying Proposition 8.6 to the above two inequalities, the results in (8.3.53) and (8.3.54) follow immediately. ■

We now make some remarks about the results obtained in Theorem 8.10. First, similar to the general convex case, the best choice for $\tilde{D}$ (cf. (8.3.52)) would be $\mathbf{V}(\mathbf{x}^0, \mathbf{x}^*)$ so that the first and the third terms in (8.3.55) are about the same order. If there exists an estimate $D_{\mathbf{X}} > 0$ satisfying (8.3.41), we can set $\tilde{D} = D_{\mathbf{X}}^2$.

Second, the complexity of the DCS method for solving strongly convex problems follows from (8.3.53) and (8.3.54). For simplicity, let us assume that X is bounded, $\tilde{D} = D_{\mathbf{X}}^2$, and $\mathbf{y}^0 = \mathbf{0}$. We can see that the total number of inter-node communication rounds and intra-node subgradient evaluations performed by each agent for finding an (ε, δ)-solution of (8.3.6) can be bounded by

$$\begin{aligned} &\mathscr{O}\left\{\max\left(\sqrt{\tfrac{\mu D_{\mathbf{X}}^2}{\varepsilon}}, \sqrt{\tfrac{\|\mathbf{L}\|}{\delta}\left(D_{\mathbf{X}} + \tfrac{\|\mathbf{L}\|\|\mathbf{y}^*\|}{\mu}\right)}\right)\right\} \text{ and} \\ &\mathscr{O}\left\{\tfrac{mM^2}{\mu}\max\left(\tfrac{1}{\varepsilon}, \tfrac{\|\mathbf{L}\|}{\mu\delta}\left(\tfrac{1}{D_{\mathbf{X}}} + \tfrac{\|\mathbf{L}\|\|\mathbf{y}^*\|}{D_{\mathbf{X}}^2\mu}\right)\right)\right\}, \end{aligned} \tag{8.3.57}$$

respectively. In particular, if ε and δ satisfy

$$\tfrac{\varepsilon}{\delta} \le \tfrac{\mu^2 D_{\mathbf{X}}^2}{\|\mathbf{L}\|(\mu D_{\mathbf{X}} + \|\mathbf{L}\|\|\mathbf{y}^*\|)}, \tag{8.3.58}$$

then the complexity bounds in (8.3.57), respectively, reduce to

$$\mathscr{O}\left\{\sqrt{\tfrac{\mu D_{\mathbf{X}}^2}{\varepsilon}}\right\} \text{ and } \mathscr{O}\left\{\tfrac{mM^2}{\mu\varepsilon}\right\}. \tag{8.3.59}$$

Third, we compare DCS method with the centralized mirror descent method (Sect. 3.2) applied to (8.3.1). In the worst case, the Lipschitz constant and strongly convex modulus of f in (8.3.1) can be bounded by $M_f \le mM$, and $\mu_f \ge m\mu$, respectively, and each iteration of the method will incur m subgradient evaluations. Therefore, the total number of subgradient evaluations performed by the mirror descent method for finding an ε-solution of (8.3.1), i.e., a point $\bar{x} \in X$ such that $f(\bar{x}) - f^* \le \varepsilon$, can be bounded by

$$\mathscr{O}\left\{\tfrac{m^2M^2}{\mu\varepsilon}\right\}. \tag{8.3.60}$$

By observing that the second bound in (8.3.59) states only the number of subgradient evaluations for each agent in the DCS method, we conclude that the total number of subgradient evaluations performed by DCS is comparable to the classic mirror descent method as long as (8.3.58) holds and hence not improvable in general for the nonsmooth strongly convex case.

8.3.3 Stochastic Decentralized Communication Sliding

In this subsection, we consider the stochastic case where only the noisy subgradient information of the functions f_i, $i = 1,\ldots,m$, is available or easier to compute. This situation happens when the function f_i's are given either in the form of expectation or as the summation of lots of components. This setting has attracted considerable interest in recent decades for its applications in a broad spectrum of disciplines including machine learning, signal processing, and operations research. We present a stochastic communication sliding method, namely the stochastic decentralized communication sliding (SDCS) method, and show that the similar complexity bounds as in Sect. 8.3.2 can still be obtained in expectation or with high probability.

8.3.3.1 The SDCS Algorithm

The first-order information of the function f_i, $i = 1,\ldots,m$, can be accessed by a stochastic oracle (SO), which, given a point $u^t \in X$, outputs a vector $G_i(u^t,\xi_i^t)$ such that

$$\mathbb{E}[G_i(u^t,\xi_i^t)] = f_i'(u^t) \in \partial f_i(u^t), \tag{8.3.61}$$

$$\mathbb{E}[\|G_i(u^t,\xi_i^t) - f_i'(u^t)\|_*^2] \le \sigma^2, \tag{8.3.62}$$

where ξ_i^t is a random vector which models a source of uncertainty and is independent of the search point u^t, and the distribution $\mathbb{P}(\xi_i)$ is not known in advance. We call $G_i(u^t,\xi_i^t)$ a *stochastic subgradient* of f_i at u^t.

The SDCS method can be obtained by simply replacing the exact subgradients in the CS procedure of Algorithm 8.7 with the stochastic subgradients obtained from SO. This difference is described in Algorithm 8.8.

Algorithm 8.8 SDCS

The projection step (8.3.25)–(8.3.26) in the CS procedure of Algorithm 8.7 is replaced by

$$h^{t-1} = H(u^{t-1},\xi^{t-1}), \tag{8.3.63}$$

$$u^t = \mathrm{argmin}_{u\in U}\left[\langle w + h^{t-1}, u\rangle + \eta V(x,u) + \eta\beta_t V(u^{t-1},u)\right], \tag{8.3.64}$$

where $H(u^{t-1},\xi^{t-1})$ is a stochastic subgradient of ϕ at u^{t-1}.

We add a few remarks about the SDCS algorithm. First, as in DCS, no additional communications of the dual variables are required when the subgradient projection (8.3.64) is performed for T_k times in the inner loop. This is because the same w_i^k has been used throughout the T_k iterations of the stochastic CS procedure. Second, the problem will reduce to the deterministic case if there is no stochastic noise associated with the SO, i.e., when $\sigma = 0$ in (8.3.62).

8.3.3.2 Convergence of SDCS on General Convex Functions

We now establish the main convergence properties of the SDCS algorithm. More specifically, we provide in Lemma 8.5 an estimate on the gap function defined in (8.3.8) together with stepsize policies which work for the general convex case with $\mu = 0$ (cf. (8.3.16)). The proof of this lemma can be found in Sect. 8.3.5.

Lemma 8.5. *Let the iterates* $(\hat{\mathbf{x}}^k, \mathbf{y}^k)$ *for* $k = 1, \ldots, N$ *be generated by Algorithm 8.8,* $\hat{\mathbf{z}}^N$ *be defined as* $\hat{\mathbf{z}}^N := \left(\sum_{k=1}^N \theta_k\right)^{-1} \sum_{k=1}^N \theta_k (\hat{\mathbf{x}}^k, \mathbf{y}^k)$*, and Assumptions* (8.3.61)–(8.3.62) *hold. If the objective* f_i, $i = 1, \ldots, m$*, are general nonsmooth convex functions, i.e.,* $\mu = 0$ *and* $M > 0$*, let the parameters* $\{\alpha_k\}$, $\{\theta_k\}$, $\{\eta_k\}$, $\{\tau_k\}$*, and* $\{T_k\}$ *in Algorithm 8.8 satisfy* (8.3.29)–(8.3.33)*, and the parameters* $\{\lambda_t\}$ *and* $\{\beta_t\}$ *in the CS procedure of Algorithm 8.8 be set as* (8.3.34)*. Then, for all* $\mathbf{z} \in \mathbf{X} \times \mathbb{R}^{md}$,

$$Q(\hat{\mathbf{z}}^N; \mathbf{z}) \le \left(\sum_{k=1}^N \theta_k\right)^{-1} \Bigg\{ \tfrac{(T_1+1)(T_1+2)\theta_1\eta_1}{T_1(T_1+3)} \mathbf{V}(\mathbf{x}^0, \mathbf{x}) + \tfrac{\theta_1\tau_1}{2} \|\mathbf{y}^0\|^2 + \langle \hat{\mathbf{s}}, \mathbf{y} \rangle \tag{8.3.65}$$

$$+ \sum_{k=1}^N \sum_{t=1}^{T_k} \sum_{i=1}^m \tfrac{2\theta_k}{T_k(T_k+3)} \left[(t+1) \langle \delta_i^{t-1,k}, x_i - u_i^{t-1} \rangle + \tfrac{4(M^2 + \|\delta_i^{t-1,k}\|_*^2)}{\eta_k} \right] \Bigg\},$$

where $\hat{\mathbf{s}} := \theta_N \mathbf{L}(\hat{\mathbf{x}}^N - \mathbf{x}^{N-1}) + \theta_1 \tau_1 (\mathbf{y}^N - \mathbf{y}^0)$ *and* Q *is defined in* (8.3.8)*. Furthermore, for any saddle point* $(\mathbf{x}^*, \mathbf{y}^*)$ *of* (8.3.7)*, we have*

$$\tfrac{\theta_N}{2}\left(1 - \tfrac{\|\mathbf{L}\|^2}{\eta_N \tau_N}\right) \max\{\eta_N \|\hat{\mathbf{x}}^N - \mathbf{x}^{N-1}\|^2, \tau_N \|\mathbf{y}^* - \mathbf{y}^N\|^2\} \tag{8.3.66}$$

$$\le \tfrac{(T_1+1)(T_1+2)\theta_1\eta_1}{T_1(T_1+3)} \mathbf{V}(\mathbf{x}^0, \mathbf{x}^*) + \tfrac{\theta_1\tau_1}{2} \|\mathbf{y}^* - \mathbf{y}^0\|^2$$

$$+ \sum_{k=1}^N \sum_{t=1}^{T_k} \sum_{i=1}^m \tfrac{2\theta_k}{T_k(T_k+3)} \left[(t+1) \langle \delta_i^{t-1,k}, x_i^* - u_i^{t-1} \rangle + \tfrac{4(M^2 + \|\delta_i^{t-1,k}\|_*^2)}{\eta_k} \right].$$

In the following theorem, we provide a specific selection of $\{\alpha_k\}$, $\{\theta_k\}$, $\{\eta_k\}$, $\{\tau_k\}$, and $\{T_k\}$ satisfying (8.3.29)–(8.3.33). Also, by using Lemma 8.5 and Proposition 8.6, we establish the complexity of the SDCS method for computing an (ε, δ)-solution of problem (8.3.6) in expectation when the objective functions are general convex.

Theorem 8.11. *Let* $\mathbf{x}^*$ *be an optimal solution of (8.3.6), the parameters* $\{\lambda_t\}$ *and* $\{\beta_t\}$ *in the CS procedure of Algorithm 8.8 be set as* (8.3.34)*, and suppose that* $\{\alpha_k\}$, $\{\theta_k\}$, $\{\eta_k\}$, $\{\tau_k\}$ *and* $\{T_k\}$ *are set to*

$$\alpha_k = \theta_k = 1, \ \eta_k = 2\|\mathbf{L}\|, \ \tau_k = \|\mathbf{L}\|, \text{ and } T_k = \left\lceil \tfrac{m(M^2+\sigma^2)N}{\|\mathbf{L}\|^2 \tilde{D}} \right\rceil, \quad \forall k = 1, \ldots, N, \tag{8.3.67}$$

for some $\tilde{D} > 0$*. Then, under Assumptions* (8.3.61) *and* (8.3.62)*, we have for any* $N \ge 1$

$$\mathbb{E}[F(\hat{\mathbf{x}}^k) - F(\mathbf{x}^*)] \le \tfrac{\|\mathbf{L}\|}{N} \left[3\mathbf{V}(\mathbf{x}^0, \mathbf{x}^*) + \tfrac{1}{2}\|\mathbf{y}^0\|^2 + 4\tilde{D} \right], \tag{8.3.68}$$

and

$$\mathbb{E}[\|\mathbf{L}\hat{\mathbf{x}}^N\|] \leq \tfrac{\|\mathbf{L}\|}{N}\left[3\sqrt{6\mathbf{V}(\mathbf{x}^0,\mathbf{x}^*)+8\tilde{D}}+4\|\mathbf{y}^*-\mathbf{y}^0\|\right]. \tag{8.3.69}$$

where $\hat{\mathbf{x}}^N = \frac{1}{N}\sum_{k=1}^N \hat{\mathbf{x}}^k$, *and* y^* *is an arbitrary dual optimal solution.*

Proof. It is easy to check that (8.3.67) satisfies conditions (8.3.29)–(8.3.33). Moreover, by (8.3.10), we can obtain

$$g(\hat{\mathbf{s}}^N,\hat{\mathbf{z}}^N) = \max_{\mathbf{y}} Q(\hat{\mathbf{z}}^N;\mathbf{x}^*,\mathbf{y}) - \left(\sum_{k=1}^N\theta_k\right)^{-1}\langle\hat{\mathbf{s}},\mathbf{y}\rangle \tag{8.3.70}$$

$$\leq \left(\sum_{k=1}^N\theta_k\right)^{-1}\left\{\tfrac{(T_1+1)(T_1+2)\theta_1\eta_1}{T_1(T_1+3)}\mathbf{V}(\mathbf{x}^0,\mathbf{x}^*)+\tfrac{\theta_1\tau_1}{2}\|\mathbf{y}^0\|^2\right.$$

$$\left.+\sum_{k=1}^N\sum_{t=1}^{T_k}\sum_{i=1}^m\tfrac{2\theta_k}{T_k(T_k+3)}\left[(t+1)\langle\delta_i^{t-1,k},x_i^*-u_i^{t-1}\rangle+\tfrac{4(M^2+\|\delta_i^{t-1,k}\|_*^2)}{\eta_k}\right]\right\},$$

where $\mathbf{s}^N = \left(\sum_{k=1}^N\theta_k\right)^{-1}\hat{\mathbf{s}}$. Particularly, from Assumption (8.3.61) and (8.3.62),

$$\mathbb{E}[\delta_i^{t-1,k}] = 0,\ \ \mathbb{E}[\|\delta_i^{t-1,k}\|_*^2] \leq \sigma^2,\ \forall i\in\{1,\ldots,m\},\ t\geq 1,\ k\geq 1,$$

and from (8.3.67)

$$\tfrac{(T_1+1)(T_1+2)}{T_1(T_1+3)} = 1+\tfrac{2}{T_1^2+3T_1} \leq \tfrac{3}{2}.$$

Therefore, by taking expectation over both sides of (8.3.70) and plugging in these values into (8.3.70), we have

$$\mathbb{E}[g(\hat{\mathbf{s}}^N,\hat{\mathbf{z}}^N)] \leq \left(\sum_{k=1}^N\theta_k\right)^{-1}\left\{\tfrac{(T_1+1)(T_1+2)\theta_1\eta_1}{T_1(T_1+3)}\mathbf{V}(\mathbf{x}^0,\mathbf{x})+\tfrac{\theta_1\tau_1}{2}\|\mathbf{y}^0\|^2\right.$$
$$\left.+\sum_{k=1}^N\tfrac{8m(M^2+\sigma^2)\theta_k}{(T_k+3)\eta_k}\right\} \leq \tfrac{\|\mathbf{L}\|}{N}\left[3\mathbf{V}(\mathbf{x}^0,\mathbf{x}^*)+\tfrac{1}{2}\|\mathbf{y}^0\|^2+4\tilde{D}\right], \tag{8.3.71}$$

with

$$\mathbb{E}[\|\hat{\mathbf{s}}^N\|] = \tfrac{1}{N}\mathbb{E}[\|\hat{\mathbf{s}}\|] \leq \tfrac{\|\mathbf{L}\|}{N}\mathbb{E}\left[\|\hat{\mathbf{x}}^N-\mathbf{x}^{N-1}\|+\|\mathbf{y}^N-\mathbf{y}^*\|+\|\mathbf{y}^*-\mathbf{y}^0\|\right].$$

Note that from (8.3.66) and Jensen's inequality, we have

$$(\mathbb{E}[\|\hat{\mathbf{x}}^N-\mathbf{x}^{N-1}])^2 \leq \mathbb{E}[\|\hat{\mathbf{x}}^N-\mathbf{x}^{N-1}\|^2] \leq 6\mathbf{V}(\mathbf{x}^0,\mathbf{x}^*)+\|\mathbf{y}^*-\mathbf{y}^0\|+8\tilde{D},$$
$$(\mathbb{E}[\|\mathbf{y}^*-\mathbf{y}^N\|])^2 \leq \mathbb{E}[\|\mathbf{y}^*-\mathbf{y}^N\|^2] \leq 12\mathbf{V}(\mathbf{x}^0,\mathbf{x}^*)+2\|\mathbf{y}^*-\mathbf{y}^0\|+16\tilde{D}.$$

Hence,

$$\mathbb{E}[\|\hat{\mathbf{s}}^N\|] \leq \tfrac{\|\mathbf{L}\|}{N}\left[3\sqrt{6\mathbf{V}(\mathbf{x}^0,\mathbf{x}^*)+8\tilde{D}}+4\|\mathbf{y}^*-\mathbf{y}^0\|\right].$$

Applying Proposition 8.6 to the above inequality and (8.3.71), the results in (8.3.68) and (8.3.69) follow immediately. ■

We now make some observations about the results obtained in Theorem 8.11. First, one can choose any $\tilde{D} > 0$ (e.g., $\tilde{D} = 1$) in (8.3.67); however, the best selection of $\tilde{D}$ would be $\mathbf{V}(\mathbf{x}^0, \mathbf{x}^*)$ so that the first and third terms in (8.3.71) are about the same order. In practice, if there exists an estimate $D_{\mathbf{X}} > 0$ satisfying (8.3.41), we can set $\tilde{D} = D_{\mathbf{X}}^2$.

Second, the complexity of SDCS method immediately follows from (8.3.68) and (8.3.69). Under the above assumption, with $\tilde{D} = D_{\mathbf{X}}^2$ and $\mathbf{y}^0 = \mathbf{0}$, we can see that the total number of inter-node communication rounds and intra-node subgradient evaluations required by each agent for finding a stochastic (ε, δ)-solution of (8.3.6) can be bounded by

$$\mathcal{O}\left\{\|\mathbf{L}\| \max\left(\tfrac{D_{\mathbf{X}}^2}{\varepsilon}, \tfrac{D_{\mathbf{X}}+\|\mathbf{y}^*\|}{\delta}\right)\right\} \text{ and } \mathcal{O}\left\{m(M^2+\sigma^2)\max\left(\tfrac{D_{\mathbf{X}}^2}{\varepsilon^2}, \tfrac{D_{\mathbf{X}}^2+\|\mathbf{y}^*\|^2}{D_{\mathbf{X}}^2\delta^2}\right)\right\}, \tag{8.3.72}$$

respectively. In particular, if ε and δ satisfy (8.3.43), the above complexity bounds, respectively, reduce to

$$\mathcal{O}\left\{\tfrac{\|\mathbf{L}\|D_{\mathbf{X}}^2}{\varepsilon}\right\} \text{ and } \mathcal{O}\left\{\tfrac{m(M^2+\sigma^2)D_{\mathbf{X}}^2}{\varepsilon^2}\right\}. \tag{8.3.73}$$

Therefore, we can show that the total number of stochastic subgradients that SDCS requires is comparable to the stochastic mirror-descent method in Sect. 4.1. This implies that the sample complexity for decentralized stochastic optimization is still optimal (as the centralized one), even after we skip many communication rounds.

8.3.3.3 Convergence of SDCS on Strongly Convex Functions

We now provide in Lemma 8.6 an estimate on the gap function defined in (8.3.8) together with stepsize policies which work for the strongly convex case with $\mu > 0$ (cf. (8.3.16)). The proof of this lemma can be found in Sect. 8.3.5.

Lemma 8.6. *Let the iterates* $(\hat{\mathbf{x}}^k, \mathbf{y}^k)$, $k = 1, \ldots, N$ *be generated by Algorithm 8.8,* $\hat{\mathbf{z}}^N$ *be defined as* $\hat{\mathbf{z}}^N := \left(\sum_{k=1}^N \theta_k\right)^{-1} \sum_{k=1}^N \theta_k(\hat{\mathbf{x}}^k, \mathbf{y}^k)$, *and Assumptions* (8.3.61)–(8.3.62) *hold. If the objective* f_i, $i = 1, \ldots, m$ *are strongly convex functions, i.e.,* $\mu, M > 0$, *let the parameters* $\{\alpha_k\}$, $\{\theta_k\}$, $\{\eta_k\}$, *and* $\{\tau_k\}$ *in Algorithm 8.8 satisfy* (8.3.30)–(8.3.33) *and* (8.3.48), *and the parameters* $\{\lambda_t\}$ *and* $\{\beta_t\}$ *in the CS procedure of Algorithm 8.8 be set as* (8.3.49). *Then, for all* $\mathbf{z} \in \mathbf{X} \times \mathbb{R}^{md}$,

$$\begin{aligned} Q(\hat{\mathbf{z}}^N; \mathbf{z}) \leq \left(\textstyle\sum_{k=1}^N \theta_k\right)^{-1} \Bigg\{ & \theta_1 \eta_1 \mathbf{V}(\mathbf{x}^0, \mathbf{x}) + \tfrac{\theta_1 \tau_1}{2}\|\mathbf{y}^0\|^2 + \langle \hat{\mathbf{s}}, \mathbf{y}\rangle \\ & + \textstyle\sum_{k=1}^N \sum_{t=1}^{T_k} \sum_{i=1}^m \tfrac{2\theta_k}{T_k(T_k+1)} \Big[t\langle \delta_i^{t-1,k}, x_i - u_i^{t-1}\rangle \\ & + \tfrac{2t(M^2 + \|\delta_i^{t-1,k}\|_*^2)}{(t+1)\mu + (t-1)\eta_k} \Big] \Bigg\}, \end{aligned} \tag{8.3.74}$$

where $\hat{\mathbf{s}} := \theta_N \mathbf{L}(\hat{\mathbf{x}}^N - \mathbf{x}^{N-1}) + \theta_1\tau_1(\mathbf{y}^N - \mathbf{y}^0)$ *and* Q *is defined in* (8.3.8). *Furthermore, for any saddle point* $(\mathbf{x}^*, \mathbf{y}^*)$ *of* (8.3.7), *we have*

$$\begin{aligned} \tfrac{\theta_N}{2}\left(1 - \tfrac{\|\mathbf{L}\|^2}{\eta_N\tau_N}\right) \max\{\eta_N\|\hat{\mathbf{x}}^N - \mathbf{x}^{N-1}\|^2, \tau_N\|\mathbf{y}^* - \mathbf{y}^N\|^2\} & \\ \le \theta_1\eta_1\mathbf{V}(\mathbf{x}^0,\mathbf{x}^*) + \tfrac{\theta_1\tau_1}{2}\|\mathbf{y}^* - \mathbf{y}^0\|^2 & \\ + \textstyle\sum_{k=1}^N \sum_{t=1}^{T_k} \sum_{i=1}^m \tfrac{2\theta_k}{T_k(T_k+1)} \Big[t\langle \delta_i^{t-1,k}, x_i^* - u_i^{t-1}\rangle & \\ + \tfrac{2t(M^2 + \|\delta_i^{t-1,k}\|_*^2)}{(t+1)\mu + (t-1)\eta_k} \Big]. & \end{aligned} \tag{8.3.75}$$

In the following theorem, we provide a specific selection of $\{\alpha_k\}$, $\{\theta_k\}$, $\{\eta_k\}$, $\{\tau_k\}$, and $\{T_k\}$ satisfying (8.3.30)–(8.3.33) and (8.3.29). Also, by using Lemma 8.6 and Proposition 8.6, we establish the complexity of the SDCS method for computing an (ε,δ)-solution of problem (8.3.6) in expectation when the objective functions are strongly convex. Similar to the deterministic case, we choose variable stepsizes rather than constant stepsizes.

Theorem 8.12. *Let* $\mathbf{x}^*$ *be an optimal solution of (8.3.6), the parameters* $\{\lambda_t\}$ *and* $\{\beta_t\}$ *in the CS procedure of Algorithm 8.8 be set as* (8.3.49), *and suppose that* $\{\alpha_k\}$, $\{\theta_k\}$, $\{\eta_k\}$, $\{\tau_k\}$, *and* $\{T_k\}$ *are set to*

$$\begin{aligned} \alpha_k &= \tfrac{k}{k+1},\ \theta_k = k+1,\ \eta_k = \tfrac{k\mu}{2},\ \tau_k = \tfrac{4\|\mathbf{L}\|^2}{(k+1)\mu},\ \text{and} \\ T_k &= \left\lceil \sqrt{\tfrac{m(M^2+\sigma^2)}{\tilde{D}}}\,\tfrac{2N}{\mu} \max\left\{\sqrt{\tfrac{m(M^2+\sigma^2)}{\tilde{D}}}\,\tfrac{8}{\mu}, 1\right\}\right\rceil, \quad \forall k = 1,\ldots,N, \end{aligned} \tag{8.3.76}$$

for some $\tilde{D} > 0$. *Then, under Assumptions* (8.3.61) *and* (8.3.62), *we have for any* $N \ge 2$

$$\mathbb{E}[F(\bar{\mathbf{x}}^N) - F(\mathbf{x}^*) \le \tfrac{2}{N(N+3)}\left[\mu\mathbf{V}(\mathbf{x}^0,\mathbf{x}^*) + \tfrac{2\|\mathbf{L}\|^2}{\mu}\|\mathbf{y}^0\|^2 + 2\mu\tilde{D}\right], \tag{8.3.77}$$

and

$$\mathbb{E}[\|\mathbf{L}\hat{\mathbf{x}}^N\|] \le \tfrac{8\|\mathbf{L}\|}{N(N+3)}\left[3\sqrt{2\tilde{D} + \mathbf{V}(\mathbf{x}^0,\mathbf{x}^*)} + \tfrac{7\|\mathbf{L}\|}{\mu}\|\mathbf{y}^* - \mathbf{y}^0\|\right], \tag{8.3.78}$$

where $\hat{\mathbf{x}}^N = \frac{2}{N(N+3)}\sum_{k=1}^N (k+1)\hat{\mathbf{x}}^k$, *and* y^* *is an arbitrary dual optimal solution.*

Proof. It is easy to check that (8.3.76) satisfies conditions (8.3.30)–(8.3.33) and (8.3.48). Similarly, by (8.3.10) and Assumptions (8.3.61) and (8.3.62), we can obtain

$$\begin{aligned} \mathbb{E}[g(\hat{\mathbf{s}}^N, \hat{\mathbf{z}}^N)] \le \left(\textstyle\sum_{k=1}^N \theta_k\right)^{-1} \Big\{ & \theta_1\eta_1\mathbf{V}(\mathbf{x}^0,\mathbf{x}^*) + \tfrac{\theta_1\tau_1}{2}\|\mathbf{y}^0\|^2 \\ & + \textstyle\sum_{k=1}^N \sum_{t=1}^{T_k} \sum_{i=1}^m \tfrac{2\theta_k}{T_k(T_k+1)} \left[\tfrac{2t(M^2+\sigma^2)}{(t+1)\mu + (t-1)\eta_k}\right] \Big\}, \end{aligned} \tag{8.3.79}$$

where $\mathbf{s}^N = \left(\sum_{k=1}^N \theta_k\right)^{-1}\hat{\mathbf{s}}$. Particularly, from (8.3.76), we have

$$\begin{aligned}
\textstyle\sum_{k=1}^N \sum_{t=1}^{T_k} \frac{4m(M^2+\sigma^2)\theta_k}{T_k(T_k+1)} \frac{t}{(t+1)\mu+(t-1)\eta_k} &= \textstyle\sum_{k=1}^N \frac{4m(M^2+\sigma^2)\theta_k}{T_k(T_k+1)\mu} \sum_{t=1}^{T_k} \frac{2t}{2(t+1)+(t-1)k} \\
&\le \textstyle\sum_{k=1}^N \frac{4m(M^2+\sigma^2)\theta_k}{T_k(T_k+1)\mu} \left(\frac{1}{2} + \sum_{t=2}^{T_k} \frac{2t}{(t-1)(k+1)}\right) \\
&\le \textstyle\sum_{k=1}^N \frac{2m(M^2+\sigma^2)(k+1)}{T_k(T_k+1)\mu} + \sum_{k=1}^N \frac{16m(M^2+\sigma^2)(T_k-1)}{T_k(T_k+1)\mu} \le 2\mu\tilde{D}.
\end{aligned}$$

Therefore, by plugging in these values into (8.3.79), we have

$$\mathbb{E}[g(\hat{\mathbf{s}}^N, \hat{\mathbf{z}}^N)] \le \tfrac{2}{N(N+3)} \left[\mu \mathbf{V}(\mathbf{x}^0, \mathbf{x}^*) + \tfrac{2\|L\|^2}{\mu}\|\mathbf{y}^0\|^2 + 2\mu\tilde{D}\right], \tag{8.3.80}$$

with

$$\begin{aligned}
\mathbb{E}[\|\hat{\mathbf{s}}^N\|] &= \tfrac{2}{N(N+3)}\mathbb{E}[\|\hat{\mathbf{s}}\|] \\
&\le \tfrac{2\|\mathbf{L}\|}{N(N+3)}\mathbb{E}\left[(N+1)\|\hat{\mathbf{x}}^N - \mathbf{x}^{N-1}\| + \tfrac{4\|\mathbf{L}\|}{\mu}(\|\mathbf{y}^N - \mathbf{y}^*\| + \|\mathbf{y}^* - \mathbf{y}^0\|)\right].
\end{aligned}$$

Note that from (8.3.75), we have, for any $N \ge 2$,

$$\begin{aligned}
\mathbb{E}[\|\hat{\mathbf{x}}^N - \mathbf{x}^{N-1}\|^2] &\le \tfrac{8}{(N+1)(N-1)}\left[\mathbf{V}(\mathbf{x}^0, \mathbf{x}^*) + \tfrac{2\|\mathbf{L}\|^2}{\mu^2}\|\mathbf{y}^0 - \mathbf{y}^*\|^2 + 2\tilde{D}\right], \\
\mathbb{E}[\|\mathbf{y}^* - \mathbf{y}^N\|^2] &\le \tfrac{N\mu}{(N-1)\|\mathbf{L}\|^2}\left[\mu\mathbf{V}(\mathbf{x}^0, \mathbf{x}^*) + \tfrac{2\|\mathbf{L}\|^2}{\mu}\|\mathbf{y}^0 - \mathbf{y}^*\|^2 + 2\mu\tilde{D}\right].
\end{aligned}$$

Hence, in view of the above three relations and Jensen's inequality, we obtain

$$\begin{aligned}
\mathbb{E}[\|\hat{\mathbf{s}}^N\|] &\le \tfrac{8\|\mathbf{L}\|}{N(N+3)}\left[3\sqrt{2\tilde{D} + \mathbf{V}(\mathbf{x}^0, \mathbf{x}^*) + \tfrac{2\|\mathbf{L}\|^2}{\mu^2}\|\mathbf{y}^0 - \mathbf{y}^*\|^2} + \tfrac{\|\mathbf{L}\|}{\mu}\|\mathbf{y}^* - \mathbf{y}^0\|\right] \\
&\le \tfrac{8\|\mathbf{L}\|}{N(N+3)}\left[3\sqrt{2\tilde{D} + \mathbf{V}(\mathbf{x}^0, \mathbf{x}^*)} + \tfrac{7\|\mathbf{L}\|}{\mu}\|\mathbf{y}^* - \mathbf{y}^0\|\right].
\end{aligned}$$

Applying Proposition 8.6 to the above inequality and (8.3.80), the results in (8.3.77) and (8.3.78) follow immediately. ■

We now make some observations about the results obtained in Theorem 8.12. First, similar to the general convex case, the best choice for $\tilde{D}$ (cf. (8.3.76)) would be $\mathbf{V}(\mathbf{x}^0, \mathbf{x}^*)$ so that the first and the third terms in (8.3.80) are about the same order. If there exists an estimate $D_{\mathbf{X}} > 0$ satisfying (8.3.41), we can set $\tilde{D} = D_{\mathbf{X}}^2$.

Second, the complexity of SDCS method for solving strongly convex problems follows from (8.3.77) and (8.3.78). Under the above assumption, with $\tilde{D} = D_{\mathbf{X}}^2$ and $\mathbf{y}^0 = \mathbf{0}$, the total number of inter-node communication rounds and intra-node subgradient evaluations performed by each agent for finding a stochastic (ε, δ)-solution of (8.3.6) can be bounded by

$$\mathcal{O}\left\{\max\left(\sqrt{\tfrac{\mu D_{\mathbf{X}}^2}{\varepsilon}}, \sqrt{\tfrac{\|\mathbf{L}\|}{\delta}\left(D_{\mathbf{X}} + \tfrac{\|\mathbf{L}\|\|\mathbf{y}^*\|}{\mu}\right)}\right)\right\} \text{ and}$$
$$\mathcal{O}\left\{\tfrac{m(M^2+\sigma^2)}{\mu}\max\left(\tfrac{1}{\varepsilon}, \tfrac{\|\mathbf{L}\|}{\mu\delta}\left(\tfrac{1}{D_{\mathbf{X}}} + \tfrac{\|\mathbf{L}\|\|\mathbf{y}^*\|}{D_{\mathbf{X}}^2\mu}\right)\right)\right\}, \tag{8.3.81}$$

respectively. In particular, if ε and δ satisfy (8.3.58), the above complexity bounds, respectively, reduce to

$$\mathcal{O}\left\{\sqrt{\tfrac{\mu D_{\mathbf{X}}^2}{\varepsilon}}\right\} \text{ and } \mathcal{O}\left\{\tfrac{m(M^2+\sigma^2)}{\mu\varepsilon}\right\}. \tag{8.3.82}$$

We can see that the total number of stochastic subgradient computations is comparable to the optimal complexity bound obtained in Sect. 4.2 for stochastic strongly convex case in the centralized case.

8.3.4 High Probability Results

All of the results stated in Sects. 8.3.3.2–8.3.3.3 are established in terms of expectation. In order to provide high probability results for SDCS method, we additionally need the following "light-tail" assumption:

$$\mathbb{E}[\exp\{\|G_i(u^t,\xi_i^t) - f_i'(u^t)\|_*^2/\sigma^2\}] \le \exp\{1\}. \tag{8.3.83}$$

Note that (8.3.83) is stronger than (8.3.62), since it implies (8.3.62) by Jensen's inequality. Moreover, we also assume that there exists $\bar{\mathbf{V}}(\mathbf{x}^*)$ s.t.

$$\bar{\mathbf{V}}(\mathbf{x}^*) := \textstyle\sum_{i=1}^m \bar{V}_i(x_i^*) := \sum_{i=1}^m \max_{x_i\in X_i} V_i(x_i^*,x_i). \tag{8.3.84}$$

The following theorem provides a large deviation result for the gap function $g(\hat{\mathbf{s}}^N,\hat{\mathbf{z}}^N)$ when our objective functions f_i, $i=1,\ldots,m$ are general nonsmooth convex functions.

Theorem 8.13. *Let x^* be an optimal solution of* (8.3.6), *Assumptions* (8.3.61), (8.3.62), *and* (8.3.83) *hold, the parameters* $\{\alpha_k\}$, $\{\theta_k\}$, $\{\eta_k\}$, $\{\tau_k\}$, *and* $\{T_k\}$ *in Algorithm 8.8 satisfy* (8.3.29)–(8.3.33), *and the parameters* $\{\lambda_t\}$ *and* $\{\beta_t\}$ *in the CS procedure of Algorithm 8.8 be set as* (8.3.34). *In addition, if X_i's are compact, then for any $\zeta > 0$ and $N \ge 1$, we have*

$$\text{Prob}\left\{g(\hat{\mathbf{s}}^N,\hat{\mathbf{z}}^N) \ge \mathcal{B}_d(N) + \zeta\mathcal{B}_p(N)\right\} \le \exp\{-\zeta^2/3\} + \exp\{-\zeta\}, \tag{8.3.85}$$

where

$$\mathscr{B}_d(N) := \left(\sum_{k=1}^N \theta_k\right)^{-1} \left[\tfrac{(T_1+1)(T_1+2)\theta_1\eta_1}{T_1(T_1+3)} \mathbf{V}(\mathbf{x}^0, \mathbf{x}^*) + \tfrac{\theta_1 \tau_1}{2} \|\mathbf{y}^0\|^2 \right. \tag{8.3.86}$$

$$\left. + \sum_{k=1}^N \tfrac{8m(M^2+\sigma^2)\theta_k}{\eta_k(T_k+3)} \right],$$

and

$$\mathscr{B}_p(N) := \left(\sum_{k=1}^N \theta_k\right)^{-1} \left\{ \sigma \left[2\bar{\mathbf{V}}(\mathbf{x}^*) \sum_{k=1}^N \sum_{t=1}^{T_k} \left(\tfrac{\theta_k \lambda_t}{\sum_{t=1}^{T_k} \lambda_t} \right)^2 \right]^{1/2} \right.$$

$$\left. + \sum_{k=1}^N \sum_{t=1}^{T_k} \sum_{i=1}^m \tfrac{\sigma^2 \theta_k \lambda_t}{\left(\sum_{t=1}^{T_k} \lambda_t\right) \eta_k \beta_t} \right\}. \tag{8.3.87}$$

In the next corollary, we show that both the primal and feasibility (or consistency) residuals associated with the output solution of SDCS decrease in the order $\mathscr{O}(1/N)$ with high probability when the objective functions are nonsmooth and convex.

Corollary 8.7. *Let* $\mathbf{x}^*$ *be an optimal solution of (8.3.6),* y^* *be an arbitrary dual optimal solution, the parameters* $\{\lambda_t\}$ *and* $\{\beta_t\}$ *in the CS procedure of Algorithm 8.8 be set as* (8.3.34)*, and suppose that* $\{\alpha_k\}$, $\{\theta_k\}$, $\{\eta_k\}$, $\{\tau_k\}$ *and* $\{T_k\}$ *are set to* (8.3.67) *with* $\tilde{D} = \bar{\mathbf{V}}(\mathbf{x}^*)$. *Under Assumptions* (8.3.61), (8.3.62), *and* (8.3.83), *we have for any* $N \geq 1$ *and* $\zeta > 0$

$$\begin{aligned} &\text{Prob}\left\{ F(\hat{\mathbf{x}}^N) - F(\mathbf{x}^*) \geq \tfrac{\|\mathbf{L}\|}{N} \left[(7+8\zeta)\bar{\mathbf{V}}(\mathbf{x}^*) + \tfrac{1}{2}\|\mathbf{y}^0\|^2\right] \right\} \\ &\quad \leq \exp\{-\zeta^2/3\} + \exp\{-\zeta\}, \end{aligned} \tag{8.3.88}$$

and

$$\begin{aligned} &\text{Prob}\left\{ \|\mathbf{L}\hat{\mathbf{x}}^N\|^2 \geq \tfrac{18\|\mathbf{L}\|^2}{N^2} \left[(7+8\zeta)\bar{\mathbf{V}}(\mathbf{x}^*) + \tfrac{2}{3}\|\mathbf{y}^* - \mathbf{y}^0\|^2\right] \right\} \\ &\quad \leq \exp\{-\zeta^2/3\} + \exp\{-\zeta\}. \end{aligned} \tag{8.3.89}$$

Proof. Observe that by the definition of λ_t in (8.3.34),

$$\begin{aligned} \sum_{t=1}^{T_k} \left[\tfrac{\theta_k \lambda_t}{\sum_{t=1}^{T_k} \lambda_t} \right]^2 &= \left(\tfrac{2}{T_k(T_k+3)} \right)^2 \sum_{t=1}^{T_k} (t+1)^2 \\ &= \left(\tfrac{2}{T_k(T_k+3)} \right)^2 \tfrac{(T_k+1)(T_k+2)(2T_k+3)}{6} \leq \tfrac{8}{3T_k}, \end{aligned}$$

which together with (8.3.87) then imply that

$$\mathscr{B}_p(N) \le \tfrac{1}{N}\left\{\sigma\left[2\bar{\mathbf{V}}(\mathbf{x}^*)\textstyle\sum_{k=1}^N \frac{8}{3T_k}\right]^{1/2} + \textstyle\sum_{k=1}^N \frac{8m\sigma^2}{\|\mathbf{L}\|(T_k+3)}\right\}$$
$$\le \tfrac{4\|\mathbf{L}\|}{N}\left\{\sqrt{\tfrac{\bar{\mathbf{V}}(\mathbf{x}^*)\tilde{D}}{3m}} + \tilde{D}\right\} \le \tfrac{8\|\mathbf{L}\|\bar{\mathbf{V}}(\mathbf{x}^*)}{N}.$$

Hence, (8.3.88) follows from the above relation, (8.3.85), and Proposition 8.6. Note that from (8.3.66) and plugging in (8.3.67) with $\tilde{D} = \bar{\mathbf{V}}(\mathbf{x}^*)$, we obtain

$$\|\hat{\mathbf{s}}^N\|^2 = \left(\textstyle\sum_{k=1}^N \theta_k\right)^{-2}\|\hat{\mathbf{s}}\|^2$$
$$\le \left(\textstyle\sum_{k=1}^N \theta_k\right)^{-2}\{3\theta_N^2\|\mathbf{L}\|^2\|\hat{\mathbf{x}}^N - \mathbf{x}^{N-1}\|^2 + 3\theta_1^2\tau_1^2\left(\|\mathbf{y}^N - \mathbf{y}^*\|^2 + \|\mathbf{y}^* - \mathbf{y}^0\|^2\right)\}$$
$$\le \tfrac{3\|\mathbf{L}\|^2}{N^2}\Bigg\{18\mathbf{V}(\mathbf{x}^0,\mathbf{x}^*) + 4\|\mathbf{y}^* - \mathbf{y}^0\|^2$$
$$+ \textstyle\sum_{k=1}^N \sum_{t=1}^{T_k}\sum_{i=1}^m \frac{12\theta_k}{T_k(T_k+3)\|\mathbf{L}\|}\Big[(t+1)\langle \delta_i^{t-1,k}, x_i^* - u_i^{t-1}\rangle$$
$$+ \tfrac{4(M^2 + \|\delta_i^{t-1,k}\|_*^2)}{\eta_k}\Big]\Bigg\}.$$

Hence, similarly, we have

$$\mathrm{Prob}\left\{\|\hat{\mathbf{s}}^N\|^2 \ge \tfrac{18\|\mathbf{L}\|^2}{N^2}\left[(7+8\zeta)\bar{\mathbf{V}}(\mathbf{x}^*) + \tfrac{2}{3}\|\mathbf{y}^* - \mathbf{y}^0\|^2\right]\right\}$$
$$\le \exp\{-\zeta^2/3\} + \exp\{-\zeta\},$$

which in view of Proposition 8.6 immediately implies (8.3.89). ■

8.3.5 Convergence Analysis

This section is devoted to the proof of the main lemmas in Sects. 8.3.2 and 8.3.3, which establish the convergence results of the deterministic and stochastic decentralized communication sliding methods, respectively. After introducing some general results about these algorithms, we provide the proofs for Lemma 8.3–8.6 and Theorem 8.13.

Before we provide proofs for Lemma 8.3-8.6, we first need to present a result which summarizes an important convergence property of the CS procedure. It needs to be mentioned that the following proposition states a general result holds for CS procedure performed by individual agent $i \in \mathscr{N}$. For notation convenience, we use the notations defined in CS procedure (cf. Algorithm 8.7).

Proposition 8.7. *If* $\{\beta_t\}$ *and* $\{\lambda_t\}$ *in the* CS *procedure satisfy*

$$\lambda_{t+1}(\eta\beta_{t+1} - \mu) \le \lambda_t(1+\beta_t)\eta,\ \forall t \ge 1. \tag{8.3.1}$$

then, for $t \geq 1$ and $u \in U$,

$$(\Sigma_{t=1}^{T}\lambda_t)^{-1}\left[\eta(1+\beta_T)\lambda_T V(u^T,u)+\Sigma_{t=1}^{T}\lambda_t\langle\delta^{t-1},u-u^{t-1}\rangle\right]+\Phi(\hat{u}^T)-\Phi(u)$$
$$\leq (\Sigma_{t=1}^{T}\lambda_t)^{-1}\left[(\eta\beta_1-\mu)\lambda_1 V(u^0,u)+\Sigma_{t=1}^{T}\tfrac{(M+\|\delta^{t-1}\|_*)^2\lambda_t}{2\eta\beta_t}\right], \tag{8.3.2}$$

where Φ is defined as

$$\Phi(u) := \langle w,u\rangle+\phi(u)+\eta V(x,u) \tag{8.3.3}$$

and $\delta^t := \phi'(u^t)-h^t$.

Proof. Noticing that $\phi := f_i$ in the CS procedure, we have by (8.3.16)

$$\begin{aligned}\phi(u^t) &\leq \phi(u^{t-1})+\langle\phi'(u^{t-1}),u^t-u^{t-1}\rangle+M\|u^t-u^{t-1}\|\\ &= \phi(u^{t-1})+\langle\phi'(u^{t-1}),u-u^{t-1}\rangle+\langle\phi'(u^{t-1}),u^t-u\rangle+M\|u^t-u^{t-1}\|\\ &\leq \phi(u)-\mu V(u^{t-1},u)+\langle\phi'(u^{t-1}),u^t-u\rangle+M\|u^t-u^{t-1}\|,\end{aligned}$$

where $\phi'(u^{t-1}) \in \partial\phi(u^{t-1})$ and $\partial\phi(u^{t-1})$ denotes the subdifferential of ϕ at u^{t-1}. By applying Lemma 3.5 in (8.3.26), we obtain

$$\begin{aligned}&\langle w+h^{t-1},u^t-u\rangle+\eta V(x,u^t)-\eta V(x,u)\\ &\leq \eta\beta_t V(u^{t-1},u)-\eta(1+\beta_t)V(u^t,u)-\eta\beta_t V(u^{t-1},u^t),\ \forall u\in U.\end{aligned}$$

Combining the above two relations, we conclude that

$$\langle w,u^t-u\rangle+\phi(u^t)-\phi(u)+\langle\delta^{t-1},u-u^{t-1}\rangle+\eta V(x,u^t)-\eta V(x,u) \tag{8.3.4}$$
$$\leq (\eta\beta_t-\mu)V(u^{t-1},u)-\eta(1+\beta_t)V(u^t,u)+\langle\delta^{t-1},u^t-u^{t-1}\rangle$$
$$+M\|u^t-u^{t-1}\|-\eta\beta_t V(u^{t-1},u^t),\ \forall u\in U. \tag{8.3.5}$$

Moreover, by Cauchy–Schwarz inequality, (8.3.12), and the simple fact that $-at^2/2+bt \leq b^2/(2a)$ for any $a>0$, we have

$$\begin{aligned}&\langle\delta^{t-1},u^t-u^{t-1}\rangle+M\|u^t-u^{t-1}\|-\eta\beta_t V(u^{t-1},u^t)\\ &\leq (\|\delta^{t-1}\|_*+M)\|u^t-u^{t-1}\|-\tfrac{\eta\beta_t}{2}\|u^t-t^{t-1}\|^2 \leq \tfrac{(M+\|\delta^{t-1}\|_*)^2}{2\eta\beta_t}.\end{aligned}$$

From the above relation and the definition of $\Phi(u)$ in (8.3.3), we can rewrite (8.3.4) as,

$$\begin{aligned}&\Phi(u^t)-\Phi(u)+\langle\delta^{t-1},u-u^{t-1}\rangle\\ &\leq (\eta\beta_t-\mu)V(u^{t-1},u)-\eta(1+\beta_t)V(u^t,u)\\ &\quad+\tfrac{(M+\|\delta^{t-1}\|_*)^2}{2\eta\beta_t},\ \forall u\in U.\end{aligned}$$

Multiplying both sides by λ_t and summing up the resulting inequalities from $t=1$ to T, we obtain

$$\begin{aligned}
&\Sigma_{t=1}^{T}\lambda_t\left[\Phi(u^t)-\Phi(u)+\langle\delta^{t-1},u-u^{t-1}\rangle\right]\\
&\le\Sigma_{t=1}^{T}\left[(\eta\beta_t-\mu)\lambda_t V(u^{t-1},u)-\eta(1+\beta_t)\lambda_t V(u^t,u)\right]\\
&\quad+\Sigma_{t=1}^{T}\tfrac{(M+\|\delta^{t-1}\|_*)^2\lambda_t}{2\eta\beta_t}.
\end{aligned}$$

Hence, in view of (8.3.1), the convexity of Φ, and the definition of $\hat{u}^T$ in (8.3.27), we have

$$\begin{aligned}
&\Phi(\hat{u}^T)-\Phi(u)+(\Sigma_{t=1}^{T}\lambda_t)^{-1}\Sigma_{t=1}^{T}\lambda_t\langle\delta^{t-1},u-u^{t-1}\rangle\\
&\le(\Sigma_{t=1}^{T}\lambda_t)^{-1}\left[(\eta\beta_1-\mu)\lambda_1 V(u^0,u)-\eta(1+\beta_T)\lambda_T V(u^T,u)\right.\\
&\left.\quad+\Sigma_{t=1}^{T}\tfrac{(M+\|\delta^{t-1}\|_*)^2\lambda_t}{2\eta\beta_t}\right],
\end{aligned}$$

which implies (8.3.2) immediately. ■

As a matter of fact, the SDCS method covers the DCS method as a special case when $\delta^t=0$, $\forall t\ge 0$. Therefore, we investigate the proofs for Lemmas 8.5 and 8.6 first and then simplify them for the proofs for Lemmas 8.3 and 8.4. We now provide a proof for Lemma 8.5, which establishes the convergence property of SDCS method for solving general convex problems.

Proof of Lemma 8.5. When f_i, $i=1,\ldots,m$, are general convex functions, we have $\mu=0$ and $M>0$ (cf. (8.3.16)). Therefore, in view of $\phi:=f_i$, and λ_t and β_t defined in (8.3.34) satisfying condition (8.3.1) in the CS procedure, Eq. (8.3.2) can be rewritten as the following,[2]

$$\begin{aligned}
&(\Sigma_{t=1}^{T}\lambda_t)^{-1}\left[\eta(1+\beta_T)\lambda_T V_i(u_i^T,u_i)+\Sigma_{t=1}^{T}\lambda_t\langle\delta_i^{t-1},u_i-u_i^{t-1}\rangle\right]+\Phi_i(\hat{u}_i^T)-\Phi_i(u_i)\\
&\le(\Sigma_{t=1}^{T}\lambda_t)^{-1}\left[\eta\beta_1\lambda_1 V_i(u_i^0,u_i)+\Sigma_{t=1}^{T}\tfrac{(M+\|\delta_i^{t-1}\|_*)^2\lambda_t}{2\eta\beta_t}\right],\ \forall u_i\in X_i.
\end{aligned}$$

In view of the above relation, the definition of Φ^k in (8.3.19), and the input and output settings in the CS procedure, it is not difficult to see that, for any $k\ge 1$,[3]

$$\begin{aligned}
&\Phi^k(\hat{\mathbf{x}}^k)-\Phi^k(\mathbf{x})\\
&\quad+(\Sigma_{t=1}^{T_k}\lambda_t)^{-1}\left[\eta_k(1+\beta_{T_k})\lambda_{T_k}\mathbf{V}(\mathbf{x}^k,\mathbf{x})+\Sigma_{t=1}^{T_k}\Sigma_{i=1}^{m}\lambda_t\langle\delta_i^{t-1,k},x_i-u_i^{t-1}\rangle\right]\\
&\le(\Sigma_{t=1}^{T_k}\lambda_t)^{-1}\left[\eta_k\beta_1\lambda_1\mathbf{V}(\mathbf{x}^{k-1},\mathbf{x})+\Sigma_{t=1}^{T_k}\Sigma_{i=1}^{m}\tfrac{(M+\|\delta_i^{t-1,k}\|_*)^2\lambda_t}{2\eta_k\beta_t}\right],\ \forall\mathbf{x}\in\mathbf{X}.
\end{aligned}$$

[2] We added the subscript i to emphasize that this inequality holds for any agent $i\in\mathcal{N}$ with $\phi=f_i$. More specifically, $\Phi_i(u_i):=\langle w_i,u_i\rangle+f_i(u_i)+\eta V_i(x_i,u_i)$.

[3] We added the superscript k in $\delta_i^{t-1,k}$ to emphasize that this error is generated at the k-th outer loop.

By plugging into the above relation the values of λ_t and β_t in (8.3.34), together with the definition of Φ^k in (8.3.19) and rearranging the terms, we have,

$$\begin{aligned}&\langle \mathbf{L}(\hat{\mathbf{x}}^k-\mathbf{x}),\mathbf{y}^k\rangle+F(\hat{\mathbf{x}}^k)-F(\mathbf{x})\\&\le \tfrac{(T_k+1)(T_k+2)\eta_k}{T_k(T_k+3)}\left[\mathbf{V}(\mathbf{x}^{k-1},\mathbf{x})-\mathbf{V}(\mathbf{x}^k,\mathbf{x})\right]-\eta_k\mathbf{V}(\mathbf{x}^{k-1},\hat{\mathbf{x}}^k)\\&+\tfrac{2}{T_k(T_k+3)}\textstyle\sum_{t=1}^{T_k}\sum_{i=1}^m\left[(t+1)\langle\delta_i^{t-1,k},x_i-u_i^{t-1}\rangle+\tfrac{2(M+\|\delta_i^{t-1,k}\|_*)^2}{\eta_k}\right],\ \forall \mathbf{x}\in\mathbf{X}.\end{aligned}$$

Moreover, applying Lemma 3.5 to (8.3.22), we have, for $k\ge 1$,

$$\langle v_i^k,y_i-y_i^k\rangle\le\tfrac{\tau_k}{2}\left[\|y_i-y_i^{k-1}\|^2-\|y_i-y_i^k\|^2-\|y_i^{k-1}-y_i^k\|^2\right],\ \forall y_i\in\mathbb{R}^d, \quad (8.3.6)$$

which in view of the definition of Q in (8.3.8) and the above two relations then implies that, for $k\ge 1$, $\mathbf{z}\in\mathbf{X}\times\mathbb{R}^{md}$,

$$\begin{aligned}Q(\hat{\mathbf{x}}^k,\mathbf{y}^k;\mathbf{z})&=F(\hat{\mathbf{x}}^k)-F(\mathbf{x})+\langle\mathbf{L}\hat{\mathbf{x}}^k,\mathbf{y}\rangle-\langle\mathbf{L}\mathbf{x},\mathbf{y}^k\rangle\\&\le\langle\mathbf{L}(\hat{\mathbf{x}}^k-\tilde{\mathbf{x}}^k),\mathbf{y}-\mathbf{y}^k\rangle+\tfrac{(T_k+1)(T_k+2)\eta_k}{T_k(T_k+3)}\left[\mathbf{V}(\mathbf{x}^{k-1},\mathbf{x})-\mathbf{V}(\mathbf{x}^k,\mathbf{x})\right]\\&-\eta_k\mathbf{V}(\mathbf{x}^{k-1},\hat{\mathbf{x}}^k)+\tfrac{\tau_k}{2}\left[\|\mathbf{y}-\mathbf{y}^{k-1}\|^2-\|\mathbf{y}-\mathbf{y}^k\|^2-\|\mathbf{y}^{k-1}-\mathbf{y}^k\|^2\right]\\&+\tfrac{2}{T_k(T_k+3)}\textstyle\sum_{t=1}^{T_k}\sum_{i=1}^m\left[(t+1)\langle\delta_i^{t-1,k},x_i-u_i^{t-1}\rangle+\tfrac{2(M+\|\delta_i^{t-1,k}\|_*)^2}{\eta_k}\right].\end{aligned}$$

Multiplying both sides of the above inequality by θ_k, and summing up the resulting inequalities from $k=1$ to N, we obtain, for all $\mathbf{z}\in\mathbf{X}\times\mathbb{R}^{md}$,

$$\begin{aligned}&\textstyle\sum_{k=1}^N\theta_kQ(\hat{\mathbf{x}}^k,\mathbf{y}^k;\mathbf{z})\le\sum_{k=1}^N\theta_k\Delta_k\\&+\textstyle\sum_{k=1}^N\sum_{t=1}^{T_k}\sum_{i=1}^m\tfrac{2\theta_k}{T_k(T_k+3)}\left[(t+1)\langle\delta_i^{t-1,k},x_i-u_i^{t-1}\rangle+\tfrac{2(M+\|\delta_i^{t-1,k}\|_*)^2}{\eta_k}\right],\end{aligned} \quad (8.3.7)$$

where

$$\begin{aligned}\Delta_k:=&\langle\mathbf{L}(\hat{\mathbf{x}}^k-\tilde{\mathbf{x}}^k),\mathbf{y}-\mathbf{y}^k\rangle+\tfrac{(T_k+1)(T_k+2)\eta_k}{T_k(T_k+3)}\left[\mathbf{V}(\mathbf{x}^{k-1},\mathbf{x})-\mathbf{V}(\mathbf{x}^k,\mathbf{x})\right]\\&-\eta_k\mathbf{V}(\mathbf{x}^{k-1},\hat{\mathbf{x}}^k)+\tfrac{\tau_k}{2}\left[\|\mathbf{y}-\mathbf{y}^{k-1}\|^2-\|\mathbf{y}-\mathbf{y}^k\|^2-\|\mathbf{y}^{k-1}-\mathbf{y}^k\|^2\right].\end{aligned} \quad (8.3.8)$$

We now provide a bound on $\sum_{k=1}^N\theta_k\Delta_k$. Observe that from the definition of $\tilde{\mathbf{x}}^k$ in (8.3.17), (8.3.29), and (8.3.31) we have

$$
\begin{aligned}
&\textstyle\sum_{k=1}^N \theta_k \Delta_k \\
&\le \textstyle\sum_{k=1}^N \left[\theta_k \langle \mathbf{L}(\hat{\mathbf{x}}^k - \mathbf{x}^{k-1}), \mathbf{y} - \mathbf{y}^k\rangle - \alpha_k \theta_k \langle \mathbf{L}(\hat{\mathbf{x}}^{k-1} - \mathbf{x}^{k-2}), \mathbf{y} - \mathbf{y}^{k-1}\rangle\right] \\
&-\textstyle\sum_{k=1}^N \theta_k \left[\alpha_k \langle \mathbf{L}(\hat{\mathbf{x}}^{k-1} - \mathbf{x}^{k-2}), \mathbf{y}^{k-1} - \mathbf{y}^k\rangle + \eta_k \mathbf{V}(\mathbf{x}^{k-1}, \hat{\mathbf{x}}^k) + \tfrac{\tau_k}{2}\|\mathbf{y}^{k-1} - \mathbf{y}^k\|^2\right] \\
&+\tfrac{(T_1+1)(T_1+2)\theta_1\eta_1}{T_1(T_1+3)} \mathbf{V}(\mathbf{x}^0, \mathbf{x}) - \tfrac{(T_N+1)(T_N+2)\theta_N\eta_N}{T_N(T_N+3)} \mathbf{V}(\mathbf{x}^N, \mathbf{x}) \\
&+\tfrac{\theta_1\tau_1}{2}\|\mathbf{y} - \mathbf{y}^0\|^2 - \tfrac{\theta_N\tau_N}{2}\|\mathbf{y} - \mathbf{y}^N\|^2 \\
&\overset{(a)}{\le} \theta_N \langle \mathbf{L}(\hat{\mathbf{x}}^N - \mathbf{x}^{N-1}), \mathbf{y} - \mathbf{y}^N\rangle - \theta_N \eta_N \mathbf{V}(\mathbf{x}^{N-1}, \hat{\mathbf{x}}^N) \\
&-\textstyle\sum_{k=2}^N \left[\theta_k \alpha_k \langle \mathbf{L}(\hat{\mathbf{x}}^{k-1} - \mathbf{x}^{k-2}), \mathbf{y}^{k-1} - \mathbf{y}^k\rangle \right.\\
&\left. +\theta_{k-1}\eta_{k-1} \mathbf{V}(\mathbf{x}^{k-2}, \hat{\mathbf{x}}^{k-1}) + \tfrac{\theta_k\tau_k}{2}\|\mathbf{y}^{k-1} - \mathbf{y}^k\|^2\right] \\
&+\tfrac{(T_1+1)(T_1+2)\theta_1\eta_1}{T_1(T_1+3)} \mathbf{V}(\mathbf{x}^0, \mathbf{x}) - \tfrac{(T_N+1)(T_N+2)\theta_N\eta_N}{T_N(T_N+3)} \mathbf{V}(\mathbf{x}^N, \mathbf{x}) \\
&+\tfrac{\theta_1\tau_1}{2}\|\mathbf{y} - \mathbf{y}^0\|^2 - \tfrac{\theta_N\tau_N}{2}\|\mathbf{y} - \mathbf{y}^N\|^2 \\
&\overset{(b)}{\le} \theta_N \langle \mathbf{L}(\hat{\mathbf{x}}^N - \mathbf{x}^{N-1}), \mathbf{y} - \mathbf{y}^N\rangle - \theta_N \eta_N \mathbf{V}(\mathbf{x}^{N-1}, \hat{\mathbf{x}}^N) \\
&+\tfrac{\theta_1\tau_1}{2}\|\mathbf{y} - \mathbf{y}^0\|^2 - \tfrac{\theta_N\tau_N}{2}\|\mathbf{y} - \mathbf{y}^N\|^2 \\
&+\textstyle\sum_{k=2}^N \left(\tfrac{\theta_{k-1}\alpha_k\|\mathbf{L}\|^2}{2\tau_k} - \tfrac{\theta_{k-1}\eta_{k-1}}{2}\right)\|\mathbf{x}^{k-2} - \hat{\mathbf{x}}^{k-1}\|^2 \\
&+\tfrac{(T_1+1)(T_1+2)\theta_1\eta_1}{T_1(T_1+3)} \mathbf{V}(\mathbf{x}^0, \mathbf{x}) - \tfrac{(T_N+1)(T_N+2)\theta_N\eta_N}{T_N(T_N+3)} \mathbf{V}(\mathbf{x}^N, \mathbf{x}) \\
&\overset{(c)}{\le} \theta_N \langle \mathbf{L}(\hat{\mathbf{x}}^N - \mathbf{x}^{N-1}), \mathbf{y} - \mathbf{y}^N\rangle - \theta_N \eta_N \mathbf{V}(\mathbf{x}^{N-1}, \hat{\mathbf{x}}^N) \\
&+\tfrac{\theta_1\tau_1}{2}\|\mathbf{y} - \mathbf{y}^0\|^2 - \tfrac{\theta_N\tau_N}{2}\|\mathbf{y} - \mathbf{y}^N\|^2 \\
&+\tfrac{(T_1+1)(T_1+2)\theta_1\eta_1}{T_1(T_1+3)} \mathbf{V}(\mathbf{x}^0, \mathbf{x}) - \tfrac{(T_N+1)(T_N+2)\theta_N\eta_N}{T_N(T_N+3)} \mathbf{V}(\mathbf{x}^N, \mathbf{x}) \\
&\overset{(d)}{\le} \theta_N \langle \mathbf{y}^N, \mathbf{L}(\mathbf{x}^{N-1} - \hat{\mathbf{x}}^N)\rangle - \theta_N \eta_N \mathbf{V}(\mathbf{x}^{N-1}, \hat{\mathbf{x}}^N) - \tfrac{\theta_1\tau_1}{2}\|\mathbf{y}^N\|^2 \\
&+\tfrac{(T_1+1)(T_1+2)\theta_1\eta_1}{T_1(T_1+3)} \mathbf{V}(\mathbf{x}^0, \mathbf{x}) + \tfrac{\theta_1\tau_1}{2}\|\mathbf{y}^0\|^2 + \langle \mathbf{y}, \theta_N \mathbf{L}(\hat{\mathbf{x}}^N - \mathbf{x}^{N-1}) + \theta_1\tau_1(\mathbf{y}^N - \mathbf{y}^0)\rangle, \\
&\overset{(e)}{\le} \left(\tfrac{\theta_N\|\mathbf{L}\|^2}{2\eta_N} - \tfrac{\theta_1\tau_1}{2}\right)\|\mathbf{y}^N\|^2 + \tfrac{(T_1+1)(T_1+2)\theta_1\eta_1}{T_1(T_1+3)} \mathbf{V}(\mathbf{x}^0, \mathbf{x}) + \tfrac{\theta_1\tau_1}{2}\|\mathbf{y}^0\|^2 \\
&+\langle \mathbf{y}, \theta_N \mathbf{L}(\hat{\mathbf{x}}^N - \mathbf{x}^{N-1}) + \theta_1\tau_1(\mathbf{y}^N - \mathbf{y}^0)\rangle,
\end{aligned}
$$

where (a) follows from (8.3.30) and the fact that $\mathbf{x}^{-1} = \hat{\mathbf{x}}^0$, (b) follows from the simple relation that $b\langle u, v\rangle - a\|v\|^2/2 \le b^2\|u\|^2/(2a), \forall a > 0$, (8.3.30) and (8.3.15), (c) follows from (8.3.32), (d) follows from (8.3.31), $\|\mathbf{y} - \mathbf{y}^0\|^2 - \|\mathbf{y} - \mathbf{y}^N\|^2 = \|\mathbf{y}^0\|^2 - \|\mathbf{y}^N\|^2 - 2\langle \mathbf{y}, \mathbf{y}^0 - \mathbf{y}^N\rangle$ and arranging the terms accordingly, (e) follows from (8.3.15) and the relation $b\langle u, v\rangle - a\|v\|^2/2 \le b^2\|u\|^2/(2a), \forall a > 0$. Using the above bound in (8.3.7) we obtain

$$
\begin{aligned}
\textstyle\sum_{k=1}^N \theta_k Q(\hat{\mathbf{x}}^k, \mathbf{y}^k; \mathbf{z}) &\le \tfrac{(T_1+1)(T_1+2)\theta_1\eta_1}{T_1(T_1+3)} \mathbf{V}(\mathbf{x}^0, \mathbf{x}) + \tfrac{\theta_1\tau_1}{2}\|\mathbf{y}^0\|^2 + \langle \hat{\mathbf{s}}, \mathbf{y}\rangle \qquad (8.3.9)\\
&+\textstyle\sum_{k=1}^N \sum_{t=1}^{T_k} \sum_{i=1}^m \tfrac{2\theta_k}{T_k(T_k+3)} \left[(t+1)\langle \delta_i^{t-1,k}, x_i - u_i^{t-1}\rangle + \tfrac{4(M^2 + \|\delta_i^{t-1,k}\|_*^2)}{\eta_k}\right],
\end{aligned}
$$

for all $\mathbf{z} \in \mathbf{X} \times \mathbb{R}^{md}$, where

$$
\hat{\mathbf{s}} := \theta_N \mathbf{L}(\hat{\mathbf{x}}^N - \mathbf{x}^{N-1}) + \theta_1\tau_1(\mathbf{y}^N - \mathbf{y}^0). \qquad (8.3.10)
$$

Our result in (8.3.65) immediately follows from the convexity of Q. Furthermore, in view of (8.3.9)(c) and (8.3.7), we can obtain the following result,

$$
\begin{aligned}
\textstyle\sum_{k=1}^N \theta_k Q(\hat{\mathbf{x}}^k,\mathbf{y}^k;\mathbf{z}) &\le \theta_N\langle \mathbf{L}(\hat{\mathbf{x}}^N-\mathbf{x}^{N-1}),\mathbf{y}-\mathbf{y}^N\rangle - \theta_N\eta_N\mathbf{V}(\mathbf{x}^{N-1},\hat{\mathbf{x}}^N)\\
&+\tfrac{\theta_1\tau_1}{2}\|\mathbf{y}-\mathbf{y}^0\|^2-\tfrac{\theta_N\tau_N}{2}\|\mathbf{y}-\mathbf{y}^N\|^2\\
&+\tfrac{(T_1+1)(T_1+2)\theta_1\eta_1}{T_1(T_1+3)}\mathbf{V}(\mathbf{x}^0,\mathbf{x})-\tfrac{(T_N+1)(T_N+2)\theta_N\eta_N}{T_N(T_N+3)}\mathbf{V}(\mathbf{x}^N,\mathbf{x})\\
&+\textstyle\sum_{k=1}^N\sum_{t=1}^{T_k}\sum_{i=1}^m \tfrac{\theta_k}{T_k(T_k+3)}\left[(t+1)\langle\delta_i^{t-1,k},x_i-u_i^{t-1}\rangle+\tfrac{4(M^2+\|\delta_i^{t-1,k}\|_*^2)}{\eta_k}\right].
\end{aligned}
$$

Therefore, in view of the fact that $\sum_{k=1}^N\theta_k Q(\hat{\mathbf{x}}^k,\mathbf{y}^k;\mathbf{z}^*)\ge 0$ for any saddle point $\mathbf{z}^*=(\mathbf{x}^*,\mathbf{y}^*)$ of (8.3.7), and (8.3.15), by fixing $\mathbf{z}=\mathbf{z}^*$ and rearranging terms, we obtain

$$
\begin{aligned}
\tfrac{\theta_N\eta_N}{2}\|\hat{\mathbf{x}}^N-\mathbf{x}^{N-1}\|^2 &\le \theta_N\langle\mathbf{L}(\hat{\mathbf{x}}^N-\mathbf{x}^{N-1}),\mathbf{y}^*-\mathbf{y}^N\rangle-\tfrac{\theta_N\tau_N}{2}\|\mathbf{y}^*-\mathbf{y}^N\|^2 \qquad (8.3.11)\\
&+\tfrac{(T_1+1)(T_1+2)\theta_1\eta_1}{T_1(T_1+3)}\mathbf{V}(\mathbf{x}^0,\mathbf{x}^*)+\tfrac{\theta_1\tau_1}{2}\|\mathbf{y}^*-\mathbf{y}^0\|^2\\
&+\textstyle\sum_{k=1}^N\sum_{t=1}^{T_k}\sum_{i=1}^m\tfrac{2\theta_k}{T_k(T_k+3)}\left[(t+1)\langle\delta_i^{t-1,k},x_i^*-u_i^{t-1}\rangle+\tfrac{4(M^2+\|\delta_i^{t-1,k}\|_*^2)}{\eta_k}\right]\\
&\le\tfrac{\theta_N\|\mathbf{L}\|^2}{2\tau_N}\|\hat{\mathbf{x}}^N-\mathbf{x}^{N-1}\|^2+\tfrac{(T_1+1)(T_1+2)\theta_1\eta_1}{T_1(T_1+3)}\mathbf{V}(\mathbf{x}^0,\mathbf{x}^*)+\tfrac{\theta_1\tau_1}{2}\|\mathbf{y}^*-\mathbf{y}^0\|^2\\
&+\textstyle\sum_{k=1}^N\sum_{t=1}^{T_k}\sum_{i=1}^m\tfrac{2\theta_k}{T_k(T_k+3)}\left[(t+1)\langle\delta_i^{t-1,k},x_i^*-u_i^{t-1}\rangle+\tfrac{4(M^2+\|\delta_i^{t-1,k}\|_*^2)}{\eta_k}\right],
\end{aligned}
$$

where the second inequality follows from the relation $b\langle u,v\rangle - a\|v\|^2/2\le b^2\|u\|^2/(2a), \forall a>0$.

Similarly, we obtain

$$
\begin{aligned}
\tfrac{\theta_N\tau_N}{2}\|\mathbf{y}^*-\mathbf{y}^N\|^2 &\le \tfrac{\theta_N\|\mathbf{L}\|^2}{2\eta_N}\|\mathbf{y}^*-\mathbf{y}^N\|^2\\
&+\tfrac{(T_1+1)(T_1+2)\theta_1\eta_1}{T_1(T_1+3)}\mathbf{V}(\mathbf{x}^0,\mathbf{x}^*)+\tfrac{\theta_1\tau_1}{2}\|\mathbf{y}^*-\mathbf{y}^0\|^2 \qquad (8.3.12)\\
&+\textstyle\sum_{k=1}^N\sum_{t=1}^{T_k}\sum_{i=1}^m\tfrac{2\theta_k}{T_k(T_k+3)}\left[(t+1)\langle\delta_i^{t-1,k},x_i^*-u_i^{t-1}\rangle+\tfrac{4(M^2+\|\delta_i^{t-1,k}\|_*^2)}{\eta_k}\right],
\end{aligned}
$$

from which the desired result in (8.3.66) follows. ■

The following proof of Lemma 8.6 establishes the convergence of SDCS method for solving strongly convex problems.

Proof of Lemma 8.6. When f_i, $i=1,\ldots,m$, are strongly convex functions, we have μ, $M>0$ (cf. (8.3.16)). Therefore, in view of Proposition 8.7 with λ_t and β_t defined in (8.3.49) satisfying condition (8.3.1), the definition of Φ^k in (8.3.19), and the input and output settings in the CS procedure, we have for all $k\ge 1$ and for all $\mathbf{x}\in\mathbf{X}$

$$\Phi^k(\hat{\mathbf{x}}^k)-\Phi^k(\mathbf{x})$$

$$+(\sum_{t=1}^{T_k}\lambda_t)^{-1}\left[\eta_k(1+\beta_{T_k}^{(k)})\lambda_{T_k}\mathbf{V}(\mathbf{x}^k,\mathbf{x})+\sum_{t=1}^{T_k}\sum_{i=1}^m\lambda_t\langle\delta_i^{t-1,k},x_i-u_i^{t-1}\rangle\right]$$
$$\le(\sum_{t=1}^{T_k}\lambda_t)^{-1}\left[(\eta_k\beta_1^{(k)}-\mu)\lambda_1\mathbf{V}(\mathbf{x}^{k-1},\mathbf{x})+\sum_{t=1}^{T_k}\sum_{i=1}^m\tfrac{(M+\|\delta_i^{t-1,k}\|_*)^2\lambda_t}{2\eta_k\beta_t}\right].$$

By plugging into the above relation the values of λ_t and $\beta_t^{(k)}$ in (8.3.49), together with the definition of Φ^k in (8.3.19) and rearranging the terms, we have

$$\langle\mathbf{L}(\hat{\mathbf{x}}^k-\mathbf{x}),\mathbf{y}^k\rangle+F(\hat{\mathbf{x}}^k)-F(\mathbf{x})$$
$$\le\eta_k\mathbf{V}(\mathbf{x}^{k-1},\mathbf{x})-(\mu+\eta_k)\mathbf{V}(\mathbf{x}^k,\mathbf{x})-\eta_k\mathbf{V}(\mathbf{x}^{k-1},\hat{\mathbf{x}}^k)$$
$$+\tfrac{2}{T_k(T_k+1)}\sum_{t=1}^{T_k}\sum_{i=1}^m\left[t\langle\delta_i^{t-1,k},x_i-u_i^{t-1}\rangle+\tfrac{(M+\|\delta_i^{t-1,k}\|_*)^2t}{(t+1)\mu+(t-1)\eta_k}\right],\ \forall\mathbf{x}\in\mathbf{X},\,k\ge1.$$

In view of (8.3.6), the above relation and the definition of Q in (8.3.8), and following the same trick that we used to obtain (8.3.7), we have, for all $\mathbf{z}\in\mathbf{X}\times\mathbb{R}^{md}$,

$$\sum_{k=1}^N\theta_kQ(\hat{\mathbf{x}}^k,\mathbf{y}^k;\mathbf{z})\le\sum_{k=1}^N\theta_k\bar{\Delta}_k$$
$$+\sum_{k=1}^N\sum_{t=1}^{T_k}\sum_{i=1}^m\tfrac{2\theta_k}{T_k(T_k+1)}\left[t\langle\delta_i^{t-1,k},x_i-u_i^{t-1}\rangle+\tfrac{(M+\|\delta_i^{t-1,k}\|_*)^2t}{(t+1)\mu+(t-1)\eta_k}\right],\tag{8.3.13}$$

where

$$\bar{\Delta}_k:=\mathbf{L}(\hat{\mathbf{x}}^k-\tilde{\mathbf{x}}^k),\mathbf{y}-\mathbf{y}^k\rangle$$
$$+\eta_k\mathbf{V}(\mathbf{x}^{k-1},\mathbf{x})-(\mu+\eta_k)\mathbf{V}(\mathbf{x}^k,\mathbf{x})-\eta_k\mathbf{V}(\mathbf{x}^{k-1},\hat{\mathbf{x}}^k)\tag{8.3.14}$$
$$+\tfrac{\tau_k}{2}\left[\|\mathbf{y}-\mathbf{y}^{k-1}\|^2-\|\mathbf{y}-\mathbf{y}^k\|^2-\|\mathbf{y}^{k-1}-\mathbf{y}^k\|^2\right].$$

Since $\bar{\Delta}_k$ in (8.3.14) shares a similar structure with Δ_k in (8.3.8), we can follow the similar procedure as in (8.3.9) to simplify the RHS of (8.3.13). Note that the only difference between (8.3.14) and (8.3.8) is in the coefficient of the terms $\mathbf{V}(\mathbf{x}^{k-1},\mathbf{x})$, and $\mathbf{V}(\mathbf{x}^k,\mathbf{x})$. Hence, by using condition (8.3.48) in place of (8.3.29), we obtain $\forall\mathbf{z}\in\mathbf{X}\times\mathbb{R}^{md}$

$$\sum_{k=1}^N\theta_kQ(\hat{\mathbf{x}}^k,\mathbf{y}^k;\mathbf{z})\le\theta_1\eta_1\mathbf{V}(\mathbf{x}^0,\mathbf{x})+\tfrac{\theta_1\tau_1}{2}\|\mathbf{y}^0\|^2+\langle\hat{\mathbf{s}},\mathbf{y}\rangle\tag{8.3.15}$$
$$+\sum_{k=1}^N\sum_{t=1}^{T_k}\sum_{i=1}^m\tfrac{2\theta_k}{T_k(T_k+1)}\left[t\langle\delta_i^{t-1,k},x_i-u_i^{t-1}\rangle+\tfrac{2t(M^2+\|\delta_i^{t-1,k}\|_*^2)}{(t+1)\mu+(t-1)\eta_k}\right],$$

where $\hat{\mathbf{s}}$ is defined in (8.3.10). Our result in (8.3.74) immediately follows from the convexity of Q.

Following the same procedure as we obtain (8.3.11), for any saddle point $\mathbf{z}^*=(\mathbf{x}^*,\mathbf{y}^*)$ of (8.3.7), we have

$$\tfrac{\theta_N\eta_N}{2}\|\hat{\mathbf{x}}^N-\mathbf{x}^{N-1}\|^2$$
$$\le\tfrac{\theta_N\|\mathbf{L}\|^2}{2\tau_N}\|\mathbf{x}^N-\mathbf{x}^{N-1}\|^2+\theta_1\eta_1\mathbf{V}(\mathbf{x}^0,\mathbf{x}^*)+\tfrac{\theta_1\tau_1}{2}\|\mathbf{y}^*-\mathbf{y}^0\|^2\tag{8.3.16}$$

$$+\sum_{k=1}^{N}\sum_{t=1}^{T_k}\sum_{i=1}^{m}\frac{2\theta_k}{T_k(T_k+1)}\left[t\langle\delta_i^{t-1,k},x_i^*-u_i^{t-1}\rangle+\frac{2t(M^2+\|\delta_i^{t-1,k}\|_*^2)}{(t+1)\mu+(t-1)\eta_k}\right],$$

$$\frac{\theta_N\tau_N}{2}\|\mathbf{y}^*-\mathbf{y}^N\|^2\le\frac{\theta_N\|\mathbf{L}\|^2}{2\eta_N}\|\mathbf{y}^*-\mathbf{y}^N\|^2+\theta_1\eta_1\mathbf{V}(\mathbf{x}^0,\mathbf{x}^*)+\frac{\theta_1\tau_1}{2}\|\mathbf{y}^*-\mathbf{y}^0\|^2$$

$$+\sum_{k=1}^{N}\sum_{t=1}^{T_k}\sum_{i=1}^{m}\frac{2\theta_k}{T_k(T_k+1)}\left[t\langle\delta_i^{t-1,k},x_i^*-u_i^{t-1}\rangle+\frac{2t(M^2+\|\delta_i^{t-1,k}\|_*^2)}{(t+1)\mu+(t-1)\eta_k}\right],$$

from which the desired result in (8.3.75) follows. ■

We are ready to provide proofs for Lemma 8.3 and 8.4, which demonstrates the convergence properties of the deterministic communication sliding method.

Proof of Lemma 8.3. When f_i, $i=1,\dots,m$ are general nonsmooth convex functions, we have $\delta_i^t=0$, $\mu=0$ and $M>0$. Therefore, in view of (8.3.9), we have, $\forall\mathbf{z}\in\mathbf{X}\times\mathbb{R}^{md}$,

$$\sum_{k=1}^{N}\theta_kQ(\hat{\mathbf{x}}^k,\mathbf{y}^k;\mathbf{z})\le\frac{(T_1+1)(T_1+2)\theta_1\eta_1}{T_1(T_1+3)}\mathbf{V}(\mathbf{x}^0,\mathbf{x})+\frac{\theta_1\tau_1}{2}\|\mathbf{y}^0\|^2+\langle\hat{\mathbf{s}},\mathbf{y}\rangle$$
$$+\sum_{k=1}^{N}\frac{4mM^2\theta_k}{(T_k+3)\eta_k},$$

where $\hat{\mathbf{s}}$ is defined in (8.3.10). Our result in (8.3.35) immediately follows from the convexity of Q. Moreover, our result in (8.3.36) follows from setting $\delta_i^{t-1,k}=0$ in (8.3.11) and (8.3.12). ■

Proof of Lemma 8.4. When f_i, $i=1,\dots,m$ are strongly convex functions, we have $\delta_i^t=0$ and μ, $M>0$. Therefore, in view of (8.3.15), we obtain, $\forall\mathbf{z}\in\mathbf{X}\times\mathbb{R}^{md}$,

$$\sum_{k=1}^{N}\theta_kQ(\hat{\mathbf{x}}^k,\mathbf{y}^k;\mathbf{z})\le\theta_1\eta_1\mathbf{V}(\mathbf{x}^0,\mathbf{x})+\frac{\theta_1\tau_1}{2}\|\mathbf{y}^0\|^2+\langle\hat{\mathbf{s}},\mathbf{y}\rangle$$
$$+\sum_{k=1}^{N}\sum_{t=1}^{T_k}\frac{2mM^2\theta_k}{T_k(T_k+1)}\frac{t}{(t+1)\mu+(t-1)\eta_k},$$

where $\hat{\mathbf{s}}$ is defined in (8.3.10). Our result in (8.3.50) immediately follows from the convexity of Q. Also, the result in (8.3.51) follows by setting $\delta_i^{t-1,k}=0$ in (8.3.16). ■

We now provide a proof for Theorem 8.13 that establishes a large deviation result for the gap function.

Proof of Theorem 8.13. Observe that by Assumptions (8.3.61), (8.3.62), and (8.3.83) on the SO and the definition of $u_i^{t,k}$, the sequence $\{\langle\delta_i^{t-1,k},x_i^*-u_i^{t-1,k}\rangle\}_{1\le i\le m,1\le t\le T_k,k\ge1}$ is a martingale-difference sequence. Denoting

$$\gamma_{k,t}:=\frac{\theta_k\lambda_t}{\sum_{t=1}^{T_k}\lambda_t},$$

and using the large-deviation theorem for martingale-difference sequence and the fact that

$$\mathbb{E}[\exp\{\gamma_{k,t}^2\langle\delta_i^{t-1,k},x_i^*-u_i^{t-1,k}\rangle^2/(2\gamma_{k,t}^2\bar{V}_i(x_i^*)\sigma^2)\}]$$

$$\leq \mathbb{E}[\exp\{\|\delta_i^{t-1,k}\|_*^2, \|x_i^* - u_i^{t-1,k}\|^2/(2\bar{V}_i(x_i^*)\sigma^2)\}]$$
$$\leq \mathbb{E}[\exp\{\|\delta_i^{t-1,k}\|_*^2/\sigma^2\}] \leq \exp\{1\},$$

we conclude that, $\forall \zeta > 0$,

$$\text{Prob}\left\{\textstyle\sum_{k=1}^N \sum_{t=1}^{T_k} \sum_{i=1}^m \gamma_{k,t}\langle \delta_i^{t-1,k}, u_i^{t-1,k} - x_i^*\rangle > \zeta\sigma\sqrt{2\bar{\mathbf{V}}(\mathbf{x}^*)\sum_{k=1}^N \sum_{t=1}^{T_k} \gamma_{k,t}^2}\right\}$$
$$\leq \exp\{-\zeta^2/3\}. \tag{8.3.17}$$

Now let

$$S_{k,t} := \frac{\theta_k \lambda_t}{\left(\sum_{t=1}^{T_k} \lambda_t\right)\eta_k \beta_t},$$

and $S := \sum_{k=1}^N \sum_{t=1}^{T_k} \sum_{i=1}^m S_{k,t}$. By the convexity of exponential function, we have

$$\mathbb{E}[\exp\{\tfrac{1}{S}\textstyle\sum_{k=1}^N \sum_{t=1}^{T_k} \sum_{i=1}^m S_{k,t}\|\delta_i^{t-1,k}\|_*^2/\sigma^2\}]$$
$$\leq \mathbb{E}[\tfrac{1}{S}\textstyle\sum_{k=1}^N \sum_{t=1}^{T_k} \sum_{i=1}^m S_{k,t}\exp\{\|\delta_i^{t-1,k}\|_*^2/\sigma^2\}] \leq \exp\{1\},$$

where the last inequality follows from Assumption (8.3.83). Therefore, by Markov's inequality, for all $\zeta > 0$,

$$\text{Prob}\left\{\textstyle\sum_{k=1}^N \sum_{t=1}^{T_k} \sum_{i=1}^m S_{k,t}\|\delta_i^{t-1,k}\|_*^2 > (1+\zeta)\sigma^2 \sum_{k=1}^N \sum_{t=1}^{T_k} \sum_{i=1}^m S_{k,t}\right\} \tag{8.3.18}$$
$$= \text{Prob}\left\{\exp\left\{\tfrac{1}{S}\textstyle\sum_{k=1}^N \sum_{t=1}^{T_k} \sum_{i=1}^m S_{k,t}\|\delta_i^{t-1,k}\|_*^2/\sigma^2\right\} \geq \exp\{1+\zeta\}\right\} \leq \exp\{-\zeta\}.$$

Combing (8.3.17), (8.3.18), (8.3.65), and (8.3.10), our result in (8.3.85) immediately follows. ■

8.4 Exercises and Notes

1. Suppose that in the gradient sliding method, the computational costs associated with gradient evaluation of ∇f and subgradient evaluation of h' are given by G and S, respectively. Use such information to refine the selection of algorithmic parameters in this method.
2. Suppose that in the communication sliding method, the computational costs associated with one communication round and subgradient evaluation of ϕ_i' are given by C and S, respectively. Use such information to refine the selection of algorithmic parameters in this method.

Notes. Lan first developed the gradient sliding method in [65]. The accelerated gradient sliding method was introduced by Lan and Ouyang in [67]. Lan, Lee, and Zhou first presented in [37] the communication sliding algorithms for decentralized optimization over networks. The basic scheme of the communication sliding algorithm was inspired by Chambolle and Pock's primal–dual method in [18] and the perturbation-based termination criterion for dual methods were first studied by Monteiro and Svaiter [84–86]. Earlier developments for decentralized algorithms can be found, e.g., in [13, 20, 49, 75, 82, 83, 90, 113, 127, 134, 135]. The seminal work on distributed optimization [134, 135] has been followed by distributed incremental (sub)gradient methods and proximal methods [13, 90, 113, 136], and more recently the incremental aggregated gradient methods and its proximal variants [14, 44, 71]. These methods are designed for networks with a central authority. To consider a more general distributed network topology without a central authority, a decentralized subgradient algorithm was first proposed in [89], and further studied in many other literature (see, e.g., [31, 87, 88, 133, 143]). Improvements on communication complexity can be obtained when the objective function (8.3.1) is smooth and/or strongly convex (see, e.g., [91, 110, 127, 128]). Besides subgradient based methods, another well-known type of decentralized algorithms relies on dual methods (see e.g., [3, 17, 82, 126, 130, 138], see also [49] for the application of mirror-prox method for solving these problems). These dual-based methods have been further studied via proximal-gradients [3, 19, 20]. Multi-step consensus has been considered in decentralized methods with smoothness assumption, and hence these methods require an increasing number of communication rounds iteratively [21, 54]. While these methods were mostly for decentralized deterministic optimization problems, earlier decentralized stochastic first-order methods have been studied in [31, 87, 112, 114, 129, 132, 139].

References

1. S.D. Ahipasaoglu, M.J. Todd, A modified Frank-Wolfe algorithm for computing minimum-area enclosing ellipsoidal cylinders: theory and algorithms. Comput. Geom. **46**, 494–519 (2013)
2. Z. Allen-Zhu, Katyusha: the first direct acceleration of stochastic gradient methods (2016). ArXiv e-prints, abs/1603.05953
3. N.S. Aybat, Z. Wang, T. Lin, S. Ma, Distributed linearized alternating direction method of multipliers for composite convex consensus optimization. IEEE Trans. Autom. Control **63**(1), 5–20 (2018)
4. F. Bach, S. Lacoste-Julien, G. Obozinski, On the equivalence between herding and conditional gradient algorithms, in *The 29th International Conference on Machine Learning* (2012)
5. A. Beck, M. Teboulle, Mirror-descent and nonlinear projected subgradient methods for convex optimization. Oper. Res. Lett. **31**, 167–175 (2003)
6. A. Beck, M. Teboulle, A conditional gradient method with linear rate of convergence for solving convex linear systems. Math. Methods Oper. Res. **59**, 235–247 (2004)
7. A. Beck, M. Teboulle, A fast iterative shrinkage-thresholding algorithm for linear inverse problems. SIAM J. Imaging Sciences **2**, 183–202 (2009)
8. Y. Bengio, Learning deep architectures for AI. Found. Trends Mach. Learn. **2**(1), 1–127 (2009)
9. A. Ben-Tal, A.S. Nemirovski, *Lectures on Modern Convex Optimization: Analysis, Algorithms, Engineering Applications*. MPS-SIAM Series on Optimization (SIAM, Philadelphia, 2000)
10. A. Ben-Tal, A.S. Nemirovski, Non-Euclidean restricted memory level method for large-scale convex optimization. Math. Program. **102**, 407–456 (2005)
11. D.P. Bertsekas, Stochastic optimization problems with nondifferentiable cost functionals. J. Optim. Theory Appl. **12**, 218–231 (1973)

G. Lan, *First-Order and Stochastic Optimization Methods for Machine Learning*, Springer Series in the Data Sciences, https://doi.org/10.1007/978-3-030-39568-1

12. D. Bertsekas, *Nonlinear Programming*, 2nd edn. (Athena Scientific, New York, 1999)
13. D.P. Bertsekas, Incremental proximal methods for large scale convex optimization. Math. Program. **129**, 163–195 (2011)
14. D.P. Bertsekas, Incremental aggregated proximal and augmented lagrangian algorithms. Technical Report LIDS-P-3176, Laboratory for Information and Decision Systems, 2015
15. D. Blatt, A. Hero, H. Gauchman, A convergent incremental gradient method with a constant step size. SIAM J. Optim. **18**(1), 29–51 (2007)
16. S. Boyd, L. Vandenberghe, *Convex Optimization*. (Cambridge University Press, Cambridge, 2004)
17. S. Boyd, N. Parikh, E. Chu, B. Peleato, J. Eckstein, Distributed optimization and statistical learning via the alternating direction method of multipliers. Found. Trends Mach. Learn. **3**(1), 1–122 (2011)
18. A. Chambolle, T. Pock, A first-order primal-dual algorithm for convex problems with applications to imaging. J. Math. Imaging Vision **40**, 120–145 (2011)
19. T. Chang, M. Hong, Stochastic proximal gradient consensus over random networks (2015). http://arxiv.org/abs/1511.08905
20. T. Chang, M. Hong, X. Wang, Multi-agent distributed optimization via inexact consensus ADMM (2014). http://arxiv.org/abs/1402.6065
21. A. Chen, A. Ozdaglar, A fast distributed proximal gradient method, in *2012 50th Annual Allerton Conference on Communication, Control, and Computing (Allerton), October* (2012), pp. 601–608
22. Y. Chen, G. Lan, Y. Ouyang, Accelerated schemes for a class of variational inequalities. Math. Program. **165**, 113–149 (2017)
23. Y. Chen, G. Lan, Y. Ouyang, Optimal primal-dual methods for a class of saddle point problems. SIAM J. Optim. **24**(4), 1779–1814 (2014)
24. K.L. Clarkson, Coresets, sparse greedy approximation, and the Frank-Wolfe algorithm. ACM Trans. Algorithms **6**(4), 63:1–63:30 (2010)
25. B. Cox, A. Juditsky, A.S. Nemirovski, Dual subgradient algorithms for large-scale nonsmooth learning problems. Math. Program. 148, 143–180 (2014). Manuscript, School of ISyE, Georgia Tech, Atlanta
26. C. Dang, G. Lan, Randomized first-order methods for saddle point optimization. Manuscript, Department of Industrial and Systems Engineering, University of Florida, Gainesville, September 2014
27. C.D. Dang, G. Lan, On the convergence properties of non-Euclidean extragradient methods for variational inequalities with generalized monotone operators. Comput. Optim. Appl. (2015). https://doi.org/10.1007/s10589-014-9673-9
28. C.D. Dang, G. Lan, Stochastic block mirror descent methods for nonsmooth and stochastic optimization. SIAM J. Optim. **25**, 856–881 (2015)
29. A. Defazio, F. Bach, S. Lacoste-Julien, SAGA: a fast incremental gradient method with support for non-strongly convex composite objectives, in *Advances of Neural Information Processing Systems (NIPS)*, vol. 27 (2014)

30. R. Dror, A. Ng, Machine learning (2018). http://cs229.stanford.edu
31. J. Duchi, A. Agarwal, M. Wainwright, Dual averaging for distributed optimization: convergence analysis and network scaling. IEEE Trans. Autom. Control **57**(3), 592–606 (2012)
32. J.C. Duchi, P.L. Bartlett, M.J. Wainwright, Randomized smoothing for stochastic optimization. SIAM J. Optim. **22**, 674–701 (2012)
33. F. Facchinei, J. Pang, *Finite-dimensional Variational Inequalities and Complementarity Problems, Volumes I and II*. Comprehensive Study in Mathematics (Springer, New York, 2003)
34. C. Fang, C.J. Li, Z. Lin, T. Zhang, Spider: near-optimal non-convex optimization via stochastic path-integrated differential estimator, in *Advances in Neural Information Processing Systems* (2018), pp. 687–697
35. M. Frank, P. Wolfe, An algorithm for quadratic programming. Naval Res. Logist. Q. **3**, 95–110 (1956)
36. R.M. Freund, P. Grigas, New analysis and results for the Frank-Wolfe method. Math. Program. **155**, 199–230 (2016)
37. S. Lee G. Lan, Y. Zhou, Communication-efficient algorithms for decentralized and stochastic optimization. Mathematical Programming (2018). Technical Report, H. Milton Stewart School of Industrial and Systems Engineering, Georgia Institute of Technology, January 15, 2017
38. S. Ghadimi, G. Lan, Optimal stochastic approximation algorithms for strongly convex stochastic composite optimization, I: a generic algorithmic framework. SIAM J. Optim. **22**, 1469–1492 (2012)
39. S. Ghadimi, G. Lan, Optimal stochastic approximation algorithms for strongly convex stochastic composite optimization, II: shrinking procedures and optimal algorithms. SIAM J. Optim. **23**, 2061–2089 (2013)
40. S. Ghadimi, G. Lan, Stochastic first- and zeroth-order methods for nonconvex stochastic programming. SIAM J. Optim. **23**(4), 2341–2368 (2013)
41. S. Ghadimi, G. Lan, Accelerated gradient methods for nonconvex nonlinear and stochastic programming. Math. Program. **156**, 59–99 (2016)
42. S. Ghadimi, G. Lan, H. Zhang, Mini-batch stochastic approximation methods for constrained nonconvex stochastic programming. Math. Program. **155**, 267–305 (2016)
43. A. Gonen S. Shalev-Shwartz, O. Shamir, Large-scale convex minimization with a low rank constraint, in *The 28th International Conference on Machine Learning* (2011)
44. M. Gurbuzbalaban, A. Ozdaglar, P. Parrilo, On the convergence rate of incremental aggregated gradient algorithms (2015). http://arxiv.org/abs/1506.02081
45. Z. Harchaoui, A. Juditsky, A.S. Nemirovski, Conditional gradient algorithms for machine learning, in *NIPS OPT Workshop* (2012)
46. T. Hastie, R. Tibshirani, J. Friedman, *The Elements of Statistical Learning: Data Mining, Inference, and Prediction*. Springer Series in Statistics, 2nd edn. (Springer, New York, 2009)

47. E. Hazan, Sparse approximate solutions to semidefinite programs, in *LATIN 2008: Theoretical Informatics*, ed. by E. Laber, C. Bornstein, L.T. Nogueira, L. Faria. Lecture Notes in Computer Science, vol. 4957 (Springer, Berlin, 2008), pp. 306–316
48. B. He, X. Yuan, On the $o(1/n)$ convergence rate of the Douglas Rachford alternating direction method. SIAM J. Numer. Anal. **50**(2), 700–709 (2012)
49. N. He, A. Juditsky, A. Nemirovski, Mirror prox algorithm for multi-term composite minimization and semi-separable problems. J. Comput. Optim. Appl. **103**, 127–152 (2015)
50. J.-B. Hiriart-Urruty, C. Lemaréchal, *Convex Analysis and Minimization Algorithms I*. Comprehensive Study in Mathematics, vol. 305 (Springer, New York, 1993)
51. M. Jaggi, Sparse convex optimization methods for machine learning. PhD thesis, ETH Zürich, 2011. https://doi.org/10.3929/ethz-a-007050453
52. M. Jaggi, Revisiting Frank-Wolfe: projection-free sparse convex optimization, in *The 30th International Conference on Machine Learning* (2013)
53. M. Jaggi, M. Sulovský, A simple algorithm for nuclear norm regularized problems, in *The 27th International Conference on Machine Learning* (2010)
54. D. Jakovetic, J. Xavier, J. Moura. Fast distributed gradient methods. IEEE Trans. Autom. Control **59**(5), 1131–1145 (2014)
55. B. Jiang, T. Lin, S. Ma, S. Zhang, Structured nonconvex and nonsmooth optimization: algorithms and iteration complexity analysis. Comput. Optim. Appl. **72**(1), 115–157 (2019)
56. R. Johnson, T. Zhang, Accelerating stochastic gradient descent using predictive variance reduction, in *Advances of Neural Information Processing Systems (NIPS)*, vol. 26 (2013), pp. 315–323
57. A. Juditsky, A.S. Nemirovski, *Large Deviations of Vector-Valued Martingales in* 2*-Smooth Normed Spaces* (Georgia Institute of Technology, Atlanta, 2008). www2.isye.gatech.edu/~nemirovs/LargeDevSubmitted.pdf
58. A. Juditsky, A.S. Nemirovski, C. Tauvel, Solving variational inequalities with stochastic mirror-prox algorithm. Stoch. Syst. **1**, 17–58 (2011). Manuscript, Georgia Institute of Technology, Atlanta
59. V. Katkovnik, Y. Kulchitsky, Convergence of a class of random search algorithms. Autom. Remote Control **33**, 1321–1326 (1972)
60. K.C. Kiwiel, Proximal level bundle method for convex nondifferentiable optimization, saddle point problems and variational inequalities. Math. Program. B **69**, 89–109 (1995)
61. G. Korpelevich, The extragradient method for finding saddle points and other problems. Ekon. Mat. Metody **12**, 747–756 (1976)
62. G. Lan, Efficient methods for stochastic composite optimization. Manuscript, Georgia Institute of Technology (2008). https://pdfs.semanticscholar.org/e8a9/331c6e3bb841ac437c8f5078fc4cd622725a.pdf
63. G. Lan, An optimal method for stochastic composite optimization. Math. Program. **133**(1), 365–397 (2012)

64. G. Lan, The complexity of large-scale convex programming under a linear optimization oracle. Manuscript, Department of Industrial and Systems Engineering, University of Florida, Gainesville, FL 32611, USA, June 2013. Available on http://www.optimization-online.org/
65. G. Lan, Gradient sliding for composite optimization. Manuscript, Department of Industrial and Systems Engineering, University of Florida, Gainesville, FL 32611, USA, June 2014
66. G. Lan, Bundle-level type methods uniformly optimal for smooth and nonsmooth convex optimization. Math. Program. **149**(1), 1–45 (2015)
67. G. Lan, Y. Ouyang, Accelerated gradient sliding for structured convex optimization. Manuscript, School of Industrial and Systems Engineering, Georgia Tech, Atlanta, GA 30332, USA, August 2016
68. G. Lan, Y. Yang, Accelerated stochastic algorithms for nonconvex finite-sum and multi-block optimization (2018). Preprint. arXiv
69. G. Lan, Y. Zhou, Conditional gradient sliding for convex optimization. Technical report, Technical Report, 2014
70. G. Lan, Y. Zhou, Random gradient extrapolation for distributed and stochastic optimization. SIAM J. Optim. **28**(4), 2753–2782 (2018)
71. G. Lan, Y. Yang, Accelerated stochastic algorithms for nonconvex finite-sum and multi-block optimization. SIAM J. Optim. **29**(4), 2753–2784 (2019)
72. G. Lan, Z. Lu, R.D.C. Monteiro, Primal-dual first-order methods with $\mathscr{O}(1/\varepsilon)$ iteration-complexity for cone programming. Math. Program. **126**, 1–29 (2011)
73. G. Lan, A.S. Nemirovski, A. Shapiro, Validation analysis of mirror descent stochastic approximation method. Math. Program. **134**, 425–458 (2012)
74. G. Lan, Z. Li, Y. Zhou, A unified variance-reduced accelerated gradient method for convex optimization, *in NeurIPS* (2019)
75. S. Lee, M. Zavlanos, Approximate projections for decentralized optimization with SDP constraints (2015). http://arxiv.org/abs/1509.08007
76. C. Lemaréchal, A.S. Nemirovski, Y.E. Nesterov, New variants of bundle methods. Math. Program. **69**, 111–148 (1995)
77. D. Leventhal, A.S. Lewis, Randomized methods for linear constraints: convergence rates and conditioning. Math. Oper. Res. **35**, 641–654 (2010)
78. H. Lin, J. Mairal, Z. Harchaoui, A universal catalyst for first-order optimization. Technical report, 2015. hal-01160728
79. D.G. Luenberger, Y. Ye, *Linear and Nonlinear Programming* (Springer, New York, 2008)
80. Z.Q. Luo, P. Tseng, On the convergence of a matrix splitting algorithm for the symmetric monotone linear complementarity problem. SIAM J. Control Optim. **29**, 1037–1060 (1991)
81. R. Luss, M. Teboulle, Conditional gradient algorithms for rank one matrix approximations with a sparsity constraint. SIAM Rev. **55**, 65–98 (2013)
82. A. Makhdoumi, A. Ozdaglar, Convergence rate of distributed ADMM over networks (2016). http://arxiv.org/abs/1601.00194

83. A. Mokhtari, W. Shi, Q. Ling, A. Ribeiro, DQM: decentralized quadratically approximated alternating direction method of multipliers (2015). http://arxiv.org/abs/1508.02073
84. R.D.C. Monteiro, B.F. Svaiter, On the complexity of the hybrid proximal extragradient method for the iterates and the ergodic mean. SIAM J. Optim. **20**(6), 2755–2787 (2010)
85. R.D.C. Monteiro, B.F. Svaiter, Complexity of variants of Tseng's modified f-b splitting and Korpelevich's methods for hemivariational inequalities with applications to saddle-point and convex optimization problems. SIAM J. Optim. **21**(4), 1688–1720 (2011)
86. R.D.C. Monteiro, B.F. Svaiter, Iteration-complexity of block-decomposition algorithms and the alternating direction method of multipliers. SIAM J. Optim. **23**(1), 475–507 (2013)
87. A. Nedić, Asynchronous broadcast-based convex optimization over a network. IEEE Trans. Autom. Control **56**(6), 1337–1351 (2011)
88. A. Nedić, A. Olshevsky, Distributed optimization over time-varying directed graphs. IEEE Trans. Autom. Control **60**(3), 601–615 (2015)
89. A. Nedić, A. Ozdaglar, Distributed subgradient methods for multi-agent optimization. IEEE Trans. Autom. Control **54**(1), 48–61 (2009)
90. A. Nedić, D.P. Bertsekas, V.S. Borkar, Distributed asynchronous incremental subgradient methods, in *Inherently Parallel Algorithms in Feasibility and Optimization and Their Applications* (2001), pp. 311–407
91. A. Nedić, A. Olshevsky, W. Shi, Achieving geometric convergence for distributed optimization over time-varying graphs (2016). http://arxiv.org/abs/1607.03218
92. A.S. Nemirovski, Optimization III (Georgia Tech, 2013). https://www2.isye.gatech.edu/~nemirovs/OPTIII_LectureNotes2018.pdf
93. A.S. Nemirovski, Prox-method with rate of convergence $o(1/t)$ for variational inequalities with Lipschitz continuous monotone operators and smooth convex-concave saddle point problems. SIAM J. Optim. **15**, 229–251 (2005)
94. A.S. Nemirovski, D. Yudin, *Problem Complexity and Method Efficiency in Optimization*. Wiley-Interscience Series in Discrete Mathematics (Wiley, XV, New York, 1983)
95. A.S. Nemirovski, A. Juditsky, G. Lan, A. Shapiro, Robust stochastic approximation approach to stochastic programming. SIAM J. Optim. **19**, 1574–1609 (2009)
96. Y.E. Nesterov, A method for unconstrained convex minimization problem with the rate of convergence $O(1/k^2)$. Doklady AN SSSR **269**, 543–547 (1983)
97. Y.E. Nesterov, *Introductory Lectures on Convex Optimization: A Basic Course* (Kluwer Academic Publishers, Boston, 2004)
98. Y.E. Nesterov, Primal-dual subgradient methods for convex problems. Core discussion paper 2005/67, CORE, Catholic University of Louvain, Belgium, September 2005

99. Y.E. Nesterov, Smooth minimization of nonsmooth functions. Math. Program. **103**, 127–152 (2005)
100. Y.E. Nesterov, Gradient methods for minimizing composite objective functions. Technical report, Center for Operations Research and Econometrics (CORE), Catholic University of Louvain, September 2007
101. Y.E. Nesterov, Efficiency of coordinate descent methods on huge-scale optimization problems. Technical report, Center for Operations Research and Econometrics (CORE), Catholic University of Louvain, February 2010
102. Y.E. Nesterov, Random gradient-free minimization of convex functions. Technical report, Center for Operations Research and Econometrics (CORE), Catholic University of Louvain, January 2010
103. L.M. Nguyen, J. Liu, K. Scheinberg, M. Takà c, A novel method for machine learning problems using stochastic recursive gradient, in *Proceedings of the 34th International Conference on Machine Learning*, vol. 70 (2017), pp. 2613–2621
104. J. Nocedal, S.J. Wright, *Numerical Optimization* (Springer, New York, 1999)
105. Y. Ouyang, Y. Chen, G. Lan, E. Pasiliao, An accelerated linearized alternating direction method of multipliers. SIAM J. Imaging Sci. **8**(1), 644–681 (2015)
106. N.H. Pham, L.M. Nguyen, D.T. Phan, Q. Tran-Dinh, Proxsarah: an efficient algorithmic framework for stochastic composite nonconvex optimization (2019). Preprint. arXiv:1902.05679
107. B. Polyak, *Introduction to Optimization*. (Optimization Software, New York, 1987)
108. B.T. Polyak, New stochastic approximation type procedures. Automat. i Telemekh. **7**, 98–107 (1990)
109. B.T. Polyak, A.B. Juditsky, Acceleration of stochastic approximation by averaging. SIAM J. Control Optim. **30**, 838–855 (1992)
110. G. Qu, N. Li, Harnessing smoothness to accelerate distributed optimization (2016). http://arxiv.org/abs/1605.07112
111. C. Qu, Y. Li, H. Xu, Non-convex conditional gradient sliding, in *Proceedings of the 35th International Conference on Machine Learning, PMLR*, vol. 80 (2018), pp. 4208–4217
112. M. Rabbat, Multi-agent mirror descent for decentralized stochastic optimization, in *2015 IEEE 6th International Workshop on Computational Advances in Multi-sensor Adaptive Processing (CAMSAP), December* (2015), pp. 517–520
113. S.S. Ram, A. Nedić, V.V. Veeravalli, Incremental stochastic subgradient algorithms for convex optimization. SIAM J. Optim. **20**(2), 691–717 (2009)
114. S.S. Ram, A. Nedić, V.V. Veeravalli, Distributed stochastic subgradient projection algorithms for convex optimization. J. Optim. Theory Appl. **147**, 516–545 (2010)
115. S.J. Reddi, S. Sra, B. Poczos, A. Smola, Stochastic Frank-Wolfe methods for nonconvex optimization (2016). Preprint. arXiv: 1607.08254

116. P. Richtárik, M. Takáč, Iteration complexity of randomized block-coordinate descent methods for minimizing a composite function. Math. Program. (2012 to appear)
117. H. Robbins, S. Monro, A stochastic approximation method. Ann. Math. Stat. **22**, 400–407 (1951)
118. R.T. Rockafellar, *Convex Analysis* (Princeton University Press, Princeton, 1970)
119. R.Y. Rubinstein, *Simulation and the Monte Carlo Method* (Wiley, New York, 1981)
120. A. Ruszczyński, *Nonlinear Optimization*, 1st edn. (Princeton University Press, Princeton, 2006)
121. M. Schmidt, N.L. Roux, F. Bach, Minimizing finite sums with the stochastic average gradient. Technical report, September 2013
122. S. Shalev-Shwartz, T. Zhang, Stochastic dual coordinate ascent methods for regularized loss. J. Mach. Learn. Res. **14**(1), 567–599 (2013)
123. S. Shalev-Shwartz, T. Zhang, Accelerated proximal stochastic dual coordinate ascent for regularized loss minimization. Math. Program. **155**, 105–145 (2016)
124. C. Shen, J. Kim, L. Wang, A. van den Hengel, Positive semidefinite metric learning using boosting-like algorithms. J. Mach. Learn. Res. **13**, 1007–1036 (2012)
125. Z. Shen, C. Fang, P. Zhao, J. Huang, H. Qian, Complexities in projection-free stochastic non-convex minimization, in *Proceedings of Machine Learning Research, PMLR 89*, vol. 89 (2019), pp. 2868–2876
126. W. Shi, Q. Ling, G. Wu, W. Yin, On the linear convergence of the admm in decentralized consensus optimization. IEEE Trans. Signal Process. **62**(7), 1750–1761 (2014)
127. W. Shi, Q. Ling, G. Wu, W. Yin, Extra: an exact first-order algorithm for decentralized consensus optimization. SIAM J. Optim. **25**(2), 944–966 (2015)
128. W. Shi, Q. Ling, G. Wu, W. Yin, A proximal gradient algorithm for decentralized composite optimization. IEEE Trans. Signal Process. **63**(22), 6013–6023 (2015)
129. A. Simonetto, L. Kester, G. Leus, Distributed time-varying stochastic optimization and utility-based communication (2014). http://arxiv.org/abs/1408.5294
130. H. Terelius, U. Topcu, R. Murray, Decentralized multi-agent optimization via dual decomposition, in *IFAC Proceedings Volumes*, vol. 44(1) (2011), pp. 11245–11251
131. P. Tseng, Convergence of a block coordinate descent method for nondifferentiable minimization. J. Optim. Theory Appl. **109**, 475–494 (2001)
132. K. Tsianos, M. Rabbat, Consensus-based distributed online prediction and optimization, in *2013 IEEE Global Conference on Signal and Information Processing, December*, pp. 807–810 (2013)

133. K. Tsianos, S. Lawlor, M. Rabbat, Consensus-based distributed optimization: practical issues and applications in large-scale machine learning, in *Proceedings of the 50th Allerton Conference on Communication, Control, and Computing* (2012)
134. J.N. Tsitsiklis, Problems in decentralized decision making and computation. PhD thesis, Massachusetts Institute of Technology, Cambridge, MA, 1984
135. J. Tsitsiklis, D. Bertsekas, M. Athans, Distributed asynchronous deterministic and stochastic gradient optimization algorithms. IEEE Trans. Autom. Control **31**(9), 803–812 (1986)
136. M. Wang, D.P. Bertsekas, Incremental constraint projection-proximal methods for nonsmooth convex optimization. Technical Report LIDS-P-2907, Laboratory for Information and Decision Systems, 2013
137. Z. Wang, K. Ji, Y. Zhou, Y. Liang, V. Tarokh, Spiderboost: a class of faster variance-reduced algorithms for nonconvex optimization (2018). Preprint. arXiv:1810.10690
138. E. Wei, A. Ozdaglar, On the $O(1/k)$ convergence of asynchronous distributed alternating direction method of multipliers (2013). http://arxiv.org/pdf/1307.8254
139. C. Xi, Q. Wu, U.A. Khan, Distributed mirror descent over directed graphs (2014). http://arxiv.org/abs/1412.5526
140. L. Xiao, T. Zhang, A proximal stochastic gradient method with progressive variance reduction. SIAM J. Optim. **24**(4), 2057–2075 (2014)
141. Y. Zhang, L. Xiao, Stochastic primal-dual coordinate method for regularized empirical risk minimization, in *Proceedings of the 32nd International Conference on Machine Learning* (2015), pp. 353–361
142. D. Zhou, P. Xu, Q. Gu, Stochastic nested variance reduction for nonconvex optimization, in *Proceedings of the 32nd International Conference on Neural Information Processing Systems, NIPS'18* (Curran Associates, New York, 2018), pp. 3925–3936
143. M. Zhu, S. Martinez, On distributed convex optimization under inequality and equality constraints. IEEE Trans. Autom. Control **57**(1), 151–164 (2012)

Index

G. Lan, *First-Order and Stochastic Optimization Methods for Machine Learning*, Springer Series in the Data Sciences, https://doi.org/10.1007/978-3-030-39568-1

GPSR Compliance
The European Union's (EU) General Product Safety Regulation (GPSR) is a set of rules that requires consumer products to be safe and our obligations to ensure this.

If you have any concerns about our products, you can contact us on

ProductSafety@springernature.com

In case Publisher is established outside the EU, the EU authorized representative is:

Springer Nature Customer Service Center GmbH
Europaplatz 3
69115 Heidelberg, Germany

www.ingramcontent.com/pod-product-compliance
Ingram Content Group UK Ltd.
Pitfield, Milton Keynes, MK11 3LW, UK
UKHW021832270726
14058UKWH00001B/107

* 9 7 8 3 0 3 0 3 9 5 7 0 4 *